Table of Contents

Glossary

abiotic factors characterized by the absence of life; include temperature, humidity, pH, and other physical and chemical influences.

absolute density the number of individuals per unit area or per unit volume.

abundance the number or biomass of organisms of a particular species in a general area.

actual evapotranspiration the actual amount of water that is used by and evaporates from a plant community over a given time period, largely dependent on the available water and the temperature.

adaptation any alteration in the structure or function of an organism by which the organism becomes better able to survive and multiply in its environment.

additive effects reproduction or mortality that simply adds or subtracts the individuals to the current population; opposite of compensatory effects.

aggregation coming together of organisms into a group, as in locusts.

aggregative response the response of predators or parasitoids to concentrate their foraging in an area of dense prey species.

Allee effects population growth rates that decrease below replacement level at low population density, potentially leading to extinction.

allele one of a pair of characters that are alternative to each other in inheritance, being governed by genes situated at the same locus in homologous chromosomes.

allelopathy organisms that alter the surrounding chemical environment in such a way as to prevent other species from using it, typically with toxins or antibiotics.

ambient energy hypothesis the idea that species diversity is governed by the amount of energy falling on an area.

apex predator in a food chain, it is the highest trophic level. Apex predators do not have other predators feeding on them within the food web.

aposematic warning coloration, indicating to a predator that this prey is poisonous or highly defended against attack.

apparent competition two species who do not share any resources but whose numbers change in relation to one another because of an indirect effect of a third species, typically a shared predator or natural enemy.

association major unit in community ecology, characterized by essential uniformity of species composition.

autotroph organism that obtains energy from the sun and materials from inorganic sources; contrast with *heterotroph*. Most plants are autotrophs.

balance of nature the belief that natural populations and communities exist in a stable equilibrium and maintain that equilibrium in the absence of human interference.

barriers any geographic feature that hinders or prevents dispersal or movement across it, producing isolation.

basal metabolic rate the amount of energy expended by an animal while at rest in a neutral temperate environment, in the post-absorptive (fasting) state; the minimum rate of metabolism.

big-bang reproduction offspring are produced in one burst rather than in a repeated manner.

biodiversity the number of species in a community or region, which may be weighted by their relative abundances; also used as an umbrella concept for total biological diversity including genetic diversity within a species, species diversity (as used here), and ecosystem diversity at the

community or ecosystem level of organization.

bioelements the chemical elements that move through living organisms.

biogeochemical cycles the movement of chemical elements around an ecosystem via physical and biological processes.

biogeography the study of the geographical distribution of life on Earth and the reasons for the patterns one observes on different continents, islands, or oceans.

biological control the reduction of pests by the introduction of predators, parasites, or pathogens; by genetic manipulations of crops or pests; by sterilization of pests; or by mating disruption using pheromones.

biomanipulation the management practice of using a trophic cascade to restore lakes to a clear water condition by removing herbivorous or planktivorous fishes or by adding piscivorous (predatory) fishes to a lake.

biomass the mass or weight of living matter in an area.

biosphere the whole-earth ecosystem, also called the *ecosphere*.

biota species of all the plants and animals occurring within a certain area or region.

biotic factors environmental influences caused by plants or animals; opposite of abiotic factors.

bottom-up model the idea that community organization is set by the effects of plants on herbivores and herbivores on carnivores in the food chain.

bryophytes plants in the phylum Bryophyta comprising mosses, liverworts, and hornworts.

Calvin-Benson cycle the series of biochemical reactions that takes place in the stroma of chloroplasts in photosynthetic organisms and results in the first step of carbon fixation in photosynthesis.

cannibalism an animal that feeds on others of the same species.

carnivores animals that eat mainly flesh from other animals; contrast with *herbivore*.

catastrophic agents term used by Howard and Fiske (1911) to describe agents of destruction in which the percentage of destruction is not related to population density; synonymous with density-independent factors.

character displacement the divergence in morphology between similar species in the region where the species both occur, but this divergence is reduced or lost in regions where the species' distributions do not overlap; presumed to be caused by competition.

climatic climax the final equilibrium vegetation for a site that is dictated by climate and toward which all successions are proceeding, according to Frederic Clements.

climax community the final equilibrium community toward which succession moves.

climax-pattern hypothesis the view that climax communities grade into one another and form a continuum of climax types that vary gradually along environmental gradients.

closed population in population estimation, a population that is not changing in size during the interval of study, having no natality, mortality, immigration, or emigration.

coarse-grained habitat from a particular species' point of view a habitat is coarse grained if it spends its life in one fragment of habitat and cannot move easily to another patch.

coevolution the evolution of two or more species that interact closely with one another, with each species adapting to changes in the other.

cohort life table a life table that follows a group of organisms from germination, birth, or hatching to the death of the last individual.

common garden an experimental design in plant ecophysiology in which a series of plants from different areas are brought together and planted in one area, side by side, in an attempt to determine which features of the plants are genetically controlled and which are environmentally determined.

community a group of populations living in the same area or habitat.

community structure the species composition of an ecological community including the abundance of all the populations in the community.

compartment any component of study for an analysis of nutrient cycling, such as a lake, a species of plant, or a functional group of nitrogen fixers, measured by its standing crop or amount of nutrient.

compartment model a type of box-and-arrow model of diseases in which each compartment contains a part of the system that can be measured and the compartments are linked by flows between them; each compartment typically has an input from some compartments and an output to other compartments.

compensation point for plants the equilibrium point at which photosynthesis equals respiration.

compensatory effects reproduction or mortality that does not add or subtract the individuals to the current population but only replaces other individuals with no change in population size; opposite of additive effects.

competition occurs when a number of organisms of the same or different species utilize common resources that are in short supply (*exploitation*) or when the organisms harm one another in the process of acquiring these resources (*interference*).

competitive exclusion principle complete competitors cannot coexist; also called Gause's hypothesis.

connectance used to describe food web complexity; the fraction of potential interactions in a food web that actually exist.

continental climates the product of weather systems over large landmasses that result in cold winters and warm summers, not influenced by the large ocean masses, typically in temperate and polar latitudes.

control in an experimental design a control is a treatment or plot in which nothing is changed so that it serves as a baseline for comparison with the experimental treatments to which something is typically added or subtracted.

cost–benefit analysis an assessment to determine whether the cost of an activity is less than the benefit that can be expected from the activity.

crassulacean acid metabolism (CAM) a form of photosynthesis in which the two chemical parts of photosynthesis are separated in time because CO_2 is taken up at night through the stomata (which are then closed during the day) and fixed to be used later in the day to complete photosynthesis carbon fixation; an adaptation used by desert plants to conserve water.

critical load the amount of a nutrient such as nitrogen that can be absorbed by an ecosystem without damaging its integrity.

cultural control the reduction of pest populations by agricultural manipulations involving crop rotation, strip cropping, burning of crop residues, staggered plantings, and other agricultural practices.

declining-population paradigm the focus of this approach is on detecting, diagnosing, and halting a population decline by finding the causal factors affecting the population.

deme interbreeding group in a population; also known as *local population*.

demographic stochasticity the random variation in birth and death rates that can lead by chance to extinction.

demographic transition the change in human populations from the two zero-population-growth states of high birth and high death rates to low birth and low death rates.

density number of organisms per unit area or per unit volume.

density-dependent rate as population density rises, births or immigration decrease or deaths or emigration increase, and consequently a graph of population density versus the rate will have a positive or negative slope.

density-independent rate as population density rises, the rate does not change in any systematic manner, so that a graph of population density versus the rate will have a slope of zero.

determinate layers birds that lay a fixed number of eggs no matter what occurs.

deterministic extinctions losses of species due to the removal of an essential resource.

deterministic models mathematical models with a fixed outcome, models that give the same answer every time they are repeatedly run with a fixed set of parameters; opposite of *stochastic model*.

detritus the plant production not consumed by herbivores.

developmental response the increasing intake rate of prey items by an organism that is growing in size as it develops.

dilution rate general term to describe the rate of additions to a population from birth and immigration.

directional selection natural selection that favors traits either above or below the average of the population, so that over time the average moves in one direction.

disease a pathological condition of an organism resulting from various causes, such as an infection, a genetic disorder, or environmental stress, with specific symptoms.

dispersal the movement of individuals away from their place of birth or hatching or seed production into a new habitat or area to survive and reproduce.

disruptive selection natural selection that favors extreme trait values rather than intermediate values so that over time extreme traits become more common.

disturbance any short-lived strong disruption to an ecological population or community, such as a fire, flood, windstorm, or earthquake.

dominant species common species of large biomass or numbers in a community.

dynamic pool models a model to predict maximum sustained yield based on detailed population information on growth rates, natural mortality, and fishing mortality; contrast with *logistic-type model*.

dynamic stability hypothesis for food chain length suggests that higher trophic levels are less stable than

lower trophic levels and past a certain point the longer chains go extinct.

dynamics in population ecology, the study of the reasons for changes in population size; contrast with *statics*.

ecological footprint the total land and water area that is appropriated by a nation or a city to produce all the resources it consumes and to absorb all the waste it generates.

ecological longevity average length of life of individuals of a population under stated conditions.

ecological specialization model a proposed explanation for Hanski's Rule, which postulates that species that exploit a wide range of resources become both widespread and common; these species are generalists; also called Brown's model.

ecosystem biotic community and its abiotic environment; the whole Earth can be considered as one large ecosystem.

ecosystem services all the processes through which natural ecosystems and the biodiversity they contain help sustain human life on Earth.

ecotone transition zone between two diverse communities (e.g., the tundra–boreal forest ecotone).

ecotype a genetic subspecies or race of a plant or animal species that is adapted to a specific set of environmental conditions such as temperature or salinity.

edaphic pertaining to the soil.

effective population size a population genetic concept of the number of breeding individuals in an idealized population that would maintain the existing genetic variability; it is typically much less than the observed population size.

Eltonian pyramid abundance or biomass of successive trophic levels of an ecosystem, illustrating the impact of energy flows through successive trophic transfers.

emigration the movement of individuals out of an area occupied by the population, typically the site of birth or hatching.

endemic phase for locusts and other organisms that show outbreaks, the phase of low numbers when indi-

viduals are difficult to find in the field.

endemic species species that occur in one restricted area but in no other.

energetic hypothesis for food chain length, postulates that higher trophic levels are restricted by the limited efficiency of energy transfer along the chain.

environment all the biotic and abiotic factors that actually affect an individual organism at any point in its life cycle.

environmental heterogeneity variation in space in any environmental parameter such as soil pH or tree cover.

environmental stochasticity variation in population growth rates imposed by changes in weather and biotic factors, as well as natural catastrophes such as floods and hurricanes.

epidemic phase for locusts and other species that show rapid increases to high density, the phase of high numbers and maximum damage; contrast with *endemic phase*.

epidemiology branch of medicine dealing with epidemic diseases.

epipelic algae algae living in or on the sediments of a body of water.

equilibrium model of community organization the global view that ecological communities are relatively constant in composition and are resilient to disturbances.

equitability evenness of distribution of species abundance patterns; maximum equitability occurs when all species are represented by the same number of individuals.

eutrophic lake a highly productive lake with dense phytoplankton, typically with green water.

eutrophic soils soils with high nutrient levels, mostly recent and often volcanic in origin.

eutrophication the process by which lakes are changed from clear water lakes dominated by green algae into murky lakes dominated by blue-green algae, typically caused by nutrient runoffs from cities or agriculture.

evapotranspiration sum total of water lost from the land by evaporation and plant transpiration.

experiment test of a hypothesis. It can be observational (observe the system) or manipulative (perturb the system). The experimental method is the scientific method.

experimental analysis an approach to studying population regulation that relies on the manipulation of populations rather than simple observation of changes used in key factor analysis.

facilitation helping another organism, providing positive feedback in a population interaction.

facilitation model the classic view that succession proceeds via one species helping the next species in the sequence to establish.

fact particular truth of the natural world. Philosophers endlessly discuss what a fact is. Ecologists make observations, which may be faulty; consequently, every observation is not automatically a fact.

facultative agents term used by Howard and Fiske (1911) to describe agents of destruction that increase their percentage of destruction as population density rises; synonymous with *density-dependent factors*.

fecundity an organism's potential reproductive capacity over a period of time, measured by the number of gametes produced.

feeding guilds organisms that eat the same general foods, such as seed-eaters.

fertility the actual number of viable offspring produced by an organism over a period of time, equivalent to realized *fecundity*.

fertility schedule the age-specific reproductive output per individual.

field metabolic rate the amount of energy used per unit of time by an organism under normal conditions of life in a natural ecosystem.

fine-grained habitat from a particular species' point of view, a habitat is fine grained if it moves freely from one patch to another at no cost.

First Principle of Population Regulation no closed population stops increasing unless either the per capita birth rate or death rate is density dependent.

fitness the ability of a particular genotype or phenotype to leave descendants in future generations, relative to other organisms.

flux rate the rate of flow of nutrients or biomass from one compartment to another.

food chain the transfer of energy and materials from plants to herbivores to carnivores.

food web a linked set of *food chains* that most often resemble a web.

frost drought for plants a shortage of water in winter when the ground is frozen so no water can be taken up by the roots and yet air temperature is high enough that plants attempt to photosynthesize.

functional group a group of species that perform the same function in a community.

functional response the change in the intake rate of a predator in relation to the density of its prey species.

fundamental niche the ecological space occupied by a species in the absence of competition and other biotic interactions from other species.

Gause's hypothesis complete competitors cannot coexist; also called the competitive exclusion principle.

gene flow the movement of alleles of genes in space and time from one population to another.

genecology study of population genetics in relation to the habitat conditions; the study of species and other taxa by the combined methods and concepts of ecology and genetics.

generalist predators predators that eat a great variety of prey species.

generalists species that eat a variety of foods or live in a variety of habitats; contrast to *specialists*.

genet a unit of genetically identical individuals, derived by asexual reproduction from a single original zygote.

genetic stochasticity any potential loss of genetic variation due to inbreeding or genetic drift (the nonrandom assortment of genes during reproduction).

genotype entire genetic constitution of an organism; contrast with *phenotype*.

genotypic under the control of the genetic endowment of an individual.

global nutrient cycles nutrient cycles that operate at very large scales over much of the Earth because the nutrients are volatile, such as oxygen.

global stability occurs when a community can recover from any disturbance, large or small, and go back to its initial configuration of species composition and abundances; compare with *neighborhood stability*.

gradocoen totality of all factors that impinge on a population, including *biotic* agents and *abiotic* factors.

grazing facilitation the process of one herbivore creating attractive feeding conditions for another herbivore so there is a benefit provided to the second herbivore.

green world hypothesis the proposed explanation for the simple observation that the world is green, that herbivores are held in check by their predators, parasites, and diseases, although other explanations have been suggested.

greenhouse effect the process in which the emission of infrared (long-wave) radiation by the atmosphere warms a planet's surface.

greenhouse gases gases present in the Earth's atmosphere that reflect infrared radiation back to Earth, thus warming it. The most important ones affected by humans are carbon dioxide, methane, nitrous oxide, and chlorofluorocarbons. Water vapor also acts as a greenhouse gas.

gross primary production the energy or carbon fixed via photosynthesis per unit time.

gross production production before respiration losses are subtracted; photosynthetic production for plants and metabolizable production for animals.

gross productivity the assimilation rate of an animal, which includes all the digested energy less the urinary waste.

group selection natural selection for traits that favor groups within a species irrespective of whether the traits favor individuals or not.

growth form morphological categories of plants, such as trees, shrubs, and vines.

guild a group of species that exploit a common resource base in a similar fashion.

habitat a particular environment in which a species lives, or broadly speaking the biotic environment occupied by an individual or population.

habitat selection the behavioral actions of organisms (typically animals) in choosing the areas in which they live and breed.

handling time the time utilized by a predator to consume an individual prey item.

Hanski's Rule the generalization that there is a positive relationship between distribution and abundance, such that abundant species have wide geographical ranges.

harvest method the measurement of primary production by clipping the vegetation at two successive times.

herbivore an animal that eats plants or parts of plants; contrast with *carnivore*.

herbivory the eating of parts of plants by animals, not typically resulting in plant death.

heterogeneity the distribution of relative abundance among the species.

heterotroph organism that obtains energy and materials by eating other organisms; contrast with *autotroph*.

homeostasis maintenance of constancy or a high degree of uniformity in an organism's functions or interactions of individuals in a population or community under changing conditions; results from the capabilities of organisms to make adjustments.

homeothermic pertaining to warm-blooded animals that regulate their body temperature; contrast with *poikilothermic*.

host organism that furnishes food, shelter, or other benefits to another organism of a different species.

hotspots of biodiversity areas of the Earth that contain many endemic species (typically 1500) and as such are of important conservation value.

hydrophyte plant that grows wholly or partly immersed in water; compare with *xerophyte* and *mesophyte*.

hypothesis universal proposition that suggests an explanation for some observed ecological situation.

hypoxia lack of oxygen, typically in lakes or parts of an ocean basin in which excessive primary production is broken down by bacteria and other decomposers, using up all the oxygen in the water.

ideal despotic distribution a theoretical spatial spread of members of a population in which the competitive dominant "aggressive" individuals take up the best resources or territories, and less competitive individuals take up areas or resources in direct relationship to their dominance status.

ideal free distribution a theoretical spatial spread of members of a population in which individuals take up areas with equal amounts of resources in relation to their needs, so all individuals do equally well (the polar opposite to the ideal despotic distribution).

immigration the movement of organisms into an area.

immunocontraception the use of genetic engineering to insert genes that stimulate the immune system of a vertebrate to reject sperm or eggs, thus causing infertility.

incidence functions the fraction of patches of a given size occupied by a breeding population of a particular species.

indeterminate layers birds that continue to lay eggs until the nest is full, thus compensating for any egg removals.

index of similarity ratio of the number of species found in common in two communities to the total number of species that are present in both.

indifferent species species occurring in many different communities; are poor species for community classification.

individual optimization hypothesis that each individual in a population has its own optimal clutch size, so that not all individuals are identical.

inducible defenses plant defense methods that are called into action once herbivore attack occurs and are nearly absent during periods of no herbivory.

inhibition model succession proceeds via one species trying to stop the next species in the sequence from establishing.

initial floristic composition the model of succession of who-gets-there-first wins, part of the inhibition model.

insect parasitoids insects that lay their eggs in or on the host species, so that the larvae enter the host and kill it by consuming it from the inside.

integrated pest management (IPM) the use of all techniques of control in an optimal mix to minimize pesticide use and maximize natural controls of pest numbers.

interactive herbivore system plant-herbivore interactions in which there is feedback from the herbivores to the plants so that herbivores affect plant production and fitness.

intermediate disturbance hypothesis the idea that biodiversity will be maximal in habitats that are subject to disturbances at a moderate level, rather than at a low or high level.

interspecific between two or more different species.

interspecific competition competition between members of different species.

intransitive competition a competitive network that never reaches a fixed endpoint because A replaces B and B replaces C but C can replace A.

intraspecific between individuals of the same species.

intrinsic capacity for increase (r) measure of the rate of increase of a population under controlled conditions, with fixed birth and death rates; also called *innate capacity for increase.*

irruption a rapid increase in a population, often after being introduced to a new area, followed by a collapse that may be rapid or prolonged and may result in a convergent oscillation to a lower equilibrium density.

isocline a contour line in graphical presentations of mathematical models in which some parameter is equal all along the line.

isotherm line drawn on a map or chart connecting points with the same temperature at a particular time or over a certain period.

key factor analysis a systematic approach using life tables to determine the factors responsible for the regulation and fluctuation of populations.

keystone species relatively rare species in a community whose removal causes a large shift in the structure of the community and the extinction of some species.

kin selection the evolution of traits that increase the survival, and ultimately the reproductive success, of one's relatives.

Krantz anatomy the particular type of leaf anatomy that characterizes C_4 plants; plant veins are encased by thick-walled photosynthetic bundle-sheath cells that are surrounded by thin-walled mesophyll cells.

***K*-selection** the type of natural selection experienced by organisms that live at carrying capacity or maximal density in a relatively stable environment.

Lack clutch size the clutch size at which productivity is maximal for the population.

Lack's hypothesis that clutch size in birds is determined by the number of young that parents can provide with food.

Leslie matrix model a method of casting the age-specific reproductive schedule and the age-specific mortality schedule of a population in matrix form so that predictions of future population change can be made.

Liebig's law of the minimum the generalization first stated by Justus von Liebig that the rate of any biological process is limited by that factor in least amount relative to requirements, so there is a single limiting factor.

life table the age-specific mortality schedule of a population.

limiting factor a factor is defined as limiting if a change in the factor produces a change in average or equilibrium density.

littoral shallow-water zone of lakes or the sea, with light penetration to the bottom; often occupied by rooted aquatic plants.

local nutrient cycles nutrient cycles that are confined to small regions because the elements are non-volatile, such as the phosphorus cycle.

local population see *deme*.

local population model a proposed explanation for Hanski's Rule, which assumes that species differ in their capacity to disperse, and if the environment is divided into patches, some species will occupy

more local patches than others as a function of their dispersal powers.

local stability occurs when communities recover from only small disturbances and return to their former configuration of species composition and abundances.

logistic equation model of population growth described by a symmetrical S-shaped curve with an upper asymptote.

logistic-type model type of optimum-yield model in which the yield is predicted from an overall descriptive function of population growth without a separate analysis of the components of mortality, recruitment, and growth; contrast with *dynamic pool model*.

log-normal distribution the statistical distribution that has the shape of a normal, bell-shaped curve when the *x*-axis is expressed in a logarithmic scale rather than an arithmetic scale.

loss rate general term to describe the rate of removal of organisms from a population by death and emigration.

Lotka-Volterra equations the set of equations that describe competition between organisms for food or space; another set of equations describes predator-prey interactions

lottery competition a type of interference competition in which an individual's chances of winning or losing are determined by who gets access to the resource first.

macroparasites large multicellular organisms, typically arthropods or helminths, which do not multiply within their definitive hosts but instead produce transmission stages (eggs and larvae) that pass into the external environment.

marine protected area a national park in the ocean where fishing is restricted or eliminated for the purpose of protecting populations from overharvesting.

match-mismatch hypothesis the idea that population regulation in many fish is determined in the early juvenile stages by food supplies, so that if eggs hatch at the same time that food is abundant, many will survive, but if eggs hatch when food is scarce, many will die.

matrix models a family of models of population change based on matrix algebra, with the Leslie matrix model being the best known.

maximum economic rent the desired economic goal of any exploited resource, measured by total revenues – total costs.

maximum reproduction the theory that natural selection will maximize reproductive rate, subject to the constraints imposed by feeding and predator avoidance.

maximum sustained yield (MSY) the predicted yield that can be taken from a population without the resource collapsing in the short or long term.

mean length of a generation the average length of time between the birth of a female and her offspring.

mechanism a biological process that explains some phenomenon.

mesic moderately moist.

mesophyte plant that grows in environmental conditions that include moderate moisture conditions.

mesopredators secondary consumers (e.g., carnivores) in a food chain that are fed upon by tertiary consumers such as apex predators.

metabolic theory of ecology an attempt to derive patterns of individual performance, population, and ecosystem dynamics from the fundamental observation that the metabolic rate of individuals is related to body size and temperature.

metapopulations local *populations* in patches that are linked together by dispersal among the patches, driven by colonization and extinction dynamics.

microparasites small pathogenic organisms, typically protozoa, fungi, bacteria, or viruses, that can cause disease.

minimum viable population (MVP) the size of a population in terms of breeding individuals that will ensure at some specified level of risk continued existence with ecological and genetic integrity.

model verbal or mathematical statement of a hypothesis.

modular organisms organisms that have an indefinite growth form, such as plants or corals.

monoclimax hypothesis the classic view of Frederic Clements that all vegetation in a region converges ultimately to a single climax plant community.

monogamy mating of an animal with only one member of the opposite sex.

morphology study of the form, structure, and development of organisms.

mortality the death of organisms in a population.

multivoltine refers to an organism that has several generations during a single season; contrast with *univoltine*.

mutualism a relationship between two organisms of different species that benefits both and harms neither.

mycorrhizae a mutually beneficial association of a fungus and the roots of a plant in which the plant's mineral absorption is enhanced and the fungus obtains nutrients from the plant.

natality birth or germination or hatching; reproductive output of a population.

natural control the limitation of pest populations by predators, parasitoids, parasites, diseases, and weather in the absence of chemical control.

natural selection the process in nature by which only the organisms best adapted to their environment tend to survive and transmit their genetic characteristics to succeeding generations while those less adapted tend to be eliminated.

neighborhood stability also called *local stability*, the ability of a community to return to its former configuration after a small disturbance.

nested subsets a sequence of habitat patches, ordered by size, is nested if all the species in the smaller patches are also included in the larger patches.

net primary production the energy (or carbon) fixed in photosynthesis minus the energy (or carbon) lost via respiration per unit time.

net production production after respiration losses are subtracted.

net reproductive rate (R_0) the average number of offspring produced per female or reproductive unit.

ͳhe the ecological space occupied by ͳecies, and the occupation of the ͳ in a community.

niche breadth a measurement of the range of resources utilized by a species.

niche overlap a measure of how much species overlap with one another in the use of resources.

nonequilibrium model of community organization the global view that ecological communities are not constant in their composition because they are always recovering from *biotic* and *abiotic* disturbances, never reaching an equilibrium.

noninteractive herbivore system plant-herbivore interactions in which there is no feedback from the herbivores to the plants.

numerical response the change in the numbers or density of a predator in relation to changes in the density of its prey species.

obligate predator or parasite that is restricted to eating a single species of prey.

oligochaetes any of a class or order (Oligochaeta) of hermaphroditic terrestrial or aquatic annelids lacking a specialized head; includes earthworms.

oligotrophic lake an unproductive, clear-water lake with a low density of phytoplankton.

oligotrophic pattern soils of very low nutrient levels that are common in tropical areas and regions with geologically old, highly eroded soils with most of the nutrients in the litter layer.

omnivore an animal that feeds on both plants and animals in a food chain.

open population in population estimation, a population that has natality, mortality, immigration, or emigration during the interval of study.

optimal defense hypothesis the idea that plants allocate defenses against herbivores in a manner that maximizes individual plant fitness, and that defenses are costly to produce.

optimal foraging any method of searching for and obtaining food that maximizes the relative benefit.

optimal foraging theory a detailed model of how animals should forage to maximize their fitness.

optimal group size the size that results in the largest relative benefit.

optimality models models that assume natural selection will achieve adaptations that are the best possible for each trait in terms of survival and reproduction.

optimum yield amount of material that can be removed from a population to maximize biomass (or numbers, or profit, or any other type of "optimum") on a sustained basis.

ordination process by which plant or animal communities are ordered along a gradient.

overcompensation hypothesis the idea that a small amount of grazing will increase plant growth and fitness rather than cause harm to the plant.

paradox of the plankton the problem of understanding how many phytoplankton species that have the same basic requirements can coexist in a community without competitive exclusion.

parasite an organism that grows, feeds, or is sheltered on or in a different organism while harming its host.

parasitoid an insect that completes larval development in another insect host.

parthenogenesis development of the egg of an organism into an embryo without fertilization.

patch any discrete area, regardless of size.

pesticide any chemical that kills a plant or animal pest.

pesticide suppression the reduction of pest populations with herbicides, fungicides, insecticides, or other chemical poisons.

Petersen method a population estimation procedure based on two periods of mark-and-recapture.

phenology study of the periodic (seasonal) phenomena of animal and plant life and their relations to the weather and climate (e.g., the time of flowering in plants).

phenotype expression of the characteristics of an organism as determined by the interaction of its genic constitution and the environment; contrast with *genotype*.

photoperiodism the physiological responses of plants and animals to the length of day.

photosynthesis the series of chemical reactions in plants that results in the fixation of carbon from CO_2 into some form of carbohydrate.

photosynthetically active radiation (PAR) that part of the solar radiation spectrum in the range 0.4 to 0.7 μm that can be used for photosynthesis by green leaves.

physiological ecology the subdiscipline of ecology that studies the biochemical, physical, and mechanical adaptations and limitations of plants and animals to their physical and chemical environments.

physiological longevity maximum life span of individuals in a population under specified conditions; the organisms die of senescence.

phytoplankton plant portion of the plankton; the plant community in marine and freshwater environments that floats free in the water and contains many species of algae and diatoms.

Plant Apparency Theory the hypothesis that herbivores attack plants that are highly visible and common, and the more apparent a plant is to herbivores, the more it must invest in defensive chemicals and structures.

plant stress hypothesis the idea that herbivores prefer to attack stressed plants, which produce leaves that are higher in nitrogen.

plant vigor hypothesis the idea that herbivores prefer to attack fast-growing, vigorous plants rather than slow-growing, stressed plants.

poikilothermic of or pertaining to cold-blooded animals, organisms that have no rapidly operating heat-regulatory mechanism; contrast with *homeothermic*.

polyandry mating of a single female animal with several males.

polyclimax hypothesis the view of Whittaker that there are several different climax vegetation communities in a region governed by many environmental factors.

polygyny mating of one male animal with several females.

pool the amount of nutrient or biomass in a compartment.

population a group of organisms of the same species occupying a particular space at a particular time.

population regulation the general problem of what prevents populations from growing without limit, and what determines the average abundance of a species.

potential evapotranspiration the theoretical depth of water that would evaporate from a standard flat pan over a given time period if water is not limiting, largely dependent on temperature.

precipitation rainfall and snowfall over a specified time period.

predation the action of one organism killing and eating another.

preemptive initial floristics model the first species at a site take over and prevent others from colonizing the site, emphasizing inhibition as the main mechanism of succession.

prey isocline the contour line of densities of predator and prey at which the prey are in equilibrium; the impact of a predator exactly balances the prey's rate of population growth, so the prey population growth rate is zero.

primary production production by green plants.

primary succession succession occurring on a landscape that has no biological legacy.

principle universal statement that we all accept because they are mostly definitions, or are ecological translations of physical–chemical laws. For example, "no population increases without limit" is an important ecological principle that must be correct in view of the finite size of the planet Earth.

probabilistic models in contrast to deterministic models, including an element of probability so that repeated runs of the models do not produce exactly the same outcome.

production amount of energy (or material) formed by an individual, population, or community in a specific time period; includes growth and reproduction only; see *primary production, secondary production, gross production, net production*.

productivity a general term that covers all processes involved in ecological *production* studies—carbon fixation, consumption, rejection, leakage, and respiration.

promiscuity a general term for multiple matings in organisms, called polyandry if multiple males are involved, or polygyny if multiple females; opposite of *monogamy*.

proximate factors the mechanisms responsible for regulating a particular trait in a physiological or biochemical manner; opposite of *ultimate factors*.

push-pull strategies management strategies that manipulate the behavior of insect pests to make the crop resource unattractive (push) and lure the pests toward an attractive source (pull) where the pests are destroyed.

quadrat a sampling frame for stationary organisms; a square, circle, or rectangle of a specified size.

ramet an individual derived by asexual reproduction from a single original zygote, which is able to live independently if separated from the parent organism. Compare with *genet*.

random colonization model succession proceeds completely randomly with no fixed sequence or fixed end point.

Rapoport's Rule the generalization that geographic range sizes decrease as one moves from polar to equatorial latitudes, such that range sizes are smaller in the tropics.

realized niche the observed resource use of a species in the presence of competition and other biotic interactions; contrast with *fundamental niche*.

reciprocal replacement two codominant plants retain their presence in the climax community by A replacing B while B replaces A.

recruitment increment to a natural population, usually from young animals or plants entering the adult population.

Red Queen Hypothesis the coevolution of parasites and their hosts, or predators and their prey, in which improvements in one of the species is countered by evolutionary improvements in the partner species, so that an evolutionary arms race occurs but neither species gains an advantage in the interaction.

Redfield ratio the observed 16:1 atomic ratio of nitrogen to phosphorus found in organisms in the open ocean by A. C. Redfield in 1934—$C_{106}N_{16}P_1$.

regulating factor a factor is defined as potentially regulating if the percentage of mortality caused by the factor increases with population density or if per capita reproductive rate decreases with population density.

Reid's paradox the observed large discrepancy between the rapid rate of movement of trees recolonizing areas at the end of the Ice Age and the observed slow dispersal rate of tree seeds spreading by diffusion.

relative benefit the difference between the costs and benefits (= net benefit).

relative density the density of a population in relation to another, specified in terms of larger/smaller without knowing the absolute density.

relay floristics the classical view of succession as specified in the facilitation model.

repeated reproduction organisms that reproduce several times over their life span.

replacement series an experimental design involving two or more species in competition in which a series of ratios are set out (such as 20:80 or 50:50) and some measure of performance is measured.

reproductive value the contribution an individual female will make to the future population.

residence time the time a nutrient spends in a given compartment of an ecosystem; equivalent to turnover time.

resilience magnitude of disturbance that can be absorbed before an ecosystem changes its structure; one aspect of ecosystem stability.

Resource Availability Hypothesis a theory of plant defense that predicts higher plant growth rates will result in less investment in defensive chemicals and structures.

resource concentration hypothesis the idea that agricultural pests are able to cause serious damage because crops are planted as monocultures at high densities.

respiration complex series of chemical reactions in all organisms by which energy is made available for use; carbon dioxide, water, and energy are the end products.

‿ection the type of natural selection experienced by populations ‿ undergoing rapid popula- ‿se in a relatively empty

safe sites for animals, sites where prey individuals are able to avoid predation; for plants, sites where seeds can germinate and plants can grow.

sampling model one proposed explanation for Hanski's Rule that the observed relationship between distribution and abundance is an artifact of the difficulty of sampling rare species and does not therefore require a biological explanation.

saprophyte plant that obtains food from dead or decaying organic matter.

scientific law universal statement that is deterministic and so well corroborated that everyone accepts it as part of the scientific background of knowledge. There are laws in physics, chemistry, and genetics, but not yet in ecology.

Second Principle of Population Regulation differences between two populations in equilibrium density can be caused by variation in either density-dependent or density-independent per capita birth and death rates.

secondary plant substances chemicals produced by plants that are not directly involved in the primary metabolic pathways and whose main function is to repel herbivores.

secondary production production by herbivores, carnivores, or detritus feeders; contrast with *primary production*.

secondary succession succession occurring on a landscape that has a biological legacy in the form of seeds, roots, and some live plants.

self-regulation process of population regulation in which population increase is prevented by a deterioration in the quality of individuals that make up the population; population regulation by adjustments in behavior and physiology within the population rather than by external forces such as predators.

self-thinning rule the prediction that the regression of organism size versus population density has a slope of −1.5 for plants and animals that have plastic growth rates and variable adult size.

senescence process of aging.

seral referring to a series of stages that follow one another in an ecological succession.

serotinous cones cones of some pine trees that remain on the trees for several years without opening and require a fire to open and release the seeds.

sessile attached to an object or fixed in place (e.g., barnacles).

shade-intolerant plants plants that cannot survive and grow in the shade of another plant, requiring open habitats for survival.

shade-tolerant plants plants that can live and grow in the shade of other plants.

Shelford's law of tolerance the ecological rule first described by Victor Shelford that the geographical distribution of a species will be controlled by that environmental factor for which the organism has the narrowest range of tolerance.

sigmoid curve S-shaped curve; in ecology, often a plot of time (x-axis) against population size (y-axis); an example is the logistic curve.

sink populations local populations in which the rate of production is below replacement level so that extinction is inevitable without a source of immigrants.

small-population paradigm the focus of this approach is on rare species and on the population consequences of rareness, and the abilities of small populations to deal with rarity.

soil drought the lack of water in the soil, less than what is needed for plant survival and growth, caused by a lack of precipitation.

source populations local populations in which the rate of production exceeds replacement so that individuals emigrate to surrounding populations.

specialist predators predators that eat only one or a very few prey species.

specialists species that eat only a few foods or live in only one or two habitats; contrast to *generalists*.

species richness the number of species in a community.

species-area curve a plot of the area of an island or habitat on the x-axis and the number of species in that island or habitat on the *y*-axis, typically done as a log-log plot and typically restricted to one taxonomic group such as plants or reptiles.

stability absence of fluctuations in populations; ability to withstand perturbations without large changes in composition.

stabilizing selection natural selection that favors the norm, the most common or average trait in a population, so the population mean stays constant.

stable age distribution the age distribution reached by a population growing at a constant rate.

stable point an equilibrium in a mathematical model to which the system converges and remains.

stage-based matrix model a type of matrix model not based on organism ages but on life history stages, such as larva, pupa, and adult.

standard error a statistical estimate of the precision of an estimate such as the mean.

static life table a life table constructed at a single point in time by doing a cross section of a population.

statics in population ecology, the study of the reasons of equilibrium conditions or average values; contrast with *dynamics*.

stationary age distribution the age distribution that is reached in a population that is constant in size over time because the birth rate equals the death rate.

steppe extensive area of natural, dry grassland; usually used in reference to grasslands in southwestern Asia and southeastern Europe; equivalent to prairie in North American usage.

sterile-insect technique the release of large numbers of sterilized males to mate with wild females and prevent the fertilization of eggs and production of viable young.

sterol any of a group of solid, mostly unsaturated polycyclic alcohols, such as cholesterol or ergosterol, derived from plants and animals.

stochastic based on probability, as in coin-flipping.

stochastic model mathematical model based on probabilities; the prediction of the model is not a single fixed number but a range of possible numbers; opposite of *deterministic model*.

stock the harvestable part of the population being exploited.

stock-recruit relationship a key graph relating how many recruits come into the exploited population from a given population of adults.

stress a condition occurring in response to adverse external influences and capable of affecting the performance of an organism, for example, in plants in a drought.

sublethal effects any pathogenic effects that reduce the well-being of an individual without causing death.

sublittoral lower division in the sea from a depth of 40 to 60 meters to about 200 meters; below the littoral zone.

succession replacement of one kind of community by another kind; the progressive changes in vegetation and animal life that may culminate in the *climax state*.

supply-side ecology the view that population dynamics are driven by immigration of seeds or juveniles from sources extrinsic to the local population, so there is no local control of recruitment processes.

sustainability the characteristic of a process that can be maintained at a certain level indefinitely, often used in an economic and environmental context. Many definitions have been suggested. The original one of the Bruntland Commission of 1987 defined sustainable development as development that meets the needs of the present without compromising the ability of future generations to meet their own needs.

symbiosis in a broad sense, the living together of two or more organisms of different species; in a narrow sense, synonymous with *mutualism*.

synecology study of groups of organisms in relation to their environment; includes population, community, and ecosystem ecology.

taiga the northern boreal forest zone, a broad band of coniferous forest south of the arctic tundra.

tannins a class of secondary compounds produced by plants (and present in tea and coffee) that reduce the digestibility of plant tissues eaten by herbivores; tannins have been used for centuries to tan animal hides.

tens rule the rule of thumb that 1 species in 10 alien species imported into a country becomes introduced, 1 in 10 of the introduced species becomes established, and 1 in 10 of the established species becomes a pest.

territory any defended area.

theory an integrated and hierarchical set of empirical hypotheses that together explain a significant fraction of scientific observations. The theory of evolution is perhaps the most frequently used theory in ecology.

thermoregulation maintenance or regulation of temperature, specifically the maintenance of a particular temperature of the living body.

theta-logistic model the modification of the original logistic equation to permit curved relationships between population density and the rate of population increase.

tillers ramets, the modular unit of construction, for example, in grasses.

time lags in population models, basing a parameter on past events, such as basing population growth rate on the density of the population last year or the year before.

tolerance model the view that plants in a successional sequence do not interact with one another in either a negative or a positive manner.

top-down model the idea that community organization is set by the effects of carnivores on herbivores and herbivores on plants in the food chain.

total fertility rate number of children a woman could expect to produce in her lifetime if the birth rate were held constant at current conditions.

total response the total losses imposed on a prey species by a combination of the numerical, functional, aggregative, and developmental responses of a predator species.

trace element chemical element used by organisms in minute quantities and essential to their physiology.

trade-offs compromises between two desirable but incompatible activities.

tragedy of the commons the inherent tendency for overexploitation of resources that have free access and unlimited demand, so that it pays the individual to continue harvesting beyond the limits dictated by the common good of sustainability.

transitive competition a linear competitive network in which A wins over B and B wins over C, so that the results of competition reach a final state of competitive exclusion.

treeline the altitude on a mountain above which no trees can survive, equivalent of timberline.

trophic cascade model the idea that a strict top-down model applies to community organization so that impacts flow down the food chain as a series of + and – impacts on successive trophic levels.

trophic efficiency net production at one trophic level as a fraction of net production of the next lower trophic level.

trophic levels classification of organisms based on their source of energy—i.e., primary producers, herbivores, carnivores, and higher carnivores.

tundra treeless area in arctic and alpine regions, varying from a bare area to various types of vegetation consisting of grasses, sedges, forbs, dwarf shrubs, lichens, and mosses.

ultimate factors the evolutionary reason for an adaptation or why a trait is maintained in a population; opposite of *proximate factors*.

umbrella species in conservation biology, species that serve as a proxy for entire communities and ecosystems, so that the entire system is conserved if they are conserved.

unitary organisms organisms appear as individual units with a definite growth form, like most animals.

univoltine refers to an organism that has only one generation per year.

unstable point an equilibrium in a mathematical model from which the system diverges and does not remain.

vector organism organism (often an insect) that transmits a pathogenic virus, bacterium, protozoan, or fungus from one organism to another.

virulence the degree or ability of a pathogenic organism to cause disease; often measured by the host death rate.

wilting point measure of soil water; the water remaining in the soil (expressed as percentage of dry weight of the soil) when the plants are in a state of permanent wilting from water shortage.

xeric deficient in available moisture for the support of life (e.g., desert environments).

xerophyte plant that can grow in dry places (e.g., cactus).

yield amount of usable material taken from a harvested population, measured in numbers or biomass.

zooplankton animal portion of the plankton; the animal community in marine and freshwater environments that floats free in the water, independent of the shore and the bottom, moving passively with the currents.

Introduction to the Science of Ecology

Key Concepts

- Ecology is the scientific study of the interactions that determine the distribution and abundance of organisms.

- Descriptive ecology forms the essential foundation for functional ecology, which asks *how* systems work, and for evolutionary ecology, which asks *why* natural selection has favored this particular solution.

- Ecological problems can be analyzed using a theoretical approach, a laboratory approach, or a field approach.

- Like other scientists, ecologists observe problems, make hypotheses, and test the predictions of each hypothesis by field or laboratory observations.

- Ecological systems are complex, and simple cause–effect relationships are rare.

...ology: *The Experimental Analysis of Distribution and Abundance*, Sixth Edition. Eugene Hecht.
...y Pearson Education, Inc. Published by Pearson Benjamin Cummings. All rights reserved.

experiment Test of a hypothesis. It can be observational (observe the system) or manipulative (perturb the system). The experimental method is the scientific method.

hypothesis Universal proposition that suggests explanations for some observed ecological situation. Ecology abounds with hypotheses.

model Verbal or mathematical statement of a hypothesis.

principle Universal statement that we all accept because they are mostly definitions, or are ecological translations of physical–chemical laws.

scientific law Universal statement that is deterministic and so well corroborated that everyone accepts it as part of the scientific background of knowledge. There are laws in physics, chemistry, and genetics, but not yet in ecology.

theory An integrated and hierarchical set of empirical hypotheses that together explain a significant fraction of scientific observations. The theory of evolution is perhaps the most frequently used theory in ecology.

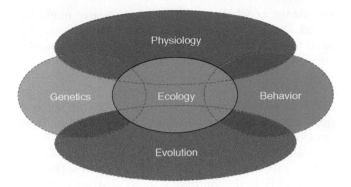

Figure 1 The four biological disciplines closely related to ecology.

Introduction to the Science of Ecology

You are embarking on a study of ecology, the most integrative discipline in the biological sciences. The purpose of this chapter is to get you started by defining the subject, providing a small amount of background history, and introducing the broad concepts that will serve as a road map for the details to come.

Definition of Ecology

The word *ecology* came into use in the second half of the nineteenth century. Ernst Haeckel in 1869 defined ecology as the total relations of the animal to both its organic and its inorganic environment. This very broad definition has provoked some authors to point out that if this is ecology, there is very little that is *not* ecology. Four biological disciplines are closely related to ecology—genetics, evolution, physiology, and behavior (**Figure 1**). Broadly interpreted, ecology overlaps each of these four subjects; hence, we need a more restrictive definition.

Charles Elton in his pioneering book *Animal Ecology* (1927) defined ecology as scientific natural history. Although this definition points out the origin of many of our ecological problems, it is again uncomfortably vague. In 1963 Eugene Odum defined ecology as the study of the structure and function of nature. This statement emphasizes the form-and-function idea that permeates biology, but it is still not a completely clear definition. A clear but restrictive definition of ecology is this: Ecology is the scientific study of the distribution and abundance of organisms (Andrewartha 1961). This definition is static and leaves out the important idea of relationships. Because ecology is about relationships, we can modify Andrewartha's definition to make a precise definition of **ecology**: *Ecology is the scientific study of the interactions that determine the distribution and abundance of organisms.*

This definition of ecology appropriately constrains the scope of our quest, and is the meaning that will be adopted in this chapter. To better understand what ecology is, we need to know what is special about scientific studies, and what is meant by distribution and abundance. **Distribution**—where organisms are found—and **abundance**—how many organisms are found in a given area—are key facts that must be determined before we can address the most difficult question: *Why* this particular distribution, *why* this abundance? We seek the cause-and-effect relationships that govern distribution and abundance.

History of Ecology

The historical roots of ecology are varied, and in this section we will explore briefly some of the origins of ecological ideas. We are not the first humans to think about ecological problems. The roots of ecology lie in natural history. Primitive tribes, for example—who depended on hunting, fishing, and food gathering—needed detailed knowledge of where and when their quarry might be found. The establishment of agriculture also increased the need to learn about the ecology of plants and domestic animals. Agriculture today is a special form of applied ecology.

Outbreaks of pests such as locusts in the Middle East and North Africa or rats in rice crops in Asia are not new problems in agriculture. Spectacular plagues of animals attracted the attention of the earliest writers. The Egyptians and Babylonians feared locust plagues (**Figure 2**), often attributing them to supernatural powers (Exodus 7:14–12:30). In the fourth century B.C., Aristotle tried to explain plagues of field mice and locusts in *Historia Animalium*. He pointed out that the high reproductive rate of field mice could produce more mice than could be reduced by their natural predators, such as foxes and ferrets, or by the control efforts of humans. Nothing succeeded in reducing these mouse plagues, Aristotle stated, except the rain, and after heavy rains the mice disappeared rapidly. And even today, Australian wheat farmers face plagues of house mice, and ask the same question: How can we get rid of these pests?

Pests are a problem for people because they violate our feeling of harmony or balance in the environment. Ecological harmony was a guiding principle basic to the Greeks' understanding of nature. The historian Frank Egerton (1968a) has traced this concept from ancient times to the modern term *balance of nature*. The concept of *providential ecology*, in which nature is designed to benefit and preserve each species, was implicit in the writings of Herodotus and Plato. A major assumption of this concept was that the number of every species remained essentially constant. Outbreaks of some populations were acknowledged, but were usually attributed to divine punishment. And since each species had a special place in nature, extinction could not occur because it would disrupt the balance and harmony in nature.

How did we get from these early Greek and Roman ideas about harmony to our modern understanding? A combination of mathematics and natural history paved the way. By the seventeenth century students of natural history and human ecology began to focus on population ecology and to construct a quantitative framework. Graunt, who in 1662 described human population change in quantitative terms, can be called the "father of demography"[1] (Cole 1958). He recognized the importance of measuring birth rates, death rates, and age structure of human populations, and he complained about the inadequate census data available in England in the seventeenth century. Graunt estimated the potential rate of population growth for London, and concluded that even without immigration, London's population would double in 64 years.

Today, human population growth is an increasing concern, but population growth was not always measured quantitatively for animals and plants. Leeuwenhoek made one of the first attempts to calculate theoretical rates of increase for an animal species (Egerton 1968b). He studied the reproductive rate of grain beetles, carrion flies, and human lice, counting the number of eggs laid by female carrion flies and calculating that one pair of flies could produce 746,496 flies in three months.

By the eighteenth century, natural history had become an important cultural occupation. Buffon, who authored *Natural History* (1756), touched on many of our modern ecological problems and recognized that populations of humans, other animals, and plants are subjected to the same processes. Buffon discussed, for example, how the great fertility of every species was counterbalanced by innumerable agents of destruction. He believed that plague populations of field mice were checked partly by diseases and scarcity of food. Buffon did not accept Aristotle's idea that heavy rains caused the decline of dense mouse populations, but thought instead that control was achieved by biological agents. Rabbits, he stated, would reduce the countryside to a desert if it were not for their predators. If the Australians had listened to Buffon before they introduced rabbits to their environment in 1859, they could have saved their rangelands from destruction (**Figure 3**). Buffon in 1756 was dealing with problems of population regulation that are still unsolved today.

Malthus, the most famous of the early demographers, published one of the earliest controversial books on demography, *Essay on Population* (1798). He calculated that although the number of organisms can increase geometrically (1, 2, 4, 8, 16, . . .), food supply can

Figure 2 A young girl looks at a dense swarm of the desert locust in North Africa.

---

[1]Demography originated as the study of human population growth and decline. It is now used as a more general term that includes plant and animal population changes.

Figure 3 European rabbit overpopulation in eastern Australia. Rabbits were introduced to Australia in 1859 and have become a serious pest because of their abundance. Their burrowing increased soil erosion, and they competed with sheep and cattle for forage.

Table 1 Total fertility rate of human populations and gross national income per person in selected countries of the globe in 2007.

Country	Total fertility rate	Gross national income per person
Sudan	4.5	2160
Gambia	5.1	1970
Niger	7.1	830
Tanzania	5.4	740
Botswana	3.1	12,240
South Africa	2.7	11,710
Canada	1.5	34,610
United States	2.1	44,260
Costa Rica	1.9	10,770
Mexico	2.4	11,330
Haiti	4.0	1490
Brazil	2.3	8800
Peru	2.5	6070
Turkey	2.2	9060
India	2.9	3800
Pakistan	4.1	2500
Indonesia	2.4	3950
China	1.6	7730
Japan	1.3	33,730
Sweden	1.9	34,780
Switzerland	1.4	40,630
Russia	1.3	11,620
Italy	1.4	29,840
Solomon Islands	4.5	2170

The total fertility rate is the average number of children a woman would have, assuming no change in birth rates. The gross national income (GNI) is in U.S. dollars per person. (Data from 2007 World Population Data Sheet.)

never increase faster than arithmetically (1, 2, 3, 4,. . .). The arithmetic rate of increase in food production seems to be somewhat arbitrary. The great disproportion between these two powers of increase led Malthus to infer that reproduction must eventually be checked by food production. What prevents populations from reaching the point at which they deplete their food supply? What checks operate against the tendency toward a geometric rate of increase? Two centuries later we still ask these questions. These ideas were not new; Machiavelli had said much the same thing around 1525, as did Buffon in 1751, and several others had anticipated Malthus. It was Malthus, however, who brought these ideas to general attention. Darwin used the reasoning of Malthus as one of the bases for his theory of natural selection. The struggle for existence results from the high reproductive output of species.

Other workers questioned the ideas of Malthus and made different predictions for human populations. For example, in 1841 Doubleday put forward the True Law of Population. He believed that whenever a species was threatened, nature made a corresponding effort to preserve it by increasing the fertility of its members. Human populations that were undernourished had the highest fertility; those that were well fed had the lowest fertility. You can make the same observations by looking around the world today (**Table 1**). Doubleday explained these effects by the oversupply of mineral nutrients in well-fed populations. Doubleday observed a basic fact that we recognize today: low birth rates occur in wealthy countries—although his explanations were completely wrong.

Interest in the mathematical aspects of demography increased after Malthus. Can we describe a mathematical law of population growth? Quetelet, a Belgian statistician, suggested in 1835 that the growth of a population was checked by factors opposing population growth. In 1838 his student Pierre-François Verhulst derived an equation describing the initial rapid growth and eventual leveling off of a population over time. This S-shaped curve he called the logistic curve. His work was overlooked until modern times, but it is fundamentally important, and we will return to it later in detail.

Until the nineteenth century, philosophical thinking had not changed from the idea of Plato's day that there was harmony in nature. Providential design was still the guiding light. In the late eighteenth and early nineteenth centuries, two ideas that undermined the idea of the balance of nature gradually gained support: (1) that many species had become extinct and (2) that resources are limited and competition caused by population pressure is important in nature. The consequences of these two ideas became clear with the work of Malthus, Lyell, Spencer, and Darwin in the nineteenth century. Providential ecology and the balance of nature were replaced by natural selection and the struggle for existence (Egerton 1968c).

The balance of nature idea, redefined after Darwin, has continued to persist in modern ecology (Pimm 1991). The idea that natural systems are stable and in equilibrium with their environments unless humans disturb them is still accepted by many ecologists and theoreticians.

Humans must eat, and many of the early developments in ecology came from the applied fields of agriculture and fisheries. Insect pests of crops have been one focus of work. Before the advent of modern chemistry, biological control was the only feasible approach. In 1762 the mynah bird was introduced from India to the island of Mauritius to control the red locust; by 1770 the locust threat was a negligible problem (Moutia and Mamet 1946). Forskål wrote in 1775 about the introduction of predatory ants from nearby mountains into date-palm orchards to control other species of ants feeding on the palms in southwestern Arabia. In subsequent years, an increasing knowledge of insect parasitism and predation led to many such introductions all over the world in the hope of controlling nonnative and native agricultural pests (De Bach 1974).

Medical work on infectious diseases such as malaria in the late 1800s gave rise to the study of epidemiology and interest in the spread of disease through a population. Malaria is still one of the great scourges of humans. In 1900 no one even knew the cause of the disease. Once mosquitoes were pinpointed as the vectors, medical workers realized that it was necessary to know in detail the ecology of mosquitoes. The pioneering work of Robert Ross (1911) attempted to describe in mathematical terms the propagation of malaria, which is transmitted by mosquitoes. In an infected area, the propagation of malaria is determined by two continuous and simultaneous processes: (1) The number of new infections among people depends on the number and infectivity of mosquitoes, and (2) the infectivity of mosquitoes depends on the number of people in the locality and the frequency of malaria among them. Ross could write these two processes as two simultaneous differential equations:

$$\begin{pmatrix} \text{Rate of increase of} \\ \text{infected humans} \end{pmatrix} = \begin{pmatrix} \text{New infections} \\ \text{per unit time} \end{pmatrix} - \begin{pmatrix} \text{Recoveries per} \\ \text{unit time} \end{pmatrix}$$
$$\downarrow$$
(Depends on number of infected mosquitoes)

$$\begin{pmatrix} \text{Rate of increase of} \\ \text{infected mosquitoes} \end{pmatrix} = \begin{pmatrix} \text{New infections} \\ \text{per unit time} \end{pmatrix} - \begin{pmatrix} \text{Death of infected} \\ \text{per unit time} \end{pmatrix}$$
$$\downarrow$$
(Depends on number of infected humans)

Ross had described an ecological process with a mathematical model, and his work represents a pioneering parasite–host model of species interactions. Such models can help us to clarify the problem—we can analyze the components of the model—and predict the spread of malaria or other diseases.

Production ecology, the study of the harvestable yields of plants and animals, had its beginnings in agriculture, and Egerton (1969) traced this back to the eighteenth-century botanist Richard Bradley. Bradley recognized the fundamental similarities of animal and plant production, and he proposed methods of maximizing agricultural yields (and hence profits) for wine grapes, trees, poultry, rabbits, and fish. The conceptual framework that Bradley used—monetary investment versus profit—is now called the "optimum-yield problem" and is a central issue in applied ecology.

Individual species do not exist in a vacuum, but instead in a matrix of other species with which they interact. Recognition of communities of living organisms in nature is very old, but specific recognition of the interrelations of the organisms in a community is relatively recent. Edward Forbes in 1844 described the distribution of animals in British coastal waters and part of the Mediterranean Sea, and he wrote of zones of differing depths that were distinguished by the associations of species they contained. Forbes noted that some species are found only in one zone, and that other species have a maximum of development in one zone but occur sparsely in other adjacent zones. Mingled in are stragglers that do not fit the zonation pattern. Forbes recognized the dynamic aspect of the interrelations between these organisms and their environment. As the environment changed, one species might die out, and another might increase in abundance. Karl Möbius expressed similar ideas in 1877 in a classic essay on the oyster-bed community as a unified collection of species.

Studies of communities were greatly influenced by the Danish botanist J. E. B. Warming (1895, 1909), one of the fathers of plant ecology. Warming was the first plant ecologist to ask questions about the composition of plant communities and the associations of species that made up these communities. The dynamics of vegetation change was emphasized first by North American plant ecologists. In 1899 H. C. Cowles described plant succession on the sand dunes at the southern end of

Lake Michigan. The development of vegetation was analyzed by the American ecologist Frederick Clements (1916) in a classic book that began a long controversy about the nature of the community.

With the recognition of the broad problems of populations and communities, ecology was by 1900 on the road to becoming a science. Its roots lay in natural history, human demography, biometry (statistical approach), and applied problems of agriculture and medicine.

The development of ecology during the twentieth century followed the lines developed by naturalists during the nineteenth century. The struggle to understand how nature works has been carried on by a collection of colorful characters quite unlike the mythical stereotypes of scientists. From Alfred Lotka, who worked for the Metropolitan Life Insurance Company in New York while laying the groundwork of mathematical ecology (Kingsland 1995), to Charles Elton, the British ecologist who wrote the first animal ecology textbook in 1927 and founded the Bureau of Animal Population at Oxford (Crowcroft 1991), ecology has blossomed with an increasing understanding of our world and how we humans affect its ecological systems (McIntosh 1985).

Until the 1970s ecology was not considered by society to be an important science. The continuing increase of the human population and the associated destruction of natural environments with pesticides and pollutants awakened the public to the world of ecology. Much of this recent interest centers on the human environment and human ecology, and is called environmentalism. Unfortunately, the word *ecology* became identified in the public mind with the much narrower problems of the human environment, and came to mean everything and anything about the environment, especially human impact on the environment and its social ramifications. It is important to distinguish ecology from environmental studies.

Ecology is focused on the natural world of animals and plants, and includes humans as a very significant species by virtue of its impact. **Environmental studies** is the analysis of human impact on the environment of the Earth—physical, chemical, and biological. Environmental studies as a discipline is much broader than ecology because it deals with many natural sciences—including ecology, geology, and climatology—as well as with social sciences, such as sociology, economics, anthropology, political science, and philosophy. The science of ecology is not solely concerned with human impact on the environment but with the interrelations of all plants and animals. As such, ecology has much to contribute to some of the broad questions about humans and their environment that are an important scientific component of environmental studies.

Environmental studies have led to "environmentalism" and "deep ecology," social movements with an important agenda for political and social change intended to minimize human impact on the Earth. These social and political movements are indeed important and are supported by many ecologists, but they are not the science of ecology. Ecology should be to environmental science as physics is to engineering. Just as we humans are constrained by the laws of physics when we build airplanes and bridges, so also are we constrained by the principles of ecology when altering the environment.

Ecological research can shed light on what will happen when global temperatures increase as a result of increasing CO_2 emissions, but it will not tell us what we *ought* to do about these emissions, or whether increased global temperature is good or bad. Ecological scientists are not policy makers or moral authorities, and should not as scientists make ethical or political recommendations. However, on a personal level, most ecologists are concerned about the extinction of species and would like to prevent extinctions. Many ecologists work hard in the political arena to achieve the social goals of environmentalism.

Basic Problems and Approaches to Ecology

We can approach the study of ecology from three points of view: descriptive, functional, or evolutionary. The descriptive point of view is mainly natural history and describes the vegetation groups of the world—such as the temperate deciduous forests, tropical rain forests, grasslands, and tundra—and the animals and plants and their interactions within each of these ecosystems. The descriptive approach is the foundation of all of ecological science, and while much of the world has been reasonably described in terms of its vegetation and animal life, some areas are still poorly studied and poorly described. The functional point of view, on the other hand, is oriented more toward dynamics and relationships, and seeks to identify and analyze general problems common to most or all of the different ecosystems. Functional studies deal with populations and communities as they exist and can be measured now. Functional ecology studies **proximate causes**—the dynamic responses of populations and communities to immediate factors of the environment. Evolutionary ecology studies **ultimate causes**—the historical reasons why natural selection has favored the particular adaptations we now see. The evolutionary point of view considers organisms and relationships between organisms as historical products of evolution. Functional ecologists ask *how*: How does the system operate? Evolutionary ecologists ask *why*: Why does natural selection favor this particular ecological solution? Since evolution not only has occurred in the past but is also going on in the present, the evolutionary ecologist must work closely with the functional ecologist to understand ecological systems (Pianka 1994). Because the environment of an organism contains all the selective forces

Science and Values in Ecology

Science is thought by many people to be value free, but this is certainly not the case. Values are woven all through the tapestry of science. All applied science is done because of value judgments. Medical research is a good example of basic research applied to human health that virtually everyone supports. Weapons research is carried out because countries wish to be able to defend themselves against military aggression.

In ecology the strongest discussions about values have involved conservation biology. Should conservation biologists be objective scientists studying biodiversity, or should they be public advocates for preserving biodiversity? The preservation of biodiversity is a value that often conflicts with other values—for example, clear-cut logging that produces jobs and wood products. The pages of the journal *Conservation Biology* are peppered with this discussion about advocacy (see, for example, *Conservation*

> *There will always be a healthy tension between scientific knowledge and public policy in environmental matters . . .*

Biology February 2007 issue, Brussard and Tull 2007, Scott et al. 2007).

Scientists in fact have a dual role. First, they carry out objective science that both obtains data and tests hypotheses about ecological systems. They can also be advocates for particular policies that attempt to change society, such as the use of electric cars to reduce air pollution. But it is crucial to separate these two kinds of activities.

Science is a way of knowing, a method for determining the principles by which systems like ecological systems operate. The key scientific virtues are honesty and objectivity in the search for truth. Scientists assume that once we know these scientific principles we can devise effective policies to achieve social goals. All members of society collectively decide on what social goals we will pursue, and civic responsibility is part of the job of everyone, scientists included. There will always be a healthy tension between scientific knowledge and public policy in environmental matters because there are always several ways of reaching a particular policy goal. The debates over public policy in research funding and environmental matters will continue, so please join in.

that shape its evolution, ecology and evolution are two viewpoints of the same reality.

All three approaches to ecology have their strengths, but the important point is that we need all three to produce good science. The descriptive approach is absolutely fundamental because unless we have a good description of nature, we cannot construct good theories or good explanations. The descriptive approach provides us maps of geographical distributions and estimates of relative abundances of different species. With the functional approach, we need the detailed biological knowledge that natural history brings if we are to discover how ecological systems operate. The evolutionary approach needs good natural history and good functional ecology to speculate about past events and to suggest hypotheses that can be tested in the real world. No single approach can encompass all ecological questions. This chapter uses a mixture of all three approaches and emphasizes the general problems ecologists try to understand.

The basic problem of ecology is to determine the causes of the distribution and abundance of organisms. Every organism lives in a matrix of space and time. Consequently, the concepts of distribution and abundance are closely related, although at first glance they may seem

quite distinct. What we observe for many species is that the numbers of individuals in an area vary in space, so if we make a contour map of a species' geographical distribution, we might get something similar to **Figure 4**.

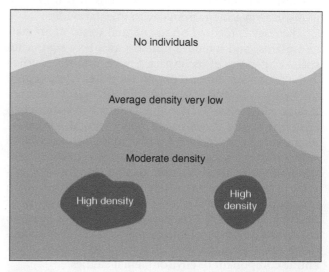

Figure 4 Schematic contour map of the abundance of a plant or animal species.

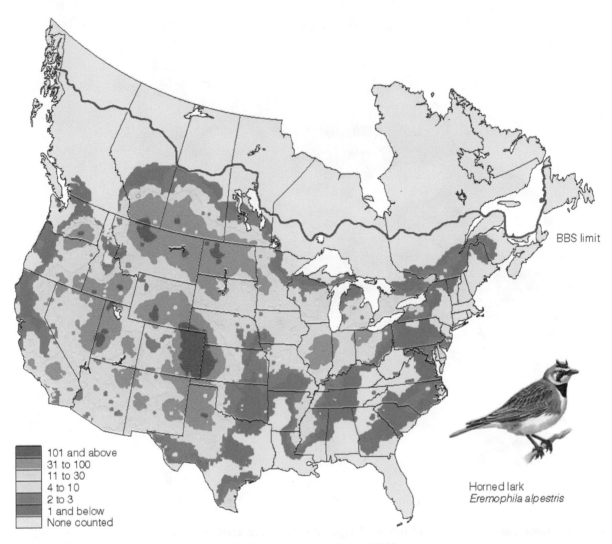

101 and above
31 to 100
11 to 30
4 to 10
2 to 3
1 and below
None counted

BBS limit

Horned lark
Eremophila alpestris

Figure 5 Abundance of the horned lark in North America from 1994 to 2003. Data are from the Breeding Bird Survey (BBS). Maximal abundance of this bird is reached in the short grass prairie of western Kansas and Nebraska and eastern Colorado. (From Sauer et al. 2005.)

Figure 5 illustrates this idea for the horned lark of North America. Horned larks are most common in the prairies of eastern Colorado and in western Kansas and Nebraska, and are absent altogether in Florida. Why should these patterns of abundance occur? Why does abundance decline as one approaches the edge of a species' geographic range? What limits the eastern and northern extension of the horned lark's range? These are examples of the fundamental questions an ecologist must ask of nature.

Similarly, the red kangaroo occurs throughout the arid zone of Australia (**Figure 6**). It is absent from the tropical areas of northern Australia and most common in western New South Wales and central Queensland. Why

are there no red kangaroos in tropical Australia? Why is this species absent from Victoria in southern Australia and from Tasmania? We can view the average density of any species as a contour map, with the provision that the contour map may change with time. Throughout the area of distribution, the abundance of an organism must be greater than zero, and the limit of distribution equals the contour of zero abundance. Distribution may be considered a facet of abundance, and distribution and abundance may be said to be reverse sides of the same coin (Andrewartha and Birch 1954). The factors that affect the distribution of a species may also affect its abundance.

The problems of distribution and abundance can be analyzed at the level of the population of a single

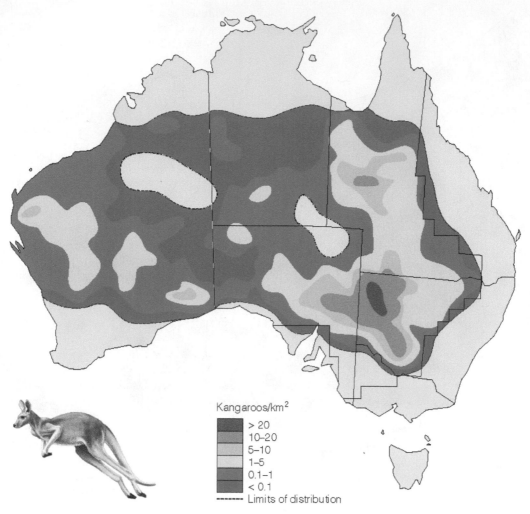

Figure 6 Distribution and abundance of the red kangaroo in Australia. Data from aerial surveys, 1980–1982. (From Caughley et al. 1987.)

species or at the level of a community, which contains many species. The complexity of the analysis may increase as more and more species are considered in a community; consequently, we will first consider the simpler problems involving single-species populations.

Considerable overlap exists between ecology and its related disciplines. Environmental physiology has developed a wealth of information that is needed to analyze problems of distribution and abundance. Population genetics and ecological genetics are two additional foci of interest that we touch on only peripherally. Behavioral ecology is another interdisciplinary area that has implications for the study of distribution and abundance. Evolutionary ecology is an important focus for problems of adaptation and studies of natural selection in populations. Each of these disciplines can become an area of study entirely on its own.

Levels of Integration

In ecology we are dealing primarily with the five starred (*) levels of integration, as shown in **Figure 7**. At one end of the spectrum, ecology overlaps with environmental physiology and behavioral studies of individual organisms, and at the other end, ecology merges into meteorology, geology, and geochemistry as we consider landscapes. Landscapes can be aggregated to include the whole-Earth ecosystem, which is called the **ecosphere** or the **biosphere**. The important message is that the boundaries of the sciences are not sharp but diffuse, and nature does not come in discrete packages.

Each level of integration involves a separate and distinct series of attributes and problems. For example, a population has a density (e.g., number of deer per square kilometer), a property that cannot be attributed

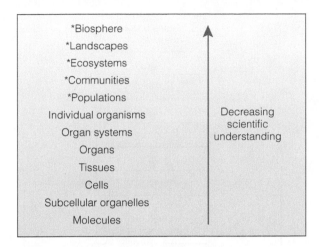

Figure 7 **Levels of integration studied in biology.**

to an individual organism. A community has biodiversity (or species richness), an attribute without meaning at the population level. In general, a scientist dealing with a particular level of integration seeks explanatory mechanisms from lower levels of integration and biological significance from higher levels. For example, to understand mechanisms of changes in a population, an ecologist might study mechanisms that operate on the behavior and physiology of individual organisms, and might try to view the significance of these population events within a community and ecosystem framework.

Much of modern biology is highly reductionistic, as it attempts to work out the physical–chemical basis of life. A good example is the Human Genome Project, an expensive and highly targeted research program to sequence all the genes on human chromosomes. The Human Genome Project is now completed, yet we do not know how many species of beetles live on the Earth, or how many species of trees there are in the Amazon basin. It should not surprise you that the amount of scientific understanding varies with the level of integration. We know an enormous amount about the molecular and cellular levels of organisms, organs and organ systems, and whole organisms, but we know relatively little about populations and even less about communities and ecosystems. This point is illustrated by looking at the levels of integration: Ecology constitutes more than one-third of the levels of biology, but no biology curriculum can be one-third ecology and do justice to current biological knowledge. The reasons for this are not hard to find; they include the increasing complexity of these higher levels and the difficulties involved in dealing with them in the laboratory.

This decrease in understanding at the higher levels has serious implications. You will not find in ecology the strong theoretical framework that you find in physics, chemistry, molecular biology, or genetics. It is

not always easy to see where the pieces fit in ecology, and we will encounter many isolated parts of ecology that are well developed theoretically but are not clearly connected to anything else. This is typical of a young science. Many students unfortunately think of science as a monumental pile of facts that must be memorized. But science is more than a pile of precise facts; it is a search for systematic relations, for explanations to problems in the physical world, and for unifying concepts. This is the growing end of science, so evident in a young science like ecology. It involves many unanswered questions and much more controversy.

The theoretical framework of ecology may be weaker than we would like at the present time, but this must not be interpreted as a terminal condition. Chemistry in the eighteenth century was perhaps in a comparable state of theoretical development as ecology at the present time. Sciences are not static, and ecology is in a strong growth phase.

Methods of Approach to Ecology

Ecology has been approached on three broad fronts: the theoretical, the laboratory, and the field. These three approaches are interrelated, but some problems have arisen when the results of one approach fail to verify those of another. For example, theoretical predictions may not be borne out by field data. We are primarily interested in understanding the distribution and abundance of organisms in nature—that is, in the field. Consequently, the descriptive ecology of populations, communities, and ecosystems will always be our basis for comparison, our basic standard.

Plant and animal ecology have tended to develop along separate paths. Historically, plant ecology got off to a faster start than animal ecology, despite the early interest in human demography. Because animals are highly dependent on plants, many of the concepts of animal ecology are patterned on those of plant ecology. Succession is one example. Also, since plants are the source of energy for many animals, to understand animal ecology we must also know a good deal of plant ecology. This is illustrated particularly well in the study of community relationships.

Some important differences, however, separate plant and animal ecology. First, because animals tend to be highly mobile whereas plants are stationary, a whole series of new techniques and ideas must be applied to animals—for example, to determine population density. Second, animals fulfill a greater variety of functional roles in nature—some are herbivores, some are carnivores, some are parasites. This distinction is not complete because there are carnivorous plants and parasitic plants, but the possible interactions are on average more numerous for animals than for plants.

During the 1960s population ecology was stimulated by the experimental field approach in which natural populations were manipulated to test specific predictions arising from controversial ecological theory. During these years ecology was transformed from a static, descriptive science to a dynamic, experimental one in which theoretical predictions and field experiments were linked. At the same time, ecologists realized that populations were only parts of larger ecosystems, and that we needed to study communities and ecosystems in the same experimental way as populations. To study a complex ecosystem, teams of ecologists had to be organized and integrated, which was first attempted during the late 1960s and the 1970s.

Modern ecology is advancing particularly strongly in three major areas. First, communities and ecosystems are being studied with experimental techniques and analyzed as systems of interacting species that process nutrients and energy. Insights into ecosystems have been provided by the comparative studies of communities on different continents. Second, modern evolutionary thinking is being combined with ecological studies to provide an explanation of how evolution by natural selection has molded the ecological patterns we observe today. Behavioral ecology is a particularly strong and expanding area combining evolutionary insights with the ecology of individual animals. Third, conservation biology is becoming a dominant theme in scientific and political arenas, and this has increased the need for ecological input in habitat management. All of these developments are providing excitement for students of ecology in this century.

Application of the Scientific Method to Ecology

The essential features of the scientific method are the same in ecology as in other sciences (**Figure 8**). An ecologist begins with a problem, often based on natural history observations. For example, pine tree seedlings do not occur in mature hardwood forests on the Piedmont of North Carolina. If the problem is not based on correct observations, all subsequent stages will be useless; thus, accurate natural history is a prerequisite for all ecological studies. Given a problem, an ecologist suggests a possible answer, which is called a **hypothesis**—a statement of cause and effect. In many cases, several answers might be possible, and several different hypotheses can be proposed to explain the observations. Hypotheses arise from previous research, intuition, or inspiration. The origin of a hypothesis tells us nothing about its likelihood of being correct.

A hypothesis makes predictions, and the more precise predictions it makes the better. Predictions follow logically from the hypothesis, and mathematical reason-

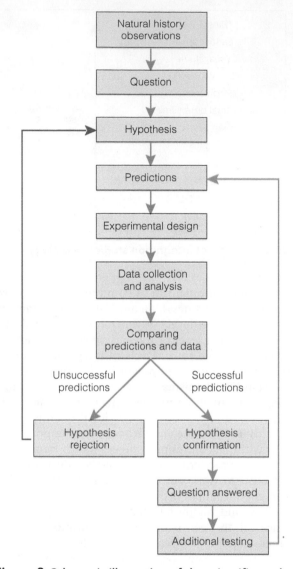

Figure 8 Schematic illustration of the scientific method as applied to ecological questions.

ing is the most useful way to check on the logic of predictions. An example of a hypothesis is that pines do not grow under hardwoods because of a shortage of light. Alternative hypotheses might be that the cause is a shortage of pine seeds, or a shortage of soil water. Predictions from simple hypotheses like these are often straightforward: If you provide more light, pine seedlings will grow (under the light hypothesis). A hypothesis is tested by making observations to check the predictions—an experiment. An **experiment** is defined as any set of observations that test a hypothesis. Experiments can be manipulative or natural. We could provide light artificially under the mature forest canopy, or we could look for natural gaps in the forest canopy. The protocol for the experiments and the data to be obtained are called

On Ecological Truth

We wish our scientists to speak the truth, and when politicians bend the truth they lose credibility. What is truth, and what in particular is the hallmark of ecological truth? The notion of truth is a profound one that philosophers discuss in detail and scientists just assume is simple.

Truth consists of correspondence with the facts. If we say that there are 23 elephants in a particular herd in the Serengeti, we are stating an ecological truth because we assume that if another person counted the elephants, he or she would get the same number. These kinds of facts are relatively simple, and scientists rarely get into arguments about them. Where arguments start is in the inferences that are drawn from whole sets of facts. For example, if we had counts of the same elephant herd over 20 years, and numbers were continually falling, we could say that this elephant population is declining in size. This statement is also an ecological truth if we have done our counting well and recorded all the data correctly.

But now suppose we wish to state that the elephant population is declining and that a disease is the cause of this decline. Is this statement an ecological truth? It is better to consider it an ecological hypothesis and to outline the predictions it makes about what we will find if we search for a disease organism in elephants dying in this particular area. We now enter a gray zone in which ecological truth is approximately equivalent to a supported hypothesis, one in which we checked the predictions and found them to be correct. But if a scientist wished to extend this argument to state that elephant populations all

over east Africa are collapsing because of this disease, this is a more general hypothesis, and before we can consider it an ecological truth we would need to test its predictions by studying many more populations of elephants and their diseases. Many of our ecological ideas are in this incomplete stage because we lack the time, money, or personnel to gather the data to decide whether the general hypothesis is correct. So ecologists, like other scientists, must then face the key question of how to deal with uncertainty when we do not know if we have an ecological truth or not.

The central idea of this principle is to do no harm to the environment, to take no action that is not reversible, and to avoid risk.

The key resolution to this dilemma for environmental management has been the precautionary principle: "Look before you leap," or "An ounce of prevention is worth a pound of cure." The precautionary principle is the ecological equivalent of part of the Hippocratic Oath in medicine: "Physician, do no harm." The central idea of this principle is to do no harm to the environment, to take no action that is not reversible, and to avoid risk. Ecological truth is never obvious in complex environmental issues and emerges more slowly than we might like, so we cannot wait for truth or certainty before deciding what to do about emerging problems in the environment, whether they concern declining elephant populations or introduced pest species.

---

the experimental design. Using the data that result from the experiments, we either accept or reject the hypothesis. And so the cycle begins again (Figure 8).

Many qualifications need to be attached to this simple scheme. Popper (1963) pointed out that we should always look for evidence that falsifies a hypothesis, and that progress in science consists of getting rid of incorrect ideas. In practice, we cannot achieve this ideal. We should also prefer simple hypotheses over complex ones, according to Popper, because we can reject simple hypotheses more quickly. This does not mean that we must be simpleminded. On the contrary, in ecology we must deal with complex hypotheses because the natural world is not simple. Every hypothesis must predict something and forbid other things from happening. The predictions of a hypothesis must say exactly what it allows and what it forbids. If a hypothesis predicts everything and forbids nothing, it is quite useless in sci-

ence. The light hypothesis for pine seedlings both predicts more seedlings if you add more light and forbids more seedlings if you add more water.

Ecological systems are complex, and this causes difficulty in applying the simple method outlined in Figure 8. In some cases factors operate together, so it may not be a situation of light *or* water for pine seedlings but one of light *and* water. Systems in which many factors operate together are most difficult to analyze, and ecologists must be alert for their presence (Quinn and Dunham 1983). The principle, however, remains—no matter how complex the hypothesis, it must make some predictions that we can check in the physical world.

All ecological systems have an evolutionary history, and this provides another fertile source of possible explanations. There is controversy in ecology about whether one needs to invoke evolutionary history to explain present-day population and community dynamics.

Evolutionary hypotheses can be tested as Darwin did, by comparative methods but not by manipulative experiments (Diamond 1986).

Ecological hypotheses may be statistical in nature, but they do not fall into the "either A or B" category of hypotheses. Statistical hypotheses postulate quantitative relationships. For example, in North Carolina forests, pine seedling abundance (per m^2) is linearly related to incident light in summer. Tests of statistical hypotheses are well understood and are discussed in all statistics textbooks. They are tested in the same way indicated in Figure 8.

Some ecological hypotheses have been very fruitful in stimulating work, even though they are known to be incorrect. The progress of ecology, and of science in general, occurs in many ways, using mathematical models, laboratory experiments, and field studies.

Review Questions and Problems

1   Discuss the connotation of the words *ecologist* and *environmentalist*. Would you like to be labeled either of these names? Where in a public ranking of preferred professions would these two fall?

2   Look up the definition of *environment* in several standard dictionaries and in the *Oxford Dictionary of Ecology* (2006), and compare them. Is it possible to measure the environment of an individual? Are other individuals part of the environment of an individual?

3   Is it necessary to define a scientific subject before one can begin to discuss it? Contrast the introduction to several ecology textbooks with those of some areas of physics and chemistry, as well as other biological areas such as genetics and physiology.

4   A plant ecologist proposed the following hypothesis to explain the absence of trees from a grassland area: Periodic fires may prevent tree seedlings from becoming established in grassland. Is this a suitable hypothesis? How could you improve it?

5   Is it necessary to study the scientific method and the philosophy of science in order to understand how science works? Consider this question before and after reading the essays by Popper (1963) and Platt (1964).

6   Discuss the application of the distribution and abundance model to microbes and viruses.

7   Quinn and Dunham (1983) argue that the conventional methods of science cannot be applied to ecological questions because there is not just one cause; one effect and many factors act together to produce ecological changes. Discuss the problem of "multiple causes" and how scientists can deal with complex systems that have multiple causes.

8   A wildlife ecologist interested in protecting large mammals by means of wolf control analyzed data from six sites at which wolves had been removed for five consecutive years. On three of the sites, the prey species (moose and caribou) had increased, and on three of the sites prey populations did not change. How would you interpret these data in light of Figure 8?

9   Plot the data in Table 1 graphically, with gross national product (*x*-axis) versus total fertility rate (*y*-axis). How tight is the relationship between these two variables? Discuss the reasons for the overall form of this relationship, and the reasons why there might be variation or spread in the data.

Overview Question

Does ecology progress as rapidly as physics? How can we measure progress in the sciences, and what might limit the rate of progress in different sciences? Will there be an "end to science"?

Suggested Readings

- Dayton, P. K. 2003. The importance of the natural sciences to conservation. *American Naturalist* 162:1–13.

- Egerton, F. N., III. 1973. Changing concepts of the balance of nature. *Quarterly Review of Biology* 48:322–350.

- Kingsland, S. E. 2004. Conveying the intellectual challenge of ecology: An historical perspective. *Frontiers in Ecology and the Environment* 2:367–374.

- Kingsland, S. E. 2005. *The Evolution of American Ecology, 1890–2000.* Baltimore: Johns Hopkins University Press.

- Krebs, C. J. 2006. Ecology after 100 years: Progress and pseudo-progress. *New Zealand Journal of Ecology* 30:3–11.

- Ludwig, D., M. Mangel, and B. Haddad. 2001. Ecology, conservation, and public policy. *Annual Review of Ecology and Systematics* 32:481–517.

- McIntosh, R. P. 1985. *The Background of Ecology: Concept and Theory*. Cambridge: Cambridge University Press.

- O'Connor, R. J. 2000. Why ecology lags behind biology. *The Scientist* 14:35.

- Paine, R. T. 2002. Advances in ecological understanding: By Kuhnian revolution or conceptual evolution? *Ecology* 83:1553–1559.

- Platt, J. R. 1964. Strong inference. *Science* 146:347–353.

- Popper, K. R. 1963. *Conjectures and Refutations*. London: Routledge & Kegan Paul.

- Thompson, J. N., et al. 2001. Frontiers of ecology. *BioScience* 51:15–24.

Credits

Illustration and Table Credits

T1 From *2007 World Population Data Sheet*. F6 Figures from *Kangaroos: Their Ecology and Management in the Sheep Rangelands of Australia* by G. Caughley et al., p. 12. Copyright © 1987 Cambridge University Press. Used with permission.

Photo Credits

Unless otherwise indicated, photos provided by the author.

CO Pete Atkinson/Getty Images. 2 Juan Medina/Reuters/CORBIS. 3 Embassy of Australia.

Evolution and Ecology

Key Concepts

- Evolution is the genetic adaptation of organisms to the environment.

- Ecology and evolution are intricately connected because evolution operates through natural selection, which is ecology in action.

- Natural selection may act by directional selection, stabilizing selection, or disruptive selection.

- Evolution results from directional selection, but for most ecological situations, stabilizing selection is most common.

- Natural selection may operate on four different levels: gametic, individual, kin, or group. Individual or Darwinian selection is probably most important in nature.

From Chapter 2 of *Ecology: The Experimental Analysis of Distribution and Abundance*, Sixth Edition. Eugene Hecht.
Copyright © 2009 by Pearson Education, Inc. Published by Pearson Benjamin Cummings. All rights reserved.

KEY TERMS

coevolution The evolution of two or more species that interact closely with one another, with each species adapting to changes in the other.

individual optimization hypothesis That each individual in a population has its own optimal clutch size, so that not all individuals are identical.

Lack clutch size The clutch size at which productivity is maximal for the population.

Lack's hypothesis That clutch size in birds is determined by the number of young that parents can provide with food.

maximum reproduction The theory that natural selection will maximize reproductive rate, subject to the constraints imposed by feeding and predator avoidance.

natural selection The process in nature by which only the organisms best adapted to their environment tend to survive and transmit their genetic characteristics to succeeding generations while those less adapted tend to be eliminated.

optimality models Models that assume natural selection will achieve adaptations that are the best possible for each trait in terms of survival and reproduction.

phenotype The observable physical characteristics of an organism.

proximate factors How a particular trait is regulated by an individual in a physiological or biochemical manner.

ultimate factors The evolutionary reason for an adaptation or why a trait is maintained in a population; opposite of *proximate factors*.

Charles Darwin was an ecologist before the term had even been coined, and is an appropriate patron for the science of ecology because he recognized the intricate connection between ecology and evolution. As we discuss ecological ideas, we will use evolutionary concepts. This chapter provides a brief survey of the basic principles of evolution that are important in evolutionary ecology. We will not discuss all aspects of evolution, which are covered in detail in books devoted to evolutionary biology (e.g., Futuyma 2005), but only those aspects that intersect directly with ecological questions of distribution and abundance.

What Is Evolution?

Evolution is change, and biological evolution might be defined as changes in any attribute of a population over time. But we must be more specific than this. Evolutionary changes often lead to adaptation and must involve a change in the frequency of individual genes in a population from generation to generation. What produces evolutionary changes?

Natural selection, said Charles Darwin and Alfred Wallace independently in 1858, is the mechanism that drives adaptive evolution. Natural selection operates through the following steps:

- Variation occurs in every group of plant and animal. Individuals of the same species are not identical in any population, as was observed in the breeding of domestic animals.

- Every population of organism produces an excess of offspring. (The high reproductive capacity of plants and animals was well known to Malthus and Buffon long before Darwin.)

- Life is difficult, and not all individuals will survive and reproduce.

- Among all the offspring competing for limited resources, only those individuals best able to obtain and use these resources will survive and reproduce.

- If the characteristics of these organisms are inherited, the favored traits will be more frequent in the next generation.

Natural selection will favor traits that allow individuals possessing those traits to leave more descendants. These individuals are said to be fitter, and evolution in general maximizes fitness. The process of natural selection is the end result of the processes of ecology in action. The environments that organisms inhabit shape the evolution that occurs. The present distribution, abundance, and diversity of animals and plants are set by the evolutionary processes of the past impinging on the environment of the present.

A simple example of natural selection is shown in **Figure 1**. The moth *Biston betularia* shows variation in the amount of black color on the wings. The typical moth is white with black speckling on the wings. The black form, *carbonaria*, was first described near Manchester in central England in 1848, and it spread over most of England during the next 50 years. When industrial pollution in central England caused lichens on tree bark to die, black-colored moths survived better because bird predators could not see them against this dark background (see Figure 1). Black wing color is in-

What Is Fitness?

Evolutionary ecologists discuss fitness in many forms, and we need to have a clear idea of what fitness means. **Fitness** is a measure of the contribution of an individual to future generations and can also be called **adaptive value**. Individuals have higher fitness if they leave more descendants. Individuals can be fitter for three reasons: They may reproduce at a high rate, they may survive longer, or both. A fish that reproduces rapidly and dies young may be fitter than another fish of the same species that lives a long time but reproduces slowly. From this definition, it should be clear that fitness is a relative term and applies to individual organisms within the same species. One individual may be fitter than another of the same species, or less fit. Ecologists tend to assume that there are traits that allow greater fitness, and that these traits have a genetic basis. Evolution will act to maximize fitness.

We should also be clear about what fitness is not:

- **Fitness is not absolute.** Measures of fitness are specific for a given environment. Individuals with genes that make them fit for cold environments may not be fit if the climate changes and they must live in warm environments.

- **Fitness cannot be compared across species.** We cannot compare the fitness of an elephant with that of an oak tree. Fitness is a measure that is defined only within a single species.

- **Fitness is not only about reproduction.** High reproductive rates may not by themselves confer high fitness if survival rates of these young are poor.

- **Fitness is not a short-term measure.** Fitness should be measured across several generations, although this is difficult for studies of long-lived plants and animals. Ecologists often study short-term measures that they hope will correlate with fitness in the long term.

- **Fitness is not about individual traits.** Evolution is a whole-organism affair. Individual traits such as large body size or fast growth rates may be components of fitness, but the test of fitness is the test of whole-organism survival and reproduction.

Figure 1 Evolution in the peppered moth _B. betularia_ in England and North America. The photo shows both phenotypes of the peppered moth. The black form, _carbonaria_, has been declining in abundance since 1950 with the decline in industrial pollution in central England. The same change has occurred in eastern North America. Differential bird predation is believed to be the major mechanism of selection. (Photo: H. B. D. Kettlewell; data from Majerus 1998.)

herited in these moths, and the result was an increase in the frequency of black moths during industrialization (Majerus 1998, Grant 2005). Because industrial pollution has decreased in England during the past 50 years, this process of natural selection is reversing (see Figure 1). The same changes have occurred in the American form of the peppered moth as air quality has improved in the eastern United States (Grant and Wiseman 2002).

Evolution through natural selection results in adaptation, and under appropriate conditions produces new species (speciation). **Adaptation** has important ecological implications because it sets limits to the life cycle traits that determine distribution and abundance.

Adaptation

Natural selection acts on **phenotypes,** the observable attributes of individuals. Different genotypes give rise to different phenotypes, but because embryological and subsequent development is affected in many ways by environmental factors, such as temperature, it is often not a direct translation. Consequently, it is simpler to observe the effect of natural selection directly on the phenotype and to ignore the underlying genotype. Ecologists, like plant and animal breeders, are primarily interested in phenotypic characters such as seed numbers or body size.

Three types of selection can operate on phenotypic characters (**Figure 2**). The simplest form is **directional selection**, in which phenotypes at one extreme are selected against. Directional selection produces genotypic changes more rapidly than any other form, so most artificial selection is of this type. Darwin's finches on the Galápagos Islands have been the best-studied example of directional selection. Peter and Rosemary Grant from Princeton University have spent more than 30 years studying these finches on the Galápagos. **Figure 3** illustrates directional selection in one of Darwin's finches, the Galápagos ground finch *Geospiza fortis*. During a prolonged drought, the birds that survived were predominantly those with large beaks that could crack large seeds (Grant and Weiner 2000). Birds with large beaks can eat both large and small seeds, while birds with small beaks can eat only small seeds. Directional selection probably accounts for many of the phenotypic changes that occur during evolution. In wild populations, resistance of pests to insecticides or herbicides is produced by directional selection.

Stabilizing selection (see Figure 2) is very common in present-day populations. In stabilizing selection, phenotypes near the mean of the population are fitter than those at either extreme; thus, the population mean value does not change. **Figure 4** illustrates stabi-

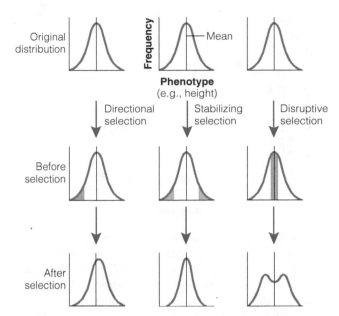

Figure 2 Three types of selection on phenotypic characters. Individuals in the colored areas are selected against. (Tamarin 1999)

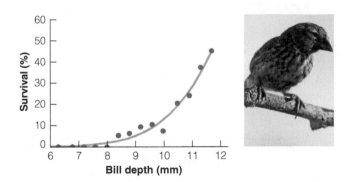

Figure 3 Directional selection for beak size in the Galápagos ground finch *Geospiza fortis*. From 1976 to 1978, a severe drought in the Galápagos Islands caused an 85 percent drop in the population, and birds with larger beaks survived better because they could crack larger, harder seeds. (Grant 1986)

lizing selection for birth weight in humans in the United States. Early mortality is lowest for babies weighing about 4.2 kilograms (kg), slightly above the observed mean birth weight of 3.4 kg for the population. Very small babies die more frequently, and very large babies are at increased risk even with modern medical care.

Figure 5 shows another example of stabilizing selection in lesser snow geese *Anser caerulescens*. Snow geese

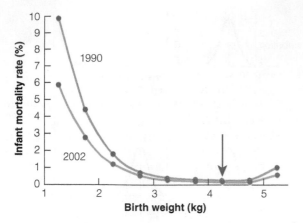

Figure 4 Stabilizing selection for birth weight in humans. Data from United States for infants, 1990 and 2002. The optimal birth weight (red arrow) is 4.25 kg, with a broad range of minimal mortality between 3.2 kg and 4.8 kg. Because of medical advances, infant mortality has been falling steadily, so the 1990 curve is higher than the 2002 curve. (Data from National Center for Health Statistics 2006).

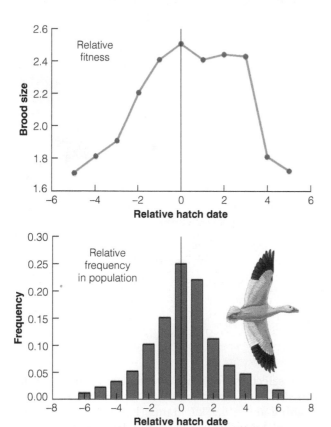

Figure 5 Stabilizing selection for hatching synchrony in lesser snow geese (*Anser caerulescens*) at La Perouse Bay in northern Manitoba, Canada. Relative hatch date is the number of days a female's eggs hatched before or after the mean date for the colony. (From Cooke and Findlay 1982.)

nest in colonies in northern Canada, and clutches hatch over a two-week period in early summer. Because predation is concentrated on whole colonies, eggs hatching synchronously confer a "safety-in-numbers" advantage against predators such as foxes. Females whose eggs hatch synchronously on or near the mean date for the colony are more likely to raise their young successfully. Nests that hatch early suffer greater predation loss, as do nests that hatch later. The result is natural selection favoring an optimum hatching time (Cooke and Findlay 1982).

In the third type of selection, **disruptive selection** (see Figure 2), the extremes are favored over the mean. But because the extreme forms breed with one another, every generation will produce many intermediate forms doomed to be eliminated. In any environment favoring the extremes, any mechanism that would prevent the opposite extremes from breeding with one another would be advantageous. Isolating mechanisms are thus an important adjunct of disruptive selection. Disruptive selection has been suggested to be important in speciation (Rueffler et al. 2006). A good illustration of how disruptive selection operates is found in three-spine sticklebacks (*Gasterosteus aculeatus*) in coastal lakes of British Columbia. Don McPhail, Dolph Schluter, and their students have shown that two forms of this small fish live in some coastal freshwater lakes (**Figure 6**). The two forms are so distinct they are effectively species. The small form lives in the open water of the lake and feeds on small plankton, while the large form lives on the bottom of the lake and feeds on insects and crustaceans that live on the bottom of the lake. These two forms seem to have originated from two separate invasions of the lakes as the sea level rose and fell during glacial periods. Competition between the earlier and the later invaders and disruptive selection have produced the two existing species that are closely related to the plankton-feeding marine ancestor species (Rundle et al. 2000).

The net result of all this selection is that organisms are adapted to their environment, and the great diversity of biological forms is a graphic essay on the power of adaptation by natural selection. But we must be careful to note that adaptation does not produce the "best" phenotypes or "optimal" phenotypes (defined as phenotypes that are theoretically the most efficient in surviving and reproducing). The "better" survive, not the "best," and the biological world can never be described as "the best of all possible worlds."

Adaptation is constrained in populations by four major forces. First, genetic forces prevent perfect adaptation because of mutation and gene flow. Mutation is always occurring, generating variation in populations, and most mutations are detrimental to organisms rather than adaptive. The immigration of individuals into an area where local environments differ will add

(a)

(b)

Figure 6 Two males of the three-spine stickleback in Paxton Lake, British Columbia. (a) The smaller male ("limnetic" species) has evolved to feed in the open water of the lake, while (b) the larger form ("benthic" species) lives and feeds on the bottom. The two forms are reproductively isolated and thus are effectively two new species that have originated from the marine ancestor species by invading coastal freshwater lakes. Both males are shown in courtship coloration. (Photos: Todd Hatfield and Ernie Taylor.)

other alleles to the gene pool and act to smooth out local adaptations. Second, environments are continually changing, and this is the most significant short-term constraint on adaptation. Third, adaptation is always a compromise because organisms have at their disposal only a limited amount of time and energy. There are trade-offs between adaptations such as wing shape in birds. A loon's wings are efficient for diving but not so efficient for flying. Fourth, historical constraints are always present because organisms have a history and change in small increments. Let us look in detail at one example of adaptation to illustrate some of these principles.

Clutch Size in Birds

Each year, Emperor penguins lay one egg; pigeons, one or two eggs; gulls, typically, three eggs; the Canada goose, four to six eggs; and the American merganser, 10 or 11 eggs. What determines clutch size in birds? We must distinguish two different aspects of this question: proximate and ultimate.

Proximate factors explain *how* a trait is regulated by an individual. Proximate factors that determine clutch size are the physiological factors that control ovulation and egg laying. **Ultimate factors** are selective factors, and ultimate explanations for clutch size differences involve evolutionary arguments about adaptations. Proximate factors affecting clutch size have to do with how an individual bird decodes its genetic information on egg laying; ultimate factors have to do with changes in this genetic program through time and with the reason for these changes (Mayr 1982). Clutch size may be modified by the age of the female, spring weather, population density, and habitat suitability. The ultimate factors that determine clutch size are the requirements for long-term (evolutionary) survival. Clutch size is viewed as an adaptation under the control of natural selection, and we seek the selective forces that have shaped the reproductive rates of birds. We shall not be concerned here with the proximate factors determining clutch size, which are reviewed by Carey (1996).

Natural selection will favor those birds that leave the most descendants to future generations. At first thought, we might hypothesize that natural selection favors a clutch size that is the physiological maximum the bird can lay. We can test this hypothesis by taking eggs from nests as they are laid. When we do this, we find that some birds, such as the common pigeon, are **determinate layers**; they lay a given number of eggs, no matter what. The pigeon lays two eggs; if you take away the first, it will incubate the second egg only. If you add a third egg, it will incubate all three. But many other birds are **indeterminate layers**; they will continue to lay eggs until the nest is "full." If eggs are removed once they are laid, these birds will continue laying. When this subterfuge was used on a mallard female, she continued to lay one egg per day until she had laid 100 of them. In other experiments, herring gull females laid up to 16 eggs (normal clutch: 2–3); a yellow-shafted flicker female, 71 eggs (normal clutch: 6–8); and a house sparrow, 50 eggs (normal clutch: 3–5) (Klomp 1970; Carey 1996). This evidence suggests that most birds under normal circumstances do not lay their physiological limit of eggs but that ovulation is stopped long before this limit is reached.

The British ornithologist David Lack was one of the first ecologists to recognize the importance of evolutionary thinking in understanding adaptations in life history traits. In 1947 Lack put forward the idea that clutch size in birds was determined ultimately by the number of young that parents can provide with food. This hypothesis stimulated much research on birds because it immediately suggested experimental manipulations. If this hypothesis is correct, the total production of young ought to be highest at the normal clutch size, and if one experimentally increased clutch size by adding eggs to nests, increased clutches should suffer greater losses because the parents could not feed the extra young in the nest.

One way to think about this problem of optimum clutch size is to use a simple economic approach. Everything an organism does has costs and benefits. Organisms integrate these costs and benefits in evolutionary time. The benefits of laying more eggs are very clear—more descendants in the next generation. The costs are less clear. There is an energy cost to make each additional egg, and there is a further cost to feeding each additional nestling. If the adult birds must work harder to feed their young, there is also a potential cost in adult survival—the adults may not live until the next breeding season. If adults are unable to work harder, there is a potential reduction in offspring quality. A cost–benefit model of this general type is shown in **Figure 7**. Models of this type are called **optimality models**. They are useful because they help us think about what the costs and benefits are for a particular ecological strategy.

No organism has an infinite amount of energy to spend on its activities. The reproductive rate of birds can

be viewed as one sector of a bird's energy balance, and the needs of reproduction must be maximized within the constraints of other energy requirements. The total requirements involve metabolic maintenance, growth, and energy used for predator avoidance, competitive interactions, and reproduction. **Lack's hypothesis** (1947)—that the clutch size of birds that feed their young in the nest was adapted by natural selection to correspond to the largest number of young for which the parents can provide enough food—has been a very fertile hypothesis in evolutionary ecology because it has stimulated a variety of experiments. According to this idea, if enough additional eggs are placed in a bird's nest, the whole brood will suffer from starvation so that, in fact, fewer young birds will fledge from nests containing larger numbers of eggs. In other words, clutch size is postulated to be under stabilizing selection (see Figure 2). Let us look at a few examples to test this idea.

In England, the blue tit normally lays a clutch of 9 to 11 eggs. What would happen if blue tits had a brood of 12 or 13? Pettifor (1993) artificially manipulated broods at hatching by adding or subtracting chicks, and found that the survival of the young blue tits in manipulated broods was poor (**Figure 8**). Blue tits feed on insects and apparently cannot feed additional young adequately, so more of the young starve. Consequently, it would not benefit a blue tit in the evolutionary sense to lay more eggs, and the results are consistent with Lack's hypothesis. Individual birds appear to produce the clutch size that maximizes their reproductive potential.

Tropical birds usually lay small clutches, and Skutch (1967) argued that this was an adaptation against nest predators. If the intensity of nest predation increases with the number of parental feeding trips away from the nest, natural selection would favor a reduced clutch size. Low clutch size and low predation rates are associated. Parents would leave more descendants if they had smaller broods and did not need to feed them as often. Exactly the same argument was used by Martin (1995) to explain the pattern of clutch size in hole-nesting birds. Hole-nesting passerine birds lay fewer eggs than comparable species that nest in the open, and predation rates are much lower for hole-nesting species (Martin 1995). So again, low clutch size and low predation rates are correlated. This suggests that a high risk of predation on the whole brood in the nest is a strong selective factor that increased clutch size in open-nesting birds. This factor also favors a shortened nesting period, independent of the ability of the parents to provide food to the nestlings. Open nesting is a gamble because of high predation rates, and passerine birds gamble on large clutches and short nesting periods.

Natural selection would seem to operate to maximize reproductive rate, subject to the constraints imposed by feeding and predator avoidance. This is called the theory

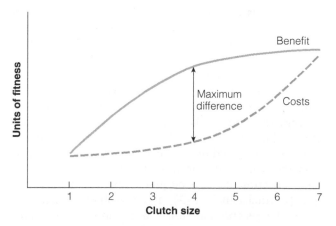

Figure 7 A cost–benefit model for the evolution of clutch size in birds. An individual benefits from laying more eggs because it will have more descendants, but it incurs costs because of increasing parental care required for larger clutches. The clutch size with the maximum difference between benefits and costs is the optimal clutch size for that individual. (The Lack clutch size, named after David Lack.)

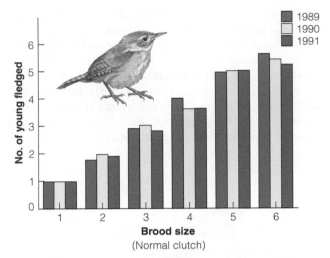

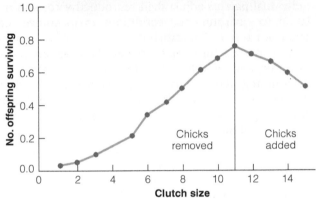

Figure 8 Production of young blue tits (*Parus caeruleus*) in relation to clutch size in Wytham Wood, Oxford, England. Only females that had laid 11 eggs in previous years are shown here, because we expect these individuals to have their highest fitness at a clutch size of 11. These results fit Lack's hypothesis because adding more chicks just after hatching does not increase fitness. (Data from Pettifor 1993, p. 136.)

Figure 9 Number of house wren chicks fledged from broods manipulated to have larger and smaller than average brood size. These results do not agree with the predictions from Lack's hypothesis that larger broods should fledge fewer young than average-sized broods. (After Young 1996.)

of **maximum reproduction**, and Lack's hypothesis is part of this theory. It is a good example of how stabilizing selection can operate on a phenotypic trait such as reproductive rate. The maximum clutch size is called the **Lack clutch size** (see Figure 7), after David Lack.

Not all manipulation experiments confirm Lack's hypothesis. Young (1996) manipulated clutch size in tropical house wrens (*Troglodytes aedon*) in Costa Rica to produce clutches ranging from one to six. House wrens in the tropics typically lay three or four eggs. **Figure 9** shows the resulting offspring produced. The number of surviving offspring per brood was maximized for broods of six eggs. Since mean brood size was 3.5, the most common clutch size was smaller than the most productive clutch size. Vanderwerf (1992) surveyed 77 experiments in which clutches had been manipulated and

found that 69 percent of these were like the house wren—the most productive clutch size was larger than the most common clutch size. Why should this be?

The presence of trade-offs is one explanation of why clutches are smaller than the Lack clutch size. Clutch size may affect the chances of the adult birds surviving to breed again. Birds may become exhausted by rearing large clutches; such exhaustion is a delayed cost of reproduction. Alternatively, laying a large clutch may postpone the next breeding attempt, leading to reduced lifetime reproduction. Laying a large clutch is energetically costly for birds, and this cost is not usually measured in brood manipulation experiments (Monaghan and Nager 1997).

The Lack clutch size may not be a constant for a species, and different individuals may vary in their parenting abilities and have a personal Lack clutch size. Clutch size is under strong genetic control in birds. One female may consistently lay three eggs and this may be best for her, while another female in the same population may consistently lay five eggs and this may be best for her. This is called the **individual optimization hypothesis**, and it explains why there is considerable variation in clutch size within a population. The individual optimization hypothesis has been the subject of several experimental tests (Pettifor et al. 2001). This hypothesis predicts that any manipulation of clutch size will reduce the fitness of the parent birds because they rear fewer young or survive less well. It also predicts that in natural broods there should be more young recruited as clutch size increases, coupled with no impairment of fitness of the parent birds. Tinbergen and Sanz (2004) did not find these predictions to be correct for a population of great tits in the Netherlands (**Figure 10**). Artificially enlarged first clutches produced more recruits, and adult survival was not affected by the manipulations either in the same year or in the following year. They rejected the individual optimization hypothesis for

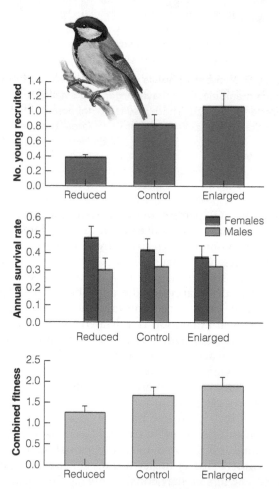

Figure 10 Test of the Individual Optimization Hypothesis for clutch size variation in the great tit (*Parus major*) in the Netherlands. These results do not agree with the predictions from this hypothesis that larger manipulated broods should fledge fewer young than unmanipulated (control) broods. The combined measure of fitness includes the number of young produced that live to the next breeding season and the survival of their parents to the next breeding season. These results would predict directional selection for increased clutch size. (Data from Tinbergen and Sanz 2004.)

this bird population. Individual birds may not be able to predict environmental variation in any given year, and food supplies may fluctuate so much that individuals cannot predict the optimal clutch size for any particular year (Török et al. 2004).

An alternative explanation of why the average clutch may be smaller than the Lack clutch size is that observed clutch sizes are a nonadaptive compromise. If gene flow occurs between two habitats, one good and one poor, clutches may be larger than optimal in poor habitats and smaller than optimal in good habitats. Blue tits and great tits in Belgium rarely breed in woodlands where they were born and show this nonadaptive compromise (Dhondt et al. 1990).

Recent work on bird reproduction investigates how individual parents adjust their reproductive costs in relation to environmental conditions to maximize the output of young. The proximate controls of reproduction operate through the energy available to reproducing birds, and the role of female condition is critical in determining reproductive effort. Reproductive effort this year may affect the chances of surviving until next year, and parents must balance the short-term and long-term costs of breeding.

Coevolution

The term *coevolution* was popularized by Paul Ehrlich and Peter Raven (1964) to describe the reciprocal evolutionary influences that plants and plant-eating insects have had on each other. **Coevolution** occurs when a trait of species *A* has evolved in response to a trait of species *B*, which has in turn evolved in response to the trait in species *A*. Coevolution is specific and reciprocal. In the more general case, several species may be involved instead of just two, and this is called diffuse coevolution (Thompson 1994).

Coevolution is simply a part of evolution, and it provides important linkages to ecology. The interactions between herbivores and their food plants have been emphasized as a critical coevolutionary interaction. Predator–prey interactions can also be coevolutionary, and in some cases can lead to "arms races" between species.

Coevolution shapes the characteristics of coevolving pairs of species, while diffuse coevolution might also occur in communities of many species. There is considerable doubt about whether whole communities of plants and animals could coevolve, and most ecologists believe that coevolution is restricted to interactions between only a few species that interact tightly (Benkman et al. 2001).

Evolution and "Arms Races"

If you look up "arms race" on the Web, you will find much discussion of military strategies and little of biological evolution. An arms race is tit-for-tat evolution—a reciprocal interaction between species—in which as species *A* evolves better adaptations to exploit species *B*, the latter fights back by evolving adaptations to thwart the improvements in species *A*. The best examples of arms races occur between hosts and parasites. The brown-headed cowbird in North America and the European cuckoo in Britain are good examples of parasitic birds that lay their eggs in the nests of other species (**Figure 11**). The host species then raise the cowbird or cuckoo chick, often to the detriment of its own young.

The brown-headed cowbird has greatly expanded its geographic range in North America because of agriculture and is invading new areas and utilizing new host species, so it has become a major conservation problem. Parasitic birds such as the cowbird often lay eggs that have the same color and pattern as the host species in order to avoid detection and the possibility that the host species will remove the parasite's eggs from their nests. The host species on the other hand should evolve the ability to discriminate cowbird eggs from its own eggs. Host individuals that discriminate more will leave more offspring (and raise fewer cowbirds). Consequently, an evolutionary arms race can develop in which both the parasite and the host are continually evolving counterstrategies in a tit-for-tat manner (Takasu 1998).

Deadly toxins and resistance to them are an evolutionary enigma and illustrate a potential difficulty in the evolution of arms races between predators and prey (Brodie and Brodie 1999). Some snakes, for example, can feed on prey that are poisonous to most other animals. There can be no natural selection for increased resistance if predators do not survive encounters with toxic prey. Similarly, deadly toxins are of no advantage to individual prey if the prey dies delivering the toxins (Williams et al. 2003). For natural selection to drive an arms race between resistant predators and lethal prey, the survivorship of individual predators must vary with their resistance. One example is the extreme toxicity of some populations of the rough-skinned newt *(Taricha granulosa)* that appear to have coevolved with resistance in its predator, the common garter snake *(Thamnophis sirtalis)* in North America (**Figure 12**). The rough-skinned newt is one of the most toxic animals known. Its skin contains a neurotoxin that is fatal to most animals in small doses. But some garter snakes feed on these newts, and have evolved resistance to their toxins. For example, San Francisco populations of garter snakes are nearly 100 times more resistant to newt neurotoxins than are garter snakes from Oregon (Brodie and Brodie 1999). There is a geographical mosaic in the amount of poison carried by the newts in their skin and the resistance shown by garter snakes, so that coevolution of this arms race has not reached the same point in all populations.

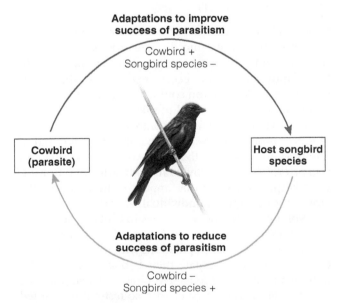

Adaptations to improve success of parasitism

Cowbird +
Songbird species –

Cowbird (parasite)

Host songbird species

Adaptations to reduce success of parasitism

Cowbird –
Songbird species +

Figure 11 Arms race. Schematic illustration of the arms race between the parasitic cowbird, which lays its eggs in other birds' nests, and the parasitized species that try to defend against this kind of parasitism by ejecting the cowbird eggs.

Units of Selection

Darwin conceived natural selection as operating through the reproduction and survival of individuals who differ in their genetic constitution. Most discussion of natural selection operates at this level of Darwinian selection, or individual selection.

But natural selection is not restricted to individuals. It can act on any biological units so long as these units meet the following criteria: (1) They have the ability to replicate; (2) they produce an excess number of units above replacement needs; (3) survival depends on some attribute (size, color, behavior); and (4) a mechanism exists for the transmission of these attributes. Three units of selection other than the individual can fulfill these criteria: gametic, kin, and group selection.

(a)

(b)

Figure 12 (a) The rough-skinned newt (*Taricha granulosa*) from western North America, an extremely toxic salamander, and (b) the garter snake (*Thamnophis sirtalis*) that preys on these newts.

Gametic Selection

Gametes (eggs and sperm) have a genetic composition that differs from the diploid organisms that produce them. Gametes are produced in vast excess and may have characteristics that they transmit through the zygote and adult organism to the next generation of gametes. Consequently, natural selection can act on a population of gametes independently from the natural selection that operates on the parent organisms. Many different characteristics of gametes could be under natural selection. Sperm mobility, for example, may be under strong selection. In plants, pollen grains that produce a faster-growing pollen tube have a better chance of releasing their sperm nuclei and fertilizing an egg. Gametic selection is an interesting and important aspect of natural selection, but it does not directly impinge on ecological relationships.

Kin Selection

If an individual is able to increase the survival or reproduction of its relatives with whom it shares some of the same genes, natural selection can operate through kin selection. Kin selection and individual selection may act together, and this action is described by the concept of inclusive fitness. Natural selection favors not only alleles that benefit an individual but also alleles that benefit close relatives of that individual because close relatives share many alleles. All relatives can help pass copies of an individual's genes to future generations.

Kin selection was recognized as one way of explaining the existence of altruistic traits such as the sounding of alarm calls. When ground squirrels sight a predator, they give an alarm call. As a result, the individual calling (1) draws attention to itself and thus may be attacked by the predator (detrimental to the individual) and (2) warns nearby squirrels to run for cover (beneficial to relatives nearby).

Kin selection has important consequences for ecological relationships because of its effects on social organization and population dynamics. Competition between individual organisms will be affected by the proximity of close relatives; thus, it can be important for an ecologist to know the degree of kinship among members of a population.

Group Selection

Group selection can occur when populations of a species are broken up into discrete groups more or less isolated from other such groups. Groups that contain less adaptive genes can become extinct, and the conditions for natural selection could occur at the level of the group, as well as at the level of the individual organism.

Group selection is highly controversial, and most biologists consider it to be rare in nature. Most of the characteristics of organisms that are favorable to groups can also be explained by individual or kin selection. Controversy erupts over traits that appear to be good for the group but bad for the individual. A classic example is the evolution of reproductive rates in birds. Group selectionists argue that many birds reproduce at less-than-maximal rates because populations with low reproductive rates will not overpopulate their habitats. Any populations with higher reproductive rates will overpopulate their habitats and become extinct, so restraint is selected for at the group level. But low reproductive rates are bad for the individual, and individual selection will act to favor higher reproductive rates, so group selection and individual selection are operating at cross purposes. In a group in which all members restrain themselves, a cheater will always be favored.

The alternative argument is that all reproductive rates are in fact maximal and have responded only to individual selection favoring individuals that leave the most offspring for future generations. Restraint does not exist, according to this view.

Group selection may occur, but at present it is not believed to be an important force shaping the adaptations that ecologists observe while trying to understand the distribution and abundance of organisms.

Summary

Organisms survive and reproduce, and because not all individuals are equally successful at these activities, natural selection occurs. The fitter individuals leave more descendants to future generations because of either higher survival or higher reproductive rates. Natural selection is ecology in action, and the ecologist asks which traits of individuals improve their chances of survival or reproduction.

The clutch size of birds is a classic problem in evolutionary ecology—why don't birds lay more eggs? David Lack suggested in 1947 that clutch size was limited by the number of chicks the adults could feed successfully. Experimental additions of eggs and chicks to nests have often shown that bird parents can in fact rear more nestlings than they usually do. This anomaly is probably due to the higher costs of reproduction for birds rearing large broods, and adults may die or lay fewer eggs in subsequent years as a cost of breeding performance in the current year. Clutch size is thus expected to be under stabilizing selection in most cases.

Coevolution can occur between interacting species. Coevolution occurs when a trait of a particular species has evolved in response to a trait of a second species, which has in turn evolved in response to the trait in the first species. Many examples of coevolution occur in plant–herbivore interactions and in predator–prey interactions. Arms races between species are a particular kind of coevolution. The best examples of arms races occur between hosts and parasites.

Individual, or Darwinian, selection is the classic form of selection on individual phenotypes, and it is the level of selection responsible for most of the adaptations we see in nature. Some adaptations may evolve by kin selection for actions that favor the survival or reproduction of close relatives carrying the same genes. Group selection might also occur if whole groups or populations become extinct because of genetic characteristics present in the group. Group selection is probably uncommon in nature.

Review Questions and Problems

1   Birds living on oceanic islands tend to have a smaller clutch size than the same species (or close relatives) breeding on the mainland (Klomp 1970, p. 85). Explain this on the basis of Lack's hypothesis.

2   Cane toads have been introduced to Australia and many of the Pacific islands. Their skin contains glands that secrete poisons that are toxic to most vertebrates. Discuss how evolution might operate on potential predators of cane toads in areas like Australia in which the predators have no prior evolutionary history of interactions with these toads. Phillips and Shine (2006) discuss this issue.

3   Ladybird beetles are distasteful to predators because of toxic chemicals they secrete, yet they also have dark melanic forms (Majerus 1998, p. 221). Melanic ladybirds have declined in frequency in central England along with the peppered moth during the past 50 years as air quality has improved. If ladybirds are not eaten by predators, how might you explain these changes in melanic frequency?

4   Figure 10 provides data that appear to contradict the Individual Optimization Hypothesis for the evolution of clutch size in birds. Are there any components of fitness in these birds that are ignored in Figure 10 and that might change the interpretation from an example of directional selection to one of stabilizing selection? Read de Heij et al. (2006) for a discussion.

5   Royama (1970, pp. 641–642) states:

Natural selection favors those individuals in a population with the most efficient reproductive capacity (in terms of the number of offspring contributed to the next generation), which means that the present-day generations consist of those individuals with the highest level of reproduction possible in their environment.

Is this correct? Discuss.

6   In many temperate zone birds, those individuals that breed earlier in the season have higher reproductive success than those that breed later in the season. If climate change is making spring weather occur at earlier dates, will this lead to directional selection for earlier breeding dates in these birds? What constraints might affect this type of directional selection?

7   Some birds such as grouse and geese have young that are mobile and able to feed themselves at hatching (precocial chicks). Discuss which factors might limit clutch size in these bird species. Winkler and Walters (1983) have reviewed studies on clutch size in precocial birds.

8   In arctic ground squirrels, adult females are more likely to give alarm calls than adult males. If alarm calls are favored by kin selection, why might this difference occur? Could alarm calls be explained by group selection? Why or why not?

9   Apply the cost–benefit model in Figure 7 to seed production in a herbaceous plant. Discuss biological reasons for the general shape of these curves. Can you apply this model to both annual and perennial plants in the same way?

10   A research scientist obtained the following data on the fitness of seven females in a small population of house sparrows:

Year	Variable	Female band number						
		A	B	C	D	E	F	G
1999	No. eggs laid	8	5	4	5	7	6	7
	No. young fledged	5	1	3	4	2	4	3
2000	No. eggs laid	5	6	4	5	7	7	9
	No. young fledged	4	2	3	5	3	3	0
2001	No. eggs laid	6	10	5	5	8	10	10
	No. young fledged	4	8	5	4	6	8	9
2002	No. eggs laid	6	6	3	5	7	8	10
	No. young fledged	2	5	3	5	3	4	4

Rank the fitness of these seven female sparrows. What data might you collect to improve on this measure of fitness for these birds?

11   Discuss how the concept of time applies to evolutionary changes and to ecological situations. Do ecological time and evolutionary time ever correspond?

12   A hypothetical population of frogs consists of 50 individuals in each of two ponds. In one pond, all of the individuals are green; in the other pond, half are green and half are brown. During a drought, the first pond dries up, and all the frogs in it die. In the population as a whole, the frequency of the brown phenotype has gone from 25 percent to 50 percent. Has evolution occurred? Has there been natural selection for the brown color morph?

Overview Question

Humans in industrialized countries increased in average body size during the twentieth century. List several possible explanations for this change, and discuss how you could decide if an evolutionary explanation is needed to interpret it. How does a physiological explanation for this change differ from an evolutionary explanation?

Suggested Readings

- Fox, C. W., D. A. Roff, and D. J. Fairbairn, eds. 2001. *Evolutionary Ecology: Concepts and Case Studies*. New York: Oxford University Press.

- Futuyma, D. J. 2005. *Evolution*. Sunderland (Massachusetts): Sinauer Associates.

- Grant, B. R., and L. L. Wiseman. 2002. Recent history of melanism in American peppered moths. *Journal of Heredity* 93:86–90.

- Grant, P. R., and B. R. Grant. 2002. Adaptive radiation of Darwin's finches. *American Scientist* 90:130–139.

- Grim, T. 2006. The evolution of nestling discrimination by hosts of parasitic birds: Why is rejection so rare? *Evolutionary Ecology Research* 8:785–802.

- Monaghan, P., and R. G. Nager. 1997. Why don't birds lay more eggs? *Trends in Ecology and Evolution* 12:271–274.

- Phillips, B. L., and R. Shine. 2006. An invasive species induces rapid adaptive change in a native predator: Cane toads and black snakes in Australia. *Proceedings of the Royal Society of London, Series B* 273:1545–1550.

- Rueffler, C., T. J. M. Van Dooren, O. Leimar, and P. A. Abrams. 2006. Disruptive selection and then what? *Trends in Ecology & Evolution* 21:238–245.

- Taylor, E. B., J. W. Boughman, M. Groenenboom, M. Sniatynski, D. Schluter, and J. L. Gow. 2006. Speciation in reverse: Morphological and genetic evidence of the collapse of a three-spined stickleback (*Gasterosteus aculeatus*) species pair. *Molecular Ecology* 15:343–355.

- Wills, C. 1996. *Yellow Fever, Black Goddess: The Coevolution of People and Plagues*. Reading (Massachusetts): Addison-Wesley.

Credits

Illustration and Table Credits

Photo Credits

Behavioral Ecology

Key Concepts

- Behavioral ecology asks how individual animals interact with other animals, plants, and their physical environments to maximize fitness.

- The consequences of decisions individual animals make will affect their survival and reproduction.

- Natural selection is assumed to have optimized the behavior of individuals to achieve maximal fitness, and the job of the behavioral ecologists is to find the mechanisms by which this is achieved.

- Foraging, antipredator, social, and mating behaviors are four critical foci of study in behavioral ecology that can be analyzed by cost–benefit models.

- Behavioral ecology is a bridge not only to evolutionary biology but also to animal population and community ecology because mechanisms driving population and community dynamics all result from the behavior of individuals.

From Chapter 3 of *Ecology: The Experimental Analysis of Distribution and Abundance*, Sixth Edition. Eugene Hecht.

KEY TERMS

cost–benefit analysis An assessment to determine whether the cost of an activity is less than the benefit that can be expected from the activity.

group selection Natural selection for traits that favor groups within a species irrespective of whether the traits favor individuals or not.

kin selection The evolution of traits that increase the survival, and ultimately the reproductive success, of one's relatives.

optimal foraging Any method of searching for and obtaining food that maximizes the relative benefit.

optimal group size The size that results in the largest relative benefit.

promiscuity A general term for multiple matings in organisms, called polyandry if multiple males are involved, or polygyny if multiple females.

relative benefit The difference between the costs and benefits (= net benefit).

territory Any defended area.

trade-offs Compromises between two desirable but incompatible activities.

The ecology of a species is ultimately determined by interactions between individuals and their environment. The environment includes other individuals of the same species as well as members of other species, such as predators. The environment also includes physical factors, such as temperature, rainfall, and wind. The ways that organisms respond to each other and to particular cues in the environment are called **behaviors**. In this chapter, we will focus on the behaviors of animals as they interact with their food resources, mates, and other members of their social group. How does a rabbit decide where to feed? How does a male lion achieve reproductive success? These are some of the questions we will address.

Behavioral ecology is a strong subdiscipline in animal ecology dealing with the ecology of individuals. Like evolutionary ecology, behavioral ecology has strong links to other sciences, in this case psychology, physiology, and developmental biology. As such, it forms an important link to understanding how populations and communities change. It is unique within ecology in that it deals almost solely with animals and largely ignores plants and microbes. Of course, plants

as well as animals respond to changes in their environment, and we shall discuss these plant responses.

All animal behaviors are generated through a complex set of physiological and neurological reactions triggered by environmental stimuli. Four questions can be asked about any behavior (Tinbergen 1963): (1) How is a behavior produced? (2) How does a behavior develop? (3) What is the adaptive value of a behavior? and (4) What is the evolutionary history of a behavior?

The first two questions are "how" questions (or "proximate" questions) that refer to the mechanisms of behavior, and the second two questions are "why" questions (or "ultimate" questions) that examine the function of behavior. The behavioral ecologist is interested in answering the last two questions, while the physiologist, neurobiologist, and developmental biologist study the first two questions. Behavioral ecologists want to understand the ecological and evolutionary contexts of behavior. They want to learn how an individual's behavior is shaped by its social and physical environment, both past and present, and how specific behaviors affect its chances of surviving and reproducing. Evolutionary questions are key to behavioral ecology.

The following is an example of the kind of questions behavioral ecologists commonly ask: "Why is promiscuity common among mammals?" Monogamy occurs in less than 3 percent of mammalian species (Kleiman 1977). **Promiscuity**, or multiple-male or multiple-female mating, is very common in mammals, and has been described in many species of mammals (Wolff and Macdonald 2004). **Figure 1** shows the frequency of multiple-male mating in the Ethiopian wolf *(Canis simiensis)*. These wolves live in packs and the males within each pack can be ranked as alpha (top male), beta, or other (lower ranking) within their pack social system. Female wolves decide which males they will copulate with, and typically solicit multiple-male matings from males that live in adjacent packs and reject matings from subordinate males within their own pack (Sillero-Zubiri et al. 1996). Why might they do this?

The basic assumption is that animals are well adapted to their environment, and hence there must be some advantage to them to behave in certain ways. Promiscuity in mammals is often an attempt to confuse paternity. For Ethiopian wolves, males from packs can attack juveniles in adjacent packs if they are not genetically related. By soliciting copulations from adjacent pack males, a female can reduce the probability of infanticide occurring because none of the males can determine the father of a litter. Much of this promiscuity seems to be an adaptation for paternity confusion (Wolff and Macdonald 2004).

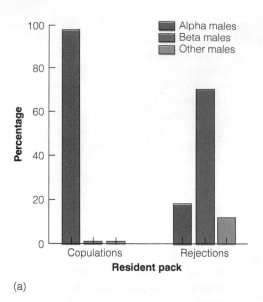

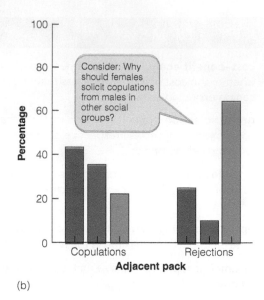

Figure 1 Percentage of copulations achieved by males of different social rank from (a) the resident pack of the female and from males in (b) adjacent packs. Wolves live in packs with well-defined territories. Alpha males are dominant males; beta males are subordinate. Female wolves copulate many times when they are in heat. (Data from Sillero-Zubiri et al. 1996.)

All Behaviors Have Costs and Benefits

We begin with the assumption that observed behaviors are beneficial, and that evolution through natural selection has molded animals to their environment. We can rarely observe the evolution of behavior because behavioral changes occur slowly in evolutionary time. And even though we can sequence the DNA in individuals, this technology will not help us understand the adaptive value of behavior because no complex behavior is under the control of a single gene. Instead, we must adopt an indirect approach to analyze why a particular kind of behavior is adaptive.

What benefits do individuals gain from behaving in certain ways? To answer this question, behavioral ecologists must examine the decisions that animals make when faced with environmental options such as where to feed, what to eat, where to live, and which individuals to mate with. An animal's decisions translate into differences in survival, fecundity, or mating success, and therefore are shaped by natural selection. Consider parental care, which is a major investment in many vertebrates. Mammals and birds in particular must divide limited resources between reproduction and other activities such as feeding. The choices involved require **trade-offs**, which are compromises between two desirable but incompatible activities.

All organisms are constrained by time, energy, and risk of injury. Time spent engaged in one activity cannot be spent on another, and energy expended in doing one thing will not be available to do something else. We can analyze some of the choices made by individuals of a given species by comparing the costs and the benefits of alternative activities. This kind of assessment, called a **cost–benefit analysis**, is commonly used in economics to determine whether the financial cost of a project is less than the economic benefit that can be expected from the project. In behavioral ecology, costs are typically measured in terms of energy consumed, the probability of injury, or the probability of being killed by a predator. Benefits are usually measured in terms of a net gain in energy or an increase in reproductive success.

Behavioral ecologists assume that natural selection favors aspects of an individual's behavior that maximize the net benefit. For example, individuals that make better decisions about where to feed should have a higher net energy intake and be in better condition. Therefore, they should be better able to avoid predators and diseases, attract mates, and produce many young. Thus, natural selection should favor any behavioral attribute that consistently leads to good feeding decisions.

Given a set of assumptions, we can construct an optimality model to predict which combination of behaviors will maximize an individual's reproductive success in a given environment. Optimality models make explicit the relationships between costs and benefits of behaviors under various conditions. They are most useful in circumstances where it is clear that making the right decision maximizes some payoff, such as survival rate, reproductive success (number of young produced), feeding efficiency (energy gained per unit time), or mating success (number of matings per unit time). The following three sections are examples of optimality models.

Territorial Defense

We can examine how an optimality model works by considering territorial defense in animals. An animal's **territory** is any defended area. Many mammals, birds, lizards, and fishes defend a feeding area against other individuals of the same species. How large a territory should an individual defend? To answer this question, we need to think about the costs and benefits of defending a territory. The costs are time, energy, and risk of injury. The total cost will increase with the size of the territory, and for simplicity, we will assume that the relationship between cost and territory size is a rising curve because larger areas are more expensive to defend (**Figure 2**). The benefit of defending a territory is exclusive access to food, and it also increases with the size of the territory but suffers from diminishing returns.

Since an individual can consume only a certain amount of food, however, the benefit curve gradually levels off as the territory becomes larger. Above a certain territory size, there is no further increase in benefit (see Figure 2). The optimal territory size is the one that maximizes the **relative benefit** or profitability, which is the difference between the costs and benefits. In the hypothetical example shown in Figure 2, the relative benefit would be greatest at the territory size indicated by the arrow. Clearly, the optimal territory size is determined by the shapes of the cost and benefit curves, which vary with the species, habitat, and an individual's age or mating status.

The benefits of defending a territory are typically thought of as obtaining exclusive use of food resources, but for some species it may be the benefit of obtaining mates, avoiding predators, or defending juvenile animals from infanticide. Typically, for birds, the main considerations seem to be food and mates. Hummingbirds that migrate defend territories even during the nonbreeding period, and the assumption is that these territories are solely about food. Hummingbirds obtain most of their food energy from the nectar in flowers. Nectar is a resource that occurs in tiny amounts in individual flowers, consists mostly of water and some dissolved sugars, and

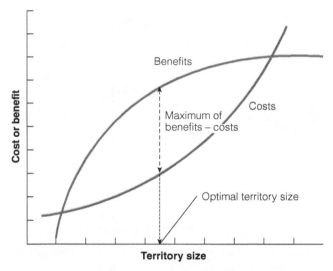

Figure 2 Hypothetical cost–benefit model for territory size in animals.

varies highly in availability. Hummingbirds have very high energy requirements for their body weight due to their small size, high body temperature, and use of hovering flight.

Rufous hummingbirds (*Selasphorus rufus*) live in western North America and migrate along the mountain chains—north to breed and south to overwinter. During their migration, they stop temporarily in mountain meadows to feed, and then move to a new site after refueling. They respond very quickly to changes in food resources—i.e., the nectar contained in flowers. Kodric-Brown and Brown (1978) showed that rufous hummingbirds adjusted their territory size to the available food supply (**Figure 3**), so that individuals always defended the same number of flowers regardless of the size of territory.

But why don't these hummingbirds defend a larger territory with more flowers? The implication is that the cost of defending a larger territory would exceed the benefits of having more food available. Carpenter et al. (1983) showed that if a hummingbird defended too large a territory, its rate of energy intake decreased because it spent too much time defending the territory and less time feeding (**Figure 4**). Diminishing returns are caused by high locomotion costs to defend more space, and a higher frequency of intrusions that reduce feeding time.

Hummingbirds are useful animals for the study of the costs and benefits of territorial defense because they can change their behavior daily and territories can change quickly in size. In many species, however, we cannot measure the costs and benefits of territorial defense at the same time, so we can see only part of the behavioral picture.

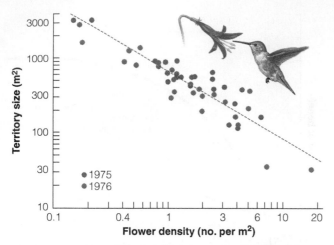

Figure 3 Territories defended by rufous hummingbirds (*S.rufus*) in relation to flower density. In this study, hummingbirds in the White Mountains of Arizona in both years defended territories with a constant number of flowers, indicating a constant food amount (indicated by the dashed line) regardless of the territory size. (Data from Kodric-Brown and Brown 1978.)

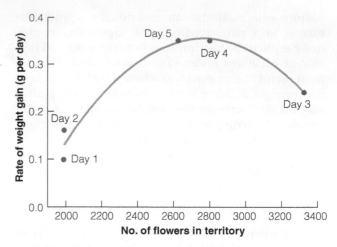

Figure 4 Daily weight change of one marked rufous hummingbird in the Sierra Nevada of California. This individual stayed in the mountain meadow for five days on its migration south. It showed the ability to adjust its territory size to an optimum in order to maximize the rate of gain of fuel for migration. (Data from Carpenter et al. 1983.)

If an animal does not behave as predicted by an optimality model, we should ask if the costs and benefits of the behavior have been correctly assessed or if additional factors should be considered. For example, the optimality model in Figure 2 assumes that cost and benefit curves are constant over time. Suppose instead that the shape of the cost curve varies from year to year. Should an animal change its territory size each year in response to these variations, or should it maintain a territory size that is optimal in average years? In ecosystems in which territories are occupied and defended year round, and the prey base fluctuates in size from year to year, individuals may adopt a territorial defense strategy that is suited to times of scarcity rather than change territory size every year. In many predators, such as the great horned owl, individuals defend territories that are larger than necessary on the basis of their food requirements (Rohner 1997).

One difficulty with optimality models is that they consider only one or two behaviors at a time, whereas individuals must simultaneously optimize all aspects of their behavior in order to survive and reproduce. We assume, however, that if a behavior such as territorial defense is directly linked to survival or reproductive success, then we should be able to detect how an individual organizes that behavior in a way consistent with the predictions of an optimality model.

Not all animals defend territories all the time, and some never defend any space. But all animals must eat

and we turn now to a more general question of foraging and how behavior can be organized to allow individuals to forage in an optimal manner.

Optimal Foraging

For all animals, food is not evenly distributed in time or in space. Consequently, acquiring food involves many behavioral decisions such as what type of food to consume, where and how to search for food, and once food is located, how much to eat and how long to keep foraging. Since animals must acquire food at a certain rate to maintain their physiological functions, the efficiency with which they can find and eat food is also important. Thus, we can assume that natural selection favors **optimal foraging**, which is any method of searching for and obtaining food that maximizes the relative benefit (the difference between costs and benefits, typically the net caloric gain per unit of time). Foraging provides an excellent opportunity to examine the factors that influence behavioral decisions because its benefits and costs are relatively easy to define, measure, and manipulate. Much of the research on foraging has been done on mammals and birds, and we begin our discussion with a simple model of optimal foraging.

Consider a predator such as an owl hunting for two kinds of prey. The prey are encountered at rates λ_1 and λ_2 prey per second during a specified time of searching, T_s seconds of searching. The two prey types yield E_1 and E_2 units of energy (measured in joules or calories), and take h_1 and h_2 seconds to handle each prey item. We define:

$$\text{Profitability of prey type 1} = E_1/h_1 \qquad (1)$$
$$\text{Profitability of prey type 2} = E_2/h_2 \qquad (2)$$

If the predator forages completely at random in T_s seconds, it will obtain on average this amount of food:

$$E = T_s(\lambda_1 E_1 + \lambda_2 E_2) \qquad (3)$$

And this foraging will take the following total amount of time (T) for searching and then handling the prey items:

$$T = T_s + T_s\,(\lambda_1 h_1 + \lambda_2 h_2) \qquad (4)$$

The overall rate of food intake of the predator is thus defined by the following equation:

$$\frac{E}{T} = \frac{(\lambda_1 E_1 + \lambda_2 E_2)}{(1 + \lambda_1 h_1 + \lambda_2 h_2)} \qquad (5)$$

Now we ask what happens if prey type 1 is more profitable to eat than prey type 2. In order to maximize the food intake (E/T), the predator should eat only prey type 1 if the rate of energy gain from prey type 1 is greater than the energy gained from eating both prey types:

$$\frac{\lambda_1 E_1}{1 + \lambda_1 h_1} > \frac{\lambda_1 E_1 + \lambda_2 E_2}{1 + \lambda_1 h_1 + \lambda_2 h_2} \qquad (6)$$

If we rearrange this equation, we obtain the following prediction: *The predator should specialize in eating only prey 1 if the equation below is true.*

$$\frac{1}{\lambda_1} < \frac{E_1}{E_2}(h_2 - h_1) \qquad (7)$$

This prediction is a threshold—eat only prey type 1 if the abundance of prey 1 exceeds this density, and eat both prey types if this inequality does not hold.

This simple model assumes there is some criterion to maximize (intake rate), some constraints to maximization (handling time), and alternative strategies (eat only prey 1 or eat both types of prey). **Table 1** lists the assumptions and the predictions of this simple optimal foraging model.

This simple optimal diet model has been very effective in stimulating research on foraging behavior in a variety of animals. In general, the results of empirical studies do not follow the model in observing a threshold change in diet. Instead, animals show partial preferences and eat the less preferred prey to some extent even when the model predicts they should eat only prey type 1 (Krebs and Davies 1993). **Figure 5** shows one example of this for the great tit. The data do not fit the model exactly because in nature birds must monitor the environment to estimate the relative abundances of the prey items, and in the process of doing this they encounter the less preferred prey occasionally and eat them in addition to the preferred prey. Animals do not

Table 1 Assumptions and predictions of the simple optimal foraging model.

Assumptions	Predictions
Prey value is measured in net energy of some single dimension	The highest-ranking prey in terms of profitability should never be ignored
Handling time is fixed for a given prey type	Low-ranking prey should be ignored according to Equation 6 above
Handling and searching cannot be done at the same time	Low-ranking prey are all or nothing in the diet, according to Equation 7 above
Prey are recognized instantly with no errors	The exclusion of low-ranking prey does not depend on their abundance (measured by λ_2)
Prey are encountered sequentially and randomly	
All prey individuals of a given prey type are identical	
Energetic costs of handling are similar for the two prey items	
Predators are maximizing the rate of energy intake	

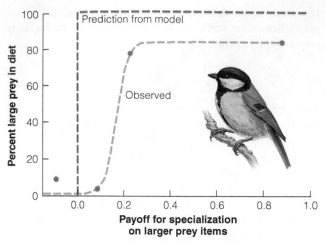

Figure 5 Test of the simple optimal foraging model for the great tit *(Parus major)*. Two sizes of worms were the prey choice in the laboratory. As more and more large prey are presented, the bird should stop eating small prey at the threshold and eat only large prey (red dashed line), according to the model. In reality the birds switch to large prey but always take some small prey (blue points). (Data from Krebs et al. 1977.)

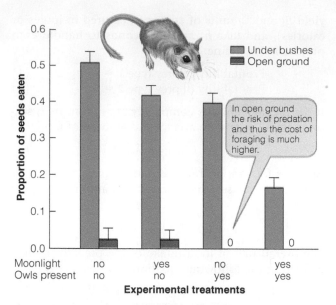

Figure 6 Proportion of seeds eaten by gerbils under bushes (green bars) and in the open (red bars) in the Negev Desert. Trays of seeds were set out in experimental enclosures in which the presence of moonlight and the presence of a predator, the barn owl, varied. Not all the seeds in the trays were eaten even in the best of circumstances in these overnight experiments. (Data from Kotler et al. 1991.)

have the perfect knowledge assumed in the simple models of foraging. Nevertheless, simple models are useful because they highlight the key processes that need studying and further analysis.

For many animals, food is distributed in a series of discrete patches across the landscape, some patches containing more food than others. If an animal is engaging in optimal foraging, it should preferentially forage in patches where the difference between benefits and costs is high. The benefits of foraging can be measured in terms of the amount of food obtained in each patch, and the costs can be measured in terms of the time taken and the probability of injury or predation. How will a forager respond when the costs of feeding in different patches are varied? We can answer this question by providing the same amount of food (a fixed benefit) in experimental patches that differ in their risk of predation or level of competition (varied costs). We can then determine how animals respond to changes in the costs of foraging by measuring how much food they eat in each patch. This approach was first used by Joel Brown in 1988 to investigate the foraging behavior of small mammals in desert habitats. He predicted that if the food levels are equal in two patches, a forager should stay longer and eat more food in the patch where the costs of foraging are lower.

One animal on which Brown's hypothesis has been tested is the gerbil *(Gerbillus* spp.). Gerbils are nocturnal, seed-eating rodents that live in sandy burrows. Their major predator is the barn owl, a rodent

specialist. Ecologists studied the foraging behavior of gerbils by placing seed trays in open areas and under bushes in experimental enclosures. Some enclosures were illuminated; others were dark. Captive, trained barn owls were released into some enclosures and not into others. If predation is the major cost of foraging by gerbils, they should eat more seeds under bushes and spend more time foraging there, especially in enclosures that are illuminated or that contain predators. This is exactly what the researchers found. As **Figure 6** shows, gerbils fed primarily at trays under bushes and reduced their overall feeding on bright nights, particularly when owls were present. They fed in open areas only when owls were absent. The results indicate that these desert rodents make choices based on the benefits of easily available food and the costs due to predation, and that the risk of predation influences their foraging behavior. If we were managing populations of gerbils, this study could tell us what kinds of habitat alterations might improve or decrease their survival and breeding success.

Simple optimal foraging models fit the observed data on many animals quite well but not perfectly, and this highlights some of the rigid assumptions of these quantitative models. Foraging models may be only partially correct (e.g., see Figure 6) because of discrimination errors (animals may confuse a prey 1 for a prey 2), simultaneous encounters (animals may see two different prey at the same time), or runs of bad luck in which animals do not encounter prey in a reasonably random manner. In spite of these problems, optimal foraging models have helped to instill quantitative rigor into studies of foraging in animals.

Optimal foraging studies support the conclusion that animals are finely adapted to searching for food in ways that achieve maximum relative benefit. Natural selection continues to favor efficient foraging traits.

Optimal Migration

Animals need information in order to make decisions, and optimality models often assume that animals are fully informed about their environment when foraging or mating. Migrating birds are a special case of the problem of decision making. Migrating animals must choose how far to move in one step, and if they cannot feed while migrating, how much fuel to carry en route. Migrating birds are a special case in decision making because they incur large locomotory costs in flight, and the strategies migratory birds use have been extensively studied (Alerstam 1990).

Migrating birds have three potential migration strategies:

- Time minimization (complete migration in the minimum possible time). The birds should optimize the overall speed of migration, which means that the birds will waste energy to achieve this goal. This would be a desirable strategy if early arrival at the destination is an important fitness advantage for the birds.

- Energy minimization (complete migration with the least energy cost). This strategy will be selected for when the risks associated with migration are relatively high and the use of energy during migration is high. This strategy would also be advantageous if energy resources along the migratory route are sparse. The net result from adopting this strategy will be some waste of energy on an annual basis. The birds are expected to minimize stopover times and increase migration speed.

- Cost of transport minimization. This is a second energy minimization strategy but focuses on the overall goal of minimizing total energy use over the entire annual cycle. The energy used in migration is only one part of the annual energy use for migratory birds, and minimizing energy in migration typically results in using more energy

over the whole annual cycle. This model optimizes migration cost but within the whole annual cycle rather than only the restricted migration period.

Because the aerodynamics and energetics of bird flight have been so well investigated, it is possible to construct optimal migration models for these three strategies (Hedenström and Alerstam 1997). We consider here only the simple case for many passerine migrants that migrate in a series of hops rather than in one long flight. At each stopover point, the birds must refuel, and there is an energy cost to finding the necessary food at the stopover points. Two variables are critical for the birds: (1) fuel deposition rate—the rate of energy accumulation by feeding before migration begins and during stopovers—which is measured by the fraction of lean body mass accumulated per day; and (2) departure load—the amount of fat and protein expressed as a fraction of lean body mass. The predicted relationship between these two variables under the three optimality models is shown in **Figure 7**. The key prediction of the third model is that the departure load of a bird will be constant and independent of the rate at which fuel can be accumulated. Both the time and the energy minimization models show an increasing relationship so that when more fuel can be accumulated, the departure load will increase.

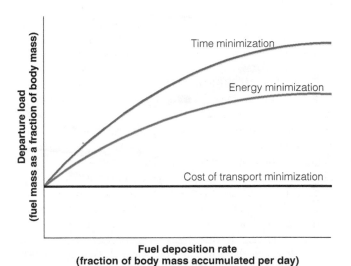

Fuel deposition rate
(fraction of body mass accumulated per day)

Figure 7 Relation between fuel deposition rate and departure load for birds migrating in a series of flights with stopovers en route. Three migration strategies are possible to minimize time or energy, and the graph shows the predictions of these three models. The costs of transport model is energy minimization on an annual basis, while the energy minimization model is energy minimization for the migratory period only. (Modified from Bayly 2006.)

We can test the models by measuring fuel deposition rate and departure loads for migrating birds. Bayly (2006) did this with reed warblers *(Acrocephalus scirpaceus)*, a trans-Saharan migrant. These birds must cross over 2500 km of desert, a feat that requires a large fuel load. Much of the migration in this species and other birds is spent in a series of fuelling phases, and migratory flight will occupy a relatively short amount of time overall. Bayly (2006) provided reed warblers with supplementary food at a site in southern England, and was able to record fuel deposition rate and departure loads, with the results shown in **Figure 8**. There is considerable variation among individual birds, but the time minimization model fit the data best. So far, most of the tests of the optimal migration model have supported the time minimization model.

The amount of energy small birds use in stopovers is typically two to three times the amount of energy used in actual migratory flights (Hedenström and Alerstam 1997). The time spent in stopovers is about 7 times that spent on flight for small birds, and much more for larger birds. These surprising results suggest that more studies need to be undertaken at stopover points for migratory birds to measure the energetics of individual birds during stopovers. Critical habitats for migratory birds are not just the endpoints (breeding and wintering areas), but also the stopover localities in between.

Many large mammals undertake seasonal migrations typically associated with seasonal food resources (Fryxell and Sinclair 1988). These migrations can have important consequences for population dynamics.

Group Living

Many animals live in groups. Grazing herbivores form large herds, fish school together, carnivores form hunting groups, birds breed in large colonies, and some animals live in extended family groups. If natural selection favors individual interests over group interests, why should animals ever associate, much less cooperate with others to hunt or raise young? We can start to understand the factors that drive the evolution of group living by evaluating its benefits and costs **(Table 2)**.

Benefits of Group Living

The two main factors affecting group size are food and predators. If food is sparsely distributed and difficult to locate, living in a group can increase an individual's foraging success by allowing it to obtain information about the location of food from successful foragers. This idea, which was first proposed by Paul Ward and Amotz Zahavi in 1973, explains why some birds nest in colonies.

Social insects are the classic example of cooperation for food gathering. Karl von Frisch discovered more than 80 years ago that when successful bee foragers returned

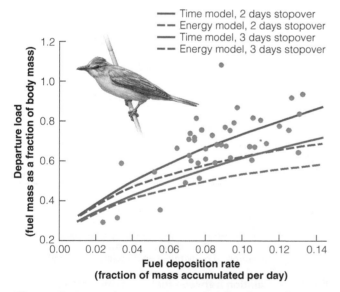

Figure 8 Optimal migration strategy for the reed warbler. The best fit to the observed data is given by the time minimization model with 3 days stopover costs. The cost of transport model, which predicts a horizontal relationship, is not supported for this species. High variability among the individual birds could be due to birds making errors when calculating the correct departure load. (Modified from Bayly 2006.)

Table 2 Potential benefits and costs of group living in animals.

Potential benefits	Potential costs
Increased foraging efficiency	Competition for food
	Increased risk of disease or parasites
Reduced predation	Attraction of predators
Increased access to mates	Loss of paternity
	Brood parasitism
Help from kin	Loss of individual reproduction

to the hive, they communicate the location of a rich food source using a waggle dance (**Figure 9**). This dance involves running through a small figure-eight pattern—a straight run followed by a turn to the right to circle back to the starting point, and then another straight run followed by a turn and circle to the left. The straight section of the dance is the most striking and the most informative part of the signaling bee's dance. While walking straight ahead, the bee waggles or vibrates its body back and forth, side to side. At the same time, the bee emits a buzzing sound. Typically, several workers cluster closely around the dancing bee, trying to maintain contact and to obtain information (von Frisch 1967).

The direction and duration of straight runs in the dance are closely correlated with the direction and distance of the patch of flowers just visited by the dancing forager (see Figure 9). The farther away the target, the longer the straight-run part of the dance. In addition to information on the direction and distance of the flower food source, the dancing bee also communicates the odor of the flowers. This communication is typically given by the pollen it carries back on its hind legs, or in the nectar it regurgitates to the surrounding bees.

There is no question that the dance of the returning honeybees gives information to the recruits, but how precise is it and over what range can it operate. Bees routinely forage up to 12 km from their nest (Seeley 1985), and it is clear that bees can recruit nestmates to forage in patches up to 10 km from home. But how precise is this recruitment? Only a small percentage of bees that closely follow the waggle dance actually find the food source. Gould (1975) described one study in which the precision of recruitment to a food source 315 m distant had an error of about 60 m in either direction. Successful recruits in several studies needed two to seven trips to find the exact food source. One suggestion is that once in the general area of a food source, bees use odors to find flowers. But it has been shown that bees in hives that are allowed to carry out dances with directional light had improved food collection compared with bees in colonies that had diffused light in the hives (which does not permit the correct dance orientation). Sherman and Visscher (2002) found that this advantage from the properly oriented dances was effective in increasing colony food collection only in those seasons of the year when the sun was at its highest.

A second potential benefit of group living is a reduced risk of predation. Group living may appear to be a benefit to the group, but it is the advantage it gives to individual animals that is the driving force in the evolution of group living. If a predator takes a single individual as prey, each individual's risk of predation would drop from 10 percent in a group of 10 to 1 percent in a group of 100, if all other factors are equal. This "dilution effect" is a passive benefit

(a)

(b)

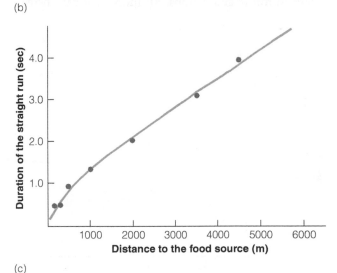

(c)

Figure 9 The waggle dance of the honey bee. (a) The patch of flowers lies 1500 m out along a line 40° to the right of the sun as the bee leaves the colony nest in the tree. (b) To advertise this target when the forager returns to the nest, the bee runs through a figure-eight pattern, vibrating her body laterally as she passes down the straight run. The straight run is oriented on the vertical honeycomb by transposing the angle shown in (a) to the angle between the straight run and the vertical. (c) Distance to the flowers is coded by the duration of the straight run. (Data from Seeley 1985.)

of larger groups. But this benefit must be balanced against the higher probability that a predator will find a large group than a small group or an individual. Animals in a group can also actively lower their risk of predation by being vigilant for predators. Increasing group size can make vigilance more effective and less costly, since many eyes increase the probability of predator detection and reduce the time each individual must spend being vigilant. Less time spent in vigilance should translate into more time for other activities, such as foraging.

Guppies *(Poecilia reticulata)* in Trinidad live in streams with differing predator densities. When predators are abundant, guppies school in more tightly spaced groups (**Figure 10**). But predators prefer to attack larger schools of guppies, which challenges the idea that it is safer in a group. The key question is whether an individual guppy is safer in a larger group. Krause and Godin (1995) tested the safety of group living in the laboratory where they could expose groups of guppies to cichlid predators for short time periods. **Figure 11** shows that predators attack larger schools more often if given a choice between a small school of guppies and a large school, but that for each individual guppy the likelihood of being captured by the predator is much higher in small groups.

From an evolutionary point of view, success is measured in terms of the number of copies of one's genes in future generations. An individual can increase its evolutionary success, or fitness, directly by producing its own young, and indirectly by increasing the survival or reproductive success of close relatives, which have some of the same genes. Helping relatives and being helped by relatives is one benefit of group living in some animals, so cooperation in these animals has an evolutionary explanation.

Belding's ground squirrels *(Spermophilus beldingi)* provide an example of apparent cooperation in group-living animals. These rodents live in burrows in alpine and subalpine meadows in western North America. Although both sexes disperse from the burrow where they were born, males move much farther than females. This difference in dispersal distance leads to neighborhoods where females are closely related but males are not. Belding's ground squirrels produce loud alarm calls when predators—chiefly coyotes, pine martens, and long-tailed weasels—are in the area. Alarm calls serve as an early warning for other ground squirrels living nearby, but they provide no immediate benefit to the caller. In fact, they may increase costs for the caller by attracting predators to it. Why then should any individual produce alarm calls? Paul Sherman addressed this question by studying a population of individually tagged Belding's ground squirrels over several years. He found that females were far more

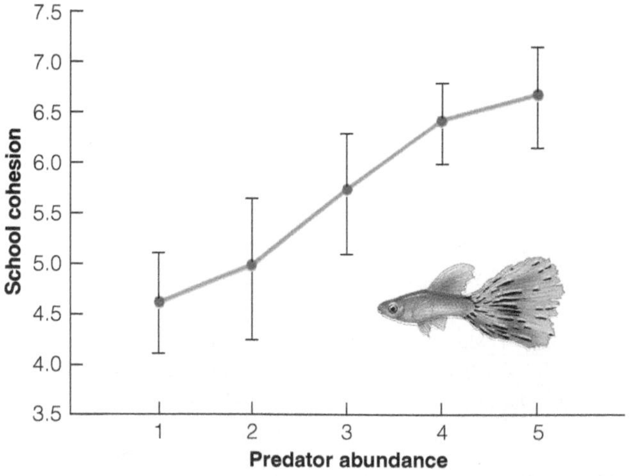

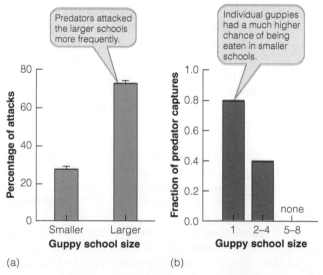

Figure 10 Guppies in Trinidad streams live in tighter schools in streams in which predators are more abundant. Each point represents a different stream, and the cohesion score is based on how much individual fish spaced from one another (with 95 percent confidence). (Data from Seghers 1974.)

Figure 11 Experimental analysis of predation risk in guppy schools. The cichlid fish *Aequidens pulcher* was used as the predator in these experiments. The overall result is that an individual guppy always benefited by joining a large school, even though large schools are attacked more often. (Data from Krause and Godin 1995.)

Do Individuals Act for the Good of the Species?

Natural selection occurs because of the reproductive advantages of some individuals. This view of the world implies that all individuals are in competition with each other and will behave to further their own interests. From a philosophical viewpoint, the idea that the world is full of selfish individuals clashes with many of the values we hold for human societies, such as cooperation, community spirit, and selflessness. Does the variety of behaviors that we observe in animals, even the apparently cooperative ones, really arise from the interactions of selfish individuals? Can traits evolve that favor the larger interests of a group or society? Does evolution lead only to selfishness? These are key questions that interest social scientists, philosophers, and biologists. Biologists do not think that individuals ever act for the good of the species, but there are many situations in which what appear to be selfish individual behaviors operate to benefit a group.

It is easy to imagine that populations of selfish individuals might overexploit the available resources and become extinct, whereas populations that have evolved social behaviors preventing overexploitation of resources might have better long-term survival prospects. Natural selection for traits that favor groups rather than individuals is termed **group selection**. The idea that groups of animals could evolve self-regulating mechanisms that prevent overexploitation of their food resources was first argued in detail in 1962 by V. C. Wynne Edwards, an ecologist at the University of Aberdeen in Scotland. Despite its intuitive appeal, group selection is not considered very important in producing changes in species traits. Group selection operates much more slowly than individual selection, making it a much weaker selective force in most circumstances.

To understand apparently cooperative behaviors that benefit the group or society, we need to look for benefits accruing to individuals.

Imagine, for example, a species of bird, such as the puffin that lives in large colonies and lays only a single egg. Could laying a single egg have evolved in puffins by group selection to limit population growth and maintain an adequate food supply for the long-term good of the puffin colony? The answer is no, because any mutation that increased the number of eggs laid would be favored only if individuals laying two eggs leave more copies of their genes to the next generation, compared with birds laying a single egg. But ecologically speaking, costs would increase as well as benefits. A puffin with two eggs would have to collect more calcium to lay two eggs and would have to fly more to feed two young. There are ecological costs to increasing the clutch size in puffins. Consequently, genes for laying two eggs would not spread through the population unless the benefits would exceed the costs. Individual selection favors the small clutch size in puffins. Short-term advantages to selfish individuals will accrue much more quickly than long-term advantages to the group, so it is difficult to see how traits favored by group selection can be maintained in a population unless they are also favored by individual selection.

But this does not mean that all behavior must be selfish and that altruism does not exist. To understand apparently cooperative behaviors that benefit the group or society, we need to look for benefits accruing to individuals. Individual selection can produce behaviors that are a benefit for the group.

Some of the best examples of individuals working for the good of the group come from the social insects. Ants and many bees live in colonies in which the individuals cooperate to rear young and defend the hive. Natural selection in the social insects operates through kin selection, and individuals in these insect colonies cooperate to further the interests of the entire colony (Queller and Strassmann 1998).

likely to give alarm calls than males (**Figure 12a**). However, females differed in the frequency with which they called: Females with nearby relatives, even young females that had no offspring of their own, called more often than females that had no relatives in the area (**Figure 12b**). Thus, Belding's ground squirrels are more likely to call when doing so may benefit the survival of their close relatives. The evolution of traits that increase the survival, and ultimately the reproductive success, of one's relatives rather than oneself is termed **kin selection**.

Costs of Group Living

Living in a group has costs, as well as benefits (see Table 2). The magnitude of these costs limits the extent to which a species forms groups and explains why some groups are larger than others. Not surprisingly, living in large groups leads to competition for resources, such as food or mates. For example, Magellanic penguins (*Spheniscus magellanicus*) form breeding colonies of up to 200,000 birds on subantarctic islands. Colony size in this species appears to be limited by competition for food,

which consists of squid and pelagic schooling fishes including anchovy. Adults and chicks in small colonies ingest more prey of high-energy content than do individuals in large colonies (**Figure 13**), and fledglings in small colonies are healthier and therefore more likely to reach adulthood. The costs of group living are related to colony size, and one of the consequences of this is population regulation by food shortage.

Breeding in large colonies can also increase the transmission of diseases and parasites, which have population consequences. Another species of penguin, the king penguin (*Aptenodytes patagonicus*), breeds in Antarctica in colonies of up to 500,000 individuals. In large colonies, adults and chicks become infested with ticks (*Ixodes uriae*). High rates of tick infestation reduce the incubation success of adults (**Figure 14**).

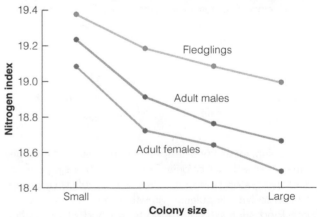

Figure 13 Relationship between nitrogen index and colony size in Magellanic penguins. The nitrogen index is based on the ratio of stable nitrogen isotopes in blood samples and is an indicator of food quality. A higher nitrogen index reflects a diet of more nutritious prey, such as anchovies. All age and sex groups suffer a poorer diet in large colonies. (Data from Forero et al. 2002.)

Figure 12 Patterns of alarm calling by Belding's ground squirrels. (a) Effect of sex on frequency of calling. (b) Effect of type of nearby relatives on frequency of calling by females. In both (a) and (b), the vertical axis is the percentage of time that squirrels produced alarm calls when a predator approached. (Data from Sherman 1977.)

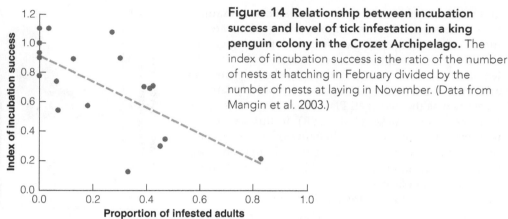

Figure 14 Relationship between incubation success and level of tick infestation in a king penguin colony in the Crozet Archipelago. The index of incubation success is the ratio of the number of nests at hatching in February divided by the number of nests at laying in November. (Data from Mangin et al. 2003.)

Another important cost of group living is loss of parentage. Breeding in a group increases the chance that an animal will raise another individual's offspring. This problem is well illustrated in cooperatively breeding birds. Splendid fairy wrens *(Malurus splendens)* in southern Australia are small songbirds that live in cooperative groups of a dominant male, a single female, and one or more auxiliaries (almost always males). All individuals in the territorial group cooperate in feeding and caring for young, and this is why dominant males tolerate auxiliary males in the group. But females engage in extra-pair copulations, both with males from another group and with auxiliary males within the group, so that about 40 percent to 70 percent of the offspring are sired by males who are not the dominant territory holder (Rowley and Russell 1997; Webster et al. 2004). The frequency of extra-pair copulations increases with the size of the cooperative group (**Figure 15**). Extra-pair copulations help to prevent inbreeding in cooperative breeders, and they explain in part the advantage that auxiliary males may gain from helping raise broods. Webster et al. (2004) found that 75 percent of the extra-pair young were fathered by the dominant male in another group, 10 percent by auxiliary males in the same group, and 14 percent by auxiliary males in another group. The potential costs and benefits of group living can vary among breeding groups with different levels of relatedness. If all members of the group are closely related, individuals will gain by helping their relatives. But if few members of the group are related, individual selection will be stronger than kin selection and the ratio of costs and benefits for an individual bird will be less favorable.

Group Living in African Lions

Ecologists have been studying the social behavior of lions *(Panthera leo)* for more than 40 years in eastern

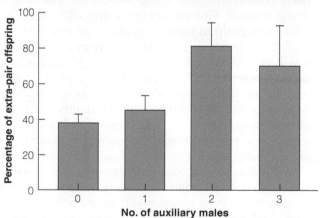

Figure 15 Percentage of offspring that were sired by extra-pair males as a function of the number of auxiliary males in the splendid fairy wren in South Australia. Bars indicate upper 95 percent confidence limits. The larger the cooperative breeding group, the less the reproductive success for the dominant male. (Data from Webster et al. 2004.)

and southern Africa, and they are now a classic example of the costs and benefits of group living. Lions are the most social member of the cat family, forming groups called prides composed of one to seven males, 2 to 18 females, and their young. Prides are relatively small in arid areas such as the Kalahari, and relatively large in areas such as the Serengeti Plains that have more abundant large prey (Packer et al. 1988). In this section, we will examine the costs and benefits of different pride sizes and try to understand the benefits of group living for lions. Why do lions live in prides, and why do pride sizes vary from place to place?

Male and female lions behave in very different ways, and these differences influence the costs and benefits of group living for each sex. Females almost never leave the pride in which they were born. They cooperate with their mothers, sisters, and other female relatives in hunting, raising young, and defending territory. In contrast, male lions are highly transient. They leave their pride of birth when two to three years old and roam widely in search of a new pride. Males that do not belong to a pride often group with related or unrelated males, forming coalitions that challenge males in existing prides for breeding positions. These challenges may result in the death of one or more of the participants. Once in a pride, the males do little hunting, and instead spend most of their time defending their territory by patrolling, scent marking, and roaring. Because of frequent challenges, males rarely retain control of a pride for more than two years.

Because of the behavioral differences between male and female lions, we will consider the benefits of male–male groups and female–female groups separately. For males, the major benefit of grouping is straightforward. Single males rarely succeed in obtaining a breeding position within a pride. Large coalitions are more likely to take over a pride and are more effective at repelling challenges from other males. Males that take over a pride kill unrelated cubs, and thus challenges must be repelled. Consequently, an individual male's reproductive success increases with the number of males in a coalition (**Figure 16**). The longer a coalition can remain in control of a pride, the more cubs those males can produce, and the greater the cubs' chances of survival. Although male reproductive success increases with coalition size, individual breeding success becomes more variable in the largest coalitions: Some males mate often, whereas others rarely mate. As a result, male–male competition for mating can act to set an upper limit on coalition size.

For female lions, the benefit of group living—as measured by reproductive success—is greatest in groups of 3 to 10 females (**Figure 17**). This appears to be the **optimal group size**, the size that results in the

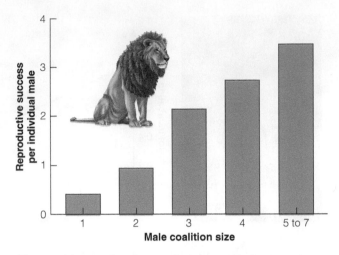

Figure 16 Benefit of group living in male lions. Males in larger coalitions have increased reproductive success. (Data from Packer et al. 1988.)

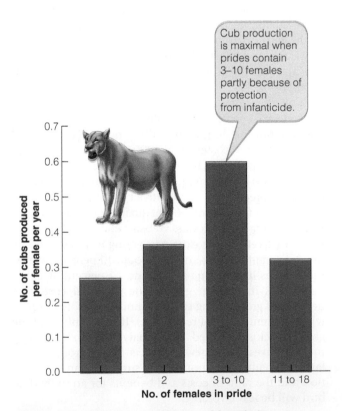

Cub production is maximal when prides contain 3–10 females partly because of protection from infanticide.

Figure 17 Reproductive success of female lions in prides of differing size. The production of cubs is maximal when prides contain 3 to 10 females. (Data from Packer et al. 1988.)

Table 3 Specific benefits and costs of forming male or female groups in African lions.

Sex	Benefits of grouping	Costs of grouping
Male	Increased ability to gain control of a pride (access to mates)	Sharing of paternity with coalition members
	Increased ability to maintain control of a pride (higher survival of offspring)	
Female	Preferential feeding of close kin (help from kin)	Lower rate of food intake
	Territorial defense (increased female and offspring survival)	

largest relative benefit. How can we explain this observation? Careful calculations have shown that very small prides (and even solitary lionesses) have the highest rates of food intake. Thus, hunting success seems to decrease as group size increases. In contrast, larger groups facilitate territorial defense, which is important in preventing male takeovers. When new males take over a pride, they typically kill all the young cubs. That causes the females in the pride to rapidly enter estrus, allowing the new males to father offspring quickly. Cub survival is higher in larger female groups because larger groups are better able to save young cubs from infanticide. Thus, the optimal group size in female lions may represent a balance between hunting success and territorial defense.

As this example of African lions illustrates, understanding which factors favor group living in a species can be complex (**Table 3**). Although we can easily identify potential costs or benefits of group living, to single out the important factors, we must determine how this behavior affects the survival and reproductive success of an individual. Doing this successfully requires detailed data on individuals from groups of different sizes, carefully designed field experiments, or both. The relative benefit of group living may vary with habitat type and other environmental conditions, making long-term studies especially important. In many species, the costs and benefits of group living differ between the sexes, which can lead to conflict between males and females over the optimal group size.

Summary

If a population is to persist, its members must obtain food, avoid predators and disease, and produce offspring. They achieve these goals through a variety of behaviors, which must be appropriate for their particular environment. Many animals must make decisions about where to forage, which individuals to mate with, how large a territory to defend, and which habitat to select for nesting. Natural selection is the force that achieves the fit between how individuals behave and their subsequent survival.

The key to understanding the behavior of individuals is to determine the costs and benefits of these decisions in terms of the number of offspring an individual produces. Optimality models assist us in understanding animal behaviors by forcing us to quantify the costs and benefits of decisions. This approach has been particularly successful for foraging behavior, and we can identify foraging rules by which animals optimize their food intake rates. A cost–benefit analysis can also help us identify the factors that affect the social structure of a species, such as its optimal group size and how large a territory it defends. Understanding the factors that influence the behavior of individuals may allow us to predict how different species will respond to conservation problems such as habitat loss.

Behavioral ecology forms a bridge to understanding the dynamics of populations and communities. Mechanisms such as climate change that affect populations and communities must ultimately relate to how individual animals adapt their behavior to a changing environment.

Review Questions and Problems

1   In 1957 Carl Haskins moved 200 guppies from a river with high predator abundance to the predator-free upper headwaters of the Oropuche River in Trinidad. What predictions would you make for the guppies moved in this transplant experiment? How would you test experimentally whether the antipredator behaviors are under genetic control or under environmental control? State two or more hypotheses for this experiment and discuss how you might test their predictions. Magurran et al. (1992) discuss this transplant experiment.

2   What assumptions underlie the cost–benefit approach to optimality models? Is it possible to test whether or not an animal is acting optimally? Could there be cases in which animals might not be well adapted? Krebs and Davies (1993) discuss these questions.

3   Altruism—personal sacrifice on behalf of others—is difficult for behavioral ecologists and evolutionary biologists to explain because natural selection favors the interests of individuals. Nevertheless, altruistic behaviors toward relatives are observed in many animal societies. Is there any way that altruism among nonrelatives can evolve in animal societies? How might altruism arise in human societies if it is based on self-interest? Gintis et al. (2003) discuss this question.

4   Many birds form groups in which only one female breeds and other birds act as helpers at the nest. Discuss the relative benefits of males and females for being a helper in such breeding groups. Why might an individual choose to stay as a helper in a group rather than move away and breed elsewhere? Heinsohn and Legge (1999) discuss this problem of cooperative breeding.

5   In Scotland, female offspring of red grouse disperse to surrounding areas, while male offspring take up a territory next to their father, if they survive. A male's territory is always occupied exclusively by one bird. Describe how the aggression associated with territorial defense might differ if a male is surrounded by his sons or by unrelated males. Mougeot et al. (2003) describe this system and some experiments on this issue.

6   Infanticide is observed in many mammals, birds, and insects. Female infanticide is surprisingly common in human cultures. Using the approaches discussed in this chapter, (a) formulate two hypotheses to explain infanticide in humans, (b) describe the data you would collect to test your hypotheses, and (c) discuss the proposition that infanticide is adaptive in humans.

Suggested Readings

- Anderson, K. G., H. Kaplan, and J. Lancaster. 1999. Paternal care by genetic fathers and stepfathers I: Reports from Albuquerque men. *Evolution and Human Behaviour* 20:405–432.

- Dornhaus, A., and L. Chittka. 2004. Why do honey bees dance? *Behavioral Ecology and Sociobiology* 55:395–401.

- Forero, M. G. et al. 2002. Conspecific food competition explains variability in colony size: A test in Magellanic penguins. *Ecology* 83:3466–3475.

- Hamilton, W. D. 1971. Geometry of the selfish herd. *Journal of Theoretical Biology* 31:295–311.

- Hedenström, A., and T. Alerstam. 1997. Optimum fuel loads in migratory birds: Distinguishing between time and energy minimization. *Journal of Theoretical Biology* 189:227–234.

- Houston, A. I. et al. 2007. Capital or income breeding? A theoretical model of female reproductive strategies. *Behavioral Ecology* 18:241–250.

- Krebs, J. R., and N. B. Davies. 1993. *An Introduction to Behavioural Ecology*. 3rd ed. Oxford: Blackwell Scientific Publications.

- Packer, C. et al. 2005. Ecological change, group territoriality, and population dynamics in Serengeti lions. *Science* 307:390–393.

- Sutherland, W. J. 1996. *From Individual Behaviour to Population Ecology*. New York: Oxford University Press.

- Wolff, J. O., and D. W. Macdonald. 2004. Promiscuous females protect their offspring. *Trends in Ecology & Evolution* 19:127–134.

Credits

Illustration and Table Credits

Photo Credits

Analyzing Geographic Distributions

Key Concepts

- All species have a limited geographical range, and the task is to discover what causes these limits.

- Transplant experiments can help to identify the potential range of a species.

- Shelford's law of tolerance can be used to define the critical environmental limits to survival and reproduction and thus the potential geographical range for a species.

- The tolerance ranges of species can change via natural selection.

control In an experimental design a control is a treatment or plot in which nothing is changed so that it serves as a baseline for comparison with the experimental treatments to which something is typically added or subtracted.

dispersal The movement of individuals away from their place of birth or hatching or seed production into a new habitat or area to survive and reproduce.

habitat selection The behavioral actions of organisms (typically animals) in choosing the areas in which they live and breed.

Liebig's Law of the Minimum The generalization first stated by Justus von Liebig that the rate of any biological process is limited by that factor in least amount relative to requirements, so there is a single limiting factor.

physiological ecology The subdiscipline of ecology that studies the biochemical, physical, and mechanical adaptations and limitations of plants and animals to their physical and chemical environments.

Shelford's Law of Tolerance The ecological rule first described by Victor Shelford that the geographical distribution of a species will be controlled by that environmental factor for which the organism has the narrowest range of tolerance.

Why are organisms of a particular species present in some places and absent from others? This is the simplest ecological question one can ask, and hence it forms a good starting point for introducing you to spatial ecology. This simple question about the distributions of species can be of enormous practical importance. Two examples illustrate why. Five species of Pacific salmon live in the North Pacific Ocean and spawn in the river systems of western North America, Asia, and Japan. Salmon are valuable fish for commercial fishermen and sport fishermen alike. Why not transplant such valuable fish to other areas—for example, to the North Atlantic or to the Southern Hemisphere? Why are five species of salmon present in the Pacific but only one species in the Atlantic? Sockeye salmon have been transplanted to France, Denmark, Mexico, and Argentina but did not survive there (Lever 1996). Coho salmon were successfully transplanted to New Zealand in 1904 (Kinnison et al. 1998). Why are some salmon introductions successful and many others not?

The African honey bee is a second example that illustrates the practical consequences of species distributions. The African honey bee (*Apis mellifera scutellata*) is a very aggressive subspecies of honey bee that was

brought to Brazil in 1956 in order to develop a tropical strain with improved honey productivity. They escaped by accident, and the spread of the African bee since 1956 is shown in **Figure 1**. Because African bees are aggressive, they may drive out established colonies of the Italian honey bee (*Apis mellifera ligustica*). In other situations, hybrids may be formed between the African and Italian subspecies.

In 1982 the African honey bee crossed the Panama Canal. It reached Mexico in 1985, and southern Texas in 1990. Moving roughly 110 km per year, it crossed the border into California in 1994 and is currently slowly spreading into the southern United States. By 2007 the African bee has reached central Oklahoma, central California and Nevada, and all of Arizona. It had hitchhiked over to southern Florida and has spread there. Unfortunately, African bees are aggressive toward humans and domestic animals, and accounts of severe stinging and even deaths have served to highlight the spread of the

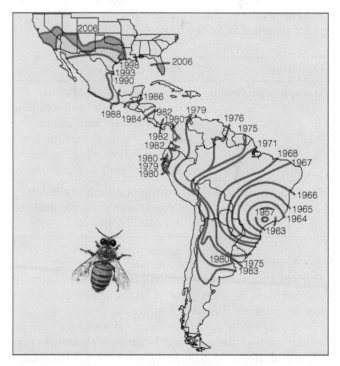

Figure 1 Spread of the African honey bee in the Americas since 1956. Southward and westward expansion in South America has been slight since 1983. Colonization of south Florida since 2001 has been the result of hitchhiking on ships or trucks. The northward movement in the southwestern United States has slowed considerably in the last few years. The dark orange area shows the spread from 1998 to 2006, and the lighter orange area the spread from 1993 to 1998. (Map data from O. R. Taylor and US Department of Agriculture, National Invasive Species Information Center.)

African bee (Schneider et al. 2004). By 2004 the African honey bee had killed 14 people in the United States, and beekeepers are understandably worried that the African bee will damage the established honey bee industry. What factors limit the distribution of the African honey bee? Will this species be able to live as far north as Oregon and North Carolina?

Transplant Experiments

To answer questions concerning distribution, we must first determine whether the limitation on distribution results from the inaccessibility of the particular area to the species. One way to determine the source of limitation is a transplant experiment. In a transplant experiment, we move individuals of a species to an unoccupied area and determine whether they can survive and reproduce successfully in the new environment. Some organisms can survive in areas but cannot reproduce there, so we should follow transplant experiments through at least one complete generation. The two possible outcomes of a transplant experiment are the following:

Outcome	Interpretation
Transplant successful	Distribution limited either because the area is inaccessible, time has been too short to reach the area, or because the species fails to recognize the area as suitable living space.
Transplant unsuccessful	Distribution limited either by other species or by physical and chemical factors.

A proper transplant experiment should have a **control**, transplants done within the distribution to provide data on the effects of handling and transplanting the individual plants or animals.

If a transplant is successful, it indicates that the potential range of the species is larger than its actual range. **Figure 2** shows this schematically for a hypothetical plant or animal. The results of successful transplant experiments thus direct our further investigations in one of two ways. If a species does not seem to occupy all of its potential range, we must determine if it can move into its potential range or if it lacks suitable means of reaching new areas. Some animal species could move into new areas but do not do so. For these species, we must study their mechanisms of **habitat selection**.

If the species cannot survive and reproduce in the transplant areas, we ask whether biotic interactions

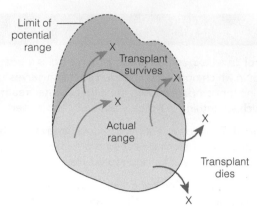

Figure 2 Hypothetical sets of transplant experiments applied to the same species. The yellow area represents the actual distribution of the population. The limit of the potential range is shown in light blue. Many separate transplant experiments are needed to define the limits of the potential range; only five are shown in this illustration.

with other species or abiotic factors exclude it from these areas. Limits imposed by other species may involve either the negative effects of predators, parasites, disease organisms, or competitors, or the positive effects of interdependent species within the actual range. We can often determine if other species are restricting distribution by conducting transplant experiments with protective devices such as cages designed to exclude the suspected predators or competitors. For example, we can transplant barnacles in mesh cages to deeper waters along the coast to see if they can survive in deep water if starfish or gastropod predators are kept away.

If other species do not set limits on the actual range, we are left with the possibility that some physical or chemical factors set the range limits. For example, many tropical plant species cannot withstand freezing temperatures, and the frost line effectively limits their distributions. Limitations imposed by physical or chemical factors have been studied extensively and are the subject of a whole discipline called **physiological ecology**.

Physiological Ecology

Physiological ecologists study the reactions of organisms to physical and chemical factors. To live in a given environment, an organism must be able to survive, grow, and reproduce, and consequently physiological ecologists must try to measure the effect of environmental factors on survival, growth, and reproduction. Victor Shelford, one of the earliest North American animal ecologists, was the first to formalize the ideas of physiological ecology with the view to understanding the distribution of species in natu-

Liebig's Law of the Minimum

Justus von Liebig (1803–1873) was a nineteenth-century German chemist. After working in organic chemistry in the first part of his life, he became interested in biochemistry and in particular how plants transform inorganic matter in the soil and atmosphere into organic matter. He postulated in 1840 what has been called Liebig's Law of the Minimum, which states that the rate of any biological process is limited by that factor in least amount relative to requirements. Crop yields were limited, according to Liebig, by a single nutrient, and if one added the limiting nutrient in fertilizer, production would increase. Liebig's work with artificial fertilizers was revolutionary in its time because of its linkage of chemistry and biology and its achievements in producing higher crop yields.

Liebig's Law has been attacked as too simplistic because it postulates that at any point in time *there is only one limiting factor* for any process. The modern view is that a number of nutrients may be limiting simultaneously, and that there may be combined enhancements from mixtures of nutrients. Although Liebig's Law is not applicable in all situations, it forms a very useful starting point for understanding what limits ecological processes. Consider one relatively simple question: What limits the distribution of African bees? The key to answering this question is experimental work in physiological ecology: Vary temperature and describe how survival and reproduction change with changing temperature. If we find that temperature is limiting in one area, however, this does not mean that moisture is not limiting in another area, or that in yet another area a combination of temperature and moisture are not the key factors. In this manner, we built complexity experimentally, but on a foundation established by a German chemist 150 years ago.

ral communities. Working at the University of Illinois, Shelford developed the major conceptual tool of physiological ecologists, **Shelford's Law of Tolerance**, which can be stated as follows: *The distribution of a species will be controlled by that environmental factor for which the organism has the narrowest range of tolerance.*

The job of physiological ecologists is to determine the tolerances of organisms to a range of environmental factors. This is not a simple chore. We could, for example, determine the range of temperatures over which a species can survive. For fish there are three methods that have been used to determine temperature tolerances (Beitinger et al. 2000) The most common method used is the critical thermal methodology **Figure 3**. Fish are acclimated to a specific temperature and then subjected to a constant linear increase or decrease in temperature until a sublethal endpoint is reached. The endpoint is observed by the behavior of the fish, which show disorganized locomotion once the critical thermal minimum or maximum is reached. In this way lethal temperatures can be estimated without actually killing fish. **Figure 4** shows the temperature tolerance polygon for the sheepshead minnow. There is an upper lethal temperature beyond which these minnows cannot survive. This upper lethal temperature is sensitive to the acclimation temperature at which the fish have been living. Clearly, the sheepshead minnow has a very broad range of temperature tolerance—in fact it has the highest critical maximum temperature of any fish yet tested. Other fish, such as the various salmon species, have very low critical temperature maxima ($\sim 20\,^\circ$C). Thus if the water temperature rises above about 20°C because of cli-mate change or human impacts, many salmon species will disappear. Critical temperature minima have not been measured for many fish species, and this information is essential to determine if introduced species can spread.

We can repeat these tolerance studies for oxygen, pH, and salinity and build up a detailed picture of the tolerance limits of any particular species of plant or animal. **Figure 5** illustrates hypothetical limitation by two factors. To these simple Shelford models we can then add complications. The stages of the life cycle may differ in their tolerance limits, and consequently we should measure the most sensitive stage when constructing these models. The young stages of both plants and animals are often most sensitive to environmental factors. These tolerance limits define the fundamental niche of the species.

Two other factors complicate the determination of tolerance limits. First, species can acclimate physiologically to some environmental factors. Figure 4 illustrates this concept: The lethal temperatures depend on the acclimation temperature, the temperature at which the fish have been living. Second, tolerance limits for one environmental factor will depend on the levels of other environmental factors. Thus in many fish, for example, pH differences will affect temperature tolerances.

Another problem arises when we try to apply these tolerance limits to situations in the real world. Animals are particularly difficult because they are mobile and can resort to a variety of tactics that help them avoid lethal environmental conditions. Both plants and animals have evolved many types of escape mechanisms. Many birds

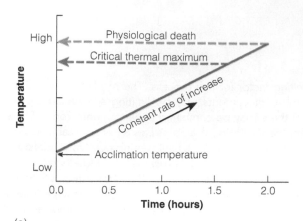

(a)

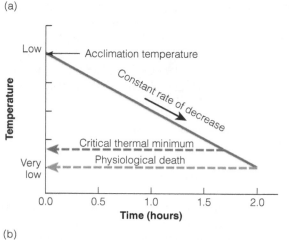

(b)

Figure 3 Determination of the critical thermal maximum and minimum temperatures in fish. (a) Fish are acclimatized at a given water temperature, and then subjected to an increasing water temperature (usually about 0.3°C per minute) until they show locomotory distress, and the trial is then stopped and the fish returned to the acclimation temperature to recover. (b) Similar methods are used to determine the critical thermal minimum temperature. This procedure is repeated for a variety of acclimation temperatures to construct the temperature tolerance polygon illustrated in Figure 4. (Modified from Beitinger et al. 2000.)

and some insects migrate from polar to temperate or equatorial regions to avoid the polar winters. Some mammals, such as the arctic ground squirrel, hibernate during the winter and thereby avoid the necessity of feeding during the cold months of the year. Plants become dormant and resistant to cold temperatures in winter, while many insects enter a cold-tolerant diapause.

Adaptation

The organisms whose distributions we study today are products of a long history of evolution, and the physio-

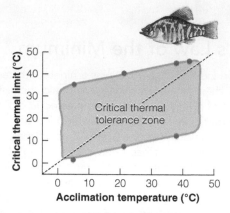

Figure 4 Temperature tolerance polygon for the sheepshead minnow (*Cyprinodon variegatus*). The dashed line indicates a 1:1 relationship between acclimation temperature and critical thermal limit. (Modified from Beitinger et al. 2000.)

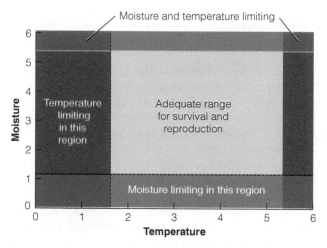

Figure 5 Idealized plot of Shelford's Law of Tolerance for two factors that might limit geographical distributions. In this hypothetical example, if temperature is below 1.7 units, or moisture level is below 1.15 units, the organism cannot survive. The geographical range must occur in the yellow zone for these two factors. Too much moisture and excessively high temperatures set the upper limits of tolerance in this simple example.

logical ecologist studies their adaptations in much the same way that we might study a single frame of a motion picture. The tolerances of species can change via the process of natural selection. Good examples are found in the adaptation of plants to heavy-metal toxicity and serpentine soils.

Heavy metals such as lead are extremely toxic to plants: 0.001% of lead and 0.00005% of copper will kill most plants within a week. In mining wastes contaminated soils often contain 1% of lead, copper, and zinc and thus should kill all plant life. But in less than 50 years the grass *Agrostis tenuis* has evolved populations that live on mine wastes in Great Britain (Antonovics et al. 1971). A few species adapt to these high concentrations of lead, zinc, and copper, but most plants from pastures will not survive on mine soil. Normal *Agrostis tenuis* populations, however, contain a few tolerant individuals. If one sows 2000 seeds on mine soil, only four or five grass plants will grow (Wu et al. 1975). On toxic mine soils only these tolerant genotypes survive.

Copper-tolerant populations of *Agrostis* have evolved by rapid natural selection acting on very rare individual grass plants that are partly tolerant to high copper levels (Wu et al. 1975). Current populations are maintained by strong disruptive selection dictated by contaminated soils. Not all plant species are able to evolve metal tolerances; many species do not have the appropriate genetic variation in their normal populations (Bradshaw and Hardwick 1989).

There is a cost to being tolerant to heavy-metal pollution. The tolerant genotypes of *Agrostis tenuis* grow poorly compared with normal genotypes when they are grown in normal soil under crowded conditions (Macnair 1987). One possible reason is that tolerant plants require more than trace amounts of heavy metals to be able to grow properly. Tolerant plants are at a selective disadvantage away from contaminated soils.

Such evolutionary changes further complicate the task of the ecologist who is trying to understand the distribution of a species. We must ask the question, What factor sets the current limitation on the geographic distribution of this species? But then we must ask further: Why for many species has natural selection not been able to increase the tolerance limits of a species and thereby expand its geographic range? If *Agrostis tenuis* has been able to increase its limits of tolerance to heavy metals, why has this not occurred in many other plant species? Is there simply a lack of genetic variability in nontolerant species so that natural selection has no impact on adaptation to tolerating heavy metals?

The grass *Agrostis tenuis* has thus increased its geographic range on a microscale by adapting to contaminated soils. When we observe geographic range changes we typically think that such shifts are due to changes in the environment, but the *Agrostis tenuis* example shows that some range shifts are caused by evolutionary changes in the physiological attributes of the individuals in a population.

Adaptive divergence in plants can be easily illustrated by reciprocal transplant experiment in which an array of strains or species are planted in a common garden (**Figure 6**). The existence of genetic variants or strains within a single species illustrates that local adaptation occurs, and these local differences can be a precursor to speciation or its reverse (Seehausen 2006).

Figure 6 Response of eight strains of *Achillea borealis* to serpentine soil (upper photo) and normal soil (lower photo). The strains were collected as seed from four different serpentine areas (142, 164, 135, and 184, red arrows), and from four sites with normal soil (125, 161, 198, and 206). (Photo courtesy of A. R. Kruckeberg.)

Figure 7 The contact between serpentine and calcareous soils on Dun Mountain in New Zealand. The *Nothofagus* beech forest (left) is unable to colonize serpentine soil (right) that contains high concentrations of Ni, Cr, Co, Mn, and Mg. Here the serpentine flora is dominated by a group of endemic plants adapted to survive under these soil conditions. Some of these plants may be able to be used to revegetate toxic mine tailings. (Photo courtesy of Brett Robinson, Swiss Federal Institute of Technology, Zurich.)

Plant adaptation to serpentine soils is a classic example of natural selection in action.

Serpentine soils are formed by the weathering of ultramafic rocks and are an excellent example of natural soils that are toxic to most plants. Serpentine soils are found around the world but are very patchy (**Figure 7**).

They contain a large number of endemic plant species. Serpentine soils contain high amounts of magnesium and low amounts of calcium, and also have high levels of heavy metals—iron, nickel, chromium, and cobalt—which are toxic to most plants (Brady et al. 2005). In serpentine soils magnesium is taken up by plant cells as a substitute for calcium, but this typically kills the cells of normal plants.

Species that can tolerate serpentine soils are often confined there because they are not able to compete in normal soils (**Figure 8**). This suggests some evolutionary trade-offs in which species evolve physiological mechanisms to live in serpentine soils, but these adaptations do not permit the species to recolonize normal soils. The key adaptations of serpentine plants are to tolerate low calcium-to-magnesium ratios, to avoid magnesium toxicity, or to have a high magnesium requirement (Macnair 1987). The physiological basis of how these adaptations are achieved is not well understood, nor is the evolutionary process by which serpentine-tolerant populations evolve.

Experiments in Geographic Ecology

Transplant experiments, such as that illustrated in Figure 2, can be disastrous when pests are introduced to new areas. It is critical that all transplant experiments be done safely, with due regard for the ecosystem. Indiscriminate transplanting of organisms contains all the seeds of ecological disaster (Ruesink et al. 1995; Pimentel et al. 2000). Most governments have stringent rules prohibiting the importation of plants and animals from other regions.

One cautionary note: We will begin by assuming that the factors affecting geographic distributions operate in isolation from one another, as Liebig first suggested in 1840. But we know this is not true from our personal experience—a spring day at 15°C will be pleasant if there is no wind, but it will seem cold if a strong wind is blowing. The effects of temperature and wind, of temperature and moisture, and of moisture and soil nutrients are not independent but often interact. We

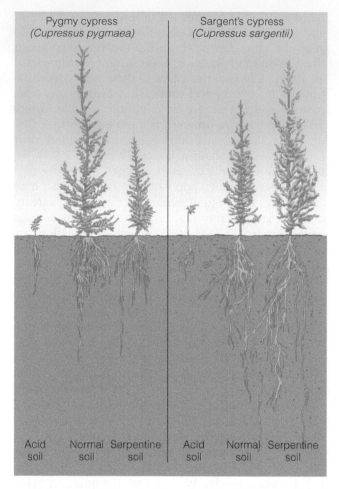

Figure 8 Schematic illustration of the growth of two tree species on acid soil, normal soil, and serpentine soil. Sargent's cypress is most common on serpentine soils in Central California, and grows less well in normal soil. The pygmy cypress by contrast grows poorly in serpentine soil and well in normal soil, and is almost never found on serpentine areas. Neither of these trees can survive in acid soils with pH less than 5.0. (Modified from McMillan 1956).

will begin simply and see how much we can understand by treating factors as separate effects, and then we will add factors together when necessary.

Summary

Why are organisms of a particular species present in some places and absent from others? This simple ecological question has significant practical consequences and thus deserves careful analysis. A transplant experiment is the major technique used to analyze the factors that limit geographic ranges. This technique leads sequentially through the hierarchy summarized in **Figure 9**.

To examine any particular problem of distribution, ecologists proceed down this chain, eliminating things one by one. We will see many examples in which part of this chain has been experimentally analyzed, but in no case has this chain been studied completely for a species.

The analytical question—What limits distribution now?—is complementary to the evolutionary question: Why has there not been more adaptation? Thus we are led to investigate the genetic variation within populations and to look for range extensions or contractions that are associated with evolutionary shifts in the adaptations of organisms to their environment.

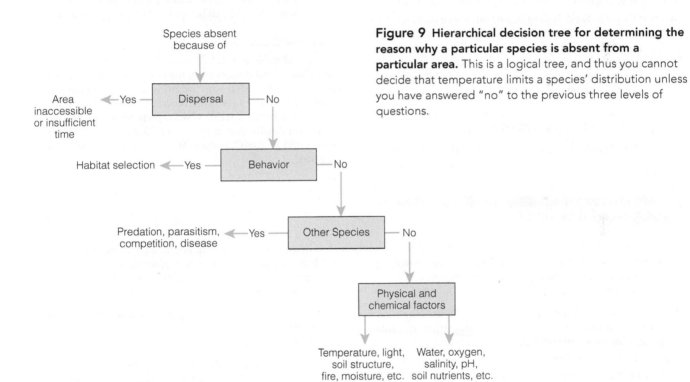

Figure 9 Hierarchical decision tree for determining the reason why a particular species is absent from a particular area. This is a logical tree, and thus you cannot decide that temperature limits a species' distribution unless you have answered "no" to the previous three levels of questions.

Review Questions and Problems

1 The northern spread of the African honey bee in North America has slowed in recent years. Discuss how you might find out the reasons for this slowdown, and how you might find out how far it might spread to the north.

2 Would you expect native fish species that have evolved in a desert climate to have higher or lower critical temperature tolerances than introduced species? Discuss the implications of both of these possible findings for species conservation. Carveth et al. (2006) provide data on this question.

3 Discuss the problem of defining exactly the "geographic distribution" of a plant or animal. Gaston (1991) reviews this problem.

4 Design a research program to introduce sockeye salmon to a New Zealand river system. You know beforehand that salmon introductions almost always fail. What studies might you undertake to increase the chances of a successful introduction? Burger et al. (2000) discuss one case history.

5 In discussing Liebig's Law of the Minimum, Colinvaux (1973, p. 278) states:

   The idea of critically limiting physical factors may serve only to obstruct a theoretical ecologist in his quest for a true understanding of nature. . . . To say that animals live where their tolerances let them live has an uninteresting sound to it. It implies that animals have been designed by some arbitrary engineer according to some preconceived sets of tolerances, and that they then have to make do with whatever places on the face of the earth will provide enough of the required factors.

   Evaluate this critique.

6 Butterflies in Europe and North America have been extending their range to the north in recent years. Discuss at least three hypotheses that could possibly explain this range extension and indicate what data you would like to have to test these hypotheses.

Overview Questions

Find a field guide to local flowers, birds, mammals, or amphibians, and discuss what the maps showing geographic ranges mean. On what scale would you map these ranges? Consult the Patuxent Wildlife Research Center of the U.S. Department of the Interior at **www.pwrc.usgs.gov** and look at the bird distribution maps for the "Breeding Bird Survey Results and Analysis" section. At what scale are these ranges mapped?

Suggested Readings

• Beitinger, T. L., W. A. Bennett, and R. W. McCauley. 2000. Temperature tolerances of North American freshwater fishes exposed to dynamic changes in temperature. *Environmental Biology of Fishes* 58:237–275.

• Brady, K. U., A. R. Kruckeberg, and H. J. D. Bradshaw. 2005. Evolutionary ecology of plant adaptation to serpentine soils. *Annual Review of Ecology, Evolution, & Systematics* 36:243–266.

• Brown, J. H., and M. V. Lomolino. 1998. Distributions of single species. Chapter 4 in *Biogeography*. Sunderland, MA: Sinauer Associates.

• Gaston, K. J. 1991. How large is a species' geographic range? *Oikos* 61:434–438.

• Hickling R., Roy D. B., Hill J. K., Fox R., and C. D. Thomas. 2006. The distributions of a wide range of taxonomic groups are expanding poleward. *Global Change Biology* 12: 450–455.

• Hierro, J. L., J. L. Maron, and R. M. Callaway. 2005. A biogeographical approach to plant invasions: The importance of studying exotics in their introduced and native range. *Journal of Ecology* 93:5–15.

• Macnair, M. R. 1987. Heavy metal tolerance in plants: A model evolutionary system. *Trends in Ecology and Evolution* 2:254–259.

• Ruesink, J. L., I. M. Parker, M. J. Groom, and P. M. Kareiva. 1995. Reducing the risks of nonindigenous species introductions. *BioScience* 45:465–477.

• Schneider, S. S., G. DeGrandi-Hoffman, and D. R. Smith. 2004. The African honey bee: Factors contributing to a successful biological invasion. *Annual Review of Entomology* 49:351–376.

Credits

PHOTO CREDITS

Factors That Limit Distributions I: Biotic

Key Concepts

- Some species do not inhabit an area because they have not yet been able to disperse there. Dispersal limitation can be tested by transplant experiments.

- Global distributions are often limited by barriers that block dispersal. On a local scale adaptations for dispersal are common and few species are limited in distribution by a failure to disperse.

- Animal species may be limited in their geographic distribution by selecting a range of habitats that is more restricted than the range they could occupy successfully.

- The presence of other organisms—predators, parasites, pathogens, or competitors—may limit the geographic distributions of many species.

- Predator limitations on prey distributions often operate on a local scale. Diseases and parasites may affect geographic distributions, but this impact has been little studied.

- Some organisms poison the environment for other species with allelochemicals, and this form of competition can affect local distributions, particularly in plants. Competition between species for food or space may affect local distributions and cause evolutionary divergence between competing species.

From Chapter 5 of *Ecology: The Experimental Analysis of Distribution and Abundance*, Sixth Edition. Eugene Hecht.

KEY TERMS

allelopathy Organisms that alter the surrounding chemical environment in such a way as to prevent other species from using it, typically with toxins or antibiotics.

barriers Any geographic feature that hinders or prevents dispersal or movement across it, producing isolation.

biogeography The study of the geographical distribution of life on Earth and the reasons for the patterns one observes on different continents, islands, or oceans.

dispersal The movement of individuals away from their place of birth or hatching or seed production into a new habitat or area to survive and reproduce.

fitness The ability of a particular genotype or phenotype to leave descendants in future generations, relative to other organisms.

gene flow The movement of alleles in space and time from one population to another.

ideal despotic distribution A theoretical spatial spread of members of a population in which the competitive dominant "aggressive" individuals take up the best resources or territories, and less competitive individuals take up areas or resources in direct relationship to their dominance status.

ideal free distribution A theoretical spatial spread of members of a population in which individuals take up areas with equal amounts of resources in relation to their needs, so all individuals do equally well (the polar opposite to the ideal despotic distribution).

Reid's paradox The observed large discrepancy between the rapid rate of movement of trees recolonizing areas at the end of the Ice Age and the observed slow dispersal rate of tree seeds spreading by diffusion.

tens rule The rule of thumb that 1 species in 10 alien species imported into a country becomes introduced, 1 in 10 of the introduced species becomes established, and 1 in 10 of the established species becomes a pest.

In this chapter we begin to unravel the first three possible explanations of what limits geographic distributions of plants and animals. Some organisms do not occupy all of their potential range, and if transplanted outside their normal range they survive, reproduce, and spread. The simplest explanation for the absence of an organism from a particular area may be the species' failure to reach the area being studied.

Dispersal Limitation on Geographic Distributions

The transport, or **dispersal**, of organisms is a vast subject that has been of primary interest not only to ecologists but also to **biogeographers**, who seek to understand the historical changes in distributions of animals and plants. Some very difficult problems are associated with the study of dispersal. For one thing, the detailed distribution is known for so few species that most dispersals are probably not noticed. Dispersal of individuals between different parts of a species' range may occur often. Second, an organism may disperse to a new area but not colonize it because of biotic or physical factors.

If colonization is successful, dispersal will result in **gene flow** and thus affect the genetic structure of a population. If the dispersing individuals are not a random sample of the population, dispersal will result in a founder effect, and the new population may be genetically quite distinct from the source population. Not all dispersing individuals survive to breed, so gene flow may be quite restricted in many species (Clobert et al. 2001). Dispersal is thus simultaneously an ecological process affecting distributions, and a genetic process affecting geographic differentiation.

The most spectacular examples of dispersal affecting distribution involve species that are introduced by humans and proliferate to occupy a new area. Other examples are exploited species recolonizing their original range. Next we look into three examples of these situations.

Zebra Mussel (*Dreissena polymorpha*)

In 1988 a fingernail-sized mussel native to the Caspian Sea of Asia was discovered in Lake St. Claire near Detroit. No one knows how these mussels got transplanted there, but the best guess is that around 1985 a ship from a freshwater port in Europe arrived at the Great Lakes, where its ballast water containing mussel larvae was dumped with no concern about what organisms it might contain. The zebra mussel quickly became a pest because it forms dense clusters on hard surfaces and grows rapidly. The mussels were noticed when they reached densities of 750,000 per m^2 in water pipes in Lake Erie, clogging the water intakes of city water systems, electrical power stations, and other industrial facilities in the Great Lakes (MacIsaac 1996).

Since 1988 zebra mussels have spread rapidly in the river systems of the central United States (**Figure 1**). While the spread in river systems has been very rapid, the colonization of small lakes in the central United

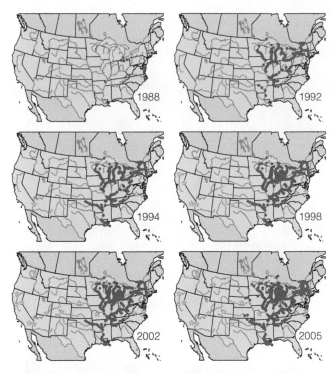

Figure 1 Expansion of the geographic range of the zebra mussel (*Dreissena polymorpha*) from its discovery near Detroit in 1988 to 2005. Yellow stars show the discovery of overland movement of zebra mussels on boats pulled in trailers. (Source: U.S. Geological Survey, Nonindigenous Aquatic Species Program, 2007.)

States has been slow. Only 8% of suitable inland lakes had been colonized up to 2003 (Johnson et al. 2006). Because they are very efficient filter feeders, zebra mussels have a positive impact on water quality, making Lake Erie, for example, much clearer than it had been previously. By feeding on phytoplankton they depress populations of zooplankton, and by making the water clearer they increase the growth of rooted aquatic plants in shallow waters. In the Hudson River in New York, phytoplankton biomass was reduced 80%–90% after zebra mussels invaded, and zooplankton that feed on phytoplankton declined by more than 70% after the invasion (Pace et al. 1998). Zebra mussels also physically smother other native clam species as they colonize all available hard surfaces, including the shells of other clams, possibly causing local extinction of some native clam species.

California Sea Otter (*Enhydra lutris*)

Sea otters were hunted by fur traders around the North Pacific to very low numbers by 1900. The few remaining small populations were protected by international treaty in 1911, and the California subpopulation of the sea otter was believed to be extinct at that time. In 1914 a small population was discovered at Point Sur in central California. Since then, otters on the central California coast have increased in numbers and expanded their geographic range to reoccupy areas from which they had been exterminated in the nineteenth century (**Figure 2**). The rate of spread of the sea otter is easy to estimate because it lives along the coastline in a linear habitat. The southern range expanded 3.1 km/year between 1938 and 1972, and the northern range expanded 1.4 km/year. These differences could result from the southern otters moving more as individuals, or from the northern otters suffering greater mortality (Lubina and Levin 1988). The recolonization of sea otter populations in the southern part of their range has been nearly completed and successful, so much so that they are considered a potential "pest" in some marine reserves (Fanshawe et al. 2003).

Cane Toad (*Bufo marinus*)

The cane toad is native to Central and South America from Mexico to Brazil. It was widely introduced during the 1930s to islands in the Caribbean and the Pacific because it was believed to control scarab beetles, an insect pest of sugarcane. It was brought into Queensland in 1935, where it failed to control any insect pests and instead became a pest itself. Cane toads have parotid glands that contain a poison that causes cardiac arrest. All forms of the toad are poisonous, and humans eating cane toad eggs have died from the toxin. Cane toads eat almost anything but mainly insects. They breed prolifically, females laying 8000–35,000 eggs at least twice a year.

Cane toads are toxic to many of their potential predators, but some species of snakes seem to be evolving resistance to the toxin (Phillips and Shine 2006). Because of their toxicity and high reproductive rate, cane toads have been moving across northern Australia since their introduction in 1935 (**Figure 3**). In 1995 Sutherst et al. (1995) predicted the possible range of the cane toad in Australia (Figure 3a), and recent data confirm their predictions closely (Figure 3b). Cane toads have been moving west at about 40 km per year (Brown et al. 2006). Individual marked toads have moved up to 1.8 km per night, primarily along roads that have served as convenient habitat corridors for rapid spread.

The Three Modes of Dispersal

These cases of colonization or spreading illustrate the important point that many organisms can spread rapidly to new areas if conditions are favorable. Before we

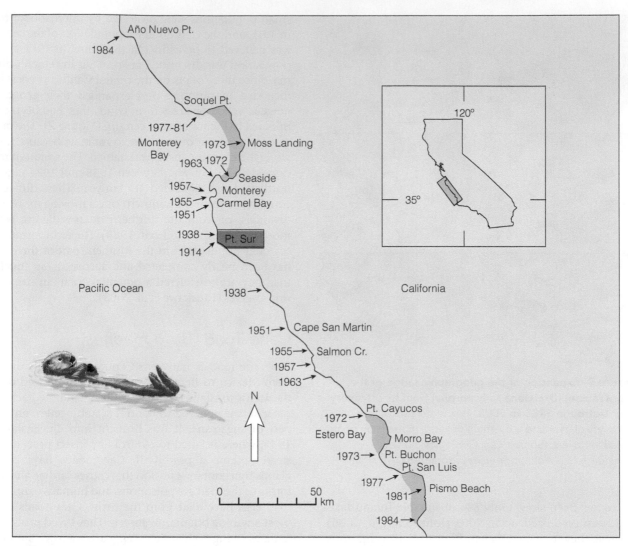

Figure 2 Expansion of the range of the California sea otter (*Enhydra lutris*) along the California coast. The current range expansion began from Point Sur (red), where 50 sea otters were rediscovered in 1914. (After Lubina and Levin 1988.)

discuss the ecological consequences of dispersal, let us define more carefully what we mean by dispersal.

The three ways in which species spread geographically, all loosely labeled as dispersal, are the following (Pielou 1979):

1. *Diffusion.* Diffusion is the gradual movement of a population across hospitable terrain for a period of several generations. This common form of dispersal is illustrated by the sea otter in California and the cane toad in Australia.

2. *Jump dispersal.* Jump dispersal is the movement of individual organisms across large distances followed by the successful establishment of a population in the new area. This form of dispersal occurs in a short time during the life of an individual, and the movement usually

occurs across unsuitable terrain. Island colonization is achieved by jump dispersal, and human introductions such as the African honey bee can be viewed as an assisted form of jump dispersal.

3. *Secular dispersal.* If diffusion occurs in evolutionary time, the species that is spreading undergoes extensive evolutionary change in the process. The geographic range of a secularly dispersing species expands over geologic time, but at the same time natural selection is causing the migrants to diverge from the ancestral population. Secular dispersal is an important process in biogeography, but, since it occurs in evolutionary time, it is rarely of immediate interest for ecologists working in ecological time.

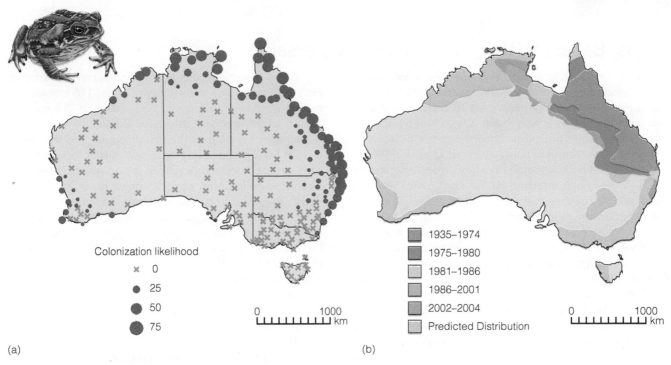

Figure 3 Cane toad distribution and expansion in Australia. (a) Potential distribution of the cane toad in Australia predicted from climate data, with climate model predictions for 2030. The larger the circles, the more likely the cane toad can survive in the area. (b) Actual expansion of the cane toad in Australia to 2005. (Source: Sutherst et al. 1996; Department of Environment and Heritage, Canberra, 2007.)

One of the most spectacular colonizations occurred at the end of the Ice Age when glaciers retreated from Europe and North America. In 1899 a British botanist, Charles Reid, raised the question of how trees recolonized the British Isles after the Ice Age. Reid (1899) identified a great discrepancy between the life history characteristics of trees before and after the Ice Age, and pointed out they spread quickly after the ice melted. From the melting of the ice about 10,000 years ago until the Romans occupied Britain about AD 50, trees such as oaks expanded their range 1000 km northwards. Reid calculated that this would take a million years. Oaks, like most deciduous temperate zone trees, mature at 10–50 years of age and drop seeds that on average fall 30 m from the parent tree. If trees migrate by simple diffusion, the migration rate is set by the following simple equation (Skellam 1951):

$$\text{Distance moved} = Dn\sqrt{\log_e R_0} \qquad (1)$$

where D = average dispersal distance
n = number of generations
R_0 = reproductive rate per generation

Thus, if a tree produces 10^7 seeds per generation over 300 generations, and seeds disperse 30 m with each generation, this simple diffusion model predicts a range extension of 36 km, far short of the observed recolonization of 1000 km since the end of the Ice Age. This discrepancy is now called **Reid's paradox**.

Paleoecologists have calculated that to repopulate Britain or northern parts of North America since the glaciers melted, trees had to migrate 100–1000 m per year (Clark 1998). How can we resolve Reid's paradox between the expected slow rates of tree diffusion and the observed rates of range expansion?

Tree seed dispersal can be mapped by putting out seed traps at different distances from the parent tree or by mapping the locations of seedlings that have been produced by isolated trees. **Figure 4** illustrates the typical pattern of the seed shadow from a deciduous tree. Most seeds fall near the parent tree, and a few are carried farther by wind or by animals. Clark (1998) measured the average seed dispersal distances of 12 species of temperate zone trees in the southern Appalachians and found a range of 4–34 m, distances far too small to account for recolonization by simple diffusion after the ice melted.

The answer to Reid's paradox seems to lie in haphazard, long-range dispersal of seeds. Even though the mean dispersal distance is small, colonization rates are driven not by the mean dispersal distance but by extreme

Ships, Ballast Water, and Marine Dispersal

Species invasions in terrestrial habitats have long been recognized as a source of environmental problems, but much less attention has been paid to marine invasions. Many marine invasions have been due to human-assisted dispersal either as organisms attached to the bottoms of ships or by the release of ballast water (Ruiz et al. 1997).

During the nineteenth century many organisms reached new ports attached to the bottoms of wooden ships, and this was the main means of human-assisted marine introductions. But the advent of metal ships, antifouling paints, and faster ships has eliminated this transport mechanism. At the same time, ballast water discharge has increased dramatically as ships have become larger. Chesapeake Bay received 10 million metric tons of ballast water discharge in 1991, mostly from ships originating in Europe and the Mediterranean. The zebra mussel is one of the best known examples in North America of a species brought in ballast water from elsewhere. A single ship can now carry 150,000 tons of ballast water to maintain trim and stability. The ballast water of five container ships entering Hong Kong contained 81 species from eight animal phyla and five protist phyla (Chu et al. 1997).

The biological results of these dispersal movements are significant. Chesapeake Bay now has 116 introduced marine species. San Francisco Bay has had 212 species added to its marine ecosystem. Some of these introduced species, such as the zebra mussel and the Asian clam in San Francisco Bay, have become dominant members of the community. Other introductions have not been studied in detail, so their impact is not known. Two health risks of introduced species have been detected. Toxic red-tide dinoflagellates are transferred worldwide in ballast water and may serve to trigger these algal outbreaks. The cholera bacterium *Vibrio cholerae* occurs in the ballast tanks of some ships and can survive for up to 240 days in seawater at 18°C (McCarthy 1996). When released into an estuary, cholera bacteria can attach to a variety of marine organisms and thus enter the human food chain.

This story contains two ecological messages: Many marine species were originally limited in their global distribution, and action to reduce the global transport of potentially harmful organisms in ballast water is urgently needed.

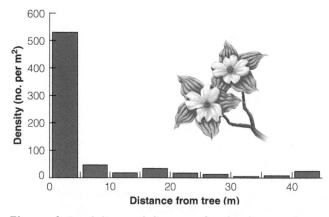

Figure 4 Seed dispersal distances for the dogwood *Cornus controversa*. This tree has fruits that are dispersed by birds, but the vast bulk of the seeds fall near the parent tree. Mean seed dispersal distance for these trees was 6.7 m. (Data from Masaki et al. 1994, Table 2.)

dispersal events (Clark et al. 2001). A few seeds are blown by wind or moved by animals a long distance from the parent tree. In the frequency diagram in Figure 4, these long-range dispersers would be off the scale to the right. Extreme long-distance events are difficult to record and measure, since less than one seed in 10,000 might be

blown a long distance by wind or moved a great distance by animals (Powell and Zimmermann 2004).

Dispersal can be affected by **barriers**. Many kinds of barriers exist for different kinds of animals and plants. But barriers are not always the factor limiting geographic ranges. On a local or global scale, dispersal may not limit distribution, because introduced species may be unable to survive. Humans have moved many species around the globe during the past 200 years, often with disastrous consequences. Long (1981) and Ebenhard (1988) list some of the early attempts to introduce nonnative birds to North America. Unfortunately, failures to establish a species are rarely studied to obtain an explanation, and accidental introductions are often recorded only when they are successful. In contrast to the global spread of weeds, few plant species introduced into continental areas can become established except in disturbed areas.

Bird introductions into continental areas are usually failures (Case 1996). **Table 1** gives some data for terrestrial and freshwater birds. In the continental United States, only 13 species of introduced birds are common, although 98 species have been introduced. In Great Britain, only 9 successful establishments of birds are recorded from 30 species introduced. About 204 species of breeding birds live around Sydney, Australia, and 50 or more bird species were introduced to this

Table 1 Historical success rates for introducing terrestrial and freshwater birds to some selected islands and mainland locations.

Location	No. of successfully introduced spp.	No. of spp. introduced	Success rate (%)*	No. of native spp. present
Australia (Victoria)	16	48	33	271
Bermuda	7	17	41	9
Continental USA	13	98	13	553
Great Britain	9	30	30	146
Hawaii (Kauai)	27	52	52	18
Mauritius	19	44	43	13
New Zealand	41	149	28	52
Tahiti	11	54	20	12
Tasmania	13	16	81	104

*The success rate is generally higher on islands than on continents.

This list includes only exotic species that increased and spread beyond the point of introduction.

SOURCE: Data from Case (1996, Table 2).

area. Only 15 species got established, and only 8 species are common. Thus continental bird introductions are successful about 10%–30% of the time.

Can we make any statistical generalizations about the success of introduced species? Williamson and Fitter (1996) have proposed the **tens rule**, which makes the statistical prediction that 1 species in 10 imported into a country become introduced, 1 in 10 of the introduced species becomes established, and 1 in 10 of the established species becomes a pest. To interpret the tens rule we need some precise definitions of these terms. Introduced species occur in four "states" and undergo three "transitions."

States	Transition	Definition
Imported		Brought into the country
↓	Escaping	Transition from imported to introduced
Introduced		Found in the wild; feral
↓	Establishing	Transition from introduced to established
Established		Has a self-sustaining population
↓	Becoming a pest	Transition from established to pest
Pest		Has a negative economic impact

In the terminology now applied to genetically engineered organisms, introduced species are *released*, whereas imported species are *contained*. The *tens rule* states that each transition in the table has a probability of about 10% (between 5% and 20%) (Williamson and Fitter 1996).

The tens rule does not apply to many taxonomic groups (Jeschke and Strayer 2005). For vertebrates introduced between Europe and North America, success at each step is nearer to 50% than to 10%. For aquatic species in Europe about 63% of introductions become established, more than expected by the tens rule. By contrast, many fewer than 10% of the imported aquatic species become introduced in Europe (Garcia-Berthou et al. 2005). For established nonnative plants in the United States, 6%–13% have invaded natural areas, which is more or less consistent with the tens rule (Lockwood et al. 2001). The first transition in introductions is the most difficult to quantify, and the general ecological message is that imported species are successful often enough to encourage strong quarantine actions for all groups.

Humans have increased dispersal on a continental scale, but on a local scale many species have good to excellent dispersal mechanisms. Plants disperse primarily by means of seeds and spores, and transport is rarely an important factor limiting distributions of plants on a local scale. Few experimental data are available to substantiate this general conclusion. Small animals often have a life cycle stage that can be transported by wind, and these species resemble plants in that their local distributions are rarely limited by lack of dispersal. Many insect species are transported by wind for long distances. Mosquitoes are a good example. The flight patterns of disease-carrying mosquitoes have been studied to enable the implementation of adequate control measures. The distances mosquitoes disperse determine the limits to which a given breeding location may allow contact with people and the area where control work must be done if a given human habitation is to be protected from diseases like malaria. Morris et al. (1991) marked 451,000 mosquitoes with fluorescent dust in central Florida in 1987 to see how far individuals would disperse into human habitations. They found that 10% of the marked individuals moved over 2.2 km within 2 days, and the maximum distance moved was 4.2 km. Malaria control zones in tropical countries typically use a 2-km barrier zone surrounding human habitations as a rule-of-thumb for control since mosquitoes rarely move that far. Wind can move mosquitoes much farther than 2 km, and there are many records of mosquitoes being carried long distances by wind (Service 1997). Salt marsh mosquitoes in Louisiana have been captured on oil rigs 74–106 km offshore, and in Australia salt marsh species have been collected 96 km inland. One marked mosquito in California was collected 61 km from the release point. Mosquitoes are serious pests in northern Canada, Alaska, and Eurasia, and local control efforts are of limited success because of dispersal. Dispersal in mosquitoes is clearly very effective in colonizing vacant areas.

Colonization and Extinction

If dispersal occurs rapidly on a local scale, one would expect areas that are cleared of organisms to be recolonized rapidly. Some large-scale colonization experiments have occurred naturally. On August 26, 1883, the small volcanic island of Krakatau in the East Indies was completely destroyed by a volcanic eruption. Six cubic miles (25 km³) of rock was blown away, and all that remained of the original island was a smaller peak covered with ashes. Two islands within a few kilometers of the volcano were buried in ashes. These sterilized islands in effect constituted a large natural experiment on dispersal. The nearest island not destroyed by the explosion was 40 kilometers away. Nine months after the eruption, only one species—a spider—could be found on the island. After only three years, the ground was thickly covered with blue-green algae, and 11 species of ferns and 15 species of flowering plants were found. Ten years after the explosion, coconut trees began growing on the island. After 25 years, 263 species of animals lived on the island, which was covered by a dense forest. Bird colonization of the islands has depended on vegetation colonization, and the flora of the islands has continued to increase (Whittaker et al. 1989). There is

some controversy about the methods of transport, but the majority of the plants and animals were probably transported by wind. Larger vertebrates probably arrived on driftwood rafts or in a few cases by swimming. The suggestion that emerges from these observations is that when there is vacant space, animals and plants are not long in finding it.

These examples suggest that dispersal may limit local distributions of a few plants and some animals, but in most cases empty places get filled rapidly. Let us now look at the other extreme and consider global distribution patterns before humans began to move organisms on a large scale.

Terrestrial mammals other than bats do not easily cross saltwater barriers (Brown and Lomalino 1998), so whole faunas can diverge if they are isolated by ocean. Marsupials, for example, became isolated in South America and in Australia early in the Tertiary period (60 million years ago). Of the placental mammals, only rodents and bats were able to colonize Australia before the arrival of humans. South America was also isolated by a water gap across Central America for most of the Tertiary and became connected to North America only during the last 2 million years. Once a land connection was established, a flood of dispersing mammals moved in both directions. The results for North America were relatively minor—the arrival of the opossum, the porcupine, and the armadillo as additions to the mammal fauna. But in South America the results of colonization were dramatic. Many South American mammals became extinct and were replaced by North American species. Carnivores from North America have completely replaced the carnivorous marsupials that previously occupied South America. Ungulates from North America have entirely replaced the unique set of South American ungulates (Darlington 1965).

The faunas and floras of oceanic islands also show in graphic detail the limitations of distribution on a global scale. New Zealand had no native marsupials or other land mammals except for two species of bats at the time Europeans first arrived. All of the plants and animals that colonize New Zealand or any oceanic island must do so across water. The unique combination of difficult access, limited dispersal powers of different species, and adaptive radiation has produced island floras and faunas of an unusual nature, such as the plants and animals of Hawaii and the species Charles Darwin found on the Galápagos Islands off Ecuador.

The antarctic beech (*Nothofagus* spp.) is a good example of how present geographic distributions are set by geologic events. Until 135 million years ago the southern continents were connected in a large landmass called Gondwana. Groups present on Gondwana now have a very disjunct distribution—*Nothofagus* is a good example (**Figure 5**). *Nothofagus* seeds are heavy and

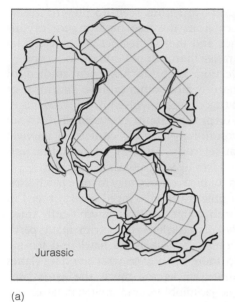
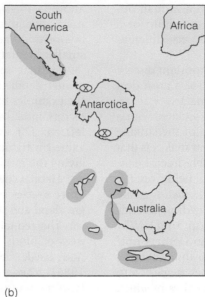

Figure 5 Effects of continental drift on geographic distributions. (a) Fit of the Gondwana continents during the Jurassic period, about 135 million years ago, before breakup. (b) Modern distribution of the genus *Nothofagus* (Antarctic beech) is outlined in dark blue. *Nothofagus* pollen of the Oligocene age (30 million years ago) has been found in Antarctica at the two sites indicated by x in part (b). (After Pielou 1979).

poorly adapted for jump dispersal. Species of *Nothofagus* have probably spread slowly overland by diffusion and have been stopped by the sea, so their present distribution is a by-product of continental drift.

Continental drift not only takes certain continents farther apart, it also brings some continents closer together. As the Australian tectonic plate, for example, drifted northward after becoming detached from Antarctica, it made contact with the Asian plate about 20 million years ago. As distances over water decreased, jump dispersal of plants between Australia and Asia has become steadily easier.

The Quaternary Ice Age is a more recent example of how geographic distributions are affected by geologic events. Chris Pielou (1991), working in Canada, has integrated much of the data on how the Ice Age affected the flora and fauna of North America. The Ice Age began about 2 million years ago. During the past 500,000 years, ice sheets in North America and Eurasia have undergone great oscillations, waxing and waning at least four times. We are now in the fourth interglacial period. At the height of the last glaciation—about 20,000 years ago—the ice volume was 77 million km³, three times the current amount. Sea level at the height of the last glaciation was 130 meters below its present level. If all the present ice melted, sea level would rise 70 meters (Pielou 1991). The biological effects of glaciations are spectacular but slow. Dropping sea levels open up migration routes for terrestrial organisms and may restrict dispersal of marine organisms.

The flora and fauna of the world today have been strongly affected both by the dispersal of species and the geological formation of barriers that prevent organisms from colonizing all of their potential range. The great sweep of evolutionary history is a prolonged essay on the role of dispersal and barrier formation in limiting species distributions.

Habitat Selection

Some organisms do not occupy all their potential range, and in the previous section we discussed cases in which limited dispersal was the reason for the absence of a species. Here we discuss cases in which organisms do not occupy all their potential range even though they are physically able to disperse into the unoccupied areas. Thus individuals "choose" not to live in certain habitats, and the distribution of a species may be limited by the behavior of individuals in selecting their habitat. We define a **habitat** as any part of the biosphere where a particular species can live, either temporarily or permanently. Habitat selection is typically thought of only with respect to animals that can in some sense choose where to live by moving among habitats. Plants show habitat preferences in quite different ways than animals because they cannot actively move from one habitat to another. Seeds or spores arrive in different habitats through dispersal, and then either survive and grow or die because of biological or physical factors.

Habitat selection is one of the most poorly understood ecological processes. If we assume that an animal cannot live everywhere, natural selection will favor the development of sensory systems that can recognize suitable habitats. What elements of the habitat do animals recognize as relevant? We must be careful here to define the perceptual world of the animal in question before we begin to postulate the mechanism of habitat selection. Areas that appear "similar" to a human observer may appear very different to a mosquito or a fish. Conversely, habitats we think are very different may be treated as the same by a bird.

Anopheline mosquitoes are often important disease vectors, and their ecology has been studied a great deal because of the practical problems of malaria eradication. Each mosquito species is usually associated with a particular type of breeding site, and one of the striking observations that a student of malaria first makes is that large areas of water seem to be completely free of dangerous mosquitoes. Large areas of rice fields on the Malay Peninsula are free of *Anopheles maculatus*, as are the majority of shallow pools in some breeding grounds of *Anopheles gambiae* (Muirhead-Thomson 1951). Why are some habitats occupied by larvae and others not? Early workers assumed that something in the water prevented the larvae from surviving, and they neglected to study the behavior of females in selecting sites in which to lay eggs. More recent work has emphasized the role of habitat selection for oviposition sites in female mosquitoes and shown that larvae can develop successfully over a much wider range of conditions than those in which eggs are laid (Bentley and Day 1989). Thus, although we presume that the female selects a type of habitat most suitable for the larvae, many of the places she avoids are suitable for larval growth and development.

In Belize, the malaria-transmitting mosquito *Anopheles albimanus* oviposits only on floating mats of blue-green algae (Rejmánková et al. 1996). In marshes with dense cattail growth, blue-green algae are shaded and do not produce mats. In cattail marshes no larval mosquitoes were found, and in oviposition tests no larvae were produced unless algal mats were present. Larval mosquitoes are not found in open waters, and in Belize *A. albimanus* is limited to marshes with algal mats.

In southern India, the mosquito *Anopheles culicifacies* (a malaria vector) does not occur in rice fields after the plants grow to a height of 12 inches (30 cm) or more, even though these older rice fields support two other *Anopheles* species. Russell and Rao (1942) could find no eggs of *A. culicifacies* in old rice fields, yet when they transplanted this mosquito's eggs into old rice fields, the larvae survived and produced normal numbers of adults. The absence of *A. culicifacies* from this particular habitat is apparently due to the selection of oviposition sites by females. In a series of simple experiments, Russell and Rao were able to show that the main limiting factor was the physical barrier posed by rice plants of a certain height. Glass rods placed vertically in small ponds also deterred female *A. culicifacies* from laying eggs, as did barriers of vertical bamboo strips. Shade did not influence egg laying. This mosquito oviposits while flying and performing a hovering dance, never touching the water but remaining 2–4 inches (5–10 cm)

above it. Physical obstructions seem to prevent the female mosquitoes from the free performance of this ovipositing dance and thereby restrict the species to a smaller habitat range than it could otherwise occupy.

Habitat selection in birds has been studied in greater detail than in most other groups, and most of the examples in this chapter involve birds. Two kinds of factors must be kept separate in discussing habitat selection: (1) evolutionary factors, conferring survival value on habitat selection, and (2) behavioral factors, giving the mechanism by which birds select areas.

Habitat cues for birds of prey may involve perch sites. Three species of buteos (broad-winged hawks) breed in grassland and shrub-steppe areas of western North America. The red-tailed hawk selects areas with many perch trees or bluffs, while the Swainson's hawk and ferruginous hawk select more open areas with few trees (Janes 1985). These three hawks eat much the same prey (ground squirrels, jackrabbits), and their habitat choice corresponds with their foraging behavior. Red-tailed hawks sit on perches and look for prey; their wings are less suited to soaring. Swainson's hawks are best at soaring and hunt from the air much more than from perches. Ferruginous hawks are intermediate in soaring abilities. Flapping flight is uncommon in all these hawks, and habitat selection is closely tied to their hunting methods.

Evolution of Habitat Preferences

Why do organisms prefer some habitats and avoid others? Natural selection will favor individuals that use the habitats in which the most progeny can be raised successfully. Individuals that choose the poorer, marginal habitats will not raise as many progeny and consequently will be selected against. Populations in marginal habitats may thus be sustained only by a net outflow of individuals from the preferred habitats. A variety of physical clues can be adopted by organisms as the proximate stimuli in choosing a particular type of habitat. Natural selection may act directly upon the behaviors that result in habitat choice, or it may select for individuals that have the capacity to learn which habitat is appropriate.

For birds, survival and reproductive success can depend on nest-site choices, and can be the bases for the evolution of nest-site preferences. When there are habitat differences between successful and unsuccessful nests, the process of natural selection can operate to ultimately change nest-site distribution (Clark and Shutler 1999). **Figure 6** illustrates how natural selection might operate to affect nest-site selection in birds. In this case there is directional selection for specific habitat features that increase the probability of successful nesting. An example of selection for nest-site habitat is given in **Figure 7**. Blue-

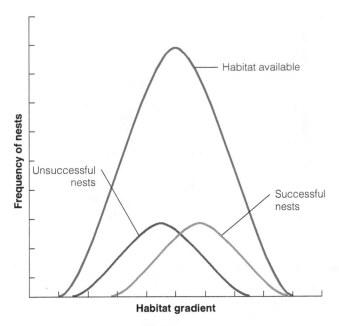

Figure 6 A hypothetical gradient showing the nest habitat available to a bird species (blue), the frequency distribution of unsuccessful nests (red), and the frequency distribution of successful nests (green). In this hypothetical example there is directional selection along the habitat gradient to the right. If this gradient is related to cover, birds would be selected to prefer areas of higher cover in the long term. (Modified after Clark and Shutler 1999.)

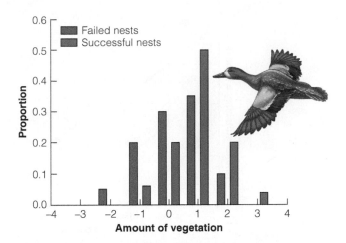

Figure 7 **Breeding success of blue-winged teal in a habitat gradient in Saskatchewan from 1983 to 1997.** Successful nests (*n* = 52) were farther from habitat edges and contained more vegetation than failed nests (*n* = 81). There is potential directional selection for nest-sites in this duck species, similar to that illustrated in Figure 6. (Data from Clark and Shutler 1999.)

winged teal (*Anas discors*) are more successful if they nest in areas that have more vegetation and are farther away from habitat edges (such as shrub/grassland borders). This differential nesting success gives rise to directional selection that should alter female nesting-habitat choice in these ducks, if this selection continues for many generations (Clark and Shutler 1999).

A simple theory of habitat selection can be used to illustrate how habitat selection may operate in a natural population (Fretwell 1972). Recall that for any particular species, we define a **habitat** as any part of the Earth where that species can live, either temporarily or permanently. Each habitat is assumed to have a suitability for that species, and in this example we assume that three habitats of different suitabilities are available to a species. Suitability is equivalent to **fitness** in evolutionary time, and we will assume that females produce more young in more suitable habitats than they do in less suitable habitats. Suitability is not constant but will be affected by many factors in the habitat, such as the food supply, shelter, and predators. But in addition, suitability in any habitat is usually a function of the density of other individuals of the species, so that overcrowding reduces suitability (**Figure 8**). We assume in this simple model that all individuals are free to move into any

habitat without any constraints, what Fretwell called the **ideal free distribution** (Fretwell 1972). As a population fills up the best habitat, it reaches a point where the suitability of the intermediate habitat is equal to that in habitat A, so individuals will now enter both habitats A and B. As these two habitats fill even more, the poor habitat finally has a suitability equal to that of habitats A and B. The prediction that arises from this simple model of habitat selection is interesting because it is counterintuitive. We would predict from the model that when density is high, good and poor habitats would have equal suitabilities (but different densities), that individuals would be crowded in the best habitats and at low density in the poor habitats (Figure 8). Is there any evidence that this might in fact be the case in natural populations?

Fretwell suggested a second model of habitat selection that could be applied to organisms that show territorial behavior: the **ideal despotic distribution**. If individuals are not free to move among all the available habitats but are constrained by the aggressive behavior of other individuals, then subordinate animals can be forced into the more marginal habitats. The ideal despotic distribution predicts that the density will not be lower in the marginal habitats and may in fact be higher if individuals are forced into these habitats. Most importantly, the ideal despotic distribution predicts that fitness will be lower in the poorer habitats.

The key to understanding habitat selection is to determine the rules by which individual animals decide which habitat to utilize. The proximate mechanisms by

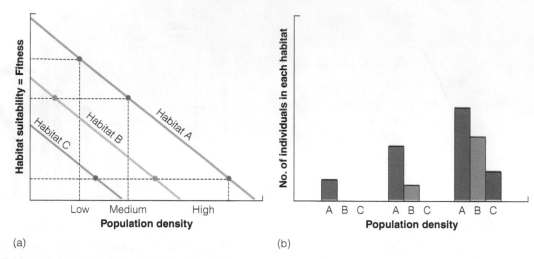

(a)   (b)

Figure 8 The ideal free distribution model of habitat selection. Three habitats are used for illustration (A = good habitat, B = intermediate habitat, C = poor habitat). Habitat suitability is measured by the fitness of individuals living in that habitat. For illustrative purposes, three levels of population density are indicated. (a) The model assumes that in all habitats fitness declines as population density increases and crowding occurs. At low density an individual can achieve the highest fitness by living in habitat A, and habitats B and C will be empty; see (b). At high density an individual can choose to live in habitat A under crowded conditions, in habitat B under less crowded conditions, or in the poorest habitat C with the least crowding. If individuals choose their habitat as in this simple model, fitnesses of individuals will be equal in all three habitats at high density. (Modified from Fretwell 1972.)

which habitats are selected are underlain by evolutionary expectations in fitness. We do not know how rapidly organisms can change the genetic and behavioral machinery that results in habitat selection.

Problems can arise whenever habitats change, and this has been a source of difficulty for many organisms since humans have modified the face of the Earth. People provide many new habitats and destroy others. Some species, but not all, have responded by colonizing *Homo sapiens*'s habitats. Other natural events, such as ice ages, cause slower habitat changes. Organisms with carefully fixed, genetically programmed habitat selection may require considerable time to evolve the necessary machinery to select a new habitat that is suitable for them. Adaptation can never be exact and instantaneous, and we must be careful not to expect perfection in organisms.

Limitation by Predators

Up to this point we have discussed cases of biotic limitation of geographical ranges in which an organism could actually live in places that it did not occupy. From now on we will be considering cases in which the organism cannot complete its full life cycle if transplanted to areas it did not originally occupy. The reason for this inability to survive and reproduce could be negative in-

teractions with other organisms, including predation, disease, and competition, or positive interactions such as mutualism or symbiosis. First we examine predation, one of the clearest interactions between species because predators eat their prey.

We begin our discussion of predation by considering the role of predators in affecting the geographical distributions of their prey. Note that we define predation very broadly. Typical predators like lions kill their prey. Herbivores prey on their food plants and usually do not kill the plants. Parasites live on or inside other organisms and again do not usually kill them.

The local distribution of some species seems to be limited by predation. Work on intertidal invertebrates has provided some classic examples of the influence of predation on distribution. Kitching and Ebling (1967) have summarized a series of studies at Lough Ine, an arm of the sea on the south coast of Ireland.

The common mussel (*Mytilus edulis*) is a widespread species on exposed rocky coasts in southern Ireland and throughout the world. Small mussels (less than 25 mm long) are abundant on the exposed rocky Atlantic coast but within Lough Ine and the more protected parts of the coast, this mussel is rare or absent. The only abundant populations are in the northern end of the lough, but these animals are typically very large (30–70 mm long).

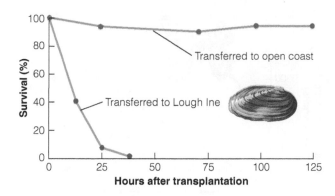

(a)

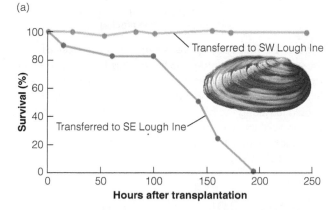

(b)

Figure 9 Percentage survival of mussels in transplant experiments in and near Lough Ine. Small mussels (a) disappear rapidly when transplanted anywhere in Lough Ine but do not disappear if transplanted to the open coast. Large mussels (b) disappear if transplanted to some parts of Lough Ine such as the southeastern part but do not disappear if transplanted to other parts of the Lough, such as the southwestern part, where they occur naturally. (After Kitching and Ebling 1967.)

Kitching and his coworkers transferred pieces of rock with *Mytilus* attached from various parts of the lough to others. (**Figure 9**) presents some typical results. Small *Mytilus* disappeared quickly from all stations to which they had been transferred within the lough, the Rapids, and the protected bays; they survived only on the open coast. The rapid loss, shown in Figure 9a, suggested that predators were responsible. Large mussels that were transplanted around the lough also disappeared rapidly from most stations (see Figure 9b), except places where they occurred naturally. Continuous observations on the transplanted mussels showed that three species of crabs and one starfish were the principal agents of mortality. By placing mussels of various sizes and crabs of the three species together in wire cages, Kitching and Ebling were able to show that one of the smaller species of crabs could not kill large *Mytilus* but that the other crabs could open all sizes of mussels. The areas of the lough where large *Mytilus* survive have few large crabs, and where the large crabs are common, *Mytilus* are scarce or absent. Predatory crabs are probably restricted in their distribution by wave action, strong currents, and low salinity. Crabs also require an escape habitat in which they spend the day.

The distribution of this mussel in the intertidal zone at Lough Ine is thus controlled as follows: on the open coast, heavy wave action restricts the size of mussels and prevents predators from eliminating small mussels. In sheltered waters, predators eliminate most of the small mussels, and *Mytilus* survive only in refuge areas safe from predators (such as steep rock faces), where they may grow to large sizes.

These kinds of experiments illustrate four criteria that must be fulfilled before one can conclude that a predator restricts the distribution of its prey (Kitching and Ebling 1967):

- Prey individuals will survive when transplanted to a site where they do not normally occur if they are protected from predators.

- The distributions of prey organisms and suspected predator(s) are inversely correlated.

- The suspected predator is able to kill the prey, both in the field and in the laboratory.

- The suspected predator can be shown to be responsible for the destruction of the prey in transplantation experiments.

In Australia, several species of small kangaroos have been driven to near extinction by predation from the introduced red fox. Rock wallabies are small kangaroos that live in rocky hill habitats throughout Australia. Their numbers have been declining for nearly a century, and numerous colonies have become extinct. Kinnear et al. (1998) tested the hypothesis that red fox predation was sufficient to limit the population size and distribution of rock wallabies in Western Australia. By poisoning red foxes around two colonies, they showed that populations of wallabies could recover dramatically in the absence of foxes (**Figure 10**). Red foxes not only kill wallabies directly but also reduce the area available for safe feeding to sites near rocky escape habitat. Without foxes in the area, wallabies ranged farther from the rocky areas to feed. This is a good example in which native species can suffer range reduction and even extinction because of introduced predators.

In the cases just discussed, the predator is believed to restrict the distribution of its prey; consequently,

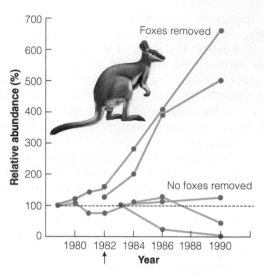

Figure 10 Changes in rock wallaby (*Petrogale lateralis*) abundance at two sites with red fox control and three sites without red fox control. Fox control on the experimental sites began in 1982 (indicated by arrow). The rock wallaby population on one of the three unmanipulated sites went extinct in 1990. (After Kinnear et al. 1998.)

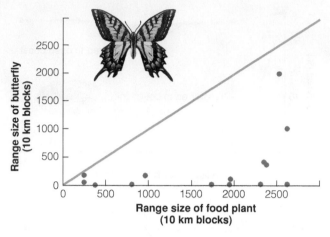

Figure 11 Relationship between the range sizes of 14 species of monophagous butterflies in Britain and the range sizes of their host plants. If butterflies were limited in their distribution by host plant distribution, the points should fall along the blue line on the diagonal. In most cases the range of these butterflies is not limited by the geographical distribution of their food plants. (Data from Quinn et al. 1998.)

the reasons for the predator's distributional limits must be sought elsewhere. In these situations, the predator may feed on a variety of prey species, and each prey species may in turn be fed upon by many predatory species. The relationship may also operate in the other direction, and the prey may restrict the distribution of its predator.

The "prey" may be a food plant and the "predator" a herbivore; alternatively, the prey may be a herbivore and the predator a carnivore. But if the prey is to restrict the predator's range, the predator must be very specialized and feed on only one or two species of prey. Such a predator is called a specialist or a monophagous predator. Many insect predators are specialists, but most vertebrate predators are not.

Insects that feed on only one host plant (monophagous insects) could be limited in their distribution by the host plant. But for the groups studied to date there is no indication that the ranges of food plants and their monophagous insect herbivores coincide (Quinn et al. 1998). **Figure 11** shows that for butterflies in Britain, no correspondence exists between food plant distributions and butterfly distributions. Even for widespread species of butterflies, the host plant occurs in many areas in which the butterfly does not. Something else must limit butterfly distributions.

Predation is a major process affecting the distribution and the abundance of many organisms.

Disease and Parasitism

In addition to predators, enemies include parasites and organisms that cause diseases. Pathogens may eliminate species from areas and thereby restrict geographical distributions. An example involves the native bird fauna of Hawaii.

A large fraction of the endemic bird species of the Hawaiian Islands has become extinct in historical times, and one possible reason for these losses is introduced diseases. Warner (1968) postulated that both avian malaria and avian pox were instrumental in causing extinctions in the Hawaiian Islands. The idea that diseases might be involved arose from the observation that native birds in Hawaii are relatively common only at elevations above 1500 m (**Figure 12**) while introduced birds occupy the lowland areas. The main malarial vector, the mosquito *Culex quinquefasciatus*, is conversely most common in the lowland areas (see Figure 12). Because the native birds are much more susceptible to malaria than the introduced species, the malaria parasite is most common at intermediate elevations (see Figure 12), where the geographical distributions of vectors and hosts overlap (Van Riper et al. 1986; Kilpatrick 2006).

The extinction of the native Hawaiian bird fauna occurred in two pulses. Before 1900 many of the low-elevation bird species disappeared coincident with extensive habitat clearing for agriculture and the introduction of rats, cats, and pigs. It is possible that other

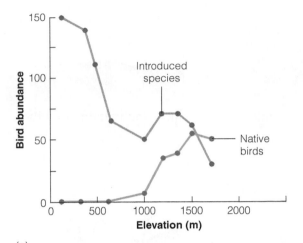

(a)

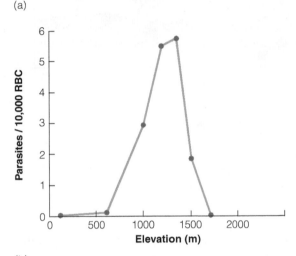

(b)

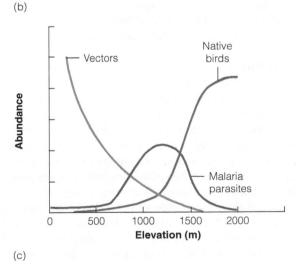

(c)

Figure 12 Effect of parasitism on the native birds of Mauna Loa, Hawaii. (a) Abundance of introduced and native birds in 1978–1979 at 16 sampling stations on Mauna Loa, Hawaii. (b) Avian malaria loads (parasites/10,000 RBCs) along this same altitudinal gradient. (c) A model of native bird abundance, malaria parasites, and mosquito vectors on Mauna Loa. (From Van Riper et al. 1986.)

introduced diseases such as avian pox played a role in the early extinctions, but avian malaria did not because it was uncommon before 1900 (Moulton and Pimm 1986; Freed et al. 2005). The second period of extinction in Hawaiian birds began in the early 1900s and was most likely the result of avian malaria. Birds that went extinct at this time lived in the mid-elevation forests where malaria parasites were most prevalent (see Figure 12b and 12c). At the same time the geographical distribution of many native birds was also reduced as they retreated to forests at the highest elevations where mosquitoes were rare. Climate change has extended the distribution of malaria-carrying mosquitoes to 1900 m elevation, with detrimental impacts on rare Hawaiian birds like honeycreepers that persist only at high elevations (Freed et al. 2005).

Diseases and parasites have always been a major factor in the ecology of humans (Diamond 1999). Their role in the geographical ecology of plants and animals has been studied far less than their potential importance would warrant.

Allelopathy

Some organisms, plants in particular, may be limited in local distribution by poisons or antibiotics, also called **allelopathic** agents. The action of penicillin among microorganisms is a classical case (Madigan et al. 2006). Interest in toxic secretions of plants arose from a consideration of soil sickness. It was observed in the nineteenth century that, as one piece of ground was continuously planted in one crop, the yields decreased and could not be improved by additional fertilizer. As early as 1832, DeCandolle suggested that the deleterious effects of continuous one-crop agriculture might be due to toxic secretions from roots. Several cases were also observed of detrimental effects of plants growing with one another—for example, grass and apple trees (Pickering 1917). Experiments of the general type shown in **Figure 13** were performed. Apple seedlings were grown with three different sources of water: tap water, water that had passed through grass growing in soil, and water that had passed through soil only. The growth of the young apple trees was apparently inhibited by something produced by the grass and carried by the water.

Agriculturalists have recognized the action of smother crops as weed suppressors. These smother crops include barley, rye, sorghum, millet, sweet clover, alfalfa, soybeans, and sunflowers. Their inhibition of weed growth was assumed to be due to competition for water, light, or nutrients. Barley, for example, is rated as a good smother crop and has extensive root growth. Wheat (*Triticum aestivum*) also has good potential for weed control in agricultural landscapes (Ma 2005). **Figure 14**

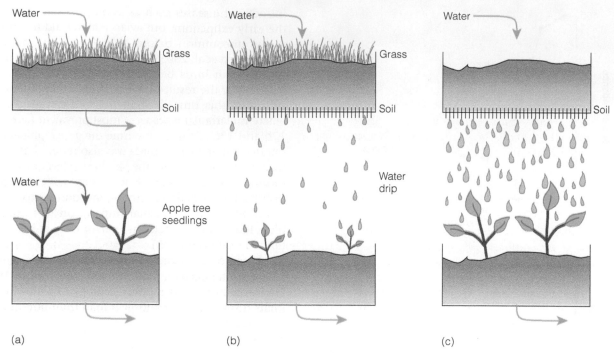

(a)　　　　　　　　　　　(b)　　　　　　　　　　　(c)

Figure 13 Experiments that demonstrated the detrimental effects of grass on apple tree seedlings. Grass and apple seedlings are grown in separate flats in a greenhouse. Water is provided either (a) independently from the tap to both grass and seedlings, (b) to grass growing in soil so that the water drips through onto the seedlings, or (c) to soil only so that the water drips onto the seedlings. Apple tree seedlings do not grow properly and often die when the water has passed through grass first (b).

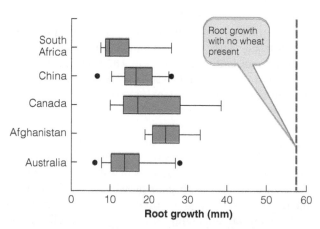

Figure 14 Allelopathic effects of 92 wheat (*Triticum aestivum*) genotypes on the root growth of annual ryegrass (*Lolium rigidum*). The box plot shows the median (line), and the green box includes the 25th and 75th percentile, and the outer lines the 10th and 90th percentiles. On average wheat seedlings reduce annual ryegrass root growth 70% from 57 mm in control assays to 17 mm when wheat seedlings are present. Phenolic acids and hydroxamic acids are the main components of the root exudates that inhibit ryegrass root development. (Data from Wu et al. 2000.)

shows how wheat seedlings reduce root growth in a common weed, annual ryegrass. There is considerable variation in the strength of allelopathy effects in different wheat varieties, but the genetic basis for the production of allelopathic chemicals is unknown (Wu et al. 2000). Part of the success of invasive plants may be their novel allelopathic effects. There is great interest among agricultural scientists in the potential uses of allelopathy for weed control in crops (Weston 1996). Since allelopathic chemicals often are highly specific, they could be used in agricultural systems in much the same manner as synthetic herbicides. This possibility is premised on our understanding the physiological mechanisms by which allelopathic chemicals operate to suppress weeds.

Whether or not allelopathy is a significant factor affecting the local distribution of plants in natural vegetation is controversial. Many plant ecologists accept the laboratory data on allelopathy but question whether or not it is effective in natural plant communities (Weidenhamer 1996). But the complex interactions going on in the root zone between different plant species suggests that allelochemicals are an integral part of biotic interactions that affect distribution and abundance (Bais et al. 2004).

Competition

The presence of other organisms may limit the distribution of some species through competition. Allelopathy is one specific type of competition for living space. But competition can occur between any two species that use the same types of resources and live in the same sorts of places. Note that two species do not need to be closely related to be involved in competition. For example, birds, rodents, and ants may compete for seeds in desert environments, and herbs and shrubs may compete for water in dry chaparral stands. Competition among animals is often over food. Plants can compete for light, water, nutrients, or even pollinators.

How can we determine whether competition could be restricting geographical distributions? One indication of competition may be the observation that when species A is absent, species B lives in a wider range of habitats. In extreme cases a habitat will contain only species A or species B and never both together. The prin-cipal difficulty in understanding these situations is that competition is only one of several hypotheses that can account for the observed distributions.

Competition between species is best studied with experiments whenever possible. In the UK two stream-dwelling carnivorous mustelids, the American mink and the Eurasian otter, show strong competition. The mink was introduced to the UK for fur farming around 1900, while the otter is a native species. Mink numbers increased dramatically from 1950 to 1980, and they expanded their geographic range, threatening a number of native species. From about 1985 mink numbers began to fall, and fewer sites were occupied by mink (Bonesi et al. 2006). To see if competition for space with otters could be responsible for the decline in mink, in 1999 Bonesi and Macdonald (2004) released 17 otters into an area occupied by mink, and followed the experimental and control populations for two years. Mink decreased rapidly in sites colonized by otters (**Figure 15**), suggesting competition for space.

ESSAY

What Is Competition?

Competition is a concept that is so familiar to us in capitalist societies that it might seem odd to ask what it means. In ecology, competition is defined as a negative interaction between two species over resources. It can take two quite different forms:

- *Resource competition*, which occurs when a number of organisms utilize common resources that are in short supply

- *Interference competition*, which occurs when the organisms seeking a resource harm one another in the process, even if the resource is not in short supply

We are concerned in this chapter with competition between two species, called *interspecific competition*. Competition can also occur among individuals of the same species, as we see every day in the business pages of the newspapers.

Competition occurs over resources, and, if competition is suspected as a mechanism affecting the local distribution of two species, we must answer two questions:

1. Does competition occur between these species? The simplest approach to answering this question is to do a removal experiment. If we remove the dominant competitor, the other species should expand its local geographic range.

2. What are the resources for which competition occurs? We tend to assume that water or nutrients are the limiting resources for plants, and food supplies the limiting resources for animals, but as always in ecology we should assume nothing without an experimental test.

Experimental studies of competition are always asking about the immediate interactions between species in ecological time, and that is our concern in this chapter. In evolutionary time we may see traces of the "ghost of competition" in adaptations that exist now because of intense competition between two species in the past and subsequent evolutionary divergence.

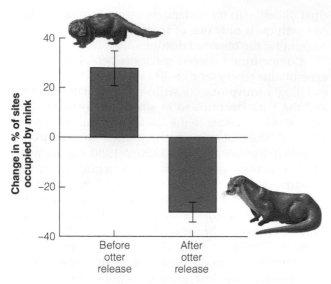

Figure 15 Competition between carnivores: Mean differences between the impact and control areas in the percentage of sites occupied by mink before and after otters were released in English river systems in 1999. In the two years after this experimental introduction, otters significantly restricted the distribution of mink in these English river systems. (Data from Bonesi and Macdonald 2004.)

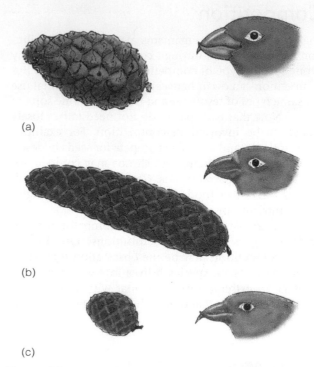

Figure 16 Heads of the three European crossbill species and the main conifer cones each species feeds on. (a) Parrot crossbill and Scotch pine cone, (b) common crossbill and spruce cone, and (c) white-winged crossbill and larch cone. (After Newton 1972.)

Otters are larger than mink, and in carnivores typically the larger species wins in competition. When otters are present, mink are excluded from stream habitats by direct aggression.

When two species compete for resources, one species will always be better than the other in gathering or utilizing the resource that is scarce. In the long run, one species must lose out and disappear, unless it evolves some adaptation to escape from competition. A species can adopt one of two general evolutionary strategies: (1) avoid the superior competitor by selecting a different part of the habitat or (2) avoid the superior competitor by making a change in diet. Let us look at an example in which possible competition is avoided by a diet shift.

Crossbills are finches that have curved crossed tips on the mandibles (**Figure 16**). Crossbills extract seeds from closed conifer cones by lateral movements of the lower jaw, and the jaw muscles are asymmetrically developed to provide the necessary leverage. Three species of crossbills live in Eurasia, and they are adapted for eating different foods (Newton 1972). The smallest crossbill is the white-winged crossbill, which has a small bill and feeds mainly on larch seeds (see Figure 16c). Larch cones are relatively soft. The medium-sized common crossbill eats mainly spruce seeds (see Figure 16b), and the larger parrot crossbill feeds on the hard cones of Scotch pine (see Figure 16a). These dietary

differences are not necessarily preserved when the species live in isolation. Thus the common crossbill has evolved a Scottish subspecies that has a large bill and feeds on pine cones, and an Asiatic subspecies that has a small bill and feeds on larch seeds. The white-winged crossbill has an isolated subspecies on Hispaniola in the West Indies that feeds on pine seeds and has a large beak. The bill adaptations of crossbills can thus be interpreted as devices for minimizing dietary overlap in regions where all three possible competitors live.

Four types of red crossbills (*Loxia curvirostra*) live in the Pacific Northwest of North America, and like their Eurasian counterparts they all represent adaptive peaks concentrating on four conifers with different seed and cone sizes (western hemlock, Douglas fir, ponderosa pine, and lodgepole pine). Benkman (1993) showed in a series of laboratory studies of foraging efficiency that the best bill size for feeding on one of these conifers was only one-half as efficient for feeding on the other conifer seeds. He postulated that disruptive selection maintained these four types of red crossbills in western North America.

Summary

A species may not occur in an area because it has not been able to disperse there. This hypothesis can be tested by artificial introductions of the organism into unoccupied habitats. Some species introduced by humans from one continent to another, such as the zebra mussel and the African honey bee have spread very rapidly. On a local scale, few introduced species, once they have become established, seem to be restricted in distribution by poor powers of dispersal.

Behavioral limitations on distribution are usually subtle and may be the most difficult to study. At present, few animal distributions are restricted on the landscape scale by behavioral reactions, but at the microhabitat scale habitat selection may be a critical limitation to local distributions. In a predictable environment, habitat selection may be very exact. When habitats change, some species are not able to adapt quickly and therefore inhabit only a portion of their potential habitat range.

Many animals and plants are limited in their local distribution by the presence of other organisms—their food plants, predators, diseases, and competitors. Experimental transfers of organisms can test for these factors, and cages or other protective devices can be used to identify the critical interactions. Predators can affect the local distribution of their prey. The converse can also occur, in which the prey's distribution determines the distribution of its predators, but this interaction does not seem to be common. Diseases and parasites may restrict geographical distributions, but few such cases have been studied in natural systems. They may play a larger role than we currently suspect in species-rich tropical communities. Much more work needs to be done on the role of disease in limiting geographic distributions.

Some organisms poison the environment for other species as a form of competition, and these chemical poisons, or allelopathic agents, may affect local distributions. Chemical interactions have been described in a variety of crop plants and in marine algae. Allelopathic interactions may have great practical importance in weed control in agriculture. Competition among organisms for resources may also restrict local distributions. Some species drive others out by aggressive interactions. Species may evolve differences in diet or habitat preferences as a result of competitive pressures.

Review Questions and Problems

1 Assume dispersal by simple diffusion (Eq. 1). How far would a plant be expected to move in 50 generations if the average dispersal distance was 100 m and the plant produced 10^3 seeds per generation? Is the distance moved more sensitive to the number of seeds produced or to the average dispersal distance? Double or triple each of these parameters and discuss the impact on the distance colonized by these life cycle changes.

2 How can natural selection maintain the particular ovipositing dance of *Anopheles culicifacies*, for example, if it results in suitable habitats being left unoccupied? Does natural selection always favor the broadest possible habitat range for a species?

3 One of the recurrent themes in studying introduced species is that introductions are more successful when more individuals are released (Green 1997; Forsyth et al. 2004). Are there cases of successful introductions by humans when only a few individuals were released? What might account for this pattern?

4 Is it possible for a transplant experiment to be successful and yet lead to the conclusion that neither dispersal nor habitat selection is responsible for range limitation? Discuss the transplant experiment of Dayton et al. (1982) on the antarctic acorn barnacle, and comment on the author's conclusions.

5 How are the predictions of the model given in Figure 8 affected if the habitat relationships are not straight lines but instead are curves? Describe a situation in which these lines in Figure 8 might cross. See Rosenzweig (1985, p. 523) for a discussion.

6 English yew (*Taxus baccata*) is an evergreen tree with an average life span of 500 years (Hulme 1996). The regeneration potential of local sites will determine the future distribution of this tree, and seed predators, seedling herbivores, or suitable microsites for germination and growth are the three factors that may limit yew distribution on a local scale. Discuss what observations could distinguish between biotic limitation and abiotic microsite limitation of yew distributions.

7 At Point Pelee National Park in Ontario, frog surveys showed that the bullfrog (*Rana catesbeiana*) disappeared in 1990, and from 1990 to 1994 the green frog (*Rana clamitans*) has increased in numbers fourfold (Hecnar and M'Closkey 1997). Suggest three possible interpretations for these natural

history observations, and indicate how you would test these hypotheses experimentally.

8 Norwegian lemmings do not live in lowland forests in Scandinavia even though they are regularly seen in these areas when their alpine populations are at high density. Suggest three hypotheses to explain the failure of lemmings to establish permanent populations in lowland forest, and discuss experiments to test these ideas. Oksanen and Oksanen (1992) discuss this question.

9 Laboratory tests for allelopathy have been criticized because the chemicals that act in the laboratory may not be effective in the field. Could this criticism be blunted by doing field experiments? Do you think that plants might evolve to produce chemical exudates that are not effective in the field?

10 Would you expect to have different factors limiting a species' geographic distribution at the northern and southern limits of its range?

11 Grizzly bears and black bears eat the same foods and live in similar places in North America. Grizzly or brown bears are much larger than black bears and more aggressive. All large islands off the coast of British Columbia and Alaska have either black bears or grizzly bears but no island has both species (Apps et al. 2006). Is this evidence for competition between these two bear species? What other evidence would you look for to show that grizzlies affect the local distribution of black bears?

Overview Question

Would you expect the same dispersal abilities in plants from tropical rain forests and from boreal conifer forests? In animals? Why or why not?

Suggested Readings

- Bonesi, L., R. Strachan, and D. W. Macdonald. 2006. Why are there fewer signs of mink in England? Considering multiple hypotheses. *Biological Conservation* 130:268–277.

- Brown, G. P., et al. 2006. Toad on the road: Use of roads as dispersal corridors by cane toads (*Bufo marinus*) at an invasion front in tropical Australia. *Biological Conservation* 133:88–94.

- Case, T. J. 1996. Global patterns in the establishment and distribution of exotic birds. *Biological Conservation* 78:69–96.

- Clark, J. S., et al. 1998. Reid's paradox of rapid plant migration. *Bioscience* 48(1):13–24.

- Diamond, J. M. 1997. *Guns, Germs, and Steel: The Fates of Human Societies.* New York: Norton.

- Freed, L. A., et al. 2005. Increase in avian malaria at upper elevation in Hawaii. *Condor* 107:753–764.

- Jeschke, J. M., and D. L. Strayer. 2005. Invasion success of vertebrates in Europe and North America. *Proceedings of the National Academy of Sciences USA.* 102:7198–7202.

- Johnson, L. E., J. M. Bossenbroek, and C. E. Kraft. 2006. Patterns and pathways in the post-establishment spread of non-indigenous aquatic species: The slowing invasion of North American inland lakes by the zebra mussel. *Biological Invasions* 8:475–489.

- Short, J., and B. Turner. 2000. Reintroduction of the burrowing bettong *Bettongia lesueur* (Marsupialia: Potoroidae) to mainland Australia. *Biological Conservation* 96:185–196.

Credits

Illustration and Table Credits

Photo Credits

Factors That Limit Distributions II: Abiotic

Key Concepts

- Temperature and moisture are the main limiting factors for both plants and animals on a global scale.

- Light, fire, pH, and other physical and chemical factors can limit distributions on a local scale.

- Species may evolve adaptations that overcome the limitations set by physical and chemical factors.

- Some of these adaptations may allow a species to extend its geographical range.

- Climatic warming in this century will have major impacts on the geographical ranges of species that are currently limited by temperature and moisture.

From Chapter 6 of *Ecology: The Experimental Analysis of Distribution and Abundance*, Sixth Edition. Eugene Hecht.

actual evapotranspiration The actual amount of water that is used by and evaporates from a plant community over a given time period, largely dependent on the available water and the temperature.

Calvin-Benson cycle The series of *biochemical* reactions that takes place in the *stroma* of *chloroplasts* in *photosynthetic organisms* and results in the first step of carbon fixation in photosynthesis.

common garden An experimental design in plant ecophysiology in which a series of plants from different areas are brought together and planted in one area, side by side, in an attempt to determine which features of the plants are genetically controlled and which are environmentally determined.

crassulacean acid metabolism (CAM) A form of photosynthesis in which the two chemical parts of photosynthesis are separated in time because CO_2 is taken up at night through the stomata (which are then closed during the day) and fixed to be used later in the day to complete photosynthesis carbon fixation; an adaptation used by desert plants to conserve water.

ecotype A genetic race of a plant or animal species that is adapted to a specific set of environmental conditions such as temperature or salinity.

Krantz anatomy The particular type of leaf anatomy that characterizes C_4 plants; plant veins are encased by thick-walled photosynthetic bundle-sheath cells that are surrounded by thin-walled mesophyll cells.

photoperiodism The physiological responses of plants and animals to the length of day.

potential evapotranspiration The theoretical depth of water that would evaporate from a standard flat pan over a given time period if water is not limiting, largely dependent on temperature.

shade-intolerant plants Plants that cannot survive and grow in the shade of another plant, requiring open habitats for survival.

shade-tolerant plants Plants that can live and grow in the shade of other plants.

Temperature and moisture are the two master limiting factors to the distribution of life on Earth, so it is not surprising that an enormous body of literature addresses the effects of temperature and moisture on organisms. Before we analyze the ecological effects of these two factors, as well as other physical-chemical limiting factors, let us look at the global temperature and moisture conditions to which organisms must adapt.

Climatology

The large temperature differentials over the Earth are a reflection of two basic variables: incoming solar radiation and the distributions of land and water. Solar radiation lands obliquely in the higher latitudes (**Figure 1**) and thus delivers less heat energy per unit of surface area. Increased day length in summer partially compensates for the reduced heat input at high latitudes, but total annual insolation is still lower in the polar regions. The amount of heat delivered to the poles is only about 40% of that delivered to the equator.

Land and sea absorb heat differently, and this effect produces more contrasts, even within the same latitude. Land heats quickly but cools rapidly as well, so land-controlled, or **continental, climates** have large daily and seasonal temperature fluctuations. Water heats and cools more slowly because of vertical mixing and a high specific heat. The net result, shown in **Figure 2**, is that annual temperature variation between summer and winter are greatest over the large continental landmasses.

Water, alone or in conjunction with temperature, is probably the most important physical factor affecting the ecology of terrestrial organisms. Land animals and plants are affected by moisture in a variety of ways. Humidity of the air is important in controlling water loss through the skin and lungs of animals. All animals require some form of water intake (in food or as drink) in order to operate their excretory systems. Plants are affected by the soil water levels as well as the humidity of the air around leaf surfaces. Cells are 85%–90% water, and without adequate moisture there can be no life.

Moisture circulates from the ocean and the land back into clouds only to fall again as rain in a continuous cycle. The global distribution of rainfall resulting

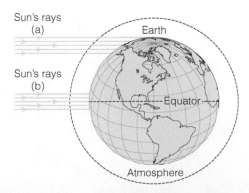

Figure 1 The sun's rays strike the Earth at an oblique angle in the polar regions (a) and vertically at the equator (b). Sunlight delivers less energy to the Earth's surface at the poles because its energy is spread over a larger surface area and because it passes through a thicker layer of absorbing, scattering, and reflecting atmosphere.

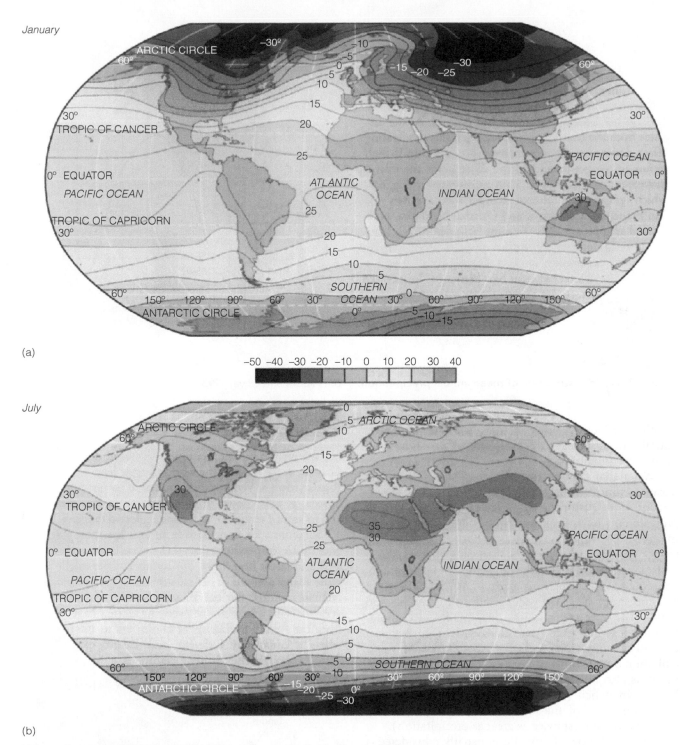

Figure 2 Average temperatures for January and July for the Earth. Temperature range (°C) from winter to summer is smallest at low latitudes and over the oceans, and largest over the continents. (From Hidore and Oliver 1993.)

from these processes is shown in **Figure 3**. A belt of high precipitation in equatorial regions is apparent in the Amazon, West Africa, and Indonesia. Low precipitation around latitude 30°N and S is associated with the distribution of deserts around the world. The distribu-

tion of continents and oceans also has a strong effect on the pattern shown in Figure 3. More rain falls over oceans than over land. The average ocean weather station for the globe records 110 cm of precipitation, compared with 66 cm for the average land weather station.

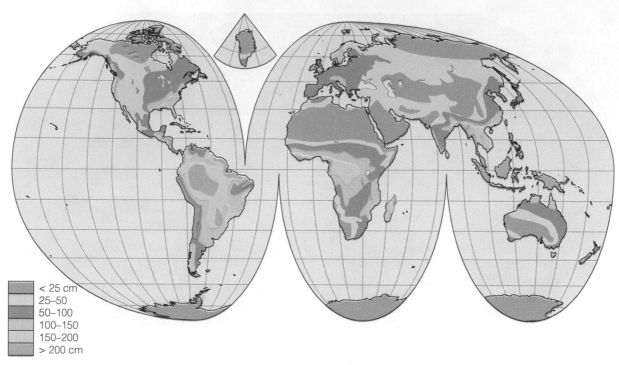

Figure 3 World distribution of mean annual precipitation. (From Hidore and Oliver 1993.)

Finally, mountains and highland areas intercept more rainfall and also leave a "rain shadow," or area of reduced precipitation, on their leeward side.

Water that falls on the land circulates back to the ocean as runoff or back to the air directly by evaporation or transpiration from plants. Only about 30% of precipitation is returned via runoff, and hence the remaining 70% must move directly back into the air by evaporation and transpiration. The rates of evaporation and transpiration depend primarily on temperature; consequently, a strong interaction between temperature and moisture affects the water relations of animals and plants. The absolute amounts of rainfall and evaporation are less important than the relationship between the two variables. Polar areas, for example, have low precipitation but are not arid because the amount of evaporation is also low. About one-third of global land area has a rain deficit (evaporation exceeds precipitation), and about 12% of the land surface is extremely arid (evaporation at least twice as great as precipitation).

The vegetation of any site is usually considered a product of the area's climate. This implies that climatic factors, temperature and moisture primarily, are the main factors controlling the distribution of vegetation (**Figure 4**). Geographers have often adopted this viewpoint and then turned it around to set up a classification of climate on the basis of vegetation. Native vegetation is assumed to be a meteorological instrument capable of measuring all the integrated climatic

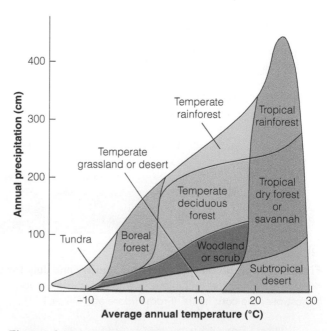

Figure 4 Terrestrial vegetation classes plotted in relation to annual precipitation and average annual temperature. Boundaries between vegetation classes are approximate. (Modified from Whittaker 1975.)

elements. Figure 4 illustrates how the world's broad vegetation groups can be mapped on temperature and precipitation averages.

Some geographers have tried to set climatic boundaries independent of vegetation. Thornthwaite (1948) developed one classification. The basis of his climatic classification is **precipitation**, which is balanced against **potential evapotranspiration**, the amount of water that would be lost from the ground by evaporation and from the vegetation by transpiration if an unlimited supply of water were available. There is no way of measuring potential evapotranspiration directly, and it is normally computed as a function of temperature. Vegetation patterns can be described more accurately if we use **actual evapotranspiration**—the evaporative water loss from a site covered by a standard crop, given the precipitation. Major vegetational types such as grassland, temperate deciduous forest, and tundra are closely associated with certain climatic types defined by the water balance (Stephenson 1990).

Temperature and Moisture as Limiting Factors

Organisms have two options in dealing with the climatic conditions of their habitat: They can simply tolerate the temperature and moisture as they are, or they can escape via some evolutionary adaptation. We begin our consideration of the effects of temperature and moisture by first examining how well organisms tolerate these two factors. Every organism has an upper and a lower lethal temperature, but these parameters are not constants for each species. Organisms can acclimate physiologically to different conditions. The resistance of woody plants to freezing temperatures is another example. Willow twigs (*Salix* spp.) collected in winter can survive freezing at temperatures below −150°C, while the same twigs in summer are killed by −5°C temperatures (Hietala et al. 1998).

Temperature and moisture may act on any stage of the life cycle and can limit the distribution of a species through their effects on one or more of the following:

- Survival

- Reproduction

- Development of young organisms

- Interactions with other organisms (competition, predation, parasitism, diseases) near the limits of temperature or moisture tolerance

If temperature or moisture acts to limit a distribution, what aspect of temperature or moisture is

relevant—maxima, minima, averages, or the level of variability? No overall rule can be applied here; the important measure depends on the mechanism by which temperature or moisture acts and the species involved. Plants (and animals) respond differently to a given environmental variable during different phases of their life cycle. For this reason, mean temperatures or average precipitation will not always be correlated with the limits of distributions, even if temperature or moisture is the critical variable.

To show that temperature or moisture limits the distribution of an organism, we should proceed as follows:

- Determine which phase of the life cycle is most sensitive to temperature or moisture.

- Identify the physiological tolerance range of the organism for this life-cycle phase.

- Show that the temperature or moisture range in the microclimate where the organism lives is permissible for sites within the geographic range, and lethal for sites outside the normal geographic range **Figure 5**.

We will now consider a set of examples that illustrates this approach and shows some of the biological complications that may occur.

The range limits of warm-blooded animals may correlate with climatic variables. Winter distributions of passerine birds in North America often correlate with

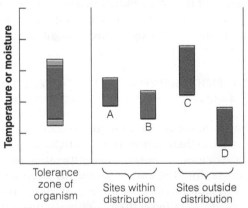

Figure 5 Hypothetical comparison of the tolerance zone of an organism and the temperature or moisture ranges of the microclimates where it lives. The tolerance zone is measured for the stage of the life cycle that is most sensitive to temperature or moisture and is subdivided into two zones, the optimal zone (dark blue) and the marginal zone (light blue). In this example the organism can live at A and B, but cannot tolerate C or D (red). The same principle can be applied to other physical-chemical factors such as pH.

minimum January temperature (Root 1988). Such climatic limitations in temperate zone birds are directly linked with the energetic demands associated with cold temperatures. It is less common for geographic range limits to coincide with rainfall contours. Few geographic distributions of animals are likely set directly by precipitation. But for plants, moisture is of direct importance, and the water relations of plants is an important area of research for plants of economic value. The water balance of plants is difficult to measure directly, and botanists usually measure the water content of plant tissues as an index of water balance. The leaves are particularly sensitive because most evaporation occurs there. Different plants vary greatly in their ability to withstand water shortages.

Drought resistance is achieved by (1) improvement of water uptake by roots; (2) reduction of water loss by stomatal closure, prevention of cuticular respiration, and reduction of leaf surface; and (3) storage of water. Rapid root growth into deeper areas of the soil is often effective in increasing drought resistance; young plants with little energy reserve will consequently suffer the worst from drought. Leaves of plants subject to poor water supply often have smaller surface areas and thicker cuticles, both of which reduce evaporation losses. By shedding their leaves in the drought season, plants have another very effective means of reducing water loss. *Xerophytes* (plants that live in dry areas) show many of these special adaptations for decreasing water loss. Additionally, leaves may be oriented vertically, which reduces the amount of absorbed radiation and resultant evaporation. Other xerophytes, such as cacti, store water in their stems and thereby overcome drought.

Interaction between Temperature and Moisture

In some cases, the moisture requirements of plants can alone restrict their geographic distributions. But in many other cases, moisture and temperature interact to limit geographic distributions, and the ecologist must consider explanations such as "both-temperature-and-moisture" rather than "either-temperature-or-moisture." Both **frost drought** and **soil drought** can be critical in determining ranges of species. Soil drought is the common notion of drought in which soil moisture is deficient (as in the desert); it can usually be described as an absolute shortage of water in the soil. Frost drought or winter drought in plants occurs when water is present but unavailable because of low soil temperatures (such as occur in the tundra in winter), and the roots are unable to take up water while the leaves continue to lose water by transpiration; it can be described as a relative shortage of water for plants. In both situations, water loss from the plant's leaves and stems is greater than water intake through the roots. Thus low temperatures can produce symptoms of drought. This fact emphasizes that water availability is the critical variable and has led to considerable research on how to measure "available" water in the soil. Many of the distributional effects attributed to temperature may in fact operate through the water balance of plants.

The simplest approach to mapping distributional limits is to combine measures of temperature and rainfall in a statistical model. Hocker (1956) described the distribution range of the loblolly pine (*Pinus taeda*) from the meteorological data available from 207 weather stations in the southeastern United States. He included values for (1) average monthly temperature, (2) average monthly range of temperature, (3) number of days per month of measurable rainfall, (4) number of days per month with rainfall over 13 mm, (5) average monthly precipitation, and (6) average length of frost-free period. Weather stations were divided into two groups, one within the natural range of the pine and the other outside the range; from the difference between these two groups it is possible to map the climatic limits for loblolly pine (**Figure 6**). There is good agreement between observed limits of range and the limits mapped from these meteorological data. Winter temperature and rainfall probably set the northern limit of this pine. The rate of water uptake in loblolly pine roots decreases rapidly at lower temperatures, and this would accentuate winter drought in more northerly areas. Hocker predicted that a northern extension of the limits of loblolly pine was not feasible under the current climate because of these basic climatic limitations. Global warming can readily change range limits set by climatic variables.

Western hemlock is a common tree in northwestern North America, and its climatic limits have been mapped by Gavin and Hu (2006). Western hemlock requires mild and humid conditions, and has a high water requirement. Its shallow root system also makes it susceptible to water deficits. Hemlock occurs in two distinct populations: in wet and mild coastal habitats, which it colonized 9000 years ago, and in the colder and drier interior valleys of western Canada, colonized only 2000–3500 years ago after the ice melted. **Figure 7** shows that for the coastal populations actual evapotranspiration is a range-limiting climatic variable, since areas outside its current geographic distribution are distinctly drier than areas within the coastal range. Interior populations by contrast show strong overlap in climatic measures (see Figure 7), which suggests that it has not yet colonized all the interior habitats where it could grow successfully (Gavin and Hu 2006). Factors limiting geographic distributions are not necessarily the same in all parts of a species' range.

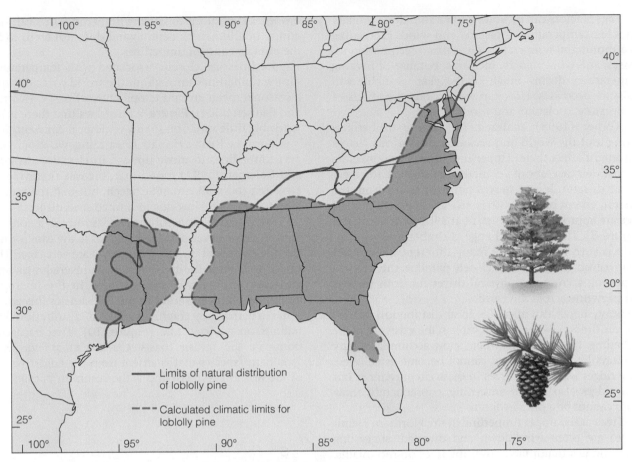

Figure 6 Natural distribution limits (solid red line) and calculated climatic limits (dashed line, blue area) of loblolly pine (*Pinus taeda*) in the southeastern United States. Winter temperature and rainfall set the northern and western limits of this pine. (After Hocker 1956.)

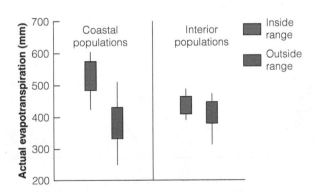

Figure 7 Climatic limitation of western hemlock (*Tsuga heterophylla*). Actual evapotranspiration (AET) is a measure that combines temperature and rainfall. For coastal population there is very little overlap for AET at geographic sites inside and outside the range. Interior populations show considerable overlap, suggesting that hemlock has not yet colonized all sites that are suitable climatically. Compare with Figure 5. (After Gavin and Hu 2006.)

As one moves up large mountains like those in the Rocky Mountains, or north or south toward the poles, one reaches the limit of trees as a vegetation type. This is called the **treeline**, or timberline, and is a particularly graphic illustration of the limitation on plant distribution imposed by the physical environment. Stevens and Fox (1991) listed nine factors that have been suggested to affect timberlines:

- Lack of soil
- Desiccation of leaves in cold weather
- Short growing season
- Lack of snow, exposing plants to winter drying
- Excessive snow lasting through the summer
- Mechanical effects of high winds
- Rapid heat loss at night
- Excessive soil temperatures during the day
- Drought

These factors can be boiled down into three primary variables: temperature, moisture, and wind. Proceeding up a mountain, temperature decreases, precipitation increases, and wind velocity increases. Because of freezing temperatures during much of the year, available soil moisture decreases. How can we separate the effects of temperature, moisture, and wind?

Körner (1998) analyzed 150 alpine timberlines throughout the world and reviewed the various factors that might affect them. Upper timberlines in temperate regions decrease about 75 meters in altitude for every degree of latitude one moves north or south from the equator, except between 30°N and 20°S, where timberlines are approximately constant at 3500 to 4000 meters (**Figure 8**). The uniform change of timberlines with latitude is surprising because many different tree species are involved. The snowline closely parallels the treeline, suggesting a common physical driver for both treeline and snowline across the Earth.

Snow depth can affect the local distribution of trees near the timberline but cannot explain the existence of the timberline. In depressions where snow accumulates early and stays late, tree seedlings cannot become established. Only ridges will support trees in these circumstances, but these ridges also have a timberline; consequently, snow depth cannot be a primary factor.

Trees at the upper timberline in the Northern Hemisphere are often windblown and dwarfed, suggesting that wind is a major factor limiting trees on mountains. Within the tropics and in the Southern Hemisphere, wind effects seem to be absent. One difficulty with the wind hypothesis is that all the evidence is relevant to old trees, whereas it is the establishment of very young seedlings that is crucial to timberline formation. Wind

has secondary effects in altering timberlines in local situations, but, like snow depth, wind does not seem to be the primary cause of timberlines.

Treelines are closely associated with temperature. Alpine timberlines throughout the world coincide with a seasonal mean ground temperature of 6.7°C (Körner and Paulsen 2004). **Figure 9** illustrates that there is remarkably little variation in this ecological constant.

Since the Earth's climate is warming, we should expect timberlines to move upward in elevation. Wardle and Coleman (1992) found that this was occurring in New Zealand with Antarctic beech, but that the rate of advance was very slow due to limited seed dispersal.

The intertidal zone of rocky coastlines is a zone of tension between sea and land, and, as is the case for the treeline, the distributional boundaries are very clear. The upper and lower limits of dominant invertebrates and algae are often very sharply defined in the intertidal zone, and on rocky shores in the British Isles this zonation is a particularly graphic example of distribution limitations on a local scale (**Figure 10**). Two barnacles dominate the British coasts. *Chthamalus stellatus* is a "southern" species that is absent from the colder waters of the British east coast and is the common barnacle of

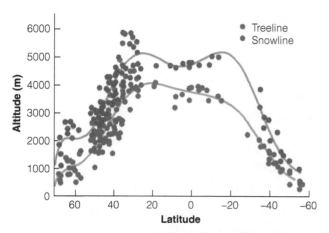

Figure 8 Alpine treeline and snowline in relation to latitude for 150 sites around the world. Snowlines mark the altitude of permanent snowfields. There is a nearly constant altitude for treelines and snowlines for nearly 50° of latitude near the equator. (Data from Körner 1998.)

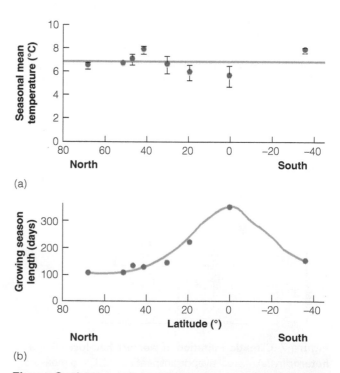

Figure 9 Alpine treelines. (a) Growing season root zone temperature at treeline across latitudes. There is a nearly constant soil temperature of 6.7°C that applies in all areas. (b) Length of the growing season at alpine treelines around the world. Mountains near the equator have a much longer growing season. (Data from Körner and Paulsen 2004.)

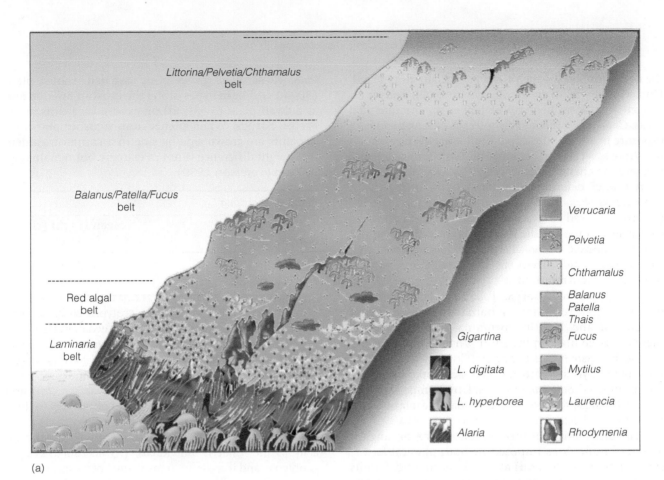

(a)

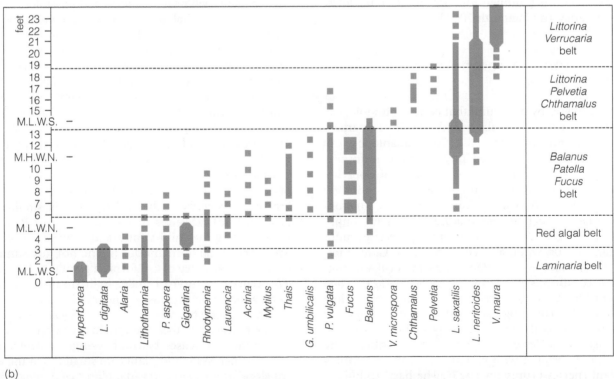

(b)

Figure 10 A very common barnacle-dominated slope on moderately exposed rocky shores of northwestern Scotland and northwestern Ireland. (a) Schematic view. (b) The distributional limits on the shore of the dominant species in this rocky intertidal. The limits of distribution are strongly defined for each species, and this results in belts of zonation that can be recognized over large areas of rocky shorelines. The width of the lines is proportional to the abundance of the species. MHWS = mean high water, spring; MHWN = mean high water, neap (i.e., minimum tide); MLWN = mean low water, neap; MLWS = mean low water, spring. (After Lewis 1964.)

the upper intertidal zone of western Britain and Ireland. Going farther north in the British Isles, one finds it restricted to a zone higher and higher on the intertidal rocks. *Chthamalus* is relatively tolerant of long periods of exposure to air, and the upper limit of its distribution on the shore is set by desiccation. This basic limitation does not seem to change over its range. Its lower limit on the shore is often determined by competition for space with *Semibalanus balanoides*, a northern species. Connell (1961b) showed that *Semibalanus* grew faster than *Chthamalus* in the middle part of the intertidal zone and simply squeezed *Chthamalus* out. He also showed that *Chthamalus* could survive in the *Semibalanus* zone if *Semibalanus* were removed.

The upper limit of *S. balanoides* is also set by weather factors, but since this barnacle is less tolerant of desiccation and high temperatures than *Chthamalus*, there is a zone high on the shore where *Chthamalus* can survive but *Semibalanus* cannot (Connell 1961a). The sensitivity of young barnacles sets this upper limit. The lower limit of *Semibalanus* is set by competition for space with algae and by predation, particularly by a gastropod, *Thais lapillus*.

The distribution of these barnacles is a striking example of limitations imposed by both physical factors (temperature, desiccation) at the upper intertidal limits and biotic factors (competition, predation) at the lower limits (Harley and Helmuth 2003).

Adaptations to Temperature and Moisture

We have begun by assuming that certain physiological tolerances are built into all the individuals of a particular species. But we know that local adaptation can occur and that genetic and physiological uniformity cannot be assumed throughout the range of a species. Darwin recognized that species could extend their distribution by local adaptation to limiting environmental factors such as temperature, but the full implications of Darwin's ideas were not appreciated until the early 1900s, when a Swedish botanist, Göte Turesson, began looking at adaptations to local environmental conditions in plants. Turesson (1922) coined the word **ecotype** to describe genetic varieties within a single species. He recognized that much of ecology had been pursued as if genetic diversity within species did not exist. In a series of publications he described some variation associated with climate and soil in a variety of plant species (Turesson 1925). The basic technique was to collect plants from a variety of areas and grow them together in field or laboratory plots at one site, a **common garden**. The type of result he obtained in this early work can be illustrated with an example.

Plantago maritima grows both as a tall, robust plant (30–40 cm) in marshes along the coast of Sweden and as a dwarf plant (5–10 cm) on exposed sea cliffs in the Faeroe Islands. When plants from marshes and from sea cliffs are grown side by side in a common garden, this height difference is not as extreme but remains significant (Turesson 1930):

Plantago maritima

Source	Mean height (cm) in garden
Marsh population	31.5
Cliff population	20.7

Turesson's early studies on ecotypes such as these helped to create a new research field of ecological genetics.

This common garden technique is an attempt to separate the **phenotypic** (environmental) and **genotypic** (genetic) components of variation. Plants of the same species growing in such diverse environments as sea cliffs and marshes can differ in morphology and physiology in three ways: (1) all differences are phenotypic, and seeds transplanted from one situation to the other will respond exactly as the resident individuals; (2) all differences are genotypic, and if seeds are transplanted between areas, the mature plants will retain the form and physiology typical for their original habitat; or (3) some combination of phenotypic and genotypic determination produces an intermediate result. In natural situations, the third case is most common. Many examples are now described in the literature, particularly in plants (Joshi et al. 2001).

A classic set of ecotypic races occurs in the perennial herb *Achillea* (yarrow), analyzed by Clausen, Keck, and Hiesey (1948) in a pioneering paper. Clausen and his colleagues studied two North American species in detail. A maritime form of *Achillea borealis* lives in coastal areas of California as a low succulent evergreen plant that grows throughout the winter. An evergreen race grows slightly farther inland that is similar but taller. A third race lives in the Pacific Coast Range; it grows during the mild winter and flowers quickly by April, becoming dormant during the hot, dry summer. In the Central Valley of California a giant race of *A. borealis* occurs that survives under high summer temperatures, a long growing season, and ample moisture.

In the Sierra Nevada, races of *Achillea lanulosa* occur. As one proceeds up these mountains, the average winter temperature decreases below freezing, so winter dormancy is necessary and plants are smaller. On the eastern slope of the Sierra Nevada, plants of *A. lanulosa* are late flowering and adapted to cold, dry conditions. Clausen, Keck, and Hiesey collected seeds from a series of populations of *A. lanulosa* across California and raised plants in a greenhouse at Stanford, with the re-

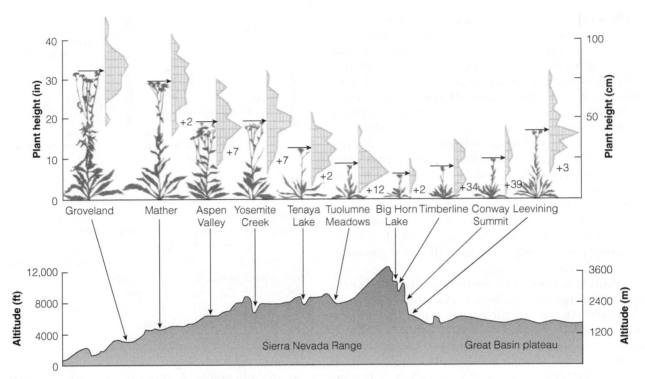

Figure 11 Representatives of populations of *Achillea lanulosa* as grown in a common garden at Stanford University, California. These originated in the localities shown in the profile of a transect across central California at approximately 38°N latitude. Altitudes are to scale, but horizontal distances are not. The plants are herbarium specimens, each representing a population of approximately 60 individuals. The frequency diagrams show variation in height within each population. The numbers to the right of some frequency diagrams indicate the number of nonflowering plants. The arrows point to the mean heights. (Modified from Clausen et al. 1948.)

sults shown in **Figure 11**. The major attributes of these races are maintained when plants are grown under uniform conditions in the same place.

Many species expanded their geographical range during the twentieth century but in nearly all cases we do not know if genotypic changes accompanied these range changes. If we could study a species in the midst of a range extension, we might obtain some insight as to how organisms can extend their tolerance limits. This opportunity may become more frequent in the future, as climatic warming occurs.

Light as a Limiting Factor

Light may be another factor limiting the local distribution of plants. Light is important to organisms for two quite different reasons: it is used as a cue for the timing of daily and seasonal rhythms in both animals and plants, and it is essential for photosynthesis in plants.

Timing, the first reason light is important, is a central issue in the life cycles of organisms. Nocturnal desert animals, for example, use light as a cue for their activity cycles. The breeding seasons of many animals and plants are set by the organisms' responses to day-length changes. The seasonal impact of day length on physiological responses, called **photoperiodism**, has been an important focus of work in environmental physiology (Eckert et al. 1997).

The second reason light is important to organisms is that it is essential for **photosynthesis**, the process by which plants convert radiant energy from the sun into energy in chemical bonds. Photosynthesis is remarkably inefficient. During the growing season, about 0.5%–1% of the incoming radiation is captured and stored by photosynthesis. In this process, carbon in the form of CO_2 is taken up from the air (or the water in the case of aquatic plants) and converted into organic compounds. We can measure the rate of photosynthesis by measuring the rate of uptake of CO_2.

Plants show a great diversity of photosynthetic responses to variations in light intensity. Some plants reach maximal photosynthesis at one-quarter full sunlight, and other species such as sugarcane never reach a maximum but continue to increase photosynthetic rate as light intensity rises. We recognize this ecologically by noting that plants in general can be divided into two groups: **shade-tolerant** species and **shade-intolerant** species. This classification is commonly used in forestry and horticulture. Plant physiologists have discovered that shade tolerance is a complex of traits, and that it is not fixed for each species but varies with plant age, microclimate, and geographical area (Kozlowski et al. 1997). Shade-tolerant plants have lower photosynthetic rates and hence would be expected to have lower growth rates than shade-intolerant species. The metabolic rate of shade-tolerant seedlings is apparently lower than that of shade-intolerant seedlings.

Plant species become adapted to live in a certain kind of habitat and in the process evolve a series of characteristics (an "adaptive syndrome") that prevent them from occupying other habitats. Grime (1979) suggests that light may be one of the major components directing these adaptations. For example, eastern hemlock seedlings are shade-tolerant and can survive in the forest understory under very low light levels. Hemlock seedlings grow slowly and have a low metabolic rate that allows them to survive low light conditions. One consequence of these adaptations is that hemlock seedlings die easily in droughts because their roots do not grow quickly enough to penetrate deep into the soil. Failure of seedlings in shaded situations is often associated with fungal attack, and part of adaptation to shade involves becoming resistant to fungal infections (Givnish 1988).

An exceedingly important principle in evolutionary ecology is that *individuals of a species cannot do everything in the best possible way.* Adaptations to live in one ecological habitat make it difficult or impossible to live in a different habitat. Thus life cycles have evolved as trade-offs between contrasting habitat requirements. Adaptations are always compromises, and there can be no superanimals or superplants.

A good illustration of this principle can be seen in the adaptations of trees to shade tolerance and drought tolerance. Niinemets and Valladares (2006) analyzed the shade, drought, and waterlogging tolerance of 806 species of North American, European, and Asian shrubs and trees to test the hypothesis that shade tolerance is negatively related to drought tolerance and waterlogging tolerance. **Figure 12** illustrates their findings for two important species groups, the oaks (11 species) and the pines (18 species). In both groups there is a strong trade-off: a species can be shade tolerant or drought tolerant but not both.

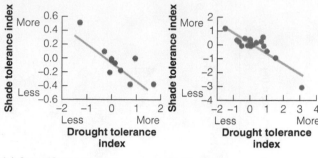

(a) Oaks (*Quercus spp.*)   (b) Pines (*Pinus* spp.)

Figure 12 **Trade-off between shade tolerance and drought tolerance for two genera of trees: (a) *Quercus* (oaks) and (b) *Pinus* (pines) from around the world.** The indices are standardized scores and the data points represent phylogenetically independent groups of species within each genus. The data illustrate graphically the trade-off principle that species are constrained in their adaptations and cannot be both drought and shade tolerant. (From Niinemets and Valladares 2006.)

An understanding of these trade-offs can be found in the physiological controls on photosynthesis in plants. One reason photosynthetic rate varies among plants is that they have evolved three photosynthetic strategies: the C_3 pathway, the C_4 pathway, and crassulacean acid metabolism. Most plants use the C_3 pathway, first described by Calvin and often called the Calvin or **Calvin-Benson cycle**. In the C_3 pathway, CO_2 from the air is first converted to 3-phosphoglyceric acid, a three-carbon molecule (hence the name C_3). Until the mid-1960s this pathway was believed to be the only important means of fixing carbon in the initial steps of photosynthesis. In 1965 sugarcane was found to fix CO_2 by first producing malic and aspartic acids (four-carbon acids), and the C_4 pathway of photosynthesis was discovered (Björkman and Berry 1973). C_4 plants have all the biochemical elements of the C_3 pathway, so they can use either method to fix CO_2.

The ecological consequences of the C_4 pathway are profound. **Figure 13** shows the rates of photosynthesis of a pair of closely related species of C_3 and C_4 plants. C_4 plants do not reach saturation light levels even under the brightest sunlight, and they always produce more photosynthate per unit area of leaf than C_3 plants. C_4 plants are thus more efficient than C_3 plants. Leaf anatomy differs in typical C_3 and C_4 plants (**Figure 14**). Chlorophyll in C_3 leaves is found throughout the leaf, but in C_4 leaves the chloroplasts are concentrated in two-layered bundles around the veins of the leaf (called **Krantz anatomy**). The bundle sheath cells in C_4 plants also have a high concentration of mitochondria. The C_4 leaf anatomy is more efficient for utilizing low CO_2 concentrations, for recycling the CO_2

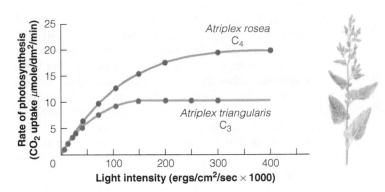

Figure 13 Comparative photosynthetic production of the C$_3$ species *Atriplex triangularis* and the related C$_4$ species *Atriplex rosea*. The plants were grown under identical controlled conditions of 25°C during the day and 20°C at night, 16-hour days, and ample water and nutrients. (After Björkman 1975.)

produced in respiration, and for rapidly translocating starches to other parts of the leaf. The biochemical reason for this anatomical difference is simple—the first step in fixing CO_2 in these two types of plants differs:

C_3: Atmospheric CO_2 + ribulose-diphosphate (RuDP)

$$\xrightarrow{\text{RuDp carboxylase}} \text{phosphoglyceric acid}$$

C_4: Atmospheric CO_2 + phospho-enolpyruvate (PEP)

$$\xrightarrow{\text{PEP Carboxylase}} \text{malic acid} + \text{aspartic acid}$$

The enzyme RuDP carboxylase is inhibited by oxygen in the air and has a lower affinity for CO_2. The enzyme PEP carboxylase is not inhibited by oxygen and has a higher affinity for CO_2. From this biochemical information we can predict that C_4 plants would be at an advantage when photosynthesis is limited by CO_2 concentration. This occurs under high light intensities and high temperatures and when water is in short supply (Epstein et al. 1997; Sage and Kubien 2003).

C_4 grasses, sedges, and dicotyledons are all more common in tropical areas than in temperate or polar areas (Hattersley 1983). **Figure 15** shows the percentage of grass species that are C_4 plants in different parts of North America and confirms the suggestion that C_4 grasses are at a selective advantage in warmer areas with high solar radiation. On Hawaiian mountains, which have small seasonal changes in temperature, C_3 grasses predominate at high elevations and C_4 grasses at low elevations (Sage and McKown 2006).

Some desert succulents, such as cacti of the genus *Opuntia*, have evolved a third modification of photosynthesis, **crassulacean acid metabolism (CAM)**. These plants are the opposite of typical plants in that they open their stomata to take up CO_2 at night, presumably

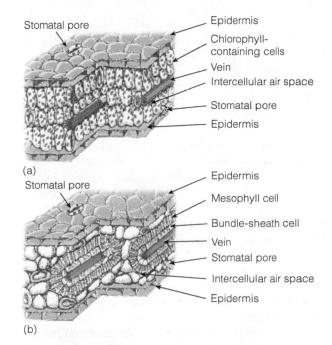

(a)

(b)

Figure 14 Leaf anatomy of C$_3$ and C$_4$ plants. (a) Leaf structure of a typical C_3 plant, *Atriplex triangularis*, in which the cells containing chlorophyll, chloroplasts (red), are of a single type and are found throughout the interior of the leaf. (b) *Atriplex rosea*, a C_4 plant, illustrating the modified leaf structure of C_4 species. The specialized leaf of *A. rosea* has nearly all its chlorophyll in two types of cells that form concentric cylinders around the fine veins of the leaf. The cells of the outer cylinder are mesophyll cells; those of the inner cylinder are bundle-sheath cells. (From Björkman and Berry 1973.)

as an adaptation for minimizing water loss through the stomata. This CO_2 is stored as malic acid, which is then used to complete photosynthesis during the day. CAM plants have a very low rate of photosynthesis and can switch to the C_3 mode during daytime. They are adapted to live in very dry desert areas where little else can grow.

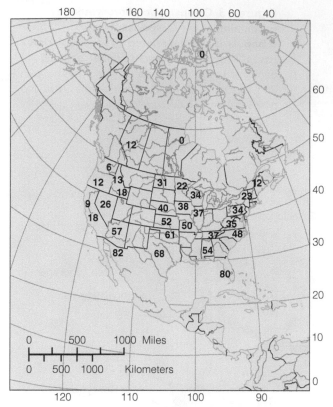

Figure 15 Percentage of C₄ species in the grass floras of 32 regions of North America. (From Teeri and Stowe 1976.)

Table 1 summarizes the main characteristics of C_3, C_4, and CAM plants.

The C_3 pathway is presumably the ancestral method of photosynthesis since no algae, bryophytes, ferns, gymnosperms, or more primitive angiosperms have the C_4 pathway or the capacity for CAM (Pearcy and Ehleringer 1984; Monson 1989). Almost half of the C_4 plant species are grasses and this pathway has apparently increased their competitive ability.

We do not know how the different photosynthetic pathways may interact with other factors to affect the geographic distribution of plant species. It is clear that the response of a plant species to temperature and moisture is strongly affected by the type of photosynthetic process it uses. Implications for animal distributions have yet to be considered. Plants possessing the C_4 pathway seem to be of lower nutritional value for herbivorous insects (Ehleringer et al. 2002). Further work is needed on the ecological consequences of the three different photosynthetic strategies, both for the plants and for the animals that depend on them.

If C_4 plants are more productive than C_3 plants, why do they not displace C_3 plants everywhere? The photosynthetic productivity plotted in Figure 13 cannot be directly translated into productivity in natural vegetation (Snaydon 1991). Competition in natural stands is not always for light, and mineral nutrients and water are often limiting to plants. Competition for soil resources can result in plants developing large root systems rather than large aboveground structures. Soil texture also affects C_3 and C_4 grasses differently; clay soils favor C_3 plants while sandy soils favor C_4 grasses in the Great Plains (Epstein et al. 1997). In spite of these differences, as climatic warming occurs in the future, C_3 grasses are predicted to shift their geographical ranges to the north in the Northern Hemisphere and to the south in the Southern Hemisphere.

Climate Change and Species Distributions

If temperature and moisture are the master limiting factors for the geographical ranges of plants and animals, the climatic warming that is now occurring will have profound effects on the Earth's biota. One way to get a glimpse of the kind of changes that may occur is to look back at the changes that have occurred in temperate regions since the end of the last Ice Age.

After the last continental glaciers began retreating in North America and Eurasia about 16,000 years ago, the northward expansion of tree distributions lagged behind the retreat of the ice. A detailed record of these migrations is captured in fossilized pollen deposited in lakes and ponds. Margaret Davis and her students have been leaders in deciphering the record left in fossilized pollen deposits (Davis 1986). In North America oaks and maples moved rapidly in a northeasterly direction from the Mississippi Valley, while hickories advanced more slowly (Delcourt and Delcourt 1987). Hemlocks and white pines moved rapidly northwest from refuges along the Atlantic Coast. The important finding of this paleoecological work is that the range of each species advanced individualistically. If you were sitting in New Hampshire, you would have seen sugar maple arrive 9000 years ago, hemlock 7500 years ago, and beech 6500 years ago (Davis 1986).

If we can determine the climatic limits of current geographical distributions, we can make predictions about how distributions will change with climatic warming. A major assumption of using this approach for plants is that seed dispersal is adequate to sustain the migrations of each species. Davis (1986) suggested that hemlock was delayed nearly 2500 years in its movement north at the end of the Ice Age, in part because of slow seed dispersal. If we use climate-change models to predict temperature and rainfall changes over the next 100 years, we can begin to estimate the size of the problem animals and plants will face over the next few centuries.

Table 1 Characteristics of photosynthesis in three groups of higher plants.

Characteristics of plants	Type of photosynthesis		
	C₃	C₄	CAM
Leaf anatomy (cross section)	Palisade and spongy mesophyll	Mesophyll compact around vascular bundles containing chloroplasts	Spongy appearance, mesophyll variable
Enzymes used in CO_2 fixation in leaf	RuDP carboxylase	PEP carboxylase and then RuDP carboxylase	Both PEP and RuDP carboxylases
CO_2 compensation point (ppm CO_2)	30–70	0–10	0–5 in dark, 0–200 with daily rhythm
Transpiration rate (water loss)	High	~25% of C₃	Very low
Maximum rate of photosynthesis (mg CO_2/dm² leaf surface/hr)	15–40	40–80	1–4
Respiration in light	High rate	Apparently none	Difficult to detect
Optimum day temperature for growth	20–25°C	30–35°C	Approx. 35°C
Response of photosynthesis to increasing light intensity at optimum temperature	Saturation about 1/4 to 1/3 full sunlight	Saturation at full sunlight or at even higher light levels	Saturation uncertain but probably well below full sunlight
Dry matter produced (t/ha/yr)	~20	~30	Extremely variable
Economically important species	wheat, rice, barley, potato	maize (corn), sugarcane, millet	

The compensation point is the CO_2 concentration at which photosynthesis just balances respiration so that there is no net oxygen generated and no net CO_2 taken up.

SOURCE: From Black (1971) and Sage (2004).

Figure 16 shows the current and potential geographical range of balsam fir (*Abies balsamea*) under anticipated climate-changes over the next 90 years. The climate change models predict that within the eastern United States there will be a 97% reduction in the geographic range of balsam fir as the species moves north with the warming climate. **Figure 17** shows a similar scenario for American beech (*Fagus grandifolia*). At present 49% of the eastern United States is occupied by beech. By 2100 as the climate warms, there will be a 90% decrease in the geographic range in the United States. The potential northern range limit of beech will move 200 km or more north in this century. If left to natural processes, beech must move at least 2 km per year to the north. By contrast, since the end of the Ice Age, beech migrated into its present range at a rate of 0.2 km per year. If these predictions are even approximately correct, slowly colonizing species like trees will require human assistance to move into their new ranges.

These effects of climate change will not appear immediately. Long-lived plant species such as trees will survive for many years as adults in inappropriate places. As the climate changes, their seed production will decline until finally they are unable to produce viable seedlings (Iverson and Prasad 1998). All of these effects are complicated by ecotypic variation within tree species. If there are specific ecotypes adapted to northern or southern climatic conditions, then the range shifts depicted in Figures 16 and 17 must be accomplished without the loss of ecotypic variation.

The geographic ranges of species are thus not static but dynamic, and as climate changes in the future, species will (if time permits) move into new areas that become climatically suitable. The major concerns of ecologists are first that the speed of climate change in the next 100 years may be too great for slowly colonizing forms to move, and second that genetic adaptation to local temperature and rainfall patterns may be lost for some species.

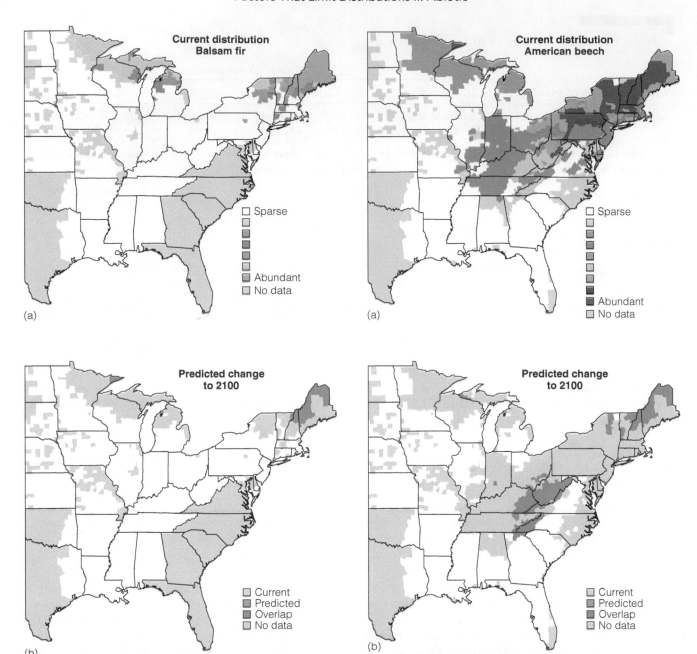

Figure 16 Predicted change in the geographic distribution of balsam fir (*Abies balsamea*) in the eastern United States as a result of predicted global climate change to 2100. (a) Current distribution. Stronger colors indicate higher abundance, white color indicates not present. (b) Predicted distribution change to 2100. Yellow outlines the current range from which balsam fir will disappear, and will remain only in the green area. (From Iverson and Prasad 1998.)

Figure 17 Predicted change in the geographic distribution of American beech (*Fagus grandifolia*) in the eastern United States as a result of predicted global climate change to 2100. (a) Current distribution. Stronger colors indicate higher abundance, white color indicates not present. (b) Predicted distribution change to 2100. Yellow outlines the current range from which American beech will disappear, and will remain only in the green area. (From Iverson and Prasad 1998.)

Summary

Temperature and moisture are the major factors that limit the distributions of animals and plants. These factors may act on any stage of the life cycle and affect survival, reproduction, or development. Temperature and moisture may also indirectly limit distributions through their joint effects on competitive ability, disease resistance, predation, or parasitism. Other physical and chemical factors, such as light and pH, can also affect the distributions of plants and animals, but they operate at a local scale.

From a global viewpoint, the distribution of plants can be associated with climate. Tropical rain forest and tundra, for example, occupy areas with different temperature and moisture regimes. The effects of climate are less clearly seen at the local level of the distribution of individual species. In only a few cases has experimental work been done in local populations, first to pinpoint the life-cycle stage affected by climate and then to describe the physiological processes involved.

Water availability is the key to moisture effects on plants, and drought occurs when adequate amounts of water are not present or are unavailable to the plant. The soil may be saturated with water, but if all of it is frozen, none may be taken up by plants, and they may suffer frost drought. Many of the distributional effects attributed to temperature may operate through the water balance of plants.

Species may adapt to temperature, moisture, or light levels phenotypically or genotypically and thereby circumvent some of the restrictions imposed by climate. Göte Turesson was one of the first to recognize the importance of ecotypes, genetic varieties within a single species. By transplanting individuals from a variety of habitats into a common garden, Turesson showed that many of the adaptations of plant forms were genotypic. Many ecotypes have now been described, particularly in plants, and these may involve adaptations to any environmental factor, including temperature and moisture. Ecotypic differentiation has often proceeded to the point where one ecotype cannot survive in the habitat of another ecotype of the same species.

Organisms have evolved an array of adaptations to overcome the limitations of high and low temperatures, drought, or other physical factors. Some adaptations might allow a species to extend its geographic range. Many species are known to have extended or reduced their geographic range in historical times, but few cases have been studied in detail.

The climatic warming that is currently under way will have strong effects on geographic distributions. The major concerns are that species will not be able to migrate fast enough to keep pace with global warming, and that genetic adaptations to local environments may be lost.

Review Questions and Problems

1 Fire ants have spread from Brazil north through Central America and Mexico into the southern United States, and they continue to spread north. Discuss how you might determine the potential geographical range of this pest species. Korzukhin et al. (2001) discuss the problem.

2 There is only one known C_4 tree species (Pearcy 1983). Explain why this is the case.

3 Cain (1944) stated:

Physiological processes are multi-conditioned, and an investigation of the effects of variation of a single factor, when all others are controlled, cannot be applied directly to an interpretation of the role of that factor in nature. It is impossible, then, to speak of a single condition of a factor as being the cause of an observed effect in an organism.

Discuss the implications of this principle—that the factors of the environment act collectively and simultaneously—with regard to methods for studying species distributions.

4 List the assumptions underlying the predictions of tree range changes illustrated in Figures 16 and 17.

5 "The frost line . . . is probably the most important of all climatic demarcations in plants" (Good 1964, p. 353). Locate the frost line in a climatological atlas, and compare the distributions of some tropical and temperate species of any particular taxonomic group with respect to this boundary.

6 Hutchins (1947) set out a simple but elegant hypothesis that the geographic limits of marine species are set by thermal tolerances of the most sensitive life history stage. Thus species are limited by intolerance to cold at the poleward limit and by

intolerance of heat at their equatorial limit. Discuss what factors might invalidate this hypothesis. Wethey (2002) discussed this issue for barnacles.

7   The British barnacle *Elminius modestus* extends higher on the shore in the intertidal zone than does the barnacle *Semibalanus balanoides* when the two species occur together. However, these two species have similar tolerances to desiccation, salinity, and temperature. The range of initial settlement of young barnacles is the same for the two species. Given these facts, can you suggest an explanation for the observation that *E. modestus* extends higher on the shore than *S. balanoides*?

8   Adult male dark-eye juncos (*Junco hyemalis*) remain farther north in winter than females and juveniles (Ketterson and Nolan 1982). Review the arguments of Root (1988) on energy balance and winter bird ranges, and suggest an explanation for these observations.

9   Fenchel and Finlay (2004) state that small organisms (less than 1 mm in length) tend to occur everywhere

around the globe, if their habitat requirements are met. In this case evolutionary history is not a possible explanation for observed geographical distributions. Is this true for larger organisms as well? Are there any microorganisms for which this statement is not correct?

10   The cane toad (*Bufo marinus*) is an introduced pest in parts of Australia. Sutherst et al. (1995) predict the geographic limits of the possible spread of this pest within Australia. Analyze their approach, and list the critical assumptions they use to make these predictions. Compare their predictions with those of Beurden (1981).

Overview Question

A herbivorous insect is to be brought into your country to control a noxious weed. Describe how you would determine the factors that limit the insect's distribution and how you might predict its new geographic range in your country.

Suggested Readings

• Iverson L. R., A. M. Prasad, and M. W. Schwartz. 1999. Modeling potential future individual tree-species distributions in the Eastern United States under a climate change scenario: A case study with *Pinus virginiana*. *Ecological Modelling* 115:77–93.

• Körner, C., and J. Paulsen. 2004. A world-wide study of high altitude treeline temperatures. *Journal of Biogeography* 31:713–732.

• Korzukhin, M. D., et al. 2001. Modeling temperature-dependent range limits for the fire ant *Solenopsis invicta* (Hymenoptera: Formicidae) in the United States. *Environmental Entomology* 30:645–655.

• Lloyd, A. H. 2005. Ecological histories from Alaskan tree lines provide insight into future change. *Ecology* 86:1687–1695.

• Sage R. F. 2004. The evolution of C₄ photosynthesis. *New Phytologist* 161:341–379.

• Somero, G. N. 2002. Thermal physiology and vertical zonation of intertidal animals: Optima, limits, and costs of living. *Integrative and Comparative Biology* 42:780–789.

• Stephenson, N. L. 1990. Climatic control of vegetation distribution: The role of the water balance. *American Naturalist* 135:649–670.

• Sutherst, R. W., R. B. Floyd, and G. F. Maywald. 1995. The potential geographical distribution of the cane toad, *Bufo marinus* L. in Australia. *Conservation Biology* 10:294–299.

• Tape, K., M. Sturm, and C. Racine. 2006. The evidence for shrub expansion in northern Alaska and the Pan-Arctic. *Global Change Biology* 12:686–702.

• Turesson, G. 1930. The selective effect of climate upon the plant species. *Hereditas* 14:99–152.

Credits

Illustration and Table Credits

F2 Hidore, J. J. and J. E. Oliver. 1993. Climatology: An atmospheric Science. Macmillan, New York. F3 Hidore, J. J. and J. E. Oliver. 1993. Climatology: An atmospheric Science. Macmillan, New York. F4 R. H. Whittaker, *Communities and Ecosystems, 2/e* © 1975 Macmillan. F10 From Lewis, J. R. (1964) The ecology of rocky shores. London: English University Press. 13 O. Björkman, "Inaugural Address" in R. Marcell, ed., *Environmental and Biological Control of Photosynthesis*. Copyright © 1975. F14 From O. Björkman and J. Berry, "High Efficiency Photosynthesis," *Scientific American*, Vol. 229, p. 86 1973. Copyright © 1973 Lorelle Raboni. Reprinted by permission of the artist. F15 From Teeri and Stowe, "Climate patterns and the distribution of C4 grasses in North America," *Oecologia*, Vol. 23:1–12, © 1976 Springer-Verlag. Reprinted by permission. F16 L. R. Iverson and A. M. Prasad, "Predicting potential future abundance of 80 tree species following climate change in the eastern United States," *Ecological Monographs* 68:465–485, 1998. Copyright © 1998 Ecological Society of America. Used with permission. F17 L. R. Iverson and A. M. Prasad, "Predicting potential future abundance of 80 tree species following climate change in the eastern United States," *Ecological Monographs* 68:465–485, 1998. Copyright © 1998 Ecological Society of America. Used with permission.

Photo Credits

Distribution and Abundance

Key Concepts

- Geographic distributions can be mapped at spatial scales from the continental to the local.

- Most species occupy small geographic areas; few are widespread.

- Polar species tend to have larger geographic ranges than tropical species in many taxonomic groups (Rapoport's Rule).

- Distribution and abundance are usually positively related such that widespread species are more abundant than species with small geographic ranges.

KEY TERMS

ecological specialization model A proposed explanation for Hanski's Rule postulating that species that exploit a wide range of resources become both widespread and common; these species are generalists; also called Brown's model.

generalists Species that eat a variety of foods or live in a variety of habitats; contrast with specialists.

Hanski's Rule The generalization that there is a positive relationship between distribution and abundance, such that abundant species have wide geographic ranges.

local population model A proposed explanation for Hanski's Rule that assumes that species differ in their capacity to disperse, and if the environment is divided into patches, some species will occupy more local patches than others as a function of their dispersal powers.

Rapoport's Rule The generalization that geographic range sizes decrease as one moves from polar to equatorial latitudes, such that range sizes are smaller in the tropics.

sampling model One proposed explanation for Hanski's Rule that the observed relationship between distribution and abundance is an artifact of the difficulty of sampling rare species and does not therefore require a biological explanation.

specialists Species that eat only a few foods or live in only one or two habitats; contrast with generalists.

We have considered the ways in which ecologists answer the two simple ecological questions *Who lives where?* and *What constrains geographic distributions?* First we will discuss the broad question of whether there is any relationship between distribution and abundance. This question was first raised in a general way by the Australian ecologists H. G. Andrewartha and L. C. Birch in their classic 1954 book *The Distribution and Abundance of Animals*. Their ideas, which tied together the two concepts of distribution and abundance, had a strong impact on ecological thinking in the past 50 years. In this chapter we focus on one aspect of this interaction between distribution and abundance— whether species that have large geographic ranges are any more or less abundant than species that have small geographic ranges.

The Spatial Scale of Geographic Ranges

We began our analysis of distribution by assuming that we can easily map the geographic range of a species, but this simple assumption breaks down as we map the detailed distribution of a species in a local area. No species occurs everywhere. **Figure 1** illustrates the range of spatial scales at which one can describe a species' geographic range. At one extreme the range of a species is defined by the worldwide extent of its occurrence, a line drawn on a map demarcating the outermost points at which the species has been observed. This is the scale of geographic range used in field guides and other regional natural history guides. At the other extreme, we could measure a much smaller area within the larger geographic range and map the location of each individual. If a particular habitat is not occupied by the species, this region would not be included in its geographic range. We would like to know the actual area occupied by each species, but this is not possible because ecologists have not collected or mapped species occurrences in this much detail for most plants and animals (Gaston 1991). The important point shown in Figure 1 is that one can measure geographic ranges at several spatial scales.

Variations in Geographic Range Size

After we have decided on a measure of geographic range, we can investigate the spread of range sizes of species within a taxonomic group. A general pattern has emerged in many separate groups: most species within a group have small geographic ranges and only a few have very large ranges. The frequency distributions of range sizes in **Figure 2** illustrate this point for the birds of North America and the vascular plants of Britain. This pattern, the "hollow curve" of Figure 2, seems to be the rule for all groups that have been studied (Gaston et al. 1998). These range-size data display other interesting patterns in addition to the "hollow curve" shape.

In 1975 the Argentinean ecologist Eduardo Rapoport suggested that within the mammals geographic range sizes decreased as one moved from polar to equatorial latitudes, such that range sizes were smaller in the tropics. This generalization has been referred to as **Rapoport's Rule** and has stimulated many studies to determine how well it describes distributions in various other groups of organisms (Stevens 1989). In North America, geographic ranges of mammals are smaller toward the tropics (see **Figure 3**). The average Canadian mammal species inhabit ranges that are an average 25 times larger than those of Mexican mammals (Pagel et al. 1991; Arita et al. 2005).

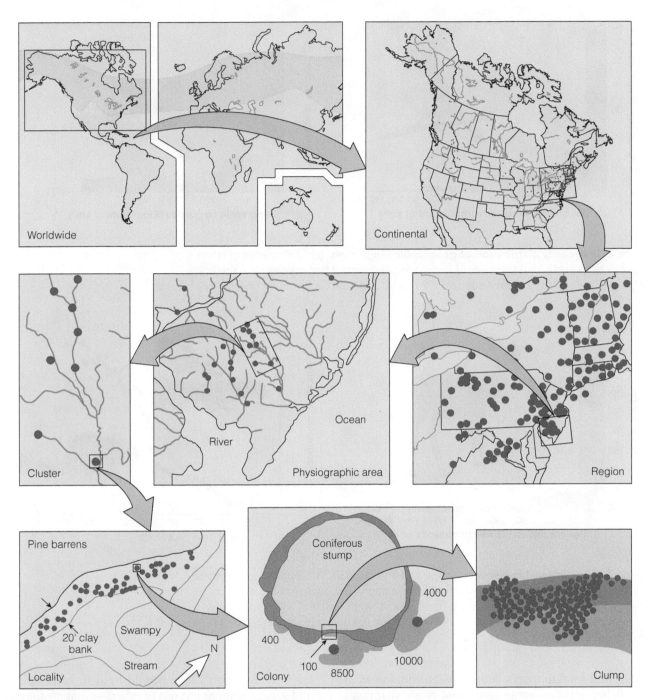

Figure 1 A hierarchy of scales for analyzing the geographic distribution of the moss
Tetraphis. The answer to the question "What limits geographic distribution?" may have
different answers when analyzed at the continental scale versus the local scale of the
individual tree stump. (After Forman 1964.)

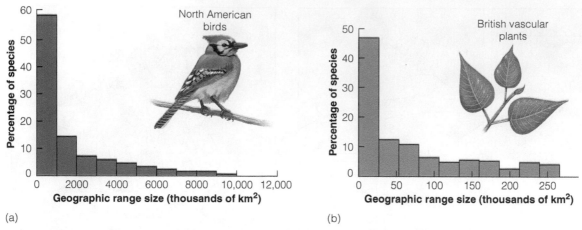

Figure 2 Frequency distribution of geographic range sizes. (a) 1370 species of North American birds and (b) 1499 species of British vascular plants. Most species have small geographic ranges. (Data from Anderson 1985 for (a) and from Gaston et al. 1998 for (b).)

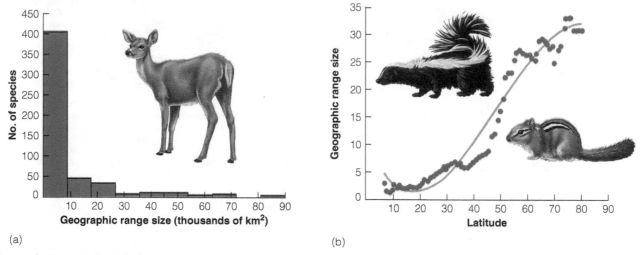

Figure 3 The relationship between range size and latitude. (a) Geographic range size for 523 species of North American mammals. (b) Relationship between range size (measured as the percentage of the total land area of North America) and latitude. Low-latitude species have smaller ranges than high-latitude species, following Rapoport's Rule. (From Pagel et al. 1991.)

Support for Rapoport's Rule has been widespread in trees, fishes, reptiles, some birds, and many mammals from all continents (Gaston et al. 1998)—but not all studies have supported Rapoport. **Figure 4** shows that geographic ranges of woodpeckers are highly variable in size and reach a minimal size at 20° N latitude. One extension of Rapoport's Rule can be made in mountain ranges, for which analogous arguments predict ranges that become larger as one moves up in altitude. Bhattarai and Vetaas (2006) tested this hypothesis for tree species in the Himalaya Range in Nepal, with the results shown in **Figure 5**. Tree range size was maximal at mid-elevation, contrary to Rapoport's Rule. We must

ask why Rapoport's Rule should hold in some species and some situations but not in others—what are the ecological mechanisms behind this pattern?

Three ecological explanations for Rapoport's Rule have been put forward. First, climatic variability is greater at high latitudes, and only organisms that have a broad range of tolerance for variable climates can live there. As a side effect of broad tolerance, these high-latitude species can occupy larger ranges. This hypothesis makes two interesting predictions. For terrestrial animals and plants, climatic tolerance should increase from tropical to polar areas. This seems to be true for amphibians (Snyder and Weathers 1975). For marine organisms more interesting

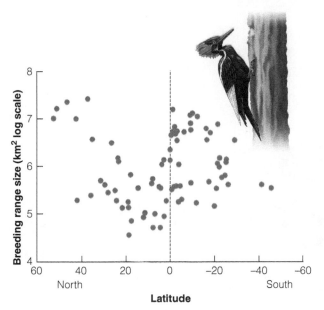

Figure 4 North and South American woodpeckers' geographic range size in relation to latitude. Each point is one species (*n* = 81). The dotted line is the equator. There is a minimum range size at 10°–20° N latitude, with much variation. (Data from Husak and Husak 2003.)

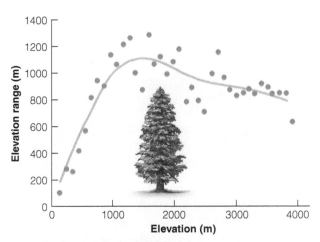

Figure 5 Elevational range sizes for 614 species of trees in the Himalaya Mountains of Nepal in relation to the midpoint of their elevation. Species are grouped into 38 elevation classes. Range size reaches a maximum at 1500 m and then falls, contrary to the predictions from Rapoport's Rule. (Data from Bhattarai and Vetaas 2006.)

patterns can be predicted. For marine fish, temperature variation is greatest in the temperate zone and much smaller in polar waters and in tropical waters. Thus the range of temperature tolerances should be minimal in both tropical and polar waters. **Figure 6** shows that this appears to be the case for shallow-water marine fish. If we go deeper in the oceans, temperature variability becomes

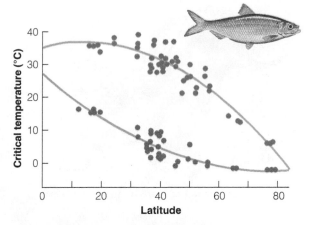

Figure 6 Critical temperature limits for marine fish from shallow waters. According to the climatic variability hypothesis, temperate fish should have the widest temperature limits. Upper critical temperatures are in blue, and lower critical temperatures are in red. These data from Brett (1970) fit this explanation well for the polar-temperate comparison and may fit the temperate-tropical comparison as well, but more data are needed from marine fish living in the 0°–10° latitude range.

minimal, and the climatic variability hypothesis would predict no relationship between latitude and range size for these deep-sea organisms. Unfortunately no data are yet available to test this prediction for the deep sea.

A second explanation of Rapoport's Rule is that it is a product of glaciation, particularly in the Northern Hemisphere (Brown 1995). When the glaciers retreated, only those species with high dispersal capacity were able to repopulate northern areas, and these species thus have large geographic ranges. The glaciation hypothesis may explain some of the patterns found in the Northern Hemisphere, but it cannot explain Rapoport's Rule in the Southern Hemisphere, where glaciation was much less prominent. Glaciation is probably a contributing factor but not the major cause for these distributional patterns.

A third explanation of Rapoport's Rule is that it arises from a lack of competition in polar communities. Because fewer species live in polar areas, the level of competition may be lower. There is no support at present for this mechanism because we do not have a simple measure of competition that can be applied across global species patterns. It remains a possibility as yet untested.

Rapoport's Rule has stimulated much work in analyzing distributions, and in doing so it has highlighted the importance of looking beyond latitude to consider the ecological mechanisms that produce the observed patterns such as that shown by the trees in Nepal (see Figure 5). Hawkins and Diniz-Filho (2006) suggest that, for birds in the Americas, range size is affected both by temperature and by topography. **Figure 7** illustrates the patterns that

Figure 7 Geographic pattern of breeding range size for 3839 species of native birds in the Americas. Island populations are excluded. The interaction of temperature and topography is illustrated in this map, explaining why Rapoport's simple rule based on latitude is incomplete. (From Hawkins and Diniz-Filho 2006.)

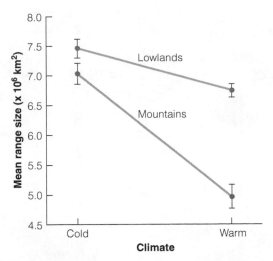

Figure 8 Mean range size for 3839 bird species from the Americas subdivided into cold and warm climates. Climates are based on the mean temperature of the coldest month above or below 0°C, and topography (lowlands = less than 1000 m elevation). Topography affects range sizes much more strongly in warmer climates, possibly because there is less habitat zonation in colder regions. (From Hawkins and Diniz-Filho 2006.)

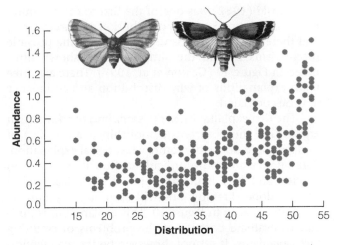

Figure 9 Relationship between distribution and abundance for 263 species of British moths. Distribution is the number of trap sites across Britain at which the species were caught. Abundance is averaged across all sites for all years. Each dot represents one species. In general, species with wider distributions are more abundant. (From Gaston 1988.)

emerge for bird ranges in the Americas. Topography affects bird ranges in tropical areas but has little effect in polar regions (**Figure 8**). The net result is that by concentrating on temperature and topography about 50% of the variation in range sizes can be explained. The remaining 50% may be associated with biotic interactions and possibly with climate and evolutionary history in relation to glaciation (Hawkins and Diniz-Filho 2006).

Range Size and Abundance

Is there any relationship between geographic range size and the abundance of a species? If a species is widespread, is it always an abundant species? Or conversely, if a species is rare or threatened, does it have a small geographic range? The data ecologists have collected on a wide variety of plants and animals reveal a correlation between distribution and abundance such that more widespread species are typically more abundant (Brown 1984; Gaston 1990). **Figure 9** shows data for 263 species of British moths. Moths were collected at light traps in 50 sites throughout Britain, and the geographic distribution was measured as the number of these sites occupied by a given species. Abundance data for each species were averaged over 6–14 years at each light trap site. (Not all light traps could be operated

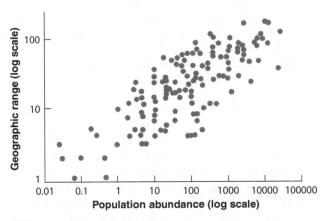

Figure 10 The geographic range-abundance relationship for geese and ducks (Anseriformes) of the world. Abundance is measured as the total estimated population for the world, and range is measured as the number of equal area grid squares of 10° longitude occupied in the world. (Data from Gaston and Blackburn 1996.)

every year.) There is much variability in these moth data, but a clear trend exists: more widespread moth species tend to be more abundant. Similar patterns can be found in birds (**Figure 10**), plants, and many other groups, such that this positive relationship between distribution and abundance is another ecological generalization that can be called **Hanski's Rule**, after Ilkka Hanski.

Hanski (1982) was one of the first to call attention to the association between distribution and abundance, and there has been extensive discussion of the possible mechanisms behind the simple pattern shown illustrated in Figure 10 (Gaston et al. 1997). There are three main explanations of why distribution and abundance may be correlated.

The first explanation is the **sampling model**, which argues that the observed relationship is an artifact of sampling and does not require a biological explanation. Rare species are more difficult to find than common species, and thus if the appropriate biological studies are not done carefully, one will automatically observe the patterns seen in Figure 10. This explanation is difficult to evaluate because of the problems of counting rare organisms. It cannot, however, be the explanation for this pattern in birds, butterflies, and mammals, which have been very well sampled and studied.

The second explanation is the **ecological specialization model**, or Brown's model, because Jim Brown first suggested it in 1984. This model argues that species that can exploit a wide range of resources become both widespread and common. These species are called **generalists** and are to be distinguished from **specialists**, which exploit only a few resources. Provided that one can determine which species are generalists and which are specialists, one should be able to test this model. A corollary of this model is that widespread generalist species should use food and habitat resources that are themselves abundant.

The third explanation is the **local population model**. In this model a population is subdivided into a series of discrete patches, or local populations,[1] that interact because animals or plants move between the patches. Since species differ in their capacity to disperse, some will occupy more local patches than others (Hanski et al. 1993). If this model is correct, we would expect species that disperse more to be more common and more widespread, when compared with less migratory species. A variant of this model is the neutral model of Bell (2001) that predicts a positive relationship between distribution and abundance for model species with identical properties that disperse between patches in a landscape.

The prediction of a strong positive relationship between distribution and abundance does not always hold. Lesica et al. (2006) tested seven species pairs in which one species of the pair was a widely distributed plant and the other was a rare endemic with a small geographic range. **Figure 11** shows the results for two species pairs. In all cases the rare species was 2 to 10 times more abun-

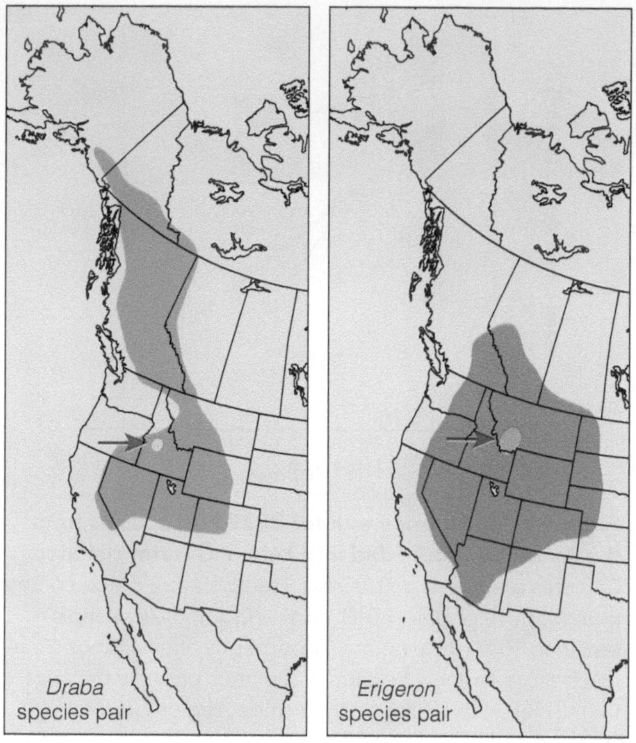

Figure 11 The geographic range of two closely related species in the two genera *Draba* (whitlowgrass) and *Erigeron* (fleabane). The arrows point to the range of the rare species and the colored area indicates the larger range of the widespread species. In *Draba* the rare species was twice as abundant as the widespread species, and in *Erigeron* the rare species was nearly five times as abundant, contrary to the expected positive relationship between geographic distribution and abundance. (Modified from Lesica et al. 2006.)

dant in plots within its range than was the common, widespread species sampled in the same size area.

The relationship between distribution and abundance is usually discussed as a pattern among many species, but it is interesting to ask if the same pattern occurs within one species. If a species is declining in abundance, does its geographic distribution also become smaller? This could be an important question for conservation biologists studying a declining species. Conversely, are species that are increasing in abundance also expanding their geographic ranges? **Figure 12** illustrates two possible trajectories for species changing in abundance. Species could follow the expected curve (Figure 12a) and decline in range size as they decline in abundance, or a different pattern may occur for rare and common species (Figure 12b).

[1]Also called *metapopulations*.

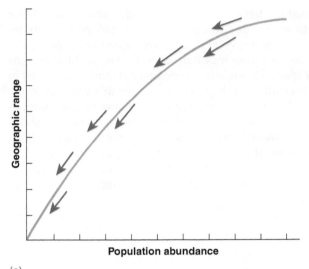

(a)

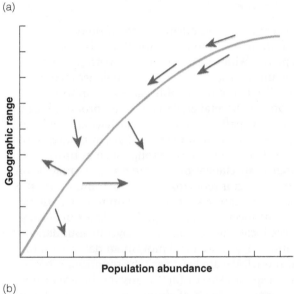

(b)

Figure 12 Potential directions of change for individual species declining in abundance in relation to the standard model of positive abundance-range size relationships.
(a) All species might move along the line, preserving the previous relationship. (b) Common species might follow the line while rare species might go in any direction. It is not yet clear which model is closer to reality. (Modified from Webb et al. 2007.)

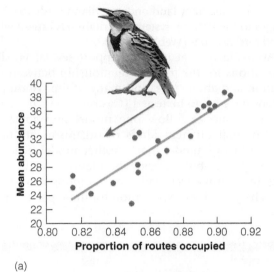

(a)

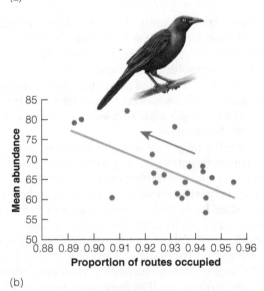

(b)

Figure 13 Changes in geographic range size for two North American birds that have been declining since 1970 in the Breeding Bird Surveys. Breeding Bird Surveys are carried out each June in a standard manner (50 stops over 24.5 miles) on more than 3700 specified routes in North America. (a) The eastern meadowlark has declined in abundance and its geographic range has also shrunk, so that there is a positive relationship between distribution and abundance in this species over the 20 years of data. (b) The common grackle has declined in abundance, but during this decline in abundance its geographic range has been increasing, so that there is a negative relationship between distribution and abundance for this bird species. The reasons for these differences are not known. (From Gaston and Curnutt 1998.)

The best data to answer this question come from bird surveys, and **Figure 13** shows two cases of birds that have declined over 20 years. Both eastern meadowlarks and common grackles declined in North America from 1970 to 1989. Meadowlarks reduced their geographic range, as predicted, but grackles increased their geographic range, contrary to prediction. The same differences have been found for British farmland and woodland birds in which changes to the farming system in Britain has caused widespread bird declines (Webb et al. 2007).

Available data at present support several possible explanations for the positive relationship between distribution and abundance, and many of the critical predictions have yet to be tested (Gaston 2003). It may be that some species follow one model and some the other, or that different kinds of organisms are more likely to fit one model than another model. Since all studies of distribution and abundance involve sampling, in an imperfect world the sampling model will always be part of the explanation for these positive rela-

tionships between distribution and abundance. More data are needed to determine the ecological attributes of successful species that are widespread and abundant. These attributes may help us to understand the reasons for species being rare or endangered, and could assist in conservation biology. We are led in this way out of the world of geographic ecology and geographic distributions into the larger and more complex world of abundance. We turn next to exploring what happens within the zone of distribution in which populations of animals and plants increase or decrease in size in response to many of the same environmental factors we have just considered.

Summary

The geographic distribution of a species is more complex than one first suspects. The scale at which a distribution is mapped can affect the answer to the simple question: What limits geographic distributions? For many species we do not know the detailed geographic range because too few data have been collected. Even for larger plants and animals in developed countries we have few details about local distributions.

Geographic ranges measured on large-scale maps show a common pattern in all groups studied—most species have small geographic ranges, and only a few species are very widespread. This relationship holds for data on fishes, birds, and mammals, as well as for plant groups. In addition, in many taxonomic groups, geographic ranges follow Rapoport's Rule, which states that polar species have larger geographic ranges than tropical species. Climatic variability, glaciation history, and competition are cited as the main causes of this pattern.

There is a broad positive correlation between distribution and abundance for all kinds of animals and plants: widespread species are typically more abundant than species that have small geographic ranges. There is considerable variability in this relationship. Several explanations are proposed for this pattern. The pattern could be an artifact arising from the problem of sampling rare species in nature, but this is unlikely for well-known groups such as birds. Ecological specialization on rare resources may be an important factor reducing abundance, and to test this second explanation we need data on resource use by abundant species and rare species. Migration among suitable local patches may affect overall abundance and distribution, and we need movement data to test this third model. We are led to ask the question: What makes a species successful? To answer this we turn to consider the problem of abundance in more detail in Part Three.

Review Questions and Problems

1   In primates there is no relation between geographic distribution and abundance at the species level but there is a clear relationship when taxonomic families are considered as the unit of analysis instead of species (Harcourt et al. 2005). Suggest why this pattern might occur.

2   Abundance can be measured as the total population size over the entire geographic range (as in Figure 10) or as density per unit of area sampled (as in Figure 11). Discuss which type of data is most appropriate

for investigating the relationship between distribution and abundance.

3   Discuss the application of Rapoport's Rule to the altitudinal distribution of species on mountains in relation to the data given in Figure 5. What predictions does this hypothesis make for mountain species? Read Fleishman et al. (1998) and Bhattarai and Vetaas (2006) and compare your analysis with theirs.

4   Discuss the implications of the relation between distribution and abundance for conservation biology.

5   The relationship between distribution and abundance is often very loose with much variability, as shown in Figure 10. How does this variability affect the interpretation of these data?

6   Would you expect species that were increasing in abundance to follow more closely the model illustrated in Figure 12a or 12b? Discuss in biological terms exactly what the arrows in Figure 12 mean with respect to distribution and abundance.

7   Geographic ranges could be mapped at a scale of 1-m, 1-km, 10-km, or 100-km squares to estimate the size of the geographic range. Would you expect the results of an abundance-range size regression (as in Figure 9) to differ if you mapped distribution at different scales? Gaston (1994) considers this issue.

Overview Question

Discuss the classification of species of plants and animals into specialists and generalists. Choose a species and describe what you would measure to convince someone that this species is either a generalist or a specialist. How does this classification assist in understanding the observed positive relationship between distribution and abundance?

Suggested Readings

- Bhattarai, K. R., and O. R. Vetaas. 2006. Can Rapoport's Rule explain tree species richness along the Himalayan elevation gradient, Nepal? *Diversity & Distributions* 12:373–378.

- Brown, J. H. 1984. On the relationship between abundance and distribution of species. *American Naturalist* 124:255–279.

- Gaston, K. J., T. M. Blackburn, and J. I. Spicer. 1998. Rapoport's Rule: Time for an epitaph? *Trends in Ecology and Evolution* 13 (2): 70–74.

- Gaston, K. J. (2003) *The Structure and Dynamics of Geographic Ranges.* Oxford: Oxford University Press.

- Harcourt, A. H., S. A. Coppeto, and S. A. Parks. 2005. The distribution-abundance (density) relationship: Its form and causes in a tropical mammal order, Primates. *Journal of Biogeography* 32:565–579.

- Hawkins, B. A., and J. A. F. Diniz-Filho. 2006. Beyond Rapoport's Rule: Evaluating range size patterns of New World birds in a two-dimensional framework. *Global Ecology & Biogeography* 15:461–469.

- Johnson, C. N. 1998. Rarity in the tropics: Latitudinal gradients in distribution and abundance in Australian mammals. *Journal of Animal Ecology* 67: 689–698.

- Lesica, P., R. Yurkewycz, and E. E. Crone. 2006. Rare plants are common where you find them. *American Journal of Botany* 93:454–459.

- Webb, T. J., D. Noble, and R. P. Freckleton. 2007. Abundance-occupancy dynamics in a human dominated environment: Linking interspecific and intraspecific trends in British farmland and woodland birds. *Journal of Animal Ecology* 76:123–134.

Credits

Illustration and Table Credits

Photo Credits

Population Parameters and Demographic Techniques

Key Concepts

- Individuals are clearly defined in *unitary* organisms such as deer, but less clearly defined in *modular* organisms of variable size, such as grasses.

- Population abundance results from an integration of the four primary population parameters of natality, mortality, immigration, and emigration.

- Age-specific natality and mortality rates for any population can be summarized quantitatively in fertility schedules and in life tables.

- The intrinsic capacity for increase (r) summarizes the natality and mortality schedules of a population and forecasts the rate of population growth implicit in these schedules.

- The age structure of a population is determined by these rates of natality and mortality.

From Chapter 8 of *Ecology: The Experimental Analysis of Distribution and Abundance*, Sixth Edition. Eugene Hecht.

KEY TERMS

absolute density The number of individuals per unit area or per unit volume.

big-bang reproduction Offspring are produced in one burst rather than in a repeated manner.

deme A population genetic unit of individuals that breed with one another; a genetic population.

emigration The movement of individuals out of an area occupied by the population, typically the site of birth or hatching.

immigration The movement of individuals into an area occupied by the population.

intrinsic capacity for increase The potential rate of increase of a population that combines the life table and fertility schedule with the speed of development.

life table The age-specific mortality schedule of a population.

mean length of a generation The average length of time between the birth of a female and her offspring.

reproductive rate (R_0) The average number of offspring produced per female or reproductive unit.

reproductive value The contribution an individual female will make to the future population.

Within their areas of distribution, animals and plants occur at varying densities. We recognize this variation when we say, for example, that black oaks are common in one woodlot and rare in another. If we are to make these statements more precise, we must quantify density. This chapter discusses some techniques used to estimate densities of animals and plants.

The Population as a Unit of Study

A **population** may be defined as a *group of organisms of the same species occupying a particular space at a particular time*. Thus, we may speak of the deer population of Glacier National Park, the deer population of Montana, or the human population of Australia. The ultimate constituents of the population are individual organisms. For sexually reproducing organisms, the population may be subdivided into local populations called **demes**, which are groups of interbreeding organisms, the smallest collective unit of a plant or animal population. Individuals in local populations share a common gene pool. The boundaries of a population both in space and in time are vague and in practice are usually fixed arbitrarily by the investigator.

Populations as units of study have received a good deal of interest from both ecologists and geneticists. Among the principles of modern evolutionary theory are the ideas that natural selection acts on the individual organism and that through natural selection populations evolve. The fields of population ecology and population genetics have much in common.

The population has group characteristics—statistical measures—that cannot be applied to individuals. The basic characteristic of a population that we are interested in is its density, and this chapter will discuss how to estimate density. The four population parameters that change density are natality (egg, seed, or spore production; births), mortality (deaths), **immigration**, and **emigration**. In addition to these attributes, one can derive secondary characteristics of a population, such as its age distribution, genetic composition, and pattern of distribution of individuals in space. Note that these population parameters result from a summation of individual characteristics.

WORKING WITH THE DATA

Calculation of Expected Population Density from the Regression Data Given in Table 2

The expected average population density of a herbivorous mammal weighing 85 g would be calculated from the coefficients given in Table 2 as follows:

$$\log (\text{population density}) = a + b \, (\log [\text{body mass}])$$
$$= 1.30 - 0.66 \, (\log [0.085])$$
$$= 1.30 + 0.707$$
$$= 2.007$$

- Population density is estimated as the antilog (2.007) or $10^{2.007}$, which is 102 individuals per km^2.
- For a seed-eating bird that also weighs 85 g, the expected average population density would be calculated as follows:

$$\log (\text{population density}) = a + b \, (\log [\text{body mass}])$$
$$= 0.22 - 0.54 \, (\log [0.085])$$
$$= 0.22 + 0.578$$
$$= 0.798$$

- Population density for this bird is estimated as the antilog (0.798) or 6.3 individuals per km^2.
- As shown in Figure 3, birds have on average smaller populations than mammals of the same mass.

Unitary and Modular Organisms

We are used to organisms in populations that come in individual units such as humans and birds. We call these unitary organisms. Deer, mice, humans, and oak trees are easy to identify as individuals. But some organisms do not come in simple units of individuals, and we call these modular organisms (**Figure 1**). Many plants are also difficult to categorize because they show great variation in size and structure. Grasses are particularly difficult to fit into anyone's definition of a single individual, and many other plants have underground connections so that what appear to be separate plants are the same genetic individual (a clone). For example, aspen trees (*Populus tremuloides*) can form clones, so that a whole stand of trees may in fact be a single genetic individual.

Organisms may thus be classified as unitary or modular organisms. Most higher animals are unitary organisms in which form is determinate. A kitten is recognizable as a cat, and a young giraffe is instantly recognized as a giraffe. In unitary organisms each easily recognizable individual is usually a separate genetic individual. By contrast, most plants are modular organisms in which the zygote or spore develops into a unit of construction (a module) that then produces additional similar modules. Modular organisms are often branched, and individuals are composed of variable numbers of modules. Modules that can exist separately are known as ramets. Most plants are modular, but many of the lower animals such as hydrozoans, corals, and bryozoans are also modular organisms. Modular organisms can have individuals that are extremely different in size.

To study populations of modular organisms, we must recognize two levels of population structure. In addition to the number of modular units, there is the number of individuals that are represented by original zygotes. These individuals are called genets, or genetic individuals (Harper 1977). In plants, an individual genet can be a single tree or a clone extending over a square kilometer. Each genet is composed of one or more modular units of construction, which vary with the type of organism. For example, grasses grow aboveground as tillers (= ramets), the modular unit of construction, and each genetic individual of a grass may have many tillers. Thus to describe a population of modular organisms, we must specify both the number of genets and the number of modular ramets, in contrast to most populations of unitary organisms, in which the individual is simultaneously the genetic unit and the modular unit. Modular organisms have added another dimension to the problem of population changes; in addition to studying whole organisms, we must measure changes in modules and measure modular "birth" and "death" rates (Harper et al. 1986).

In some of these cases we can circumvent the problem by measuring biomass (weight) instead of counting numbers. Foresters are not interested in the number of oak trees in a woodlot but rather the sizes of the trees. In some species we will be able to apply population methodology only by making some arbitrary decision about what units to measure or count. Fortunately, in many cases, individuals and populations are easy to recognize and study.

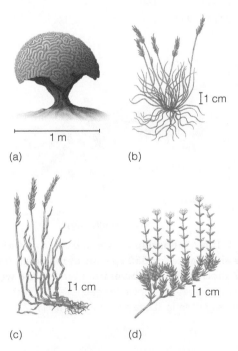

Figure 1 Some examples of modular organisms. (a) Brain coral from the Great Barrier Reef. This large structure is actually a colony consisting of thousands of individual coral polyps, each a module budded from the original coral polyp.
(b) Fescue grass (*Festuca brachyphylla*), which forms a tussock of tightly packed modules. (c) Wheatgrass (*Agropyron boreale*), in which tillers spread laterally. (d) Sandwort (*Arenaria laricifolia*) with many tillers coming off a spreading stolon.

Estimation of Population Parameters

The population attributes concerned with changes in abundance are interrelated as shown in **Figure 2**. When we ask why population density has gone up or gone down in a particular species, we are asking which one (or more) of these primary population parameters has changed. In this section we examine briefly the methods employed in estimating these vital statistics. We can

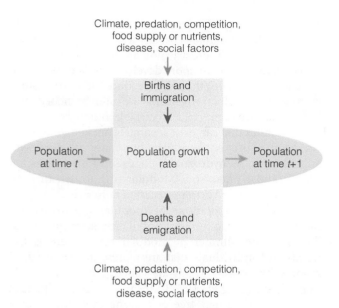

Figure 2 The dynamics of populations centers on understanding the population growth rate. In this chapter we will describe the four primary population parameters (pink box) that together determine how population density will change.

Table 1	Observed density of small to large organisms in natural populations.	
	Density in conventional units	**Density per m² (or m³)**
Diatoms	5,000,000/m³	5,000,000
Soil arthropods	500,000/m²	500,000
Barnacles (adult)	20/100 cm²	2000
Trees	500/ha	0.0500000
Field mice	250/ha	0.0250000
Woodland mice	10/ha	0.0010000
Deer	4/km²	0.0000040
Human beings		
Netherlands	395/km²	0.0003950
United States	31/km²	0.0000310
Canada	3.2/km²	0.0000032

1 ha = 10,000 m² = 2.47 acres

1 km² = 100 ha = 0.386 sq. mile

appreciate the problems involved in estimating density by considering the approximate densities of organisms in nature listed in **Table 1**. Given such a wide range of figures, covering more than a dozen orders of magnitude, it is clear that techniques for estimating density that work nicely with deer cannot be applied to bacteria or protozoa. The two fundamental attributes that affect our choice of techniques for population estimation are the size and mobility of the organism with respect to humans.

Small animals are usually more abundant than large animals. **Figure 3** shows this trend for 350 species of mammals and 552 species of birds and allows us to predict the approximate density for a species of bird or mammal of given size. Similar plots can be constructed for other species groups (Peters 1983). **Table 2** presents the regression estimates for several groups of animals, and the Working with the Data box "Calculation of Per Capita Rates" illustrates the use of those regressions to estimate average population density for animals of a particular size.

The systematic differences that exist among groups are clear in Figure 3. Birds, for example, are less abundant than mammals of equivalent size. If we were to study a 1-kg bird, we should expect a density of approx-

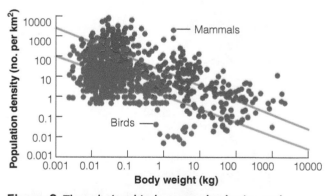

Figure 3 The relationship between body size and average abundance for 350 species of mammals (red) and 552 species of birds (blue) from around the world. Average trend lines are shown for each group. Note that the scales are logarithmic. (Data from Silva et al. 1997.)

imately 1 per sq. km. For a similar-sized mammal, we should expect about 100 per sq. km.

In most cases we cannot rely on these estimates of average abundance because we need to know if, for example, a population of fish is declining from overfishing, or if a population of endangered plants is increasing or

Table 2 Predictive equations for the relationship of average population density to body size for various animal groups.

Group	Intercept of regression line (a)	Slope of regression line (b)	Sample size
Mammals	1.316	−0.688	364
Herbivores	1.30	−0.66	98
Carnivores	1.69	−1.02	25
Birds	−0.045	−0.604	564
Insect eaters	−0.05	−0.64	277
Seed eaters	0.22	−0.54	80
Fishes	1.81	−0.77	11
Aquatic invertebrates	5.37	−0.58	56
Terrestrial invertebrates	3.48	−0.69	106

The equation for each group has the following general form:

log (population density) = a + b (log [body mass]),

where all logs are base 10, population density is the number per km², and body mass is in kg.

SOURCE: Data are from Silva et al. (1997) and Peters (1983).

WORKING WITH THE DATA

Calculation of Per Capita Rates

Demographers are usually interested in per capita rates of birth and death. You can see why with a simple example. If you were told that 400 ducks had been killed in a disease outbreak, your reaction would be that this number is difficult to evaluate without knowing the size of the duck population. If you were then told that the duck population of this region was 250,000 individuals, you would be able to evaluate this mortality as a per capita rate:

$$\text{Per capita death rate} = \frac{\text{No. of deaths}}{\text{Population at risk}}$$

$$q_x = \frac{d_x}{n_x} = \frac{400}{250,000} = 0.0016$$

or a death rate of 0.16%, a tiny figure. Similarly, if you were told that nine marmots died in a severe storm, you would need to know that the total marmot population was 26 individuals to know that this is a high per capita mortality rate:

$$q_x = \frac{d_x}{n_x} = \frac{9}{26} = 0.346$$

and that more than one-third of the population died.

Natality rates should also be expressed as per capita rates for the same reasons:

$$\text{Per capita birth rate} = \frac{\text{No. of births}}{\substack{\text{Size of the reproductive} \\ \text{population}}}$$

Plotting population data on a logarithmic scale is a simple way to emphasize per capita rates. A numerical example shows this. If one-half of a population dies, we obtain:

Starting population size	No. dying	Per capita death rate	Final population size
1000	500	0.5	500
500	250	0.5	250
250	125	0.5	125

On a logarithmic scale, all these decreases are equal (base 10 logs):

log(1000) − log(500) = log(500) − log(250)

3.00 − 2.70 = 2.70 − 2.40

Even though *numbers* lost differ greatly, the *per capita death rates* are the same.

decreasing over time. How can we estimate population density? Ecologists have developed an array of techniques to estimate population density. Here we will only scratch the surface of the methods used, but it is important to understand in general how these estimates of populations are made for two reasons: we can begin to see how ecologists quantify nature, and we will come to appreciate how difficult it is to obtain reliable estimates of populations. Reliable quantitative methods are the backbone of science, and one of the great triumphs of ecologists in the past 60 years has been the development of these methods to a high degree of precision (Krebs 1999; Sutherland 2006).

There are two broad approaches to estimating population density. In many cases we need to know the **absolute density** of a population (for example, number of individuals per hectare or per square meter) to make management decisions or conservation recommendations. In other cases we may find it adequate to know the relative density of the population (that is, for two areas of equal size, area *x* has more organisms than area *y*). This division of approaches is reflected in the techniques developed for measuring density.

Measurements of Absolute Density

Ecologists go about determining absolute density in two ways: by making total counts and by using sampling methods.

Total Counts

The most direct way to find out how many organisms are living in an area is to count them. One good example of this is a human population census. Other examples come from populations of plants and from vertebrate animals. With trees one can easily count all the individuals in a given area. With territorial birds one can count all the singing males in an area, or with bobwhite quail one can count the number of birds in each covey. Other animals, such as the northern fur seal, may be counted when they are all gathered in breeding colonies. Few invertebrates, however, can be counted in total, the exceptions being barnacles and other sessile invertebrates such as some rotifers. Large animals on small areas can sometimes be counted in total or photographed and counted in the photos, but in general direct counts are possible for very few organisms.

Sampling Methods

Usually investigators must be content to count only a small proportion of the population and to use this sample to estimate the total. There are two general sampling techniques: the use of quadrats and the capture-recapture method.

Use of quadrats

The general procedure in this technique is to count all the individuals on several quadrats of known size and then to extrapolate the average count to the whole area. A quadrat is a sampling area of any shape. Although the word literally describes a four-sided figure, it has been used in ecology for areas of all shapes, including circles. An example will illustrate this estimation procedure: if you counted 19, 21, 17, and 19 individuals of a beetle species in four soil samples of 10 cm by 10 cm, you could extrapolate this to 1900 beetles per square meter of soil surface.

Achieving reliable estimates using this technique requires three things: (1) The population of each quadrat examined must be determined accurately, (2) the area of each quadrat must be known, and (3) the quadrats counted must be representative of the whole area. This last condition is usually achieved by random sampling procedures; students acquainted with statistics will find a good discussion of this problem in Zar (1999). The population of each quadrat may be counted without error in some organisms but only estimated in other species. Many special techniques have been developed for applying quadrat-sampling techniques to different kinds of animals and plants in terrestrial and aquatic systems. Next we examine one example of the use of quadrats.

Wireworms are click beetle larvae (*Elateridae*) that live in the soil. Some species feed on seeds and seedlings and damage the roots of agricultural crops. To estimate populations of a wireworm root pest (*Agriotes* spp.), Salt and Hollick (1944) devised a technique of extracting larvae from soil samples. This technique involved breaking the lumps of soil, separating the very coarse and very fine material using sieves, and separating the wireworms from other organic material by benzene flotation (insects accumulate at the benzene-water interface; the plant matter stays in the water). Exhaustive tests were made at each step in this process to see if larvae were lost. The investigators sampled soil by using a corer that removed a cylinder of soil 10 cm in diameter and 15 cm deep. In one pasture near Cambridge, England, they collected 240 random samples that contained a total of 3742 larvae of wireworms., an average of 15.6 larvae per 10-cm core, or an infestation of 19.3 million larvae per hectare. Variation among the samples can be used to construct confidence limits for this density estimate. Salt and Hollick were able to show by this careful work that wireworm populations were about three times higher in English pastureland than people had previously supposed.

Quadrats have been used extensively in plant ecology; indeed it is the most common method for sampling plants. There is an immense literature on the problems of sampling plants with quadrats; some articles, for example, deal with the relative efficiency of round, square, and rectangular quadrats. We will not go into these detailed problems of methodology; interested students should refer to Krebs (1999, Chapter 4).

Capture-recapture method

The technique of capture, marking, release, and recapture is an important one for mobile animals, because it allows not only an estimate of density but also estimates of birth rate and death rate for the population being studied.

Several models can be used for capture-recapture estimation. Basically, they all depend on the following line of reasoning: if you capture animals, mark them, and then release them, the proportion of animals marked in subsequent samples taken from this population should be representative of the proportion marked in the entire population. This is illustrated here for a simple example that includes two capture sessions in **Figure 4**.

This simple type of population estimation is known as the Petersen method[1] because it was developed by the Danish fisheries scientist C. G. J. Petersen in 1898. It involves only two sampling periods: capture, mark, and release at time 1 and capture and check for marked animals at time 2. The time interval between the two samples must be short because this method assumes a closed population with no recruitment of new animals into the population between times 1 and 2 and no losses of marked animals. We assume that a sample, if random, will contain the same proportion of marked animals as that in the whole population:

$$\frac{\text{Marked animals in}}{\substack{\text{second sample} \\ \text{Total caught in} \\ \text{second sample}}} = \frac{\text{Marked animals in}}{\substack{\text{first sample} \\ \text{Total population} \\ \text{size}}}$$

---

[1] Also called the *Lincoln method* by wildlife ecologists, because it was first used by F. C. Lincoln on ducks in 1930.

For this simple example, using N for total population size:

$$\frac{5}{20} = \frac{16}{N}$$

or $N = 64$.

One of the first to use the Petersen method was Dahl (1919) who marked trout (*Salmo fario*) in small Norwegian lakes to estimate the size of the population that was subject to fishing. He marked and released 109 trout, and in a second sample a few days later caught 177 trout, of which 57 were marked. From these data we estimate:

$$\frac{\text{Proportion of total}}{\text{population marked}} = \frac{\substack{\text{Number marked} \\ \text{in sample}}}{\substack{\text{Total number} \\ \text{caught}}}$$

$$= \frac{57}{177} = 0.322 \qquad (1)$$

The number of fish marked in the first sample was 109, and therefore

$$\frac{\text{Total population}}{\text{size}} = \frac{\substack{\text{Size of marked} \\ \text{population}}}{\substack{\text{Proportion of} \\ \text{population marked}}}$$

$$= \frac{109}{0.322} = 338 \text{ trout} \qquad (2)$$

To estimate density with capture-recapture methods, two situations must be considered: closed and open populations. For population estimation a population is defined as closed if it is not changing in size during the period of capture, marking, and recapturing; a population is defined as open if it is changing in size during the study period. Real populations are clearly open, unless we sample them over a very brief period. Population estimation for open populations is more complex because we need to take account of all the parameters shown in Figure 2. Details are given in Seber (1982), Pollock et al. (1990), and Krebs (1999).

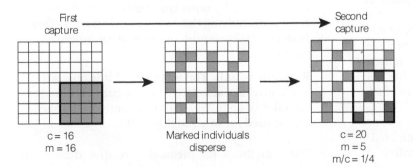

First capture → Second capture

c = 16
m = 16

Marked individuals disperse

c = 20
m = 5
m/c = 1/4

Figure 4 Schematic illustration of a capture-recapture sampling session with two capture periods. Each square represents one individual, each light blue square represents an individual that has been marked (m) once, each dark blue square indicates individuals that have been marked twice, and bold lines indicate the individuals that have been caught (c).

All capture-recapture models make three crucial assumptions:

- Marked and unmarked animals are captured randomly.

- Marked animals are subject to the same mortality rate as unmarked animals. The Petersen method assumes there is no mortality during the sampling interval.

- Marked animals are neither lost nor overlooked.

All these assumptions have caused trouble at one time or another. For example, field mice may become trap-happy or trap-shy and thus violate the first assumption. Fish tagged on the high seas may be weakened by the nets and the tagging procedure (being held out of water) such that they suffer increased mortality just after release. In some cases, fishermen have not returned tags recovered from marked fish because they considered them good-luck charms. Leg bands may wear thin and be lost from long-lived birds. Numerous variations of the techniques of marking and recapture analysis have been designed to very cleverly circumvent some of these problems.

The capture-recapture technique has been used mainly on larger forms, such as butterflies, snails, beetles, and many vertebrates, that can be readily marked.

Note that when we have estimated the population size for a mobile species we still have the problem of estimating density because we must determine the area occupied by the population. When we are studying an island, or an isolated patch of forest, the area occupied is easy to determine. When habitats are continuous, the extent of the area occupied is less clear, and some correction must be made to take into account the movements of the animals being studied.

Indices of Relative Density

The characteristic feature of all methods for measuring relative density is that they depend on the collection of samples that represent some relatively constant but unknown relationship to total population size. These methods provide an index of abundance that is more or less accurate. When an index of abundance (such as tracks in sand plots) is 4.0 on area x and 8.0 on area y, we can conclude that area y has a higher density of animals than area x. You cannot conclude that area y has twice the density of area x, because it may be that there are only 40% more animals on area y, but they are much more active. There are a great many indices of relative density, and here we will list only a few:

1. **Traps.** We previous noted that traps are often used in capture-recapture studies to estimate absolute density. The number of individuals caught per day per trap may also be used as an index of relative density. The traps could include mousetraps spread across a field, light traps for night-flying insects, pitfall traps in the ground for beetles, suction traps for aerial insects, and plankton nets. The number of organisms trapped depends not only on the population density but also the animals' activity and range of movement, and on the researcher's skill in placing traps, so these techniques provide only a rough idea of abundance.

2. **Number of fecal pellets.** This technique has been used for snowshoe hares, deer, field mice, and rabbits. If we know the average rate of defecation, the number of fecal pellets in an area can provide an index of population size. Similar methods are used for defoliating caterpillars by estimating the amount of frass falling from trees.

3. **Vocalization frequency.** The number of bird calls heard per 10 minutes in the early morning has been used as an index of the size of bird populations. The same method can be used for frogs, crickets, and cicadas.

4. **Pelt records.** The number of animals caught by trappers has been used to estimate population changes in several mammals such as Canada lynx; some records extend back 300 years.

5. **Catch per unit fishing effort.** This measure can be used as an index of fish abundance, for example, number of fish per 100 hours of trawling.

6. **Number of artifacts.** This count can be used for organisms that leave evidence of their activities, for example, mud chimneys for burrowing crayfish, tree-squirrel nests, and pupal cases from emerged insects.

7. **Questionnaires.** Questionnaires can be sent to sportsmen or trappers to get a subjective estimate of population changes. This technique is useful only for detecting large changes in population among animals large enough to be noticed.

8. **Cover.** The percentage of the ground surface covered by a plant as a measure of relative density has been used by botanists and by invertebrate ecologists studying the rocky intertidal zone. This is an especially important method for modular organisms.

9. **Roadside counts.** The number of birds observed while driving a standard distance has been used as an index of abundance, and the same technique can be used for other highly visible organisms.

These methods for measuring relative density all need to be viewed skeptically until they have been carefully evaluated (Anderson 2003). They are most useful as a supplement to more direct census techniques and for detecting large changes in population density.

We conclude our discussion of techniques for measuring density by noting that detailed, accurate census information is obtainable for few animals. In many cases we must be content with an order-of-magnitude estimate. Because of this, a disproportionate amount of work has been done on the more easily censused forms, particularly butterflies, birds, and mammals, and species of economic importance. This fact introduces an obvious bias into the following discussions of the mechanisms determining population density.

Natality

A major factor in population increase is natality, a broad term covering the production of new individuals by birth, hatching, germination, or fission.

Two aspects of reproduction must be distinguished. The concept of fecundity is a physiological notion that refers to an organism's potential reproductive capacity. Fertility is an ecological concept that is based on the number of viable offspring produced during a period of time. We must distinguish between *realized fertility* and *potential fecundity*. For example, the realized fertility rate for a human population may be only one birth per 15 years per female in the childbearing ages, whereas the potential fecundity rate for humans is one birth per 10 to 11 months per female in the childbearing ages.

Natality rate may be expressed as the number of organisms produced per female[2] per unit time, and is synonymous with the realized fertility rate. The magnitude of the natality rate is highly dependent on the type of organism being studied. Some species breed once a year, some breed several times a year, and others breed continually. Some produce many seeds or eggs, others few. For example, a single oyster can produce 55–114 million eggs; fish commonly lay eggs in the thousands; frogs produce eggs in the hundreds; birds usually lay between 1 and 20 eggs; and mammals rarely have litters of more than 10 offspring and often have only one or two. Fecundity is usually inversely related to the amount of parental care given to the young.

Mortality

Biologists are interested not only in why organisms die but also why they die at a given age. Mortality—or its converse, survival—can be looked at from several per-

spectives. Longevity focuses on the age of death of individuals in a population. Two types of longevity can be recognized: *potential longevity* and *realized longevity.* Potential longevity, the maximum life span attainable by an individual of a particular species, is a limit set by the physiology of the organism, such that it simply dies of old age. Another way of describing potential longevity is the average longevity of individuals living under optimum conditions. But organisms in nature rarely live in optimum conditions; most animals and plants die from disease, or are eaten by predators, or succumb to any one of a number of natural hazards. Realized longevity is the actual life span of an organism. Realized longevity can be averaged for all the individuals in a population living under real environmental conditions, and this average longevity can be measured in the field, whereas potential longevity can be measured only in the laboratory or in a zoo or botanical garden.

Two examples will illustrate these distinctions. The European robin has an average life expectation of one year in the wild, whereas it can live at least 11 years in captivity (Lack 1954). In ancient Rome, the average life expectation at birth for human females was about 21 years, and in England in the 1780s it was about 39 years (Pearl 1922). Realized longevity in humans has risen dramatically during the twentieth century. In the United States in 2007, females at birth could expect to live 81 years on average. Potential longevity in humans is around 100 years. Some individuals in Rome and preindustrial England did live to be 80 or more, but very few. Low longevity in human populations is due to high mortality in infants and children. The simplest measurements of mortality in plants and animals are done directly. Mortality rates are estimated by marking a series of organisms and observing how many survive from time t to time $t + 1$.

Immigration and Emigration

Dispersal—immigration and emigration—is seldom measured in a population study. In most cases it is either assumed that the two components are equal or else work is done in an island type of habitat, where dispersal is presumably of reduced importance. Both assumptions are highly questionable. The capacity to disperse is an essential part of the life cycle of most organisms; it is the ecological process that produces gene flow between local populations and thus helps to prevent inbreeding. Dispersal can set limits on geographic distributions, and it affects community composition. Some populations sustain a net emigration and thus export individuals; others are sustained only by a net immigration. One example is small songbirds in woodlots in the eastern United States. Small woodlots are not productive for birds because of heavy nest predation, and these populations can be sustained

---

[2]For asexual organisms that reproduce by fission or budding, all individuals would be included in estimating natality rates. The same would apply to bisexual (monoecious) plants.

only by immigration (Tittler et al. 2006). Dispersal may be a critical parameter in population changes.

Dispersal can be measured if individuals can be marked in a population. The use of radio-telemetry has revolutionized the study of animal movements, particularly for larger organisms (Millspaugh and Marzluff 2001). The major technical problem in studying dispersal is the scale of the movements involved. Because animals move distances greater than the size of study areas, information on "long distance" dispersal can be lost. One of the major unsolved problems of conservation biology is how to facilitate immigration and emigration from populations in isolated parks or refuges in a fragmented landscape.

Demographic Techniques

One of the great strengths of population ecology is that it is quantitative. If the survival rate of adult bald eagles decreased 2% per year, would their populations decline? If we could increase the survival of juvenile salmon 0.5% in their first year, how many more adults would reach maturity and be available for fishermen? It is possible to answer these questions precisely with some simple mathematics. Population mathematics is not difficult, but it is sufficiently different to merit some of your attention if you wish to achieve a more precise understanding of how and why populations change. This chapter and the next provide a quantitative background for population ecology.

Life Tables

Mortality is one of the four key parameters that drive population changes, as we saw in Figure 2. We need a technique to summarize how mortality is occurring in a population. Is mortality high among juvenile organisms? Do older organisms have a higher mortality rate than younger organisms? We can answer these kinds of questions by constructing a **life table**, a convenient format for describing the mortality schedule of a population. Life tables were developed by human demographers, particularly those working for life insurance companies, which have a vested interest in knowing how long people can be expected to live. There is a correspondingly immense literature on human life tables, but less data are available on other animals or on plants.

Plant and animal populations may be composed of several types of individuals, and in any given analysis a demographer may group them together or may keep them separate. A life insurance company offers to males a policy different from the one they give to females for good demographic reasons, and thus it may be useful for some purposes to classify individuals by sex or age.

A life table is an age-specific summary of the mortality rates operating on a cohort of individuals. A cohort may include the entire population, or it may include only males, or only individuals born in a given year. An example of a cohort life table for song sparrows is given in **Table 3**.

Table 3 Cohort life table for the song sparrow on Mandarte Island, British Columbia.[a]

Age in years (x)	Observed no. of birds alive (n_x)	Proportion surviving at start of age interval x (l_x)	No. dying within age interval x to x + 1 (d_x)	Rate of mortality (q_x)
0	115	1.0	90	0.78
1	25	0.217	6	0.24
2	19	0.165	7	0.37
3	12	0.104	10	0.83
4	2	0.017	1	0.50
5	1	0.009	1	1.0
6	0	0.0	—	—

[a] Males hatched in 1976 were followed from hatching until all had died six years later.
SOURCE: From Smith (1988).

The columns of this life table are assigned the following symbols, which are consistently used in ecology:

x = age

n_x = number alive at age x

l_x = proportion of organisms surviving from the start of the life table to age x

d_x = number of individuals dying during the age interval x to $x + 1$

q_x = per capita rate of mortality during the age interval x to $x + 1$

To set up a life table, we must decide on age intervals in which to group the data. For humans or trees the age interval may be five years; for deer, birds, or perennial plants one year, and for annual plants or field mice one month. By making the age interval shorter, we increase the detail of the mortality picture shown by the life table at the price of needing more data.

Note that if you are given any one of the columns of the life table, you can calculate the rest. Put another way, there is nothing "new" in each of the three columns l_x, d_x, and q_x; they are just different ways of summarizing one set of data. The columns are related as follows:

$$n_{x+1} = n_x - d_x \tag{3}$$

$$q_x = \frac{d_x}{n_x} \tag{4}$$

$$l_x = \frac{n_x}{n_0} \tag{5}$$

For example, from Table 3,

$$n_3 = n_2 - d_2 \qquad q_2 = \frac{d_2}{n_2} \qquad l_4 = \frac{n_4}{n_0}$$

$$= 19 - 7 = 12 \qquad = \frac{7}{19} = 0.37 \qquad = \frac{2}{115} = 0.017$$

The rate of mortality q_x is expressed as a rate for the time interval between successive census stages of the life table. For example, q_x for the song sparrows in Table 3 is 0.78 for the interval between egg and one year, or per year. Thus, 78% of the birds are lost in the nest or during their first year of life.

The most frequently used part of the life table (see Table 3) is the n_x column, the number of survivors at age x. This is often expressed from a starting cohort of 1000, but some human demographers prefer a starting cohort of 100,000. Other workers prefer to plot the l_x column to show the proportion surviving. The n_x (or l_x) data are

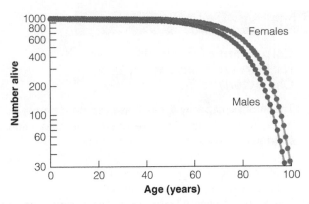

Figure 5 Survivorship curve for all males (red) and females (blue) in the United States in 2003 for a starting cohort of 1000 individuals. Life expectancy at birth was 75 years for males and 80 years for females. (Data from the U.S. National Center for Health Statistics, 2006.)

plotted as a survivorship curve; **Figure 5** presents the survivorship curves for the human population of the United States in 2003. Note that the n_x values are plotted on a logarithmic scale. Population data should be plotted this way when one is interested in per capita rates of change rather than absolute numerical changes.

The life table was introduced to ecologists in 1921 by Raymond Pearl, one of the most important population ecologists in the United States during the first four decades of the twentieth century. Pearl (1928) described three general types of survivorship curves (**Figure 6**). Type 1 curves are characteristic of populations with low per capita mortality for most of the life span and then high losses of older organisms. The linear survivorship curve (type 2) implies a constant per capita rate of mortality independent of age. Type 3 curves indicate high per capita mortality early in life, followed by a period of much lower and relatively constant loss.

No population has a survivorship curve exactly like these idealized ones, and real curves are composites of the three types. In developed nations, for example, humans tend to have a type 1 survivorship curve (except for the first few days of life). Many birds have a type 2 survivorship curve, and a large number of populations would fall in the area intermediate between types 1 and 2. Often a period of high loss in the early juvenile stages alters these ideal type 1 and 2 curves. Type 3 curves occur in many fishes, marine invertebrates, and parasites.

Now that we have seen what a life table looks like, how do we get the data to construct one? The answer is: it depends, because there are two very different ways of gathering data for life tables, and they produce two different types of life tables: the cohort life table (which

WORKING WITH THE DATA

Calculation of the Intrinsic Capacity for Increase from Lotka's Characteristic Equation

The intrinsic capacity for increase can be determined more accurately by solving the characteristic equation, a formula derived by Lotka (1907, 1913):

$$\sum_0^\infty e^{-rx}l_x b_x = 1$$

This equation cannot be solved explicitly for r because it cannot be rearranged to have r on one side and all else on the other. By substituting trial values of r, we can solve this equation iteratively, by trial-and-error. Our hypothetical animal (see Figure 11) can be used as an example. For our estimate of $r = 0.824$, we get

x	$l_x b_x$	$e^{-0.824x}$	$e^{-0.824x}l_x b_x$
0	0.0	1.00	0.000
1	2.0	0.44	0.880
2	1.0	0.19	0.190
3	0.0	0.08	0.000
4	0.0	0.04	0.000

$$\sum_0^\infty e^{-rx}l_x b_x = 1.070$$

If the sum is too large (as it is here), then the estimate of $r = 0.824$ is too low. We repeat with $r = 0.85$, and after several trials we find that for this hypothetical organism, $r = 0.881$ provides

$$\sum_0^\infty e^{-rx}l_x b_x = 1.004$$

which is a close enough approximation. Carey (1995) works out another example in detail. The intrinsic capacity for increase is an instantaneous rate and can be converted to the more familiar finite rate by the formula

$$\text{Finite rate of increase} = \lambda = e^r \qquad (6)$$

For example, if $r = 0.881$, then $\lambda = 2.413$ per individual per year in our hypothetical organism. Thus for every individual present this year, 2.413 individuals will be present next year.

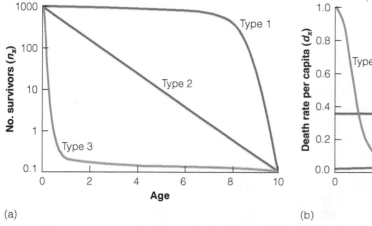
(a)

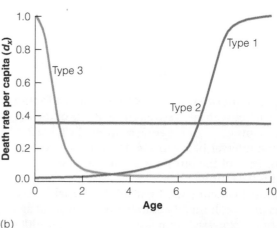
(b)

Figure 6 Types of survivorship curves. (a) Hypothetical survivorship curves (n_x). (b) Mortality rate (d_x) curves corresponding to these hypothetical survivorship curves. Type 2 (red) curves show constant survival rate with respect to age. Type 1 (blue) curves show increasing mortality late in life, and Type 3 (green) curves show the highest mortality early in life. (After Pearl 1928.)

we have already seen in Table 3) and the static life table. These two life tables are different in form, except under unusual circumstances, and are always quite different in meaning (Caughley 1977).

The static life table (also called a stationary, time-specific, current, or vertical life table) is calculated on the basis of a cross section of a population at a specific time. **Table 4** is a static life table composed from the census data and mortality data for human females in Canada in 2006. A cross section of the female population in 2006 provides the number of deaths (d_x) in each age group and the number of individuals in that age group. This allows us to estimate a set of mortality rates (q_x) for each age group, and the q_x values can be used to calculate a complete life table in the way outlined previously, if we assume that the population is stationary.

Table 4 Static life table for the human female population of Canada, 2006.

Age group (yr)	No. in each age group	Deaths in each age group	Mortality rate per 1000 persons (1000 q_x)
0–4	829,300	911	1.10
5–9	899,500	70	0.08
10–14	1,016,500	136	0.13
15–19	1055500	317	0.30
20–24	1,100,200	370	0.34
25–29	1,101,200	377	0.34
30–34	1,101,100	511	0.46
35–39	1,168,400	853	0.73
40–44	1,341,700	1481	1.10
45–49	1,336,900	2364	1.77
50–54	1,193,800	3338	2.80
55–59	1,054,000	4775	4.53
60–64	805,500	5729	7.11
65–69	636,800	7253	11.39
70–74	554,300	10,210	18.42
75–79	490,800	15,221	31.01
80–84	389,200	21,236	54.56
85–90	227,900	22,256	97.66
90 and above	125,300	38,742	309.19

SOURCE: Statistics Canada (2007).

The cohort life table (also called a generation or horizontal life table) is calculated on the basis of a cohort of organisms followed throughout life. For example, we could, in principle, get all the birth records from New York City for 1931 and trace the history of all these people throughout their lives, following those that move out of town—a very tedious task. We could then tabulate the number surviving at each age interval. Very few data like these are available for human populations.[3] This procedure would give us the survivorship curve directly, and we could calculate the other life-table functions, as previously described.

These two types of life table will be identical if and only if the environment does not change from year to year and the population is at equilibrium. But normally birth rates and death rates do vary from year to year, and consequently large differences exist between the two forms of life table. These differences can be illustrated most easily for human populations. For example, a static life table for humans born in 1900 in the United States would show what the survivorship curve would have been if the population had continued surviving at the rates observed in 1900. But of course the human population did not retain these same 1900 rates. The continual improvement in medicine and sanitation in the past 100 years has increased survival rates and life expectancy by more than 15 years, and the people born in 1900 had a cohort or generation survivorship curve unlike that of any of the years through which they lived. Static life tables assume static (stationary) populations.

Insurance companies would like to have data from cohort life tables covering the future, but these data are obviously impossible to get. Insurers are definitely not interested in cohort life tables covering the past—the life table for the 1900 cohort would be of little use for predicting mortality patterns today. So insurers use static life tables and correct them at each census. These predictions will never be completely accurate but will be close enough for their purposes.

Life tables from nonhuman populations are more difficult to come by. In general, ecologists use three types of data to construct life tables:

- *Survivorship directly observed.* The information on survival *(l$_x$)* of a large cohort born at the same time, followed at close intervals throughout its existence, is the best to have, since it generates a

---

[3]For human populations, unlike those of other animals and plants, it is possible to construct cohort life tables indirectly from mortality rate (q_x) data. To construct a cohort life table for the 1931 New York City cohort, we can obtain the mortality statistics for the 0- to 1-year-olds for 1931, the 1- to 5-year-olds for 1932–1935, the 6- to 10-year-olds for 1936–1940, and so on, and use these q_x rates to estimate the life-table functions.

cohort life table directly and does not involve the assumption that the population is stable over time. A good example of data of this type is that of Connell (1961a) on the barnacle *Chthamalus stellatus* in Scotland. This barnacle settles on rocks during the autumn. Connell did several experiments in which he removed a competing barnacle, *Semibalanus balanoides*, from some rocks but not from others, and then about once a month counted the *Chthamalus* surviving on these defined areas (**Figure 7**). Barnacles that disappeared had certainly died; they could not emigrate.

- *Age at death observed.* Data on age at death may be used to estimate the life-table functions for a static life table. In such cases we must assume that the population size is constant over time and that the birth and death rates of each age group remain constant. A good example of this type of data comes from the work of Bronikowski et al. (2002) on the baboons of Amboseli National Park in Kenya, East Africa. In this park primate researchers were able to follow female groups from 1971 to 1999 and identify individual females, so that age at death could be directly observed. The Amboseli baboons live in a semiarid environment and are subjected to considerable predation, so that

mortality accelerates rapidly after age 5, as shown in **Figure 8**.

- *Age structure directly observed.* Ecological information on age structure, particularly of trees, birds, and fishes, is considerable and in some cases can be used to construct a static life table. In these cases, we can often determine how many individuals of each age are living in the population. For example, if we fish a lake, we can get a sample of fish and determine the age of each from annular rings on the scales. (The same type of data can be obtained from tree rings.) The difficulty is that to produce a life table from such data, we must assume a constant age distribution, something that is rare for many populations. Consequently, data of this type are not always suitable for constructing a life table.

Attempts to gather life-table data on organisms other than humans and to establish a general theory of senescence have suggested that, except for early ages when mortality is high, mortality rates (q_x) increase inexorably with age, so that for all organisms the mortality curve is roughly U-shaped, as illustrated in Figure 8. Humans and primates fit this pattern of mortality increasing with age (Bronikowski et al. 2002). The mortality rate doubling time for baboons was estimated at 3.5–4.8 years, compared with the current human estimate of 8 years for U.S. females. Humans have an extended life span beyond the reproductive years, and this is probably an evolutionary adaptation because post-reproductive women can make a contribution to the fitness of their children and grandchildren (Reznick et al. 2006).

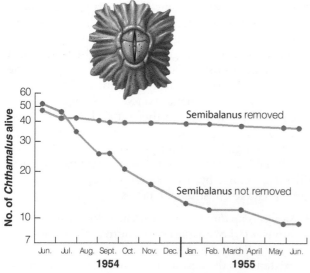

Figure 7 Survivorship curves of the barnacle *Chthamalus stellatus*, which had settled naturally on the shore at Millport, Scotland, in the autumn of 1953. The survival of *Chthamalus* growing without contact with *Semibalanus* is compared with survival in an area with both species. *Semibalanus* crowds out *Chthamalus* when the two species are side by side. (Data from Connell 1961a and personal communication.)

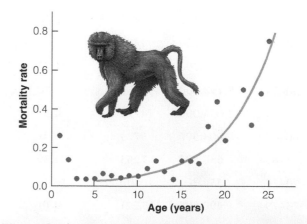

Figure 8 Mortality rate per year for each one-year age interval for 274 female baboons (*Papio hamadryas*) of Amboseli National Park, Kenya. Age at death was determined by direct observation of marked animals. (Data from Bronikowski et al. 2002.)

Intrinsic Capacity for Increase in Numbers

A life table summarizes the mortality schedule of a population, and we have just seen several examples. We must now consider the reproductive rate of a population and techniques by which we can combine reproduction and mortality estimates to determine net population changes. Students of human populations were the first to appreciate and solve these problems. One way of combining reproduction and mortality data for populations utilizes a demographic parameter called the **intrinsic capacity for increase** derived by Alfred Lotka in 1925.

Any population in a particular environment will have a mean longevity or survival rate, a mean natality rate, and a mean growth rate or speed of development of individuals. The values of these means are determined in part by the environment and in part by the innate qualities of the organisms themselves. These qualities of an organism cannot be measured simply because they are not a constant, but by measuring their expression under specified conditions we can define for each population its intrinsic capacity for increase (also called the Malthusian parameter), a statistical population characteristic that depends on environmental conditions.

Environments in nature vary continually. They are never consistently favorable or consistently unfavorable but fluctuate between these two extremes, for example, from winter to summer. When conditions are favorable, numbers increase; when conditions are unfavorable, numbers decrease. It is clear that no population goes on increasing forever. Darwin (1859, Chapter 3) recognized the contrast between a high potential rate of increase and an observed approximate balance in nature. He illustrated this problem by asking why there were not more elephants, given his estimate that two elephants could give rise to 19 million elephants in 750 years.

Therefore, in nature we observe an actual rate of population change that is continually varying from positive to negative in response to changes within the population in age distribution, social structure, and genetic composition, and in response to changes in environmental factors. We can, however, ask what would happen to a population if it persisted in its current configuration of births and deaths. This abstraction is the ecologist's version of the perfect vacuum of introductory physics: we ask what would happen in terms of population increase if conditions remained unchanged for a long time in a particular environment.

An organism's innate or intrinsic capacity for increase depends on its fertility, longevity, and speed of development. For any population, these processes are integrated and measured by the natality rate and the death rate. When the natality rate exceeds the death rate, the population will increase. If we wish to estimate quantitatively the rate at which the population increases or decreases, we need to describe how both the natality rate and the death rate vary with age.

How can we express the variations of natality and mortality rates with age? We have just discussed the method of expressing survival rates as a function of age. The life table includes a table of age-specific survival rates. The portion of the life table needed to compute the capacity for increase is the l_x column, the proportion of the population surviving to age x. Similarly, the natality rate of a population is best described by an age schedule of births, seed production, egg production, or fission. This is a table that gives (for sexual species) the number of female offspring produced per female aged x to $x + 1$ and is called a fertility schedule, or b_x function. Usually only females are counted, and the demographer typically views populations as females giving rise to more females. **Table 5** gives the survivorship table, the l_x schedule with which we are familiar, and the fertility schedule for women in the United States in 2007. In this case, the great majority of women live through the childbearing ages. The fertility schedule gives the expected number of female offspring for each woman living through the five years of each age group. For example, slightly fewer than three women in 10 between the ages of 25 and 29 will, on average, have a female baby.

Given these data, we can obtain a useful statistic, the **net reproductive rate (R_0)**. If a cohort of females lives its entire reproductive life at the survival and fertility rates given in Table 5, what will this cohort or generation leave as its female offspring? We define the net reproductive rate as follows:

$$\begin{matrix} \text{Net} \\ \text{reproductive} \\ \text{rate} \end{matrix} = R_0 = \frac{\begin{matrix}\text{Number of} \\ \text{daughters produced} \\ \text{in generation } t + 1\end{matrix}}{\begin{matrix}\text{Number of} \\ \text{daughters produced} \\ \text{in generation } t\end{matrix}} \quad (7)$$

R_0 is thus the multiplication rate per generation[4] and is obtained by multiplying together the l_x and b_x schedules and summing over all age groups, as shown in Table 5:

$$R_0 = \sum_0^\infty l_x b_x \quad (8)$$

Thus, we temper the natality rate by the fraction of expected survivors to each age. If survival were 100%, R_0 would just be the sum of the b_x column. In this example

---

[4]A generation is defined as the mean period elapsing between the birth of parents and the birth of offspring; see Figure 12.

Table 5 Survivorship schedule (l_x) and fertility schedule (b_x) for women in the United States, 2007.

Age group	Midpoint or pivotal age x	Proportion surviving to pivotal age l_x	No. female offspring per female aged x per 5-year period (b_x)	Product of l_x and b_x
0–9	5.0	0.9945	0	0.0000
10–14	12.5	0.9939	0.0020	0.0020
15–19	17.5	0.9929	0.1432	0.1422
20–24	22.5	0.9913	0.2855	0.2830
25–29	27.5	0.9896	0.2863	0.2833
30–34	32.5	0.9878	0.2160	0.2134
35–39	37.5	0.9851	0.0918	0.0904
40–44	42.5	0.9809	0.0175	0.0172
45–49	47.5	0.9743	0.0075	0.0073
50 +	—	—	0.0	0.00

$$R_0 = \sum_0^\infty l_x b_x = 1.0388$$

SOURCE: *Statistical Abstract of the United States 2007.*

(see Table 5), if the human population of the United States continued at these 2007 rates, it would multiply 1.039 times in each generation. If the net reproductive rate is 1.0, the population is replacing itself exactly; when the net reproductive rate is below 1.0, the population is not replacing itself; and if the rates in the example continue for a long time, the population will increase about 3.9% each generation in the absence of immigration or emigration. The net reproductive rate is illustrated in **Figure 9**.

Given these two schedules expressing the age-specific rates of survival and fertility, we may inquire at what rate a population subject to these rates would increase, assuming (1) that these rates remain constant and (2) that no limit is placed on population growth. Because these survival and fertility rates vary with age, the actual natality and mortality rates of the population will depend on the existing age distribution. If the whole population were over 50 years of age, it would not increase. Similarly, if all females were between 20 and 25, the rate of increase would be much higher than if they were all between 35 and 39. Before we can calculate the population's rate of increase, it would seem that we must specify (1) age-specific survival rates (l_x), (2) age-specific natality rates (b_x), and (3) age distribution.

This intuitive conclusion is not correct. Contrary to intuition, we do not need to know the age structure of

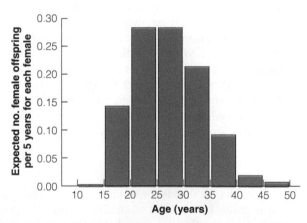

Figure 9 Expected number of female offspring per five-year period for each female in the United States in 2007. Data are from the final column in Table 5. The area under the histogram is the net reproductive rate R_0. (Data from the *Statistical Abstract of the United States 2007*.)

the population. Lotka (1922) showed that a population that is subject to a constant schedule of natality and mortality rates will gradually approach a fixed or stable age distribution, whatever the initial age distribution may have been, and will then maintain this age distribution indefinitely. This theorem is one of the most important discoveries in mathematical demography. When the population has reached this stable age distribution, it will increase in numbers according to the differential equation

$$\frac{dN}{dt} = rN \qquad (9)$$

or, as rewritten in integral form:

$$N_t = N_0 e^{rt} \qquad (10)$$

where N_0 = number of individuals at time 0

N_t = number of individuals at time t

e = 2.71828 (a constant)

r = intrinsic capacity for increase for the particular environmental conditions

t = time

This equation describes the curve of geometric increase in an expanding population (or geometric decrease to zero if r is negative).

A simple example illustrates this equation. Let the starting population (N_0) be 100 and let $r = 0.5$ per female per year. The successive populations would be:

Year	Population size
0	100
1	$(100)(e^{0.5}) = 165$
2	$(100)(e^{1.0}) = 272$
3	$(100)(e^{1.5}) = 448$
4	$(100)(e^{2.0}) = 739$
5	$(100)(e^{2.5}) = 1218$

This hypothetical population growth is plotted in **Figure 10**. Note that on a logarithmic scale the increase is linear, but on an arithmetic scale the curve swings upward at an accelerating rate.

To summarize to this point: (1) Any population subject to a fixed age schedule of natality and mortality will increase in a geometric way, and (2) this geometric increase will dictate a fixed and unchanging age distribution called the stable age distribution.

Let us invent a simple hypothetical organism to illustrate these points. Suppose that we have a parthenogenetic animal that lives three years and then dies. It produces two young at exactly one year of age, one young at exactly two years of age, and no young at year 3. The life table and fertility table for this hypothetical animal are thus extremely simple:

x	l_x	b_x	$l_x b_x$	$(x)(l_x)(b_x)$
0	1	0	0	0
1	1	2	2	2
2	1	1	1	2
3	1	0	0	0
4	0	–	–	–

$$R_0 = \sum_0^4 l_x b_x = 3$$

If a population of this organism starts with one individual at age 0, the population growth will be as shown in **Figure 11**, or, in tabular form, as follows:

	Number at ages				Total population size	% Age 0 in total population
Year	0	1	2	3		
0	1	0	0	0	1	100.0
1	2	1	0	0	3	66.7
2	5	2	1	0	8	62.50
3	12	5	2	1	20	60.00
4	29	12	5	2	48	60.42
5	70	29	12	5	116	60.34
6	169	70	29	12	280	60.36
7	408	169	70	29	676	60.36
8	985	408	169	70	1632	60.36

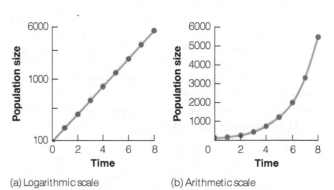

(a) Logarithmic scale (b) Arithmetic scale

Figure 10 Geometric growth of a hypothetical population when $N_0 = 100$ and $r = 0.5$, according to Equation (10). (a) On a logarithmic scale, geometric population growth appears as a straight line. (b) On an arithmetic scale, geometric population growth is a curve that rises more rapidly with time.

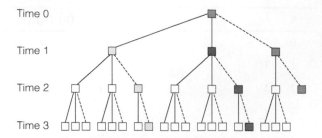

Figure 11 Population growth of a simple hypothetical organism that is parthenogenetic. Start at the top of the diagram with one green individual (each box represents one individual). At time 1 this individual gives birth to two young (yellow, red), so that there are now three individuals at time 1. At time 2 the two young individuals (red and yellow) give birth to two young each and the old green individual gives birth to one young, so at time 2 there are now eight individuals. The green individual then dies and the others reproduce, so that at time 3 there are 19 individuals. Solid lines indicate reproduction and dashed lines indicate the aging of individuals from one time to the next. Three of the individuals are color-coded to show their presence through time.

Note that the age distribution quickly becomes fixed or stable with about 60% at age 0, 25% at age 1, 10% at age 2, and 4% at age 3. This demonstrates Lotka's (1922) conclusion that a population growing geometrically develops a stable age distribution.

We may also use our hypothetical animal to illustrate how the intrinsic capacity for increase r can be calculated from biological data. The data of the l_x and b_x tables are sufficient to allow the calculation of r, the intrinsic capacity for increase in numbers. To do this, we first need to calculate the net reproductive rate (R_0), explained earlier. For our hypothetical animal, $R_0 = 3.0$, which means that the population can triple its size each generation. But how long is a generation? The **mean length of a generation** (G) is the mean period elapsing between the production or "birth" of parents and the production or "birth" of offspring. This is only an approximate definition, because offspring are produced over a period of time and not all at once. The mean length of a generation is defined approximately as follows (Dublin and Lotka 1925):

$$G_c = \frac{\sum l_x b_x x}{\sum l_x b_x} = \frac{\sum l_x b_x x}{R_0} \qquad (11)$$

For our model organism, $G = 4.0/3.0 = 1.33$ years. **Figure 12** uses the metaphor of a balance to illustrate the approximate meaning of generation time for a human population. Leslie (1966) has discussed some of the difficulties of applying the concept of generation time to a continuously breeding population with over-

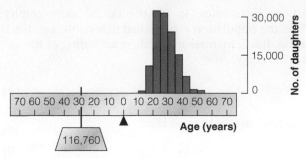

Figure 12 A mechanical balance to illustrate the idea of the mean length of one generation. Histogram of daughters from a cohort of 100,000 mothers starting life together (right side) is balanced by sum of total daughters (116,760) at exactly 28.46 years from the fulcrum. The mean length of a generation (G_c) is thus 28.46 years for these data. (Data from the U.S. population of 1920, $R_0 = 1.168$.)

lapping generations. For organisms such as annual plants and many insects with a fixed length of life cycle, the mean length of a generation is simple to measure and to understand.

Knowing the multiplication rate per generation (R_0) and the length of a generation (G), we can now determine r directly as an instantaneous rate:

$$r = \frac{\log_e(R_0)}{G} \qquad (12)$$

For our hypothetical organism,

$$r = \frac{\log_e(3.0)}{1.33} = 0.824 \text{ per individual per year}$$

Because the generation time G is an approximate estimate,[5] this value of r is only an approximate estimate when generations overlap.

The capacity for increase is an instantaneous rate and can be converted to the more familiar finite rate[6] by the formula

$$\text{Finite rate of increase} = \lambda = e^r \qquad (13)$$

The essay "Demographic Projections and Predictions" gives some examples of the utility and the difficulties of calculating the intrinsic capacity for increase of real world situations.

---

[5]Generation time has also been defined by Caughley (1977) as:

$$G_M = \frac{\sum (l_x b_x x e^{-rx})}{\sum (l_x b_x e^{-rx})}$$

This will not give exactly the same value for generation time as defined in Equation (11); see Gregory (1997).

Demographic Projections and Predictions

How much will the whooping crane population grow in the next two years? What will the AIDS epidemic do to the population of Africa between now and 2050? To answer questions such as these, we can use the demographic methods outlined in this chapter, but in doing so it is crucial that we make one subtle but important distinction: these methods can provide *projections*—that something will happen *if* conditions *a* and *b* are met—but not *predictions* that something will happen, period. (Scientists cannot predict the future; if you want a prediction, consult an astrologer or a Ouija board.) A *demographic projection* is a statement of what will happen to a population if certain assumptions are met, and demographic projections are correct only under very specific assumptions. A demographer can project population changes into the future, for example, on the assumption that the age-specific birth and death rates will remain constant. But in the real world the simple assumption that things will remain as they are now is rarely a correct one. Thus projections on the effects of AIDS on a population are most difficult because they require some uncertain assumptions about future death rates. Moreover, unpredictable changes such as catastrophic environmental events are especially damaging to demographic projections. No demographer can foresee the mortality to one flock of 18 whooping cranes caught in an episode of severe weather in Florida in February 2007.

In spite of the fact that they cannot predict the future, it is still useful for conservationists and resource managers to make projections of what will happen if specific assumptions are fulfilled. Such projections can limit our optimism and pessimism alike.

It should now be clear why the intrinsic capacity for increase in numbers cannot be expressed quantitatively except for a particular environment. Any component of the environment, such as temperature, humidity, or rainfall, might affect the natality and mortality rates and hence *r*.

Charles Birch, working at the University of Sydney, did some of the classic early research applying these quantitative demographic techniques to insects. One illustration of the effect of the environment on the capacity for increase was developed by Birch (1953a) in his work on *Calandra oryzae*, a beetle pest that lives in stored grain. The capacity for increase in this species varied with the temperature and with the moisture content of the wheat, as shown in **Figure 13**. The practical implications of these results are that wheat should be stored where it is cool and dry to prevent losses from *C. oryzae*.

In general, the intrinsic capacity for increase is not correlated with the abundance of species: species with a high *r* are not always common, and species with a low *r* are not always rare. Some species, such as the bison in North America, the elephant in central Africa, and the periodical cicadas, are (or were) quite common and yet have a low *r* value. Many parasites and other invertebrates with a high capacity for increase are nevertheless quite rare. Darwin (1859) pointed this out in *The Origin of Species*. From a conservation viewpoint species with a high *r* can recover more quickly from disturbances, and these calculations will permit us to calculate exactly how fast they might recover.

We can calculate how certain changes in the life history of a species would affect its capacity for increase in numbers. In general, three factors will increase *r*: (1) reduction in age at first reproduction, (2) increase in number of progeny in each reproductive event, and (3) increase in number of reproductive events (increased longevity). In many cases when *r* is large, the most profound effects are

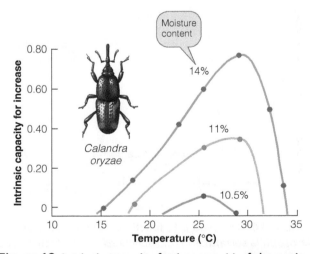

Figure 13 Intrinsic capacity for increase (*r*) of the grain beetle *Calandra oryzae* living in wheat of different moisture contents and at different temperatures. The higher the moisture content of the wheat, the more rapidly these beetles can increase in numbers. (After Birch 1953a.)

achieved by changing the age at first reproduction. For example, Birch (1948) calculated for the grain beetle *C. oryzae* the number of eggs needed to obtain $r = 0.76$ according to the age at first reproduction:

Age at which breeding begins (weeks)	Total no. eggs that must be laid to produce $r = 0.76$
1	15
2	32
3	67
4	141 (actual life history)
5	297
6	564

The earlier the peak in reproductive output, the larger the *r* value, as a rule. Lewontin (1965) provides an excellent example to illustrate this in *Drosophila serrata* (**Figure 14**). The Rabaul race of this fruit fly survives poorly and lays fewer eggs than the Brisbane race, but because it begins to reproduce at an earlier age (11.7 days compared with 16.0 days) and has a shorter generation length, its capacity for increase is equal to that of the longer-living, more fertile Brisbane race.

To conclude: The concept of an intrinsic capacity for increase in numbers is an oversimplification of na-

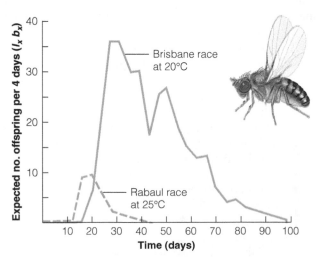

Figure 14 Observed $l_x b_x$ functions for two races of *Drosophila serrata*. Both $l_x b_x$ functions give the same value of the innate capacity for increase (*r*) because of the overriding importance of earlier reproduction and shorter generation length of the Rabaul race. Brisbane females lay an average of 546 eggs at 20°C, while Rabaul females lay only 151 eggs during their life span. (After Lewontin 1965.)

ture. In nature, we do not find populations with stable age distributions or with constant age-specific mortality and fertility rates. The actual rate of increase we observe in natural populations varies in more complex ways than the theoretical constant *r*. The importance of *r* lies mostly in its use as a model for comparison with the actual rates of increase we see in nature. The actual rate of increase along with its components in the life table and fertility table can be used in the diagnosis of environmental quality because they are sensitive to environmental conditions.

Reproductive Value

We can use life tables and fertility tables to determine the contribution to the future population that an individual female will make. We call this the **reproductive value** of a female aged *x* (Williams 1966), and this is most easily expressed for a population that is stable in size as follows:

$$\text{Reproductive value at age } x = V_x = \sum_{t=x}^{w} \frac{l_t b_t}{l_x} \quad (14)$$

where *t* and *x* are age and *w* is the age of last reproduction. Note that as defined here, reproductive value at age 0 is the same as net reproductive rate (R_0) as defined earlier in this chapter.

Reproductive value can be partitioned into two components (Pianka and Parker 1975):

Reproductive value at age *x* = present progeny + expected future progeny

$$V_x = b_x + \sum_{t=x+1}^{w} \frac{l_t b_t}{l_x} \quad (15)$$

We call the second term residual reproductive value, because it measures the number of progeny on average that will be produced in the rest of an individual's life span.

Reproductive value is more difficult to define if the population is not stable (Roff 1992; Fox et al. 2001). In this case we must discount future reproduction if population growth is occurring because the value of one progeny is less in a larger population. **Figure 15** illustrates the change of reproductive value with age in a red deer population in Scotland. Red deer stags defend harems, and their effective breeding span is three to five years between the ages of six and 11 years. By contrast, red deer hinds start to produce calves at age 3 and breed until they are 15 years old or older. These differences in reproductive biology explain the shapes of the reproductive value curves in Figure 15.

Reproductive value is important in the evolution of life-history traits. Natural selection acts more strongly

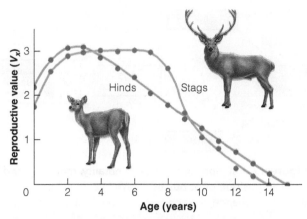

Figure 15 Reproductive value for red deer stags (males) of different ages, compared with that of hinds (females) on the island of Rhum, Scotland. Reproductive value is calculated in terms of the number of female offspring surviving to one year of age that parents of different ages can expect to produce in the future. (From Clutton-Brock et al. 1982.)

on age classes with high reproductive values and very weakly on age classes with low reproductive values. Predators will have a greater effect on a population if they prefer individuals of high reproductive value.

Age Distributions

We have already discussed the idea of age distribution in connection with the intrinsic capacity for increase. We noted that a population growing geometrically with constant age-specific mortality and fertility rates would assume and maintain a stable age distribution. The stable age distribution can be calculated for any set of life tables and fertility tables. The stable age distribution is defined as follows:

C_x = proportion of organisms in the age category x to $x + 1$ in a population increasing geometrically

Mertz (1970) has shown that:

$$C_x = \frac{\lambda^{-x} l_x}{\sum\limits_{i=0}^{\infty} \lambda^{-i} l_i} \tag{16}$$

where $\lambda = e^r$ = finite rate of increase
l_x = survivorship function from life table
x, i = subscripts indicating age

Let us go through these calculations with our hypothetical organism from Figure 11:

$\lambda = e^r = e^{0.881} = 2.413$			
Age (x)	l_x	λ^{-x}	$\lambda^{-x} l_x$
0	1.0	1.0000	1.00001
1	1.0	0.4144	0.4144
2	1.0	0.1717	0.1717
3	1.0	0.0711	0.0711
4	0.0	0.0295	0.0000

$$\sum_{x=0}^{4} \lambda^{-x} l_x = 1.6572$$

Thus to calculate C_0, the proportion of organisms in the age category 0 to 1 in the stable age distribution, we have

$$C_x = \frac{\lambda^{-0} l_0}{\sum\limits_{i=0}^{4} \lambda^{-i} l_i} = \frac{(1.0)(1.0)}{1.6572} = 0.6035$$

For C_1, we have

$$C_1 = \frac{\lambda^{-1} l_1}{\sum\limits_{i=0}^{4} \lambda^{-i} l_i} = \frac{(0.4144)(1.0)}{1.6572} = 0.250$$

In a similar way,

$$C_2 = 0.104$$
$$C_3 = 0.043$$

Compare these calculated values with those obtained empirically earlier. Carey (1993) illustrates another method of calculating the stable age distribution for a set of l_x and b_x schedules.

Populations that have reached a constant size, in which the fertility rate equals the mortality rate, will also assume a fixed age distribution, called a stationary age distribution (or *life-table age distribution*) and will maintain this distribution. The stationary age distribution is a hypothetical one and illustrates what the age composition of the population would be at a particular set of mortality rates (q_x) if the fertility rate were exactly equal to the mortality rate. **Figure 16** contrasts the stable and stationary age distributions for the short-tailed vole in a laboratory colony.

A constant age structure in a population is attained only if the l_x and b_x distributions are fixed and unchanging. This typically occurs in only two situations: (1) When the age-specific fertility and mortality rates are fixed and unchanging and the population grows exponentially, the population assumes a constant age structure called the stable age distribution; and (2) when the fertility rate

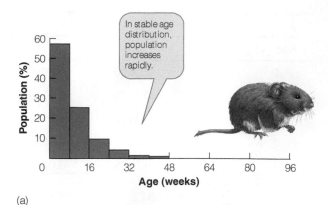

(a)

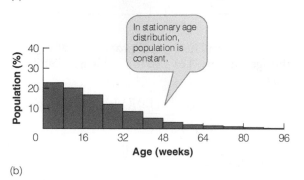

(b)

Figure 16 Age Distributions. (a) Stable age distribution and (b) stationary age distribution for the vole *Microtus agrestis* in the laboratory. The stable age distribution should be observed when populations are growing rapidly, and the stationary age distribution when populations are constant in size. (After Leslie and Ranson 1940.)

exactly equals the mortality rate and the population does not change in size over time, the population assumes a constant age structure called the stationary age distribution, which has the same form as the l_x distribution. Under any other circumstances, the population's age structure is not a constant but changes over time. In natural populations, the age structure is thus almost constantly changing. We rarely find a natural population that has a stable age structure because populations do not increase for long in an unlimited fashion. Nor do we often find a stationary age distribution because populations are rarely in a stationary phase for long. We illustrate these relationships in **Figure 17**.

With proper care, information on age composition can be used to judge the status of a population. Increasing populations typically have a predominance of young organisms, whereas constant or declining populations do not (see Figure 16). The same principles apply to human populations. In 2006 the Sudan's population was growing at 2.6% per year and had 44% of its population less than 15 years old, and 2% over 65 years of age. The comparable figures for Canada (growing at 0.3% per year) were 18% less than 15 years, and

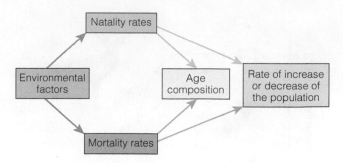

Figure 17 Relationships between natality, mortality, and age composition of populations. When either or both of these rates change, the age composition must also change.

13% over 65 years. The age structure of human populations has been analyzed in detail because of its economic and sociological implications (Weeks 1996). A country with a high fertility rate and a large proportion of children (such as the Sudan) has a much greater demand for schools and other child services than do countries such as Canada and the United States, with a lower proportion of people under age 15.

In populations of plants and animals, even more variation in age composition is apparent. In long-lived species such as trees and fishes, one may find dominant year-classes. **Figure 18** illustrates this for Engelmann spruce and subalpine fir trees of the Rocky Mountains, in which some year-classes may be 100 times as numerous as others. In these situations, the age composition can change greatly from one year to the next. Eberhardt (1988) discusses the use of age composition information in the management of wildlife populations, and Walters and Martell (2004) discuss this problem in exploited fish populations.

Evolution of Demographic Traits

We can use the demographic techniques just described to investigate one of the most interesting questions of evolutionary ecology: Why do organisms evolve one type of life cycle rather than another? Only certain kinds of l_x and b_x schedules are permissible if a population is to avoid extinction. How does evolution act, within the framework of permissible demographic schedules, to determine the life cycle of a population?

Pacific salmon grow to adult size in the ocean and return to fresh water to spawn once and die. We may call this **big-bang reproduction**.[7] Oak trees may

---

[6]Big-bang reproduction = semelparity, and repeated reproduction = iteroparity, for those who prefer the more classical terms derived from Greek roots.

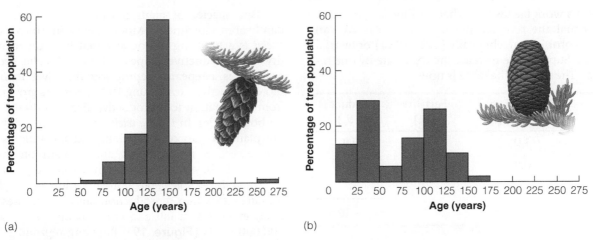

Figure 18 Age structure. (a) Engelmann spruce and (b) subalpine fir in a forest stand at 3150 m elevation in northern Colorado. Neither of these tree species has an age distribution like those shown in Figure 16 for stable or stationary age distributions. (Data from Aplet et al. 1988.)

become mature after 10 or 20 years and drop thousands of acorns for 200 years or more. We call this **repeated reproduction**. How have these life cycles evolved? What advantage might be gained by salmon that breed more than once, or by oak trees that drop only one set of seeds and then die?

The population consequences of life cycles were first explored by Cole (1954), who asked a simple question: What effect does repeated reproduction have on the intrinsic capacity for increase (r)? Assume that we have an annual species that produces offspring at the end of the year and then dies, has a simple survivorship of 0.5 per year, and has a fertility rate of 20 offspring. The life table for this species is as follows:

Age (x)	Proportion surviving (l_x)	Fertility (b_x)	Product ($l_x b_x$)
0	1.0	0	0
1	0.5	20	10
2	0.0	–	0
			$R_0 = 10.0$

The net reproductive rate (R_0) is 10.0, which means that the species could increase 10-fold in one generation (= 1 year). We can determine r from the characteristic equation of Lotka:

$$r = (\log_e R_0)/G \qquad (17)$$

where R_0 is the net reproductive rate defined in Equation (7) and G is generation time in Equation (11). From this equation we determine that $r = 2.303$ per year for the annual species with big-bang reproduction. What ad-

vantage could this species gain by continuing to live and reproduce at years 2, 3, ... ∞. Let us assume the most favorable condition, no mortality after age 1 and survival to age 100. The life table now becomes the following:

Age (x)	Proportion surviving (l_x)	Fertility (b_x)	Product ($l_x b_x$)
0	1.0	0.0	0.0
1	0.5	20	10.0
2	0.5	20	10.0
3	0.5	20	10.0
4	0.5	20	10.0
5	0.5	20	10.0
–	–	–	–
–	–	–	–
–	–	–	–
99	0.5	20.0	10.0
100	0.0	0.0	0.0
$R_0 = \sum l_x b_x = 990.0$			

In the manner outlined above, we determine that $r = 2.398$ for the perennial species with repeated reproduction. If we adopt repeated reproduction in our hypothetical organism, we raise the intrinsic capacity for increase only about 4%:

$$\frac{2.398}{2.303} = 1.04$$

Now let us work backward. What fertility rate at year 1 would equal the r of the perennial (2.398)? We can solve this problem algebraically (Cole 1954) or by trial and error. Suppose we increase the birth rate by one individual. The annual life table is now:

Age (x)	Proportion surviving (l_x)	Fertility (b_x)	Product ($l_x b_x$)
0	1.0	0.0	0.0
1	0.5	21.0	10.5
2	0.0	–	0
			$R_0 = 10.5$

This is almost the gain achieved by repeated reproduction. If we increase the fertility rate by two individuals, we get $r = 2.398$ per year, equal to the r for the perennial. This is obviously an ideal case, because we assume no mortality after age 1 in the perennial form. Cole (1954) generalized this ideal case to a surprising conclusion: for an annual species, the maximum gain in the intrinsic capacity for increase (r) that could be achieved by changing to the perennial reproductive habit would be equivalent to adding one individual to the effective litter size $(l_x b_x$ for age 1$)$. Cole assumed for his ideal case perfect survival to reproductive age (Charnov and Schaffer 1973). In our hypothetical example we assumed that half of the organisms die before reaching reproductive age.

This simple model for the evolution of big-bang reproduction is unrealistic because it is a "cost-free" model: present reproduction is assumed to have no effect on future reproduction or future survival (Roff 1992). Let us assume that an organism can "decide" how much of its resources it will devote to reproduction. If it uses all its resources to reproduce, it will die and thus be a big-bang reproducer. Big-bang reproduction will be favored if the greater benefits of reproduction come only at high levels of reproductive effort; conversely, if good reproductive success can be achieved at low levels of effort, organisms will be selected to be repeat reproducers. A trade-off between reproductive effort and reproductive success is implied in the reproductive effort model. The key demographic effect of big-bang reproduction is higher reproductive rates. Plants that reproduce only once typically produce 2 to 5 times as many seeds as closely related species that reproduce repeatedly (Young 1990). Repeated reproduction can also be favored when adult survival rates are high and juvenile survival is highly variable. The critical division between big-bang reproduction and repeated reproduction is set by the survival rate of the juvenile stages. If survival of juveniles is very poor or unpredictable, selection will usually favor repeated reproduction (Roff 1992). Let us look at one example to illustrate this theory.

Two species of giant rosette plants occur abundantly above treeline on Mount Kenya in Africa. *Lobelia telekii* is a big-bang reproducer that lives on relatively dry, less productive slopes, whereas *Lobelia deckenii keniensis* is a repeated reproducer that lives in moist, more productive sites (Young 1990). Rosettes grow slowly from germination to reproductive size over 40–60 years for both species. In *Lobelia telekii* the resources of the entire plant go into reproduction, and the inflorescence may exceed 3 m in height and contain on average 500,000 seeds. After reproduction the entire plant dies. In *Lobelia deckenii keniensis* only a portion of the plant's resources goes into reproduction, and the inflorescence rarely exceeds 1 m tall and contains on average about 200,000 seeds (**Figure 19**). Big-bang reproduction in

Figure 19 Two species of *Lobelia* from Mount Kenya. *Lobelia telekii* is a big-bang reproducer that grows over 2 m tall, seeds once, and dies. *Lobelia deckenii* is a smaller plant that produces seeds several times throughout its life.

Lobelia telekii is favored by a high mortality rate of adult plants in the dry sites where they live so that it is unlikely after flowering that an individual plant will survive 10 or more additional years for a second flowering season, and there is an evolutionary payoff in the greater fecundity of big-bang reproduction.

Some of the best examples of the evolution of life history strategies come from studies within a single species. Capelin are a good example because males are big-bang reproducers while females are repeated reproducers. Capelin, small (15–25 g), sardine-like, pelagic fish with a circumpolar arctic distribution, form an important part of the food chain for seals, seabirds, and other fish such as cod. Males have adopted the big-bang strategy because each male can mate with several females during a spawning season and because male mortality is very high after spawning (Huse 1998). Female capelin are limited by the number of eggs they can carry, and they can improve their reproductive success only by spawning several times at yearly intervals.

Much interest in life history evolution has centered on determining the costs of reproduction. Reproductive effort at any given age can be associated with a biological cost and a biological profit. The biological cost derives from the reduction in growth or survival that occurs as a consequence of using energy to reproduce. For example, the more seeds a meadow grass (*Poa annua*) plant produces in one year, the less it grows the following year (Law 1979). Fruit fly (*Drosophila melanogaster*) females that mate often typically live shorter lives than females that mate less often (Fowler and Partridge 1989). The biological profit associated with reproduction is measured in the number of descendants left to future generations, which will be affected by the survival rate and the growth rate. The hypothetical organism must in effect ask at each age: Should I reproduce this year, or would I profit more by waiting until next year? Obviously, if the mortality rate of adults is high, it would be best to reproduce as soon as possible. But if adult mortality is low, it

may pay for an organism to put its energy into growth and wait until the next year to reproduce.

Many organisms do not reproduce as soon as they are physiologically capable of doing so. The key quantity that we must measure to predict the optimal age at maturity is the potential fecundity cost (Bell 1980). Individuals that reproduce in a given year will often be smaller and less fecund in the following year than an individual that has previously abstained from reproduction. This is best established in poikilotherms, such as fishes, that show a reduction in growth associated with spawning. Potential fecundity costs also occur in homeotherms (Clutton-Brock et al. 1989), and the period of lactation in mammals is energetically very expensive for females (**Figure 20**). Social behavior associated with reproduction can produce great differences in the costs associated with breeding in the two sexes and thus cause differences in the optimal age at maturity for males and females. Red deer stags, for example, defend harems and attain a breeding peak after seven years of age through their fighting ability. Females mature at three years and live longer than males.

Repeated reproducers must "decide" in an evolutionary sense to increase, decrease, or hold constant their reproductive effort with age. In every case analyzed so far, reproductive effort increases with age but only to the age of senescence (Berube et al. 1999; Weladji et al. 2002). **Figure 21** illustrates two examples of a senescent decline in reproduction in large mammals. The senescence hypothesis seems to apply equally well to reproduction as it does to mortality (cf. Figure 8).

Why do species expend the effort to have repeated reproduction? The answer seems to be that repeated reproduction is an adaptation to something other than achieving maximum reproductive output. Repeated reproduction may be an evolutionary response to uncertain survival from zygote to adult stages (Roff 1992). The greater the uncertainty, the higher the selection for a longer reproductive life. This may involve channeling

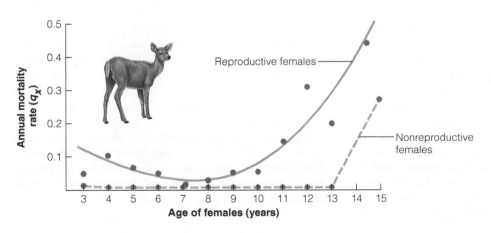

Figure 20 Cost of reproduction in female red deer on the island of Rhum in Scotland. Mortality in winter is always higher in females that reproduced during the previous summer, no matter the age of the female. (After Clutton-Brock et al. 1982.)

145

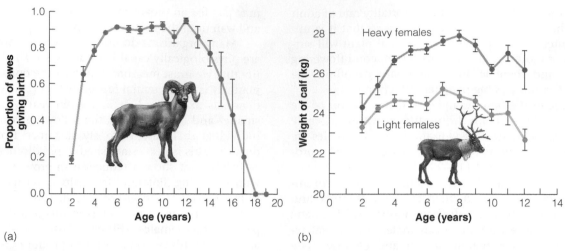

Figure 21 Age-related reproductive effort in (a) Rocky Mountain bighorn sheep and (b) Norwegian reindeer. In both species an age of senescence is clearly demarcated by a decline in reproductive effort. In this sheep population, reproduction declined after age 12. In reindeer reproduction declined after age 8 in larger females and age 7 in lighter females. (Data from Berube et al. 1999 and Weladji et al. 2002.)

more energy into growth and maintenance, and less into reproduction. Thus, we can recognize a simple scheme of possibilities:

	Long life span	Short life span
Steady reproductive success	?	Possible
Variable reproductive success	Possible	Not possible

We now believe that the advantage of repeated reproduction is that it spreads the risk of reproducing over a longer time period and thus acts as an adaptation that thwarts environmental fluctuations.

Limitations of the Population Approach

Two fundamental limitations restrict the methods used for studying populations. First, how can we determine what constitutes a population for any given species? What are the boundaries of a population in space? In some situations the boundaries are clear. Wildebeest populations in the Serengeti area of East Africa form five herds that rarely exchange members (Sinclair 1977). The largest herd is highly migratory and moves seasonally between plains and woodlands following the rainy season. The four smaller herds are less migratory and breed in different areas and at slightly different times of the year.

But in many other cases, organisms are distributed in a continuum, and no boundaries are evident. White

spruce trees grow in northern coniferous forests from Newfoundland to Alaska. Do all these white spruce trees belong to one population? Most population biologists would answer no to this question, but the reasons for their answer would likely differ. Part of the definition of a population should involve the probability of genetic exchange between members of a given population, but no one is able to specify this probability in a rigorous manner. Moreover, some species are asexual, and we are left with the same general problem that troubles systematists when they seek to determine what constitutes a species. One pragmatic answer we can give is that a population is a group of individuals that a population biologist chooses to study. To say this is only to say that we may have to start our study by making a completely arbitrary decision on what to call a population. To remember this decision after the population study is completed is a mark of ecological wisdom.

The second limitation is that populations do not exist as isolates but are imbedded in a community matrix of associated species. When we study a population, we assume that we can abstract from the whole community a single species and the small number of species that interact with it. Whether this abstraction can be done effectively is controversial, but for the moment we will proceed with the assumption that a population can be isolated from the complex tapestry of a biological community. We must keep in mind that just as communities are made of populations, populations are made of individuals.

Composition of Populations

Populations are indeed composed of individuals, but not of a series of identical individuals; yet we tend to forget this heterogeneity when considering population density. Three major variables distinguish individuals in many populations: sex, age, and size. The composition of many populations deviates from the expected sex ratio of 50% of each gender, so we cannot assume a constant 1:1 (50%) sex ratio. Populations of the common lizard *Lacerta vivipara* in France average 39% males (Le Galliard et al. 2005). Sex ratios of many vertebrate species are adjusted by breeding females in relation to the available resources (Johnson et al. 2001). The sex ratio of a population clearly affects the potential reproductive rate and the level of inbreeding, and it may affect social interactions in many vertebrates (Wolff et al. 2002).

Age is a significant variable in human populations, and age effects are common to many species. Older individuals are frequently larger, and changes in size may be the main mechanism by which age effects occur. Larger fish lay many more eggs than smaller fish, and larger plants produce more seeds. Young mammals may be prone to diseases that older animals can resist. Age and size are very significant individual attributes in all animals that have social organization because they help to specify an individual's social position. Old individuals in some species may be post-reproductive, as they are in humans.

Size is a particularly significant variable in modular organisms. Size and age may be correlated in plants and animals with indeterminate growth, but much individual variation in size is independent of age. For modular plants and animals size is usually the ecologically relevant variable that defines their importance in ecological processes.

Other, secondary variables may distinguish individuals in some populations. Color is one obvious trait. Social insects such as ants have castes with distinctive individual morphology. Many phenotypic traits can affect survival, reproduction, or growth and be important to a population.

Because of these individual differences, population ecology must be seen to have a split personality. While we view populations as aggregates of individuals and calculate population density and other population parameters as averages over all individuals, we also recognize that to explain population processes we must understand the individuals that make up the population, and the mechanisms by which they reproduce, move, and die. Not all individuals are equal in a population, and individual variability affects the mechanisms behind population processes. We will flip back and forth between this dual view of populations as statistical averages and as mixtures of heterogeneous individuals as we try to understand population dynamics.

Summary

Populations are composed of organisms that may be unitary or modular. Most animals are unitary with determinate form—cats all have four legs. In most unitary organisms the individual is easily recognized, and each individual is an independent genetic unit. Most plants are modular with repeated units of construction (modules)—oak trees can have any number of leaves. Individuals may be difficult to recognize in modular organisms, and a genetic individual (*genet*) may be a clone with many separate modules (*ramets*). Abundance in unitary organisms is typically a count of the number of individuals, while in modular organisms abundance is often measured by biomass or cover.

Every aspect of the problem of abundance comes down to one crucial question: How can we estimate population abundance? Absolute abundance can be estimated by total counts or by sampling methods using quadrats or capture-recapture methods. Relative abundance can be estimated by many techniques, depending on the species studied. Once we obtain estimates of population size, we can investigate changes in numbers by analyzing the four primary demographic parameters of natality, mortality, immigration, and emigration.

Population changes can be analyzed with a set of quantitative techniques first developed for human population analysis. A life table is an age-specific summary of the mortality rates operating on a population. Life tables are necessary because mortality does not fall equally on all ages, and in most species the very young and the old suffer high mortality.

A fertility schedule that summarizes reproduction with respect to age can describe the reproductive component of population increase. The **intrinsic capacity for increase** of a population is obtained by combining the life table and the fertility schedule for specified environmental conditions. This concept leads to an important demographic principle: a population that is subject to a constant schedule of mortality and natality rates will (1) increase in numbers geometrically at a rate equal to the capacity for increase (r), (2) assume a fixed or stable age distribution, and (3) maintain this age distribution indefinitely. The age distribution of a population is constant and unchanging only as long as the life table and the fertility table remain constant.

Demographic techniques are useful for exploring quantitatively the consequences of adopting an annual life cycle versus a perennial one. Very little gain in potential for population increase occurs in species that reproduce many times in each generation, and repeated reproduction seems to be an evolutionary response to conditions in which survival from zygote to adult varies unpredictably from good to poor. An organism thus "hedges its bets" by reproducing several times.

Review Questions and Problems

1. Canada lynx are now listed as a threatened species in the contiguous 48 United States. Given an average body mass of 9.7 kg, calculate what population density you would expect for lynx from the global relationship for mammalian carnivores in Table 2.

2. One technique for estimating the springtime abundance of sheep ticks in Scotland is by dragging a wool blanket over the grass. (Ticks will cling to anything that brushes against them during the spring.) Does this technique measure absolute density or relative density? How might you determine this?

3. Compare the definition of *population* presented here with that used in statistics texts (see Sokal and Rohlf 1995, p. 9; Zar 1999, p. 15) and in evolution texts (see Futuyma 2005).

4. Suggest three hypotheses to explain why, in Figure 3, birds should in general exist at a lower density than mammals of the same size. Make predictions from each hypothesis and discuss how the predictions could be tested. Silva et al. (1997) discuss this general problem.

5. What are the problems of using the oldest living human to estimate the upper limit to human longevity? What other approaches might you use to estimate potential longevity? Is there much scope for increasing human longevity in the developed countries? Read Litzgus (2006) for an analysis of this question.

6. In human populations in developed countries women generally outlive men by a margin of 5–10 years. Is this advantage in female longevity a general characteristic of nonhuman animal species as well? What might explain such a pattern in humans and other animals? Cohen (2004) reviews this question, and Carey et al. (1995) provide data on fruit flies to test this generalization.

7. What additional data, if any, are required to determine the stable age distribution for the human population described in Table 5?

8. The life table and the seed production of the winter annual plant *Collinsia verna* for 1983–84 was as follows (Kalisz 1991):

Life cycle	Age interval (months)	Number alive n_x	Average no. seeds produced per plant b_x
Seed	0–5	23,061	0
Seedling	5–7	6019	0
Overwintering plants	7–12	4617	0
Flowering plants	12–13	2612	0
Fruiting plants	13–14	692	10.754

Calculate the net reproductive rate for these plants and discuss the biological interpretation of this rate.

9. Forest ecologists usually measure the *size* structure of a forest and less often make use of the annual rings of temperate-zone trees to get the age structure of the forest. What might one learn from determining age structure in addition to size structure in a forest stand?

10. Can the reproductive value of males and females at a given age differ? Discuss the data presented on red deer in Clutton-Brock et al. (1982, p. 154) as an example.

Overview Question

A life table and a fertility schedule are available for a species of threatened plant. If you were in charge of a management plan for this plant species, what could you conclude from these two tables, and what further demographic information would you want to have?

Suggested Readings

- Boyce, M. S., et al. 2006. Demography in an increasingly variable world. *Trends in Ecology & Evolution* 21:141–148.

- Bronikowski, A. M., et al. 2002. The aging baboon: Comparative demography in a non-human primate. *Proceedings of the National Academy of Sciences of the USA* 99:9591–9595.

- Caughley, G. 1977. *Analysis of Vertebrate Populations*. Chapters 7–9. New York: Wiley.

- Cohen, A. A. 2004. Female post-reproductive lifespan: A general mammalian trait. *Biological Reviews* 79:733–750.

- Erickson, G. M., P. J. Currie, B. D. Inouye, and A. A. Winn. 2006. Tyrannosaur life tables: An example of nonavian dinosaur population biology. *Science* 313:213–217.

- Roff, D. A. 1992. *The Evolution of Life Histories: Theory and Analysis*. New York: Chapman and Hall.

- Sandercock, B. K., K. Martin, and S. J. Hannon. 2005. Life history strategies in extreme environments: comparative demography of arctic and alpine ptarmigan. *Ecology* **86**:2176–2186.

- Sibly, R. M., J. Hone, and T. H. Clutton-Brock. 2003. *Wildlife population growth rates.* Cambridge: Cambridge University Press.

- Sutherland, W. J., ed. 2006. *Ecological Census Techniques: A Handbook.* Cambridge: Cambridge University Press.

- Weladji, R. B., A. Mysterud, Ø. Holand, and D. Lenvik. 2002. Age-related reproductive effort in reindeer (*Rangifer tarandus*): Evidence of senescence. *Oecologia* 131:79–82.

Credits

Ilustration and Table Credits

Photo Credits

Population Growth

Key Concepts

- Population growth can be described with simple mathematical models both for organisms with discrete generations and for organisms with overlapping generations.

- Simple models for discrete generations can lead to complex dynamics from a stable equilibrium to cycles and chaotic fluctuations in numbers.

- For species with overlapping generations, the logistic equation is a simple mathematical description of population growth to an asymptotic limit.

- Natural populations often grow rapidly, but then density fluctuates widely rather than remaining at an equilibrium density.

- More complex and more realistic models of population growth incorporate time lags and chance into population growth models.

From Chapter 9 of *Ecology: The Experimental Analysis of Distribution and Abundance*, Sixth Edition. Eugene Hecht.

competition Occurs when the number of organisms of the same or different species utilize common resources that are in short supply or when the organisms harm one another in the process of acquiring these resources.

disease A pathological condition of an organism resulting from various causes, such as an infection, a genetic disorder, or environmental stress, with specific symptoms.

herbivory The eating of parts of plants by animals, not typically resulting in plant death.

Leslie matrix model A method of casting the age-specific reproductive schedule and the age-specific mortality schedule of a population in matrix form so that predictions of future population change can be made.

logistic model A specific population growth model based on the logistic equation that predicts an S-shaped population growth curve.

matrix models A family of models of population change based on matrix algebra, with the Leslie matrix model being the best known.

population regulation The general problem of what prevents populations from growing without limit, and what determines the average abundance of a species.

predation The action of one organism killing and eating another.

probabilistic models In contrast to deterministic models, including an element of probability so that repeated runs of the models do not produce exactly the same outcome.

theta-logistic model The modification of the original logistic equation to permit curved relationships between population density and the rate of population increase.

Population growth is a central process of ecology. But no population goes on growing forever, and this leads us to the problem of **population regulation**. Because species interactions such as **predation**, **competition**, **herbivory**, and **disease** affect population growth, and population growth produces changes in community structure, it is important to understand how population growth occurs.

The demographic techniques described earlier are useful because they permit us to project future changes in population density in a precise manner. In this chapter we will apply these demographic parameters to the description of population growth and explore some of the difficulties of analyzing the growth of natural populations. To illustrate these methods' utility, we will use them to address a practical problem in conservation biology.

Mathematical Theory

A population that has been released into a favorable environment will begin to increase in numbers. What form will this increase take, and how can we describe it mathematically? We start by considering a simple case in which generations are separate, as in univoltine insects (one generation per year) or annual plants.

Growth in Populations with Discrete Generations

Consider a species with a single annual breeding season and a life span of one year. Let each reproductive unit in the population on average produce R_0 offspring that survive to breed in the following year. Then

$$N_{t+1} = R_0 N_t \qquad (1)$$

where N_t = population size at generation t
N_{t+1} = population size at generation $t + 1$
R_0 = net reproductive rate, or population growth rate per generation

Note that R_0 is the net reproductive.

What happens to this population will very much depend on the value of R_0. Consider two cases:

1. **Multiplication rate constant.** Let R_0 be a constant. If $R_0 > 1$, the population increases geometrically without limit; if $R_0 < 1$, the population decreases to extinction. For example, let $R_0 = 1.5$ and $N_t = 10$ when $t = 0$:

Generation	Population size (N_t)
0	10
1	$15 = (1.5)(10)$
2	$22.5 = (1.5)(15)$
3	$33.75 = (1.5)(22.5)$

Figure 1 shows some examples of geometric population growth with different R_0 values.

2. **Multiplication rate dependent on population size.** Populations do not normally grow with a

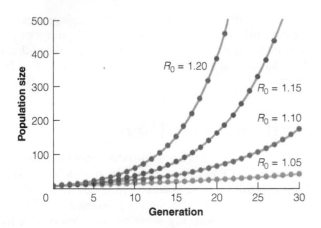

Figure 1 Geometric or exponential population growth, discrete generations, population growth rate (R_0) constant. Starting population is 10. Equation (1).

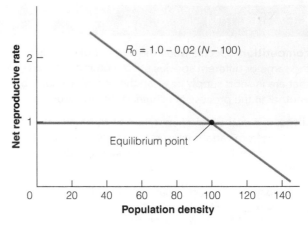

Figure 2 Net reproductive rate (= population growth rate) (R_0) as a linear function of population density (N) at time t. In this hypothetical example, equilibrium density is 100 (black dot). The red line marks the equilibrium line at which $R_0 = 1.0$ and the population remains constant. The blue line is an example of the relationship given in Equation (3).

constant multiplication rate as in Figure 1. If we look at the trajectory of a species population through time, we observe a variety of dynamics, including populations that fluctuate very little, others that fluctuate chaotically, and still others that fluctuate in cycles. How can we explain this variety of dynamic behavior?

The simplest way is to assume that the multiplication rate changes as population density rises and falls. At high densities, birth rates will decrease or death rates will increase from a variety of causes, such as food shortage or epidemic disease. At low densities birth rates will be high and losses from diseases and natural enemies low. We need to express the way in which the multiplication rate slows down as density increases. The simplest mathematical model is linear: Assume that there is a straight-line[1] relationship between the density and multiplication rates such that the higher the density, the lower the multiplication rate (**Figure 2**). The point where the line crosses $R_0 = 1.0$ is a point of equilibrium in population density at which the birth rate equals the death rate. It is convenient to measure population density in terms of deviations from this equilibrium density, expressed as

$$z = N - N_{eq} \tag{2}$$

where z = deviation from equilibrium density
N = observed population size
N_{eq} = equilibrium population size (where $R_0 = 1.0$)

---

[1] A straight line is described by the equation $y = a + bx$, where b is the slope and a is the y-intercept (the y value when $x = 0$).

The equation of the straight line shown in Figure 2 is thus

$$R_0 = 1.0 - B(N - N_{eq}) \tag{3}$$

where R_0 = net reproductive rate or rate of population growth per generation
$(-)B$ = slope of line

In Figure 2, $B = 0.02$ and $N_{eq} = 100$. Equation (1) can now be written

$$N_{t+1} = R_0 N_t = (1.0 - Bz_t)N_t \tag{4}$$

The properties of this equation depend on the equilibrium density and the slope of the line. Let us work out a few examples to illustrate this. Consider first a simple example in which $B = 0.011$ and $N_{eq} = 100$. Start the population at $N_0 = 10$:

$$N_1 = [1.0 - 0.011(10 - 100)]10$$
$$= (1.99)(10) = 19.9$$
$$N_2 = [1.0 - 0.011(19.9 - 100)]19.9$$
$$= (1.881)(19.9) = 37.4$$

Similarly,

$$N_3 = 63.2$$
$$N_4 = 88.8$$
$$N_5 = 99.7$$

and the population density converges smoothly toward the equilibrium point of 100. A second example is worked out in **Table 1**, and three additional examples are plotted in **Figure 3**.

Table 1 Growth of a hypothetical population with discrete generations and net reproductive rate that is a linear function of density.

Population sizes are calculated from Equation (4) using $B = 0.025$, $N_{eq} = 100$, and a starting density of 50 individuals.

General formula: $N_{t+1} = [1.0 - 0.025(N_t - 100)]N_t$

$$N_1 = [1.0 - 0.025(50 - 100)]50$$
$$= (2.25)(50) = 112.5$$

$$N_2 = [1.0 - 0.025(112.5 - 100)]112.5$$
$$= (0.6875)(112.5) = 77.34$$

$$N_3 = [1.0 - 0.025(77.34 - 100)]77.34$$
$$= (1.5665)(77.34) = 121.15$$

$$N_4 = [1.0 - 0.025(121.15 - 100)]121.15$$
$$= (0.4712)(121.15) = 57.09$$

Similarly,

$$N_5 = 118.33$$
$$N_6 = 64.10$$
$$N_7 = 121.63$$
$$N_8 = 55.86$$

The population continues to oscillate in a stable two-generation cycle.

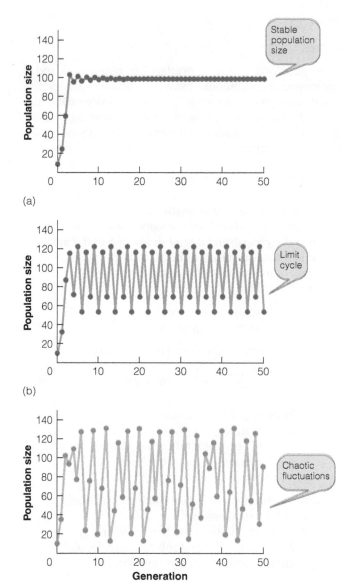

(a)

(b)

(c)

Figure 3 Examples of population growth with discrete generations and multiplication rate as a linear function of population density as in Figure 2. Starting density is 10 and equilibrium density is 100. Three examples with different slopes are shown: (a) For $B = 0.018$, the population shows convergent oscillations to equilibrium density at 100. (b) For $B = 0.025$, the population oscillates in a two-generation limit cycle. (c) For $B = 0.029$, the population fluctuates chaotically in an irregular pattern that never repeats itself.

The behavior of this simple population model is very surprising because it generates many different patterns of population changes. If we define $L = BN_{eq}$, then:

- If L is between 0 and 1, the population approaches the equilibrium without oscillations.

- If L is between 1 and 2, the population undergoes oscillations of decreasing amplitude to the equilibrium point (*convergent oscillations*) (see Figure 3a).

- If L is between 2 and 2.57, the population exhibits stable limit cycles that continue indefinitely (see Figure 3b).

- If L is above 2.57, the population fluctuates chaotically showing what appear to be random changes, depending on the starting conditions (Maynard Smith 1968; May 1974a) (see Figure 3c).

Much of this mathematical theory of population growth was clarified and elaborated by the mathematical ecologist Robert May working at Princeton University and later at Oxford University. The fact that such a simple population model can produce such a diversity of population growth trajectories is one of the most surprising results found by twentieth-century mathematical ecologists. This model, in which the net reproductive rate decreases in a linear way with density, is the discrete-generation version of the logistic equation described in the next section, in which we consider populations with overlapping generations.

Growth in Populations with Overlapping Generations

In populations that have overlapping generations and a prolonged or continuous breeding season, we can describe population growth more easily by using differential equations. As earlier, we will assume for the moment that the growth of the population at time t depends only on conditions at that time and not on past events of any kind.

1. **Multiplication rate constant.** Assume that, in any short time interval Δt (usually written as dt), an individual has the probability $b\,dt$ of giving rise to another individual. In the same time interval, it has the probability $d\,dt$ of dying. If b and d are instantaneous rates[2] of birth and death, the instantaneous rate of population growth per capita will be

$$\text{instantaneous rate of population growth} = r = b - d \qquad (5)$$

and the form of the population increase is given by

$$\frac{dN}{dt} = rN = (b - d)N \qquad (6)$$

where N = population size
t = time
r = per capita rate of population growth
b = instantaneous birth rate
d = instantaneous death rate

This is the curve of geometric increase in an unlimited environment.

Note that we can use the geometric growth model to estimate the doubling time for a population growing at a certain rate:

$$\frac{N_t}{N_0} = e^{rt} \qquad (7)$$

But if the population doubles, $N_t/N_0 = 2$. Thus

$$\frac{N_t}{N_0} = 2 = e^{rt}$$

or

$$\log_e(2) = rt \text{ or, } \frac{0.69315}{r} = t \qquad (8)$$

where t = time for population to double its size
r = realized rate of population growth per capita

A few values for this relationship are given for illustration:

r	t
0.01	69.3
0.02	34.7
0.03	23.1
0.04	17.3
0.05	13.9
0.06	11.6

Thus if a human population is increasing at an instantaneous rate of 0.0300 per year (finite rate = 1.0305), its doubling time would be about 23 years, if geometric increase prevails.

2. **Multiplication rate dependent on population size.** Populations, however, do not show continuous geometric increase. When a population is growing in a limited space, the density gradually rises until eventually the presence of other organisms reduces the fertility and longevity of the individuals in the population. This reduces the rate of increase of the population until eventually the population ceases to grow. The growth curve defined by such a population is *sigmoid*, or S-shaped (**Figure 4**). The S-shaped curve differs from the geometric curve in two ways: It has an upper asymptote (that is, the curve does not exceed a certain maximal level), and it approaches this asymptote smoothly, not abruptly.

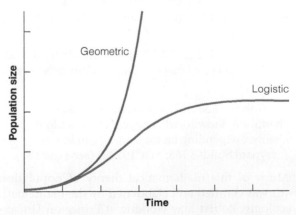

Figure 4 Population growth: geometric growth in an unlimited environment, and logistic (sigmoid) growth in a limited environment.

The simplest way to produce an S-shaped curve is to introduce into our geometric equation a term that will smoothly reduce the rate of increase as the population builds up. We can do this by making each individual added to the population reduce the rate of increase an equal amount. This produces the equation

$$\frac{dN}{dt} = r N \left(\frac{K - N}{K} \right) \qquad (9)$$

where N = population size
t = time
r = intrinsic capacity for increase
K = upper asymptote or maximal value of N ("carrying capacity")

This equation states that

$$\begin{pmatrix} \text{Realized rate} \\ \text{of population} \\ \text{increase per} \\ \text{unit time} \end{pmatrix} = \begin{pmatrix} \text{Potential rate} \\ \text{of population} \\ \text{growth per} \\ \text{capita} \end{pmatrix} \times \begin{pmatrix} \text{Population} \\ \text{size} \end{pmatrix} \times \begin{pmatrix} \text{unutilized} \\ \text{opportunity for} \\ \text{population growth} \end{pmatrix}$$

and is the differential form of the equation for the logistic curve. Verhulst first suggested this curve to describe the growth of human populations in 1838. Pearl and Reed (1920) independently derived the same equation as a description of the growth of the population of the United States.

Note that r is the potential rate of population growth per individual in the population.

The integral form of the logistic equation can be written as follows:

$$N_t = \frac{K}{1 + e^{a - rt}} \qquad (10)$$

where N_t = population size at time t
t = time
K = maximal value
e = 2.71828 (base of natural logarithms)
a = a constant of integration defining the position of the curve relative to the origin
r = intrinsic capacity for increase

Let us look for a minute at the factor $(K - N)/K$, also called the "unutilized opportunity for population

WORKING WITH THE DATA

What Is Little-r, and Why Is It So Confusing?

Unfortunately ecologists have used r to mean two quite different things: r the intrinsic capacity for increase, and r the realized population growth rate per capita.

When populations are growing geometrically, these two meanings are identical:

$$\frac{dN}{dt} = rN$$

$$\frac{dN}{dt\,N} = r = \text{per capita rate of population growth}$$

This is good mathematics, but it becomes confusing when we deal with population growth that is not geometric. To keep matters clear we define two concepts:

r = potential per capita population growth rate = intrinsic capacity for increase

dN/dtN = realized per capita population growth rate

The distinction between these two concepts is easily seen in the following two ways of writing the logistic equation:

$$\frac{dN}{dt} = r N \left(\frac{K - N}{K} \right)$$

$$\frac{dN}{dt\,N} = r \left(\frac{K - N}{K} \right)$$

The realized population growth rate per capita (dN/dtN) is not equal to the potential growth rate (r). Ecologists in the field measure the realized growth rate, which depends (in the logistic model) on the intrinsic capacity for increase r, the carrying capacity K, and the existing population size N.

It is important to keep these two concepts clear. The intrinsic capacity for increase can be considered a constant for a particular population and thus is always a positive number. The realized population growth rate can be negative when a population is declining and then becomes positive when the population grows. Even though in ideal situations the population grows geometrically, and the potential growth rate and the realized rate are the same, but this is rarely the case in the real world.

growth." To demonstrate that this factor does in fact put the brakes on the basic geometric growth pattern, we consider a situation like the following:

$$K = 100$$

$$r = 1.0$$

$$N_0 = 1.0 \text{ (starting density)}$$

Very early in population growth, there is little difference between the curves for the logistic and the geometric equations (see Figure 4). As we approach the middle segments of the curves, they diverge more. As we approach the upper limit of the logistic curve, the curves diverge much farther, and when we reach the upper limit, the population stops growing because $(K - N)/K$ becomes zero. The following calculations demonstrate this:

r	Population size [N]	Unutilized opportunity for population growth $[(K - N)/K]$	Rate of population growth (dN/dt)
1.0	1	99/100	0.99
1.0	50	50/100	25.00
1.0	75	25/100	18.75
1.0	95	5/100	4.75
1.0	99	1/100	.99
1.0	100	0/100	0.00

Note that the addition of one animal has the same effect on the rate of population growth at both the low and high ends of the curve (in this example, 1/100).

Two attributes of the logistic curve make it attractive: its mathematical simplicity and its apparent reality. The differential form of the logistic curve contains only two constants, r and K. Both these mathematical symbols can be translated into biological terms. The constant r is the per capita (or per individual) potential rate of population increase (the intrinsic capacity for increase). It seems reasonable to attribute to K a biological meaning—the density at which the space being studied becomes "saturated" with organisms, the "carrying capacity" of the environment.

There are two ways of viewing the logistic curve. The more general, more flexible viewpoint is to consider it an empirical description of how populations tend to grow in numbers when conditions are initially favorable. The other way is to view the logistic curve as an implicit strict theory of population growth, as a "law" of population growth.

Does the logistic curve fit the facts? One way to find out is to rear a colony of organisms in a constant space with a constant supply of food. From this information we can calculate a logistic curve. If the data fit the sig-

moid pattern of the logistic, we can confirm this model of population growth for that organism. We look into this approach next.

Laboratory Tests of the Logistic Theory

Many populations have been observed in the laboratory as they increase in size. Let us consider a few relatively simple organisms first. Gause (1934) studied the growth of populations of *Paramecium aurelia* and *P. caudatum*. He began his experiments with 20 *Paramecium* in a tube with 5 milliliters (mL) of a salt solution buffered to pH 8. Each day Gause added a constant quantity of bacteria, which served as food and could not multiply in the salt solution. The cultures were incubated at 26°C, and every second day they were washed with fresh salt solution to remove any waste products. Thus, Gause had a *constant environment in a limited space*; the temperature, volume, and chemical composition of the medium were constant, waste products were removed frequently, and food was added in uniform amounts each day. The growth of some of Gause's *Paramecium* populations is shown in **Figure 5**. In general, the fit of these data to the logistic curve was quite good. Under these conditions the asymptotic density (K) was approximately 448 individuals per mL for *P. aurelia* and 128 individuals per mL for *P. caudatum*.

Populations of organisms with more complex life cycles may also increase in an S-shaped curve. Raymond Pearl (1927) fitted a logistic curve to the growth of *Drosophila melanogaster* populations he maintained in bottles with yeast as food. The fit of the data was fairly good (**Figure 6**), and Pearl ushered in the "logistic

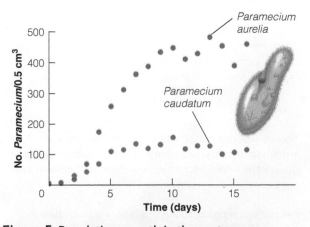

Figure 5 Population growth in the protozoans *Paramecium aurelia* and *P. caudatum* at 26°C in buffered Osterhout's medium, pH 8.0, with "one loop" of bacteria added as food. (Data from Gause 1934.)

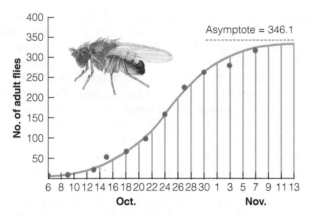

Figure 6 Growth of an experimental laboratory population of the fruit fly _Drosophila melanogaster_. The circles are observed census counts of adult flies, and the smooth curve is the fitted logistic equation. (After Pearl 1927.)

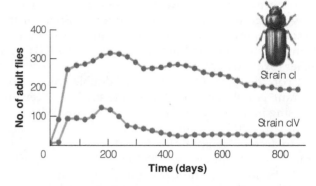

Figure 7 Population growth of two genetic strains of the flour beetle _Tribolium castaneum_ at 29°C and 70% relative humidity in 8 grams of flour. Considerable variation in population growth occurs among different genetic strains of this flour beetle. (Data from Park et al. 1964.)

era" when he proclaimed the logistic curve to be the universal law of population growth. But Sang (1950) criticized the application of the logistic curve to _Drosophila_ populations by identifying complexities in the _Drosophila_ cultures that Pearl did not recognize. First, the flies did not receive a constant amount of food because the yeast that was the source of food was itself a growing population. Also, the composition of the yeast varied as the cultures aged. Second, because the fruit fly has several stages in its life cycle, it is not clear just which stage should be used in measuring "population size." Pearl counted only the adult flies, but to some extent adults and larvae feed on the same thing.

Beetles that live in flour (_Tribolium_) and wheat (_Calandra_) have been also used very often for experimental population studies. These beetles are preferable to _Drosophila_ because, even though they have as complex a life cycle (involving eggs, larvae, pupae, and adults), their food source is nonliving, so their medium can be precisely controlled. Chapman (1928), one of the first to use _Tribolium_ for laboratory studies in ecology, found that colonies of these beetles grew in a logistic fashion. Most workers stopped their cultures as soon as they reached the upper asymptote. Thomas Park, however, reared populations of _Tribolium_ for several years and obtained the results shown in **Figure 7**. The upper asymptote of the logistic is imaginary—the density does not stabilize after the initial sigmoid increase but rather shows a long-term decline. When Birch (1953b) did similar studies on _Calandra oryzae_, he found logistic growth initially, followed by large fluctuations in density with no indication of stabilization around an asymptote.

It is important to note that these populations of a single species of beetle living in a constant climate with

constant food supply show wide fluctuations in numbers. These fluctuations are brought about by the influence of the animals on each other completely independent of any fluctuations in temperature, food, predators, or disease. No cases have as yet been demonstrated in which the population of any organism with a complex life history comes to a steady state at the upper asymptote of the logistic curve. For these reasons the logistic "law" of population growth has been rejected as a general model of how populations increase in size (Kingsland 1995).

Interestingly population ecologists have historically focused on the logistic model for overlapping generations and have largely overlooked the simpler discrete models, with their much richer dynamic behavior (May 1981). Insects constitute a large fraction of animal species, and many insects have nonoverlapping generations that are described well by the simpler discrete models. This change of focus away from the logistic equation as a model for population growth has been highlighted by data on laboratory populations and has been demonstrated even more graphically by data on field populations.

Field Data on Population Growth

Population growth does not occur continuously in field populations. Many species living in seasonal environments show population growth during the favorable season each year. Long-lived organisms may show population growth only rarely, and few populations in nature fill up a vacant habitat the way they do in the laboratory. Some populations have been released from hunting pressure, and we have good records of how they subsequently increased in numbers. Some of the best examples are from birds and mammals recovering from overhunting or from mortality due to DDT and other toxic chemicals.

For the past 30 years many double-crested cormorant populations in North America have been increasing in abundance (Ridgway et al. 2006; Wires and Cuthbert 2006). From the early 1950s cormorants began decreasing rapidly because of reproductive failure from toxic chemicals. By 1973 only 125 nesting pairs could be found in the Great Lakes. Since then colony counts have been increasing rapidly in most colonies (**Figure 8**). There are now about 38,000 cormorant pairs nesting in the Great Lakes, and concerns have been raised by the fishing industry that cormorants are eating too many game fish in the Great Lakes. Population growth in these cormorant colonies has not followed a smooth sigmoid pattern, and an upper limit appears to have been reached in some colonies in the mid-1990s, possibly due to nest site limitation or local food depletion. Whether this upper limit will be stable for different colonies is not yet known.

The ibex lives on steep high-mountain slopes in Europe and Northern Africa. It was hunted to near extinction so that by the early 1700s there was a single population remaining in part of the Italian Alps. Ibex populations had declined steadily since the 1500s because of overhunting and poaching in spite of stringent laws protecting the species. They were hunted for food,

skins, their magnificent horns, and the perception of the curative value of their body parts. At its low point only about 100 individuals survived in northwestern Italy, where the population received full protection in 1821. A captive breeding program was begun in Switzerland between 1906 and 1942, and animals from this program were used in reintroductions. The ibex was reintroduced in 1919 into the Swiss National Park and has recovered in numbers since then to become a conservation success story. **Figure 9** shows the population recovery, which is well described by the logistic equation until 1960, when it began to decline slightly (Sæther et al. 2002). The ibex is a high-altitude goat-like animal that feeds on a great variety of forbs and bushes, and in protected areas its populations may be restricted by the food supplies available in alpine areas.

The whooping crane (*Grus americana*) is another good example of an endangered species now recovering from the brink of extinction. Only 47 whooping cranes existed when this species was first protected in 1916, and only 15 birds were still alive in 1941. The whooping crane breeds in the Northwest Territories of Canada and migrates to overwinter on the Texas coast at the Aransas National Wildlife Refuge. Counts of the entire population on the wintering grounds in Texas since 1938 have yielded the population growth curve shown in **Figure 10**. Population growth has been irregular in the whooping crane. Binkley and Miller (1983, 1988) found that the rate of increase (*r*) changed around 1956, at which time the population began to recover more rapidly than before. Moreover, a 10-year cycle, possibly a spin-off of predation from the 10-year

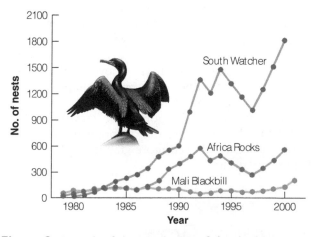

Figure 8 Growth of three colonies of the double-crested cormorant (*Phalacrocorax auritus*) on Lake Huron in the Great Lakes, 1978–2003. The drop in active nests during the mid-1990s was possibly due to climatic events, with some birds not nesting. Colony size differs, partly because of competition for nest sites and some colonies have stopped growing. (Data from Ridgway et al. 2006.)

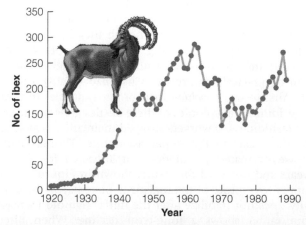

Figure 9 Growth of the ibex (*Capra ibex*) population in Swiss National Park in southeastern Switzerland from its introduction in 1919 to 1990. The ibex went extinct in Switzerland and in most of the Alps at the beginning of the nineteenth century. (Data from Sæther et al. 2002.)

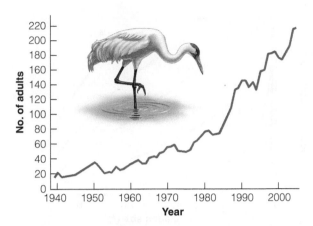

Figure 10 Population growth of the whooping crane, an endangered species that has recovered from near extinction in 1941. Counts of adults are made annually on the wintering grounds at Aransas, Texas. (Data from Johns 2005.)

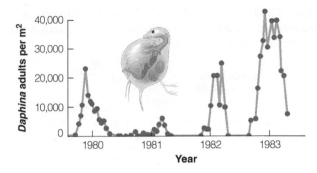

(a) Eunice Lake

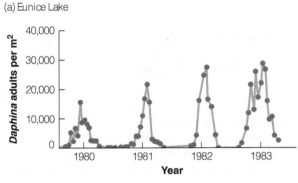

(b) Katherine Lake

Figure 11 Density of the cladoceran *Daphnia rosea* in Eunice Lake and Katherine Lake, British Columbia, from 1980 to 1983. Because these temperate lakes—(a) Eunice Lake and (b) Katherine Lake—show strong seasonal dynamics that vary from year to year, the population growth curve cannot be described by a simple equation like the logistic equation. (Data from Walters et al. 1990.)

cycle of snowshoe hares in the breeding areas, is superimposed on the population growth curve (Nedelman et al. 1987). Hare predators like coyote and lynx may turn to whooping crane nests and chicks once hares begin to decline. The whooping crane population has continued to increase in a sigmoid fashion and is another conservation success story (Johns 2005).

Many organisms show strong annual fluctuations in density, and thus the pattern of population growth can be observed once a year. The cladoceran *Daphnia*, common in the plankton of many temperate lakes and ponds, shows a spring increase in numbers that varies dramatically from year to year (Walters et al. 1990). These cladocerans increase in numbers in an almost exponential manner (**Figure 11**), remain abundant for a variable amount of time in midsummer, and then decline in autumn, possibly because of reduced algal density in the lake water. The maximum density reached varies greatly in different years, so there is no constant carrying capacity (K) for these planktonic organisms.

Very often, field data on population growth are too crude to show definitely whether or not the logistic curve is a good representation of the data. The cases we illustrated here suggest that the logistic curve only approximately describes field population increases.

We conclude from this analysis that population growth may sometimes be sigmoid in natural populations and thus fit the logistic model, but often it is not. Natural populations almost never achieve the asymptotic stable density of the logistic curve, and hence the logistic model has serious drawbacks as a general model of population growth. What can be done about this? Work on population growth models has proceeded along four lines. One has been to generalize the

logistic growth equation by allowing curvilinear relationships between population growth rate and population density. A second approach has been to analyze the effect of time lags on the logistic model, because the assumption of no time lags in the logistic model is most clearly at odds with the biological realities of complex organisms. A third approach has been to construct **probabilistic (stochastic) models** of population growth. The fourth approach has been to use more specific models based on age or size to project population changes (**Leslie matrix models**). Next we look briefly at these four approaches.

Theta (θ) Logistic Model of Population Growth

One of the first attempts to generalize the logistic growth model was to relax the assumption of a linear decrease in the population growth rate as density increases (see

Figure 2). The easiest way to do this mathematically is to add one parameter (θ) to the simple logistic equation. We rewrite Equation (9) as:

$$\text{change in numbers per unit time} = \frac{dN}{dt} = rN\left(\frac{K-N}{K}\right)$$

$$\text{population growth rate per capita} = \frac{dN}{dt\,N} = r\left(\frac{K-N}{K}\right) = r\left(1 - \frac{N}{K}\right) \qquad (11)$$

The theta logistic modifies this as follows:

$$\text{population growth rate per capita} = \frac{dN}{dt\,N} = r\left(\frac{K-N}{K}\right)^{\theta} = r\left(1 - \left(\frac{N}{K}\right)^{\theta}\right)$$

or in terms of changes in population size:

$$N_{t+1} = N_t e^{r\left(1 - \left(\frac{N_t}{K}\right)^{\theta}\right)} \qquad (12)$$

where: N = population size
N_t = population size at time t
K = carrying capacity
θ = scaling parameter defining the shape of the relationship of population growth rate to population size

Figure 12 shows some theoretical θ-logistic curves. When θ is > 1 the curve is convex, and when θ is < 1 the curve is concave (Gilpin and Ayala 1973). When $\theta = 1$ the equation simplifies to the normal logistic Equation (9).

The introduction of the θ-logistic raised the question about how often it was a better fit than the normal

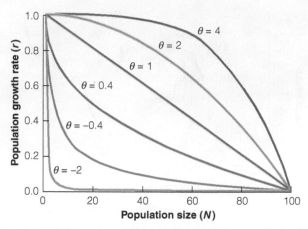

Figure 12 Population growth described by the theta (θ) logistic equation [Equation (12)]. For all of these curves, K = 100 and r = 1.0. The red line is identical to that shown in Figure 2 and represents the normal logistic equation assumption of a linear decline of population growth rate with increasing population size.

logistic. Sibley et al. (2005) analyzed time series of population growth for 3269 sets of data and fitted the θ-logistic to each series. **Figure 13** shows the frequency distributions of the resulting θ-values. For a majority of populations of birds, mammals, fish, and insects for which we have adequate long-term data, θ-values are less than 1. A concave relationship in the θ-logistic means that the population growth rate is low

Figure 13 Population growth described by the theta (θ) logistic equation for 3269 time series of population data in the Global Population Dynamics Database. For all of these curves, 78% of the population growth curves had $\theta < 1$, suggesting a concave regression (see Figure 12). The black arrows point to $\theta = 1$, the value expected if the normal logistic equation fitted the observed data. Each species is represented only once in the data. Note that the scale of theta values is linear between −1 and 2 and then logarithmic. (Data from Sibley et al. 2005.)

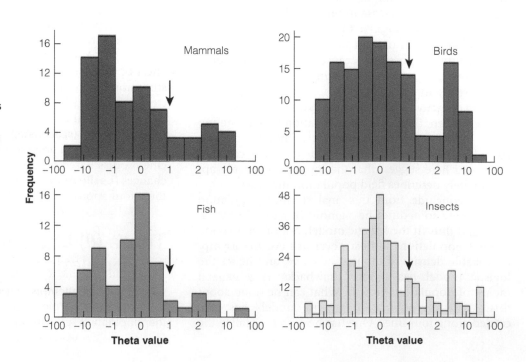

when populations are below carrying capacity (K), low relative to what would be predicted if a normal logistic was fitted in which θ is assumed to be 1. Populations that have a concave θ-logistic regression would recover from a disturbance more slowly than one might predict.

Time-Lag Models of Population Growth

Animals and plants do not respond instantly to changes in their environment, and this leads us to consider what effect time lags might have on population growth models. We can look at this problem in the simplest way by changing our assumption that the reproductive rate at generation t depends on density not in the same generation but instead on density in the last generation ($t - 1$) (see Figure 2). The Working with the Data box "A Simple Time-Lag Model of Population Growth" gives the details of how to do these calculations, and **Figure 14** shows the results. A delay in feedback of only one generation can change a stable population growth pattern into an unstable one. Maynard Smith (1968, p. 25) has shown that, defining $L = BN_{eq}$:

If $0 < L < 0.25$, then stable equilibrium with no oscillation

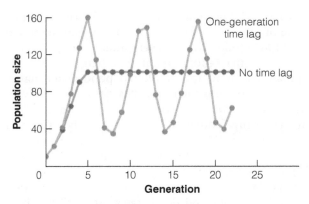

Figure 14 Hypothetical population growth with and without a time lag, with discrete generations, and with reproductive rate a linear function of density. Starting density = 10, slope of reproductive curve = 0.011, equilibrium density = 100.

If $0.25 < L < 1.0$, then convergent oscillation

If $L > 1.0$, then stable limit cycles or divergent oscillation to extinction

Compare the results of this time-lag model with those obtained without any time lags.

WORKING WITH THE DATA

A Simple Time-Lag Model of Population Growth

Consider a simple population growth model with discrete generations. Assume that the reproductive rate at generation t depends on density in a linear manner but that, instead of depending on density at generation t (as in Figure 2), it depends on density at the previous generation ($t - 1$). We measure density as a deviation from the equilibrium point:

$$z = N - N_{eq} \qquad (13)$$

where z = deviation from equilibrium density

N = observed population size

N_{eq} = equilibrium population size (where $R_0 = 1.0$)

The reproductive rate is described in Figure 2 as a straight line, $R_0 = 1.0 - Bz$. The population growth model can thus be written as:

$$N_{t+1} = R_0 N_t$$

$$= \left(1 - Bz_{t-1}\right)N_t \qquad (14)$$

which is similar to the preceding treatment except that the reproductive rate is now defined by the density of the previous generation. The properties of this equation depend on the equilibrium density and the slope of the line.

Let us work out a hypothetical case with a time lag to illustrate a simple model of time-lag population growth:

$$B = 0.011 \qquad\qquad N_{eq} = 100$$

Start a population at $N_0 = 10$ (and use $N = 10$ for first-generation calculation of the time-lag term). From Equation (14):

$$N_1 = [1.0 - 0.011(10 - 100)]10 = 19.9$$

$$N_2 = [1.0 - 0.011(10 - 100)]19.9 = 39.6$$

$$N_3 = [1.0 - 0.011(19.9 - 100)]39.6 = 74.4$$

These results are plotted in Figure 14. This population oscillates more or less regularly, with a period of six or seven generations between peaks in numbers, in contrast to the smooth approach to equilibrium density that occurred in the absence of a time lag.

Laboratory populations of *Daphnia* are a good example of the effect of time lags on population growth. Pratt (1943) followed the development of *Daphnia* populations in the laboratory at two temperatures. The populations, in 50 mL of filtered pond water, started with two parthenogenetic females each. *Daphnia* were counted every two days and transferred to a fresh culture. The only food used was a green alga, *Chlorella*. Populations at 25°C showed oscillations in numbers, whereas those at 18°C were approximately stable (**Figure 15**). Oscillations that occurred at 25°C resulted from a delay in the depressing effect of population density on birth rates and death rates. At 25°C, the birth rate is affected first by rising density, and only later is the death rate increased. This causes the *Daphnia* population to continually "overshoot" and then "undershoot" its equilibrium density. Note that these oscillations are intrinsic to the biological system and are not caused by external environmental changes.

The biological mechanisms in *Daphnia* that account for these time lags are now well understood (Goulden and Hornig 1980). *Daphnia* store energy in the form of oil droplets, mainly as triacylglycerols, when food is superabundant. They use these energy reserves once the food supply has collapsed, so the effects of low food supply are not instantaneous but delayed. Females can thus continue to produce offspring even after food has become scarce. After these energy reserves are exhausted, *Daphnia* starve and die, producing the oscillations shown in Figure 15.

So we see that the introduction of time lags into simple models of population growth permits three possible alternatives: (1) a converging oscillation toward equilibrium, (2) a stable oscillation around the equilibrium level, or (3) a smooth approach to equilibrium density. In addition, some configurations of time lags will produce a divergent oscillation that is unstable and leads to extinction of the population. These outcomes are clearly more realistic models of what seems to occur in natural populations (Forsyth and Caley 2006).

Stochastic Models of Population Growth

The models we have discussed so far are **deterministic** models, which means that given certain initial conditions, each model predicts one exact outcome. But biological systems are probabilistic, not deterministic. Thus, we speak of the probability that a female will have a litter in the next unit of time, or the probability that there will be a cone crop in a given year, or the probability that a predator will kill a certain number of animals within the next month. The realization that population trends are thus the joint outcome of many individual probabilities has led to the development of probabilistic, or stochastic, models.

We can illustrate the basic nature of stochastic models very simply. Recall the geometric growth equation, a deterministic model we previously developed for discrete generations [Equation (1)]:

$$N_{t+1} = R_0 N_t$$

Consider an example in which the net reproductive rate (R_0) is 2.0 and the starting density is 6:

$$N_{t+1} = (2.0)(6) = 12$$

The deterministic model thus predicts a population size of 12 at generation 1. In constructing a stochastic model for this, we might assume that the probabilities of reproduction are as follows:

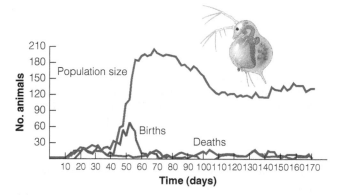

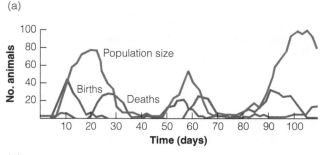

Figure 15 Population growth in the water flea (*Daphnia magna*) in 50 mL of pond water (a) at 18°C and (b) at 25°C. The numbers of births and deaths have been doubled to make them visible at this scale. (Modified from Pratt 1943.)

	Probability
One female offspring	0.50
Three female offspring	0.50

Clearly, on the average, two female parents will leave four female offspring, so $R_0 = 2.0$. Let us use coin tosses to construct some numerical examples. If the coin comes up heads, one offspring is produced; if tails, three offspring.

	Outcome			
Parent	Trial 1	Trial 2	Trial 3	Trial 4
1	(h)1	(t)3	(h)1	(t)3
2	(t)3	(h)1	(t)3	(h)1
3	(h)1	(t)3	(h)1	(h)1
4	(t)3	(t)3	(t)3	(t)3
5	(t)3	(t)3	(t)3	(h)1
6	(t)3	(t)3	(h)1	(h)1
Total population in next generation	14	16	12	10

Some of the outcomes are above the expected value of 12, and some are below it. If we continued doing this many times, we could generate a frequency distribution of population sizes for this simple problem; an example is shown in **Figure 16**. Note that populations starting from exactly the same point with exactly the same biological parameters could, in fact, finish one generation later with either as few as six or as many as 18 members.

The population growth of species with overlapping generations can also be described by stochastic models.

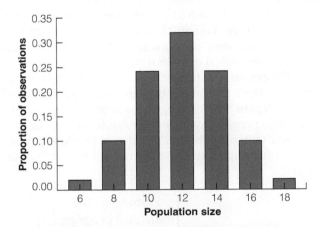

Figure 16 Stochastic population growth. Frequency distribution of the size of the female population after one generation for the example of stochastic population growth discussed in the text. $N_0 = 6$, $R_0 = 2.0$, probability of having one female offspring = 0.5, probability of having three female offspring = 0.5.

Geometric growth in this case follows the differential equation

$$\frac{dN}{dt} = r N = (b - d)N \qquad (15)$$

where b = instantaneous birth rate
d = instantaneous death rate

In the simplest case (the pure birth process), we assume that $d = 0$, so no organisms can die. If we assume a simple binary fission type of reproduction, the probability that an organism will reproduce in the next short time interval dt is $b\,dt$, in which b is the instantaneous birth rate. Consider an example where $b = 0.5$ and $N_0 = 5$ (starting population). In one time interval, according to the deterministic model (Equation 7)

$$N_t = N_0 e^{rt}$$
$$N_1 = (5)e^{(0.5)(1)} = 8.244$$

For the stochastic equivalent of this simple model, we must determine two things from the instantaneous rate of birth:

Probability of not reproducing in one time interval = $e^{-b} = 0.6065$

Probability of reproducing at least once in one time unit = $1.0 - e^{-b} = 0.3935$

Thus for five organisms, the chance that none of the five will reproduce in the next unit of time is
$(0.6065)(0.6065)(0.6065)(0.6065)(0.6065) = 0.082$

so, in approximately one trial out of 12, no population change will occur in the unit of time ($N_1 = 5$). We could laboriously count up all the other possibilities, remembering that each individual may undergo fission more than once in each unit of time. Or we may follow a mathematician's application of probability theory to this problem (Pielou 1969, p. 9). The key point is that probability values inject uncertainty into the predicted outcome, so that there is much variation in final population size when births and deaths are considered in a probabilistic manner. **Figure 17** illustrates these principles of stochastic models of population growth and contrasts them with deterministic model predictions.

If we use probabilistic models and allow both births and deaths to occur randomly, there is a chance that a population will become extinct. What is the chance of extinction for a population starting with N_0 organisms and undergoing stochastic changes in size with average instantaneous birth rate b and death rate d, as in Figure 17? Pielou (1969, p. 17) discussed two cases:

1. **Average birth rate greater than average death rate.** These populations should increase geometrically but may by chance drift to extinction, particularly during the first few time

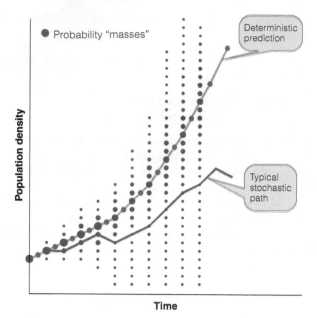

● Probability "masses"

Deterministic prediction

Typical stochastic path

Figure 17 Stochastic model of geometric population growth for continuous overlapping generations. Population predictions cannot be represented by a single value in stochastic models, as they can with deterministic models, and the uncertainty of the prediction increases over time. (After Skellam 1955.)

periods if population size is small. The probability of extinction at some time is given by

$$\text{Probability of extinction} = \left(\frac{d}{b}\right)^{N_0} \quad (16)$$

For example, if $b = 0.75$ and $d = 0.25$ for $N_0 = 5$, we have:

$$\text{Probability of extinction} = \left(\frac{0.25}{0.75}\right)^5 = 0.0041$$

But if $b = 0.55$, $d = 0.45$, and $N_0 = 5$, then

$$\text{Probability of extinction} = \left(\frac{0.45}{0.55}\right)^5 = 0.367$$

Thus, the larger the initial population size and the greater the difference between birth and death rates, the greater chance a population has of staying in existence. The effects of random fluctuations in birth and death rates on individuals is called demographic stochasticity, or demographic uncertainty (Soulè 1987). The important principle is that, even when the birth rate exceeds the death rate on average, there is a finite probability of a population going extinct.

2. **Average birth rate equals average death rate.** These populations are stationary in numbers, fluctuating around constant densities, as is typical of the real world on the average, and by Equation (16):

$$\text{Probability of extinction} = \left(\frac{d}{b}\right)^{N_0} = (1.0)^{N_0} = 1.0$$

as time approaches infinity. Thus, when births equals deaths on average, extinction is a certainty for any population subject to stochastic variations in births and deaths, if we allow a long enough time span.

Stochastic models of population growth thus introduce the important idea of biological variation into the consideration of population changes. The probability approach to these ecological problems is consequently more realistic. The price we must pay for the greater realism of stochastic models is the greater difficulty of the mathematics. The variation inherent in stochastic models becomes more important as population size becomes smaller just as predictions about the change in size of an individual family from one year to the next are much less certain than predictions about the change in size of the world's population. If all populations were in the millions, stochastic models could be eliminated, and deterministic models would be adequate.

Population Projection Matrices

One realistic way of estimating population growth was pioneered by Patrick Leslie (1945), who calculated population changes from age-specific birth and survival rates. Such an age-classified model is called a Leslie matrix. Leslie, who worked closely with Charles Elton's ecologists at the Bureau of Animal Population in Oxford, was responsible for many applications of mathematics to ecological questions (Crowcroft 1991).

The essential feature of Leslie matrix models is that the organism's life cycle is broken down into a series of stages (**Figure 18**). Each age class is one stage in a simple Leslie matrix. Organisms survive from one stage to the next with probability P_x, and they produce a number of offspring F_x. In the conventional life table notation

$$P_x = \frac{l_{x+1}}{l_x} = (1 - q_x) = \left\{ \begin{array}{c} \text{Probability that an} \\ \text{individual of age group} \\ x \text{ will survive to enter} \\ \text{age group } x + 1 \text{ at the} \\ \text{next time interval of} \\ \text{the life table} \end{array} \right\} (17)$$

What Is a "Good" Population Growth Model?

Raymond Pearl in the 1920s had a vision of the logistic equation as a universal model of population growth. We now know that this was too optimistic and that many if not most episodes of population growth do not fit this model. In the real world most population growth patterns are so highly variable that they defy description by a simple model. Nevertheless, applied ecologists are often faced with a need to forecast how the population of an endangered species might increase if it were protected, or how a fish population might recover from overharvesting. So, what can we do? Our choice of approach depends very much on how much we already know about the species in question:

1. *Considerable background knowledge.* For some species we know the approximate birth rate, the number of eggs they lay, the approximate generation time, and their life expectancy. For these species we can use the Leslie matrix models or stage-based matrix models to make a simple forecast of short-term changes in population size.

2. *Little background knowledge.* For many species we know almost nothing about the vital demographic parameters. These species are probably best treated by the use of simple models such as the logistic equation or the geometric growth equation for short-term forecasts. We know these simple models are not precise but they are better than nothing, and ecologists must often follow the old adage that a poor model is better than no model.

We must keep in mind that mathematical models should not be classified as *right* or *wrong*, or as *valid* or *invalid.* All models are wrong but some are useful. Models must be evaluated primarily by their *utility* in helping to answer a question. All models simplify reality to help us understand it, and to help us explore *what if* questions in population dynamics. So even though we may well conclude that the geometric model of population growth is a poor general model for describing population growth, we may still decide to use it to forecast the short-term path of recovery of an endangered frog species. *Utility* is the key to deciding which models are valuable to ecologists.

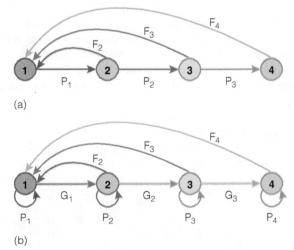

(a)

(b)

Figure 18 Population projection matrices. (a) The Leslie matrix, or age-classified life cycle. Four age classes are shown in this example, with different fecundities (F_x) for each age class and different probabilities (P_x) of surviving from one age class to the next. (b) A size- or stage-based life cycle, in which the only added complication is that an individual has probability P_x of remaining in the same life cycle stage in one time period and a probability G_x of surviving and moving on into the next stage of the life cycle. (After Caswell 2001.)

$$F_x = b_x s_x = \left\{ \begin{array}{l} \text{Number of female offspring born} \\ \text{in one time interval per female} \\ \text{aged } x \text{ to } x + 1; \text{ these offspring} \\ \text{must survive to enter age group} \\ 0 \text{ at the next time interval} \end{array} \right\} \quad (18)$$

where l_x = number of individuals alive at start of age interval x

b_x = number of births in one time interval per adult female aged x to $x + 1$

s_x = proportion of the b_x offspring that are alive at the start of the next time interval

q_x = probability that an individual of age group x will not survive to enter age group $x + 1$ at the next time interval

Begin with a population having specified age structure at time t:

N_0 = number of organisms between ages 0 and 1

N_1 = number of organisms between ages 1 and 2 (and so on to the oldest age class)

N_k = number of organisms between ages k and $k + 1$ (oldest organisms)

Time units for age in a Leslie matrix are often one year but can be any fixed time unit, depending on the organism. Usually only the female population is considered for sexually reproducing species.

If we assume no emigration and no immigration, the population's age structure at the next time interval is defined as follows:

New age structure
↓
{Number of new organisms at time $t + 1$}

$$= F_0N_0 + F_1N_1 + F_2N_2 + F_3N_3 + ... + F_kN_k \quad (19)$$

$$= \sum_{x=0}^{k} F_xN_x$$

Number of age 1 organisms at time $t + 1 = P_0N_0$
Number of age 2 organisms at time $t + 1 = P_1N_1$
Number of age 3 organisms at time $t + 1 = P_2N_2$

and so on.

Leslie (1945) recognized that this problem could be cast as a simple matrix problem if one defined a transition matrix **M** as follows:

$$\mathbf{M} = \begin{bmatrix} F_0 & F_1 & F_2 & F_3 & F_4 & F_5... & F_{k-1} & F_k \\ P_0 & 0 & 0 & 0 & 0 & 0... & 0 & 0 \\ 0 & P_1 & 0 & 0 & 0 & 0... & 0 & 0 \\ 0 & 0 & P_2 & 0 & 0 & 0... & 0 & 0 \\ 0 & 0 & 0 & P_3 & 0 & 0... & 0 & 0 \\ 0 & 0 & 0 & 0 & P_4 & 0... & 0 & 0 \\ ... & ... & ... & & & & & \\ ... & ... & ... & & & & & \\ 0 & 0 & 0 & 0 & 0 & 0 & ...P_{k-1} & 0 \end{bmatrix} \quad (20)$$

where $F_x \geq 0$ and P_x ranges from 0 to 1. By casting the present age structure as a column vector, we get

$$\vec{N}_t = \begin{bmatrix} N_0 \\ N_1 \\ N_2 \\ N_3 \\ N_4 \\ . \\ . \\ N_k \end{bmatrix} \quad (21)$$

Leslie showed that the age distribution at any future time could be found by premultiplying the column vector of age structure by the transition matrix **M**:

$$\mathbf{M}\vec{N}_t = \vec{N}_{t+1}$$
$$\mathbf{M}\vec{N}_{t+1} = \vec{N}_{t+2} \quad (22)$$

Students who are familiar with matrix algebra will benefit from the discussion of the properties of this matrix in Leslie (1945) and in Caswell (2001).

Lefkovitch (1965) realized that the Leslie matrix was a special case of a more general stage-based matrix, in which life history stages replace ages. Such a stage-based or size-based model is illustrated in Figure 18b. One new complexity is added to the age-based model: Whereas all individuals of age x move to age $x + 1$ after 1 unit of time, in a stage- or size-based model some individuals will remain in the same life cycle stage. We thus have two probabilities associated with each stage:

P_x = probability that an individual will survive and remain in stage- or size-class x in the next time unit

G_x = probability that an individual will survive and move up to the next stage- or size-class $x + 1$ in the next time unit

Note that in stage-based matrices we set the time unit such that it is impossible for the organism to jump up two or more stages in one time step.

All of this seems uncomfortably abstract, so let us look at an example of a size-based matrix model. Crouse et al. (1987) analyzed the dynamics of the loggerhead sea turtle (*Caretta caretta*), an endangered species from the Atlantic Ocean off the southeastern United States. Sea turtles have a long life span that can be broken down into seven stages based on size. **Table 2** lists these stages along with the size and approximate age of turtles in each stage. Survivorship varies with size, and only individuals over 87 cm long are sexually mature.

The population projection matrix based on this life history takes the following form:

$$\mathbf{M} = \begin{bmatrix} P_1 & F_2 & F_3 & F_4 & F_5 & F_6 & F_7 \\ G_1 & P_2 & 0 & 0 & 0 & 0 & 0 \\ 0 & G_2 & P_3 & 0 & 0 & 0 & 0 \\ 0 & 0 & G_3 & P_4 & 0 & 0 & 0 \\ 0 & 0 & 0 & G_4 & P_5 & 0 & 0 \\ 0 & 0 & 0 & 0 & G_5 & P_6 & 0 \\ 0 & 0 & 0 & 0 & 0 & G_6 & P_7 \end{bmatrix} \quad (23)$$

The best estimates of the parameters of this matrix are given in **Table 3**.

Given this model of population growth for the loggerhead sea turtle, we can ask some interesting questions about how to reverse the population decline of this endangered species. By holding all but one of the life history pa-

Table 2 Stage-based life table and fecundity table for the loggerhead sea turtle.[a]

Stage number	Class	Size (carapace length) (cm)	Approximate age (yr)	Annual survivorship	Fecundity (eggs/yr)
1	Eggs, hatchlings	<10	<1	0.6747	0
2	Small juveniles	10.1–58.0	1–7	0.7857	0
3	Large juveniles	58.1–80.0	8–15	0.6758	0
4	Subadults	80.1–87.0	16–21	0.7425	0
5	Novice breeders	>87.0	22	0.8091	127
6	First-year remigrants	>87.0	23	0.8091	4
7	Mature breeders	>87.0	24–54	0.8091	80

[a] These values assume a population declining at 3% per year.

SOURCE: Data from Crouse et al. (1987).

Table 3 Stage-class population matrix for the loggerhead sea turtles.[a]

0	0	0	0	127	4	80
0.6747	0.7370	0	0	0	0	0
0	0.0486	0.6610	0	0	0	0
0	0	0.0147	0.6907	0	0	0
0	0	0	0.0518	0	0	0
0	0	0	0	0.8091	0	0
0	0	0	0	0	0.8091	0.8089

[a]Estimates based on the life table presented in Table 2, with the survival estimates broken down into survival within the same stage and survival and movement into the next stage.

SOURCE: Data from Crouse et al. (1987).

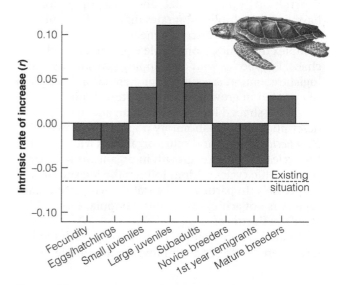

Figure 19 Hypothetical changes in the rate of population increase of loggerhead sea turtle populations off the southeastern United States resulting from either simulated increases of 50% in fecundity or simulated increases in survival to 100% for the different stages of the life cycle. For the four stages of the life cycle shown in red, improving survival or fecundity would still leave the population in decline. The greatest improvement for this endangered turtle would occur by improving the survival of the large juveniles (blue). (From Crouse et al. 1987.)

rameters constant, we can investigate quantitatively the impact of conservation efforts. **Figure 19** shows the results of either increasing fecundity 50% or improving survival in each stage of the life cycle. Improving fecundity 50% still leaves the population declining. Maximum improvement is achieved by improving the survival of juvenile turtles. Many sea turtle conservation efforts have been focused on protecting the eggs on beaches, even though 20–30 years of protecting nests on beaches has produced no increase in sea turtle abundance (Crouse et al. 1987). In fact this is exactly what the model in Figure 19 would predict. What is needed for conservation is an improvement of juvenile turtle survival at sea. Much juvenile loss is caused when turtles are caught in shrimp nets and drown, so

shrimp trawlers are now being fitted with a device to prevent the capture and drowning of sea turtles (Crowder et al. 1994).

Stage-based or size-based matrix models have been used extensively for plant populations in which size is a more useful measure of an individual than is age (Caswell 2001). Matrix models also permit plants to grow or decrease in size, a biologically useful feature. The solution to matrix population growth models based on age- or size stages is just as complex as those previously illustrated for simple population growth models. Populations may increase or decrease geometrically or may show oscillations. Because these models assume a constant schedule of survival and reproduction, they can be applied to natural populations only for the short time periods for which this assumption is valid.

Summary

The growth of a population can be described with simple mathematical models for organisms with discrete generations and for those with overlapping generations. If the multiplication rate is constant, geometric population growth occurs. Populations stabilize at finite sizes only if the multiplication rate depends on population size, and large populations have lower multiplication rates than small populations. Simple models for discrete generations can lead to complex dynamics, from stable equilibria to cycles and chaos. For species with overlapping generations, the logistic equation is a simple mathematical description of population growth to an asymptotic limit.

The S-shaped logistic curve is an adequate description for the laboratory population growth of *Paramecium*, yeast, and other organisms with simple life cycles. Population growth in organisms with more complex life cycles seldom follows the logistic curve very closely. In particular, the stable asymptote of the logistic is not achieved in natural populations and numbers fluctuate.

Four different types of population growth models have been developed to improve on the simple logistic model. The **theta (θ) logistic model** relaxes the assumption of a linear relationship between population growth rate and population size in favor of potential nonlinear relationships. Time-lag models have been used to analyze the effects of different time lags on the population growth curve. The introduction of time lags into the simple models of population growth can produce oscillations in population size instead of a stable asymptotic density. Stochastic models of population growth introduce the effects of chance events on populations. Populations starting from the same density and having the same average birth and death rates may increase at different rates because of chance events, which can lead to extinction and are particularly important in small populations. **Matrix models** of population growth can be age- or size-based and thus can be used for plants and animals alike. Matrix models are ideally suited to asking hypothetical questions about the contribution of specific life table parameters to population growth and to exploring the consequences of alternative management plans for endangered species or pests.

Review Questions and Problems

1. List for plants and animals six reasons why the assumption that population growth at a given point in time depends only on conditions at that time and not on past events might be incorrect.

2. Determine the population growth curve for 10 generations for an annual plant with a net reproductive rate of 6 and a starting density of 35. Assume a constant reproductive rate [Equation (1)].

3. African elephant numbers in Addo National Park, South Africa, have increased as the table on the following page indicates (Gough and Kerley 2006). What shape of population growth curve is shown by these data? Could one fit a logistic equation to these data? Why or why not? Calculate the average instantaneous rate of increase (r) for this elephant population.

Year	Total no. elephants
1976	94
1977	96
1978	96
1979	98
1980	103
1981	111
1982	113
1983	120
1984	128
1985	138
1986	142
1987	151
1988	160
1989	170
1990	181
1991	189
1992	199
1993	205
1994	220
1995	232
1996	249
1997	261
1998	284
1999	315
2000	324
2001	336
2002	377
2003	388

4   Determine the population growth curve for the annual plant in question 2 if reproductive rate is a linear function of density of the form $R_0 = 1.0 - 0.01z$ (where z = deviation from equilibrium density), equilibrium density is 3000, and starting density is 35 [Equation (4)]. Repeat under the assumption that there is a one-generation time lag in changing reproductive rate [Equation (14)].

5   Determine the doubling time for the following human populations (2006 data) from Equation (8):

Country	Realized instantaneous rate of population growth
Sudan	0.026
Niger	0.034
Canada	0.003
Argentina	0.011
United Kingdom	0.002
Ireland	0.008
Russia	−0.006

What assumptions must one make to predict these doubling times?

6   Use the theta logistic Equation (12) to contrast the population growth of two insect populations that both have $K = 100$, $r = 0.5$, starting population 15, and differ only in $\theta = 1.0$ for population A and $\theta = 0.4$ for population B. Which population grows faster?

7   The snapping turtle (*Chelydra serpentina*) has the following life history parameters shown in the table below (Cunnington and Brooks 1996). Calculate the course of population growth for a snapping turtle population that begins with eight adults. Compare the demographic rates of this turtle with those of the loggerhead sea turtle (see Table 3).

Stage of life cycle	Annual probability of survival for this stage	Probability of remaining in this stage for next year	Fecundity (No. of eggs per year)
Eggs	0.0635	0.0	0
Small juveniles	0.0554	0.6985	0
Large juveniles	0.0554	0.6985	0
Subadults	0.0554	0.6985	0
Novice breeders	0.9660	0.0	15.63
Second-year breeders	0.9660	0.0	15.63
Mature breeders	0.9660	0.9638	15.63

8 Discuss how the logistic pattern of population growth might be changed if K and r are not constant but vary over time. May (1981, pp. 24–27) discusses some simple examples.

9 The giant lobelia *Lobelia deckenii keniensis* on Mount Kenya produces on average 250,000 seeds and flowers every eight years. Average adult survival is 0.984 per year. Plants do not begin setting seed until they are 50 years old. Assuming for simplicity a two-stage life cycle (seeds, adult plants), calculate what survival rate of seeds would produce a stable population ($\lambda = 1.0$). How would this survival rate change if the plants flowered every year instead of only once every eight years?

10 A feral house mouse population can increase at $r = 0.0246$ per day. At this rate of increase, how many days are needed for the population to double?

11 Discuss the current projections for the human population of the Earth for 2050. These projections can be obtained from the Web site of the Population Reference Bureau, the U.S. Census Bureau, or the United Nations Population Division. To what variables are these projections most sensitive?

12 If the human population instantly adopted zero population growth ($R_0 = 1.0$), the population would continue to grow until it reached the stationary age structure. Keyfitz (1971) showed that such a population would increase by demographic momentum, as follows:

$$Q = \frac{be}{ra}\left[\frac{R_0 - 1.0}{R_0}\right]$$

where Q = finite rate of population change
 (1.0 = no change)
 b = crude birth rate per 1000 persons
 e = life expectancy at birth in years
 r = current rate natural increase per 1000 persons
 a = average age at first reproduction in years
 R_0 = current net reproductive rate

A human population growing at these rates would increase Q times before it reached equilibrium, if zero population growth was instantly adopted.

 Calculate how much the human population of the Earth would increase from current levels if zero population growth happened overnight. In 2006 the human population parameters were: $b = 21$, $e = 67$ years, $r = 12$, $a = 22$ years, and $R_0 = 1.13$.

 How sensitive is this estimate to changes in the birth rate? To changes in average age at reproduction?

Overview Question

Most analyses of population growth describe processes applicable to unitary organisms. Plants and other modular organisms also undergo population growth. Discuss the application of the models discussed in this chapter to population growth in modular organisms.

Suggested Readings

- Berryman, A. A. 1981. *Population Systems: A General Introduction.* Chapter 2. New York: Plenum Press.

- Caswell, H. 2001. *Matrix Population Models: Construction, Analysis, and Interpretation.* 2nd ed. Sunderland, MA: Sinauer Associates.

- Dennis, B., R. A. Desharnis, J. M. Cushing, and R. F. Costantino. 1997. Transitions in population dynamics: Equilibria to periodic cycles to aperiodic cycles. *Journal of Animal Ecology* 66:704–729.

- Gaillard, J. M., M. Festa-Bianchet, and N. G. Yoccoz. 1998. Population dynamics of large herbivores: Variable recruitment with constant adult survival. *Trends in Ecology and Evolution* 13:58–63.

- Jonzen, N., A. R. Pople, G. C. Grigg, and H. P. Possingham. 2005. Of sheep and rain: large-scale population dynamics of the red kangaroo. *Journal of Animal Ecology* 74:22–30.

- Madsen, T., B. Ujvari, R. Shine, and M. Olsson. 2006. Rain, rats and pythons: Climate-driven population dynamics of predators and prey in tropical Australia. *Austral Ecology* 31:30–37.

- Pearl, R. 1927. The growth of populations. *Quarterly Review of Biology* 2:532–548.

- Rankin, D. J., and H. Kokko. 2007. Do males matter? The role of males in population dynamics. *Oikos* 116:335–348.

- Silvertown, J. W., and D. Charlesworth. 2001. *Introduction to Plant Population Biology.* 4th ed. Oxford: Blackwell Science.

- Turchin, P. 2003. *Complex Population Dynamics: A Theoretical/Empirical Synthesis.* Princeton, NJ: Princeton University Press.

- Watkinson, A. R. 1997. "Plant population dynamics." Pages 359–400 in M. J. Crawley, ed. *Plant Ecology,* 2nd ed. Oxford: Blackwell Science.

Credits

Ilustration and Table Credits

T2 From D. T. Crouse et al., "A stage-based population model for loggerhead sea turtles and implications for conservation," *Ecology* 68:1412–1423, 1987. Copyright © 1987. Reprinted by permission of The Ecological Society of America. T3 From D. T. Crouse et al., "A stage-based population model for loggerhead sea turtles and implications for conservation," *Ecology* 68:1412–1423, 1987. Copyright © 1987. Reprinted by permission of The Ecological Society of America. F19 D. Crouse et al., "A stage-based population model for loggerhead sea

turtles and implications for conservation," *Ecology* 68:1412–1423, 1987. Copyright © 1987 Ecological Society of America. Used with permission.

Photo Credits

Unless otherwise indicated, photos provided by the author.

CO Jim Watt/Pacific Stock.

Species Interactions I: Competition

Key Concepts

- Competition between species can result from exploitation of resources that are in short supply or from interference in gaining access to needed resources.

- Competition between species can be analyzed with simple mathematical models based on the logistic growth equation.

- Competition is common in natural populations of plants and animals, and is particularly strong among herbivores.

- In natural populations, competition over evolutionary time leads to niche differentiation, observed as character displacement, which acts to minimize competition between species.

- To understand the effects of competition we need to study the mechanisms by which it operates and the resources that are being utilized.

From Chapter 10 of *Ecology: The Experimental Analysis of Distribution and Abundance*, Sixth Edition. Eugene Hecht.

KEY TERMS

character displacement The divergence in morphology between similar species in the region where the species both occur, but this divergence is reduced or lost in regions where the species' distributions do not overlap; presumed to be caused by competition.

fundamental niche The ecological space occupied by a species in the absence of competition and other biotic interactions from other species.

Gause's hypothesis Complete competitors cannot coexist; also called the competitive exclusion principle.

Lotka-Volterra equations The set of equations that describe competition between organisms for food or space; another set of equations describes predator-prey interactions.

niche The ecological space occupied by a species, and the occupation of the species in a community.

realized niche The observed resource use of a species in the presence of competition and other biotic interactions; contrast with fundamental niche.

r-selection The type of natural selection experienced by populations that are undergoing rapid population increase in a relatively empty environment.

Organisms do not exist alone in nature but instead in a matrix of other organisms of many species. Many species will be unaffected by the presence of one another in an area, but in some cases two or more species will interact. The evidence for this interaction is quite direct: populations of one species change in the presence of a second species.

Classification of Species Interactions

Interactions between populations can be classified on the basis of either the mechanism of the interaction or the effects of the interaction (Abrams 1987). Ecologists use both of these classifications and often combine them. In categorizing interactions on the basis of mechanism, we can identify six interactions between individuals of different species:

- *Competition.* Two species use the same limited resource, or seek that resource, to the detriment of both.

- *Predation.* One animal species eats all or part of a second animal species.

- *Herbivory.* One animal species eats part or all of a plant species.

- *Parasitism.* Two species live in a physically close, obligatory association in which the parasite depends metabolically on the host.

- *Disease.* An association between a pathogenic microorganism and a host species in which the host suffers physiologically.

- *Mutualism.* Two species live in close association with one another to the benefit of both.

Some authors do not distinguish parasitism from disease, or predation from herbivory, and there is great variability in how loosely these terms are used in the ecology literature.

In categorizing interactions on the basis of effects, the most common effects studied are on population growth. Odum (1983) categorized effects as 0, +, and −. A zero indicates no effect of one species on the other, a plus indicates that the population has benefited at the expense of the other, and a minus indicates that the population has been adversely affected by the other. This system has fatal flaws for classifying interactions because it does not specify a time frame for recognizing effects, and more importantly it cannot describe many indirect interactions (Abrams 1987) that result when one species affects a second species, which in turn affects a third species. Ecological communities are composed of many species linked in complex food webs, and a simple two-species interaction such as (+,−) cannot adequately summarize the possible interactions in a web. For this reason, we will define species interactions based on their mechanism and then explore the variety of effects these mechanisms can produce in populations of plants and animals.

In this chapter we discuss the interactions between two species that result from competition. There are two different types of competition, defined as follows (Birch 1957):

- *Resource competition* (also called *scramble* or *exploitative competition*). Occurs when a number of organisms (of the same or of different species) utilize common resources that are in short supply.

- *Interference competition* (also called *contest competition*). Occurs when the organisms seeking a resource harm one another in the process, even if the resource is not in short supply.

In scramble competition, all individuals are equally affected; there are no "winners" or "losers." In contest competition, some individuals acquire resources at the expense of other individuals, so there are "winners" and "losers." Note that competition may be interspecific (between two or more different species) or intraspecific (between members of the same species). In this chapter, we discuss interspecific competition only.

Competition occurs for resources, and a variety of resources may become the center of competitive interactions. Light, nutrients, and water may be important resources for plants, but plants may also compete for pollinators or for space. Water, food, and mates are possible sources of competition for animals. In some animals, competition for space may involve many types of specific requirements, such as nesting sites, wintering sites, or resting sites that are safe from predators. Species must share a common interest in one or more resources before they can be potential competitors.

Several aspects of the process of competition must be kept clear. First, animals need not see or hear their competitors. A species that feeds by day on a plant may compete with a species that feeds at night on the same plant if the plant is in short supply. Second, many or most of the organisms that an animal sees or hears will not be its competitors. This is true even if resources are shared by the organisms. Thus, even though oxygen is a resource shared by most terrestrial organisms, there is no competition for it among these organisms because this resource is superabundant. Third, competition in plants usually occurs among individuals rooted in position and therefore differs from competition among mobile animals. The spacing of individuals is thus more important in plant competition.

Theories on Competition for Resources

Mathematical models have been used extensively to build hypotheses about what happens when two species live together, either sharing the same food, occupying the same space, or preying on or parasitizing the other. The classical models of these phenomena are the **Lotka-Volterra equations**, which were derived independently by Lotka (1925b) in the United States and Volterra (1926) in Italy. More mechanistic models by

Tilman (1982, 1990) have provided another important perspective on competition theory.

Mathematical Model of Lotka and Volterra

Lotka and Volterra each derived two different sets of equations: One set applies to predator-prey interactions, the other set to nonpredatory situations involving competition for food or space. We are concerned here only with their second set of equations for nonpredatory competition.

The Lotka-Volterra equations, which describe competition between organisms for food or space, are based on the logistic curve. We have seen that the logistic curve is described by the following simple logistic equations: for species 1,

$$\frac{dN_1}{dt} = r_1 N_1 \left(\frac{K_1 - N_1}{K_1} \right) \tag{1}$$

and for species 2,

$$\frac{dN_2}{dt} = r_2 N_2 \left(\frac{K_2 - N_2}{K_2} \right) \tag{2}$$

where N_1 = population size of species 1
t = time
r_1 = intrinsic capacity for increase of species 1
K_1 = asymptotic density or "carrying capacity" for species 1

and these variables are similarly defined for species 2.

We can visualize two species interacting—that is, affecting the population growth of each other—with the following simple analogy illustrated in **Figure 1**.

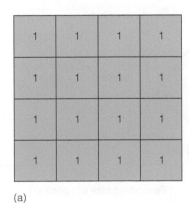
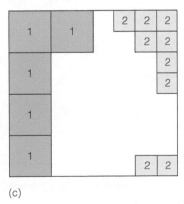

(a)   (b)   (c)

Figure 1 Schematic illustration of the resources two species utilize in competition.
(a) Species 1 has a high utilization rate, and only 16 individuals can be supported in this habitat. (b) Species 2 uses much less of this resource per individual, and 64 individuals can be supported. (c) In competition these two species vie for the common resource. The resource might be nitrogen in the soil for two competing plant species, or a particular food source for two animal species. The size of the box represents the amount of the resource that is available for both species.

Consider the environment to contain a certain amount of a limiting resource, such as nitrogen in the soil. Species 1 uses this resource, and the environment will hold K_1 individuals of this species (shown in green) when all the resource is being monopolized. But some of this resource can also be used by a competitor, species 2 (shown in yellow), which in this example needs much less of the resource to support one individual.

In most cases, the amount of resource used by one individual of species 2 is not exactly the same as that used by one individual of species 1, as illustrated in Figure 1. For example, species 2 may be smaller and require less of the critical resource that is contained in the environment. For this reason, we need a factor to convert species 2 individuals into an equivalent number of species 1 individuals. For this competitive situation, we define

$$\alpha N_2 = \text{equivalent number of species 1 individuals} \qquad (3)$$

where α is the conversion factor for expressing species 2 in units of species 1. This is a very simple assumption, which states that under all conditions of density there is a constant conversion factor between the competitors. We can now write the competition equation for species 1 as

$$\frac{dN_1}{dt} = r_1 N_1 \left(\frac{K_1 - N_1 - \alpha N_2}{K_1} \right) \qquad (4)$$

This equation is mathematically equivalent to the simple analogy we just developed. **Figure 2** shows this graphically for the equilibrium conditions, when dN_1/dt is zero.

The two extreme cases are shown at the ends of the diagonal line in Figure 2. All the "space" for species 1 is used (1) when there are K_1 individuals of species 1, or (2) when there are K_1/α individuals of species 2. Populations of species 1 below this line will increase in size until they reach the diagonal line, which represents all points of equilibrium and is called the isocline. Note that we do not yet know where along this diagonal we will finish, but it must be somewhere at or between the points $N_1 = K_1$ and $N_1 = 0$.

Now we can retrace our steps and apply the same line of argument to species 2. We now have a volume of K_2 spaces to be filled by N_2 individuals but also by N_1 individuals. Again we must convert N_1 into equivalent numbers of N_2, and we define

$$\beta N_1 = \text{equivalent number of species 2 individuals} \qquad (5)$$

where β is the conversion factor for expressing species 1 in species 2 units.[1] We can now write the competition equations for the second species, as follows:

$$\frac{dN_2}{dt} = r_2 N_2 \left(\frac{K_2 - N_2 - \beta N_1}{K_2} \right) \qquad (6)$$

Figure 3 shows this equation graphically for the equilibrium conditions when dN_2/dt is zero.

---

[1] α and β can be written more generally as a_{ij}, the effect of species j on species i. Thus $\alpha = a_{12}$ and $\beta = a_{21}$.

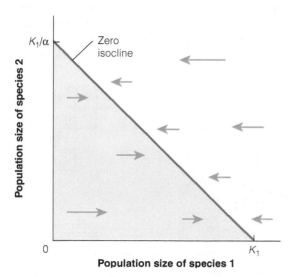

Figure 2 Changes in population size of species 1 when competing with species 2. Populations in the yellow area will increase in size and will come to equilibrium at some point on the blue diagonal line. The sizes of the arrows indicate the approximate rate at which the population will move toward the blue diagonal line. The blue diagonal line represents the zero growth isocline, all those points at which $dN_1/dt = 0$.

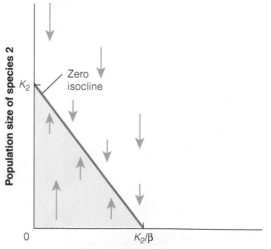

Figure 3 Changes in population size of species 2 when competing with species 1. Populations in the yellow area will increase in size and will come to equilibrium at some point on the blue zero growth isocline, all those points at which $dN_2/dt = 0$. The sizes of the arrows indicate the approximate rates at which the population will move toward the isocline.

Now if we put these two species together, what might be the outcome of this competition? Only three outcomes are possible: (1) Both species coexist, (2) species 1 becomes extinct, or (3) species 2 becomes extinct. Intuitively, we would expect that species 1, if it had a very strong depressing effect on species 2, would win out and force species 2 to become extinct. The converse would apply for the situation in which species 2 strongly affected species 1. In a situation in which neither species has a very strong effect on the other, we might expect them to coexist. These intuitive ideas can be evaluated mathematically in the following way.

Solve the following simultaneous equations at equilibrium:

$$\frac{dN_1}{dt} = 0 = \frac{dN_2}{dt} \tag{7}$$

This can be done by superimposing figures (such as Figures 2 and Figure 3) and adding the arrows by vector addition. **Figure 4** shows the four possible

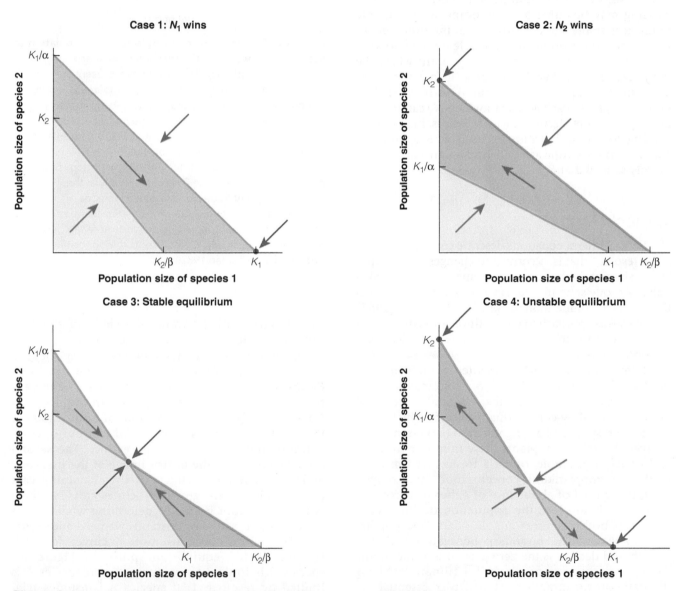

Figure 4 Four possible outcomes of competition between two species. Blue arrows indicate direction of change in populations, and red dots and red arrows indicate the final equilibrium points. In the yellow zone, both species can increase; in the green zone, only species 1 can increase; in the orange zone, only species 2 can increase; and in the white zone both species must decrease.

geometric configurations. In each of these, the vector arrows have been abstracted, and the results can be traced by following the arrows. Species 1 will increase in yellow and green areas, and species 2 will increase in yellow and orange areas. There are a number of principles to keep in mind in viewing these kinds of curves. First, there can be no equilibrium of the two species unless the diagonal curves cross each other. Thus, in cases 1 and 2, there can be no equilibrium, because one species is able to increase in a zone in which the second species must decrease. These cases lead to the extinction of one competitor. Second, if the diagonal lines cross, the equilibrium point represented by their crossing may be either a stable point or an unstable point. It is stable if the vectors about the point are directed toward the point, and unstable if the vectors are directed away from it. In case 4, the point where the two lines cross is unstable because if in response to some small disturbance the populations move slightly downward, they reach a zone in which N_1 can increase but N_2 can only decrease, which results in species 1 coming to an equilibrium by itself at K_1. Similarly, slight movement upward will lead to an equilibrium of only species 2 at K_2.

Tilman's Model

The Lotka-Volterra equations describe competition only by its results—that is, according to changes in the population sizes of the two competing species. In the Lotka-Volterra models, no mechanisms are specified by which the effects of competition are produced. Tilman (1987) criticized this approach to competition and emphasized that we need to study the mechanisms by which competition occurs.

Tilman (1977, 1982) presented a mathematical model of competition based on resource use. We begin our examination of the essential features of Tilman's model by considering **Figure 5**, which illustrates the response of an organism to two essential resources; for terrestrial plants these might be nitrogen and light, for example, or for a freshwater fish these might be zooplankton concentration and oxygen level. If the level of abundance of either resource 1 or resource 2 is too low, the population declines; conversely, if both resources are abundant, the population increases. The boundary between population growth and decline is the zero growth isocline of this species. A second key parameter for Tilman's model is the rate of consumption of the two essential resources. Each species will consume resources at different rates. For example, a plant might utilize water more rapidly than it utilizes nitrogen. These rates of

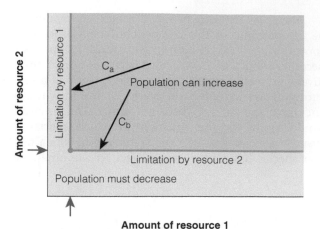

Figure 5 The response of a single species population to variations in two essential resources (such as nitrogen and water, for plants). The blue lines represent the zero growth isoclines, the lower one set by resource 2 and the left one set by resource 1 (red arrows). Above these isoclines in the blue shaded area, the population can increase in size; below these isoclines in the gray area, the population will decline. In the left side of the gray area, resource 1 is limiting; in the bottom side of the gray area, resource 2 is limiting. Only at the intersection point (blue dot) are both resources simultaneously limiting. At the hypothetical consumption vectors C_a the organism uses resource 1 more rapidly and resource 2 more slowly; C_b represents the opposite case. (Modified from Tilman 1982.)

consumption will determine the slope of the consumption vectors illustrated in Figure 5.

If we repeat this analysis for a second species, we can superimpose the two zero growth isoclines. **Figure 6** shows the possible outcomes of competition for the two competing species. In the first case (Figure 6a), species B needs more of both resources than species A. Thus species A will win out in competition, and species B will go extinct. The second case (Figure 6b) is the mirror image of the first case, and species A goes extinct. In the remaining case (Figure 6c) the zero growth isoclines cross, so there is an equilibrium point. To determine whether this equilibrium is stable or unstable, we need additional information on the consumption curves for each species. At the equilibrium point in Figure 6c, species A is limited by resource 2, and species B is limited by resource 1. If species A consumes relatively more of resource 1 than does species B, the equilibrium point is unstable, and one species or the other will go extinct. To apply Tilman's model to a

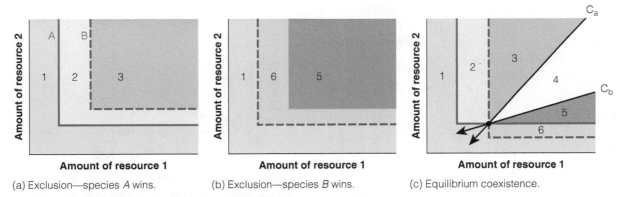

(a) Exclusion—species *A* wins.   (b) Exclusion—species *B* wins.   (c) Equilibrium coexistence.

Figure 6 Tilman's model of competition for two essential resources. The zero isoclines for species A (blue) and species B (red, dashed line) are shown, along with the consumption rate vectors for each species (C_a and C_b). For all three cases the regions are labeled and colored as follows: 1 (gray) = neither species can live; 2 (yellow) = only species A can live; 3 (blue) = species A wins out in competition; 4 (white) = stable coexistence; 5 (orange) = species B wins out in competition; 6 (green) = only species B can live. • = stable equilibrium point. (From Tilman 1982.)

particular environment, we must know the rate of supply of the limiting resources to the populations (a function of the habitat) and the rates of consumption of these resources by each species (represented by the vectors in Figure 5).

Tilman's model provides the same final predictions as the Lotka-Volterra model (compare Figure 6 with Figure 4), but Tilman's model can be extended to make community-level predictions about species diversity and succession (Tilman 1986, 1990). The strength of Tilman's model is in its emphasis on mechanism, and because of this it can help us understand more precisely how species interact over limited resources.

Three important ideas have come from these mathematical models of two competing species:

1. Competition can lead to one species winning and the second species going extinct.

2. Some competitive interactions can lead to coexistence.

3. We can understand competitive interactions only by knowing the resources involved and the mechanisms by which species compete.

Now that we have these mathematical formulations and some simple hypotheses of competitive interactions, we must see if they are an adequate representation of what happens in actual biological systems.

Competition in Experimental Laboratory Populations

One of the first and most important investigations of competitive systems was conducted by a Russian microbiologist named Georgyi Frantsevich Gause working at Moscow University. Gause (1932) studied in detail the mechanism of competition between two species of yeast, *Saccharomyces cervisiae* and *Schizosaccharomyces kephir*.[2] In the first aspect of his investigations, concerning the growth of these two species in isolation, he found that the population growth of both species of yeast was sigmoid and could reasonably be fitted by the logistic curve.

Gause then asked: What are the factors in the environment that depress and stop the growth of the yeast population? Richards (1928) had previously shown that when the growth of yeast stops under anaerobic conditions, a considerable amount of sugar and other necessary growth substances remain in the cultures. Because growth ceases before the reserves of food and energy are exhausted, something else in the environment must be responsible for the restriction of population increase. The decisive factor seems to be the accumulation of ethyl alcohol, which is produced by the breakdown of sugar for energy under anaerobic conditions

---

[2]These organisms' scientific names have changed since Gause's studies.

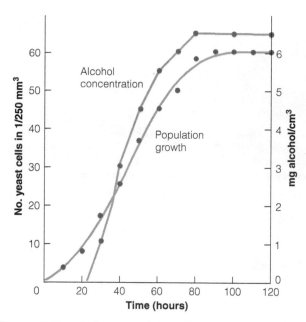

Figure 7 Population growth (purple) and ethyl alcohol accumulation (red) in a population of yeast (*Saccharomyces*). (After Richards 1928.)

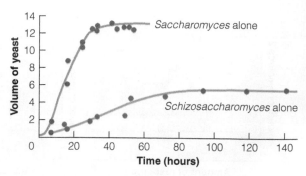

Figure 8 Population growth of pure cultures of two yeasts, *Saccharomyces* and *Schizosaccharomyces*. (After Gause 1932.)

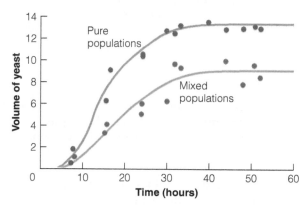

Figure 9 Growth of populations of the yeast *Saccharomyces* in pure cultures and in mixed cultures with *Schizosaccharomyces*. (After Gause 1932.)

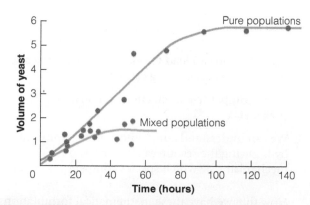

Figure 10 Growth of populations of the yeast *Schizosaccharomyces* in pure cultures and in mixed cultures with *Saccharomyces*. (After Gause 1932.)

(**Figure 7**). High concentrations of alcohol kill the new yeast buds just after they separate from the mother cell. Richards showed that the yeast growth could be reduced by artificially adding alcohol to cultures, and changes in the pH of the medium were of secondary importance. Thus with yeast we apparently have a quite simple relationship, with the population in test tube cultures being limited principally by one factor: ethyl alcohol concentration.

When grown separately, the two yeast species reacted as shown in **Figure 8**. From these curves, Gause calculated logistic curves (calculated in units of volume):

	Saccharomyces	Schizosaccharomyces
K	13.00	5.80
r	0.22	0.06

Gause then investigated what would happen when the two yeast species were grown together, and he obtained the results shown in **Figure 9** and **Figure 10**. Gause assumed that these data fit the Lotka-Volterra equations, and using the equations on the data from the mixed cultures, he obtained the following data:

| Age of Culture (hr) | Competition coefficients | |
	α Saccharomyces	β Schizosaccharomyces
20	4.79	0.501
30	2.81	0.349
40	1.85	0.467
Mean value	3.15	0.439

The influence of *Schizosaccharomyces* on *Saccharomyces* is measured by α, and this means that, in terms of competition, *Saccharomyces* can fill its K_1 spaces according to the equivalence

1 volume of *Schizosaccharomyces* = 3.15 volumes of *Saccharomyces*

Note that the α and β values tend to change with the age of the culture, but as a first approximation we can assume α and β to be constants.

If alcohol concentration is the critical limiting factor in these anaerobic yeast populations, Gause argued, then we should be able to determine the competition coefficients α and β by measuring the alcohol production rate of the two yeasts. He found:

	Alcohol production (% EtOH/mL yeast)
Saccharomyces	0.113
Schizosaccharomyces	0.247

Gause then argued that since alcohol was the limiting factor of population growth, the competition coefficients, α and β, should be determined by a direct ratio of these alcohol production figures:

$$\alpha = \frac{0.247}{0.113} = 2.18$$

$$\beta = \frac{0.113}{0.247} = 0.46$$

These independent physiological measurements agree in general with those obtained from the population data given previously. Gause attributed the differences in the α values to the presence of other waste products affecting *Saccharomyces*. Gause assumed that the competition coefficients would be the reciprocals of each

other, but this assumption need not apply to all cases of competition.

In many laboratory experiments, a species can do well when raised alone but can be driven to extinction when raised in competition with another species. When Birch (1953b) raised the grain beetles *Calandra oryzae* and *Rhizopertha dominica* at several different temperatures, he found that *Calandra* would invariably eliminate *Rhizopertha* at 29°C (**Figure 11**) and that *Rhizopertha* would always eliminate *Calandra* at 32°C (**Figure 12**). Birch found that he could predict these results from the intrinsic capacity for increase; for example,

	r	Temperature	Winner
Calandra	0.77	29.1°C	*Calandra*
Rhizopertha	0.58		
Rhizopertha	0.69	32.3°C	*Rhizopertha*
Calandra	0.50		

Thus we could change the outcome of competition by changing only one component of the environment, temperature, by only 3°C.

In all the grain beetle experiments just discussed, one species or the other died out completely. All these situations fall under cases 1 or 2 in our treatment of the Lotka-Volterra equations. What about case 3, in which the species coexist? Yeasts coexisted in Gause's experiments; does coexistence ever occur in grain beetles?

Under the conditions of extreme crowding in laboratory experiments, it is possible for two species to live together indefinitely if they differ even slightly in their

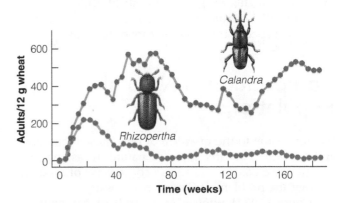

Figure 11 Population trends of adult grain beetles (*Calandra oryzae* and *Rhizopertha dominica*) living together in wheat of 14% moisture content at 29.1°C. *Calandra* eliminates *Rhizopertha* in competition at this temperature. (After Birch 1953b.)

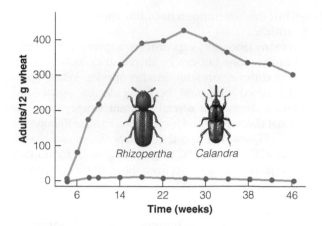

Figure 12 Population trends of adult grain beetles (*Calandra oryzae* and *Rhizopertha dominica*) living together in wheat of 14% moisture content at 32.3°C. *Calandra* goes extinct and *Rhizopertha* wins in competition at this temperature. (After Birch 1953b.)

requirements. For example, Crombie (1945) reared the grain beetles *Rhizopertha* and *Oryzaephilus* in wheat and found that they would coexist indefinitely. The larvae of *Rhizopertha* live and feed inside the grain of wheat; the larvae of *Oryzaephilus* live and feed outside the grain. (The adults of both species have the same feeding behavior, feeding outside the wheat grain.) Apparently these larval differences were sufficient to allow coexistence.

Gause (1934) found that *Paramecium aurelia* and *P. bursaria* would coexist in a tube containing yeast. *P. aurelia* would feed on the yeast suspension in the upper layers of the fluid, whereas *P. bursaria* would feed on the bottom layers. This difference in feeding behavior allowed these species to coexist.

Thus by introducing only very slight differences in the environment, or given very slight differences in species habits, coexistence can occur between competing animal species under laboratory conditions.

Competition in Natural Populations

We now come to the question of how these theoretical and laboratory results apply to nature. In asking this question, we come up against a controversy of modern ecology, the problem of Gause's hypothesis.

Gause (1934) wrote: "As a result of competition two similar species scarcely ever occupy similar niches, but displace each other in such a manner that each takes possession of certain peculiar kinds of food and modes of life in which it has an advantage over its competitor" (p. 19). Gause referred to Elton (1927), who had defined **niche** as follows: "The niche of an animal

means its place in the biotic environment, its relations to food and enemies" (p. 64). Thus Elton used the term niche to describe the role of an animal in its community, so one could speak (for example) of a broad herbivore niche, which could be further subdivided.

Gause went on to say that the Lotka-Volterra equations do "not permit any equilibrium between the competing species occupying the same 'niche,' and [lead] to the entire displacing of one of them by another. . . . Both species survive indefinitely only when they occupy different niches in the microcosm in which they have an advantage over their competitors" (p. 48). Gause identifies case 3 (stable coexistence) with the situation of "different niches" and cases 1, 2, and 4 with the situation of "same niche."

Gause himself never formally defined what is called **Gause's hypothesis**. In 1944 the British Ecological Society held a symposium on the ecology of closely related species. An anonymous reporter (who turned out to be David Lack) wrote that year in the *Journal of Animal Ecology* that "the symposium centered about Gause's contention (1934) that two species with similar ecology cannot live together in the same place . . ." (p. 176).

As is usual, several workers immediately searched out and found earlier statements of "Gause's hypothesis." Monard, a French freshwater biologist, had expressed the same idea in 1920, and Grinnell, a California biologist, had written much the same thing in 1904. Darwin apparently had the same idea but never clearly expressed it. The solution to this has been to drop the use of names and call this idea the competitive exclusion principle, which Hardin (1960) states succinctly: "Complete competitors cannot coexist." The competitive exclusion principle encapsulates the conclusions of the Lotka-Volterra models for competition.

The concept of the niche is intimately involved with the competitive exclusion principle, and so we must clarify this concept first. The term niche was almost simultaneously defined to mean two different things. Joseph Grinnell, who in 1917 was one of the first to use the term niche, viewed it as a subdivision of the habitat: Each niche was occupied by only one species. Elton in 1927 independently defined the niche as the "role" of a species in the community. These vague concepts were incorporated into Hutchinson's redefinition of the niche in 1958. If we consider just two environmental variables, such as temperature and precipitation, and determine for each species the range of values that allow the species to persist, we can produce an analysis like that in **Figure 13**. This ecological space in which the species can survive is defined as the realized niche of that species. We could measure other environmental variables, such as pH or soil nutrients for plants, until all the ecological factors relative to the species have been measured. In an ideal world we could measure the ecological

Geographical space

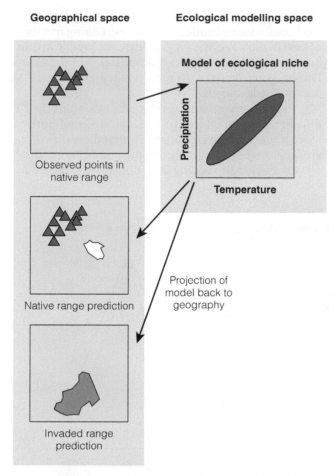

Observed points in
native range

Native range prediction

Projection of
model back to
geography

Invaded range
prediction

Ecological modelling space

Model of ecological niche

Figure 13 **Schematic illustration of how we can define the realized niche of a species in which the key limiting resources are temperature and precipitation.** First, we determine the environmental characteristics of the geographic area occupied by the species. Only two environmental variables are used for illustration, but this approach could be extended to three or more variables. Given these data, we define the realized ecological niche of the species (red ellipse). We can now project this ecological niche model back into geographic space to predict both the native range the species could occupy and the possible geographic range it might occupy in a newly invaded landscape.

space occupied by the species in the absence of competition and other biotic interactions, and this ecological space would define the **fundamental niche** of the species—the set of resources the species can utilize in the absence of other organisms.

This idea of a fundamental niche has some practical difficulties. It has an infinite number of dimensions, and thus we cannot completely determine the fundamental niche of any organism. The fundamental niche is thus an abstract concept, and we can measure only the **realized niche** of a species, as illustrated in Figure

13. The realized niche is the observed resource use of a species in the presence of competition.

The fundamental niche, which describes a species' role in the absence of competition and other interactions, can be measured for some species in the laboratory. When species are deliberately or accidentally introduced into new regions, they often leave behind their competitors and predators, so that they occupy more of their fundamental niche.

Given that we have now defined a realized niche, we can next ask whether two species in the same community can exist in a single niche. Does competitive exclusion occur in natural communities? Before answering this question, we must realize that every hypothesis has its limits, and thus we should be careful to set down at the start some situations in which competitive exclusion would *not* be expected to occur. These situations are (1) unstable environments that never reach equilibrium and are occupied by colonizing species, (2) environments in which species do not compete for resources, and (3) fluctuating environments that reverse the direction of competition before extinction is possible (Hutchinson 1958).

Field naturalists were the first to question Gause's hypothesis. They pointed out that one might see in the field many examples of closely related species living together and apparently in the same habitat. Anyone who has made field collections of plants or insects will attest to the great number of species living in close association. This observation brings us to the ecological paradox of competition: How can we reconcile the frequent extinction of closely related species in laboratory cultures with the apparent coexistence of large numbers of species in field communities?

Ecologists have developed two simple views in attempting to answer this question. One holds that competition is rare in nature, and since species are not competing for limited resources, there is no need to expect evidence of competitive exclusion in natural communities. The other view holds that competition has been very common throughout the evolutionary history of communities and has resulted in adaptations that serve to minimize competitive effects.

How common is competition in nature? Much investigation has centered on closely related species on the assumption that taxonomic similarity should promote competition. Robert MacArthur of Princeton University was instrumental in bringing the study of competition to the fore in North America because of his theoretical and empirical work on birds. His classic research was on a group of closely related birds in the boreal forests of New England. Five warbler species of the genus *Dendroica* coexist in these forests, and all of these warblers are insect eaters and about the same size. Why does one species not exterminate the others by competitive exclusion, if Gause's hypothesis is correct? MacArthur (1958) showed that these warblers feed in different positions in the canopy (**Figure 14**), feed in different manners, move in different directions through the trees, and have slightly different nesting dates. The feeding-zone differences seem sufficiently large to explain the coexistence of the blackburnian, black-throated green, and bay-breasted warblers. The myrtle warbler is uncommon and less specialized than the other species. The Cape May warbler is different from these other species because it depends on occasional irruptions of forest insects to provide a superabundant food source for its continued existence. During irruptions of insects, the Cape May warbler increases rapidly in numbers and obtains a temporary advantage over the others. During years between irruptions, they are reduced in numbers to low levels.

Thus closely related species of birds either live in different sorts of places or else use different sorts of foods. One possible explanation is that these differences arose because of competition in the past between closely related species. In keeping with Gause's hypothesis and

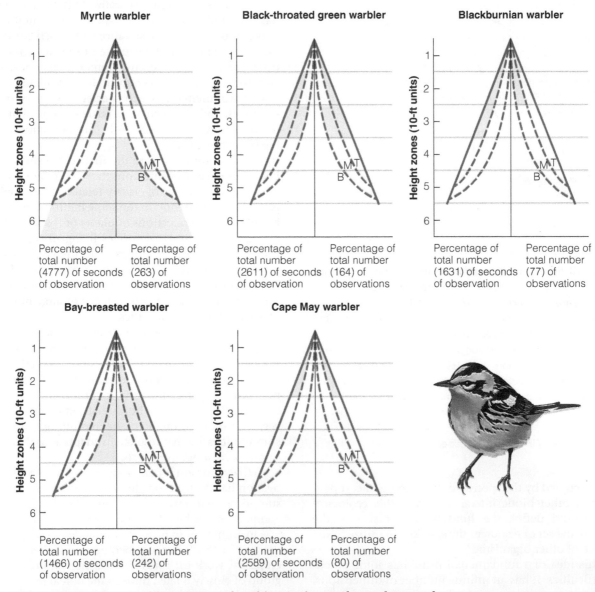

Figure 14 Feeding positions of five species of warblers in the coniferous forests of the northeastern United States. The zones of most concentrated feeding activity are shaded. B = base of branches, M = middle of branches, T = terminal portions of branches. The blackburnian warbler is illustrated. (After MacArthur 1958.)

its associated selection pressure, species either "moved" to different places and so avoided competition, or they changed their feeding behavior to avoid competition. What we observe now is the "ghost of competition past" (Connell 1980). This explanation may be correct but there is a logical difficulty in testing the hypothesis that competition has caused two species to differ (Simberloff and Boecklen 1981). Two species are always somewhat different from each other as a by-product of speciation, and consequently observing differences between species does not necessarily mean that competition caused the differences. This fundamental difficulty means that descriptive studies of species differences by themselves are not useful for understanding the importance of competition in natural populations. Experimental work is needed on populations that are possible competitors, and the important issue is the significance of competition between species at the present time.

Animal ecologists have attributed the coexistence of many different species to these species' abilities to specialize in their diet: herbivores feed on different plant species or different parts of plants, and carnivores feed on different animal species, or eat both plants and animals. Thus many food resources are available to animals. But all plants need only a few resources, and it has never been clear how all the plant species we see can coexist in natural communities (Grace 1995). One possible explanation is that plant communities are not in equilibrium, so competition can never reach the end point of competitive exclusion. How might this work? One way to approach this dilemma has been described as the paradox of the plankton.

The phytoplankton of marine and freshwater environments consists of a large number of autotrophic species that utilize a common pool of nutrients and undergo photosynthesis in a relatively unstructured environment. How can all these species coexist especially given that, because natural waters are often deficient in nutrients, competition should be strong and competitive exclusion should be common? This dilemma, the paradox of the plankton, has been aptly described by Hutchinson (1961) as a possible exception to the competitive exclusion principle. Hutchinson suggested that these species could coexist because of environmental instability; before competitive displacement could have time to occur, seasonal changes in the lake or the sea would occur. The phytoplankton may thus be viewed as a nonequilibrium community of competing species and thus are not an exception to the principle of competitive exclusion.

All vascular plants require water, light, and nutrients, and consequently competition between plants over essential resources is common. Plant ecologists have developed several methods for studying the effect of one plant species on a competing species. The most common approach is through the use of replacement series either in the field or in the greenhouse. Replacement series were pioneered by the Dutch ecologist C. T. de Wit nearly 50 years ago (de Wit 1960). A replacement series can be viewed schematically as an array of plots with different combinations of the two species. **Figure 15** illustrates a replacement series for two species. In this series, the density of plants is kept constant, and only the percentage composition is changed. The variables of interest are the yield of each species and the combined yield of both species. Competition between the two species, as well as competition among individuals of the same species, determines the yield. **Figure 16** illustrates the results from one replacement series involving perennial ryegrass (*Lolium perenne*) and white clover (*Trifolium repens*) (Jolliffe 2000). The total yield was maximal when clover comprised about 65%–75% of the mixture, indicating that the mixture was more productive than either single-species monoculture in this experiment. This study is then repeated with combined densities greater than 24 plants per plot to explore how

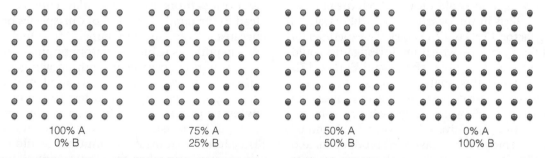

100% A
0% B

75% A
25% B

50% A
50% B

0% A
100% B

Figure 15 Replacement series for the study of plant competition. Schematic illustration of four plots in which the density of plants is held constant and the composition varied from a monoculture of species A (green dots) to various mixtures and a monoculture of species B (red dots). (Data from Jolliffe 2000.)

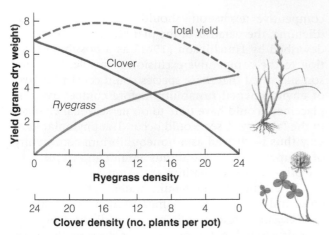

Figure 16 Results from replacement series for the study of plant competition. An example of a replacement series for perennial ryegrass (green, top diagram) and white clover (red, lower diagram). At this particular plant density there is evidence of competition between these two species because yields decline in the presence of the competing species. (Data from Jolliffe 2000.)

these relationships vary with overall plant density. Since replacement series are short-term experiments, it is rare to observe competitive exclusion of the inferior competitor.

Plant interactions can have a positive effect on other plants, a phenomenon called facilitation. The focus of ecologists on negative effects has tended to obscure positive effects that might be common, and in natural ecosystems both positive and negative effects occur. In arid environments, competition among roots for water should be expected, and range managers have long been interested in improving arid zone grazing by replacing shrubs such as sagebrush with grasses or legumes. Holzapfel and Mahall (1999) tested for the effects of competition and facilitation between a desert shrub (*Ambrosia dumosa*, burroweed) and annual grasses and herbs in the Mojave Desert of California. By eliminating shrubs the investigators could measure both the positive and negative effects in this desert system, where water is the limiting factor. **Figure 17a** shows the experimental design and the means by which the investigators could separate positive and negative effects, and **Figure 17b** shows the results of this analysis. Annual grasses and herbs had strong negative effects on *Ambrosia*, measured by water balance and by shrub growth. By contrast, *Ambrosia* had strong positive effects on all the annual plant species, improving biomass, seed production, and survival. Shading by shrubs lowers ambient temperature and improves water availability to annuals, and this is the mechanism behind facilitation. Plant-plant interactions are not always negative.

Interspecific competition has been analyzed in a wide variety of plants and animals during the past 50 years, and we can now ask how frequently competition occurs between species in nature, and how strong its effects are. Gurevitch et al. (1992) have tabulated the results of 218 competition experiments. To compare different groups of plants and animals, they defined effect size in the usual statistical manner:

$$\text{Effect size} = \frac{\overline{X}_e - \overline{X}_c}{s} \qquad (8)$$

where $\overline{X}_c$ = mean biomass of the control group
(with competition)
$\overline{X}_e$ = mean biomass of the experimental group
(without competition)
s = standard deviation of both groups pooled

Since the experimental treatments involve the removal of potential competitors, a positive effect size means that competition is reducing the density or biomass of the species. A negative effect size implies facilitation, a higher density or biomass under conditions of interaction.

Competition had a strong overall effect in 218 studies covering 93 species (average effect size = 0.8; Gurevitch et al. 1992). **Figure 18** shows the average effect sizes in four categories of organisms. Several general trends are apparent in this figure. Plants and carnivores showed relatively small effects of interspecific competition, compared with herbivores. Considerable variation in competitive effects is apparent within herbivores. Some herbivores like frogs and toads show strong effects of interspecific competition, while other herbivores like marine molluscs show only moderate effects. The overall conclusion of this survey of interspecific competition is that it occurs frequently (but not always) in natural populations, and that more information is needed on the mechanisms by which competition operates in nature and how large the effects of competition might be in different species groups.

Gause's hypothesis would seem to predict that if two competitors are very similar, competition would lead to the rapid extinction of one species or the other because of very strong competition. One way to test this hypothesis using laboratory populations is to use strains of microorganisms with known genetic differences. Kashiwagi et al. (1998) used mutants of the bacterium *Escherichia coli* to test for competition between two strains that differ at a single genetic locus, the smallest possible difference between competitors. They found that these two mutants would coexist in a chemostat, even when the starting population sizes of the two strains were varied. These results suggest that Gause's hypothesis should be rejected as a general ecological model for competition, since even the

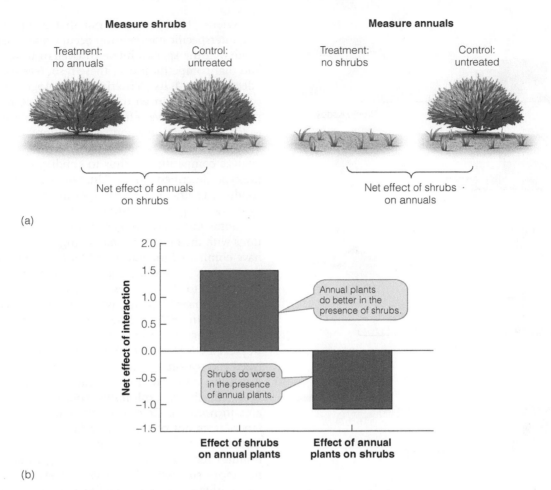

(a)

(b)

Figure 17 The analysis of competition and facilitation between the desert shrub *Ambrosia dumosa* (burroweed) and annual grasses and herbs in the Mojave Desert. (a) The experimental design to measure the net effect of annual plants on shrubs and shrubs on annual plants. (b) The average results of this interaction, which includes growth, seed production, water balance, and mortality. (Data from Holzapfel and Mahall 1999.)

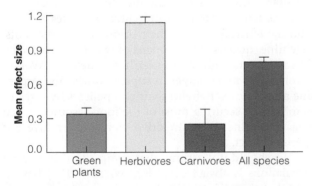

Figure 18 Relative strength of competition between species as measured by mean effect size on biomass in 93 species of organisms in diverse groups. Mean effect size is demonstrated in Equation (8). Ninety-five percent confidence limits are shown for each category. (Data from Gurevitch et al. 1992, Table 1.)

smallest differences can permit coexistence of closely related forms.

The role of competition in natural populations can be analyzed in several ways (Wiens 1989). We can search for patterns in resource utilization to determine how much different species overlap in their resource use. One good example comes from the study of the diets of five species of terns on Christmas Island in the Pacific Ocean (**Figure 19**). Terns on Christmas Island are ecologically segregated according to their diets, and these data are consistent with the idea that competition has favored ecological divergence in diets. Schoener (1986b) compiled data from many studies of this sort that show ecological segregation. Even if such segregation occurred by evolutionary changes in the past, it is still an open question whether interspecific competition is operating today in these populations.

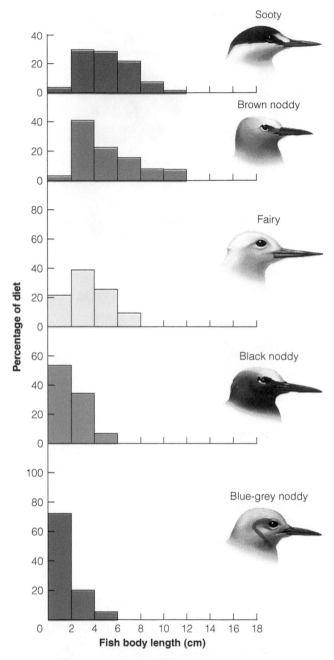

Figure 19 Resource partitioning in five species of terns on Christmas Island, Pacific Ocean, as seen in frequency distributions of prey size. Terns are arranged in order of size from the largest species at the top to the smallest at the bottom. The two largest terns are nearly the same size and eat very similar sizes of fish, but the sooty tern feeds at sea several hundred kilometers from land and the brown noddy tern feeds within 100 km of land. (From Ashmole 1968.)

Wiens (1989) pointed out that in order to show that interspecific competition occurs, one must demonstrate that the species involved overlap in resource use and that competition over these resources has negative effects. **Table 1** lists criteria that ecologists use to become more convinced that interspecific competition is producing negative effects in modern populations. Much of the data on resource utilization, such as Figure 19, satisfy criteria 1 and 2 only. A second, better way to analyze competition is thus to conduct experiments of the type described earlier. Humans have inadvertently conducted some of the best experiments in competition by introducing species into new areas.

Some introduced ants have extended their distributions with the help of human beings and in the process have eliminated the native ant fauna through competition. Relatively few species of ants have shown a striking ability to displace resident species. The Argentine ant (*Linepithema humile*), which was first discovered in California in 1907 and has been spreading ever since, is displacing native ants in many temperate and subtropical areas (Holway 1999). Holway (1999) has analyzed the reasons for the strong competitive ability of Argentine ants. He found that Argentine ants were more effective at exploitative competition than native ants in northern California. Baits set out at fixed distances from ant colonies were found within four minutes by Argentine ants, in contrast to the 10–35 minutes required by native ants. Argentine ants were not more successful in direct aggression—sometimes native ants won, and sometimes Argentine ants won. The situation concerning chemical defensive compounds is similar: both native ants and Argentine ants produced chemical repellents that worked equally well against other species. The key to Argentine ant success seems to be in pure numbers—they are more numerous than native ants, because they form supercolonies. Whereas native ants form individual colonies and defend the territory around their colony from other ants, Argentine queens and workers move freely between different nests without any territorial defense. Worker numbers are much larger in supercolonies, and Argentine ants thus overwhelm their competitors by force of numbers. By securing most of the food resources in an area, Argentine ants can drive competing native ant species extinct.

We should not assume that competition in natural populations is always occurring. Wiens (1977) has argued that competition may be rare in some populations because of high environmental fluctuation. According to this argument, populations are typically below the carrying capacity of their environment, and thus resources

Table 1 Criteria for establishing the occurrence of interspecific competition, listed according to the strength of the evidence for its occurrence in natural populations.

Criteria	Strength of evidence
1. Observed checkerboard patterns of distribution consistent with predictions	Weak
2. Species overlap in resource use	↓
3. Intraspecific competition occurs	Suggestive
4. Resource use by one species reduces availability to another species	↓
5. One or more species is negatively affected	Convincing
6. Alternative process hypotheses are not consistent with patterns	

SOURCE: Wiens (1989), p. 17.

are plentiful. Occasionally a "crunch" occurs, a period of scarcity in which competition does occur. This may happen only once every five or ten years, or even less frequently, implying that competition may be difficult to detect in most short-term studies.

Evolution of Competitive Ability

If two species are competing for a resource that is in short supply, both would benefit by evolving differences that reduce competition. The benefit involved is a higher average population size for each species, and presumably a reduced possibility of extinction. But in many cases it will be impossible to evolve differences that reduce competition. Consider, for example, food size as a limiting resource. If species A evolves such that it uses smaller food items than species B, it still may encounter a third species, C, that also feeds on small-sized food. Thus species may be constrained by a web of other possible competitors, such that the option of evolving to avoid competition is not always feasible. If a species cannot avoid competition, it must evolve competitive ability. Competitive ability is one element of the more general problem of the evolution of life history strategies.

Ecologists have used two general approaches to the general question of the evolution of competitive ability as an element in life history strategies. Animal ecologists have utilized the theory of r-selection and K-selection first proposed by MacArthur and Wilson (1967), while plant ecologists have utilized a related theory of plant strategies, the C-S-R model developed by Grime (1979).

Theory of r-Selection and K-Selection

The idea of competitive ability in animals is an ecological concept that is intuitively clear but difficult to define, and to understand how competitive ability might evolve we need to look at life history strategies more broadly. To understand life history evolution, we can begin with the Lotka-Volterra equations for competition, which are based on the logistic curve for each competing species. Two parameters characterize the logistic curve of each competing species: r (rate of increase) and K (saturation density). We can characterize organisms by the relative importance of r and K in their life cycles.

In some stable environments, organisms exist near the asymptotic density (K) for much of the year, and these organisms are subject to K-selection. In other unstable or unpredictable habitats the same organisms may rarely approach the asymptotic density but instead remain on the rising portion of the curve for most of the year; these organisms are subjected to **r-selection**. MacArthur and Wilson (1967) defined r-selection and K-selection to be density-dependent natural selection. As a population initially colonized an empty habitat, r-selection would predominate for a time, but ultimately the population would come under K-selection.

Species that are r-selected seldom suffer much pressure from interspecific competition, and hence they evolve no mechanisms for strong competitive ability (**Table 2**). Species that are K-selected exist under both intraspecific and interspecific competitive pressures. The pressures of K-selection should thus push organisms to use their resources more efficiently.

If K-selection is a complete description of competitive ability, we should be able to predict the outcome

Table 2 Characteristics of *r*-selected species and *K*-selected species. Many species will have characteristics intermediate between these two extreme life history strategies.

r-selected life history	*K*-selected life history
Small-sized organisms	Large-sized organisms
Many small reproductive units (seeds, spores, offspring)	Few larger reproductive units
Little energy used per reproductive unit	Much energy used to produce one reproductive unit
Early maturity	Late maturity and often parental care
Short expectation of life	Long life expectancy
Single reproductive episode (semelparous)	Many reproductive episodes (iteroparous)
Type 3 survival curve (Figure 6)	Type 1 or 2 survival curve

of competition in laboratory situations by knowing the *K* values for the two competing species. We cannot do this, however, because of the third parameter in the Lotka-Volterra equations for competition—the competition coefficients α and β. Species can evolve competitive ability by the process of α-selection (Gill 1974). Any mechanism that prevents a competitor from gaining access to limiting resources will increase α (or β) and thereby improve competitive ability. Most types of interference competition fall into this category. Territorial behavior in mammals and birds, and allelopathic chemicals in plants, are two examples of interference attributes that keep competing species from using resources.

One major evolutionary problem with α-selection is that the strategy of interference often affects members of the same species as well as members of competing species, such that competitive ability is achieved only at the expense of a reduction in the species' own values of *r* and *K*. An example is a shrub that produces chemicals that retard the germination and growth of competing plants but that also induces autointoxication after several years (Rice 1984). An individual's negative effects on members of its own species present no evolutionary problem so long as the affected individuals do not include the individual itself or its kin.

Alpha-selection for interference attributes can also operate when organisms are at low density. In animals, the evolution of a broad array of aggressive behaviors

has been crucial in substituting ability in mock combat for ability to utilize resources in competition in many situations (MacArthur 1972), and we can recognize an idealized evolutionary gradient:

Low density − colonization and growth (*r* selection)
$\downarrow$
High density − resource competition (*K* selection)
$\downarrow$
High density − interference mechanisms (α selection)
prevent resource competition

Populations may exist at all points along this evolutionary gradient because competition for limiting resources is only one source of evolutionary pressure that molds the life cycles of plants and animals (Roff 1992).

Grime's Theory of Plant Strategies

Vascular plants face two broad categories of factors that affect their growth and reproduction. One category includes shortages of resources such as light, water, or nitrogen; temperature stresses; and other physical-chemical limitations. This category Grime (1979) called stress. A second category includes all the factors classified as disturbances, including grazing, diseases, wind storms, frost, erosion, and fire. Grime examined the four possible combinations of these two categories and recognized that, for one combination, no strategy was possible:

| | *Intensity of stress* | |
Intensity of disturbance	Low	High
Low	Competitive (*K*) strategy	Stress-tolerant strategy
High	Ruderal (weed or *r*) strategy	None possible

If stress and disturbance are too severe, no plant can survive. Grime (1979) suggested that the other three strategies formed the primary focus of plant evolution, and that individual plant species have tended to adopt one of these three life history models.

The three strategies can be diagrammed as a triangle (**Figure 20**), which emphasizes that these three strategies represent trade-offs in life history traits. A plant cannot be good at all three strategies but must trade off one set of traits against another (Wilson and Lee 2000). Competitive plants show characteristics of *K*-selection—dense leaf canopies, rapid growth rates, low levels of seed production, and relatively short life

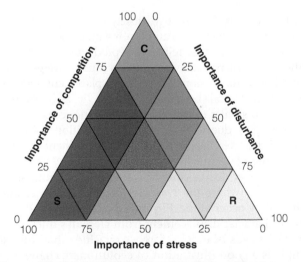

Figure 20 Grime's triangle model of plant life history strategies. Plant attributes evolve within this life history space depending on the relative importance of three factors: competition (*C*), stress (*S*), and disturbance (represented by *R*, for ruderal or weed strategy). It is for these factors that this model is also called the *C-S-R* model. (Modified from Grime 1979.)

spans—and many perennial herbs, shrubs, and trees show these characteristics. Stress-tolerant plants often have small leaves, slow growth rates, evergreen leaves, low seed production, and long life spans. Ruderals are weeds that thrive on disturbance; they exhibit small size, rapid growth, are often annual plants, and devote much of their resources to seed production. Ruderals are the *r*-strategists of the plant world.

The evolution of competitive ability, although viewed differently by botanists and zoologists, has achieved a convergence of ideas. Both *r*- and *K*-selection theory and Grime's theory describe well the trade-offs organisms must face in evolutionary time. Organisms cannot become good at everything, and adaptations are always a compromise between conflicting goals.

Westoby's Leaf-Height-Seed Theory of Plant Strategies

One problem with the C-S-R triangle theory of Grime (1979) is that it has been difficult to place a particular species at a point within the triangle by measuring some ecological traits. Mark Westoby (1998) proposed an alternative scheme of plant strategies that is empirically based on three measures of plants that can be readily taken in the field. The three axes of Westoby's scheme are:

- Specific leaf area—the light-capturing area deployed by the plant per unit of dry mass allocated to leaves

- Height of the plant canopy at maturity

- Seed mass

Both plant height and seed mass are readily measured for plants. Specific leaf area is somewhat more difficult to conceptualize. It measures the light-catching area deployed by a plant, and is analogous to the expected rate of return on investment. A high specific leaf area allows the plant to obtain a shorter payback time on a gram of leaf matter invested in a leaf. A plant species employing a low specific leaf area achieves a longer life span through higher structural strength and sometimes by way of defensive chemicals such as tannins.

Plants may produce many small seeds or fewer large seeds, and we expect ruderal or *r*-selected species of plants to produce many small seeds and competitive or

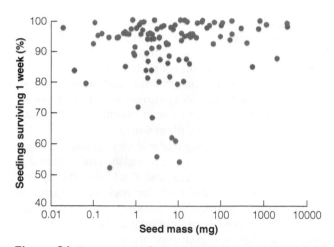

Figure 21 Percentage of plant seedlings surviving the first week after germination in relation to seed size. Data for 112 species from tropical to temperate habitats were used. Seed mass is on a logarithmic scale. There is a significant positive relation, indicating that larger seeds are more likely to survive the first week after emergence, but there is wide scatter from different species, so that not all small seeds survive poorly. (Data from Moles and Westoby 2004.)

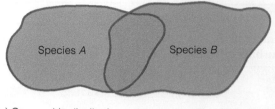

(a) Geographic distribution

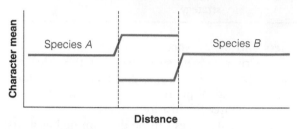

(b) Character changes

Figure 22 Schematic view of character displacement arising from interspecific competition in the zone of overlap of two species. The character measured must be one that is critical in competition between the species. This scheme is inferred as an explanation of the observations illustrated in Figure 23.

K-selected plants to produce fewer smaller seeds. If seed survival was constant over all seed sizes, we would predict that small seed producers would win out. But this clearly does not happen, and one hypothesis is that large-seeded species make up for their low seed production by increased survival during seedling establishment. There is no relationship between seed mass and survival to the newly emerged seedling stage, but survival to one week of age is higher in plant species with larger seeds (**Figure 21**). Moles and Westoby (2004) concluded that this survival advantage of larger seeds was not large enough to permit larger-seeded species to outcompete smaller-seeded plants unless larger-seeded species had a longer reproductive life span. There is a clear trade-off between a plant producing many small seeds, each with a lower chance of establishment, and producing fewer larger seeds, each with a high chance of successful establishment (Westoby et al. 2002). However, the exact quantitative trade-offs need to be measured more carefully to decide which of these strategies will be favored by natural selection in a particular competitive environment.

Character Displacement

One evolutionary consequence of competition between two species has been the divergence of the species in areas where they occur together. This sort of divergence is called **character displacement (Figure 22)** and can arise

for two reasons. Because two closely related species must maintain reproductive isolation, some differences between them may evolve that reinforce reproductive barriers. In other cases, interspecific competition causes divergence in critical niche dimensions. Character displacement is an important ecological hypothesis because it assumes that species too similar to one another could not coexist without diverging due to interspecific competition. Observations of character displacement are thus consistent with the predictions of Gause's hypothesis.

Character displacement is often inferred from studies in areas where the two species occur together and where they occur alone. **Figure 23** gives a classic example of character displacement from Darwin's finches on the Galápagos Islands. Before we conclude that this example is a good illustration of evolutionary changes in competing populations, we must satisfy six criteria (Schluter and McPhail 1992):

1. The pattern observed could not have occurred by chance.

2. The observed phenotypic differences should have a genetic basis.

3. The trait differences should result from actual evolutionary changes.

4. The morphological differences should reflect differences in resource use.

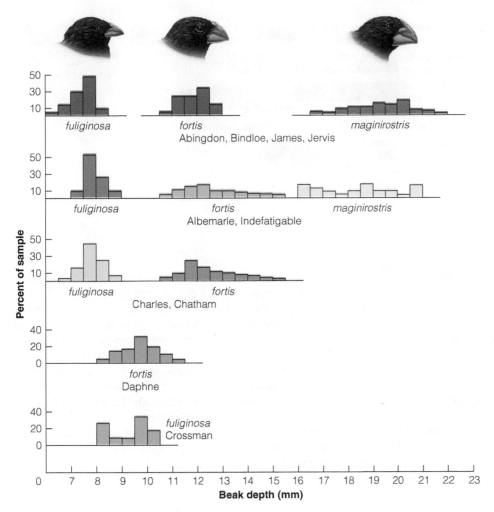

Figure 23 Character displacement in beak size in Darwin's finches from the Galápagos Islands. Beak depths are given for *Geospiza fortis* and *G. fuliginosa* on islands where these two species occur together (upper three sets of islands) and alone (lower two islands). *Geospiza magnirostris* is a large finch that occurs on some islands. (After Lack 1947.)

5. The sites of sympatry and allopatry should not differ greatly in environmental factors that affect the phenotype.

6. There must be independent evidence for competition between the species.

For Darwin's finches, all these criteria are satisfied (Grant and Grant 2006). The change in beak size in *Geospiza fortis* in isolation on Daphne Island, for example, is much greater than one would predict from observed variation on any of the other islands. Beak characters in *G. fortis* have a very high heritability (off-spring resemble parents), which suggests that the variation in beak depth in *Geospiza* shown in Figure 23 is largely genetic in origin. There is good observational evidence of competition for food in Darwin's finches.

Many examples of character displacement in the feeding morphology of carnivores have been measured. One example comes from three closely related small carnivorous marsupials that live on Tasmania—the spotted-tailed quoll (*Dasyurus maculatus*), the eastern quoll (*Dasyurus viverrinus*), and the Tasmanian devil (*Sarcophilus laniarius*). Because these carnivores are sexually dimorphic (males are about twice the size of females), the sexes must be considered separately. **Figure 24** shows the even spacing of body mass in these three carnivores that is consistent with the hypothesis of character displacement in the past (Jones 1997).

There are now many cases in which character displacement has been conclusively demonstrated (Dayan and Simberloff 2005). These cases all attest to the important role of interspecific competition in the evolution of species traits.

Apparent Competition and Indirect Effects

Competition between species is usually thought of in terms of two species directly interacting over limited resources. But organisms of different species may interact

Figure 24 Character displacement in body size in the carnivorous marsupials of Tasmania. There is a regular progression of body sizes that is reflected in their hunting behavior and prey eaten, and this character displacement is most readily explained by interspecific competition for food in the past. (Data from Jones 1997.)

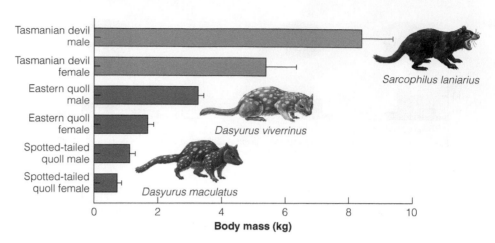

directly or indirectly (**Figure 25**). Interference competition occurs by direct effects in which, for example, two species of birds vie for access to tree holes for nesting. Exploitative competition involves indirect effects because the two species have no interactions with each other but interact only through a third species or a shared resource. For example, if buffalo and grasshoppers eat the same grass, exploitative competition may occur even though buffalo have nothing directly to do with grasshoppers. Indirect effects are often surprising and can take on a variety of forms (Abrams 1987). Holt (1977) pointed out how indirect effects could produce apparent competition. Consider two herbivores, such as

rabbits and pheasants, that do not eat any of the same foods or compete for any essential resources. If these two species have a common predator, an increase in the abundance of rabbits could increase the abundance of the predator, which might then eat more pheasants and reduce their numbers. In systems like this, one could easily be fooled into thinking that two species were competing because when one increased in numbers, the other decreased, and vice versa. The important idea here is that we should try to understand the mechanisms behind interactions between species and not simply describe how numbers may go up or down without knowing why.

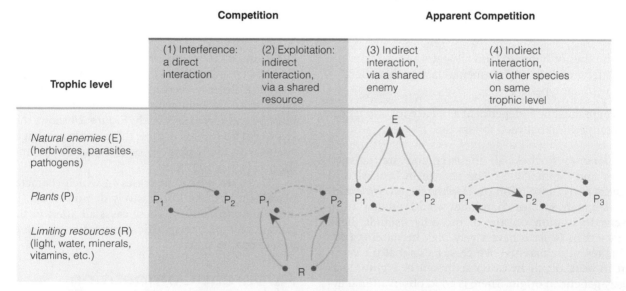

Figure 25 Illustration of possible pathways of interspecific competition, in this case for plants. Solid lines are direct interactions, dashed lines are indirect ones. An arrowhead indicates a positive effect, a circle indicates a negative effect. A similar type of interaction scheme can be applied to animals. Species can affect the abundance of other species without direct interactions. (After Connell 1990.)

What Is a Phase Plane, and What Is an Isocline?

The dynamics of two interacting populations can be illustrated graphically in two ways:

1. *Time series.* This is the usual way that population data are plotted, with population size on the *y*-axis and time on the *x*-axis. For two or more interacting populations there will be two or more lines on the graph.

2. *Phase plane.* By taking cross sections of the time series plot, one can abstract time from the diagram and plot species 1 numbers (*x*-axis) directly against species 2 numbers (*y*-axis). This type of diagram is called a phase plane. Figure 2 is an example, and phase-plane diagrams will be used in this chapter to illustrate the changes of one species population relative to another.

Phase planes are most useful for plotting models that have an equilibrium as a final solution. For population growth we are interested in the equilibrium that represents zero population growth $dN_1/dt = 0$. The line connecting all points that have zero population growth is called the *zero growth isocline* for that population. A simple thought-experiment will illustrate how the zero growth isocline can be constructed. Consider Figure 2 with respect to species 1 numbers:

1. Start a population in the upper right corner of the graph, with high numbers of species 1 and species 2 individuals. Species 1 will decrease in numbers because it is above carrying capacity in this simple model. From your starting point put an arrow on the graph pointing to the left for species 1.

2. Now start another population in the upper left side of the graph, with low numbers of species 1 and high numbers of species 2. Species 1 will now increase in numbers because it is below carrying capacity. Put another arrow on the graph pointing to the right from this starting point.

3. Somewhere between these extremes you could start a population that would not change in numbers for species 1 because it was exactly at carrying capacity for the fixed number of species 2. This point would be on the zero growth isocline.

4. Repeat these thought-experiments many times all over the area of this graph (the "phase plane"), placing arrows in the direction of movement of the species 1 population. Eventually you would define the blue diagonal line shown in Figure 2.

You have now constructed a phase-plane diagram with a zero growth isocline.

Summary

Competition is a negative interaction between species that occurs when both species strive to obtain resources that each needs. Theoretical models of competition indicate that, in cases of competition between two similar species, one species may be displaced, or both may reach a stable equilibrium. The possibility of displacement has given rise to the *competitive exclusion principle*, which states that complete competitors cannot coexist. Under simple laboratory conditions, one species often becomes extinct but sometimes coexists with another species. Natural communities show many examples of the coexistence of similar species, and this must be reconciled with the principle of competitive exclusion. One approach to solving this paradox is to suggest that natural communities are in a constant state of flux so that competition is interrupted in nature, and hence final ecological displacement is not observed. Another approach is to suggest that competition has occurred and that the interrelations we now see are the outcome of competition, displacement, and subsequent evolution in the past, the "ghost of competition past." Organisms evolve competitive ability by becoming more efficient resource users and by developing interference mechanisms that keep competing species from using scarce resources.

Interspecific competition is common and can exert a major influence on population size in many natural populations. Experimental work suggests that the effects of competitive interactions in field populations are greater in herbivores than in plants or carnivores. Detailed studies of the mechanisms of competition between species are needed to understand multispecies systems and to predict patterns in natural and agricultural communities.

Character displacement, or the evolution of morphological difference between competing species, is commonly observed in closely related species that live in the same area. Competition theory predicts that species will shift in the morphological traits that relate to the way in which competition occurs. Character

displacement thus follows as a prediction from the competitive exclusion principle of Gause.

The evolution of competitive ability can be evaluated within a broad framework of the evolution of life history traits. Weedy species colonize quickly and avoid competition and are often referred to as *r*-selected species, while species in stable communities are under evolutionary pressure to minimize competition by niche differentiation and specialization and are often referred to as *K*-selected species. Life history evolution in all organisms involves trade-offs in many dimensions, of which competitive ability is only one.

Review Questions and Problems

1. The introduced house sparrow (*Passer domesticus*) competes with the native house finch (*Carpodacus mexicanus*) in the western United States for nesting sites, and the house finch seems to lose out more frequently in interference competition both at feeders and at nest sites, even if nesting sites are not limited. In 1940 the house finch was introduced into the eastern United States. Discuss the potential impact of this eastern introduction of the house finch on the house sparrow, and list the observations and experiments you would like to do to investigate this species interaction. Bennett (1990) summarizes data on these species.

2. Black bears and grizzly bears in North America are presumed to be in competition. Discuss the resources for which they might be competing, and, following Table 1, design field experiments that would determine if they are competing and what the mechanisms of competition are. Apps et al. (2006) discuss one approach to this question.

3. Charles Darwin in *The Origin of Species* (1859, Chapter 3) states:

   As the species of the same genus usually have, though by no means invariably, much similarity in habits and constitution, and always in structure, the struggle will generally be more severe between them, if they come into competition with each other, than between the species of distinct genera.

   Discuss.

4. This chapter has discussed interspecific competition. What should be the relationships between interspecific competition and intraspecific competition? How could one measure the relative strengths of these two types of competition for a plant or animal species?

5. Both Adelie penguins and minke whales feed on crystal krill (*Euphausia crystallorophias*) and Antarctic

silverfish (*Pleuragramma antarcticum*) in the Western Ross Sea of Antarctica. How could you determine if there is competition between these two species in a large-scale system in which no possible experimental manipulation can be performed? Ainley et al. (2006) discuss this competitive interaction.

6. Many trees such as oaks and spruces are long-lived and form extensive mature forests but still produce many small seeds frequently throughout their long life. Discuss why this mixture of *K*-selection and *r*-selection traits (Table 2) might evolve.

7. Analyze the yeast results of Gause (1932) by the use of Lotka-Volterra plots (as in Figure 4), and predict the outcome of this competition from the estimates of α, β, K_1, and K_2.

8. Fruiting plants may compete with birds that disperse their seeds. If this competition occurs, it would benefit plants to evolve a sequence of fruit ripening times that do not overlap and thereby avoid interspecific competition. How could you test for character displacement in fruiting times in woody plants? Burns (2005) discusses this problem.

9. Competition for light in trees should produce an immediate benefit for individuals that are taller than their neighbors. Discuss the factors that may affect the height to which trees grow in terms of the costs and benefits of being tall. Koch (2004) discusses this problem.

10. Where in Grime's triangle (see Figure 20) would one expect to find annual plants? Trees? Cacti? What characteristics of plants might one use to quantify these three axes?

Overview Question

Sheep, rabbits, eastern grey kangaroos, and red kangaroos are possible competitors for food in the rangelands of eastern Australia. Design an interactive flowchart for testing the presence and intensity of competition among these herbivores.

Suggested Readings

- Abrams, P. A. 1998. High competition with low similarity and low competition with high similarity: Exploitative and apparent competition in consumer-resource systems. *American Naturalist* 152:114–128.

- Apps, C. D., B. N. McLennan, and J. G. Woods. 2006. Landscape partitioning and spatial inferences of competition between black and grizzly bears. *Ecography* 29:561–572.

- Chase, J. M., and M. A. Leibold. 2003. *Ecological Niches: Linking Classical and Contemporary Approaches.* Chicago: University of Chicago Press.

- Dayan, T., and D. Simberloff. 2005. Ecological and community-wide character displacement: The next generation. *Ecology Letters* 8:875–894.

- Grace, J. B. 1995. In search of the Holy Grail: Explanations for the coexistence of plant species. *Trends in Ecology & Evolution* 10:263–264.

- Gurevitch, J., J. A. Morrison, and L. V. Hedges. 2000. The interaction between competition and predation: A meta-analysis of field experiments. *American Naturalist* 155:435–453.

- Krasnov, B. R., N. V. Burdelova, I. S. Khokhlova, G. I. Shenbrot, and A. Degen. 2005. Larval interspecific competition in two flea species parasitic on the same rodent host. *Ecological Entomology* 30:146–155.

- Park, S. E., L. R. Benjamin, and A. R. Watkinson. 2003. The theory and application of plant competition models: An agronomic perspective. *Annals of Botany* 92:741–748.

- Tannerfeldt, M., B. Elmhagen, and A. Angerbjörn. 2002. Exclusion by interference competition? The relationship between red and arctic foxes. *Oecologia* 132:213–220.

- Wilson, J. B., and W. G. Lee. 2000. *C-S-R* triangle theory: Community-level predictions, tests, evaluations of criticisms, and relation to other theories. *Oikos* 91:77–96.

Credits

Illustration and Table Credits

Photo Credits

Species Interactions II: Predation

Key Concepts

- Predator-prey interactions can be analyzed with simple models for one predator–one prey systems.

- Simple models of predation often lead to predator-prey cycles rather than a stable equilibrium.

- Laboratory systems rarely lead to stable interactions between predators and prey, but they show the importance of prey refuges and spatial heterogeneity.

- Predation can be broken down into components—numerical, functional, developmental, and aggregative responses of predators to prey—to aid our understanding of the predation process.

- Multiple predator-multiple prey systems lead to more complex dynamics, and show the importance of predation to the evolution of escape behavior and warning coloration in animals.

From Chapter 11 of *Ecology: The Experimental Analysis of Distribution and Abundance*, Sixth Edition. Eugene Hecht.
Copyright © 2009 by Pearson Education, Inc. Published by Pearson Benjamin Cummings. All rights reserved.

KEY TERMS

aposematic Warning coloration, indicating to a predator that this prey is poisonous or highly defended against attack.

coevolution The mutual evolutionary influence between two species; each party in a coevolutionary relationship exerts selective pressures on the other, thereby affecting each others' evolution, back and forth.

environmental heterogeneity Variation in space in any environmental parameter such as soil pH or tree cover.

functional response The change in the intake rate of a predator in relation to the density of its prey species.

generalist predators Predators that eat a great variety of prey species.

handling time The time utilized by a predator to consume an individual prey item.

numerical response The change in the numbers or density of a predator in relation to changes in the density of its prey species.

optimal foraging theory A detailed model of how animals should forage to maximize their fitness.

prey isocline The contour line of densities of predator and prey at which the prey are in equilibrium; the impact of a predator exactly balances the prey's rate of population growth, so the prey population growth rate is zero.

safe sites For animals, sites where prey individuals are able to avoid predation; for plants, sites where seeds can germinate and plants can grow.

In addition to competing for food or space, species may interact directly via predation. Predation in the broad sense occurs when members of one species eat those of another species. Often, but not always, this involves the killing of the prey. Humans are now one of the major predators of the Earth's ecosystem. We prey on fishes in the oceans, and hunt grouse, geese, and deer for sport. In this chapter we explore how predation operates and what we need to know to understand its effects.

Five specific types of predation may be distinguished. Herbivores are animals that prey on green plants or their seeds and fruits; often the plants eaten are not killed but may be damaged. Typical predation occurs when carnivores prey on herbivores or on other carnivores. Insect parasitoids are a type of predator that

lay eggs on or near the host insect, which is subsequently killed and eaten. Parasites are plants or animals that live on or in their hosts and depend on the host for nutrition. They do not consume their hosts and thus differ little in their effects from herbivores. Finally, cannibalism is a special form of predation in which the predator and the prey are members of the same species. All these processes can be described initially with the same kind of mathematical models, and we will begin by considering them together as "predation" in the broad sense.

Predators do not interact only with their prey species: they can also interact with one another via competition (**Figure 1**). Competition between predators may be indirect when both predator species eat the same prey species that is in short supply, or it may be indirect via prey species that themselves compete for space or food. The important point is that predation in nature goes on within a context of other biotic interactions, including competition.

Predation is an important process from three points of view. First, predation on a population may restrict distribution or reduce abundance of the prey. If the affected

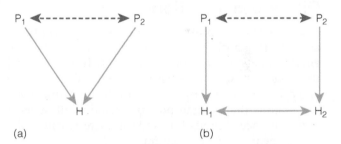

Figure 1 Schematic diagram of possible indirect effects (dotted arrows) between two predator species P_1 and P_2 that eat herbivores H_1 and H_2. (a) Indirect effects via exploitation. Two predators that share a common prey species may interact indirectly through exploitative competition of the common prey species such that there could be an indirect effect of P_1 on P_2. (b) Indirect effects without competition. Two predators eat two different prey species and do not interfere with one another and do not compete for food. But competition between the two prey species can cause effects on either or both predators indirectly. For example, if H_1 increases, P_1 will increase, H_2 will decrease because of competition with H_1, and because H_2 decreases, P_2 will also decrease. This indirect effect between the two predators is called apparent competition because an examination of predator numbers alone suggests that P_1 increases and P_2 decreases as a result of these interactions. The important point is that food web linkages can produce effects between two predators that ecologically do not interact directly.

animal is a pest, we may consider predation useful. If the affected animal is a valuable resource like caribou or domestic sheep, we may consider the predation undesirable. Second, along with competition, predation is another major type of interaction that can influence the organization of communities. Third, predation is a major selective force, and many adaptations we see in organisms, such as warning coloration, have their explanation in predator-prey coevolution.

We begin our analysis of the predation process by constructing some simple models. All these models have the underlying assumption that we can isolate in nature a system consisting of one predator species and one prey species.

Mathematical Models of Predation

The models we discuss in this section are of two types: those for organisms with discrete generations and those for organisms with continuous generations.

Discrete Generations

First we explore a simple model of predator-prey interactions using a discrete generation system. In seasonal environments, many insect parasitoids (predator) and their insect hosts (prey) have one generation per year and can be described by a model of the following type.

Assume that a small prey population will increase in the absence of predation, and this increase can be described by the logistic equation:

$$N_{t+1} = (1.0 - B z_t)N_t \qquad (1)$$

where N_1 = population size
t = generation number
B = slope of reproductive curve
$z_t = (N_t - N_{eq})$ = deviation of present population size from equilibrium population size in the absence of the predator

In the presence of a predator, we must subtract from this equation a term accounting for the individuals eaten by predators, and this could be done in a number of ways. All the prey above a certain number (the number of **safe sites**) might be killed by predators, or each predator might eat a constant number of prey. If, however, the abundance of the prey is determined by the abundance of

predators, the whole predator population must eat proportionately more prey when prey are abundant and proportionately less prey when prey are scarce. They could do this by becoming more abundant when prey are abundant or by being very flexible in their food requirements. We subtract a term from the prey's logistic equation:

$$N_{t+1} = (1.0 - B z_t)N_t - CN_tP_t \qquad (2)$$

where P_t = population size of predators in generation t
C = a constant measuring the efficiency of the predator

What about the predator population? We assume that the reproductive rate of the predators depends on the number of prey available. We can write this simply as

$$P_{t+1} = QN_tP_t \qquad (3)$$

where P_t = population size of predator
N = population size of prey
t = generation number
Q = a constant measuring the efficiency of utilization of prey for reproduction by predators

Note that if the prey population (N) were constant, this equation would describe geometric population growth for the predator.

To put these two equations together and interpret them, we must first obtain the maximum reproductive rates of both predator and prey. When predators are absent and prey are scarce, the net reproductive rate of the prey will be, approximately,

$$N_{t+1} = (1.0 - B N_{eq})N_t \qquad (4)$$

or

$$R = \frac{N_{t+1}}{N_t} = 1.0 - B N_{eq} \qquad (5)$$

where R = maximum finite rate of population increase of the prey

For the predator, when the prey population is at equilibrium and predators are scarce, predators will increase at

$$P_{t+1} = QN_{eq}P_t \qquad (6)$$

or

$$S = \frac{P_{t+1}}{P_t} = QN_{eq} \qquad (7)$$

where S is the maximum finite rate of population increase of the predator.

Let us now work out an example. Let the maximum rate of increase of the prey (R) = 1.5 and N_{eq} = 100, so that the absolute value of the slope of the reproductive curve $B = 0.005$. Assume that the constant C measuring the efficiency of the predator is 0.5. Thus

$$N_{t+1} = (1.0 - 0.005 z_t) N_t - 0.5 N_t P_t \qquad (8)$$

Assume that under the best conditions, the predators can double their numbers each generation ($S = 2.0$), so that the constant Q is

$$S = Q N_{eq}$$

$$2.0 = Q(100) \qquad (9)$$

or

$$Q = 0.02 \qquad (10)$$

Consequently, the second equation is

$$P_{t+1} = 0.02 N_t P_t \qquad (11)$$

Start a population at $N_0 = 50$ and $P_0 = 0.2$:

$$N_1 = \left([1.0 - 0.005(50 - 100)]50 \right) - [(0.02)(50)(0.2)]$$
$$= 62.5 - 5.0 = 57.5$$
$$P_1 = (0.02)(50)(0.2)$$
$$= 0.2$$

For the second generation,

$$N_1 = \left([1.0 - 0.005(57.5 - 100)]57.5 \right)$$
$$\quad - [(0.5)(57.5)(0.2)]$$
$$= 69.72 - 5.75 = 63.97$$
$$P_1 = (0.02)(57.5)(0.2)$$
$$= 0.2$$

These calculations can be carried over many generations to produce the results shown in **Figure 2**—a cycle of predator and prey numbers.

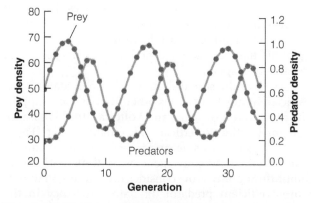

Figure 2 Population changes in a hypothetical predator-prey system with discrete generations. For the prey population $N_{eq} = 100$, $B = 0.005$, and $C = 0.5$. For the predator, $Q = 0.02$.

A stable oscillation in the numbers of predators and prey is only one of four possible outcomes; the others are stable equilibrium with no oscillation, convergent oscillation, and divergent oscillation leading to the extinction of either predator or prey. Maynard Smith (1968) has shown that the range of variables for a stable equilibrium without oscillation is very restricted. An example will illustrate this solution. Let $N_{eq} = 100$, $B = 0.005$, and $C = 0.5$ for the prey, while $Q = 0.0105$ ($S = 1.05$) for the predator. For the first generation, from a starting population of 50 prey and 0.2 predators:

$$N_1 = \left([1.0 - 0.005(50 - 100)]50 \right) - [(0.5)(50)(0.2)]$$
$$= 62.5 - 5.00 = 57.50$$
$$P_1 = (0.0105)(50)(0.2)$$
$$= 0.105$$

Similarly,

	N	P
Second generation	66.70	0.063
Third generation	75.70	0.044
Fourth generation	83.20	0.035
Fifth generation	88.70	0.031

The populations show decreasing small oscillations and gradually stabilize around a level of 95.2 for the prey and 0.048 for the predator.

Discrete generation predator-prey models show a variety of dynamic behaviors much like those seen in discrete population growth models.

Continuous Generations

Many predators and prey have overlapping generations, with births and deaths occurring continuously; vertebrate predators provide many examples. For the continuous-generation case, Lotka (1925) and Volterra (1926) independently derived a set of equations to describe the interaction between populations of predators and prey. Vito Volterra, a professor of physics in Rome, became interested in population fluctuations in 1925 when his daughter became engaged to a young marine biologist who was studying the effects of World War I on fish catches in the Adriatic. The early models of Lotka and Volterra were unrealistic, and other models that are capable of greater biological realism have replaced them (Berryman 1992). The best general models were developed by Rosenzweig and MacArthur (1963) as graphic models.

Consider first the population growth of a prey species in relation to predator and prey abundance (**Figure 3**). Now do a hypothetical experiment: construct a series of populations at different predator and prey densities, and at each point measure whether the prey increase or decline. For example, at point A in Figure 3 there are many predators, and prey will certainly decline. At point B, there are few predators, and prey will increase. At point C there are many predators and many prey, and excessive predation will drive prey numbers down. By following this process for a series of points we can divide the area of the graph into a zone of prey increase and a zone of prey decrease. This fixes the **prey isocline**, the boundary between these two zones at which the rate of increase of the prey population is zero. At equilibrium the prey population must exist somewhere on this line. Lotka and Volterra made the simple assumption that the prey isocline was a horizontal line (see Figure 3), but in the more realistic Rosenzweig-MacArthur model the prey isocline always has a "hump." What ecological factors cause a hump-shaped prey curve? The key process is that, as prey numbers build up, prey begin to limit their own rate of increase because of food shortage, disease, or social interactions. To the left of the hump, the dominant limitation on the prey is from the predators. Above the isocline hump, at higher prey numbers, the prey isocline curve falls off because the dominant limitation on the prey rate of increase comes from prey intraspecific competition, and predator limitation on the prey becomes less and less significant. The exact shape of the prey curve will depend on the demographic characteristics of the prey and the carrying capacity of the environment, which sets an upper limit to prey abundance.

Now consider the population changes of a predator that is food-limited at low prey densities and eats only a single prey species. When prey numbers are high, predator numbers should increase. But at high predator density, predators stop increasing because of other limitations, such as territorial behavior in wolves or a shortage of burrow sites for predatory crabs. The resulting predator isocline is shown in **Figure 4**. The predator isocline will not always be this shape, and not all predators will have exactly the same shape of isocline. A key point to note is that the more efficient the predator, the more the predator isocline is positioned to the left in Figure 4.

By superimposing the two isoclines in Figures 3 and 4, we get a graphic model of a predator-prey interaction. In this case, by examining the vectors around the equilibrium point, we can see that this is a stable equilibrium for both predator and prey (**Figure 5a**). In the lower right quadrant (C in Figure 5), the prey is decreasing but the predator is increasing, so the vector points upward and inward. The lower left quadrant (B) represents increasing prey and decreasing predators. The upper left quadrant represents both species decreasing. This model, the Rosenzweig-MacArthur model of predator-prey interactions, is useful because we can explore in a graphic manner the effects of simple changes to the predator-prey system.

Consider the situation in which the predator is not restricted by any limitations other than its food supplies (the assumption of Lotka and Volterra). In this case the predator isocline is vertical and remains linear (Figure 5a). This system is stable, and if disturbed from equilibrium, it will show convergent oscillations back to the equilibrium point. Now consider this same system with a more efficient predator. Predator efficiency in this graphic presentation means that the predators can subsist on lower prey numbers, so the predator isocline is moved to the left on the graph (Figure 5b). When the predator isocline intersects the prey isocline to the left of the hump, there is no point equilibrium for the sys-

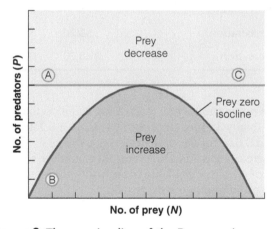

Figure 3 The prey isocline of the Rosenzweig-MacArthur model for a predator-prey interaction. In the purple zone the prey can increase in abundance. The simple model of a horizontal prey isocline assumed by Lotka and Volterra is shown in blue. The hump-shaped Rosenzweig-MacArthur prey isocline is more realistic than the Lotka-Volterra isocline because as prey numbers increase more predators can be supported but at a diminishing rate. As prey numbers build up, prey begin to limit their own increase because of food shortage, disease, or social interactions. At the hump of the isocline a maximum number of predators can be supported. Above the hump, at higher prey numbers, the prey isocline curve falls off because the dominant limitation on the prey rate of increase comes from prey intraspecific competition, and fewer predators are needed to hold down the prey rate of increase. (After Rosenzweig and MacArthur 1963.)

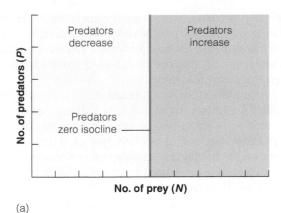

(a)

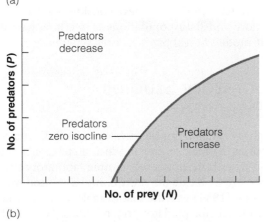

(b)

Figure 4 The predator isocline of the Rosenzweig-MacArthur model for predator-prey interaction. In the pink-colored zones the predator can increase in abundance. (a) In the simple model of a vertical isocline assumed by Lotka and Volterra, there is a single prey density above which predator populations can grow, and below which they decrease. (b) A more realistic predator isocline, which bends to the right because as predators increase in number they compete with one another for breeding sites and other resources. Not all predators will have the same shape of isocline.

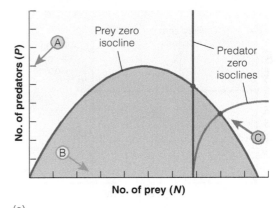

(a)

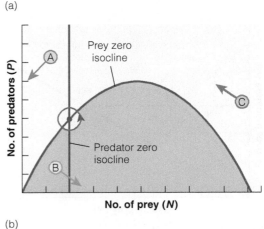

(b)

Figure 5 The predator and prey isoclines superimposed for the Rosenzweig-MacArthur model for predator-prey interaction. The equilibrium points are indicated by red dots, and the vectors from points A, B, and C indicate the direction of movement in the phase plane. (a) When the predator isocline intersects the prey isocline to the right of the hump, there is a stable equilibrium point, regardless of the exact shape of the predator isocline. (b) When the predator isocline intersects the prey isocline to the left of the hump, limit cycles like those in Figure 2 arise. Depending on the exact slopes of the lines, these cycles may be large enough to lead to the extinction of the predator, the prey, or both. The key point is that predator-prey systems that intersect to the left of the hump in the prey zero isocline are unstable compared with those that intersect to the right of the hump.

tem, and populations endlessly follow a stable cycle around the hypothetical equilibrium point. The farther the equilibrium point is from the hump of the prey isocline, the larger will be the amplitude of the resulting cycles and the greater the possibility of extinction.

The Rosenzweig-MacArthur model of predator-prey interactions thus reveals a wide variety of dynamic behavior, from stability to strong oscillations. This model provides a focus for asking simple questions about predator-prey systems, such as, What would happen if prey became less abundant? The model also serves as an entry point into understanding the more complex real world.

All these predator-prey models make a series of simplifying assumptions about the world, including a

homogeneous world in which there are no refuges for the prey or different habitats, and that the system is one predator eating one prey. Relaxing these assumptions leads to more complex and more realistic Rosenzweig-MacArthur models (Hastings 1997).

The classical form of the Rosenzweig-MacArthur model uses a vertical predator isocline, as in **Figure 6**, which implies that the rate of increase of the predator population

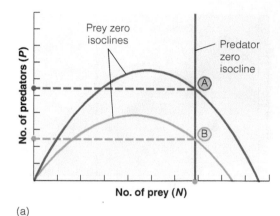

(a)

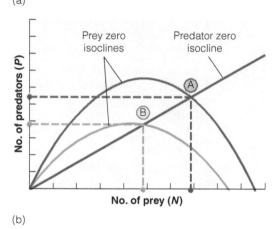

(b)

Figure 6 Predator-prey isoclines in (a) the classical Rosenzweig-MacArthur model and (b) the ratio-dependent model.

Two prey isoclines are shown for less productive (blue) and more productive (purple) habitat. Increasing prey productivity changes only the equilibrium predator abundance from A to B in the classical model, but changes both predator and prey abundance in the ratio-dependent model. The equilibrium intersection points are shown by dotted lines, and the resulting equilibrium numbers of predators and prey by the dots on each axis.

is controlled completely by the density of the prey. In this model the exact equilibrium for the prey species depends only upon the predator's characteristics. In particular, if the productivity of the prey population increases, the equilibrium density of the prey does not change (Figure 6a). All the gain in prey productivity goes to the predators, which increase in abundance. An alternative model, the ratio-dependent model suggested by Arditi et al. (1991), postulates a predator isocline that runs diagonally upward (Figure 6b). The ratio-dependent model assumes that the predation rate depends on the ratio of predators to prey, rather than just on prey numbers alone (Arditi et al. 1991; Akcakaya et al. 1995). These two models make quite different predictions about the relationship between prey abun-

dance and predator abundance. In the ratio-dependent model, as prey productivity is increased, predator and prey equilibria both rise. In some biological systems the classical theory may be adequate, but in other systems the ratio-dependent theory fits better.

The simple models of predation that we have just discussed are interesting in that they indicate that oscillations may be an outcome of a simple interaction between one predator species and one prey species in an idealized environment. In discrete generation systems, the outcome of a simple predation process may be stable equilibrium, oscillations, or extinction. Discrete systems are more likely to lead to extinction in a fluctuating environment (Gotelli 1998). We next consider evidence from laboratory and field populations to see how well these simple models fit real predator-prey systems.

Laboratory Studies of Predation

Laboratory systems can be set up in which the major assumptions of predator-prey models can be met, and then we can investigate how these simple laboratory systems work before we tackle the more complex natural world.

Gause (1934) was the first to make an empirical test of the models for predator-prey relations. He reared the protozoans *Paramecium caudatum* (prey) and *Didinium nasutum* (predator) together in an oat medium. In his initial experiments, *Didinium* always exterminated *Paramecium* and then died of starvation—that is, the system went to extinction (**Figure 7a**)—and this is not very interesting biologically. Extinction occurred under all the circumstances Gause used for this system—making the culture vessel very large, introducing only a few *Didinium*, and so on. The conclusion was that the *Paramecium-Didinium* system did not show either a stable equilibrium or a stable limit cycle. Gause thought that stability could not be achieved because of a biological peculiarity of *Didinium*: It was able to multiply very rapidly even when prey were scarce, the individual *Didinium* becoming smaller and smaller in the process.

Gause then introduced a complication into the system: To the oat medium he added sediment, which constituted a refuge for the prey. *Paramecium* in the sediment were safe from *Didinium*, which never entered it. In this system, the *Didinium* again eliminated the *Paramecium*, but only from the clear-fluid medium; *Didinium* then starved to death, and the *Paramecium* hiding in the sediment emerged to increase in numbers (Figure 7b). The experiment ended with many prey and no predators. The system had reached a stable point predicted by the mathematical model, but it was a biologically uninteresting system with the predators extinct. In doing these experiments Gause added an

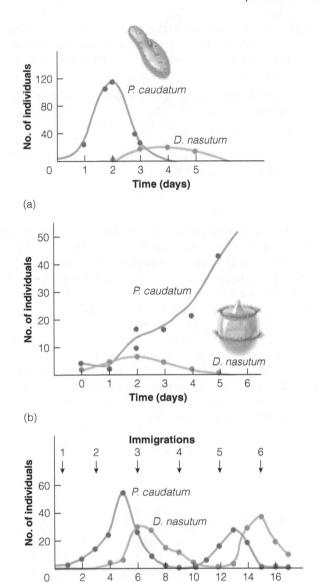

Figure 7 Predator-prey interactions between the protozoans *Paramecium caudatum* and *Didinium nasutum* in three microcosms. (a) Oat medium without sediment, (b) oat medium with sediment, and (c) oat medium without sediment and with immigration. (After Gause 1934.)

important idea to our understanding of predation: the potential importance of refuges for prey species.

Gause, quite determined, tried yet another system, introducing **immigration** into the experimental setup. Every third day he added one *Paramecium* and one *Didinium*, which produced the results shown in Figure 7c. Gause concluded that in *Paramecium* and *Didinium* stable oscillations in predator and prey numbers are not a property of the predator-prey interaction itself, as

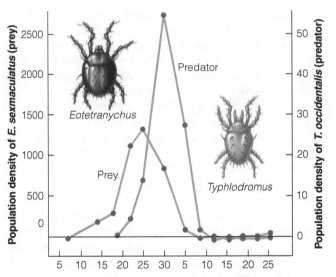

Figure 8 Densities (per unit area of orange) for the prey mite *Eotetranychus sexmaculatus* and the predator mite *Typhlodromus occidentalis*, with 40 oranges, 20 of which provided food for the prey (good habitat) alternating with 20 foodless (covered) oranges (poor habitat). (After Huffaker 1958.)

some models predict, but apparently are a result of constant interference from outside the system.

Carl Huffaker, working at Berkeley on the biological control of insect pests, completed a classic set of experiments on predator-prey dynamics that had important implications for predator-prey theory. Huffaker (1958) questioned Gause's conclusions that the predator-prey system was inherently self-annihilating without some outside interference such as immigration. He claimed that Gause had used too simple a microcosm. Huffaker studied a laboratory system containing a phytophagous mite, *Eotetranychus sexmaculatus*, as prey, and a predatory mite, *Typhlodromus occidentatis*, as predator. The prey mite infests oranges, so Huffaker used these fruits for his experiments. When the predator was introduced onto a single prey-infested orange, it completely eliminated the prey and died of starvation (like Gause's *Didinium*). Huffaker gradually introduced more and more spatial heterogeneity into his experiments. In some cases he placed 40 oranges on rectangular trays similar to egg cartons and partly covered some oranges with paraffin or paper to limit the available feeding area; in other cases he used rubber balls as "substitute oranges" so that he could either disperse the oranges among the rubber balls or place all the oranges together. In still other cases, he added whole new trays that included artificial barriers of petroleum jelly, which the mites could not cross.

All of Huffaker's simple systems eventually resulted in extermination of the populations. **Figure 8** illustrates a population that became extinct in a moderately complex environment containing 40 oranges. Finally, Huffaker produced the desired oscillation in a 252-orange universe with a complex series of petroleum-jelly barriers; in this system, the prey were able to colonize oranges in "hop, skip, and jump" fashion and keep one step ahead of the predator, which exterminated each little colony of the prey it found (**Figure 9**). The predators died out after 70 weeks, and the experiment was terminated.

Huffaker concluded that he could establish an experimental system in which the predator-prey relationship would not be inherently self-destructive. He admitted, however, that his system was dependent on local emigration and immigration, and that a great deal of environmental heterogeneity was necessary to prevent immediate annihilation of the system. The important idea that Huffaker's work added to our perspective on predator-prey theory is the concept of **environmental heterogeneity**. The world is not a uniform environment but consists of a variety of patches that are either good or bad for predator and prey alike. The addition of the simple idea of environmental heterogeneity into our thinking about ecological systems has had a revolutionary effect on our thinking about ecological communities, as we will see in our discussions of community dynamics.

Laboratory studies of predator-prey systems have carried us a long way from our starting point. What might we look for in predator-prey systems in the field? We must consider four aspects of predator-prey dynamics that have been simplified in both theoretical and laboratory studies:

- Multiple prey species being eaten by multiple predator species
- Refuges for the prey
- Spatial heterogeneity in habitat suitability for both the predator and the prey
- Evolutionary changes in predator and prey characteristics

We have assumed so far that predators have a strong effect on the abundance of their prey and vice versa, and we should consider whether this generalization holds for field situations. We can look for evidence of population oscillations that might result from predator-prey interactions in field populations. Associations of predator and prey might show evolutionary changes, and these evolutionary changes could be looked for in species that have recently come into contact in the field. Highly efficient predators introduced into a new ecosystem might cause the extinction of vulnerable prey. The richness of predator-prey theory should map onto the richness of interactions in field populations.

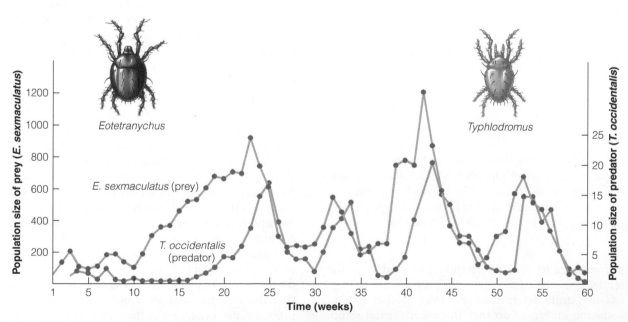

Figure 9 Predator-prey interaction between the prey mite *Eotetranychus sexmaculatus* and the predator mite *Typhlodromus occidentalis* in a complex laboratory environment consisting of a 252-orange system in which one-twentieth of each orange was exposed for possible feeding by the prey. (After Huffaker et al. 1963.)

Field Studies of Predation

How can we find out whether predators have a strong effect on the abundance of their prey? The obvious experiment is to remove predators from the system and to observe its response. Few such direct experiments have been properly conducted with adequate controls. An alternative is to use natural experiments in which selected areas differ in their predator fauna. Let us examine some case studies.

Woodland caribou in North America have been declining in abundance for the past 50 years, particularly in the southern part of their distribution along the Canada-U.S. border. Two reasons have been suggested for this decline: habitat loss leading to food limitation or increased predation from wolves and bears. **Figure 10** shows the kind of natural experiment that suggests that predators are the chief cause of the decline. On the north shore of Lake Superior, Pukaskwa National Park occupies about 2000 km² of nearly undisturbed boreal forest with an intact predator-prey system of caribou, moose, wolves, black bears, and lynx. On the Slate Islands there are no predators of caribou. Predation holds the average density of caribou in the park at the low density of 0.06 caribou per km², and on the predator-free Slate Islands caribou are about 100 times as abundant. Island caribou populations appear to be limited by food shortage (Bergerud and Elliot 1998).

ESSAY

Laboratory Studies and Field Studies of Predation

Predator-prey dynamics can be studied either in the laboratory or in the field. What are the advantages and disadvantages of these two kinds of studies for analyzing biological interactions? Can we directly apply the results of laboratory studies to field situations? These critical questions are not easy to answer.

Ecological laboratory studies are done in model systems or *microcosms*—small ecosystems housed in containers. Microcosms can range from simple two-species systems to complex communities of many different species. Although most microcosms are small, some—such as Biosphere 2 in Arizona or the Ecotron in England—are very large. We have already seen good examples of microcosms in Gause's work on predation in this chapter. Many of our ideas about competition and predation have come from microcosm research.

Laboratory studies of microcosms are controlled, and in the classical laboratory study only one or two factors are manipulated. In his predator-prey studies, Gause could vary the number of prey and predators introduced to start the cultures. Other factors that may affect the system, such as temperature or the size of the containers, are held constant. Replication is relatively easy to achieve, particularly with small organisms. In some but not all cases, results are obtained in a short time period. Costs of doing experiments are relatively low. Small-sized containers are typically used in laboratory microcosms.

Field studies are uncontrolled, and even in experimental field studies in which one or two factors are manipulated, all other factors are left to vary naturally. Consequently because there are warm years and cold years, wet years and dry years, all this natural variation can impinge on the results obtained. At first sight this would seem to be a great disadvantage of field studies. But in fact this variability is part of the real world, and the results of field studies are thus robust with natural variation in uncontrolled environmental factors. The greatest advantages of field studies relate to scale; some processes are too large spatially to study in the laboratory. An example would be turbulence in lakes or the oceans as it affects predation of larval fish, or dispersal and territorial social organization of wolves as it affects prey consumption rates. Another difference between field and microcosm studies is duration. Microcosm studies are typically of short duration, and many mistakes in ecology have been prompted by microcosm studies of too short duration (Carpenter 1996; Drenner and Mazumder 1999). Field studies of longer duration often uncover more of the complexity of ecological relationships that require additional study. Of course, field experiments are usually more expensive and require a long time commitment to obtain the results.

The best model for ecological studies is to use laboratory and field experiments together (Srivastava et al. 2004). Microcosms can suggest hypotheses and mechanisms that can be tested in longer-term field manipulations. Microcosms are not suited to studies of more than a few years in duration, but by combining their statistical power and experimental rigor with long-term field experiments, ecologists can have the best of both worlds (Fraser and Keddy 1997).

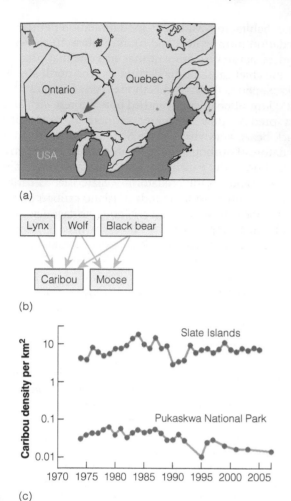

(a)

(b)

(c)

Figure 10 Impact of wolf predation on woodland caribou numbers along the north shore of Lake Superior.
(a) Pukaskwa National Park, indicated by the red arrow on the map, occupies about 2000 km² of undisturbed boreal forest. The Slate Islands (36 km²) lie offshore in Lake Superior, about 35 km west of the park. (b) The partial predator-prey food web for Pukaskwa National Park is illustrated. (c) There are no predators on the Slate Islands, and this lack of predators is correlated with a nearly 100-fold difference in average caribou density on the islands. Note that caribou densities are graphed on a logarithmic scale. Caribou are nearly extinct in the park. (Data from Bergerud et al. 2007.)

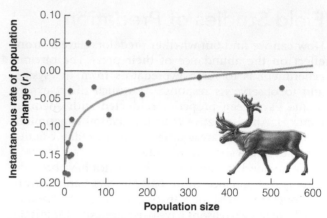

Figure 11 Average rates of increase of 15 populations of woodland caribou in British Columbia from 1992 to 2002 in relation to population size in 2002.
These results are at variance with the general belief that the smaller the population, the larger the rates of population increase. These results are consistent with the predation hypothesis and not with food limitation. (Data from Wittmer et al. 2005.)

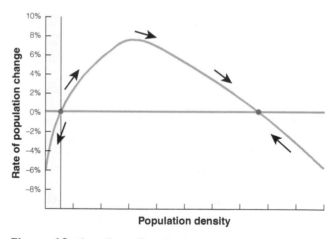

Figure 12 The Allee effect for small populations.
The Allee effect (beige zone) describes a region of low population density in which rates of population change are negative and the population heads to extinction. The arrows indicate the direction of population change. This can be viewed as a variant of the simple model of population change given and is a good description of the problem of woodland caribou declines in southern Canada.

Woodland caribou show puzzling population responses with changes in density (**Figure 11**). We expect in general that as population density falls, the rate of population growth should go up, exactly the opposite of what appears to be happening with woodland caribou (Wittmer et al. 2005). The ecologist W. C. Allee described this possibility in 1931 (Allee 1931) and it is illustrated schematically in **Figure 12**. The mechanism responsible for the Allee effect in woodland caribou appears to be excessive predation from wolves. Wolf populations are pri-

marily limited by their primary prey, moose, and they treat caribou as secondary prey items in their diet. Woodland caribou thus suffer from apparent competition with moose, as illustrated in the third diagram because of shared predators.

Predation losses do not always translate into reduced prey populations, and this puzzling result can be understood only by analyzing the details of population interactions. Paul Errington studied muskrats (*Ondatra zibethicus*) in the marshes of Iowa for 25 years to determine the effects of predation on muskrat populations. He questioned the common assumption that if a predator kills a prey animal, the prey population must then be one animal lower than it would have been without predation. You cannot study the effects of predation, Errington argued, by counting the numbers of prey killed; one must determine the factors that condition predation, the factors that make certain individuals vulnerable to predation while others are protected. Mink predation on muskrats was indeed a primary cause of death in Iowa marshes, but Errington contended that mink were removing only surplus muskrats that were doomed to die for other reasons. The territorial hostility of muskrats toward one another determined their numbers, and the muskrats driven out by this hostility over space were doomed to die—if not from predators, then from disease or exposure. Predators were merely acting as the "executioners" for animals excluded by the social system (Errington 1963). Errington introduced the important idea that in some systems predation may remove from populations only the "doomed surplus," and that predator effects should be inferred only from proper experiments involving both predator reduction areas and unmanipulated control areas.[1]

The role of predators in limiting the abundance of mammals is controversial. When a proper experimental design is used involving either natural experiments or manipulative experiments, the question can be clearly answered. The Serengeti Plains of eastern Africa contain a suite of large mammals and their predators, but the predators—lions, leopards, cheetahs, wild dogs, and spotted hyenas—seem to have little effect on their large mammal prey (Sinclair and Arcese 1995). Most of the prey individuals taken by predators are doomed surplus—older, injured, or diseased animals. Also, the vast majority of the prey species are migratory, whereas most of the predators are resident. Lions, for example, seem to be limited in numbers by the resident prey species available in the dry season, when the migratory ungulates are elsewhere.

Without detailed studies we cannot answer the general question of whether predators limit the abundance of their prey in field populations. Spectacular examples of the influence of predators have occurred where humans have accidentally introduced a new predator. A striking example is the virtual elimination of the lake trout fishery in the Great Lakes by the sea lamprey (*Petromyzon marinus*). The marine lamprey lives on the Atlantic coast of North America and

migrates into fresh water to spawn. Adult lampreys have a sucking, rasping mouth by which they attach themselves to the sides of fish, rasp a hole, and suck out body fluids. Only a few fish attacked by lampreys survive. Niagara Falls presumably blocked the passage of the lamprey to the upper Great Lakes before the Welland Canal was built in 1829. The first sea lamprey was found in Lake Erie in 1921, in Lake Michigan in 1936, in Lake Huron in 1937, and in Lake Superior in 1938 (Applegate 1950). Lake trout catches decreased to virtually zero within about 20 years of the lamprey invasion (**Figure 13**). Control efforts to reduce the lamprey population have been implemented since 1951, and lamprey are now reduced in abundance. Attempts to rebuild the Great Lakes fishery have been made by releasing trout bred in hatcheries. Lake trout have increased reasonably well in Lake Superior but are still rare in all the other Great Lakes (Krueger et al. 1995). Restoration of lake trout in these lakes has been hampered by a loss of genetic diversity, loss of spawning areas, chemical contaminants, and the introduction of new exotic species such as Pacific salmon (Holey et al. 1995). Lake trout in Lake Superior continue to recover toward their historical population size and composition, and have become a major restoration success story driven by lamprey control, hatchery releases, and native fish recovery (Bronte et al. 2003, Sitar and He 2006).

We conclude that in some but not all cases, the abundance of predators does influence the abundance of their prey in field populations. This raises an important question: What is it about certain predators that makes them effective in controlling populations of their prey? Can we find some type of system by which we can effectively classify predators? This question has great economic implications both in the management of fish and wildlife populations and in agricultural pest control. It is, of course, possible to proceed in a case-by-case manner and to investigate each individual predator-prey system on its own, but this is clearly inefficient, and we would rather attempt to reach some generalizations that apply to many individual cases.

The first approach to this problem was outlined by Solomon (1949) who recognized two components of predation. (1) **Functional response**, defined as the response of an average predator to the abundance of the prey. The key question here is whether an individual predator eats more prey individuals when prey are abundant. (2) **Numerical response**, defined as the response of a predator population to a change in prey density. The key question here is whether the density of predators will change as prey numbers increase. These two components of predation were extended by C. S. Holling, working at the Canadian Forest Research Laboratory at Sault Ste. Marie, Ontario, in the late 1950s. Holling (1959) defined four possible responses in predator-prey interactions: (1) a functional response, in which the number of prey eaten by individual predators

---

[1]Control here is used in the experimental design sense to mean an unmanipulated area. It should not be confused with the use of control to mean animal or plant removal, as in "pest control."

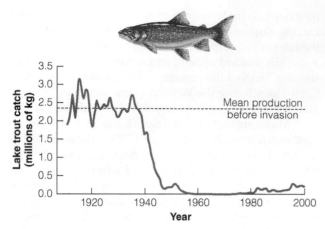

(a) Lake Huron

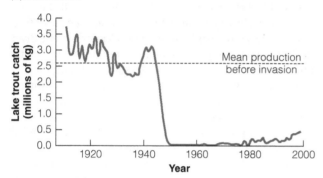

(b) Lake Michigan

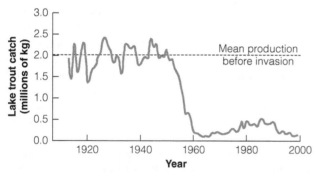

(c) Lake Superior

Figure 13 Effect of sea lamprey introduction on the lake trout fishery of the upper Great Lakes of North America. Lampreys were first seen in (a) Lake Huron in 1937, (b) Lake Michigan in 1936, and (c) Lake Superior in 1938. Commercial fish production from 1978 to 2000 is shown. These data are not actual population estimates, and in particular during the past 20 years, stocks have recovered (particularly in Lake Superior) while commercial catches have been tightly restricted. (Data from the Great Lakes Fishery Commission, 2005.)

changes; (2) a *numerical response*, in which the density of predators in a given area increases by reproduction; (3) an *aggregative response*, in which individual predators move into and concentrate in certain areas within the study area; and (4) a *developmental response*, in which individual predators eat more or fewer prey as predators grow toward maturity. The combination of these four components of predation is called the *total response*. Considerable theoretical and practical work has been done on the numerical, functional, aggregative, and developmental responses since the early analyses by Solomon (1949) and Holling (1959).

The functional response measures for each individual predator how many prey it eats in a given time period. Three general types of functional responses are recognized (**Figure 14**). The functional response of many predators rises to a plateau as prey density increases, so that over some range of prey density each individual predator eats more prey, but at some high prey density the predator becomes satiated and will not eat more. **Figure 15** shows one example of a Type 2 functional response for Canada lynx (*Lynx canadensis*) preying on snowshoe hares (*Lepus americanus*) in North America. The upper plateau of these functional responses is fixed by **handling time**, the time it takes for a predator to catch, kill, and eat a prey organism. The curve rises rapidly when the searching capacity of the predator is high. Note that the exact shape of the functional response curve observed for field populations will depend on the range of prey densities observed. If only low prey densities occur, the functional response may be a rising straight line; if only high prey densities occur, the functional response may be a horizontal line

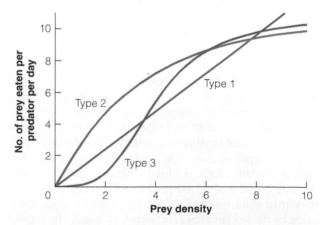

Figure 14 Three types of possible functional responses for predators to changes in prey abundance. Type 1 responses show a constant consumption of prey, with no satiation; Type 2 and Type 3 responses reach saturation at high prey densities.

(a)

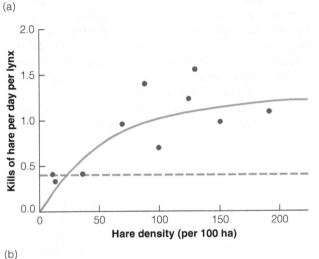

(b)

Figure 15 Functional response of Canada lynx to the abundance of their main prey, snowshoe hares. (a) The snowshoe hare is the main prey of the Canada lynx. (b) Lynx show a Type 2 functional response to hares (cf. Figure 14). The dashed red line shows the estimated daily energy needs of a lynx, and kill rates above this line could be labeled as "surplus killing." (Data from O'Donoghue et al. 1997.)

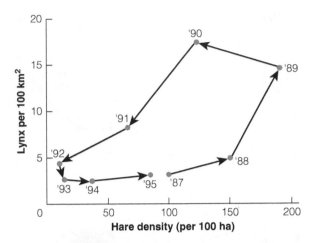

Figure 16 The numerical response of Canada lynx to changes in snowshoe hare density from 1987 to 1995 at Kluane Lake, Yukon Territory. Lynx respond to rising hare numbers by increasing in density but with a time lag, resulting in a counterclockwise spiral, indicated by the arrows. (Data from O'Donoghue et al. 1997.)

with no relationship between prey density and the number of prey eaten per predator per day.

A numerical response of predators can occur because of reproduction by the predator, and an aggregative response results from the movements or concentration of predators in areas of high prey density. Predators are usually mobile, and they do not search at random but instead concentrate on patches of high prey density. **Figure 16** illustrates the numerical response of Canada lynx to snowshoe hares in North America. When hares increase in abundance, lynx increase in numbers, and this is a common observation for many predator-prey systems. The ability of predators to reproduce more in areas of high prey density and to aggregate to patches of high prey abundance is a critical element in determining how effective the predator can be at limiting prey populations.

The developmental response occurs because predators are often growing and maturing during laboratory and field studies of predation. **Figure 17** illustrates the effects of a functional response and the additional effects of the developmental response on the number of mosquito larvae eaten by backswimmers (*Notonecta hoffmanni*) in the laboratory. Backswimmers grow more rapidly at higher food levels, and this explains the rise in the curve for total consumption in Figure 17 (Murdoch and Sih 1978).

Much of the work on predator-prey models has been conducted on laboratory populations and has proved difficult to translate to field populations (Sih et al. 1998). Thus we cannot give more than a vague answer to our general question about what makes some predators effective in controlling their prey. Because much of the theoretical work has concentrated on single-species systems, there is a need to consider the more complex cases in which predators feed on several prey species (Pech et al. 1995).

If we can measure the functional, numerical, developmental, and aggregative responses for a predator-prey system, we can determine the total response of the predators, as illustrated by the simple flowchart in **Figure 18**.

The total response gives the percentage of prey organisms eaten per unit time by the entire predator population, plotted against prey density. If the total response increases as prey density increases, the predator may limit the density of the prey. By contrast, if the total response remains constant or falls as prey density

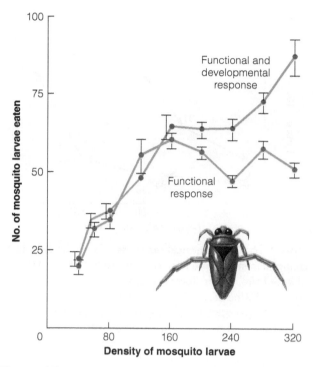

Figure 17 Functional and developmental responses of the predatory insect _Notonecta_ feeding on mosquito larvae in the laboratory. The prey consumption rate is measured by the number of mosquito larvae eaten per day. At high food levels _Notonecta_ grow larger faster, and thus the combined functional and developmental response curve accelerates upward. (From Murdoch and Sih 1978.)

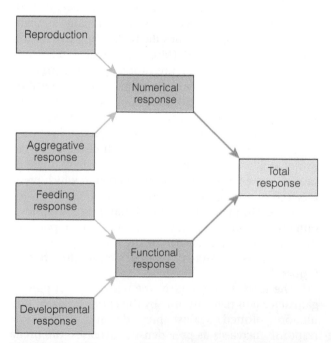

Figure 18 The components of predation that combine to give the total response of predators to changes in the density of their prey species.

increases, the predator cannot limit prey numbers. The key question is always whether percent mortality imposed by the predator on the prey increases as prey density increases. For many predator-prey systems, there may be a threshold or tipping point of prey density at which predators can no longer limit prey population growth. At high prey densities, some predators will exert no controlling influence on the prey because they will be swamped by prey numbers. This threshold of prey density above which prey escape from being limited by predators may be important in the conservation of endangered species (Sinclair et al. 1998).

One important general implication of our analysis is that predators may have effects on prey abundance that are important when prey populations are low but become unimportant when prey densities are high. Populations of this sort can exist in two different phases, a low-density endemic phase and a high-density epidemic (outbreak) phase. Some insect pests, such as the desert locust (Belayneh 2005) and the spruce budworm (Royama et al. 2005), show such biphasic densities, and the key to the endemic or low phase may be the action of predators at low insect densities.

How Do Prey Persist?

The most general question we can ask about predator-prey systems is how they continue to persist. There is a general assumption that there is a dynamic equilibrium in predator-prey interactions that results in the continuing existence of both predator and prey species. There are two general mechanisms for achieving this dynamic equilibrium. First, the prey species persists because it has a refuge in which it is safe from predators. This refuge could be spatial or temporal. For example, there could be habitats in which predators cannot effectively find their prey. Alternatively, there could be diurnal or seasonal periods in which predation is ineffective. Second, the predators may switch their hunting to other species as the original prey falls to low abundance. The behavior of predators in choosing prey is part of **optimal foraging theory**. When a predator has a choice of two or more different foods, the situation becomes more complex than the simple functional response we discussed above. How should a predator decide what items to eat? What is an optimal diet for an animal faced with many different prey items?

In natural foraging situations predators typically prefer some prey over others, and we can classify these as primary prey species and secondary prey. Depending on the profitability of each prey type, predators may switch from eating mainly one prey type to eating other prey as the relative abundance of the different prey species changes (Murdoch and Oaten 1975). Switching can be important in predators that feed on several types of prey because it could act to stabilize the density fluctuations of the prey species. As one prey species increases in abundance relative to the others, the predator would concentrate its feeding on the more abundant prey species and possibly restrict that prey's population growth. Conversely, switching to alternative foods may help a prey population to recover if it falls to a low level. Switching behavior could thus be a benefit to the predator by allowing it to maintain a stable population size (Elliott 2004).

But not all predators switch from eating rare prey, and in the process can drive their prey to extinction. The usual assumption that prey persist because predators stop eating them when they are rare may be wrong in some predator-prey systems (Matter and Mannan 2005). Two examples will illustrate this situation. Nile perch (*Lates niloticus*) were introduced to Lake Victoria in central Africa in the 1950s. They increased in abundance and by 1990 about 200 species of the 500 endemic cichlid fishes in Lake Victoria were driven extinct by this voracious predator (Witte et al. 2000). In Australia Short et al. (2002) describe nine examples of red foxes (an introduced predator) preying on native Australian species and killing them in excess of their immediate food needs. Some cases of local extinction resulted from this surplus killing. Short et al. (2002) attributed this killing behavior to an introduced predator interacting with prey which have no antipredator adaptations and no refuges. The key point is that in some cases the predation rate may not fall when prey become scarce.

Much of predation theory has been directed toward understanding how predators might stabilize prey populations (Gotelli 1998). It is clear from many field studies that predators do not necessarily stabilize prey numbers. Predators can be loosely classified into **generalist predators** and **specialist predators.** Generalist predators eat a great variety of prey and do not heavily depend on one species; specialist predators, by contrast, depend on only one or two species for the majority of their diet. The effects of specialist and generalist predators on prey populations differ:

- Generalist predators tend to stabilize prey numbers.

- Specialist predators tend to cause instability in prey numbers.

These generalizations are not ironclad and should be treated as hypotheses rather than facts. The Lotka-Volterra model and other simple predator-prey models all deal with specialist predators eating one prey species, and as such they may be of limited value in un-

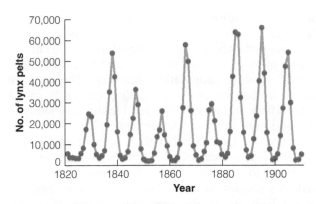

Figure 19 Canada lynx fur returns of the Northern Department, Hudson's Bay Company, 1821–1913. Canada lynx are specialist predators of snowshoe hares, and both hares and lynx oscillate in numbers in a 9- to 10-year cycle. (After Elton and Nicholson 1942.)

derstanding predator-prey systems in which the prey are fed on by a variety of predator species (Sih et al. 1998).

The inference from the Lotka-Volterra predation model that predator-prey interactions can result in oscillations (see Figure 2) appears to be strikingly applicable to some biological systems. The Canada lynx eats snowshoe hares and both species show dramatic cyclic oscillations in density with peaks every 9 to 10 years (**Figure 19**). Charles Elton analyzed the records of furs traded by the Hudson's Bay Company in Canada and showed that the cycle is a real one that has persisted unchanged for at least 200 years (Elton and Nicholson 1942). This lynx-hare cycle has been interpreted as an example of an intrinsic predator-prey oscillation, but more recent experimental studies have suggested that both food shortage and predation are involved in generating cycles. Lynx depend on snowshoe hares as primary prey, and are thus food-limited, whereas hares are affected by both food limitations and predators (Krebs et al. 2001). The time lag inherent in the numerical response of lynx to hare numbers induces the density cycle of hares (see Figure 16).

Prey populations persist because of refuges from predators and because of adaptations that have evolved toward the hunting behavior of predators, and this turns our attention to the evolutionary dimension of predator-prey interactions.

Evolution of Predator-Prey Systems

One of the striking features of the simple models of predator-prey interactions is that these models are often unstable. Oscillations are common in many predator-prey models (see Figure 2), but although they occur, they are not common in the real world. One way to explain the stability of real predator-prey systems is to postulate that natural selection has changed the characteristics of predators and prey alike such that their interactions produce population stability. Evolutionary change in two or more interacting species is called **coevolution**, and we are concerned in this case with the coevolution of predator-prey systems.

If one predator is better than another at catching prey, the first individual will probably leave more descendants to subsequent predator generations. Thus predators should be continually selected to become more efficient at catching prey. The problem, of course, is that by becoming too efficient, the predator will exterminate its prey and then suffer starvation. The prey at the same time are being selected to be better at escaping predation. Because of the conflicting adaptive goals of predator and prey, many evolutionists have described predator-prey evolution as an "arms race" (Dawkins and Krebs 1979). Predators may have an inherent disadvantage in this arms race because of the "life-dinner" principle, which states that selection will be stronger on the prey than on the predator because a prey individual that loses the race loses its life, whereas the unsuccessful predator loses only a meal. Dawkins (1982) suggested that the inherent disadvantage of a predator could be offset if the predator is rare and the prey is common. In this case the predator will be only a minor selective agent on the prey population as a whole.

Abrams (1986) criticized the "arms race" analogy for predator-prey coevolution, citing many theoretical situations in which the arms race would not occur, but he recognized that some asymmetry in the evolutionary responses of predators and prey was common. Prey should always increase their investment in escape mechanisms, if predators invest in becoming more efficient. But the reverse is often not true—predators do not always respond to prey investment—and whether or not predators will respond depends on the details of the specific predator-prey

system. The "arms race" analogy may not be correct in many particular cases of predator-prey coevolution.

Two obvious constraints operate in systems having several species of predators and prey. The existence of several species of predators feeding on several species of prey places limits on predator efficiency (Brodie and Brodie 1999). For example, one prey species may escape by hiding under rocks, while a second species may run very fast. Clearly, a predator is constrained by conflicting pressures either to get very good at turning rocks over or to get very good at running, and it is difficult to be good at both these activities. Conversely, we can imagine that the prey population is always being selected for escape responses. Faced with several predators with different types of hunting strategy, prey will not be able to evolve a specific escape behavior suitable to all species of predators.

One persistent belief about predator-prey systems is that predators typically capture substandard individuals from prey populations, so that weak, sick, aged, and injured prey are culled from prey populations. Temple (1987) tested this idea by flying a trained red-tailed hawk (*Buteo jamaicensis*) and measuring the outcome of its attacks on eastern chipmunks, cottontail rabbits, and gray squirrels. **Table 1** shows that the more difficult the prey is to catch, the higher the fraction of substandard individuals that are caught. Gray squirrels are particularly difficult for red-tailed hawks to catch, and squirrels taken were in markedly poorer condition than squirrels in the general population. The same generalization seems to hold for other vertebrate predators—substandard individuals are captured disproportionately when the type of prey is difficult to catch but not when it is easy to capture. Wirsing et al. (2002) found that substandard red squirrels (*Tamiasciurus hudsonicus*) in poor condition were taken more often by weasels but avian predators showed no such discrimination, taking the strong and the weak in equal proportions. Hyenas in Africa take wildebeest in poor condition because wildebeest are difficult to

Table 1 Evaluation of prey quality in predation by a trained red-tailed hawk on individuals of three prey species in Wisconsin. Results based on 447 attacks.

Species of prey	Difficulty of prey capture	Percentage of attacks that failed	Percentage of substandard individuals in hawk kills—percentage of substandard individuals in the population
Eastern chipmunk	Easy	72	8
Cottontail rabbit	Moderate	82	21
Gray squirrel	Difficult	88	33

SOURCE: Modified from Temple (1987).

capture, but take gazelles at random because they are relatively easy to catch (Kruuk 1972).

The coevolution of predator-prey systems occurs most tightly when the predators strongly affect the abundance of the prey. In some predator-prey systems the predator does not determine the abundance of the prey, so the evolutionary pressures are considerably reduced. In some cases the prey has refuges available where the predator does not occur, or the prey may have certain size classes that are not vulnerable to the predator. In other cases the predators have developed territorial behavior that restricts their own density so that they cannot easily respond to excessive numbers of prey animals (Sinclair and Arcese 1995).

Much of the stability we see in the natural world may result from the continued coevolution of predators and prey. Predators that do not have prudence forced on them by their prey may exist for only a short time in the evolutionary record, and we are left today with a residue of highly selected predator-prey systems.

Predators need not limit the density of their prey species to play an important role in the evolution of prey characteristics. Two antipredator defense strategies that are common in animals—warning coloration and group living—illustrate how evolutionary pressures can affect predator-prey systems. We discuss here the antipredator strategy of warning coloration.

Warning Coloration

Many animals have conspicuous coloration that advertises their presence, and this would appear to be a risky strategy, making them highly visible to prospective predators. But these animals either contain chemical toxins or possess physical defenses that deter predators once they have learned about the warning coloration. For example, many butterflies and other insects that are brightly colored contain poisons that are distasteful to predators. The theory of warning (or **aposematic**) coloration is usually put forward as an explanation of this correlation (Marples et al. 2005).

Mechanisms of prey defense using warning coloration must evolve by increasing the chances of survival of the individuals in which they are found. But for distasteful species, the predator must first sample one individual before the predator learns to avoid other prey of similar color. If the prey are gregarious and nearby individuals are closely related, kin selection would operate to favor the warning coloration. If only a few siblings are sampled from a large brood and the predator learns to avoid other individuals of the group, an allele for distastefulness can increase in frequency by kin selection. Predators do in fact seem to learn very quickly to avoid

Figure 20 The strawberry poison-dart frog (*Dendrobates pumilio*) of Costa Rica. This brightly colored frog, about 20 mm in length, contains alkaloids in its skin that are toxic to potential predators.

distasteful insects (Brower 1988). Three examples of warning coloration illustrate these ideas.

The strawberry poison dart frog (*Dendrobates pumilio*) is native to tropical rain forests of Nicaragua, Costa Rica, and Panama (**Figure 20**). The bright coloration of these frogs is a good example of aposematic coloration, and predators avoid these frogs because of the bitter, toxic, alkaloid secretions in their skin (Walls 1994) Toxins are accumulated by these frogs from their diet of arthropod prey, particularly ants and mites. A great variety of alkaloids can be present in these frogs from their diet (Saporito et al. 2006).

Monarch butterflies (*Danaus plexippus*) are brightly colored both as a caterpillar and as an adult (**Figure 21**). Monarch butterflies are distasteful and toxic. When they are caterpillars, they feed on milkweed plants, which contain a host of toxins. The caterpillars sequester the toxins within parts of their bodies, where they cause no harm, and these toxins stay in the animals when they become adult monarchs. The adults are strikingly colored, and after a bird has tried to eat one it will typically spit the butterfly out and will avoid them thereafter. Monarch butterflies are famous for their annual migrations. In late summer, monarchs from the eastern two-thirds of the United States and Canada migrate south to overwinter in a small area of pine forests high in the mountains of Mexico. When spring arrives, the individuals that have survived begin a flight northward to complete the annual migration. Not all predators are deterred by the warning coloration of monarchs. The introduced Asian lady beetle (*Harmonia axyridus*) is now judged to be a risk to monarch butterflies because it preys on both monarch eggs and small larvae in corn and soybean fields where milkweeds are common (Koch et al. 2006).

Figure 21 Monarch butterfly caterpillar and adult. The larvae concentrate toxins from their milkweed food plants and these are stored in the adults, making them poisonous to potential bird predators. The bright colors serve to warn off predators.

Coral snakes are brightly colored with red, yellow, and black bands. All of the 120 species of coral snakes in tropical America are extremely poisonous. Many other nonpoisonous snakes have evolved color patterns to mimic the appearance of coral snakes (**Figure 22**). These nonpoisonous snakes are called Batesian mimics because they mimic the color patterns of unrelated poisonous species.[2] Birds that live in areas occupied by coral snakes have an innate tendency to avoid snakes with these color patterns (Brodie and Janzen 1995), so a predator need not have a lethal encounter to avoid the poisonous species (Pough 1988). The mimic species can profit from resembling a poisonous or unpalatable prey species, and this coevolution has been particularly well developed in tropical species groups.

Many of the most striking characteristics of animal morphology and behavior are adaptations related to predation.

Figure 22 Geographic variation in color pattern in poisonous coral snakes and their nonvenomous mimics in Central America. The poisonous models (*Micrurus*) are shown on the left, and the mimics (*Pliocercus*) on the right, for five different areas (A–D, F). In E, simultaneous mimicry of two models is shown. The colubrid snake *Pliocercus elapholdes* (center) combines elements of the patterns of *Micrurus diastema* (left) and *Micrurus elegans* (right). (From Greene and McDiarmid 1981. Illustration copyright by the artist Frances J. Irish; used with permission.)

[2]A Batesian mimic could be likened to a sheep in wolf's clothing.

Summary

One species interaction involves predation. Simple mathematical models can be used to describe this interaction. When generations are discrete, simple models can produce stable equilibria of predator and prey, but usually produce oscillations in the numbers of both species. When generations are continuous, graphic models developed by Rosenzweig and MacArthur can be used to evaluate the equilibrium levels and the stability of predator-prey systems. Both stable equilibria and cyclic oscillations may occur. All these simple models make the assumption that the world is homogeneous (one habitat), that there are no prey refuges, and that only one predator species eats one prey species. Relaxing these assumptions leads to more complex models.

In laboratory systems of predators and prey, cyclic oscillations are produced only in complex environments, and most simple systems do not reach stability but instead are self-annihilating. The importance of refuges and spatial heterogeneity can be illustrated readily in laboratory systems, and these factors are even more critical in field populations of predators and prey.

Field populations can be models of predator-prey systems only if predators have a strong effect on the abundance of their prey. This assumption can be tested by predator-removal experiments. In some but not all cases studied, the abundance of predators does influence the abundance of prey. The properties of effective predators can be described in a general manner, but we cannot yet predict which predators will be good agents of prey control without actually doing field tests. Both predator and prey species are affected by many other factors in the environment, and consequently the population trends predicted by simple predator-prey models are rarely found in field populations.

Predator-prey systems always involve a coevolutionary race in which prey are selected for escape and predators for hunting ability. These systems stabilize most easily when several species are involved, when prey have safe refuges from predators, and when predators take old animals of little reproductive value. Many characteristic structures and behavior patterns of animals are adaptations related to predation.

Review Questions and Problems

1   One of the long-standing controversies in predator-prey limitation involves the wolf-moose interaction in North America. Eberhardt (1998, 2000) and Messier and Joly (2000) present alternative views on how much this interaction affects moose abundance. Evaluate the data they present and their arguments for population control of moose by wolves.

2   Calculate the population changes from Equations (2) and (3) for ten generations in a hypothetical predator-prey system with discrete generations in which the parameters for the prey are $B = 0.03$, $N_{eq} = 100$, $C = 0.5$, and starting density is 50 prey, and for the predators, $Q = 0.02$ (or $S = 2.0$) and starting density is 0.2. How would the prey population change in the absence of the predators?

3   When (if ever) would it be adaptive for a predator to engage in surplus killing of prey? Evaluate the exact definition of "surplus killing" in Short et al. (2002) on arctic fox predation on goose eggs.

4   Buckner and Turnock (1965) studied bird predation on the larch sawfly in Manitoba. They obtained the following data for the chipping sparrow (*Spizella passerina*):

	Plot I		Plot II	
Year	Sparrows per acre	Sawfly larvae per acre	Sparrows per acre	Sawfly larvae per acre
1954	—	—	3.2	235,000
1956	—	—	2.9	33,400
1957	1.4	2,138,700	2.3	40,000
1958	0.5	879,400	2.5	41,200
1959	0.4	437,800	2.2	27,300
1960	0.2	354,300	2.2	54,600
1961	0.5	199,900	2.3	15,000
1962	1.1	191,800	5.0	3200
1963	0.2	366,800	0.3	3900

Plot the numerical response of chipping sparrows to changes in sawfly larval abundance for each of the two study plots, and discuss the differences between plots I and II for this predator-prey system.

5   The collapse of lake trout populations in the Great Lakes coincided with a general increase in commercial fishing of the lakes. Discuss the hypothesis that the collapse of fish stocks in the Great Lakes (see Figure 13) was caused more by overfishing (human predation) than by the introduction of the sea lamprey. Coble et al. (1990) and Bronte et al. (2003) give references.

6   How does the predation by herbivores on green plants differ from either the predation of insect parasitoids on their hosts or the predation of carnivores on herbivores? Make a list of similarities and differences, and discuss how they affect the simple models of predation discussed in this chapter.

7   In the northern Gulf of St. Lawrence, cod (*Gadus morhua*) numbers have been steadily declining for the past 25 years, while one of their main prey species, northern shrimp (*Pandanus borealis*), have been increasing (Worm and Myers 2003). How strong is this observation as a test of the hypothesis of predator limitation? What other data would you like to have to test this hypothesis? Worm and Myers (2003) discuss this question.

8   Tammar wallabies (*Macropus eugenii*), a small 6–10 kg macropod, were introduced to New Zealand about 130 years ago. In their native Australia, they have been subject to predation by a variety of predators such as the marsupial lion (now extinct) and the marsupial tiger (also extinct), as well as large lizards and now the introduced dingo and red fox. New Zealand has none of these predators. Discuss how tammar wallabies might evolve in the absence of predation. Under what conditions would you expect antipredator behaviors to disappear from the New Zealand population of tammar wallabies? Blumstein et al. (2004) provide data on this issue.

9   Wildebeest in the Serengeti area of east Africa have a very restricted calving season. All females give birth within a space of three weeks at the start of the rainy season (Sinclair and Arcese 1995). How would you test the hypothesis that this restricted calving season is an adaptation to reduce predation losses of calves?

10   The graphic model of Rosenzweig and MacArthur (see Figure 5) predicts that the predator-prey system will become unstable when nutrients are added to the prey population (the paradox of enrichment). Evaluate the evidence for the occurrence of the paradox of enrichment in laboratory and field populations. Does the same prediction follow from ratio-dependent predation theory? Jensen and Ginzburg (2005) provide background references and a discussion of the problem.

11   The birds, lizards, and mammals of Guam in the western Pacific Ocean have been driven to extinction or to low numbers by the introduced brown tree snake (*Boiga irregularis*). How could this happen? Is it adaptive for a predator to drive its prey to extinction? Read the discussion in Rodda et al. (1997) and evaluate the uniqueness of this situation.

Overview Question

When populations of moose, caribou, or deer decline in Alaska or Canada, a great public pressure to instigate wolf control programs typically ensues. What data would you collect to describe and understand the dynamics of this predator-prey system? List the alternative hypotheses you would test and their management implications.

Suggested Readings

- Barbosa, P., and I. Castellanos, eds. 2005. *Ecology of Predator-Prey Interactions*. New York: Oxford University Press.

- Boveng, P. L., L. M. Hiruki, M. K. Schwartz, and J. L. Bengtson. 1998. Population growth of Antarctic fur seals: Limitation by a top predator, the leopard seal? *Ecology* 79:2863–2877.

- Creel, S. and D. Christianson. 2008. Relationships between direct predation and risk effects. *Trends in Ecology & Evolution* 23:194–201.

- Johnson, C. N., J. L. Isaac, and D. O. Fisher. 2007. Rarity of a top predator triggers continent-wide collapse of mammal prey: dingoes and marsupials in Australia. *Proceeding of the Royal Society of London, Series B* 274:341–346.

- Lima, S. L. 1998. Nonlethal effects in the ecology of predator-prey interactions. *Bioscience* 48:25–34.

- Marples, N. M., D. J. Kelly, and R. J. Thomas. 2005. The evolution of warning coloration is not paradoxical. *Evolution* 59:933–940.

- Matter, W. J., and R. W. Mannan. 2005. How do prey persist? *Journal of Wildlife Management* 69:1315–1320.

- Short, J., J. E. Kinnear, and A. Robley. 2002. Surplus killing by introduced predators in Australia—evidence for ineffective anti-predator adaptations in native prey species. *Biological Conservation* 103:283–301.

- Sih, A., G. Englund, and D. Wooster. 1998. Emergent impacts of multiple predators on prey. *Trends in Ecology and Evolution* 13:350–355.

- Valkama, J., E. Korpimäki, B. Arroyo, et al. 2004. Birds of prey as limiting factors of gamebird populations in Europe: A review. *Biological Reviews* 80:171–203.

- Wittmer, H. U., A. R. E. Sinclair, and B. N. McLellan. 2005. The role of predation in the decline and extirpation of woodland caribou. *Oecologia* 144:257–267.

Credits

Illustration and Table Credits

Photo Credits

Species Interactions III: Herbivory and Mutualism

Key Concepts

- That the world is green implies that herbivores are prevented from completely destroying their food sources, either by their own behavior, by their enemies, or by plant defense strategies.

- The Resource Availability Hypothesis predicts that plants growing slowly in poor habitats should invest most in plant defense because they have the most to lose from herbivory.

- Herbivores may not achieve a stable interaction with their food plants, and many ungulates undergo irruptions with subsequent oscillations in numbers.

- Models of predator-prey dynamics can be applied to grazing systems to determine the kinds of plant-herbivore interactions that might lead to stability.

- Not all plant-herbivore interactions are detrimental to plants. Mycorrhizal fungi grow on most plant roots to the advantage of both the plants and the fungi. Many mutualistic interactions have evolved in which both the plants and the herbivores gain from their association.

KEY TERMS

grazing facilitation The process of one herbivore creating attractive feeding conditions for another herbivore so there is a benefit provided to the second herbivore.

inducible defenses Plant defense methods that are called into action once herbivore attack occurs and are nearly absent during periods of no herbivory.

mutualism A relationship between two organisms of different species that benefits both and harms neither.

mycorrhizae A mutually beneficial association of a fungus and the roots of a plant in which the plant's mineral absorption is enhanced and the fungus obtains nutrients from the plant.

optimal defense hypothesis The idea that plants allocate defenses against herbivores in a manner that maximizes individual plant fitness, and that defenses are costly to produce.

overcompensation hypothesis The idea that a small amount of grazing will increase plant growth and fitness rather than cause harm to the plant.

plant stress hypothesis The idea that herbivores prefer to attack stressed plants, which produce leaves that are higher in nitrogen.

plant vigor hypothesis The idea that herbivores prefer to attack fast-growing, vigorous plants rather than slow-growing, stressed plants.

resource availability hypothesis A theory of plant defense that predicts higher plant growth rates will result in less investment in defensive chemicals and structures.

secondary plant substances Chemicals produced by plants that are not directly involved in the primary metabolic pathways and whose main function is to repel herbivores.

Plant-animal interactions are the focus of many population interactions. **Herbivory** is a major interaction in which animals prey on plants, and herbivory is traditionally considered a profit for the animals and a loss for the plants. Many examples of **mutualism** involve plant-animal interactions that are by contrast a gain for both species. In this chapter we discuss herbivory and mutualism to assess their effects on abundance and their evolutionary origins.

Herbivory is a special kind of predation because the herbivore does not kill the plant but eats only part of it. Over half of the macroscopic species on Earth are plants, and consequently a major part of species interactions involve plant-herbivore interactions. In this

chapter we will examine some of the specific relationships between herbivores and plants. The uniqueness of these relationships is often only a reflection of the simple fact that most plants cannot move, so "escape" from herbivores can be achieved only by some clever adaptations. Herbivores can be important selective agents on plants, and the evolutionary interplay between plants and animals is a major theme in this chapter.

Defense Mechanisms in Plants

The world is green, and there are three possible explanations for this. First, some herbivore populations may evolve self-regulatory mechanisms that hold their own numbers in check and prevent them from destroying their food supply. Or second, other control mechanisms, such as predation or disease, may hold herbivore abundance down so that plants escape being totally eaten. Third, not all that is green may be edible. Plants have evolved an array of defenses against herbivores, and this has set up a coevolutionary contest between plants and herbivores in evolutionary time.

Plants may discourage herbivores by structural adaptations, as anyone who has tried to prune a rosebush will attest, but they may also use a variety of chemical weapons that we are only now starting to appreciate. Plants contain a variety of chemicals that have always puzzled plant physiologists and biochemists (Feeny 1992). These chemicals, called **secondary plant substances**, are found only in some plants and not in others and are by-products of the primary metabolic pathways in plants. **Figure 1** gives a simplified view of the biochemical origins of some of the major chemical groups of secondary plant substances. A number of these substances are familiar to us already. One acetogenin, juglone, is produced by walnut trees as an allelopathic chemical. Among the phenylpropanes found in

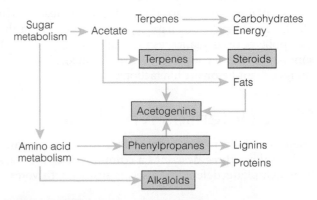

Figure 1 Relationships of the major groups of secondary plant substances (shown in boxes) to the primary metabolic pathways of plants. (After Whittaker and Feeny 1971.)

some trees are the spices cinnamon and cloves. Familiar terpenoids include peppermint oil and catnip. Well-known alkaloids include nicotine, morphine, and caffeine.

Nearly 50 years ago ecologists suggested that secondary plant substances were specifically evolved by plants to thwart herbivores (Fraenkel 1959; Ehrlich and Raven 1964). This view assumes that secondary plant substances are actively produced at a metabolic cost to the plant. Such a chemical variety exists in different plant groups only because plants are eaten by an array of different herbivores, and the chemicals that deter one herbivore species may not deter another. If all animals could be removed from the community, plants would not produce secondary substances because they are costly to make.

This plant-defense view argues that herbivores have a strong effect on plant fitness, and that well-defended plants are fitter. If plant defense characteristics are inherited, then all the elements needed for natural selection are present.

If plant defense has a cost in terms of plant fitness, we can make four general predictions:

- Plants evolve more defenses if they are exposed to much damage, and fewer defenses if the cost of defense is high.

- Plants allocate more defenses to valuable tissues that are at risk.

- Defense mechanisms are reduced when enemies are absent, and increased when plants are attacked.

- Defense mechanisms are costly and cannot be maintained if plants are severely stressed by environmental factors.

The cost of defense is due to the diversion of energy and nutrients from other needs. Much evidence to support these four predictions has now accumulated, and the hypothesis that secondary substances have an ecological role as deterrents to herbivory has become a fruitful and exciting area of research (Stamp 2003). Secondary substances in plants are not static but have rapid turnover rates in the metabolic pool as plants respond to herbivory. **Table 1** summarizes the key messages that have emerged from the study of plant-herbivore interactions.

Plant Defense Hypotheses

There are three major plant defense hypotheses, and each of these three describes a component of the ways by which plants defend themselves against herbivores.

The Optimal Defense Hypothesis

The oldest hypothesis of plant defense is the **Optimal Defense Hypothesis**, which arose in the 1970s from re-

Table 1 Six key concepts about plant-herbivore interactions that are illustrated in this chapter.

1. Defensive chemicals are widespread among plant species

2. Individual plants or species have an array of defenses, rather than only one defense against herbivores

3. Many plants have dynamic defenses against herbivores, so they can respond chemically or physically once they are attacked

4. Characteristics of the environment (or resource availability) affects the ability of plants to mount defenses against herbivores

5. There is geographic variability in the interactions between plants and herbivores, so that not all populations of a species have the same defenses

6. Plant adaptations and herbivore feeding specialization reflect their evolutionary history

SOURCE: Stamp (2005).

search by plant physiologists and plant evolutionary ecologists. The basic hypothesis states that organisms allocate defenses in a manner that maximizes individual inclusive fitness, and that defenses are costly in terms of fitness. The assumptions of this hypothesis are that there is genetic variation in plants for secondary compounds, that herbivory is the primary selective agent for these secondary compounds, and that these defenses reduce herbivory. There are many studies that support these general assumptions, and the critical issue is to make these ideas more specific so they can be tested.

Paul Feeny, working at Cornell University, was one of the first to recognize the importance of plant defenses and to suggest a general theory. Feeny's Plant Apparency Theory (1976) suggested that plants can be divided into two classes: "apparent" plants are those found easily by herbivores; "unapparent" plants are hard for herbivores to find because they are small or rare or short lived. The major premise of the Plant Apparency Theory is that the type of defenses the plant uses depends on how easily a herbivore can find the plant. Plants that are easily found by herbivores evolve chemical defenses of a different type from those used by plants that are difficult for herbivores to locate. Short-lived plants may be able to escape the attention of herbivores by developing so quickly that their herbivores are unlikely to discover them—such plants are "unapparent" to their herbivores. In contrast, "apparent" plants are sure to be found by herbivores because they are long-lived.

Some defense mechanisms in plants are quantitative, and secondary compounds may vary in concentration. For example, tannins and resins in leaves may occupy up to 60% of the dry weight of a leaf (Feeny 1976). Quantitative plant defenses should be used by "apparent" plants. Other defense mechanisms may be qualitative or +/− defenses because the compounds involved are present in very low concentrations (less than 2% dry weight). Examples are alkaloids and cyanogenic compounds in leaves. Qualitative defenses are poisons that protect plants against generalized herbivores that are not adapted to cope with the toxic chemicals, but they do not stop specialized herbivores that have evolved detoxification mechanisms in the digestive system. Qualitative defenses should be used by "unapparent" plants (Feeny 1976).

Both the type of defense and the amount of defense plants use depend on the vulnerability of the plant tissues (Rhoades and Cates 1976). Growing shoots and young leaves are more valuable to plants than mature leaves, so plants typically invest more heavily in the defense of growing tips and young leaves. Tannins, resins, alkaloids, and other defense chemicals are concentrated at or near the surface of the plant, thereby increasing their effectiveness.

The Plant Apparency Theory stimulated a great deal of work on plant defense during the 1970s and 1980s, and it was soon found to be inadequate as an explanation of plant defense. Some "apparent" plants have qualitative, toxic defenses, and some "unapparent" species use quantitative defenses. In some cases it is difficult for researchers to decide whether or not plants are apparent to their herbivores and not just to humans. It became clear that a more general theory was needed (Stamp 2003).

The Resource Availability Hypothesis

One element missing from the Optimal Defense Hypothesis is the ability of a plant to defend itself, given the resources available to it. Coley et al. (1985) proposed the **Resource Availability Hypothesis** to take into account the plant's ability to replace tissues taken by herbivores. This hypothesis is also called the *Growth Rate Hypothesis*. Both fast-growing and slow-growing plants have a suite of physiological characteristics that are summarized in **Table 2**. Herbivores prefer fast-growing plants and tend to avoid slow-growing plants. Because each leaf represents a greater investment for a slow-growing plant, slow-growing plants stand to lose more to herbivores and

Table 2 Characteristics of inherently fast-growing and slow-growing plant species.

Variable	Fast-growing species	Slow-growing species
Growth characteristics		
Maximum growth rates	High	Low
Maximum photosynthetic rates	High	Low
Dark respiration rates	High	Low
Leaf protein content	High	Low
Responses to pulses in resources	Flexible	Inflexible
Leaf lifetimes	Short	Long
Successional status	Often early	Often late
Antiherbivore characteristics		
Expected rates of herbivory	High	Low
Amount of defense metabolites	Low	High
Type of defense	Qualitative (alkaloids)	Quantitative (tannins)
Turnover rate of defense	High	Low
Flexibility of defense expression	More flexible	Less flexible

SOURCE: Coley et al. (1985).

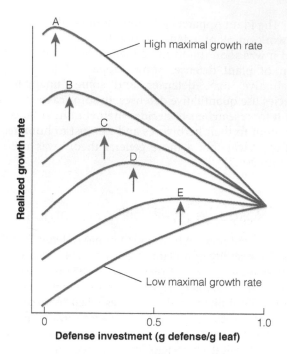

Figure 2 The Resource Availability Hypothesis of plant defense. Each curve represents a plant species (A to E) with a different maximal growth potential. Levels of defense that maximize growth rate are indicated by an arrow. Slow-growing plants should invest much more in defense because losses to herbivores are more difficult to replace. (From Ryther and Dunstan 1971.)

thus invest more in defensive chemicals. **Figure 2** summarizes a conceptual model of the Resource Availability Hypothesis. A prediction of this hypothesis is that the higher the plant's growth rate, the lower is the predicted investment in defense. Coley (1987) tested this prediction in 47 species of rain forest trees in Panama (**Figure 3**).

There was a strong negative correlation between growth rate and defense levels, as predicted by the Resource Availability Hypothesis. The key idea that this hypothesis adds to plant-herbivore interactions is that growth rates in plants are strongly affected by the available resources—such as nitrogen, phosphorus, and water—and plant defenses are well correlated with plant growth potential.

Plants can either defend themselves at all times, or only when attacked by a herbivore. Structural defenses are clearly more permanent than chemical defenses. By utilizing **inducible defenses**, plants can avoid the cost of producing defensive chemicals when they are not needed. Once a plant is attacked by herbivores, it can then activate its defenses. Induction times for defensive reactions by plants have been studied for only a few species and vary from 12 hours to one year or more (Tollrian and Harvell 1999). Rapid defensive responses in plants were unexpected but are now being found in more and more species. If defenses are costly, we would expect a relaxation of defenses after a herbivore's attack, but few measurements have yet been made (Karban and Baldwin 1997). If induced responses are occurring in a plant, we need to answer two questions to determine if these responses are antiherbivore defense responses:

1. Do the induced changes affect herbivore foraging or herbivore distributions?

2. Do plants suffer less damage and have greater fitness as a result of induced changes in leaf chemistry?

The brown seaweed *Ascophyllum nodosum* is a common seaweed in the intertidal zone of rocky shores of the North Atlantic. It is grazed by herbivorous gastropods,

Figure 3 Relationship of growth rate and defense investment in 47 species of Panamanian forest trees. The negative relationship fits a prediction of the Resource Availability Hypothesis illustrated in Figure 2. Growth rate is measured as annual increase in height. Defense investment is a combined measure of physical and chemical defenses against herbivores. (Data from Coley 1987.)

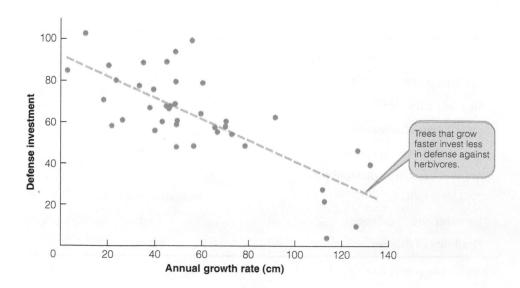

and it responds to grazing by increasing the production of secondary metabolites (phlorotannins) (Toth et al. 2005). Phlorotannins are polyphenols produced as secondary chemicals by these seaweeds. Toth et al. (2005) showed in laboratory aquarium studies that *Ascophyllum* plants exposed to two weeks of grazing by gastropods increased their phlorotannin content (**Figure 4**). Induction of tannins was greatest in the most important tissues for plant fitness: the basal shoots that support all the vegetative and reproductive tissues of this seaweed. These data support one of the key assumptions of the Optimal Defense Hypothesis that the most valuable tissues of a plant will be defended most heavily. Moreover, Toth et al. (2005) showed that the induced tannins in these seaweeds reduced the number of viable eggs laid by the herbivorous gastropods that fed on previously grazed plants.

Herbivores do not, of course, sit idly by while plants evolve defense systems (Thompson 1999). Herbivores circumvent plant defenses either by evolving enzymes to detoxify plant chemicals or by altering the timing of their life cycle to avoid the noxious chemicals of the plants (as in the next example of tannins in oak leaves). The coevolution of animals and plants can thus occur, and we will examine three cases to illustrate this.

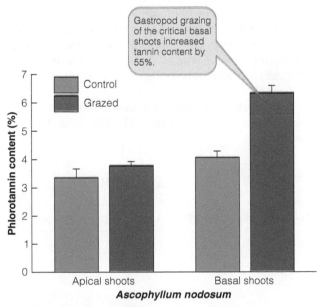

Figure 4 Induction of phlorotannins in the brown seaweed *Ascophyllum nodosum* by exposing plants to grazing by the gastropod *Littorina obtusata* for two weeks in the laboratory. Grazing increased the secondary chemical content by 12% in the less critical apical shoots and by 55% in the critical basal shoots that support the plant in the rocky intertidal zone. (Data from Toth et al. 2005.)

Tannins in Oak Trees

The common oak (*Quercus robur*), a dominant tree in the deciduous forests of western Europe, is attacked by the larvae of over 200 species of Lepidoptera, more species of insect attackers than any other tree in Europe withstands. The attack of insects is concentrated in the spring with a smaller peak of feeding in the fall. Among the most common of these insects is the winter moth, whose larvae feed on oak leaves in May and drop to the ground to pupate late in that month. Why is insect attack concentrated in the spring? One possibility is that oak leaves become less suitable insect food as they age (Feeny 1970.)

Winter moth larvae fed "young" oak leaves grow well, but if larvae are fed slightly "older" leaves, they grow very poorly:

Winter moth larvae diet	Mean peak larval weight (mg)
May 16: "young" oak leaves	45
May 28–June 8: "old" oak leaves	18

No adults emerged from the larvae fed older leaves. Thus some change occurs very rapidly in oak leaves in the spring to make them less suitable for winter moth larvae. The most obvious changes in oak leaves during the spring are a rapid darkening and an increase in toughness. The thin oak leaves of May become thick and more difficult to tear by early June (**Figure 5**). If leaf toughness is a sufficient explanation for the feeding pattern of oak insects in the spring, then ground-up older leaves should provide an adequate diet. But if chemical changes have occurred as well, ground-up older leaves should still be inadequate as a larval diet. Ground-up leaves seem to be an adequate diet, at least until early June:

Larvae fed ground-up leaves	Mean peak larval weight (mg)
May 13: "young" leaves	37
June 1: "old" leaves	35

If mature oak leaves can provide an adequate diet, why has natural selection not favored insect mouthparts able to cope with tough leaves? Some Lepidoptera do feed on summer oak leaves, so it is possible to feed on tough leaves. If mature oak leaves later in summer are relatively poor nutritionally compared with young spring leaves from May and early June, this would produce natural selection toward early feeding.

Two related chemical changes in oak leaves seem to be significant for feeding insects: the amount of tannins

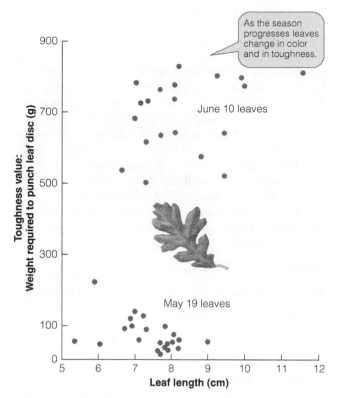

Figure 5 Toughness of "young" oak (*Quercus robur*) leaves collected May 19 and "old" oak leaves collected **June 10.** Toughness is one way in which plants can make their leaves less palatable to herbivores. (After Feeny 1970.)

in the leaves increases from spring to fall (especially after July), and the amount of protein decreases from spring to summer and remains low from June onward. Tannins, which are secondary plant substances that may reduce palatability, may act in oak leaves by tying up proteins in complexes that insects cannot digest and utilize. Larval

weights of winter moths are significantly reduced if their diet contains as little as 1% oak-leaf tannin.

Nevertheless, some insects have evolved ways of minimizing the effect of tannins. Insects that feed on oak leaves in the summer and fall tend to grow very slowly, which may be an adaptation to a low-nitrogen diet. **Table 3** shows that many of the late-feeding insects on oak overwinter as larvae and complete their development on the spring leaves. Many others are leaf miners, which may avoid tannins by feeding on leaf parts that contain little tannin.

Thus the oak tree has defended itself against herbivores by the use of tannins as a chemical defense and altered leaf texture (toughness) as a structural defense. Herbivores have compensated by concentrating feeding in the early spring on young leaves and by altering life cycles in the summer and fall.

Ants and Acacias

A mutualistic system of defense has coevolved in the swollen-thorn acacias and their ant inhabitants in the tropics of Central and South America and in Africa. The ants depend on the acacia tree for food and a place to live, and the acacia depends on the ants for protection from herbivores and neighboring plants. Not all of the approximately 700 species of acacias (*Acacia* spp.) depend on the ants in the New World tropics, and not all the acacia ants (*Pseudomyrmex* spp.), 150 species or more, depend completely on acacia. In a few cases a high degree of mutualism has developed, described in detail by Janzen (1966). Some of the species of ants that inhabit acacia thorns are obligate acacia ants and live nowhere else.

Swollen-thorn acacias have large, hollow thorns in which the ants live (**Figure 6**). The ants feed on

Table 3 Larval feeding habits of early-feeding and late-feeding *lepidoptera* species on leaves of the common oak in Britain.

Feeding habit	Early-feeding species[a] (%)	Late-feeding species[b] (%)
Larvae complete growth on oak leaves in one season	92	42
Larvae complete growth on low herbs after initial feeding on oak leaves	3	11
Larvae overwinter and complete growth in following year	4	38
Larvae bore into leaf parenchyma (leaf miners)	3	26

[a]Early-feeding larvae are in May and June; total of 111 species.

[b]Total of 90 species.

NOTE: Some species exhibit more than one of the feeding habits, so the columns do not add to 100%.

SOURCE: After Feeny (1970).

(a)   (b)   (c)

Figure 6 Ant-Acacia Coevolution. (a) *Acacia collinsii* growing in open pasture in Nicaragua. This tree had a colony of about 15,000 worker ants and was about 4 m tall. (b) Area cleared over 10 years around a growing *Acacia collinsii* in Panama by ants chewing on all vegetation except the acacia. The machete in the photo is 70 cm long. The area was not disturbed by other animals. (c) Swollen thorns of *Acacia cornigera* on a lateral branch. Each thorn is occupied by 20–40 immature ants and 10–15 worker ants. All the thorns on the tree are occupied by ants belonging to a single colony. An ant entrance hole is visible in the left tip of the fourth thorn up from the bottom.

modified leaflet tips called Beltian bodies, which are the primary source of protein and oil for the ants, and also on enlarged extrafloral nectaries, which supply sugars. Swollen-thorn acacias maintain year-round leaf production, even in the dry season, providing food for the ants. If all the ants are removed from swollen-thorn acacias, the trees are quickly destroyed by herbivores and crowded out by other plants. Janzen (1966) showed that acacias without ants grew less and were often killed:

	Acacias with ants removed	Acacias with ants present
Survival rate over 10 months (%)	43	72
Growth Increment		
May 25–June 16 (cm)	6.2	31.0
June 16–August 3 (cm)	10.2	72.9

Swollen-thorn acacias have apparently lost (or never had) the chemical defenses against herbivores found in other trees in the tropics.

The acacia ants continually patrol the leaves and branches of the acacia tree and immediately attack any herbivore that attempts to eat acacia leaves or bark. The ants also bite and sting any foreign vegetation that touches an acacia, and they clear all the vegetation from the ground beneath the acacia tree. As a result the swollen-thorn acacia often grows in a cylinder of space virtually free of all competing vegetation (see Figure 6b).

Similar interactions between ants and acacias have been described for Africa (Stapley 1998). The ant-acacia system is thus a model system of the coevolution of two species in an association of mutual benefit. By reducing herbivore destruction and competition from adjacent plants, the ants serve as a living defense mechanism for the acacias.

Spines and Thorns in Terrestrial Plants

Thorns, spines, and prickles occur widely on terrestrial plants, and even though everyone assumes that they act as physical defenses against large herbivores, there is remarkably little evidence that this is true (Myers and Bazely 1991). A variety of observations are consistent with this

Figure 7 Percentage of the cactus *Opuntia stricta* having spines on three islands off the coast of Queensland, Australia. Cattle were present on one island and grazing damage was observed, but cattle were absent from the other two islands. (After Myers and Bazely 1991.)

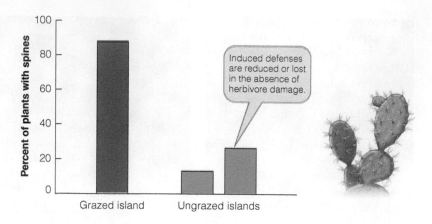

idea. For example, the cactus *Opuntia stricta* has more spiny individuals on Australian islands on which cattle graze than on islands with no grazing (**Figure 7**). If thorns and spines are herbivore defense mechanisms, they could be used to test ideas about plant defenses. The resource availability hypothesis (see Figure 2) predicts that plants growing in nutrient-poor soils should invest more in plant defense than plants on rich soils. This is not the case for the fynbos vegetation of South Africa (Campbell 1986). Fynbos is a shrubland of sclerophyllous, evergreen plants growing on very poor soils. Only 4% of the total plant cover in fynbos has spines, compared with 13% of the plant cover in nutrient-richer areas that lack fynbos. Campbell (1986) suggests that the fynbos vegetation is so poor that no large herbivores can live on it, and consequently there is no selection for physical plant defenses such as thorns.

Acacias in Africa and the Mediterranean region grow spines and thorns that appear to be an adaptation against large herbivores. In Tanzania *Acacia tortilis* trees protected from grazing do not grow spines (Gowda 1996). Goats feeding on these acacias induce spines on the trees, and Gowda (1996) found that the more spines on individual plants, the fewer shoots they lost to goat browsing on branches and leaves. In the Negev Desert of Israel, Rohner and Ward (1997) compared acacias on fenced areas that excluded large herbivores for more than 10 years with acacias on open areas. They found that browsed acacias increased the numbers of spines and thorns, but they did not consistently increase their chemical defenses compared with controls.

One way to test whether thorns are an induced response to herbivory is to exclude the herbivores and monitor thorn production. The Mediterranean shrub *Hormathophylla spinosa* is heavily browsed by ungulates in southern Spain, and typically 80% of the flowers and fruits are eaten each year. Thorns are grown anew each year in this shrub, since they grow only on the flower stems. When ungulates were excluded from 20 shrubs for three years, thorn production decreased by about 40% (**Figure 8**). Thorn production is costly to these plants, and up to 58%

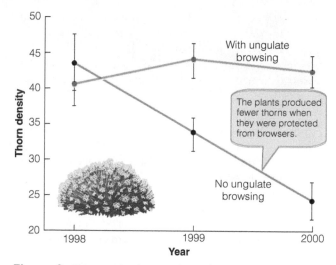

Figure 8 Changes in the number of thorns on the Mediterranean shrub *Hormathophylla spinosa* in fenced (black) and unfenced (red) plots. There is a trade-off between thorn production and fruiting in this plant: the fewer thorns it produces, the more fruit it can bear. (After Gomez and Zamora 2002.)

more fruits were produced when thorns were removed experimentally. A trade-off between thorn production and fruiting occurs in this shrub: the fewer thorns it produces, the more fruit it can bear (Gomez and Zamora 2002).

Thorns do not prevent all herbivore damage but they are an effective partial deterrent to many large herbivores. For some plants such as *Hormathophylla spinosa* thorns are plastic traits that can be induced by heavy herbivore damage and relaxed in herbivore-free environments.

Herbivores on the Serengeti Plains

Because of the many defense mechanisms of plants, all that is green is not necessarily edible and herbivores may be more food-limited than they appear. The result is that herbivores may still compete for food plants. But in some cases herbivores may cooperate in the harvesting of plant matter. The grazing system of ungulates on the Serengeti Plains of east Africa is an excellent illustration of how herbivores may interact over their food supply. The Serengeti Plains contain the most spectacular concentrations of large mammals found anywhere in the world. A million wildebeest (**Figure 9**), 600,000 Thomson's gazelles, 200,000 zebras, and 65,000 buffaloes occupy an area of 23,000 km² (9000 mi²), along with undetermined numbers of 20 other species of grazing animals (Sinclair and Arcese 1995).

The dominant grazers of the Serengeti Plains are migratory and respond to the growth of the grasses in a fixed sequence (**Figure 10**). First, zebras enter the long-grass communities and remove many of the longer stems. Zebras are followed by wildebeest, which migrate in very large herds and trample and graze the grasses to near the ground. Wildebeest are in turn followed by Thomson's gazelles, which feed on the short grass during the dry season (Bell 1971).

Figure 9 Blue wildebeest grazing on the Serengeti Plains of East Africa. An estimated 1 million migratory wildebeest inhabit the Serengeti region.

Different grazers in the Serengeti system do not select different species of grasses but instead select different parts of the grass plant during different seasons (**Figure 11**). Zebras eat mostly grass stems and sheaths and almost no grass leaves. Wildebeest eat more sheaths and leaves, and Thomson's gazelles eat grass sheaths and a large fraction of herbs not touched by the other two ungulates. These feeding differences have significant consequences for the ungulates because grass stems are very low in protein and high in lignin, whereas grass leaves are relatively high in protein and low in lignin, such that leaves provide more energy per gram of dry weight. Herb leaves typically contain even more protein and energy than grass leaves (Gwynne and Bell 1968). So zebras seem to have the worst diet and Thomson's gazelles the best.

How can zebras cope with grass stems as the major part of their diet during the dry season? Most of the ungulates in the Serengeti are ruminants, which have a specialized stomach containing bacteria and protozoa that break down the cellulose in the cell walls of plants. But the zebra is not a ruminant and is similar to the horse in having a simple stomach. Zebras survive by processing a much larger volume of plant material through their gut than ruminants do, perhaps roughly twice as much. So even though a zebra cannot extract all the protein and energy from the grass stems, it eats more and compensates by volume. Zebras also have an advantage of being larger than wildebeest and Thomson's gazelles, and larger animals need less energy and less protein per unit of weight than smaller animals. The net result of these factors is that in times of dietary stress, large animals are able to tolerate low food quality better than small animals can.

Competition for food may occur between wildebeest and Thomson's gazelles because they eat the same parts of the grass. Wildebeest have what appears to be a devastating effect on the grassland as they pass through in migration. Green biomass was reduced by 85% and average plant height by 56% on sample plots. By establishing fenced areas as grazing exclosures, McNaughton (1976) was able to follow the subsequent changes both in grassland areas subject to wildebeest grazing and in areas protected from all grazing. Grazed areas recovered after the wildebeest migration had passed and produced a short, dense lawn of green grass leaves. As gazelles entered the area during the dry season, they concentrated their feeding on areas where wildebeest had previously grazed and avoided areas of grassland that the wildebeest herd had missed.

Grass production was reduced by both wildebeest and gazelles, but no signs of competition were found. The Serengeti ungulate populations show possible evidence of **grazing facilitation**, in which the feeding activity of one herbivore species improves the food supply available to a second species. Heavy grazing by

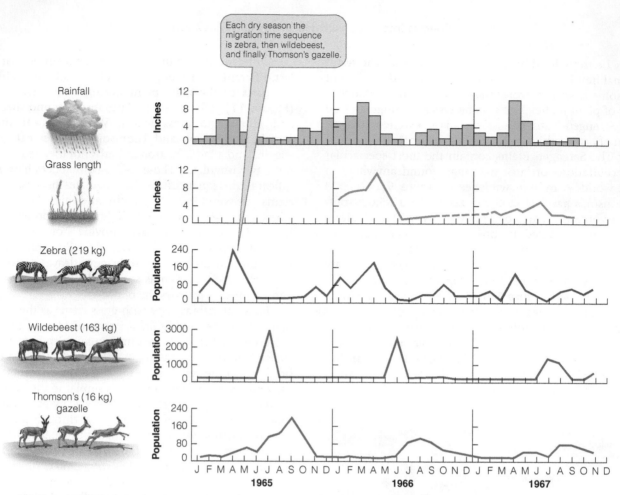

Figure 10 Populations of three migrating ungulates in relation to rainfall and grass length on the Serengeti Plains of East Africa. The figures were obtained in the western Serengeti by a series of daily transects in a strip approximately 3000 m long and 800 m wide. Successive peaks during each year mark the passage of the main migratory species in the early dry season. (Estate holds rights.)

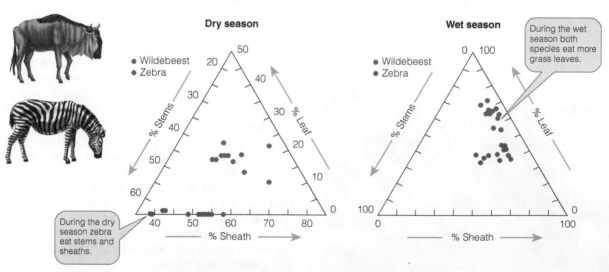

Figure 11 Frequency of the three structural parts of grass in the diets of wildebeest and zebras during the dry season and the wet season, Serengeti Plains, East Africa. The diets of these two ungulates differ more in the dry season than in the wet season. (After Gwynne and Bell 1968).

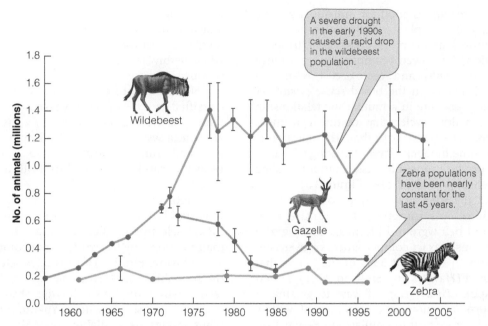

Figure 12 **Population changes in migratory ungulates in the Serengeti Plains of East Africa from 1959 to 2006.** Estimates were obtained by aerial census. If grazing facilitation is strong in this system, wildebeest should not increase in numbers unless zebra also increase. This has not occurred. (Data from Mduma et al. (1999) and A.R.E. Sinclair, personal communication.)

wildebeest prepares the grass community for subsequent exploitation by Thomson's gazelles in the same general way that zebra feeding improves wildebeest grazing. Thus potential competition may be replaced by mutualism. Feeding systems of this type may be severely upset by the selective removal of one herbivore in the sequence.

Grazing facilitation may have strong or weak effects on populations of these ungulates. By comparing population trends of wildebeest, zebras, and Thomson's gazelles in the Serengeti, Sinclair and Norton-Griffiths (1982) could test if this facilitation was strong or weak. If grazing facilitation is strongly mutualistic and obligatory, wildebeest numbers should not increase if zebra numbers do not increase, and if wildebeest numbers increase, gazelle numbers should also increase. **Figure 12** shows that this has not happened. Wildebeest numbers more than doubled during the 1970s while gazelle numbers fell slightly and zebra numbers remained constant. Predation may hold zebra numbers down, and these three ungulates apparently are not as closely linked as Bell (1971) suggested. Grazing facilitation has little effect on population changes of these ungulates (Arsenault and Owen-Smith 2002).

Competition for grass in the Serengeti region may occur between very different types of herbivores (Sinclair and Arcese 1995). In addition to the large ungulates, 38 species of grasshoppers and 36 species of rodents consume parts of the grasses and herbs. In the Serengeti Plains, most of the plant material consumed by herbivores is consumed by the large ungulates, but in some plant communities within the Serengeti, grasshoppers consumed nearly half as much grass as did the ungulates. The grazing system of the Serengeti is thus even more complex than is suggested in Figure 10. In any grazing system, herbivores of greatly differing size and taxonomy may be affecting one another positively or negatively.

Can Grazing Benefit Plants?

Herbivores eat parts of plants, and at first view this action would appear to be detrimental to the individual plant, a negative interaction from the plant's viewpoint. But could grazing or browsing in fact be beneficial to a plant so that it is a win-win situation for both the plant and the herbivore? On a more practical level of public policy, should public grazing land be protected from sheep and cattle grazing, or should we encourage cattle and sheep production? If cattle and sheep grazing is good for plants, then we would have a clear, ecologically based reason to support current land management policies in the western United States and Australia. What is the ecological evidence?

The idea that grazing is good for individual grasses, good for cattle, good for plant communities (to retard succession to shrubs), and good for ecosystems (to speed up decomposition) have been promoted by both range managers (Savory 1988) and ecologists (Owen and Wiegert 1981). This idea in the broad sense postulates that grazers and grasses are in a mutualistic relationship in which both gain. But it is clear that too much grazing is detrimental—everyone agrees with that. The question is whether or not some moderate level of grazing will stimulate plants to produce more biomass, a proposal called the **overcompensation hypothesis (Figure 13)**.

To evaluate the idea that grazing could improve plant production, it is important to measure both aboveground and belowground biomass. Grazing at a moderate level causes exact compensation or overcompensation in growth for about 35% of the plants that have been studied (Hawkes and Sullivan 2001). Different species respond in a variety of ways to grazing or browsing. **Figure 14** shows the response of willow shrubs and black spruce trees to simulated browsing. Fast-growing willows in Sweden are being grown for bioenergy, and the response of willow plants to shoot removal is critical for management (Guillet and Bergström 2006). Figure 14a shows that total biomass of these willows under clipping never exceeded the biomass production of unclipped shrubs. By contrast, black spruce trees showed overcompensation for shoot removals during the first growing season but exact compensation one year later (Bast and Reader 2003). Plants respond to grazing by regrowth but they never recover completely from the losses caused by moderate to severe grazing. The prevailing view of the plant-herbivore interaction

for grazing systems that it is a predator-prey type of interaction in which the herbivore gains and the plant loses is correct for some but not all plants, particularly when herbivores remove only small amounts of plant production.

Not all plant-herbivore interactions can thus be classified as +/− or negative—some plant-animal interactions are mutualistic, or +/+, as we saw for the ant-acacia system, in which both parties gain. Pollination and fruit dispersal are two additional interactions that can be beneficial for both the plant and the herbivore.

Herbivores are commonly thought to be "lawn mowers," but it is important to recognize that they are highly selective in their feeding. This selectivity is a major reason why the world is not completely green for an herbivore. **Figure 15** illustrates selective feeding in the snowshoe hare (*Lepus americanus*). Snowshoe hares feed in winter on the small twigs of woody shrubs and trees. In the southwestern Yukon, only three main plant species are available above the snow, and hares clearly prefer dwarf birch (*Betula glandulosa*) over willow (*Salix glauca*). These preferences may be caused by plant secondary substances such as phenols (Sinclair and Smith 1984).

Dynamics of Herbivore Populations

There are two basic types of plant-herbivore systems. We have discussed one type, called an interactive herbivore system because the vegetation affects the herbivore

Figure 13 The overcompensation hypothesis of grazing. Grazing is postulated to be favorable for plant production up to some optimum level of grazing pressure (green arrow). The classical view of grazing (undercompensation) is shown by the red line, in which plants cannot completely replace grazed tissues so that all grazing has a negative effect on plant production. The horizontal black line indicates exact compensation of lost tissues by a grazed plant. (Modified after Belsky 1986.)

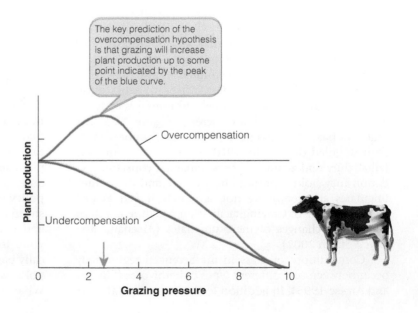

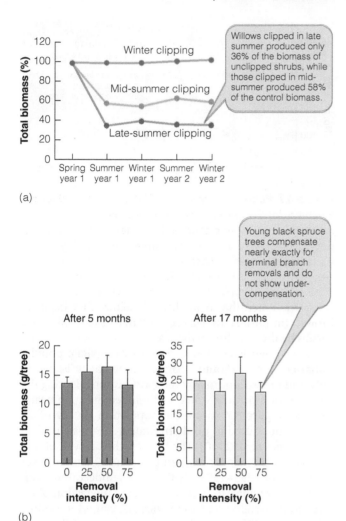

(a)

(b)

Figure 14 Tests of the overcompensation hypothesis. (a) Willows (*Salix viminalis*) grown for biofuel in Sweden. Winter clipping of these shrubs shows exact compensation, while summer clipping shows undercompensation. (b) Black spruce (*Picea mariana*) trees five years of age were clipped at three intensities. A nonsignificant hint of overcompensation can be seen after five months, but one year later all treatments showed exact compensation. (Data from Guillet and Bergström 2006, and Bast and Reader 2003.)

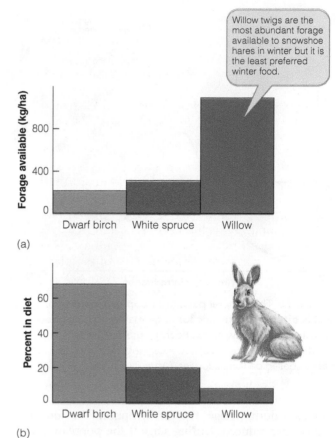

(a)

(b)

Figure 15 Selective feeding of snowshoe hares on woody plants during winter. (a) Biomass of available forage. (b) Percentage of forage type in diet. Hares do not eat a random sample of the plants available but are highly selective and prefer dwarf birch. (Data from Kluane Lake, Yukon, winter 1979–1980; unpublished data from A. R. E. Sinclair and J. N. M. Smith.)

population and the herbivores influence the rate of growth and the subsequent fate of the vegetation:

vegetation production ↔ herbivore abundance

This feedback is critical for the dynamics of the plant-herbivore system. Other herbivore systems, called noninteractive herbivore systems, show no relationship between herbivore population density and the subsequent condition of the vegetation because there is no feedback:

vegetation production → herbivore abundance

These systems will be discussed below.

Many herbivore systems are interactive, with feedback occurring between herbivores and plants. Serengeti ungulates provide many examples, and most grazing systems are of this type. Next we look at an example of each type to compare and contrast the two ways in which animal populations react to their food plants.

Interactive Grazing: Ungulate Irruptions

Many large mammals introduced into a new region increase dramatically to high densities and then collapse to lower levels. The increase and subsequent collapse is called an irruption. Introduced reindeer populations have provided several examples. **Figure 16** illustrates the stages of an irruption. Irruptions commonly occur when the introduced population has an excellent food supply and no natural predators. As the population

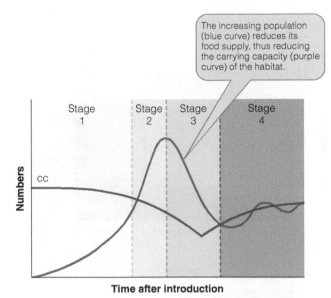

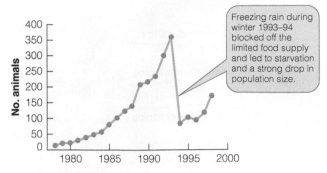

Figure 17 Population irruption of introduced reindeer on Svalbard. In 1978 twelve individuals were introduced to this island area that was free of other herbivores. This isolated population showed a classic irruptive sequence (cf. Figure 16). (Data from Aanes et al. 2000.)

Figure 16 The general pattern of population irruptions that is often characteristic for large mammal populations introduced into new areas. Four general stages of the irruption can be recognized. cc = carrying capacity of the habitat based on food supplies. (After Riney 1964.)

increases during stage 1 of an irruption, the food resources are reduced. During stage 2 the population exceeds the carrying capacity of the habitat, and food plants are overutilized and damaged. In stage 3 the population collapses because of food shortage, often aggravated by severe weather. This collapse may continue to near-extinction, or the population may stabilize at lower numbers (Leader-Williams 1988).

Graeme Caughley studied irruptions of the Himalayan thar in New Zealand (Caughley 1970). The Himalayan thar, a goatlike ungulate of Asia, was introduced into New Zealand in 1904 and has since spread over a large region of the Southern Alps. As its density increased, its birth rate fell only slightly and its death rate increased, primarily because of increased juvenile mortality. After a period of high density the population declined due to a combination of reduced adult fecundity and a further increase in juvenile losses.

What caused these population changes? Caughley (1970) suggested that grazing by the thar both reduced its food supply and changed the character of the vegetation. The link between ungulates and their food plants is critical in these irruptions. The most conspicuous effect of thar grazing was found in the abundance of snow tussocks (*Chionochloa* spp.), evergreen perennial grasses that were the dominant vegetative cover where thar were absent but were scarce where thar had become common. Snow tussocks were believed to be important as food in late winter and cannot tolerate even moderate grazing

pressures. When thar reach high densities, they begin to browse on shrubs in winter and may even kill some shrubs by their feeding activities.

Norwegian whalers introduced two separate populations of reindeer to South Georgia, a subantarctic island, in 1911 and in 1925, primarily for sport hunting. During the 1950s whales became scarce and reindeer hunting nearly stopped. **Figure 17** shows the population history of 12 reindeer introduced to northern Svalbard (Spitsbergen) in 1978 (Aanes et al. 2000). The herd reached a peak of 360 animals in 1993 and collapsed to about 80 animals over the winter of 1993–1994. By this time reindeer were overgrazing their winter food plants, and the interaction of severe winter weather and food shortage caused the initial collapse. By 1998 this population was rising rapidly again, as predicted in the irruption model shown in Figure 16. This simple island system, which lacks predators and other grazing animals, clearly illustrates the interplay between plants and herbivores in an interactive grazing system.

A general picture of an ungulate irruption emerges: A small number of animals is introduced onto a range with superabundant food, and a gradual increase in animal density and decrease in plant density occurs until the animals have reduced or eliminated their best forage. Animal numbers then decline until a new, lower density is reached, at which the herbivores and their plants may stabilize or continue to fluctuate. Irruptions occur in both native and introduced species of large herbivores (Forsyth and Caley 2006).

This sequence of events is similar to that predicted by some simple predator-prey models (Noy-Meir 1975; Caughley 1976b). The Rosenzweig-MacArthur predator-prey model can also be applied to a simple grazing system. Plant growth is a simple function of plant biomass for most plants, and the logistic equation can describe plant growth in the absence of grazing. **Figure 18** illustrates some

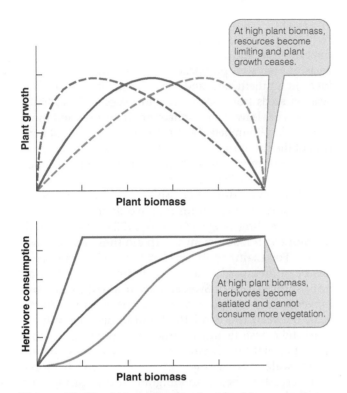

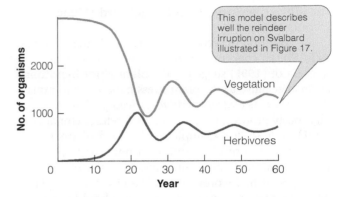

Figure 19 Simple model of an interactive herbivore-plant grazing system, which is similar to an ungulate irruption in showing a damped oscillation. The model used is a modified version of the Lotka-Volterra predator-prey model in which vegetation increases logistically instead of exponentially. (After Caughley 1976a.)

Figure 18 Simple models of a plant-herbivore grazing system. (a) Plant growth as a function of plant biomass. The logistic growth function is shown in red, and two other possible shapes of growth functions are shown in blue and green. (b) Herbivore consumption as a function of plant biomass. These are functional response curves like those for predator-prey relations. The blue curve is Type 1, the red curve is Type 2, and the green sigmoid curve is Type 3. (Modified from Noy-Meir 1975.)

possible models for plant growth and herbivore consumption in a grazing system. If we combine the logistic plant growth model with the Type 2 consumption curve, we can generate the dynamics shown in **Figure 19**, a simple herbivore-vegetation model that mimics the reindeer irruption shown in Figure 17 (Caughley 1976b). The behavior of simple herbivore-vegetation models is highly dependent on the rates of increase of the plants and herbivores alike, and also on the feeding rates of the herbivores. Caughley (1976b) showed that such simple model systems oscillate in cycles if the grazing pressure tends to hold the amount of vegetation below about half the amount present in the ungrazed state. If the herbivore is a very efficient grazer, such simple systems can collapse completely (Noy-Meir 1975).

Many insect populations experience irruptions that damage their food plants (Myers 1993, 2000). The spruce budworm, for example, periodically irrupts to epi-

demic proportions in the coniferous forests of eastern Canada (Royama et al. 2005). Budworms eat the buds, flowers, and needles of balsam fir trees. Outbreaks occur every 35 to 40 years, in association with the maturing of extensive stands of balsam fir, and during budworm outbreaks many balsam fir trees are defoliated and killed. Populations of the large aspen tortrix moth irrupt in interior Alaska at intervals of 10–15 years; during these irruptions quaking aspen trees are severely defoliated (80%–100%) for two to four years (Brandt et al. 2003).

Many herbivorous insect populations may be held at low densities by a protein deficiency in their food plants. White (1993) has suggested that most plant material is not suitable food for insects because of nitrogen deficiencies. When plants are physiologically stressed—by water shortage, for example—they often respond by increasing the concentration of amino acids in their leaves and stems. Some larval insects may survive much better when more amino acids are available, and thus the stage is set for an insect irruption. This hypothesis, called the **plant stress hypothesis**, postulates that plants under abiotic stress become more suitable as food for herbivorous insects (Larsson, 1989; Huberty and Denno 2004). Laboratory tests of this hypothesis have consistently failed to validate the plant stress hypothesis, and the problem seems to be that continuous plant stress increases nitrogen in the leaves but also reduces their water content, so that insects feeding on sap in leaves cannot gain access to this nitrogen. The increased nitrogen is available to these herbivores only if there is intermittent stress. The key to understanding insect irruptions seems to lie in the interaction between water stress and nitrogen availability in plants, and the simple plant stress hypothesis needs to be modified (Huberty and Denno 2004).

A common observation of foresters is that insect attack is often concentrated on young, vigorously growing trees. Price (1991) suggested the **plant vigor hypothesis** as an alternative to the plant stress hypothesis to explain these outbreaks. Many herbivores feed preferentially on vigorously growing plants or plant modules (Inbar et al. 2001). Moose, for example, prefer to feed on rapidly growing shoots of birch and seem immune to plant defense chemicals (Danell and Huss-Danell 1985). Different types of herbivores feed on rapidly growing plants, and others prefer stressed plants, so we should not expect all species to fit only one of these two hypotheses. Understanding patterns of herbivore attacks on plants will likely require a diversity of hypotheses.

Noninteractive Grazing: Finch Populations

European finches feed on the seeds of trees and herbs, and their feeding activities do not in any way affect the subsequent production of their food plants. These species form a good example of a noninteractive system in which controls operate in only one direction:

Food plant production → herbivore density

In these systems herbivore abundance has no effect at all on the production of food plants. There is no direct feedback from the herbivores to the plants, in contrast to the grazing systems discussed in the previous section. This lack of interaction has important consequences for plant-herbivore interactions.

Among two groups of British finches, one group feeds on the seeds of herbs, and their populations are quite stable (Newton 1972); a second group feeds on the seeds of trees, and their populations fluctuate greatly. Population dynamics in these finches are determined by fluctuations in seed crops from year to year. Herbs in the temperate zone produce nearly the same numbers of seeds from one year to another, but trees do not. Most trees require more than one year to accumulate the reserves necessary to produce fruit. Spruce trees in Europe, for example, have moderate to large cone crops every two to three years in central Europe, every three to four years in southern Scandinavia, and every four to five years in northern Scandinavia. Good weather is also needed when the fruit buds are forming during the year before the seed crop is produced. The net result is that trees in a given geographic region usually fruit in synchrony. Various geographic regions may or may not be in synchrony with each other, depending on local weather conditions.

Finches that depend on tree seeds experience great irruptions in population density. They exist only by being opportunistic and moving large distances to search for areas of high seed production. All the "irruptive" finches breed in northern areas and rely at some critical part of the year on seeds from one or two tree species. Periodically these finches leave their northern breeding areas and move south in large numbers. **Figure 20** shows the years of invasion of the common crossbill into southwestern Europe. A major invasion of crossbills into western Europe occurred in 1990 and a smaller invasion in 1999 and 2002.

Mass emigration of crossbills and other finches is presumably an adaptation that avoids food shortages on the breeding range (Newton 1972). But crop failure alone is not sufficient to explain these mass movements. For example, in Sweden the spruce cone crop has been measured in all districts since 1900. Not all poor spruce crops in Sweden have resulted in crossbill movements. Very poor spruce crops occurred in 14 years between 1900 and 1963, but in only six of these years did crossbills move. Other evidence suggests that high population density may be necessary before larger scale movements can be triggered. In some years, crossbills began to emigrate in the spring, even before the new cone crop was available. Crossbills also put on additional fat before they emigrate, in the same way that migratory birds do. The suggestion is that high crossbill density is a prerequisite for large-scale movements and that emigration occurs in response to the first inadequate cone crop once high bird densities are present.

Why emigrate? Mass emigration presumably is advantageous to the birds that stay behind, provided they find sufficient food. Emigration, by contrast, is often considered suicidal, and the question arises as to how such an adaptation could exist. Crossbill emigrants might have two potential advantages: They could colonize new habitats in the south and thereby leave descendants. However it is more likely they obtain an advantage by migrating back north again after the food crisis has passed. Newton (1972) described four common crossbills that were banded in Switzerland during an irruption and were recovered a year later in northern Russia. Thus some birds return north, even though many die in the south during the irruption.

A close correlation exists between crossbill breeding densities and the size of their food supply, the conifer cone crop. This correspondence is obtained by having great mobility such that populations can concentrate their nesting in areas with good cone crops. How random mobility within the normal breeding becomes a unidirectional emigration in years of irruption is not understood.

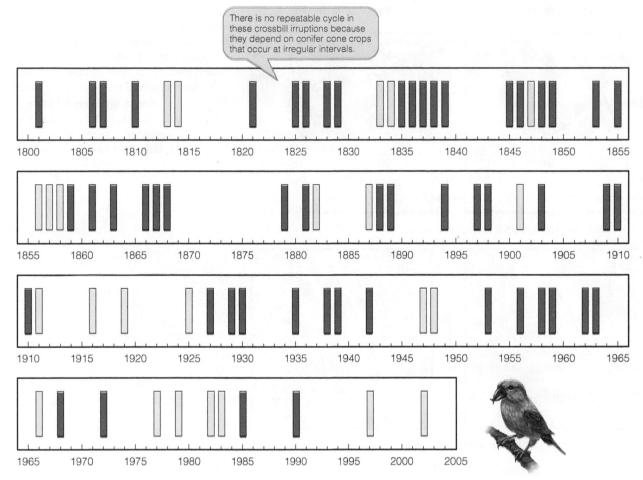

Figure 20 Years of invasions of the common crossbill into southwestern Europe from 1800 to 2005. Red blocks indicate large invasions; yellow blocks indicate small invasions. (Data from Ian Newton 1972 and personal communication.)

Mycorrhizae: An Example of Mutualism

Not all interactions between species are detrimental to plants; one good example of mutualism involves mycorrhizal fungi in the soil. Plants must take up nutrients from the soil to grow, and almost all plants have fungi called **mycorrhizae** growing on or in their roots. These fungi help the plant by taking up inorganic nutrients such as phosphorus from the soil and donating these nutrients to the plant in exchange for carbohydrates like sugars that the fungi obtain from the plant roots. **Figure 21** illustrates schematically the interactions between plant roots and mycorrhizal fungal hyphae. These kinds of win-win interactions are called mutualisms because they benefit both of the species involved.

Ecologists and agricultural scientists discovered the importance of mycorrhizae by observing what happens to plants that do not have mycorrhizae. In Oregon, a Douglas fir tree nursery was started in the Willamette Valley in 1961 on old agricultural fields. Because the foresters were concerned about root diseases and reducing weeds, they fumigated the sandy soil before sowing the first crop and killed all the soil organisms, good as well as bad. The photo in **Figure 22** shows the Douglas fir seedlings in their third growing season. Most seedlings are stunted, off color, and deficient in all nutrients, especially phosphorus. However, some tree seedlings got inoculated with mycorrhizal fungi, presumably by airborne spores, and started growing normally. Once the mycorrhizae were established, the fungi grew out from the first root system to colonize adjacent seedlings, which then began to grow. The fungi spread through the growing season, resulting in patches with the largest seedlings in the middle. The soil in the nursery was heavily fertilized when the stunting syndrome

237

No mycorrhizal fungi　　　**Mycorrhizas present**

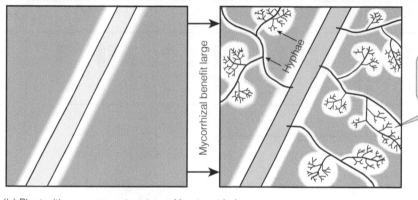

(a) Plant with a fine root system and long root hairs.

(b) Plant with a coarse root system without root hairs.

Available P　　　Non-mycorrhizal roots

P depletion zone　　　Roots with mycorrhizae

Figure 21 Schematic view of the interaction between soil phosphorus and plant roots with and without mycorrhizae. (a) Plants with fine root hairs gain little from having mycorrhizae that can bring in nutrients from distant parts of the soil. (b) Plants without fine root hairs gain a large benefit from mycorrhizae. This schematic illustrates the reason for the variation in tree growth shown in Figure 22. (Modified from Brundrett et al. 1996.)

Nutrients from the white zone are picked up by the fungal hyphae and moved into the plant roots, benefiting the plant.

Figure 22 Douglas fir seedlings growing in a nursery in Oregon. The soil was sterilized, which killed all of the mycorrhizal fungi, and tree seedlings were then planted. A few of the seedlings had mycorrhizal fungi colonize their roots (probably from airborne spores), but most did not. The differences in growth are striking. (Photo courtesy of Jim Trappe, Oregon State University.)

appeared, but the seedlings could not pick up the soil phosphorus without the mycorrhizal fungi, and without the phosphorus they could not extend their root systems to acquire other nutrients.

The mycorrhizal association is a true mutualism because both the plant and the fungus benefit. In infertile soils, nutrients taken up by the mycorrhizal fungi lead to improved plant growth. Plants with mycorrhizae are more competitive in infertile soils and better able to tolerate environmental stresses than are plants without mycorrhizae. There are many different species of mycorrhizal fungi, and any single species of plant may be colonized by a variety of different fungal species (van der Heijden and Sanders 2002). Ecologists are just now beginning to appreciate how this complex community of soil organisms interacts with trees, shrubs, grasses, and herbs. Mycorrhizal fungi are essential for modern agriculture and forestry but go largely unnoticed because they are hidden from view in the soil.

Complex Species Interactions

Species interactions are rarely one-on-one in natural communities, and the untangling of complex sets of species interactions is an important focus in ecology today. Complex interactions illustrate how difficult it can be to determine whether a single species exerts a beneficial or a harmful influence on another species.

Interactions between homopterous insects and the ants that either tend them or prey on them have been described for nearly a century (Buckley 1987). These plant-homopteran-ant interactions are significant because many of the world's major plant pests are homopteran insects, and many of the worst crop diseases are transmitted by homopterans. The manipulation of ant assemblages to control homopteran pests has been practiced in China since at least AD 300 (Needham 1986).

Ants that tend homopterans provide a positive benefit for the homopterans, including protection from predators or parasitoids, sanitation in honeydew removal, and transportation to new feeding sites. Ants may also remove dead individuals from homopteran populations and provide nest sites. For the plants, plant-homopteran-ant interactions impose many costs and confer few if any benefits. Homopterans consume phloem sap and tax the plant by removing metabolites, damaging plant tissues, and increasing water loss. Homopterans may also transmit plant pathogens. For the ants that tend homopterans, the main benefit is the food they obtain in the form of sugars from the honeydew secreted by homopterans (Buckley 1987), and ant colonies that feed on honeydew have higher populations than colonies with no honeydew source.

Pinyon pine trees in northern Arizona are attacked by a stem-boring moth (*Dioryctria albovittella*) with a cascade of effects on the trees, their cone crops and seed harvest by birds, and their mycorrhizal fungi (Whitham and Mopper 1985; Brown et al. 2001). Pinyon pine growing on volcanic cinder soils with low nutrients are subject to attack by this moth, but while 80% of the pine trees are attacked, others are resistant to the moth (**Figure 23**). Resistance to the moth has a genetic basis. The result is pinyon pine that appear like shrubs because of the moth attack on their terminal shoots and normal conical, upright pines not subject to attack. By experimentally removing moth larvae from some susceptible trees for 18 years, Whitlam was able to show that moth removal reversed the herbivore impact on tree shape. These effects flowed on to other components of the ecological community (**Table 4**). Cone production was much lower on susceptible pine trees, providing fewer seeds to avian seedeaters like Clark's nutcracker. Moreover, only one-third of the harvested seeds were from the cones of susceptible trees compared with resistant trees. Finally, mycorrhizae colonized the resistant trees more readily than the susceptible trees, which had fewer coarse roots, giving them a growth advantage in nutrient uptake from the poor volcanic soils (Gehring and Whitham 1994). Herbivory in this example has consequences not only for the pines being eaten but also for other animals that feed on the seeds of this pine and the soil fungi that assist the pines in nutrient uptake.

Figure 23 Pinyon pines (*Pinus edulis*) growing in the Sunset Crater of northern Arizona on black cinder soil. Pines susceptible to the stem-boring moth *Dioryctria albovittella* are reduced to tall shrubs by this herbivore, while resistant pines grow into a normal tree. By removing the moth over 18 years Thomas Whitham and his students were able to change the growth form of susceptible trees back to normal growth. Further consequences of this plant-herbivore interaction are given in Table 4. (Photo by T. G. Whitham.)

Table 4 Levels of herbivory on moth-resistant, moth-susceptible, and moth-removal pinyon pines in northern Arizona, and the consequences of this herbivory for other components of the community.

Tree type	Shoot mortality caused by moths	Cone production[a]	Cone harvest by birds	Mycorrhizal colonization
Resistant: 150-year-old trees	9%	2440	100%	37%
Resistant: 60-year-old trees	2%	165	38%	51%
Susceptible: 150-year-old trees	24%	172	38%	25%
Susceptible: 60-year-old trees	16%	13	19%	34%
Experimental moth removal: 60-year-old trees	1%	147	36%	55%

[a]Data from all trees for the cone production year of 1994.

SOURCE: After Brown et al. (2001).

In Hawaii an insect called green scale (*Coccus viridis*) is tended by the ant *Pheidole megacephala* on the host plant Indian fleabane (*Pluchea indica*). Bach (1991) analyzed this system using removal and addition experiments to measure the strength of these interactions. When she removed ants from plants, the number of parasitized green scales increased, and the mortality of scales to other predators and diseases also increased (**Figure 24**). Ants removed ladybird beetle larvae introduced onto plants as possible scale

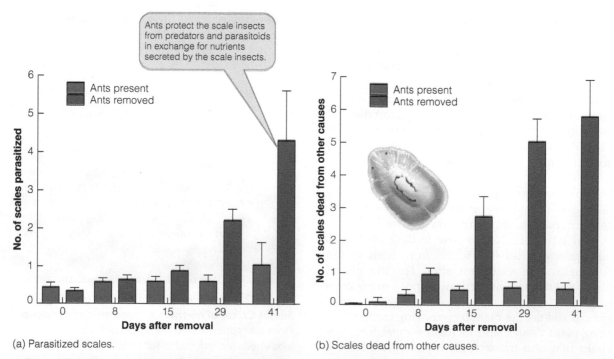

(a) Parasitized scales.

(b) Scales dead from other causes.

Figure 24 Number of green scales (*Coccus viridis*) per leaf on Indian fleabane plants (*Pluchea indica*) with and without ants (*Pheidole megacephala*). Mortality of the insects increased dramatically when ants were removed. (After Bach 1991.)

predators. On plants without ants, honeydew from the green scale accumulates, and a sooty mold grows on the leaves. This mold reduces the photosynthetic rate of the leaf, and leaves infested with mold were often shed by the plant. Ants thus indirectly benefited the plant by removing the honeydew. Ants may also remove other herbivorous insects such as moth larvae from plants and thus protect the plants from additional losses (Bach 1991).

Complex interactions become more difficult to unravel as the number of species involved grows. Interaction webs may involve herbivory, mutualism, predation, and competition, and emphasize the need to look at both direct interactions and indirect effects in ecological systems in a quantitative way (Brown et al. 2001).

Herbivory, Economics, and Land Use

Ecological ideas about herbivory meet economic ideas about land use in the grazing lands of the western United States. About 70% of the land area in the western states is grazed by livestock, including wilderness areas, wildlife refuges, national forests, and some national parks (Fleischner 1994). Ecologists ask two questions about the effects of grazing: (1) What are the ecological costs of grazing these areas? And (2) is grazing in its current form sustainable? Economists ask about the balance of costs and benefits of grazing, but in doing so rarely consider the ecological costs, which almost never have dollar values attached to them. The result has been an ongoing and acrimonious controversy over land use in the West, a controversy with multiple dimensions.

The experimental ecologist would like to look at comparable grazed and ungrazed land in order to measure the ecological effects of grazing on populations of plants and animals. But almost no ungrazed land is available for such comparisons. Much of the land left ungrazed is on steep slopes or in rocky areas that differ dramatically from the surrounding habitats. One solution to this problem is to use livestock exclosures to study effects. But this approach also has problems because most exclosures are small in area and were previously grazed. Small exclosures do not include all the species in a community, especially the rare

ones. And if the initial grazing effects are the most severe ones, historical carryover will affect even long-term exclosure studies on sites that were previously grazed. As such, exclosures will underestimate the true effects of grazing on plants and animals.

Some ecologists and land managers argue that grass needs grazing and that livestock are thus essential for the ecological health of western grazing lands. Some of the justification for this has come from the *overcompensation* or herbivore optimization hypothesis, which suggests that plant productivity may increase if plants are grazed. There is little ecological evidence for this hypothesis in western rangelands, but the idea keeps coming up as but one justification for the current grazing system.

The use of public lands for grazing must be balanced with the needs of conservation and recreation. In particular, all those concerned need to work out sustainable land-use practices that will achieve these diverse economic and ecological goals. The western rangelands should not be all national parks, nor should they be all overgrazed plant communities. There must be cooperation among all interested groups to achieve the goal of sustainable land use, and good science conducted to show us what policy goals can be achieved by good land management (Brown and McDonald 1995).

Summary

Herbivory is a form of predation. Because plants are modular organisms, herbivores usually eat only part of the plant, and typical herbivory thus differs from typical predation. Plants have a variety of structural and chemical defenses that discourage herbivores from eating them. Many secondary plant substances stored in plant parts discourage herbivores, which have responded to these evolutionary challenges by timing their life cycle to avoid the chemical threats or by evolving enzymes to detoxify plant chemicals. Several theories have attempted to specify the strategies of plant defense but no general theory has yet proven possible because of the diverse methods of plant defense against herbivores. The Resource Availability Hypothesis is currently the best model for plant defense. It emphasizes the differential costs and benefits of defense for slow-growing and fast-growing plants, and predicts when plants ought to invest in either chemical or physical defense.

Some herbivores can affect the future density and productivity of their food plants. A very efficient herbivore can thus drive itself to extinction unless it has some constraints that prevent overexploitation of its food plants. Most herbivore-plant systems seem to exist in a fluctuating equilibrium. Some herbivore populations track their food supply, and large fluctuations in food supply are often translated into large fluctuations in herbivore densities.

Not all herbivory is detrimental. Mycorrhizal fungi in the soil take nutrients from plant roots but in exchange extract nutrients like phosphorus from the soil and provide these nutrients to the roots. Plants without mycorrhizae often grow poorly, and this mutualism has major consequences for plant-plant competition.

Mutualism occurs in many plant-animal interactions. Pollination and seed dispersal are two examples of processes that benefit both plant and animal species. Ants may form mutualistic relationships with plants, particularly in tropical areas or with herbivorous insects such as homopterans. Many plant-animal interactions are complex and involve many species in a web of relationships that are not easily categorized as positive or negative overall.

Review Questions and Problems

1. Large mammals in the Serengeti utilize grazing facilitation (Figure 10) but their populations change independently of one another (Figure 12). Suggest at least two hypotheses that might explain this discrepancy, and discuss what data would be needed to test these hypotheses.

2. Discuss the advantages and disadvantages of physical versus chemical defenses in plants.

3. Plants endemic to islands without large mammalian herbivores are believed to be vulnerable to damage and possible extinction because they have no evolutionary history of being grazed. How would you test this hypothesis that island plants lack defenses against herbivores? Compare your approach with that of Bowen and Van Vuren (1997), who studied the plants of Santa Cruz Island off California.

4. An early view of plant-herbivore interactions was that plants and their insect herbivores are engaged in an evolutionary arms race, and for many interactions pairwise coevolution was the dominant explanation for the observed patterns of plant defense. This view is not widely held now (Stamp 2003). Discuss why the evolutionary arms race analogy might not hold for plant-herbivore interactions.

5. Wildlife managers and range ecologists both speak of the "carrying capacity" of a given habitat for a herbivore population and try to prevent "overgrazing." Write an essay on the concepts of carrying capacity and overgrazing, how they can be measured, and how the concepts can be applied to agricultural and natural situations. McLeod (1997), Price (1999), and Mysterud (2006) discuss the definition and use of these terms.

6. Sap-feeding insects do more poorly on water-stressed plants, while leaf-chewing insects on average are not affected by water-stressed plants. Discuss why this difference in response might occur. Huberty and Denno (2004) discuss these results.

7. Alkaloids are plant defense chemicals, but not all plants contain alkaloids. Among annual plants, the incidence of alkaloids is nearly twice that among perennial plants. Tropical floras also contain a much higher fraction of species with alkaloids than temperate floras, and this is true for both woody and nonwoody plants. Suggest why these patterns might exist, and how to test your ideas. Compare your ideas with those of Levin (1976).

8   Caughley and Lawton (1981) suggest that the growth of many plant populations will be close to logistic. Review the assumptions of the logistic equation, and discuss why this suggestion might be true or false.

9   Eucalyptus trees in Australia have high rates of insect attack on leaves, with 10%–50% of the leaves eaten every year, even though these trees also contain very high concentrations of essential oils and tannins (Gras et al. 2005). Discuss how this situation could occur if eucalyptus oils and tannins are defensive chemicals.

10   Large herbivorous mammals are not always present in habitats dominated by spiny plants. Why might this be? Janzen (1986) reviews the vegetation of the Chihuahuan Desert of north-central Mexico and interprets the abundance of spiny cacti as reflecting the "ghost of herbivory past." Read Janzen's analysis and discuss how one might test his ideas.

11   Three species of crossbills in northern Europe tend to irrupt together. But two species concentrate on larch and spruce cones, which mature in one year, while the third species feeds on pine cones, which mature in two years. Poor flowering seems to occur at the same time in pine, spruce, and larch. How can you explain this puzzle? Suggest an experiment to test your hypothesis, and compare your ideas with those of Newton (1972, p. 239).

12   If a plant-homopteran-ant interaction has a net negative effect on the individual plants occupied by the homopterans (see Figure 23), why are these plants not selectively eliminated from the population?

Overview Question

Under what conditions might herbivory benefit a plant, such that a plant-herbivore interaction could be mutualistic? How would you test the overcompensation model for a grassland grazing system like the Serengeti in Africa or the Great Plains in North America?

Suggested Readings

- Allen, M. F., W. Swenson, J. I. Querejeta, L. M. Egerton-Warburton, and K. K. Treseder. 2003. Ecology of mycorrhizae: A conceptual framework for complex interactions among plants and fungi. *Annual Review of Phytopathology* 41:271–303.

- Brown, J. H., T. G. Whitham, S. K. M. Ernest, and C. A. Gehring. 2001. Complex species interactions and the dynamics of ecological systems: Long-term experiments. *Science* 293:643–650.

- Caughley, G., and J. H. Lawton. 1981. Plant-herbivore systems. In *Theoretical Ecology*, ed. R. M. May, 132–166. Oxford: Blackwell.

- Coley, P. D., J. P. Bryant, and F. S. Chapin III. 1985. Resource availability and plant antiherbivore defense. *Science* 230:895–899.

- Forsyth, D. M., and P. Caley. 2006. Testing the irruptive paradigm of large-herbivore dynamics. *Ecology* 87:297–303.

- Huberty, A. F., and R. F. Denno. 2004. Plant water stress and its consequences for herbivorous insects: A new synthesis. *Ecology* 85:1383–1398.

- Huntzinger, M., R. Karban, T. P. Young, and T. M. Palmer. 2004. Relaxation of induced indirect defenses of acacias following exclusion of mammalian herbivores. *Ecology* 85:609–614.

- Mysterud, A. 2006. The concept of overgrazing and its role in management of large herbivores. *Wildlife Biology* 12:129–141.

- Palmer, T. M., M. L. Stanton, T. P. Young, J. R. Goheen, R. M. Pringle, and R. Karban. 2008. Breakdown of an ant-plant mutualism follows the loss of large herbivores from an African savanna. *Science* 319:192–195.

- Stamp, N. 2003. Out of the quagmire of plant defense hypotheses. *Quarterly Review of Biology* 78:23–55.

- White, T. C. R. 1993. *The Inadequate Environment: Nitrogen and the Abundance of Animals*. New York: Springer-Verlag.

Credits

Ilustration and Table Credits

F1 Reprinted with permission from R. H. Whittaker and P. P. Feeny, "Allelochemics: Chemical interactions between species," *Science*, Vol. 171, p. 760, 1971. Copyright © 1971 American Association for the Advancement of Science. T1 N. Stamp, "The problem with the messages of plant-herbivore interactions in ecology textbooks," *Bulletin of the*

Ecological Society of America 86:27–31. Copyright © 2005 Ecological Society of America. Used with permission. T2 Reprinted with permission from P. D. Coley et al., "Resource availability and plant antiherbivore defense," *Science*, Vol. 230 , p. 895, 1985. Copyright © 1985 American Association for the Advancement of Science. F2 Reprinted with permission from J. H. Ryther and W. M. Dunstan, "Nitrogen, phosphorus, and eutrophication in the coastal marine environment," *Science*, Vol. 171, pp. 1008–1009, 1971. Copyright © 1971 American Association for the Advancement of Science. F5 From P. P. Feeny, "Seasonal changes in oak leaf tannins and nutrients as a cause of spring feeding by winter moth caterpillars" *Ecology*, Vol. 51, pp. 565–581. Copyright © 1970 Ecological Society of America. Reprinted by permission. T3 From P. P. Feeny, "Seasonal changes in oak leaf tannins and nutrients as a cause of spring feeding by winter moth caterpillars," *Ecology* 51:565–581, 1970. Copyright © 1970. Reprinted by permission of The Ecological Society of America. F10 Estate holds rights. F11 From M. D. Gwynne and R. H. V. Bell, "Selection of vegetation components by grazing ungulates in the Serengeti National Park," *Nature*, Vol. 220., p. 391 Copyright © 1968 Macmillan Magazines Ltd. Reprinted by permission. T4 Reprinted with permission from J. H. Brown et al., "Complex species interactions and the dynamics of ecological systems: Long-term experiments," *Science* 293:643–650, 2001. Copyright © 2001. Reprinted with permission from AAAS. F24 From C. E. Bach, "Direct and indirect interactions between ants, scales, and plants," *Oecologia*, Vol. 87, p. 235 © 1991 Springer-Verlag. Reprinted by permission.

Photo Credits

Species Interactions IV: Disease and Parasitism

Key Concepts

- Disease is one of the major causes of debilitation and death of animals and plants, and the interactions between parasites and disease agents and their hosts are important for individuals and populations.

- Simple host-parasite models can predict extinction, stability, or host-parasite cycles. Stable interactions of host and parasite are rather rare in most disease models.

- Diseases and parasites can affect reproductive output or mortality rates, but only in a few cases do we understand the effects of disease on the host population.

- Parasites and diseases do not necessarily coevolve to become more benign, but instead face an arms race in which each is attempting to maximize fitness in evolutionary time.

- While human disease has been a major preoccupation of medical scientists, we know much less about the role of disease in ecological systems. Diseases introduced to new hosts have caused major effects on population dynamics.

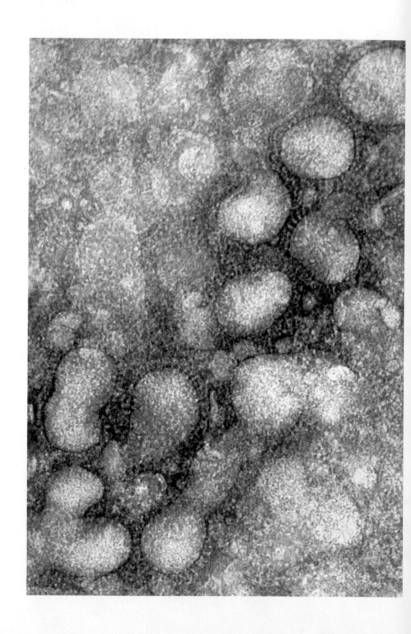

From Chapter 13 of *Ecology: The Experimental Analysis of Distribution and Abundance*, Sixth Edition. Eugene Hecht.

KEY TERMS

compartment model A type of box-and-arrow model of diseases in which each compartment contains a part of the system that can be measured and the compartments are linked by flows between them; each compartment typically has an input from some compartments and an output to other compartments.

disease An interaction in which a disease organism lives on or within a host plant or animal, to the benefit of the disease agent and the detriment of the host.

macroparasites Large multicellular organisms, typically arthropods or helminths, which do not multiply within their definitive hosts but instead produce transmission stages (eggs and larvae) that pass into the external environment.

microparasites Small pathogenic organisms, typically protozoa, fungi, bacteria, or viruses, that can cause disease.

parasite An organism that grows, feeds, or is sheltered on or in a different organism while having a negative impact on the host.

Red Queen Hypothesis The coevolution of parasites and their hosts, or predators and their prey, in which improvements in one of the species is countered by evolutionary improvements in the partner species, so that an evolutionary arms race occurs but neither species gains an advantage in the interaction.

sublethal effects Any pathogenic effects that reduce the well-being of an individual without causing death.

virulence The degree or ability of a pathogenic organism to cause disease; often measured by the host death rate.

Disease is an important interaction between organisms, ranking with competition, predation, and herbivory as one of the four agents of population change. Disease has been one of the great preoccupations of humans through all recorded history, and our history books are replete with tales of the Black Death of the fourteenth century, the smallpox scourge of the nineteenth century, and the Spanish flu pandemic of 1918–1919. Today we are occupied with the AIDS epidemic, drug-resistant tuberculosis, and mad-cow disease. Generations of children come down with everyday diseases like measles, and each winter we suffer another flu epidemic, but both diseases are rarely more than inconveniences, given modern medicine.

Disease is defined as an interaction in which a disease organism lives on or within a host plant or animal, to the benefit of the disease agent and the detriment of

the host. Disease agents are typically bacteria or viruses, but may be pathogenic fungi or prions (protein bodies); these agents are called **microparasites**. Parasitism has much in common with disease as a biotic interaction, and differs from disease mainly because **parasites** are often large, multicellular organisms such as tapeworms; these large agents are called **macroparasites**. But there is a middle ground of parasites, like the spirochetes responsible for syphilis, that are really disease organisms, and so these two interactions can be treated together. We tend to think of parasites as inflicting nonlethal harm on their hosts, but many diseases are also rarely lethal.

The **virulence** of a pathogen depends on the intensity of the disease it causes and is measured by host mortality. Although people are often very concerned about the lethal effects of pathogens and parasites, the **sublethal effects**—any effects that reduce well-being without causing death—are probably more important for plants and animals in ecological settings. Infected or parasitized animals may produce fewer offspring, be captured more easily by predators, or be less tolerant of temperature extremes. Disease and parasitism can thus interact with competition and predation in affecting population dynamics. Almost every individual of every plant and animal species harbors both pathogens and parasites.

We begin our analysis of disease by constructing some simple models, all of which have the underlying assumption that we can isolate in nature a system consisting of one host species and one disease organism. We will restore the system's complexities later.

Mathematical Models of Host-Disease Interaction

Human epidemiology is a focus of much disease research and has been the source of mathematical models that explore the host-disease interaction. In contrast to models of competition and predation, disease models have traditionally been continuous time models that use differential equations. These models are applicable to many ecological systems in which birth, death, and infection processes are continuous in time. Roy Anderson of Oxford University has been a world leader in bringing mathematical models of disease into ecology during the past 20 years; it is his work, and that of his colleague Robert May, on which much of the following analysis is based.

Many types of models have been used in the study of disease epidemiology (Anderson and May 1978). **Compartment models** are box-and-arrow models that include simplified population dynamics, and they are a good starting point for learning to think about epidemics. In their simplest form they assume a constant host population, and because this fits the human situation in the short term, they have been used extensively for exploring human disease problems. More complex models can be developed that allow both host and parasite populations to vary in size and have been used to explore the dynamics of the entire ecological system of hosts and parasites. Let us explore some simple examples of these models.

Compartment Models with Constant Population Size

We begin by considering microparasites such as viruses and bacteria that are directly transmitted between hosts and reproduce within the host. Microparasites like the influenza virus typically are very small, have a short generation time, and thus have very high replication rates. Hosts that recover from infections typically acquire some immunity against reinfection, sometimes for life. In many cases the duration of the infection is short relative to the life span of the host, and we think of microparasitic infections as transient for the host.

For microparasitic infections we can divide the host population into three parts: susceptible, infected, and recovered. **Figure 1** illustrates a simple compartment model for a microparasitic disease. The host population is characterized by the relative sizes of the three compartments and the instantaneous rates of birth (b) and death (d). The effects of the disease agent are summarized by

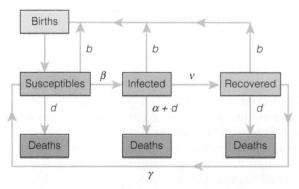

Figure 1 Compartment model for a directly transmitted microparasitic infection such as measles. Hosts are divided into susceptibles, infected, and recovered (= now immune). The parameters controlling this simple model are in the natural birth (b) and death (d) rates of the host, and the parameters of the disease agent: disease-induced deaths (α), recovery rate (ν), transmission rate (β), and rate of loss of immunity (γ). (From Anderson and May 1979.)

four parameters: the per capita rate of disease mortality (α), the per capita recovery rate of hosts (ν), the transmission rate (β), and the per capita rate of loss of immunity (γ). This is a relatively simple compartment model because it does not take into account either the abundance of the disease agent in the host (individuals are either infected or not infected) or individual differences in susceptibility due to genetic or nutritional effects.

Compartment models are useful for answering questions about the stability of the host-disease interaction. Will the disease persist in a population or will it die out? How do the proportions of susceptible and infected individuals change through time as the disease goes through a population? We can answer these questions by converting the compartment model into a mathematical model, as follows (Heesterbeek and Roberts 1995).

Consider first the susceptibles in the population. We can estimate their rate of change by the differential equation

$$\begin{bmatrix} \text{Rate of change in} \\ \text{the susceptible} \\ \text{population} \end{bmatrix} = \begin{bmatrix} \text{Rate of transmission} \\ \text{of disease from} \\ \text{infected} \\ \text{to susceptibles} \end{bmatrix}$$

$$\frac{dX}{dt} = \frac{-\beta XY}{N} \tag{1}$$

where X = number of susceptibles
  Y = number of infected individuals
  N = total number of individuals = $X + Y + Z$
  Z = number of recovered and immune individuals
  β = transmission rate per encounter

In this simple model we assume that the population size (N) and the transmission rate (β) are constants. The number of infected individuals (Y) is given by

$$\begin{bmatrix} \text{Rate of change in} \\ \text{the infected} \\ \text{population} \end{bmatrix} = \begin{bmatrix} \text{Rate of transmission} \\ \text{of disease from} \\ \text{infected} \\ \text{to susceptibles} \end{bmatrix} - \begin{bmatrix} \text{Rate of recovery} \\ \text{of infected} \\ \text{individuals} \end{bmatrix}$$

$$\frac{dY}{dt} = \frac{\beta XY}{N} - \gamma Y \tag{2}$$

where γ = recovery rate, and all other terms are as previously defined. The term $\dfrac{\beta XY}{N}$ is called the transmission term. In this simple model it is assumed that disease transmission is proportional to the product of the number of susceptible individuals (X) and the proportion of the population that is infected (Y/N). For simplicity we assume that the recovery rate (γ) is a constant. Finally, the dynamics of the recovered individuals (Z) can be written as

$$
\begin{bmatrix} \text{Rate of change in} \\ \text{the recovered} \\ \text{population} \end{bmatrix} = \begin{bmatrix} \text{Rate of recovery} \\ \text{of infected} \\ \text{individuals} \end{bmatrix}
$$

$$
\frac{dZ}{dt} = \gamma Y \tag{3}
$$

where all terms are as previously defined above.

Compartment models are named after the types of compartments used, so this model is sometimes called an *SIR model* (susceptible, infected, recovered). If there were no recovery from the disease (as with untreated rabies), then $\gamma = 0$ and we would have an SI model.

What use can we make of this simple model? The first question we can ask is whether or not an epidemic develops when a small number of infected individuals enters a large population of susceptibles. The answer to this question depends on the value of a critical epidemiological parameter, the basic reproductive rate of the disease organism, called R_0. We define the basic reproductive rate as

R_0 = the average number of secondary infections produced by one infected individual

For this simple model,

$$
R_0 = \frac{\beta}{\gamma} \tag{4}
$$

On average, one infected individual meets and infects β susceptible individuals per unit of time, and it does this for a time period of average length $1/\beta$ until it recovers.[1] An epidemic can develop only if $R_0 > 1$, which ensures a chain reaction of infection. In this simple model R_0 is a constant. Note that the basic reproductive rate of these disease models is analogous to the net reproductive rate of population growth models.

The course of the disease under this simple model is illustrated in **Figure 2**. The number of infected individuals rises steadily to a peak and then declines to zero, and the infection dies out. The susceptible population becomes too small after a certain time for an infected individual to encounter a susceptible one in order to cause new disease cases. In this simple model, organisms become immune and the epidemic dies out.

---

[1] For instantaneous rates of death or recovery, the average duration until the event occurs is 1/rate. Thus, for a death rate of d, the average life span will be $1/d$, and for a recovery rate of γ, the average time of being infective before recovery will be $1/\gamma$.

WORKING WITH THE DATA

How Can We Determine R_0? A Mathematical Excursion

One of the critical parameters in simple disease models is the net reproductive rate of the disease agent, R_0. When the net reproductive rate is > 1, the disease will spread, and when it is < 1, the disease dies out. We can derive this parameter with a bit of algebra. Begin with Equation (2) for the simple model:

$$
\frac{dY}{dt} = \frac{\beta XY}{N} - \gamma Y
$$

where Y = Number of infected individuals
       X = Number of susceptible individuals
       N = Total number of individuals
       β = Transmission rate per encounter
       γ = Recovery rate per capita

By definition, at equilibrium the rate of change in the number of infected individuals is zero, so

$$
\frac{dY}{dt} = 0 \text{ and consequently, } 0 = \frac{\beta XY}{N} - \gamma Y
$$

We can divide all these terms by Y to obtain

$$
0 = \frac{\beta X}{N} - \gamma \text{ and thus, } \gamma = \frac{\beta X}{N}
$$

By rearranging terms we can obtain

$$
\frac{\gamma}{\beta} = \frac{X}{N} \text{ or taking reciprocals, } \frac{\beta}{\gamma} = \frac{N}{X}
$$

At equilibrium, N is carrying capacity K, and X, the number of susceptibles, is defined to be the threshold population density K_T. But at equilibrium the net reproductive rate is defined as

$$
R_0 = \frac{K}{K_T} = \frac{\text{equilibrium population density}}{\text{threshold population density}}
$$

Thus, since X is equal to the threshold density K_T, and $N = K$, we can put these two relationships together to obtain Equation (4):

$$
R_0 = \frac{\beta}{\gamma}
$$

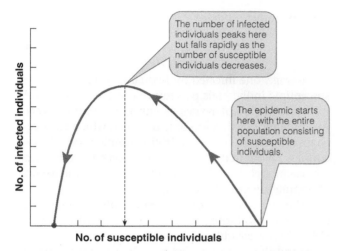

Figure 2 Trajectory for the simple epidemic described by Equations (1) through (3). Begin with a large number of susceptible individuals and no infected individuals. The number of infected individuals rises to a peak, indicated by the dotted line, when $x = \gamma/\beta$ and then falls to zero at the equilibrium marked by the red dot. (Modified after Heesterbeek and Roberts 1995.)

One interesting question that we can ask about this simple model is what happens if there is a steady influx of new susceptible individuals. Populations are often growing during the breeding season, or immigrants may move into an area. The resulting disease model becomes more realistic and more complicated if there is influx. In some cases the host-disease system reaches an equilibrium at a density of the susceptible population at which the basic reproductive rate is 1.0 (Anderson and May 1991). In other cases the populations will oscillate around an equilibrium point, or there may be no steady state for the system.

How might we control a disease described by this simple model? If we vaccinate individuals or cull susceptibles from the population, how many must we treat or remove to eradicate the disease? If we make a fraction c of the susceptible population immune, then a fraction $(1 - c)$ remains susceptible. Thus from Equation (4) we can calculate:

$$R_0 = \frac{(1 - c)\beta}{\gamma} \tag{5}$$

which is equivalent to

$$c > 1 - \frac{\gamma}{\beta} = 1 - \frac{1}{R_0} \tag{6}$$

Thus, for example, if R_0 is 4, then we would have to vaccinate or cull 75% of the population to control the disease.

Compartment Models with Variable Population Size

We can add more realism to this first compartment model by allowing the population size of the host to vary over time. Second, we can allow the contact rate to be a function of population size, so that disease transmission increases with population density. To keep the model simple, we assume a host-disease system in which there is no recovery from the disease, so that we construct an SI model (susceptibles and infecteds only). Infecteds must die in this model.

Models of this type allow us to ask an important ecological question: Does the disease affect population size of the host? We modify the simple models described in Equations (1) and (2) to allow for changes in host numbers. For the susceptibles we have

$$\frac{dX}{dt} = bN - dX - \frac{c\beta XY}{N} \tag{7}$$

where X = number of susceptibles
Y = number of infected individuals
N = total number of individuals = $X + Y$
b = instantaneous birth rate of the host (constant)
d = instantaneous death rate of the host in the absence of disease (constant)
c = contact rate, a function of population density N (Figure 3)
β = transmission rate per encounter

For the infected individuals we get

$$\frac{dY}{dt} = \frac{c\beta XY}{N} - (\alpha + d)Y \tag{8}$$

where α = increase in host mortality due to disease

and all other terms are as previously defined (**Figure 3**).

We can solve these equations at equilibrium

$$\left(\frac{dX}{dt} = 0 = \frac{dY}{dt}\right)$$

to get the following solutions (Heesterbeek and Roberts 1995):

$$Y^* = \left(\frac{b - d}{\alpha}\right)N^* \tag{9}$$

$$c_{N^*} = \frac{\alpha(\alpha + d)}{\beta(\alpha + d - b)} \tag{10}$$

where Y^* = number of infectives at equilibrium
N^* = total population size at equilibrium
c_{N^*} = contact rate at equilibrium population density

and all other terms are as previously defined.

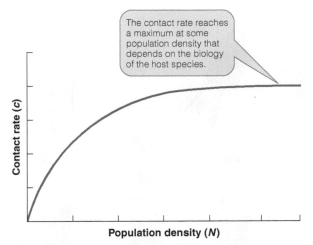

Figure 3 **The expected relationship between contact rate and population density.** At higher densities contact rates will increase, and disease transmission will be facilitated, but at some density contacts reach a maximum and do not increase. (Modified after Heesterbeek and Roberts 1995.)

This equilibrium will have a solution only if the contact rate (see Figure 3) can possibly be as high as that given by Equation (10). If there is a solution to the equilibrium, the disease will reduce population size when

$$\alpha > b - d \qquad (11)$$

that is, if disease-induced mortality is greater than the potential rate of population growth. Equation (8) can be used to investigate the possibility of eradicating a disease from a wild population by culling (or by vaccination). Assume a fixed per capita culling rate δ. Heesterbeek and Roberts (1995) showed that culling can eradicate a disease from a population if

$$\delta > \beta c - (\alpha + d) \qquad (12)$$

where
- δ = culling rate per capita
- c = contact rate at the equilibrium population size after culling
- d = natural death rate at the equilibrium population size after culling
- β = transmission rate of the disease
- α = increase in host mortality due to disease

We will discuss later in this chapter cases in which culling has been used to reduce wildlife losses due to disease.

If infected animals can be identified in the field and killed, this action will effectively increase the mortality rate caused by the disease (α), and reduce the amount of culling needed to achieve eradication.

These simple models can be elaborated to account for the specific details of particular diseases (Grenfell

and Dobson 1995). There is a large literature on epidemiological models, and to explore it further consult Bailey (1975) or Busenberg and Cooke (1993). We next explore the effects diseases have on individuals and populations of animals and plants.

Effects of Disease on Individuals

Individual hosts are effectively islands for a disease agent, and from the viewpoint of the disease organism these islands or patches of habitat must be colonized for the disease to spread. We begin by looking at these individual hosts and ask how a disease agent might affect them as individuals.

Effects of disease and parasitism on individual organisms are relatively easy to study (Gulland 1995). We have relatively few data on wild animals and plants compared to the large amount of data on domestic animals and humans. One reason why few studies have been done in the wild is that parasites and diseases are thought to coevolve with their hosts such that they become relatively harmless, and consequently one would not expect to find strong effects. But this may not be correct, and more studies are finding significant effects on reproduction, survival, and growth of infected organisms.

Effects on Reproduction

Because organisms have a limited amount of available energy, it is not surprising that parasite and disease infections can reduce reproductive output. A good illustration of these effects can be seen in lizards. Even though malaria parasites infect many vertebrates, including humans (four species of *Plasmodium*), a majority of the 125 malaria species attack lizards. In California, about 25% of western fence lizards (*Scleroporus occidentalis*) are infected with lizard malaria, and infected females have smaller clutch sizes than uninfected females (**Figure 4**). Clutches are about 20% smaller in malaria-infected lizards compared with uninfected individuals (Schall 1983). The cause of this reduction in reproductive effect is that individuals store less fat in a given summer and thus females have less energy available the following spring to lay eggs.

Bird chicks are often attacked by nest parasites that suck blood from the chicks, and if parasite infestation is severe, reproductive output can be reduced. Birds that repeatedly use the same nest are particularly susceptible to ectoparasites of this type. Barn swallows (*Hirundo rustica*) typically raise two or three broods in southern Europe (de Lope and Møller 1993). To test whether

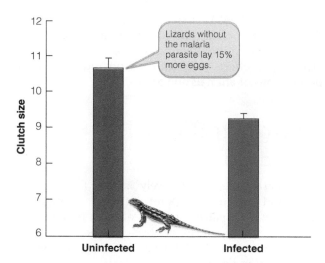

Figure 4 Effect of lizard malaria (*Plasmodium mexicanum*) on clutch size of western fence lizards (*Scleroporus occidentalis*). Data averaged from 1978 to 1982. (Data from Schall 1983.)

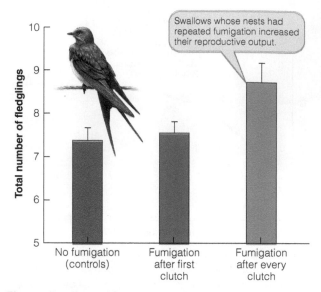

Figure 5 Effect of fumigation of nests for ectoparasites on the production of young in the swallow *Hirundo rustica* in Spain. Swallows have several broods, and pyrethrin was used to fumigate nests just after clutches were completed to reduce the abundance of blood-sucking ectoparasites that live in the nests. Fumigation after the first clutch increased reproductive output only 2%, but repeated fumigation increased output 19% over untreated nests. (Data from de Lope and Møller 1993.)

ectoparasites in swallow nests reduced reproductive output, de Lope and Møller (1993) fumigated nests immediately after the eggs were laid (eggs were removed during fumigation), and obtained the results shown in **Figure 5**. When nests were fumigated, more birds with treated nests added a third clutch when compared to unmanipulated birds. There was also a significant increase in the mass of nestlings in treated nests, so chicks had more energy to use for growth when less energy was lost to ectoparasites.

One way for a swallow to avoid nest parasites would be to change nests after each clutch but only 14% of swallows in Spain built new nests for their second clutch. But pairs that built new nests did not have fewer ectoparasites than birds that used the same nest twice, so while this strategy would appear to be attractive, in fact it does not alleviate the ectoparasite problem.

There are many examples now that show a reduction of reproductive output for organisms that carry large parasite loads. But populations differ in their parasite loads, and not all populations have high levels of infection, so that demographic impacts are variable (Marzal et al. 2005).

Effects on Mortality

No one doubts that diseases kill animals, and there are numerous cases in which veterinary examinations of dead animals suggest that a parasite or disease was the immediate cause of death. The population ecologist, however, needs to know more. What fraction of mortality is disease-caused? This is a more difficult question to

answer, and while we have much data of this type for humans, we have very little for natural populations of animals or plants. One example will illustrate the problems of obtaining good information.

In the spring of 1988, harbor seals in the North Sea began to die in large numbers. Dead seals were first noted in the central Baltic off Denmark, and mortality spread around the Baltic, to the Dutch coast, to Britain, and as far as Ireland by August 1988 (**Figure 6**). Harbor seals occur throughout the North Atlantic, and before the epizootic approximately 50,000 harbor seals lived in European waters (Swinton et al. 1998).[2] An estimated 60% of the total seal population in the Baltic Sea died from this epizootic, and the deaths occurred very rapidly (**Figure 7**). During the outbreak the exact cause of death was not clear, but a viral disease was suspected because the dying seals had symptoms that resembled those of canine distemper.

One characteristic of the disease is that it caused pregnant female seals to abort. Osterhaus and Vedder (1988) identified the infective agent as a morbillivirus similar to canine distemper virus, and they named it

---

[2]An epizootic is a disease epidemic among wild animals.

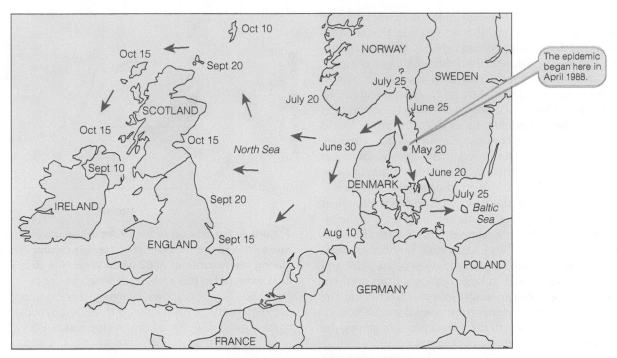

Figure 6 Map of the spread of the phocine distemper virus epizootic among the harbor seal populations of northern Europe in the summer of 1988. The epizootic began in a seal colony in the central Baltic in April. This outbreak, the first well-documented epizootic among free-ranging marine animals, had a very rapid spread and a high rate of mortality. (Data from Swinton et al. 1998.)

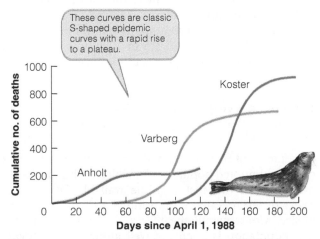

Figure 7 Cumulative number of harbor seal deaths recorded in the central Baltic Sea in the summer of 1988. The epizootic of phocine distemper virus started at the small Anholt seal colony in April, spread to the larger Varberg colony in mid-May, and reached the Koster colony in mid-June. On average an estimated 60% of the seals were killed at each site. (Data from Heide-Jørgensen and Härkönen 1992.)

phocine distemper virus. Within two weeks of infection seals developed the symptoms and typically died of pneumonia with secondary bacterial and viral infections.

The incidence of infection for this seal epizootic could not be measured directly, but Heide-Jørgensen and Härkönen (1992) estimated that 95% of the harbor seals were infected with the virus. Deaths from phocine distemper virus seemed to be more common in males than females, although both were infected. There was no indication that the epizootic was affected by the number of seals in each colony, and the main predictor of the spread was distance between colonies. Harbor seal colonies in northern Norway and Iceland escaped the epizootic, presumably because no infected seals dispersed to these distant colonies.

The key question from the harbor seal's viewpoint is whether or not this viral disease could persist in the population. Infected individuals that recover are immune for life, but since births occur each year there is a continual source of susceptibles in the population. In the Baltic, seal pups are about 20%–22% of the total population in any given year. Swinton et al. (1998) used these estimates to construct a compartment model of the 1988 epizootic. Seal colonies are discrete population patches, and the transmission rate (β) between individuals seems to be constant at 0.005 per day. The net reproductive rate (R_0) for this viral disease is approximately 2.8. The critical additional variable needed for the types of simple compartment models previously discussed (single-population models) is the rate of spread of the virus from seal colony to seal colony. Dispersal of infected seals between colonies must have been frequent to enable the rapid spread shown in Figure 7. Given these estimates for a model, Swinton et al. (1998) showed that phocine distemper virus could not be maintained in harbor seal populations as a persistent infection. This conclusion relates to the origin of the disease in the first place. Phocine distemper virus is found in both grey seals and harp seals in the Atlantic and seems to be a relatively innocuous disease in harp seals (Harwood 1989). Harp seals are northern seals and are normally rare in southern waters. In 1987 and 1988 harp seals moved in large numbers from northern Norway and Spitzbergen south into the North Sea. The phocine distemper virus may have crossed species boundaries at this time to set off the 1988 epizootic among the more susceptible harbor seal population.

In spite of all the harbor seal deaths in 1988, the seal populations of western Europe were only temporarily affected and quickly recovered to their former numbers. The seal epizootic of 1988 raises the general question of how often a disease can exert a long-term effect on a population, a question that also arises for the current outbreak of West Nile virus in North America.

West Nile virus is an RNA virus closely related to the Japanese encephalitis complex of viruses. It infects mainly birds but also is known to attack humans, horses, dogs, cats, skunks, and various rodents. The main route of infection is through a mosquito bite. Birds are amplifying hosts, and infected birds can pass the virus on to other mosquitoes. By contrast, mammals do not amplify the virus and are dead-end infections. The West Nile virus has been known since 1937 when it was detected in Uganda and then found to be common in Africa and the Mediterranean region. It first appeared in North America in 1999 in the New York City area.

West Nile virus causes high mortality particularly in crows and other members of the family Corvidae (ravens, magpies, jays), and the presence of dead birds in cities has been an early indicator that the virus has spread. There have been few studies of the direct impact of West Nile virus on crow populations. Yaremych et al. (2004) reported on one epidemic in American crows that were radio-tagged so that precise data on individuals could be obtained. **Figure 8** shows that 68% of the radio-tagged crows died from West Nile virus infection over one summer. In laboratory studies with crows, 100% of the infected birds died within six to seven days of being infected. Some infected crows survive in the field and develop antibodies to West Nile virus, and consequently there will be strong selection for genetic resistance to this

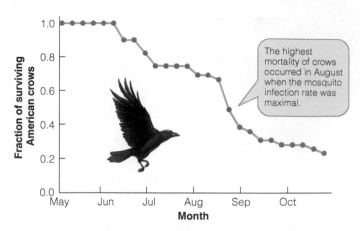

Figure 8 Survival curve for 39 American crows that were marked with radio-transmitters in central Illinois during the summer of 2002 when West Nile virus was first detected in this area. In a normal crow population about 5%–10% of the population might be lost over the summer months, a great contrast to the loss of two-thirds of these birds in 2002. (Data from Yaremych et al. 2004.)

disease. The recovery time for crow populations that have been severely reduced by West Nile is not yet known.

The numbers of human cases from West Nile virus infection have been rising. The key to epidemics of this disease lies in the feeding behavior of the mosquitoes that transmit the virus (Kilpatrick et al. 2006). The mosquito *Culex pipiens* shifts its feeding behavior in late summer from birds to humans, thus driving the human epidemics of recent years. The Centers for Disease Control and Prevention in Atlanta reported for the United States that there were 100 human fatalities in 2004, 119 in 2005, 177 in 2006, and 115 in 2007. It is not known if these numbers will continue to increase.

Effects of Disease on Populations

Few studies of plant or animal diseases have included a closely monitored population in which each individual's history is known. Most often the available data are estimates of seroprevalence from individuals of known age or size.[3] Consequently the effect of a disease on a particular population is often not well known. Most disease studies have concentrated on the effects on humans or on agriculture, and there is a need to bring ecologists and epidemiologists together to measure population effects (Mills 1999). The following three examples illustrate the

---

[3]Seroprevalence is the percentage of individuals in the host population with antibodies to a particular disease agent. It measures how widespread a disease has been in a population.

range of problems faced in trying to measure the effects of disease on populations.

Brucellosis in Ungulates

Brucellosis is a highly contagious disease of ungulates caused by a bacterium (*Brucella abortus*). Prevalent in cattle throughout the world, it manifests itself in females by abortion, so its common name is "contagious abortion." Much effort has been expended by the livestock industry to eradicate brucellosis in cattle, but the possible transmission of infection from wild ungulates to cattle has caused much controversy in the western United States, where brucellosis is endemic in bison and elk (Aguirre and Starkey 1994; Rhyan et al. 2001). **Figure 9a** illustrates the age pattern of seroprevalence to brucellosis of bison in Yellowstone National Park. Seroprevalence increases with age in bison, so that about 60% of older adults have antibodies to the *Brucella* bacterium. There is considerable controversy over whether or not brucellosis is a native disease of bison or whether it was introduced

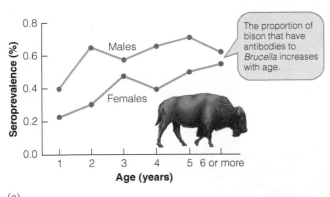

(a)

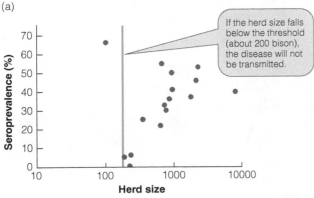

(b)

Figure 9 Brucellosis in bison as measured by seroprevalence. (a) Relationship between seroprevalence and age of male and female bison in Yellowstone National Park in the winter of 1990–1991. (Data from Pac and Frey 1991.) (b) Relationship between seroprevalence and size of bison herds in six national parks in Canada and the western United States. (After Dobson and Meagher 1996.)

into North America by cattle (Meagher and Meyer 1994). Most probably it was not present in bison before 1917 and was contracted from domestic cattle.

A simple model of the interaction between brucellosis and bison in Yellowstone National Park was constructed by Dobson and Meagher (1996) to determine whether brucellosis could be eliminated by a culling program. Brucellosis has a sharply defined threshold for establishment (**Figure 9b**), and the proportion of bison infected rises smoothly with population density. These data illustrate one of the important principles of epidemiology: the critical threshold. Most diseases have a threshold host population density that is needed for the continued presence of the pathogen. In this case, brucellosis will persist in bison populations of 200 or larger, a low number. Bison in Yellowstone now number about 4000 animals. Whereas it is possible to cull bison down to this low density, this action is unacceptable because it would put them in danger of extinction (and would be politically unacceptable to a variety of people). So it is unlikely that culling will be a viable strategy for eliminating brucellosis in bison in Yellowstone National Park (Dobson and Meagher 1996). Note that brucellosis could infect bison populations in very small herds, but once it passed through a small population it would fail to maintain itself and would die out.

Rabies in Wildlife

Rabies is one of the oldest known diseases, and one of the most terrifying diseases for humans. Around 500 BC, Democritus recorded a description of rabies, and 200 years later Aristotle wrote about rabies in his *Natural History of Animals*. Rabies is a directly transmitted viral infection of the central nervous system, and all mammals are susceptible. The disease is particularly common in foxes, wolves, coyotes, skunks, raccoons, jackals, and bats, but domestic dogs most frequently transmit it to humans. Rabies virus, present in saliva, is transmitted directly by the bite of an infected animal. A few cases of aerosol transmission from bats in caves have been reported (Krebs et al. 1995). Once rabies is contracted, death is inevitable: there is no cure. Rabies is widespread in the world (**Figure 10**) and only a few countries are free of this disease. Worldwide the incidence of rabies in humans is low; about 55,000 people a year are victims, mostly in India and the Far East (Knobe et al. 2005). The incubation period in humans is highly variable, ranging from less than ten days to more than six years. Malaria and tuberculosis are much more significant causes of human deaths globally, but no disease is as feared as rabies.

Rabies is caused by a number of different viruses belonging to the *Lyssavirus* genus in the Family Rhabdoviridae. Carnivorous mammals are the essential hosts for the virus. In Europe the red fox is the main reservoir

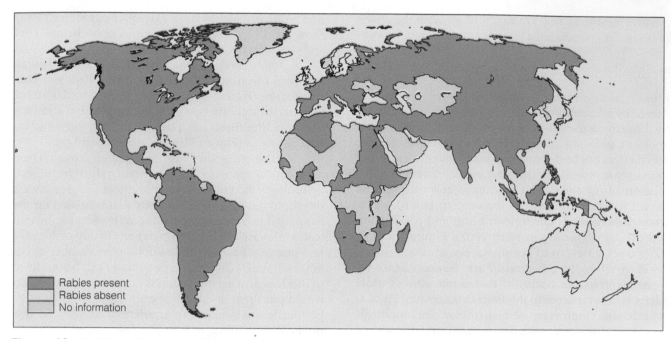

Figure 10 World distribution of rabies. There are only a few countries in which rabies is absent. The annual number of deaths worldwide caused by rabies is estimated to be 55,000, mostly in rural areas of Africa and Asia. An estimated 10 million people receive postexposure treatments each year after being exposed to animals suspected to be infected with rabies. (Data from World Health Organization for 2001–2006.)

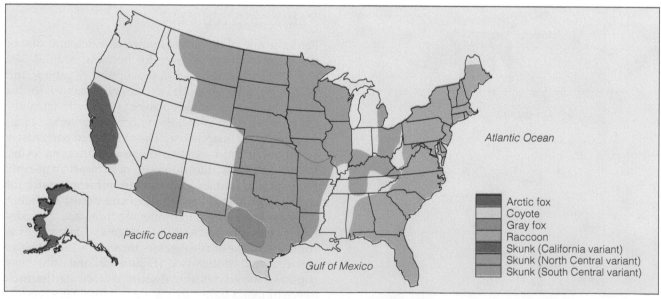

Figure 11 Main rabies reservoirs in different regions of the United States. Other mammals serve as minor reservoirs of the disease in each region. The geographic ranges of these five species are much wider than the areas shown here. There are several variants of the rabies virus that are spread by specific mammals. The light green areas have no major rabies problem. (Modified from Krebs et al. 2005.)

for rabies (Anderson et al. 1981); in North America raccoons, skunks, foxes, and bats are the main reservoirs, and in 1997 wild animals represented 93% of the reported cases. The main vectors of rabies differ in differ-ent regions of the United States (**Figure 11**). These vectors carry a diverse set of rabies virus genotypes. The raccoon is a keystone host of rabies in the southeastern United States and a majority of the recorded cases in wild animals in the United States are now from raccoons (**Figure 12**). The striped skunk currently represents about 25% of rabies reports, although it was more important as a host in the 1970s. The reason for these host shifts in rabies incidence is completely unknown.

An epizootic of rabies in eastern North America began around 1970 in Virginia and has been spreading for 30 years (**Figure 13**). This epizootic probably began from diseased raccoons brought into the area by humans. Rabies in raccoons has since spread north to Ontario, crossing the border in 1998, and has also spread south to meet another epizootic moving north from Florida. Because raccoons are so common, particularly around human habitations, rabies in raccoons has been particularly targeted by control agencies in the United States and Canada in recent years.

A recent attempt has been made to reduce rabies in raccoons by vaccination of wild raccoons, using a recombinant virus vaccine approved in April 1997. A raccoon bait, a small cube of fish oil and wax polymer, contains the oral rabies vaccine. Millions of baits are distributed annually to immunize susceptible raccoons and foxes. In addition, raccoons can be easily live trapped, injected with the vaccine, and released. This vaccination program has been used in Massachusetts, New York, New Jersey, Vermont, and Ontario, but its effect on the incidence of rabies in raccoons is not yet clear.

In many parts of the world, rabies reaches humans through domestic dogs, but in North America and Europe vaccination of dogs has cut this link to humans. From 1997 to 2003, 26 people died in the United States from rabies, and 90% of the confirmed cases have been caused by rabies virus variants carried by bats (Krebs et al. 1998). In 2006 two deaths occurred in the United States and one in Canada, with bats as the suspected carriers. Little is known about either the incidence of rabies in bats or the impact of rabies on bat populations.

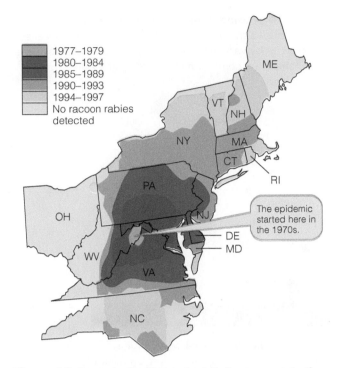

Figure 13 Spread of rabies epizootic in raccoons in the eastern United States since 1977. The epizootic began in Virginia and moved as far north as southern Canada by 1998. It has moved south as well and has met another rabies epizootic in raccoons spreading from Florida. The outbreak in Virginia was probably caused by human translocation of infected raccoons from the southeastern states during the 1970s. (From Centers for Disease Control and Prevention, courtesy John W. Krebs.)

In Eastern Europe a major epidemic of rabies began in Poland in 1939 and gradually moved 1400 km westward at a rate of 20–60 km per year. The epidemic reached the Atlantic coast in northern France in the late 1980s and stopped. The main carrier has been the red fox, with over 70% of the reported cases in Europe (Anderson et al. 1981). After extensive culling programs failed to stop rabies or reduce its incidence, most European countries began to use oral vaccination of foxes in baits to stop the spread of the disease. Vaccination via baits has proven to be highly successful in Europe. By 1999 rabies was much reduced in western Europe, and Switzerland had reached the status of rabies free as a result of this extensive vaccination program.

Figure 14 gives a simple compartment model for rabies. Many attempts have been made to model a rabies outbreak (Barlow 1995). Anderson et al. (1981) presented a simple model of rabies that captures much of the ecology of this disease. From this model we can ask a critical management question: Can we eliminate

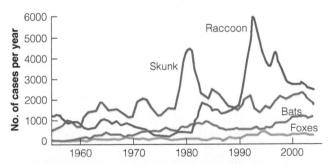

Figure 12 Number of rabies cases reported to the Centers for Disease Control in the United States from 1955 to 2004. The rise in the number of raccoon rabies cases since 1980 has resulted from an epidemic that spread through the eastern United States. (Data from Krebs et al. 2005.)

Figure 14 A compartment model for rabies. All infected animals die, so there is no recovery compartment. Because animals can be vaccinated artificially as a control measure, a vaccination rate compartment is added. (Modified from Bacon 1985.)

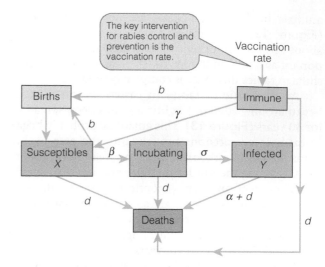

WORKING WITH THE DATA

A Simple Rabies Model

Anderson et al. (1981) have presented the following model as a representation of rabies in red foxes in Europe. Figure 14 shows the compartment model visually. The model contains susceptible (X), incubating (I), and infected (Y) foxes, and it assumes logistic growth for the fox population without rabies. Transmission for rabies is assumed to be proportional to the product of the number of susceptible (X) foxes and the number of infected (Y) foxes. The equations for the model are as follows:

$$\frac{dX}{dt} = rX - \gamma XN - \beta XY \qquad (13)$$

$$\frac{dI}{dt} = \beta XY - (\sigma + d + \gamma N)I \qquad (14)$$

$$\frac{dY}{dt} = \sigma t - (\alpha + d + \gamma N)Y \qquad (15)$$

where X = number of susceptible foxes
I = number of incubating foxes
Y = number of infected foxes
N = total number of foxes = $X + I + Y$
t = transmission rate per encounter
r = population growth rate per capita in absence of disease = $b - d$
d = death rate of foxes per capita in absence of rabies (life expectancy =$1/d$)
b = birth rate of foxes per capita in absence of rabies
γ = r/K where K is the fox carrying capacity

σ = rate of incubation (incubation period =$1/\sigma$)
α = death rate of rabid foxes (life expectancy of rabid foxes = $1/\alpha$)

Anderson et al. (1981) estimated these parameters for the red fox in Europe to be as follows:

Parameter	Definition	Estimated value
b	birth rate per capita	1 per year
d	death rate per capita	0.5 per year
r	population growth rate = $b - d$	0.5 per year
σ	rate of incubation	13 per year
α	death rate of rabid foxes	73 per year
β	transmission coefficient	80 km^2 per year
K	fox carrying capacity	1 to 4 per km^2

This model with these parameters produces cycles in fox numbers with a three-to-five-year period, in agreement with the data that is currently available. From this model the basic reproductive rate R_0 is given by

$$R_0 = \frac{\sigma \beta K}{(\sigma + b)(\alpha + b)} \qquad (16)$$

When R_0 is less than 1, rabies will die out in the fox population. The threshold density at which rabies will be maintained in the population in this model is estimated to be around 1 fox per km^2.

rabies from the fox population by culling or by vaccination? Attempts to control the spread of rabies in Europe and in North America by culling have been unsuccessful despite heroic efforts. Foxes have high reproductive rates and high dispersal rates, and these two parameters combine to make culling attempts unsuccessful at controlling the disease unless the foxes are in poor habitat or the rate of culling is extremely high.

Vaccination directly reduces the size of the susceptible pool and is much more effective in the control of rabies. **Figure 15** shows that the proportion of foxes that would need to be vaccinated varies with the density of the fox population. If foxes are at a density of 2 per km², the model predicts that vaccinating about 50% of the foxes would break the transmission cycle and eradicate the disease. Extensive programs of vaccination of wild foxes using baits have been carried out in Switzerland (since 1978), Austria, Hungary, France, Belgium, and Germany (Pastoret and Brochier 1999). These vaccination programs have been successful in eliminating rabies from wildlife reservoirs in large areas and thus in reducing the health risk.

At present we have no data at all on the effects of rabies on mammal host populations. Most of the effort has been directed at the public health aspects of this disease, and on preventative measures to reduce damage to humans and domestic animals. The most

critical issues involve the assumption that for mammalian hosts the transmission rate (β) of rabies is a constant at all host densities, and that the threshold for persistence of the disease is also a constant and thus identical in both good and poor host habitats (McCallum et al. 2001).

Myxomatosis in the European Rabbit

The European rabbit (*Oryctolagus cuniculus*) was introduced into Australia in 1859 and increased to very high densities within 20 years. After World War II, an attempt was made to reduce rabbit numbers by releasing a viral disease called myxomatosis. Myxomatosis originated in the South American jungle rabbit *Sylvilagus brasiliensis*. In its original host, myxomatosis is a mild disease that rarely kills its host. The disease agent is the myxoma virus, a pox virus of the genus *Leporipoxvirus*. Transmission of myxomatosis occurs via biting arthropod vectors, principally mosquitoes and fleas. Transmission is passive, and the virus does not replicate in the vector.

Myxomatosis was highly lethal to European rabbits when it was introduced into Australia in 1950, killing over 99% of individuals infected. **Figure 16a** shows the precipitous crash in rabbit numbers that followed the introduction of myxomatosis in one area in southeastern Australia in 1951. Myxomatosis was also introduced to France in 1952, from where it spread throughout western Europe, reaching Britain in 1953. In Britain 99% of the entire nation's rabbit population was killed in the first epizootics from 1953 to 1955 (Ross and Tittensor 1986).

Very soon after its introduction, weaker myxoma virus strains were detected in England and in Australia (Fenner and Ratcliffe 1965). Frank Fenner, working at the Australian National University, was instrumental in studying the changes that have occurred in the myxoma virus in Australia. Since myxomatosis was introduced into Britain and Australia, evolution has been going on in both the virus and the rabbit. The virus has become attenuated such that it kills fewer and fewer rabbits and takes longer to cause death. Because mosquitoes are a major vector of the disease, the time period between exposure and death is critical to viral spread. **Table 1** summarizes changes that have occurred in the virus. These data were obtained by testing standard laboratory rabbits against the virus, so they measure viral changes while holding rabbit susceptibility constant. Since 1951 less-virulent grades of virus have replaced more-virulent grades in field populations.

Rabbits have also become more resistant to the virus (**Figure 16b**). By challenging wild rabbits with a constant laboratory virus source, we can detect that natural selection has produced a growing resistance of rabbits to this introduced disease.

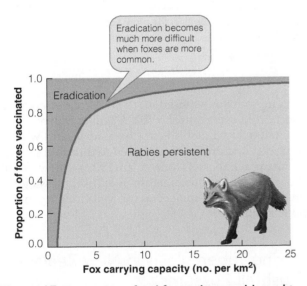

Figure 15 Proportion of red foxes that would need to be vaccinated to eliminate rabies in Europe in relation to the carrying capacity of the habitat. The simple model in Working with the Data "A Simple Rabies Model" predicts that if fox carrying capacity is relatively low, only a small proportion of the foxes would need to be vaccinated to eradicate the disease from Europe. (Modified from Anderson et al. 1981.)

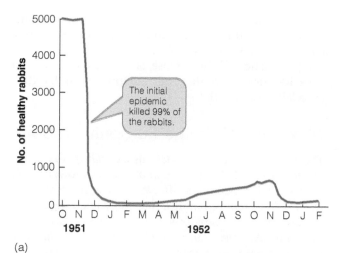

(a)

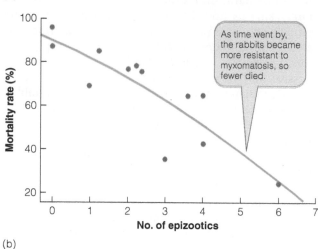

(b)

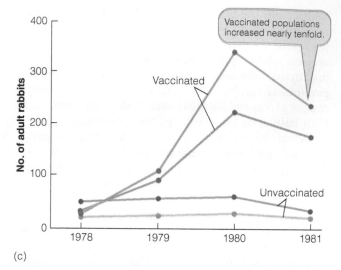

(c)

Figure 16 Effects of myxomatosis and vaccination on European rabbits. (a) Crash of the rabbit population at Lake Urana, New South Wales, after the myxoma virus was introduced in 1951. Numbers of healthy rabbits were counted on standardized transects. (After Myers et al. 1954.) (b) Decline in mortality rates of wild rabbits near Lake Urana as a function of time since the myxoma virus was introduced. Mortality was measured after infection with a virulent strain of the virus. (After Fenner and Myers 1978.) (c) Effect of vaccination on the numbers of adult rabbits in four fenced areas in southeastern Australia. Rabbits in two areas (blue and red) were vaccinated with an attenuated strain of the myxoma virus that produced immunity to virulent strains. Rabbits in the other two areas (purple and green) were inoculated with a virulent strain. (Data from Parer et al. 1985.)

What is the net effect of these changes in the virus and in the genetic resistance of the rabbits to the population dynamics of the host? Because over time myxomatosis has caused less and less mortality, it is tempting to assume that the disease was having little effect on rabbit numbers. One way to test the idea that myxomatosis was no longer effective in rabbit control is to compare rabbit populations with and without exposure to myxomatosis. This is difficult to do technically because it is impossible to find a field population of rabbits that does not already have myxomatosis. The only method possible is to reduce the effect of myxomatosis by making rabbits immune or by cutting the transmission by vectors. Two such attempts have been made. In Australia, Parer et al. (1985) compared four fenced populations of rabbits, two inoculated with an attenuated strain of the virus (to produce immunity with little mortality) and two inoculated with a virulent strain of the virus. Figure 16c illustrates the effects of this experiment on the numbers of rabbits. Populations protected

from myxomatosis-caused mortality increased eightfold and 12-fold over control levels. A similar experiment in England reduced rabbit fleas (the main vector) with insecticides, and produced a twofold to threefold increase in rabbit numbers (Trout et al. 1992). These results show clearly that myxomatosis is still suppressing rabbit populations, in spite of its reduced virulence in field populations.

Bovine Tuberculosis in New Zealand Brushtail Possums

Tuberculosis is a chronic disease affecting humans and many animal species including cattle and deer. In New Zealand, brushtail possums (*Trichosurus vulpecula*) are the main vector of bovine tuberculosis, which affects animals in about 38% of the country. Possums transmit TB to cattle in areas where pastures are bordered by forest containing infected possums. Infected cattle lose

Table 1 Virulence of field myxoma virus in laboratory rabbits in Australia, Great Britain, and France after the introduction of myxomatosis to these three countries between 1949 and 1951.

	Virulence type—grade					
	I	II	IIIA	IIIB	IV	V
Mean survival of rabbits (days)	<13	14–16	17–22	23–28	29–50	—
Mortality rate (%)	>99	95–99	90–95	70–90	50–70	<50
Australia						
1950–1951	100	—	—	—	—	—
1958–1959	0	25.0	29	27	14	5
1963–1964	0	0.3	26	33	31	9
Great Britain						
1953	100	—	—	—	—	—
1968–1970	3	15	48	23	10	1
1971–1973	0	3	37	57	3	0
France						
1953	100	—	—	—	—	—
1962	11	19	35	21	13	1
1968	2	4	14	21	59	4

Values in the table are the percentages of virus samples collected in the field that were classified as each virulence type. These studies measure changes in the virulence of the virus to a standardized host, the laboratory rabbit. Viruses collected in all three countries in three different time periods show the rapid change brought about by selection for less-virulent virus strains.

SOURCE: After Fenner and Myers (1978) and Anderson and May (1982).

weight and must be culled to prevent infecting other cattle, so the costs of this disease transmission are about $50 million annually to farmers. International markets for beef demand a certification of TB-free status, and consequently this has stimulated a major effort in New Zealand to rid the country of bovine TB.

Three ecological questions immediately arise for this disease system. First, do possums transmit TB to other wildlife species as well as to cattle? Second, can the transmission of TB from possums be stopped by reducing the density of possums? And three, what effect does TB have on the possum population?

Tuberculosis bacteria have been found in ferrets, stoats, deer, rodents, and rabbits in New Zealand. The ferret (*Mustela furo*) has been a chief suspect, since it also is common in pasture areas. The important question is whether the ferret can be part of the transmission cycle to cattle, because if it is, control efforts must be directed against it as well as against possums. Caley and Hone (2004) investigated the transmission of TB

between possums and ferrets by experimentally reducing possum numbers and measuring the prevalence of TB in ferrets. Possums were controlled for three years from 1998 to 2000 in two sites. There was a dramatic decline in the prevalence of TB in ferrets when possums were removed from the two experimental populations (**Figure 17**). This can occur only if possums were transmitting TB to ferrets. There was evidence of transmission of TB between ferrets but only at high ferret densities. There is little doubt now that brushtail possums are the main vector for TB, spreading it both to ferrets and to cattle.

The stimulus behind the studies of bovine TB in New Zealand is the requirement of the World Organization for Animal Health to reach disease-free status in order to be able to sell meat on the international market. This requirement is set at 0.2% of cattle herds infected, which at present means about 50 cattle herds. From an economic point of view, a key question in this disease system is whether reducing the density of brushtail

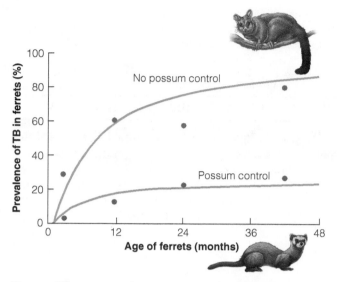

Figure 17 Age-specific prevalence of bovine TB infection in ferrets from areas without control of brushtail possum numbers and from areas in which possums were reduced to very low numbers over three years. Each data point represents the average of two New Zealand sites. (Data from Caley and Hone 2004.)

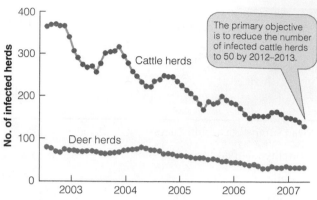

Figure 18 Incidence of bovine TB in New Zealand cattle and deer herds, 2002–2007. The World Organization for Animal Health sets an international benchmark of 0.2% of cattle herds infected to be recognized as officially free of bovine TB. As of June 2007, 0.5% of cattle herds were infected with bovine TB. (Data from New Zealand Animal Health Board, 2007.)

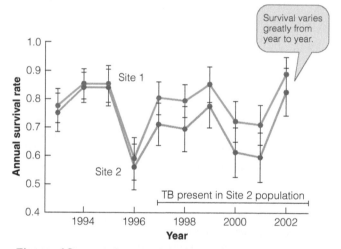

Figure 19 Annual survival rates estimated from live trapping of brushtail possums on two intensive study sites on the North Island of New Zealand.
Both sites were free of bovine TB until the disease turned up in 1997 at Site 2. Site 1 has been free of TB since 1980. The subsequent drop in survival rates at Site 2 can be attributed to mortality caused by bovine TB. (Data from Arthur et al. 2004.)

possums can stop the transmission of TB to cattle. The National Bovine Tuberculosis Pest Management Strategy of New Zealand aims to minimize transmission by reducing populations of the main vector of TB—the brushtail possum—by poisoning programs, by slaughtering cattle and deer that have bovine TB, and by controlling the movement of cattle and deer. This strategy is working well (**Figure 18**), and the plan is on target to achieve its goals by 2012–2013. At least some of this achievement can be placed on the ability of wildlife managers to reduce the density of vectors like brushtail possums in buffer zones around cattle herds.

To test the impact of bovine TB on populations of brushtail possums, Arthur et al. (2004) made use of a natural experiment in which bovine TB entered one of two intensive study sites partway through a study of their population dynamics. Survival fell about 10% on average after TB was detected (**Figure 19**). If birth and movements had remained constant, Arthur et al. (2004) estimated the possum population should have declined by about 30% over the last four years of the study, but in fact the population did not decrease in spite of this additional mortality. Juvenile survival or immigration appeared to compensate for the additional mortality caused by TB, so the consequences of this disease were strong at the individual survival level but absent at the population level.

Consequently not all diseases have population consequences. We have seen that myxomatosis is a good example of a strong effect that a disease can have on a wild population. The fact that myxomatosis was transferred between species by humans raises the broad question of how disease organisms and their hosts coevolve in evolutionary time. Do diseases gradually evolve to be benign for their hosts? This is a critical question to which we now turn.

Evolution of Host-Parasite Systems

One of the striking features of the simple models of host-parasite interactions is that these models are often unstable. Oscillations are common in many host-parasite models, as they are in predator-prey models: diseases may explode or go to extinction in simple models. But even though in real disease systems some diseases disappear, most seem to persist. One way in which we can explain the stability of real host-parasite systems is to postulate that natural selection has changed the characteristics of both hosts and disease organisms so that their interactions produce population stability. In par-

ticular, the conventional wisdom about host-parasite evolution is that virulence is selected against, so that diseases and parasites become less harmful to their hosts and thus persist. Thus the well-adapted parasite is a benign parasite (Ewald 1995). If this traditional view of peaceful coexistence is correct, we would expect to see diseases and parasites becoming less harmful over evolutionary time. But does natural selection work that way with host-parasite systems? What can we say about the evolution of virulence?

Natural selection does not necessarily favor peaceful coexistence of hosts and parasites and the view that the well-adapted parasite is benign has now been completely rejected (Walther and Ewald 2004). To maximize

ESSAY

What Is the Transmission Coefficient (β), and How Can We Measure It?

All host-parasite models have within them a difficult parameter called the transmission coefficient (β), which measures the rate at which a disease or parasite moves from infected individuals to susceptibles. The transmission coefficient enters simple models as a mass-action term depending only on the numbers of susceptibles (X) and infecteds (Y). For example, Equation (2) states that

$$\begin{pmatrix} \text{Change in number} \\ \text{of infected per} \\ \text{unit time} \end{pmatrix} = \begin{pmatrix} \text{Per capita contact} \\ \text{rate between infected} \\ \text{and susceptibles} \end{pmatrix} - \begin{pmatrix} \text{Rate of recovery} \\ \text{of infected} \end{pmatrix}$$

$$\frac{dY}{dt} = \frac{\beta XY}{N} - \gamma Y$$

In this simple model the transmission coefficient is the probability that a single contact between a susceptible host and an infected one will result in disease transmission. The transmission coefficient is thus a dimensionless number, a probability between 0 and 1.

How can we estimate β from empirical data? Hone et al. (1992) used one method for a model of swine fever in wild pigs. The critical data are the number of deaths on each day of the epidemic. For example, a swine fever epidemic in Pakistan gave the following detailed results for an initial population of 465 pigs: the rate at which deaths from the disease accumulate is clearly related to the transmission coefficient β as well as to the disease-induced death rate α (**Figure 20**). Given that we can get estimates of all the parameters in the disease model and that we know the starting population size, we can select an arbitrary value of β and then run the model to see if it fits the data shown in the curve. We can keep doing this until we zero in on the

value of β that gives the best fit to the cumulative number of deaths curve. If we select a value of β that is too large, the deaths will happen too fast; if we select a value of β that is too small, the deaths will happen too slowly. For the previous data on swine fever, Hone et al. (1992) obtained an estimate of β of 0.001 per day, which means that a single daily contact of an infected pig with a susceptible one has a probability of only 1 in 1000 of transmitting the infection.

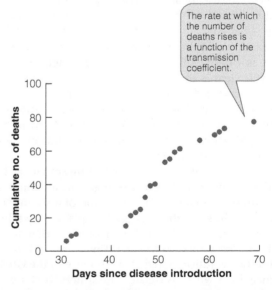

Figure 20 **Cumulative deaths curve for a swine fever epidemic in wild pigs in Pakistan.** (Data from Hone et al. 1992.)

fitness a parasite or a disease agent must optimize the trade-off between virulence and other fitness components such as transmissibility. If the host did not evolve, the parasite should be able to reach this optimal balance of host exploitation. But hosts do evolve, and this produces an arms race between the host and the parasite. If hosts are genetically variable, the parasite or disease agent will be on average less virulent than if the hosts are uniform (Ebert 1999). The evolutionary time scales of the host organism and the disease agent are typically greatly different. Hosts evolve slowly; bacteria and viruses evolve quickly.

One way to study the evolution of host-parasite systems involves serial passage experiments in the laboratory (Ebert 1998). In serial passage experiments, disease organisms or parasites are transferred from one host to another, holding host properties constant so that the evolutionary changes in the disease organisms can be monitored. Because the disease organisms are propagated under defined laboratory conditions, their biological attributes can be compared with those of the ancestral organism at the outset of the experiment. Although serial passage was developed for vaccine studies, it can be used very effectively in studies of the evolution of virulence. **Figure 21** shows a serial passage experiment in laboratory mice with the mouse typhoid bacterium *Salmonella typhimurium*.

Diseases become more virulent with passage in artificial serial passage experiments in their native host species (Ebert 1998), and this appears to be a general result with many different viral, bacterial, fungal, and protozoan disease agents. One explanation of this is the **Red Queen Hypothesis**, which states that genetic variation is beneficial because it hinders parasite and disease adaptation. In laboratory serial passage experiments the host is often clonal or of limited genetic variability. What is clear is that the increase in virulence of disease agents observed in serial passage experiments in the laboratory does not occur in most natural disease systems, and host genetic variability is believed to be the principal reason that such runaway evolution does not often happen in nature.

The coevolution of rabbits and myxomatosis is one of the best empirical studies of host-parasite interactions in natural populations. The evolution of resistance to the myxoma virus in rabbits is easily explained by selection operating at the individual level—rabbits that are more resistant leave more offspring. It is more difficult to explain the evolution of reduced virulence in the virus. Virulence in a virus is related to fitness because more virulent viruses make more copies of themselves. But if more virulent viruses kill rabbits more quickly, less time will be available for transmission of the virus through

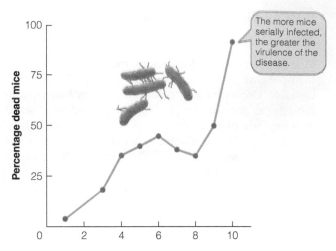

Figure 21 Change in virulence of the mouse typhoid bacterium *Salmonella typhimurium* after serial passage in laboratory mice. A constant source of laboratory mice was used in these studies so that no evolution of the host could occur. Virulence increased rapidly over time as the *Salmonella* adapted to its host. (Data from Ebert 1998.)

mosquitoes or fleas. The result for the myxoma virus is group selection that operates to reduce virulence to a moderate level (Levin and Pimentel 1981). The basic reproductive rate (R_0) of the virus is highest at intermediate virulence. Group selection occurs because less-virulent viral strains are favored over more-virulent viral strains because they take longer to kill the host rabbit (see Table 1). Host-parasite systems may be ideal candidates for group selection along these lines.

We do not know if the rabbit-myxoma system has reached a stable equilibrium, or whether continuing evolution will allow the rabbit population to slowly recover to its former levels. There is some evidence that the rabbit-myxomatosis interaction in Britain is changing, and the population size of rabbits in Britain seems to be slowly increasing (Trout et al. 1992). Evolutionary changes in the rabbit-myxoma system in Australia have been complicated by the accidental release in 1997 of a second viral disease, rabbit hemorrhagic disease, that has further reduced the rabbit's average density (Mutze et al. 2002).

The evolution of virulence differs in pathogens that are carried by vectors and those such as tuberculosis that are transmitted directly between individuals with no intermediate vectors. In particular, nonvector pathogens should utilize a sit-and-wait strategy of transmission, in which susceptible hosts pick up the infective parasites by moving around and contacting infected hosts. If these

What Is the Red Queen Hypothesis?

Lewis Carroll's *Alice in Wonderland* has a scene in which Alice and the Red Queen must run as fast as they can to get nowhere because the world is running by at the same speed. Van Valen (1973) used this metaphor to illuminate biotic evolution. Any evolutionary adjustment that a particular species makes can be countermanded by natural selection acting on all other species in the community. For example, if a prey evolves to run faster to escape its predators, the predators can also evolve to run faster to catch the prey. Thus disease-host systems, plant-herbivore systems, and predator-prey systems may show consistent evolutionary change, not to increase adaptedness but simply to maintain it. The species run, run, run but get nowhere. Increasing fitness in one species is always balanced by decreasing fitness in all other species.

Rates of evolution can be much faster in disease agents and parasites that have short generation times relative to their hosts. The Red Queen Hypothesis predicts a continuing evolutionary battle between hosts and parasites, with the important implication that because parasites evolve faster, the main selection pressures will come from the most common host genotypes. By changing genotypes over time, the host can present a moving target that the parasite or disease cannot catch. This is one possible reason for the evolution of sex, in which recombination at each generation presents a new array of host genotypes to the coevolving array of diseases and parasites. The Red Queen Hypothesis thus predicts continually changing evolutionary dynamics between parasites and hosts, not a stable equilibrium.

pathogens are highly virulent, they must be long-lived in the external environment, and the sit-and-wait hypothesis predicts a positive correlation between virulence and durability of the parasite. Walther and Ewald (2004) tested the sit-and-wait hypothesis for human respiratory pathogens (**Figure 22**). The pattern found is exactly what is predicted by the sit-and-wait hypothesis. Thus the tuberculosis bacterium, which produces a mortality rate of 5% per infection, survives in an infectious state on a standard glass plate for 244 days while the common influenza virus, which kills 0.002% of infections, survives only 1.3 days in the external environment.

Insect pests in agriculture can be controlled with directly transmitted pathogens that follow the sit-and-wait strategy. For example, the nuclear polyhedrosis virus that infects many insects remains viable in the soil for at least six years. These pathogens combine the traits of high virulence, long durability after application, and host specificity, traits most useful for the control of injurious pests.

Similarly, some of the most dangerous hospital pathogens have long survival times in the external environment. For example, golden staph (*Staphylococcus aureus*) can survive for months on fabric or on dust particles (Walther and Ewald 2004). High durability of pathogens may be linked genetically to high virulence, and measuring durability may be one way to identify potentially hazardous pathogens. We should not expect evolution of pathogens to move in the direction of less virulence.

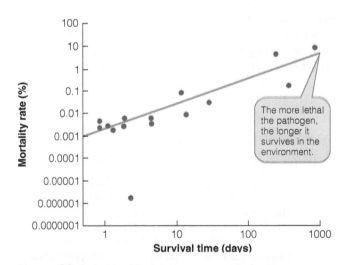

Figure 22 Relationship between average human mortality rate and survival time of the pathogen in a standardized external environment for 16 human pathogens that are not transmitted by vectors. Pathogens that kill many of their hosts have evolved a sit-and-wait strategy of durability to retain their infectivity for a long time in the external environment. (Data from Walther and Ewald 2004.)

Summary

Disease, one of the four major interactions between species, is an interaction between organisms in which the host loses and the parasite gains. Disease has been one of the major preoccupations of humans throughout history, and much of our understanding of disease dynamics has its roots in the efforts of epidemiologists and medical scientists to understand the dynamics of human diseases such as malaria.

Mathematical models of host-parasite systems utilize compartment models to represent the interactions. The host population is usually broken down into susceptible, infected, and recovered individuals, and can be considered to be either constant (as in many human disease models) or variable in size (with birth and death rates). These simple models are characterized by a few parameters that define the outcome of the interaction. The most critical parameter is the *basic reproductive rate* (R_0) of the disease organism—the number of new infections produced by the average infected individual over its life span. If the reproductive rate is 1 or more, the disease will propagate, and if it falls below 1 the disease disappears.

Simple models of host-parasite systems all show a *threshold density* below which the disease or parasite will die out. The objective of much of the study of applied disease ecology is to determine how best to move the host population below threshold density. In general, culling of animals has not been very successful in achieving eradication or even control of diseases of wild animals, and vaccination may be a better general strategy for practical control.

Diseases and parasites can affect the reproductive rate or the mortality rate of their hosts. Even though many studies show effects on mortality, few measure how large these effects are in nature or show whether a disease or parasite can reduce the average density of the host species. Rabies is used to illustrate these concepts, and while we know much about the transmission of rabies to humans, we know little about its effects on the foxes, skunks, coyotes, and bats that are the main carriers. The best studies of disease in nature have been done on myxomatosis, a viral disease introduced into Australia and Europe to control European rabbits, and on bovine tuberculosis in New Zealand.

Our view of the evolution of virulence has progressed from the conventional wisdom that well-adapted parasites and diseases are benign, to a more dynamic view in which diseases and their hosts are locked in an arms race, with each group evolving to maximize its fitness. Virulence will increase in evolutionary time if the parasite or disease organism can increase its fitness by harming the host more and producing more copies of itself. One of the main factors limiting disease virulence is host genetic variability, and monocultures of crops or clonal populations are particularly susceptible to virulent disease outbreaks. Selection for higher virulence needs to be carefully studied and understood to manage human diseases such as AIDS and tuberculosis.

Review Questions and Problems

1   By treating house martins (*Delichon urbica*) with antimalarial drugs, Marzal et al. (2005) were able to show that the malarial blood parasites in Spain reduced production of young birds by about 40%. In Denmark house martins do not carry this malarial parasite. Would you expect the population density of these birds to be higher in Denmark? Why or why not?

2   Calculate the population changes from Equations (1) to (3) in a hypothetical host-parasite system. The parameters for the interaction are: $\beta = 0.025$ (transmission rate) and $\gamma = 0.01$ (recovery rate). Start the population with 500 susceptibles and 5 infecteds, and investigate how the dynamics would change if β increased to 0.040 or 0.060.

3   About 20 million waterfowl die each year in North America from avian cholera, which is caused by the bacterium *Pasteurella multocida* (Blanchong et al. 2006). Over 100 species have been known to be infected. Epizootics are typically explosive and involve hundreds and sometimes thousands of birds. There is high variation from year to year in the incidence of this disease. Plan a research program to determine the effects of avian cholera on a species of duck. What are the key questions you need to answer to be able to control this disease?

4   One resolution to emerging human health problems with diseases is to use evolutionary thinking to manage virulence. The suggestion is that with appropriate public health measures and treatment

protocols, we could reduce disease and cause the parasites to become less virulent. In this way we could engineer the AIDS virus, for example, to become like the common cold. How might we drive evolution to manage virulence in human diseases? Ebert and Bull (2003) discuss this approach to virulence management.

5   Simple models of host-parasite systems do not have any spatial component. What advantages might be gained by constructing a spatial model of disease? Rabies is an example of a disease with interesting spatial spread patterns (see Figure 13). Foxes defend discrete, nonoverlapping territories. How might territorial behavior affect the spatial dynamics of rabies spread in foxes?

6   Why do not all pathogens evolve to become highly virulent and durable so that they survive a long time in the external environment? Is it possible to design a perfect pathogen?

7   Barlow (1995) showed that the vaccination rate required to eliminate a disease will always be greater than the culling rate required for elimination, given the standard SIR host-parasite model. If this is correct, why might we still prefer vaccination as a strategy for disease control in wild animals?

8   One of the controversies in disease ecology is whether the parasitic nematode Trichostrongylus tenuis has a strong effect on red grouse populations in Scotland and northern England. Review this controversy and evaluate the experiments that have been done to resolve the different points of view. Hudson et al. (1998), Moss and Watson (2001), and Redpath et al. (2006) discuss the differing points of view.

9   Anderson and May (1980) suggested that fluctuations in forest insect populations could be explained as host-parasite interactions, because simple disease models could generate population cycles or outbreaks of the host insect species. Review the subsequent history of this suggestion from the papers in Berryman (2002) and the discussions in Turchin (2003).

10 Anthrax, a bacterial disease caused by Bacillus anthracis, is lethal to most mammalian herbivores. Within a few months during 1983–1984 an anthrax epizootic wiped out 90% of the impala population in Lake Manyara National Park in Tanzania. How is it possible for an epizootic of this type to suddenly appear in a population and then disappear for decades? Discuss the biological mechanisms that might permit this type of phenomenon. Prins and Weyerhaeuser (1987) discuss this particular impala epizootic.

Overview Question

Snowshoe hares in Canada and Alaska are hosts to many species of internal parasites (nematodes and tapeworms) and external parasites such as ticks and fleas. Outline a research program to determine the effects of parasites on individual hares and on their population dynamics.

Suggested Readings

- Anderson, R. M. 1991. Populations and infectious diseases: Ecology or epidemiology? *Journal of Animal Ecology* 60:1–50.

- Barlow, N. D. 2000. Non-linear transmission and simple models for bovine tuberculosis. *Journal of Animal Ecology* 69:703–713.

- Caley, P. 2006. Bovine tuberculosis in brushtail possums: Models, dogma and data. *New Zealand Journal of Ecology* 30:25–34.

- Davis, S., et al. 2004. Predictive thresholds for plague in Kazakhstan. *Science* 304:736–738.

- Ebert, D. 1998. Experimental evolution of parasites. *Science* 282:1432–1435.

- Grenfell, B. T., and A. P. Dobson, eds. 1995. *Ecology of Infectious Diseases in Natural Populations*. Cambridge: Cambridge University Press.

- Juliano, S. A., L. P. Lounibos, and J. M. Chase. 2005. Ecology of invasive mosquitoes: Effects on resident species and on human health. *Ecology Letters* 8:558–574.

- Krkosek, M., J. S. Ford, A. Morton, S. Lele, R. A. Myers, and M. A. Lewis. 2007. Declining wild salmon populations in relation to parasites from farm salmon. *Science* 318: 1772–1775.

- Marra, P. P., et al. 2004. West Nile virus and wildlife. *BioScience* 54:393–402.

- Ostfeld, R. S. 1997. The ecology of Lyme-disease risk. *American Scientist* 85:338–346.

- Smith, M. J., A. White, J. A. Sherratt, S. Telfer, M. Begon, and X. Lambin. 2008. Disease effects on reproduction can cause population cycles in seasonal environments. *Journal of Animal Ecology* 77:378–389.

- Walther, B. A., and P. W. Ewald. 2004. Pathogen survival in the external environment and the evolution of virulence. *Biological Reviews* 79:849–869.

Credits

Ilustration and Table Credits

F1 Roy M. Anderson and Robert M. May, "Population biology of infectious diseases: Part I," *Nature* 280:361–367, 1979. Copyright © 1979 Nature Publishing Group. T1 R. M. Anderson and R. M. May. 1982. Coevolution of hosts and parasites. *Parasitology* 85:411–426.

Photo Credits

Unless otherwise indicated, photos provided by the author.

CO NIBSC/Photo Researchers, Inc.

Regulation of Population Size

Key Concepts

- Two questions are central to population dynamics: (1) What stops population growth? and (2) What determines average abundance?

- To stop population growth, natality, mortality, or movement rates must change with population density. Population regulation requires density dependence.

- Biotic agents such as predators and diseases can limit or regulate populations, as can climatic and physical factors such as temperature, water, and nutrients.

- Individual differences in physiology, genetics, or behavior can limit or regulate populations through intraspecific competition for resources.

- Some populations may be subdivided into local populations or metapopulations that may go extinct and be recolonized by dispersing individuals. Local populations may be unstable while the entire metapopulation is stable.

- Local populations may be source populations exporting emigrants or sink populations importing immigrants. Sink populations go to extinction if they are too isolated.

From Chapter 14 of *Ecology: The Experimental Analysis of Distribution and Abundance*, Sixth Edition. Eugene Hecht.

KEY TERMS

Allee effects Population growth rates that decrease below replacement level at low population density, potentially leading to extinction

balance of nature The belief that natural populations and communities exist in a stable equilibrium and maintain that equilibrium in the absence of human interference.

density-dependent rate As population density rises, births or immigration decrease or deaths or emigration increase, and consequently a graph of population density versus the rate will have a positive or negative slope.

density-independent rate As population density rises, the rate does not change in any systematic manner, so that a graph of population density versus the rate will have a slope of zero.

limiting factor A factor is defined as limiting if a change in the factor produces a change in average or equilibrium density.

metapopulations Local populations in patches that are linked together by dispersal among the patches, driven by colonization and extinction dynamics.

regulating factor A factor is defined as potentially regulating if the percentage of mortality caused by the factor increases with population density or if per capita reproductive rate decreases with population density.

self-thinning rule The prediction that the regression of organism size versus population density has a slope of −1.5 for plants and animals that have plastic growth rates and variable adult size.

sink populations Local populations in which the rate of production is below replacement level so that extinction is inevitable without a source of immigrants.

source populations Local populations in which the rate of production exceeds replacement so that individuals emigrate to surrounding populations.

We have often asked the question about whether predation, disease, or competition could affect the population dynamics of a particular animal or plant. How can we decide that? If a predator kills a prey individual, does that automatically affect the population level of the prey? If we kill pests with insecticides, will they necessarily become less abundant? The answer to these questions is *no*, and in this chapter we explore why simple concepts of population arithmetic can be misleading. These questions are at the core of conservation,

land management, fisheries, and pest control issues that occupy our news media daily. For that reason it is important that our understanding of population regulation is correct.

We can make two fundamental observations about populations of any plant or animal. The first observation is that abundance varies from place to place; there are some "good" habitats where the species is, on the average, common and some "poor" habitats where it is, on the average, rare. The second observation is that no population goes on increasing without limit, and the problem is to find out what prevents unlimited increase in low- and high-density populations. This is the problem of explaining fluctuations in numbers. **Figure 1** illustrates these two problems, which are often confused in discussions of population regulation.

Prolonged controversies spanning more than 50 years have arisen over the problems of the regulation of populations. The idea of the **balance of nature** has been a background assumption in natural history since the time of the early Greeks and underlies much of the controversy about population regulation (Egerton 1973). The simple idea of early naturalists was that the numbers of plants and animals were fixed and in equilibrium, and observed deviations from equilibrium, such as the locust plagues described in the Bible, were the result of a punishment sent by divine powers. Only after Darwin's time did biologists try to specify how a

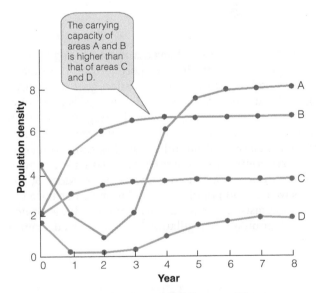

Figure 1 Hypothetical annual censuses of four populations of the same species occupying different types of habitat. Two questions may be asked about these populations: (1) Why do all populations fail to go on increasing indefinitely? (2) Why are there more organisms on the average in the good (red) habitats A and B compared with the poor (blue) habitats C and D? (After Chitty 1960.)

Why Is Population Regulation So Controversial?

The controversies over population regulation are legendary in the history of ecology. During the 1950s and 1960s highly charged exchanges in the literature and strong public verbal attacks at scientific meetings were the order of the day. While most of the vituperative attacks have stopped as time has passed, exchanges still occur in scientific journals (Murray 1999; Turchin 1999). It is interesting to ask why this subject has been so controversial.

There are two aspects to any such controversy, one scientific and one personal. The scientific issue behind the population regulation controversy has been focused on the identification of density-dependent regulating factors as biotic agents—predators, diseases, parasites, food supplies—and density-independent nonregulating factors such as weather and other physical factors. The side issue was always that density-dependent factors are important and density-independent factors are not important, which we now know is not correct (see Figure 3). The difficulty of identifying density-dependent effects in real-world data has greatly prolonged the arguments (Wolda and Dennis 1993). The conclusion after all the controversy was that regulation is an empirical question for each population,

and that one cannot a priori assign factors such as predators or weather to one category or another. The critical thing is to measure what effect a particular factor is having on a particular population, preferably in an experimental setting with proper controls. The realization that intrinsic processes could impinge on regulation, and that mortality could be compensatory rather than additive, also made the original 1950s controversy obsolete.

The personal element to scientific controversy is fascinating because many leading scientists are forceful personalities with large egos. This element is not so easily captured in the written word, but it is apparent at scientific meetings in which proponents of differing paradigms come face to face. Controversy galvanizes people, and population ecologists are indeed human. Population ecology has had an array of fascinating scientists that historians are now beginning to evaluate (Kingsland 2005). The important message is that not every scientist, no matter how distinguished, is right about everything, and in science we should appeal not to authority or personality but to experiment and observation, to empirical tests of ideas, not dogmatic assertions, no matter how articulate the speaker.

---

balance of nature was achieved and how it might be restored in areas where it was upset. Before 1900 many authors had noted that no population goes on increasing without limit, that there are many agents of destruction that reduce the population. During the twentieth century researchers attempted to analyze these facts more formally. The stimulus for this came primarily from economic entomologists, who had to deal with both introduced and native insect pests. Most of the ideas we have on population regulation can be traced to entomologists. Their ideas specifying the basic principles of population regulation can be derived from a simple model.

A Simple Model of Population Regulation

If populations do not increase without limit, what stops them? We can answer this question with a simple graphic model similar. A population in a closed system[1]

will increase until it reaches an equilibrium point at which

Per capita birth rate = per capita death rate

Figure 2 illustrates three possible ways in which this equilibrium may be defined. As population density goes up, birth rates may fall or death rates may rise, or both changes may occur.[2] To determine the equilibrium population size for any field population, we need only determine the curves shown in Figure 2. Note that this simple model in no way depends on the shapes of the curves, provided that they rise or fall smoothly. In particular, these curves need not be straight lines.

We now introduce a few terms to describe the concepts shown in Figure 2. The per capita death rate is said to be **density dependent** if it increases as density

---

[1] A closed system has no immigrants and no emigrants, so the population dynamics are driven solely by births (**natality**) and deaths.

[2] In all discussions of population regulation, "birth rates" always refers to per capita birth rates, and death rates always refers to per capita death rates.

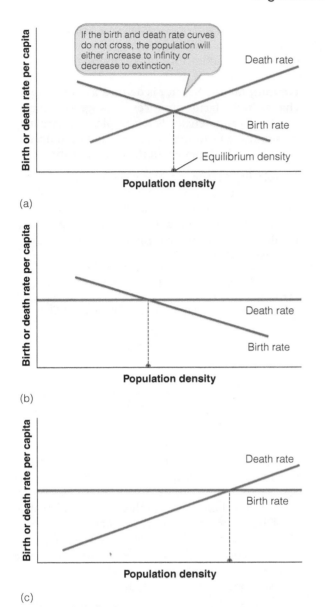

(a)

(b)

(c)

Figure 2 Simple graphic model to illustrate how equilibrium population density may be determined. Population density comes to an equilibrium only when the per capita birth rate equals the per capita death rate, and this is possible only if birth or death rates are density dependent. Note that these relationships need not be straight lines. (Modified from Enright 1976.)

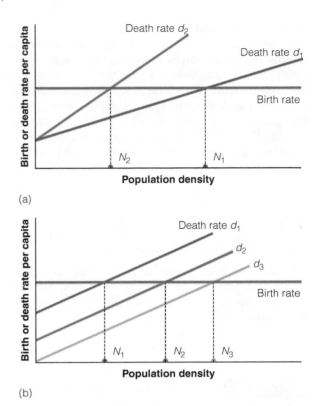

(a)

(b)

Figure 3 Simple graphic model to illustrate how two populations may differ in average abundance. In this example the birth rate is density independent and the death rate is density dependent. In (a) the two populations differ in the amount of density-dependent mortality because the slopes of the lines differ. In (b) the populations differ in the amount of density-independent mortality (the slopes of the lines do not differ). Dashed lines indicate the equilibrium population densities. (Modified from Enright 1976.)

increases (see Figure 2a and c). Similarly, the per capita birth rate is called density dependent if it falls as density rises (see Figure 2a and b). Another possibility is that the birth or death rates do not change as density rises; such rates are called **density-independent rates**.

Note that Figure 2 does not include all logical possibilities. Birth rates might, in fact, *increase* as population density rises, or death rates might *decrease*. Such

rates are called *inversely density-dependent rates* because they are the opposite of directly density-dependent rates. Inversely density-dependent rates are not shown in Figure 2 because they can never lead to an equilibrium density. Figure 2 can be formalized into the **First Principle of Population Regulation**: *No closed population stops increasing unless either the per capita birth rate or death rate is density dependent.*

We can extend this simple model to the case of two populations that differ in equilibrium density to answer the question of why abundance varies from place to place (**Figure 3**). Consider first the simple case of populations with a constant (density-independent) birth rate. Equilibrium densities vary for two reasons: (1) Either the slope of the mortality curve changes (see Figure 3a), or (2) the general position of the mortality curve is raised or lowered (see Figure 3b). In the first case, the density-dependent rate is changed because the slopes of the lines differ, but in the second case, only the density-independent mortality rate is changed. From this graphic model we can arrive at the **Second Principle of Population Regulation**: *Differences between two populations in equilibrium density can be caused by variation in either density-dependent or density-independent per capita birth and death rates.* This principle seems simple: it states that anything that alters birth or death rates can affect equilibrium density. Yet this principle was in fact denied by many population ecologists for 40 years (Enright 1976; Sinclair 1989).

A Synthesis of Population Regulation

There has been a great deal of controversy in ecology over the concepts of population regulation (Sinclair 1989), and we need to highlight the areas of agreement and disagreement.

The definition of terms has always plagued discussions about population regulation. Let us start with clear operational definitions of two confusing terms:

1. **Limiting factor.** A factor is defined as limiting if a change in the factor produces a change in average or equilibrium density. For example, a disease may be a limiting factor for a deer population if deer abundance is higher when the disease is absent.

2. **Regulating factor.** A factor is defined as potentially regulating if the percentage mortality caused by the factor increases with population density.[3] For example, a disease may be a potential regulating factor only if it causes a higher fraction of losses as deer density increases.

---

[3]Or alternatively, if the reproductive rate declines as population density rises.

ESSAY

Definitions in Population Regulation

The lack of clear definitions has plagued debates and discussions of population regulation for decades. We begin by separating two problems:

1. *Population limitation.* What factors and processes can change average density?

2. *Population regulation.* What processes halt population increase?

If we keep these two problems separate, we will solve about half of the confusion in terminology. Answering the first question does not answer the second question.

Population limitation implies a before and after or experimental-control type of comparison. For example, European rabbits in Australia were at high density before myxomatosis and at low density after this disease was introduced. Myxomatosis limits rabbit density.

Population regulation implies some form of negative feedback between increasing density and factors such as predation, disease, food shortage, or territoriality. The effects of a regulating factor must be density dependent, as defined in Figure 2. But the problem is that not all density-dependent processes will achieve population regulation; they may not be quantitatively large enough. A predator that eats one lizard out of a total population of 1000 and

three lizards out of 2000 is inflicting mortality that is density dependent, but it is also quantitatively trivial for population regulation in this species. Population regulation can be inferred only from a comprehensive model that includes all the factors affecting a population.

Compensation can complicate inferences about population regulation. Compensation occurs when a change in one factor produces the opposite change of identical magnitude in another factor, such that their combined effects on the population remain unchanged. One factor can essentially take the place of another factor. The opposite of compensation is *additivity*. Compensation is most easily seen experimentally by comparing, say, mortality rates with and without a particular factor. For example, measure overwinter mortality rates in two populations:

Population A: disease and food shortage

Population B: disease and no food shortage (food supplemented)

If the overwinter mortality rates are identical, food shortage and disease are completely compensatory. If processes are compensatory, population regulation is *either-or*, rather than *both-and*.

The distinction between a potential regulating factor and an actual regulating factor is quantitative. Unless the change in mortality is large enough, a regulating factor will not stop population growth. Regulation is much more difficult to study than limitation. Most experimental manipulations of populations involve studies of limitation, and most practical problems in population ecology are problems of limitation, not regulation.

Factors that influence population size can be subdivided into intrinsic and extrinsic factors (**Figure 4**). Extrinsic factors impinge on populations by the actions of other species such as predators, and by physical-chemical factors such as climate or nutrient supplies. Intrinsic factors are internal to the population and result from the interactions of the individuals making up the population. All individuals in a population are not identical—they differ in sex, age, size, behavior, and in a variety of physiological and genetic traits. The key point is that population regulation results from the interaction of extrinsic and intrinsic factors. For example, predators typically take individuals of certain age groups or preferentially take females over males. The social environment of insects and vertebrates may affect populations. For example, many bird and mammal species defend territories, and the size of the territory defended sets a limit on population density (Durell and Clarke 2004; Packer et al. 2005).

The simple model of population regulation shown in Figure 2 is critically focused on the concept of equilibrium, and we must begin by asking whether natural populations can be equilibrium systems. Recent work on ecological stability has given us a more comprehensive view of the factors that affect stability (**Figure 5**). There is no reason to expect all populations to show the stable equilibria expected under the balance of nature model. Strong environmental fluctuations in weather can produce instability, but biotic interactions may also promote instability. We have seen examples of predator-prey interactions that are unstable. Time lags can also affect population stability. We should expect

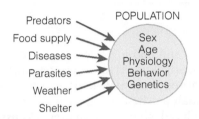

Figure 4 Schematic relationship of population regulation processes that are extrinsic to the population and those that are intrinsic. Extrinsic processes (for example, disease) interact with the properties of individuals that make up the population (intrinsic processes), so that population regulation results from interplay between these two kinds of factors.

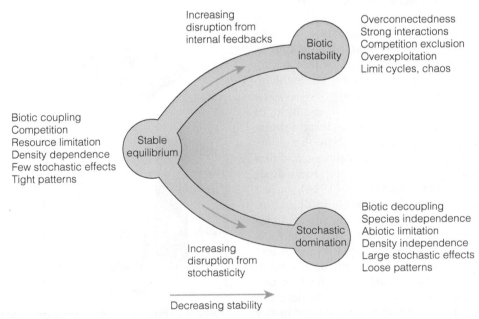

Figure 5 Schematic representation of ecological systems along a continuum from stable to unstable. Unstable or fluctuating populations can result from either biotic instability, caused by internal feedbacks, or by stochastic domination caused by strong environmental fluctuations, or by a combination of both kinds of disruption. (From DeAngelis and Waterhouse 1987.)

275

Figure 6 Hypothetical metapopulation dynamics. Closed circles represent habitat patches; dots represent individual plants or animals. Arrows indicate dispersal between patches. Over time the regional metapopulation changes less than each local subpopulation.

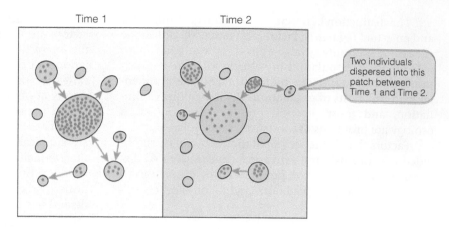

Time 1   Time 2

Two individuals dispersed into this patch between Time 1 and Time 2.

real-world populations to fall along the continuum from stable, equilibrium dynamics to unstable, nonequilibrium dynamics. The simple model shown in Figure 2 will be difficult to detect in a real population that shows unstable dynamics.

The spatial scale is critical in considerations of stability. If we study a very small local population on a small area, it may fluctuate widely and even go extinct; at the same time, a large regional population may be stable in density. The important concept here is that some local populations are linked together through dispersal into **metapopulations (Figure 6)**. To study population regulation, we must know if a population is subdivided and, if so, how the patches are linked (Hanski 1998). Ensembles of randomly fluctuating subpopulations, loosely linked by dispersal, will persist if irruptions at some sites occur at the same time as extinctions at other sites. The result can be that at a regional level the population appears stable while the individual subpopulations fluctuate greatly.

Butterflies on islands are a good example of metapopulations. To show that a set of local populations is a metapopulation, we must show that some metapopulations go extinct in ecological time, and that these can be recolonized by dispersing individuals from nearby populations. Hanski et al. (1996) studied 1502 small populations of the Glanville fritillary butterfly (*Melitaea cinxia*) on islands in the Åland Archipelago between Finland and Sweden. This butterfly is an endangered species that has recently become extinct on mainland Finland and now exists only on islands in the Åland Archipelago. Larval caterpillars feed on two host plants and spin a web, which is easy to detect in field surveys. These butterfly populations ranged in size from 1 to 65 larval groups per meadow, but most populations are small, averaging four larval groups per patch (corresponding to about 5–50 butterflies). From 1991 to 1993 an average of 45% of these local populations went extinct; smaller patches supported smaller

populations and had a greater chance of going extinct (**Figure 7**). Small populations went extinct more often for two reasons. First, male and female butterflies tend to leave small patches, in which they presumably perceive a reduced chance of mating (Kuussaari et al. 1998). **Figure 8** shows the residence time for female butterflies in populations of different sizes, and the fraction of mated females. Small butterfly populations suffered reduced population growth rates, the exact opposite of what is predicted by the simple density-dependent model (shown in Figure 2).

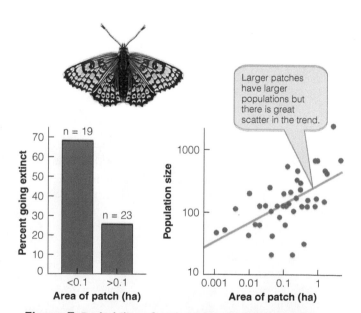

Larger patches have larger populations but there is great scatter in the trend.

Figure 7 Probability of extinction over three years in relation to the patch area for metapopulations of the Glanville fritillary butterfly in the Åland Archipelago, Finland. Small patches are much more likely to go extinct, and small patches tend to have smaller populations of this endangered butterfly. (Data from Hanski et al. 1994, 1995.)

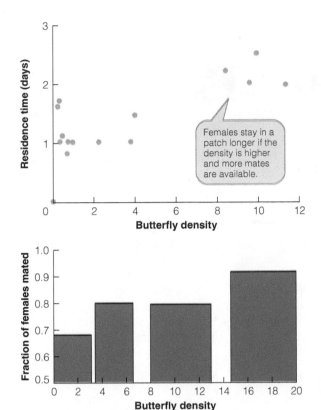

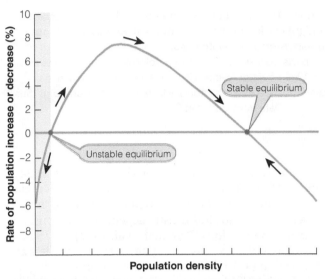

Figure 9 Schematic illustration of the Allee effect, which could have important consequences for endangered species driven to low population densities. The standard population regulation model (see Figure 2) assumes that things get better for populations as density falls to a low level—the death rate goes down and the birth rate goes up. But if an Allee effect (or negative density dependence) occurs, things get worse for a population as density falls. In the yellow zone the rate of population change is negative, and if a population falls below the density indicated by the blue dot, extinction is inevitable. The horizontal red line indicates a stable population (zero population growth).

Figure 8 Effects of population density on the residence time of females and on the fraction of females mated in Glanville fritillary butterflies in the Åland Archipelago, Finland. The net result of these processes is that small populations suffer decreased population growth rates. (Data from Kuussaari et al. 1998.)

Allee Effects and Compensation

Small populations can suffer reduced population growth rates, an effect called the Allee effect, first described by W. C. Allee in 1931. Allee effects produce instability in populations and may contribute to local extinctions. For that reason they are a focus of great interest in conservation biology. **Allee effects** are defined as inverse density dependence at low density (**Figure 9**). Allee (1931) pointed out that undercrowding could be as harmful to social species as overcrowding. If species become too rare, mates may become difficult to locate or group defenses against predators may become ineffective. The key point for populations is that there is a critical threshold density below which a social group or an entire population may go extinct.

A good example of an Allee effect is shown by shearwaters nesting on New Zealand coastal areas and islands. Shearwaters are small petrels that nest in burrows and lay a single egg. Hutton's shearwater is classified as an endangered species because its populations have been in decline due to predation by weasels (stoats) and pigs—predators introduced to New Zealand. **Figure 10** shows that shearwater colonies suffer from an Allee effect in which smaller colonies have poor breeding success and high chick mortality. A minimum colony size of 600 birds is required before Allee effects disappear (Cuthbert 2002).

Allee effects can be widespread in many plant and animal species, and in particular can arise when predation is a major source of mortality (Gascoigne and Lipcius 2004). The important point is that these effects occur below a threshold population size or density, and that once below this threshold, extinction is likely. As more and more examples of Allee effects are being uncovered, the simple view of density-dependent regulation of population size shown in Figure 2 is being replaced by more realistic models (Gilchrist 1999; Dulvy et al. 2004).

An additional complication for the analysis of population regulation is that real-world populations rarely show smooth curves like those in Figure 2. A more usual observation is of a cloud of points, such that density

Figure 10 Impact of predation on (a) breeding success and (b) chick mortality in Hutton's shearwater and sooty shearwaters in New Zealand. Smaller colonies of these seabirds do less well than larger colonies, thus demonstrating an Allee effect or inverse density-dependence. The main predators are weasels (stoats, *Mustela erminea*) and pigs (*Sus scrofa*). (Data from Cuthbert 2002.)

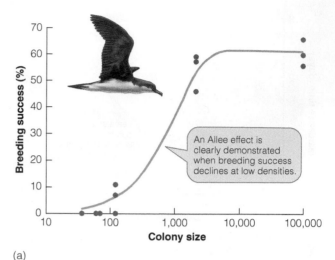

(a)

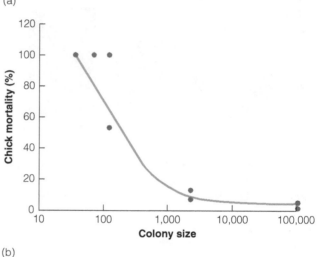

(b)

dependence is either "vague" or absent (Strong 1984). **Figure 11** illustrates the type of density-dependent relationships that might be observed in the real world. It may be very difficult to find density-dependent relationships in natural populations (Berryman et al. 2002).

If a population does not continue to increase, it is axiomatic that births, deaths, or movements must change at high density. The first step is to ask which of these parameters changes with increasing population density (Sinclair 1989; Sibly et al. 2003). Does reproductive rate decline at high density, or does mortality increase (or both)? If mortality increases, does this fall more heavily on younger or on older animals, on males or on females? The first step to understanding population regulation in animals, then, is to see whether these patterns of changing reproduction and mortality with changing population density occur in a variety of populations.

The second step is to determine the reason for the changes in reproduction or mortality. Determining the cause of death of plants or animals in natural populations is not always simple. If a fox or a bat has rabies, a fatal disease, the cause of death is clear; a caterpillar with a

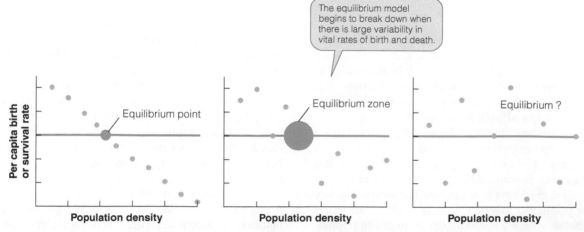

Figure 11 Types of density-dependent relationships for survival rates likely to be observed in real-world populations. For simplicity, a density-independent birth rate is shown (blue line). Increasingly, scattered points make it difficult to determine if there is a stable equilibrium point, or an equilibrium zone, or any equilibrium at all. (Modified after Strong 1984.)

tachinid parasite is certain to die from this infection. But as we examine more complex cases, decisions about causes of death are not clear. If a moose has inadequate winter food and the snow is deep, it may be killed by wolves (Peterson 1999). Is predation the cause of death? Yes, but only in the immediate sense. Malnutrition and deep snow have increased the probability of the moose's death. Because many components of the environment can affect one another and not be independent, mortality can be **compensatory**, as distinguished from **additive**. The concepts of compensatory and additive mortality are crucial to our understanding of population regulation.

Additive mortality is applicable to the agriculture model of population arithmetic. If a farmer keeps sheep and a coyote kills one of them, the farmer's flock is smaller by one. In this model, deaths are additive, and to measure their total effect on a population, we simply add them up. But in natural populations, in which several causes of death operate, the arithmetic is not so simple. Consider, for analogy, a sheep population in which winter food is limiting such that starvation will kill many individuals by the end of winter. In this case, any sheep a coyote kills may have been doomed to die anyway from starvation, and the number of sheep left at the end of winter will be the same, whether predation occurs or not (in this hypothetical scenario). In this case, predation mortality is not additive but is compensatory, and simple arithmetic does not work.

Figure 12 illustrates how additive and compensatory effects can be recognized. Consider, for example,

what happens if wolf predation increases elk calf mortality from 10% per year to 20% per year. If this mortality is additive, total elk calf mortality will increase from 45% to 55% per year (in this hypothetical example). If this mortality is compensatory, total elk calf mortality will remain unchanged at 45% per year. Clearly, if mortality from predation is very high, compensation is not possible, as shown on the right side of Figure 12.

Compensatory mortality is the reason behind many ecological anomalies that puzzle the average person. If we kill pests, they will not necessarily become less abundant. Compensatory mortality has practical consequences when it occurs.

In natural populations, mortality agents will rarely be completely additive or completely compensatory. We can determine if a particular cause of mortality is compensatory only by doing an experiment in which total losses are measured with and without the particular cause of death. **Figure 13** shows the results of this kind of experiment on bobwhite quail. Six study areas were harvested during the early winter at 60% of the birds present, and other areas were not harvested at all. If compensation to hunting mortality is occurring, we would expect to have equal overwinter survival of radio-tagged birds. This did not occur, and Williams et al. (2004) concluded that hunting mortality was

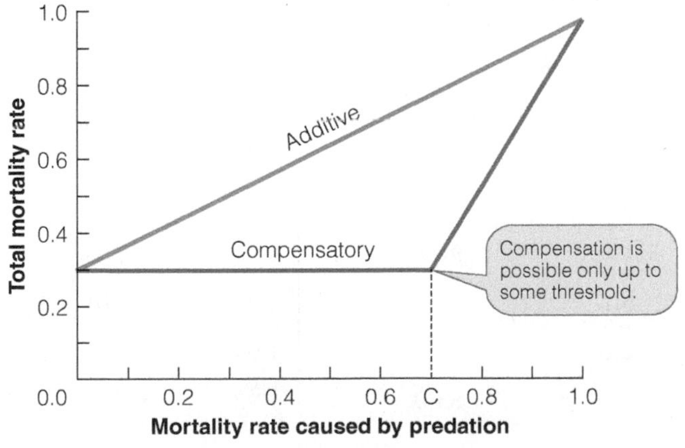

Figure 12 Schematic illustration of additive and compensatory mortality for losses due to predation. The additive hypothesis (green) predicts that for any increase in predation mortality, total mortality increases by a constant amount. The compensatory hypothesis (blue) predicts that, below a threshold mortality rate C, any change in predation losses has no effect on total mortality. This model can be applied to any mortality agent—predation, disease, starvation, or hunting. (Modified after Nichols et al. 1984.)

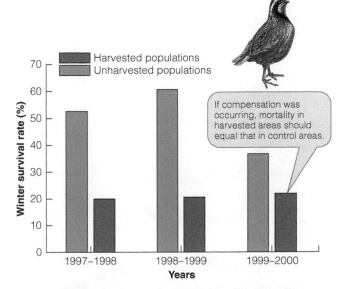

Figure 13 Winter survival of bobwhite quail in Kansas. Six areas (259 ha each) were harvested at 60% of the resident birds in November, and six areas were left as controls with no harvesting. The average survival over winter was 48% for control birds (not hunted) and 21% for treated (hunted) birds. (Data from Williams et al. 2004.)

additive and not compensatory. Similar conclusions have come from studies of hunting mortality in mallards in North America (Pöysä 2004). We should not assume that compensation will occur without adequate studies, particularly in populations that are harvested.

If birth rates change with population density, it is important to identify the factors that cause reproduction to change. Food supply is usually the first hypothesis to be tested for animals; nutrient availability is the first to be tested for plants. But other factors may cause birth rates to change as well. Social interactions can inhibit reproduction in vertebrates (Ishibashi et al. 1998), and risk of predation can change the behavior of animals such that they can gather less energy and thus produce fewer offspring (Lima 1998). These factors can most easily be identified experimentally by manipulations of field populations, or by careful descriptive studies of processes in unmanipulated populations.

The bottom line is that inferences about population limitation and population regulation are both important and difficult to come by. Given these problems, how might one develop a systematic approach to answer these key questions of population dynamics?

Two Approaches to Studying Population Dynamics

There are two competing paradigms about how best to study population dynamics to uncover the causes of population change. **Key factor analysis** is a method of analyzing populations through the preparation of life tables and a retrospective analysis of year-to-year changes in mortality and reproduction. **Experimental analysis** forms a second method of analyzing population changes that approach questions of limitation and regulation directly. Let us consider the advantages and disadvantages of each of these two approaches.

Key Factor Analysis

Morris (1957) developed key factor analysis as a technique for determining the cause of population irruptions in the spruce budworm, which periodically defoliates large areas of balsam fir forests in eastern Canada. Varley and Gradwell (1960) improved Morris's method, and their approach is now used.

Key factor analysis begins with a series of life tables of the type shown in **Table 1**. The life table data are

Table 1 Life table for the winter moth in Wytham Woods, near Oxford, England, 1955–1956.

	Percentage killed in previous stage	No. killed (per m²)	No. alive (per m²)	Log no. alive (per m²)	k Value
Adult Stage					
Females climbing trees, 1955				4.39	
Egg Stage					
Females × 150			658.0	2.82	
Larval Stage					$0.84 = k_1$
Full-grown larvae	86.9	551.6	96.4	1.98	$0.03 = k_2$
Attacked by *Cyzenis*	6.7	6.2	90.2	1.95	$0.01 = k_3$
Attacked by other parasites	2.3	2.6	87.6	1.94	$0.02 = k_4$
Infected by microsporidian	4.5	4.6	83.0	1.92	
Pupal Stage					$0.47 = k_5$
Killed by predators	66.1	54.6	28.4	1.45	$0.27 = k_6$
Killed by *Cratichneumon*	46.3	13.4	15.0	1.18	
Adult Stage					
Females climbing trees, 1956				7.5	

NOTE: The figures in bold are those actually measured. The rest of the life table is derived from these.

SOURCE: After Varley et al. (1973).

most easily obtained for organisms with one discrete generation per year. The life cycle is broken down into a series of stages (eggs, larvae, pupae, adults) on which a sequence of mortality factors operate. We define for each drop in numbers in the life table:

$$k = \log_e(N_s) - \log_e(N_e) \qquad (1)$$

where k = instantaneous mortality coefficient[4]
       N_s = number of individuals starting the stage
       N_e = number of individuals ending the stage

For example, from Table 1 we see that 83.0 winter moth larvae entered the pupal stage in 1955, and of these, 54.6 were killed by pupal predators (shrews, mice, beetles) during late summer, which reduced the population to 28.4 per m². Thus

$$k_5 = \begin{bmatrix} \text{instantaneous mortality} \\ \text{coefficient for pupal} \\ \text{predation} \end{bmatrix}$$

$$= \log_e(83.0) - \log_e(28.4) = 0.47$$

We do these calculations in logarithms to preserve the additivities of the mortality factors. Thus we can define *generation mortality K* as

$$K = k_1 + k_2 + k_3 + k_4 + k_5 + \cdots \qquad (2)$$

Key factor analysis assumes that all mortality factors are additive and ignores compensatory mortality, and this is an important limitation to this method. For our sample data in Table 1,

$$K = \log_e(658) - \log_e(15) = 1.64$$

$$\text{(no.eggs) (no. adults of both sexes)}$$

which is identical to:

$$K = 0.84 + 0.03 + 0.01 + 0.024 + 0.47 + 0.27$$
$$= 1.64$$

Note that since the k values are instantaneous rates, they may take on any value between zero and infinity. Larger k values represent higher mortality rates. Varley et al. (1973) give a detailed description of these calculations.

Given a series of life tables like Table 1 over several years, we can proceed to the second step of key factor analysis, in which we ask an important question: *What causes the population to change in density from year to year?* Simple visual inspection of **Figure 14** shows that k_1 (winter disappearance) is the *key factor* causing population fluctuations. A *key factor* is defined as the component of the life table that causes the major fluctuations in population size. An implication of this definition is

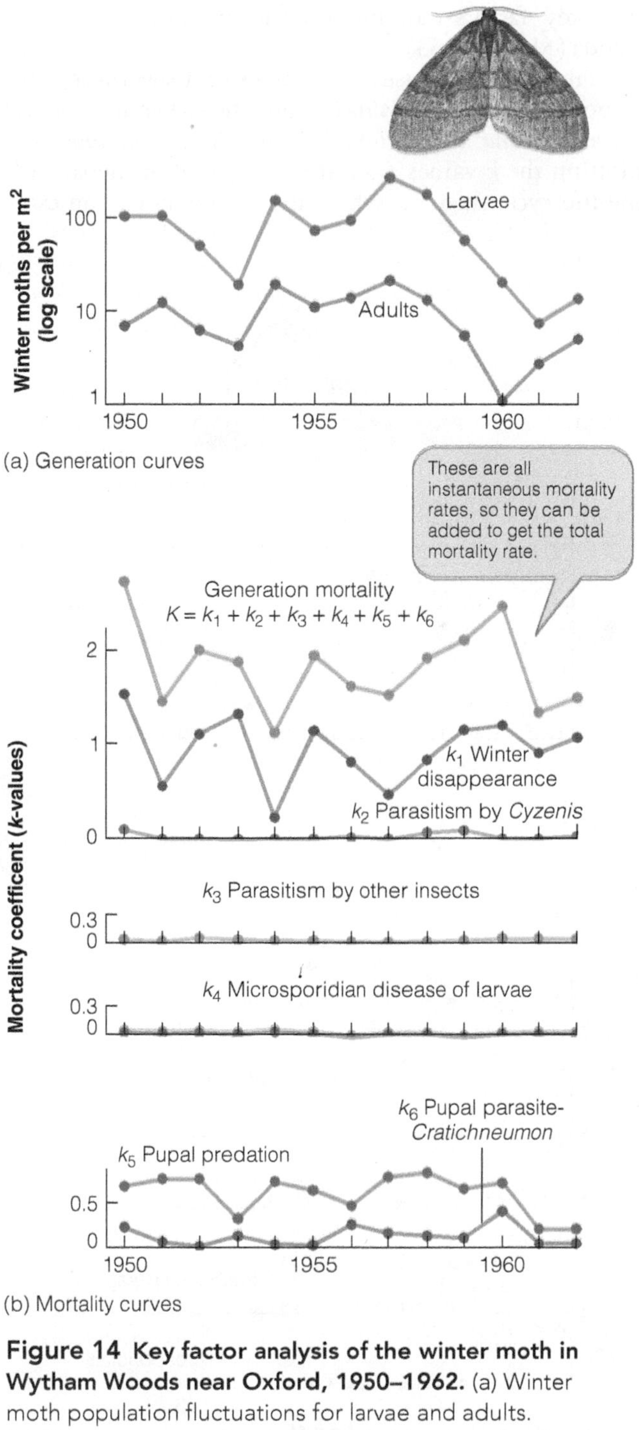

---

[4]Note that these k values are the same as the instantaneous mortality rate without the minus sign.

(a) Generation curves

These are all instantaneous mortality rates, so they can be added to get the total mortality rate.

(b) Mortality curves

Figure 14 Key factor analysis of the winter moth in Wytham Woods near Oxford, 1950–1962. (a) Winter moth population fluctuations for larvae and adults. (b) Changes in mortality, expressed as k values, for the six mortality factors listed in Table 1. The biggest contribution to change in the generation mortality K comes from changes in k_1, winter disappearance, which is the key factor for this population. (After Varley et al. 1973.)

that key factors can be used to predict population trends (Morris 1963).

Finally, we can use the *k* values to answer a second important question: *Which mortality factors are density dependent and thus might halt population increase?* By plotting the *k* values against the population density of the life cycle stage on which they operate, we can esti-

mate density dependence. **Figure 15** shows these data for the winter moth, and **Figure 16** shows the idealized types of curves that can arise from this type of key factor analysis. Note that the key factor need not be density dependent and need not be involved in population regulation. In this example for the winter moth, winter disappearance is the key factor, but pupal predation is the major density-dependent factor.

Key factor analysis has been widely applied to insect populations (Varley et al. 1973; Casanova and do Prado 2002), but it has some important limitations. It cannot be applied to organisms with overlapping generations, including birds and mammals. Mortality factors may be difficult to separate into discrete effects that operate in a linear sequence, do not overlap, and are

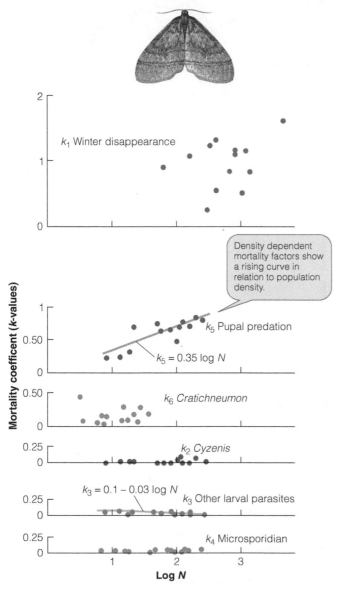

Figure 15 Relationship of winter moth mortality coefficients to population density. The *k* values for the different mortalities are plotted against the population densities of the stage on which they acted. k_1 and k_6 are density independent and quite variable, k_2 and k_4 are density independent but constant, k_3 is inversely density dependent, and k_5 is strongly density dependent. Compare these data with the idealized curves in Figure 16. (After Varley et al. 1973.)

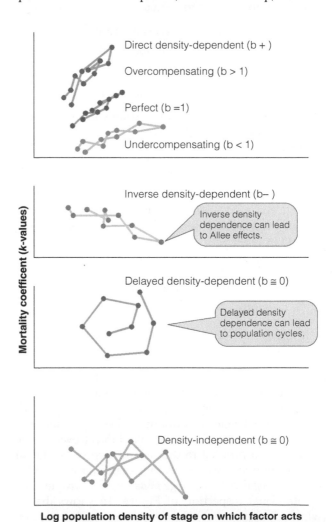

Log population density of stage on which factor acts

Figure 16 Idealized forms of the possible relationships between *k* values determined from key factor analysis and population density. The points are connected in a time sequence, and *b* is the slope of the regression line. Compare with Figure 15. (After Southwood and Henderson 2000.)

completely additive (Åström et al. 1996). In addition, key factor analysis is sensitive to the number of stages in the life cycle that are lumped together into one k value. For example, in the winter moth data, winter disappearance (k_1) includes many distinct mortality processes that are grouped together in this stage of the life cycle (Royama 1996). Such groupings may blur the interpretation of key factors.

Finally, density dependence may be difficult to detect if the equilibrium density (see Figure 3) varies greatly from year to year (Moss et al. 1982). Nevertheless, key factor analysis has provided for some populations a reliable quantitative framework within which the problems of natural regulation can be discussed.

Experimental Analysis

An alternative approach to population regulation is to ask the empirical question: *What factors limit population density during a particular study?* This approach does not utilize the density-dependent paradigm because density dependence is often impossible to demonstrate with field data. Instead we try to identify *limiting factors* and study them with manipulative experiments, an approach that has been called the mechanistic paradigm (Hone and Sibly 2002; Krebs 2002). A population may be held down by one or more limiting factors, and these factors can be recognized empirically by a manipulation—by adding to or reducing the relevant factor. If we suspect that food is limiting a population, we can increase the food supply and see if population size increases accordingly. Alternatively, we can observe changes in population density and the supposed limiting factors over several years and see if they vary together. This is another way of testing hypotheses about limitation, but gathering the relevant data may take a long time. Observations of this type, however, always provide weaker evidence than manipulations involving experimental and control populations.

The experimental approach uses the most direct and empirical techniques for answering the two central questions of regulation—what determines average abundance, and what stops population growth? If we think that parasites reduce the average abundance of pheasants, we can increase or reduce parasite loads and observe the changes in pheasant numbers. If we think that food shortage halts population growth in cabbage aphids, we can manipulate the food resources and measure aphid population growth. It is important to realize that more than one factor may be involved in population limitation. Perhaps both parasite levels and food supplies affect average abundance, so if we have shown one factor to be significant, we cannot assume that only one factor is involved.

Often it is not possible to manipulate a suspected factor, either because it is physically impossible (for example, the weather) or because it is not possible biologically or politically. Experimental analysis can be carried out without manipulations if one sets up a hypothesis and makes a prediction about what can be observed. A good example of this approach has been taken to understand population changes in the western tent caterpillar.

In the northern United States and Canada two species of tent caterpillars fluctuate cyclically in numbers. The western tent caterpillar (*Malacosoma californicum pluviale*) numbers rise and fall in a 6- to 11-year cycle that is synchronized over a broad geographic area (**Figure 17**) (Myers 2000). Tent caterpillars have one generation per year, and larvae hatch in the spring as the leaves of their deciduous host trees begin to form. Each female lays only one egg mass, and the resulting colony of larvae make conspicuous silk tents in which they congregate between bouts of feeding. Population size is measured by counting these highly visible tents in shrubs and trees during the spring. The key question for this tent caterpillar is what factors cause these large fluctuations in abundance.

Three hypotheses have been suggested. The weather could produce runs of good and bad years that are reflected in population changes. Second, insect parasites and predators may attack these populations with a delayed density-dependent action, generating a cycle. Or

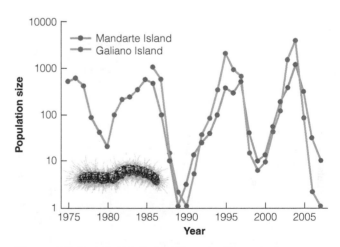

Figure 17 Population fluctuations in the western tent caterpillar on two islands near Vancouver, British Columbia. Population size is measured by the number of silk tents in deciduous trees in the same sites each spring. Populations fluctuate in synchrony on these two islands with peaks in 1985–86, 1995–97 and 2003–04. Populations change through almost four orders of magnitude from a low of 1 to a high of 4000 larval tents. (Data from Myers 2000 and Myers, personal communication.)

finally, virus diseases may spread in high populations and maintain high mortality through the population decline. The strongest support has been for the disease hypothesis to explain tent caterpillar population fluctuations (Myers 1988). The main pathogen that attacks western tent caterpillars is a baculovirus in the nucleopolyhedrovirus (NPV) group, extremely small viruses that are composed of double-stranded DNA (Cory and Myers 2003). Each type of NPV is species specific, and insect baculoviruses must be eaten by the host to produce an infection, which is typically fatal. **Figure 18** illustrates how NPV infection rate follows population density changes in the western tent caterpillar. These data are consistent with the disease hypothesis suggesting that the tent caterpillar cycles are caused by outbreaks of virus disease, and models that include virus disease can successfully mimic insect population fluctuations (Dwyer et al. 2004). But NPV disease by itself cannot be the entire explanation for these fluctuations, since some populations collapse with little evidence of NPV disease (Myers pers.comm.).

Experimental analysis is oriented toward testing hypotheses about regulation mechanisms. Key factor analysis is retrospective looking and is confined to a descriptive analysis of a population. Both methods should converge to provide an understanding about population changes, many of which have important economic consequences.

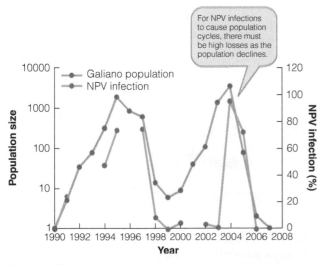

Figure 18 Population changes and NPV infection rates in the western tent caterpillar on Galiano Island near Vancouver, British Columbia. NPV infections are measured from samples of larvae brought into the laboratory. (Data from Myers 2000 and Myers personal communication.)

Plant Population Regulation

Because most plants are modular organisms, population regulation in plants must be discussed as the regulation of biomass rather than of numbers. Plant ecologists have not usually addressed the problem of population regulation in the same way as have animal ecologists (Crawley 1990, 1997), but the same principles can be applied. As a plant population increases in numbers and biomass, either reproduction or survival will be reduced by a shortage of nutrients, water, or light; by herbivore damage; by parasites and diseases; or by a shortage of space. Because plants are typically fixed in one location, competition for light or nutrients is often implicated in population regulation. This competition has been described by the −3/2 power rule (also called *Yoda's law* or the **self-thinning rule**).

The self-thinning rule describes the relationship between individual plant size and density in even-aged populations of a single species. Mortality, or "thinning," from competition within the population is postulated to fit a theoretical line with a slope of −3/2:

$$\log(\overline{m}) = -\frac{3}{2}(\log N) + K \qquad (3)$$

where $\overline{m}$ = average plant weight (g)
N = plant density (individuals/m²)
K = a constant

This line has been suggested as an ecological law (Hutchings 1983; Westoby 1984) that applies both within any given plant species and among different plant species. **Figure 19** illustrates the −3/2 power rule. The self-thinning rule highlights the trade-offs that can occur in organisms having plastic growth, such that the size of an individual can become smaller as population density increases.

Evaluations of the self-thinning rule have found many exceptions to it (Weller 1987, 1991). However, the principle of a trade-off between average plant size and total plant population density is supported by all plant studies. The self-thinning rule has been replaced by a more general "−3/2 boundary rule," which postulates the self-thinning line as an upper limit for the relationship between plant size and population density in monocultures (Hamilton et al. 1995). The self-thinning rule expresses competition between plants for essential resources. If this competition is largely for light, the self-thinning rule should predict that leaf area should remain constant during thinning. This type of formulation would have practical consequences for stand densities of forest trees in plantations (Newton 2006). The conclusion is that we should view the self-thinning rule as a boundary rule rather than an absolute thinning law for all plants. The slope of the thinning line is variable but gives us

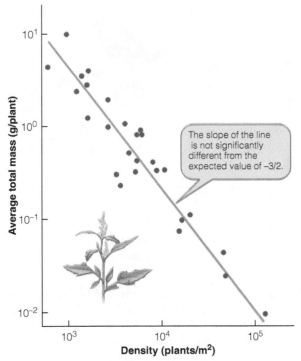

Figure 19 **The self-thinning line for the herb**
Chenopodium album. The slope of this line is −1.37, close
to the theoretical –3/2 of the self-thinning rule. Populations
started at densities to either side of this line would be
expected to move to the line and then reach equilibrium
along the line. (Data from Yoda et al. 1963.)

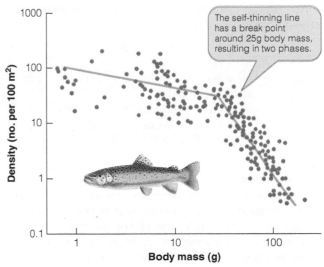

Figure 20 **Self-thinning in populations of stream-living
brown trout (*Salmo trutta*) in Spain.** There is a strong
trade-off between population density and the size of these
fish in streams, but these trout follow a two-phase self-
thinning line rather than a one-phase line that seems to
occur in most plant species. (Data from Lobon-Cervia and
Mortensen 2006.)

further insight into species differences under strong
competition for light and nutrients.

The self-thinning rule has been applied to animals as well. Animals with plastic growth rates, such
as fish, can respond to changes in population density
by changing growth rates and body size (Lobon-
Cervia and Mortensen 2006). Animals of larger body
size use more energy, and when populations are food
limited or space limited, a trade-off can occur between average size and population density. Salmon
and trout fingerlings living in streams are a good example (**Figure 20**), and for these kinds of animals
with plastic growth, the self-thinning rule is a useful
empirical description of these trade-offs between
body size and population density.

Source and Sink Populations

Local populations can be classified as **source populations**, in which there is a net excess of reproduction
over mortality, and **sink populations**, in which there
is a net excess of mortality over reproduction. Left to

themselves, source populations would grow to infinity, and sink populations would shrink to extinction.
But in practice, source populations do not increase
forever, but are regulated. This regulation may involve
a net export of animals via dispersal, such that in a
source population emigration exceeds immigration.
Sink populations may indeed go to extinction, so we
would not necessarily know about them, but more
typically they are in negative balance in that immigration exceeds emigration. Sink populations continue to
exist only if they attract immigrants from nearby
source populations.

Sources and sinks have become more important
in human-impacted landscapes, such as formerly large
continuous areas of forest or grassland that have been
dissected by modern agriculture into a series of small
fragments (Pulliam 1988). Source and sink dynamics
are thus often part and parcel of habitat fragmentation. Forests in agricultural landscapes have been particularly fragmented, and there is much concern that
fragmentation can turn source populations into sink
populations.

To identify source and sink populations we need to
measure reproduction, mortality, and movements
among a whole set of local populations. Much of the
concern about source and sink populations has concerned migratory birds in North America (Robbins et al.
1989). For the simplest model of population change for

birds, we can estimate the finite rate of population increase from three parameters:

$$\lambda = P_A + P_F\beta \qquad (4)$$

where λ = finite rate of population growth ($\lambda = 1$ for stable population)

P_A = adult survival rate during the year

P_F = juvenile survival rate during the year

β = number of juveniles produced per adult by the end of the breeding season

and assuming an equal sex ratio of males to females.

If we can estimate these three parameters for any population, we can determine if that population is a source ($\lambda > 1$) or a sink ($\lambda < 1$). For example, a population of house sparrows on an island off the coast of Norway produced 6.33 fledglings per female in 1993 (or 3.165 fledglings per adult bird), and the nonbreeding-period finite survival rate was 0.579 for juveniles and 0.758 for adults (Sæther et al. 1999). Assuming a 1:1 sex ratio, from Equation (4) we obtain for this population

$$\lambda = P_A + P_F\beta$$

$$= 0.758 + (0.579)\left(\frac{6.333}{2}\right)$$

$$= 2.59$$

Given these demographic rates, this population will more than double each year and must be a source population.

Source-sink dynamics may be characteristic of particular metapopulations, or they may be a product of variation in weather from year to year. Sæther et al. (1999) studied house sparrows on four islands off Norway to measure variation in population growth rate among the islands and over time. **Figure 21** shows that some islands on average were much more productive than others, but that all islands are sink populations in particularly severe years. Populations on each of the four islands remained nearly constant from 1993 to 1996, with immigration boosting the sinks and emigration evening the source populations. The dynamics of source-sink populations are graphic illustrations of how immigration and emigration can be just as important as reproduction and mortality as agents of population change.

Evolutionary Implications of Population Regulation

How are systems of population regulation affected by evolutionary changes? We have already discussed some of the problems involved in coevolution of predator-prey systems and herbivore-plant systems. In many of these interactions, evolutionary changes operate very slowly and are difficult to detect. But recent work in ecological genet-

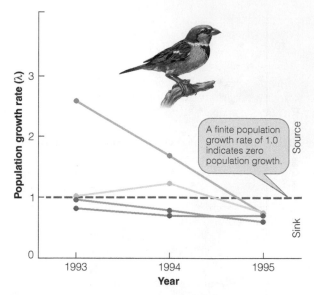

Figure 21 Population growth rates of house sparrows (*Passer domesticus*) on four islands off northern Norway from 1993 to 1995. All populations were sinks in 1995, and one island (Indre Kvarøy, green) was particularly productive on average. Two islands were always sink populations because of poor juvenile survival. (From Sæther et al. 1999.)

ics (Futuyma 2005) has shown that evolutionary changes may occur very rapidly, such that the evolutionary time scale approaches the ecological time scale. Natural selection may thus impinge upon population regulation in some organisms.

Many changes in average abundance can be attributed to changes in extrinsic factors such as weather, disease, or predation. But some changes in abundance are the result of changes in the genetic properties of the organisms in a population. Such evolutionary changes can be produced by natural selection. Pimentel (2002) catalogs some spectacular examples of genetic changes playing a role in population limitation. For example, the population of the herbivorous Hessian fly was reduced drastically in Kansas after 1942 when genetically altered, fly-resistant varieties of wheat were introduced. Another example is the myxomatosis-rabbit interaction in Australia in which evolutionary changes occurred in both the virus and the rabbit.

Genetic changes in populations can affect the interspecific interactions that limit abundance. The coevolution of interacting populations of predator and prey, disease and host, and food plant and herbivore may have implications for population dynamics. The important point is that we should not assume that the ecological traits of species are constant and unchanging in ecological time. In particular, an evolutionary perspective on questions of population limitation and regulation serves as a warning with respect to the continual introduction of new species into ecological communities of distant areas.

Populations in which intrinsic processes regulate abundance present yet another problem in evolutionary ecology. Under what conditions should we expect a population to be regulated by intrinsic processes involving spacing behavior in the broad sense, including territoriality, dispersal, and reproductive inhibition? We might expect that vertebrates, with their relatively complex behavior, would be the most obvious species to show intrinsic regulation. Wolff (1997) has suggested a conceptual model that predicts which vertebrates have the potential for intrinsic regulation. He discusses mammals in particular, but similar arguments could be made for birds and other vertebrates. The key to Wolff's model is that territoriality in female mammals has evolved as a counterstrategy to infanticide committed by strange females. Infanticide is a mechanism of competition by which intruders usurp the breeding space of residents and increase their fitness by killing the offspring of resident females.

Female mammals should evolve territorial behavior to defend their young from infanticide only if young are not mobile at birth. Females with precocial young, which have their eyes open and can move very soon after birth, will not be susceptible to infanticide and will not defend territories. These predictions from Wolff's model are consistent with most of what is known about mammalian social systems. For example, hares have precocial young while rabbits have altricial young.[5] Infanticide is unknown in hares but is known to occur in rabbits. Many carnivores (for example, lions) have altricial young, are subject to infanticide, and are territorial. By contrast, kangaroos have altricial young but carry them around in a pouch so that they are not vulnerable to infanticide. None of the kangaroo species are territorial.

Another feature of self-regulation in mammals is reproductive suppression of juveniles (Wolff 1997). If juveniles do not disperse from their natal area, they risk the possibility of breeding with close relatives. Selection against inbreeding has molded the dispersal pattern of mammals such that male juveniles will emigrate while female offspring remain near their natal site (Clobert et al., 2001; Lambin et al. 2001). But high density may make dispersal costly due to aggressive encounters such that all juveniles stay near the birthplace. At high density, adults may suppress sexual maturation of their offspring through pheromones in order to prevent inbreeding, especially if space for breeding is limited. The result can be that a large fraction of the population is not breeding, as has been observed in many rodents, primates, and wolves. This reproductive suppression of

juveniles at high density acts as a density-dependent factor to potentially regulate the population.

Figure 22 summarizes Wolff's model for the evolution of intrinsic regulation in mammals. Many mammal species and many other vertebrates will not be subject to potential infanticide, and these species would be expected to be subject to extrinsic regulation by predators, food shortage, disease, or weather. Note that intrinsic regulation is not in itself an evolved strategy. What evolves are behavioral strategies such as territoriality, dispersal, and reproductive inhibition, and these individual strategies can result in population regulation at the level of the population. Evolution works at the level of the individual, in most cases, and not at the population level.

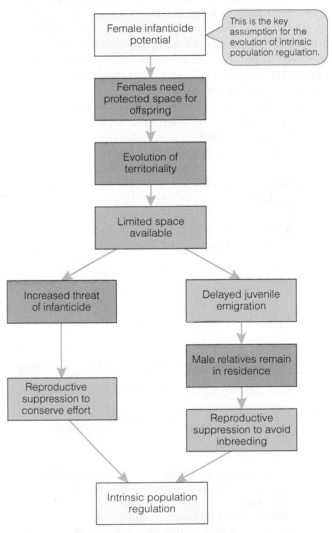

Figure 22 Wolff's hypothesis for the evolution of intrinsic population regulation in mammals. Spacing behavior is the key mechanism to evolve in species that compete for space free from infanticidal individuals. The demographic attributes that contribute to population regulation are shown in blue boxes. (From Wolff 1997.)

---

[5]Altricial young are typically blind, naked, and cannot move around at birth.

Summary

Populations of plants and animals do not increase without limits but show more or less restricted fluctuations. Two general questions may be raised for all populations: (1) What stops population growth? (2) What determines average abundance?

Two principles encapsulate the answers to these questions. The First Principle of Population Regulation is that no closed population stops increasing unless either the per capita birth rate or death rate is density dependent. The Second Principle of Population Regulation is that differences between two populations in average abundance can be caused by variation in either density-dependent or density-independent per capita birth and death rates. The key to understanding population dynamics lies in mapping the changes observed in populations onto the driving factors that can be divided into extrinsic factors (weather, food supplies, nutrients, predators, pathogens, and shelter) and intrinsic factors (behavioral, physiological, and genetic changes associated with social interactions within the population).

Population regulation theory has focused on equilibrium conditions, and many ecologists now emphasize nonequilibrium concepts and ask what factors reduce stability for populations. The spatial scale of a study affects conclusions about stability, and if a population is subdivided into local populations, stability may be increased for the entire population. *Metapopulations*, or clusters of local populations, are critical foci for conservation as habitats are broken up into small, isolated blocks linked by dispersal.

Plant population dynamics follow the same general principles as those for animals, but because of plastic growth, changes in biomass are more significant than changes in numbers of individuals. The self-thinning rule describes a consistent trade-off between individual size and population density in both plants and animals with plastic growth rates.

Local populations may be sources or sinks. Source populations export emigrants to other local populations, while sink populations continue to exist only because of immigration. Source and sink dynamics may change from year to year, and if the metapopulation structure becomes too broken up, sink population can go to extinction.

The theories of population regulation are not mutually exclusive but overlap, and a synthesis of several approaches may be most useful in attempting to answer practical questions. The limitation and regulation of populations are critical areas of theoretical ecology because they are central to many questions of community ecology and because they have enormous practical consequences.

Review Questions and Problems

1   Morris (1957, p. 49), in discussing the interpretation of mortality data in population studies, states:

We tend to overlook the fact that these mortality estimates do not represent an ultimate objective in population work. Long columns of percentages, which are sometimes presented only with the conclusion that high percentages indicate important mortality factors and low percentages indicate unimportant ones, contribute little to our understanding of population dynamics.

Discuss this claim.

2   Density-dependent relationships can be looked for by studying different local populations living in different patches (spatial density dependence) or by following one local population over several years (temporal density dependence). Discuss the interpretation of these two types of data with regard to the problem of regulation.

3   Singer et al. (1997) reported on the population dynamics of elk in Yellowstone National Park, with the following data from 1975–1991. Calf recruitment is the number of calves per adult female in autumn; survival rates are finite annual rates. Population estimates are for autumn of each year, and calf data are from the following summer and winter. There is a gap in the data between 1978 and 1982.

Year	Summer calf recruitment rate	Summer calf survival rate	Winter calf survival rate	Population size
1975–76	0.18	0.22	0.36	15,797
1976–77	0.30	0.38	1.00	13,305
1977–78	0.27	0.32	0.59	15,350
1982–83	?	?	0.76	19,523
1983–84	?	?	0.52	20,837
1984–85	0.52	0.80	0.28	21,115
1985–86	0.37	0.61	0.54	22,115
1986–87	0.44	0.69	0.38	19,825
1987–88	0.26	0.40	0.32	21,706
1988–89	0.20	0.31	0.17	20,619
1989–90	0.30	0.37	0.35	17,843
1990–91	0.78	?	1.00	17,950

Are any of these three measures of recruitment or mortality density dependent? What can you conclude about population regulation in Yellowstone elk? Compare your conclusions with those of Singer et al. (1997).

4 If you wished to increase the abundance of a threatened species like Hutton's shearwater that shows an Allee effect (see Figure 10), what management actions might you recommend?

5 Mourning doves (*Zenaida macroura*) are hunted in eastern North America. McGowan and Otis (1998) reported the following data for two populations of doves in South Carolina:

Area	Recruitment per adult bird	Adult survival rate (annual)	Juvenile survival rate (annual)
Bennettsville	3.400	0.359	0.118
Eutawville	2.325	0.359	0.118

Calculate the finite rate of population growth for these two dove populations from Equation (4), and discuss what management action these results might indicate.

6 Spatial synchrony is relatively common in forest insect pests that have outbreaks (Liebhold and Kamata 2000). Suggest three possible mechanisms that could produce synchrony among local populations, and discuss what data could test among these alternative hypotheses.

7 The saguaro is a prominent columnar cactus of the Sonoran Desert of Arizona and northern Mexico. Saguaro are long-lived perennials, and individuals may reach 150–200 years of age. Pierson and Turner (1998) reported the following data from a long-term study of four populations in an ungrazed desert preserve:

Plot	Census			
	1964	1970	1987	1993
North	284	265	232	221
South	1308	1316	—	1087
East	1367	1394	—	1277
West	603	586	—	459

What would you conclude about the population dynamics of these cacti from these data? What additional data would you like to have to predict future population trends?

8 Can a population persist without regulation? How could you determine if a population was persisting without regulation? Read Strong (1984) and Reddingius and den Boer (1970) and discuss.

9 The autumnal moth *Epirrita autumnata* shows population outbreaks at 9- to 10-year intervals in the mountain birch forests of Fennoscandia. Scandinavian ecologists have data on these outbreaks going back 112 years (Nilssen et al. 2007). The sunspot cycle has a periodicity of about 11 years, but both these cycles are somewhat variable. How could you test the hypothesis that the sunspot cycle causes the periodicity in autumnal moth populations through its effect on weather? Nilssen et al. (2007) discuss this issue.

10 What general guidelines would you recommend as to how many generations should be analyzed in order to complete a key factor analysis of a population? Would you expect that a new study of the key factors affecting winter moth populations (see Figure 14) would reach the same conclusions?

11 Is the dispersal rate in mammals density dependent? Read Wolff's (1997) arguments and discuss the conditions under which emigration might regulate population density in mammals.

Overview Question

Local populations can be classified as source populations ($\lambda > 1$) or sink populations ($\lambda < 1$). How would you determine for a metapopulation of plants which local populations were sources and which were sinks? Discuss the application of population regulation theories to a metapopulation of plants containing sources and sinks.

Suggested Readings

- Dwyer, G., J. Dushoff, and S. H.Yee. 2004. The combined effects of pathogens and predators on insect outbreaks. *Nature* 430:341–345.

- Gascoigne, J. C., and R. N. Lipcius. 2004. Allee effects driven by predation. *Journal of Applied Ecology* 41:801–810.

- Gough, K. F., and G. I. H. Kerley. 2006. Demography and population dynamics in the elephants *Loxodonta africana* of Addo Elephant National Park, South Africa: Is there evidence of density dependent regulation? *Oryx* 40:434–441.

- Hone, J., and R. M. Sibly. 2002. Demographic, mechanistic and density-dependent determinants of population growth rate: A case study in an avian predator. *Philosophical Transactions of the Royal Society*, Series B 357:1171–1177.

- Mougeot, F., S. B. Piertney, F. Leckie, S. Evans, R. Moss, S. M. Redpath, and P. J. Hudson. 2005. Experimentally increased aggressiveness reduces population kin structure and subsequent recruitment in red grouse *Lagopus lagopus scoticus. Journal of Animal Ecology* 74:488–497.

- Myers, J. H. 2000. Population fluctuations of the western tent caterpillar in southwestern British Columbia. *Population Ecology* 42:231–242.

- Newey, S., F. Dahl, T. Willebrand, and S. Thirgood. 2007. Unstable dynamics and population limitation in mountain hares. *Biological Reviews* 82:527–549.

- Pierson, E. A., and R. M. Turner. 1998. An 85-year study of saguaro (*Carnegiea gigantea*) demography. *Ecology* 79:2676–2693.

- Pulliam, H. R. 1988. Sources, sinks, and population regulation. *American Naturalist* 132:652–661.

- Sibly, R. M., Barker, M. C. Denham, J. Hone, and M. Pagel. 2005. On the regulation of populations of mammals, birds, fish, and insects. *Science* 309:607–610.

- Wolff, J. O. 1997. Population regulation in mammals: An evolutionary perspective. *Journal of Animal Ecology* 66:1–13.

Credits

Illustration and Table Credits

Photo Credits

Applied Problems I: Harvesting Populations

Key Concepts

- Any harvested population must decline in abundance, and the losses due to harvesting must be compensated for by increased growth, increased reproduction, or decreased natural mortality.

- Simple population growth models such as the logistic equation can be used to estimate the maximum sustainable yield of a harvested population.

- All simple yield models assume equilibrium conditions and fail when there are changes in ocean conditions, weather, predators, or diseases.

- Uncertainty in our knowledge of population dynamics and the variable effects of weather argue for more conservative harvesting goals than maximum yield.

- Incessant social and political pressure for increased harvests coupled with uncertainty concerning biological information has produced many examples of overfishing and overharvesting of renewable natural resources. Sustainability can be achieved by a combination of good ecological knowledge and good governance.

From Chapter 15 of *Ecology: The Experimental Analysis of Distribution and Abundance*, Sixth Edition. Eugene Hecht.
Copyright © 2009 by Pearson Education, Inc. Published by Pearson Benjamin Cummings. All rights reserved.

KEY TERMS

dynamic pool models Models to predict maximum sustained yield based on detailed population information on growth rates, natural mortality, and fishing mortality.

logistic models Models to predict maximum sustained yield by the use of sigmoid curves of population increase modified by fishing removals.

marine protected area A national park in the ocean where fishing is restricted or eliminated for the purpose of protecting populations from overharvesting.

match/mismatch hypothesis the idea that population regulation in many fish is determined in the early juvenile stages by food supplies, so that if eggs hatch at the same time that food is abundant, many will survive, but if eggs hatch when food is scarce, many will die.

maximum economic rent The desired economic goal of any exploited resource, measured by total revenues – total costs.

maximum sustained yield (MSY) The predicted yield that can be taken from a population without the resource collapsing in the short or long term.

stock The harvestable part of the population being exploited.

stock-recruit relationship A key graph relating how many recruits come into the exploited population from a given population of adults.

tragedy of the commons The inherent tendency for overexploitation of resources that have free access and unlimited demand, so that it pays the individual to continue harvesting beyond the limits dictated by the common good of sustainability.

yield Amount of usable material taken from a harvested population, measured in numbers or biomass.

To manage a population effectively, we must have some understanding of its dynamics. The list of populations destroyed by inadequate management throughout human history should serve as both a warning and as a stimulus for us to achieve a better understanding of harvesting principles. The central problem of economically oriented fields such as forestry, agriculture, fisheries, and wildlife management is how to produce the largest crop without endangering the resource being harvested. The problem may be illustrated with a simple example from forestry. If you were managing a forest woodlot that was growing to maturity, you obviously would not cut the trees when they were saplings because this would yield little wood

production and less profit. At the other extreme, you would not let the trees grow too old and begin to rot because you would get little timber to sell. Somewhere between these two extremes is some optimum point to harvest the trees, and the problem is how to identify it.

Next to forestry and agriculture, the greatest amount of work on the problem of optimum harvesting has been done in fishery biology, especially because of the tremendous economic importance of marine fisheries in particular. Many marine fisheries have dwindled in size since the 1920s because of overfishing, and this has stimulated a great deal of research on "the overfishing problem."

For any harvested population, the important unit of measure is the crop or **yield**. The yield may be expressed in *numbers* or in *biomass* of organisms, and it always involves some unit of time (often a year). We are interested in obtaining the optimum yield from any harvested population. We will begin by defining *optimum yield* very specifically, and at the end of the chapter we will reconsider other ways of defining *optimum*. The concept of **maximum sustained yield** has been the basis of scientific resource management since the 1930s (Larkin 1977; Mace 2001). Let us consider first the simple situation in which maximum yield in biomass is defined as the optimum yield. Implicit in this concept is the idea of a sustained yield over a long time period.

Russell (1931) was one of the first to deal in detail with the harvesting problem in fisheries. In any exploited fish population, there is usually a portion of the population that cannot be caught by the type of gear used or is purposely not harvested. The harvestable sector of the population is called the **stock**. For a fishery, interest normally centers on yield in weight, so instead of individuals we will deal in biomass units. Russell pointed out that two factors decrease the weight of the stock during a year: natural mortality and fishing mortality. Similarly, two factors increase the weight of the stock: growth and recruitment.[1] Consequently, one can write the following simple equation to describe this relationship:

$$S_2 = S_1 + R + G - M - F \qquad (1)$$

where S_2 = biomass of the stock at the end of the year
S_1 = biomass of the stock at the start of the year
R = biomass of new recruits
G = growth in biomass of fish remaining alive
M = biomass of fish removed by natural deaths
F = yield in biomass to fishery

If we wish to balance the biomass of the fish population, $S_1 = S_2$, and hence

$$R + G = M + F \qquad (2)$$

---

[1]Recruitment in fisheries is usually measured when the fish reach a certain size or age. Recruitment thus includes natality and early life history survival and growth.

This means that in an unexploited stage ($F = 0$), in which the stock biomass remains approximately constant from one year to the next, all growth and recruitment is on the average balanced by natural mortality. When exploitation begins, the size of the exploited population is usually reduced, and the loss to the fishery is made up by compensatory changes such as (1) greater recruitment rate, (2) greater growth rate, or (3) reduced natural mortality. In some populations, none of these three occurs, and the population is exploited to extinction because the right side of Equation (2) always exceeds the left side.

Note that stability at *any* level of population density is described by the equation:

Recruitment + growth = natural losses + fishing yield

Thus a crucial question arises: What level of population stabilization provides the greatest weight of catch to the fishery? One of the first attempts to solve this problem was made by Graham (1935), who proposed the *sigmoid-curve theory*.

Start by considering a very small stock of fish in an empty area of the sea, said Graham. At what rate will such a stock increase in size? Graham suggested that the growth of this population would follow a sigmoid curve like the one described by the logistic equation (**Figure 1**). Initially, the population grows more slowly in absolute size, reaches a maximum rate of increase near the middle of the curve, and grows slowly again as it approaches the asymptote of maximal density. We can use the terminology of the logistic equation to show that two factors interact to determine the amount of increase per year. For simplicity, let $K = 200$ biomass units and $r = 1.0$:

Point on curve	Population size	$\dfrac{K - N}{K}$	rN	Amount of increase per year
S_1	20	0.90	20	18
S_2	50	0.75	50	38
S_3	100	0.50	100	50
S_4	150	0.25	150	38
S_5	180	0.10	180	18

According to the logistic equation, the amount of population increase depends on the carrying capacity (K), the intrinsic rate of increase (r), and the current population size (N):

$$\frac{dN}{dt} = rN\left(\frac{K - N}{K}\right) \qquad (3)$$

and this is maximal at the midpoint of the curve (S_3).

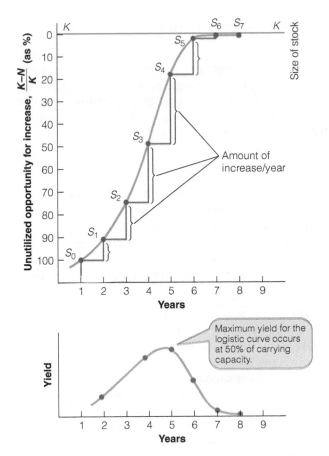

Figure 1 Use of sigmoid curve theory to describe the growth of a population that could be exploited. The amount of increase per year is the maximum yield that could be taken sustainably by the fishery. According to this model maximum yield is obtained by keeping the population at one-half of carrying capacity. (After Graham 1939.)

If we wish to maintain the maximal yield from such a population, Graham pointed out, we should keep the stock around point S_3 of the curve. The important point here is that the highest production from such a population is not near the top of the curve, where the fish population is relatively dense, but at a lower density. This can be expressed as the first rule of exploitation: *Maximum yield is obtained from populations at less than maximum density.*

All the vital statistics of an exploited population—recruitment, growth, and natural mortality—may be a function of population density and also of age composition. Because in most fisheries we do not know how these vital statistics relate to density or age, we make some simplifying assumptions. Two alternative approaches have been developed for determining optimum yield: **logistic models** and **dynamic pool models**. We next discuss each of them in turn.

Logistic Models

In logistic models,[2] we do not distinguish among growth, recruitment, and natural mortality but instead combine them into a single measure, *rate of population increase*, which is a function of population size. Graham's *sigmoid-curve theory* is a classic example of this type of model. The general case can be written as:

$$\frac{\text{Rate of}}{\text{population increase}} = f(\text{population size}) - \frac{\text{amount of}}{\text{fishing losses}}$$

If we specify that the function of population size in this equation is a simple linear function,

$$f(\text{population size}) = r\left(\frac{K - N}{K}\right) = r - \left(\frac{rN}{K}\right) \qquad (4)$$

we obtain the logistic equation modified for fishing losses:

$$\frac{dN}{dt} = rN\left(\frac{K - N}{K}\right) - qXN \qquad (5)$$

where
N = population size
t = time
r = per capita rate of population growth
K = asymptotic density (in absence of fishing)
q = catchability (a constant)
X = amount of fishing effort (so qX = fishing mortality rate)

The ecological assumptions of logistic models are that no time lags operate in the system, that age structure has no effect on the rate of population increase, and that catchability remains constant at all densities of fish. **Figure 2** illustrates how an exploited population is postulated to respond to a series of fishing episodes in this model. This model, although crude, may be useful for populations that are in approximately steady states in the absence of fishing and that do not change greatly from year to year. Because of their simplicity, logistic models can be used on fisheries with relatively few data available. The following example illustrates how this can be done.

The Peruvian anchovy (*Engraulis ringens*) is restricted in distribution to areas of upwelling of cool, nutrient-rich water along the coasts of Peru and northern Chile. The upwelling causes very high productivity in the coastal zone. The Peruvian anchovy is a short-lived fish, spawning first at about one year of age and rarely living beyond three years. It is a small fish, about 12 cm in length at one year and seldom reaching 20 cm in length. Young anchovies enter the fishery at only five months of age (8–10 cm). Anchovies occur in schools and are caught near the surface.

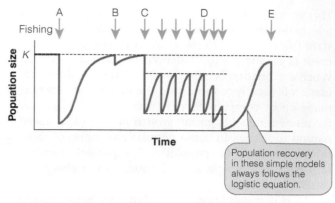

Figure 2 Schematic diagram of the assumed response of a fish population to exploitation according to logistic-type models. Periodic fishing of different intensity and frequency is indicated at the top of the diagram. At point A fishing is intensive, but the time interval is long enough that the fish population recovers to carrying capacity (K). At C a moderate intensity fishery is operating, and at D this fishing intensity is applied more frequently, causing the stock to collapse. At E excessive fishing drives the stock to extinction. Note that during every recovery phase the fish population increases logistically.

Population recovery in these simple models always follows the logistic equation.

The Peruvian anchovy fishery was the largest fishery in the world until 1972, when it collapsed. From 1955, when the major fishery first began, the anchovy catch doubled every year until 1961. In 1970, 12.3 million metric tons were harvested, and this single-species fishery constituted 18% of total global harvest of fish. **Figure 3** shows the total catch and the total fishing effort. These two parameters were used to fit a logistic model to the fishery (Boerema and Gulland 1973). The logistic model predicts a parabolic relationship between fishing effort and total catch, with an optimal catch at half the carrying capacity (K) for the simple logistic ($\theta = 1$). Anchovy are taken both by fishermen and by large colonies of seabirds, and these two were combined to measure the total "catch." Figure 3 indicates a maximum sustained yield between 10 million and 11 million metric tons, which, after subtraction of the birds' share, left about 9 to 10 million tons for the fishery. From 1964 to 1971 the catch was close to the supposed maximum indicated in the figure. Note that the estimate of maximum sustainable yield in Figure 3 assumes a population at equilibrium with average environmental conditions.

In 1972 average environmental conditions disappeared, and the Peruvian anchovy fishery collapsed. Early in 1972 the upwelling system off the coast of Peru weakened, and warm tropical water moved into the area. This phenomenon—known as "El Niño" (The Child) because it often happens around Christmas—occurs about every five years and greatly changes the regional ecosystem (Mysak 1986). The productivity of

---

[2]Also called *surplus yield models*, *stock production models*, or *Schaefer models*.

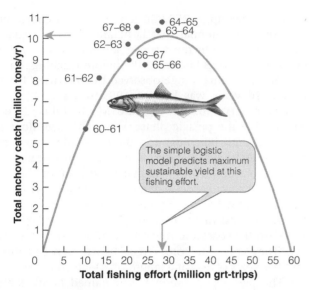

Figure 3 **Relation between total fishing effort and total catch for the Peruvian anchovy fishery, 1960–1968.** The effects of humans and seabirds are combined in these data. The parabola represents the logistic model fitted to these data, as in Equation (5). Arrow indicates maximum sustained yield and appropriate fishing effort. During the mid-1960s catches were near what was believed to be maximum sustained yield. (After Boerema and Gulland 1973.)

the sea drops, seabirds starve, and anchovies move south to cooler waters and may congregate. In early 1972 very few young fish were found; the spawning of 1971 had been very poor, only one-seventh of normal. Adult fish were highly concentrated in cooler waters in early 1972, and these concentrations produced large catches for the fishermen. By June 1972 the anchovy stocks had fallen to a low level, catches had declined drastically, and no young fish were entering the population. Seabird numbers fell from about 16 million birds to about 3 million (Jahncke et al. 2004).

The fishery was suspended to allow the stocks to recover, but from 1972 to 1985 there was little sign of a return of the anchovy to its former abundance. Catches fell to low levels and began to recover only during the 1990s, after 20 years of low catches (**Figure 4**). The economic consequences of the fishery collapse of 1972 were very great, and some of them might have been avoided if the fishery had been closed a few months earlier or if the fishing intensity had been slightly less than the maximum shown in Figure 3.

The Peruvian anchovy has become a model case of overfishing and has raised the important question of how to manage fisheries in a sustainable manner. The important message the collapse of the Peruvian anchovy fishery conveys is the fragility of the assumptions that fish populations are in a state of equilibrium and that average conditions never change.

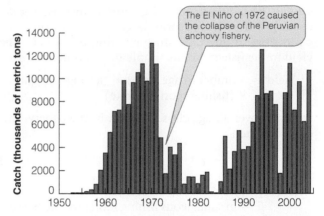

Figure 4 **Total catch for the Peruvian anchovy fishery, 1950–2005.** This fishery was the largest in the world until it collapsed in 1972 during an El Niño event. In spite of reduced fishing, it took 20 years for the fishery to recover. (Data from FAO FishStat.)

Dynamic Pool Models

Dynamic pool models of harvested populations are more biologically explicit because they include estimates of growth, recruitment, and mortality for the population being harvested. These models originated in a classic fisheries book by Beverton and Holt in 1957 and represented a biologically realistic approach to fisheries management that appealed strongly to fishery scientists. Ray Beverton and Sydney Holt revolutionized fishery science in the 10 years following World War II by applying mathematics to the problem of defining the optimum yield from a fishery. In the simplest form of these models, various assumptions are made. Natural mortality rate is assumed to be constant, independent of density, and the same for all ages of harvested fish. Growth rates are assumed to be age specific but unrelated to population density. Fishing mortality (effort) is assumed to act just like natural mortality—to be independent of density and constant for all ages of harvested fish. These assumptions are unrealistic, but they are useful as a starting point, and they can be relaxed later in the analysis. The object is to determine what yield a given level of fishing mortality will produce. In this simple model, the population size of R recruits after t years in the fished population is given by the formula for geometric decrease:

$$N_t = Re^{-(F+M)t} \tag{6}$$

where N_t = number of recruits alive at t years after entering fishery

   t = time in years since recruits

   R = number of original recruits

   F = instantaneous fishing mortality rate

   M = instantaneous natural mortality rate

This is the familiar curve of geometric increase (or decrease). If $R = 1$, this formula gives the fraction of recruits alive at any time since entering the fishery. The yield to the fishery in this simple model is defined as

Yield = (number in age class) × (average weight)
    × (fishing mortality rate)

summed over all age classes caught in the fishery. This can be written

$$Y = \sum_{t=t_c}^{\infty} FN_tW_t \qquad (7)$$

where Y = yield in weight for a year
    F = instantaneous fishing mortality rate per year
    N_t = population size of age t fish
    W_t = average weight of age t fish
    t_c = age at which fish enter the fishery

Let us illustrate this simple dynamic pool model with a classic example used by Beverton and Holt (1957)—the plaice (*Pleuronectes platessa*) fishery in the North Sea. The plaice is a shallow-water flatfish that is an important commercial species in the North Sea. Plaice spawn in midwinter when females are five to seven years old and males are four to six years old. Fe-

males can lay up to 350,000 fertile eggs, an enormous reproductive potential that is balanced by an equally high mortality. On the average, all but 10 fish out of every million eggs laid must die before reaching maturity, and the actual range observed by Beverton (1962) during 26 years was between 999,970 and 999,995 dying for every 1 million eggs laid. Much of this loss occurs during the pelagic phase, when the eggs float as plankton until hatching, and the larval plaice are carried about by water currents in the North Sea. After about two months the larval plaice settle out on nursery areas off the sandy coasts of the Netherlands, Denmark, and Germany. There the young plaice remain until between two and three years of age, when they begin to move off the coast and toward the middle of the North Sea. They enter the commercial fishery between three and five years of age, at a length of 20–30 cm.

The plaice population has remained fairly stable, with the exception of the periods during the world wars, when fishing was reduced and stocks increased. We can illustrate a dynamic pool model most easily in this type of near-equilibrium condition. First, we must determine growth rate with respect to age in the plaice, and we can do that with samples from the fishery (**Figure 5**). We

Figure 5 Dynamic pool model of equilibrium yield for plaice in the North Sea. The components of individual growth, recruitment of young fish, and survivorship of adults combine to produce the dynamic pool model of the fishery. The fishing intensity before World War II is indicated by the dashed line ($F = 0.73$) and was about twice the level at which maximum yield would be obtained. (After Beverton and Holt 1957.)

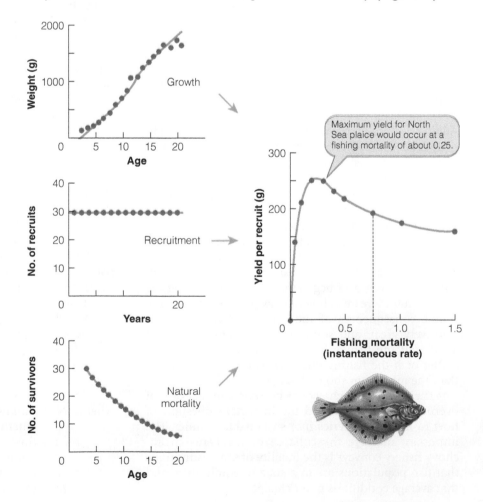

assume in this simple model that growth does not depend on population density. Second, we need to specify recruitment of young fish into the population, and we assume a constant number of recruits each year. For the plaice, this is not an unreasonable first approximation (see Figure 5). Third, we must determine the natural mortality rate of adults. We can do this by mark-and-recapture techniques or by indirect means. We assume that in the simple case natural mortality is constant at all ages and at all population densities. For the plaice, Beverton and Holt (1957) estimated the instantaneous mortality rate $M = 0.10$, and Figure 5 shows how a cohort of recruits would decline according to predictions of natural mortality *only*.

Because recruitment is assumed constant, we can express the yield as yield per recruit, and by combining the three factors we obtain the yield curve shown in Figure 5. An example of how this yield for plaice was calculated for a fishing mortality of 0.5 is given in **Table 1**. The yield per recruit was then calculated for several values of fishing mortality to obtain the curve in Figure 5. This is only an approximate calculation because we should use calculus instead of finite summation to find the yield (details in Beverton and Holt 1957). Figure 5 also shows the pre-World War

II fishing intensity ($F = 0.73$), which was clearly not at the point of optimum yield.

We have treated fishing mortality in the same way that we have treated natural mortality, using humans as just another "predator" in the system. This fishing mortality rate must be converted into fishing effort before the results of a yield analysis, such as that in Figure 5, can be applied to an operating fishery. This application is a complex problem that revolves around the types of equipment used in the fishery, the equipment's efficiency, the interactions between different units of equipment, the spatial and seasonal patterns of exploitation, and the area occupied by the stock (Winters and Wheeler 1985). Walters and Martell (2004) discuss these problems in some detail, and the analysis depends on the type of fishery operation.

Once we have built a dynamic pool model of a fishery, we can test it by regulating the fishery accordingly. Thus for the North Sea plaice we would predict from Figure 5 that an increased yield would result from lowering fishing mortality to one-third or one-half the prewar level of 0.73. This is the critical test of any model: Does it predict accurately? The North Sea plaice is one of the success stories of fisheries management.

Table 1 Calculation of equilibrium yield per recruit for North Sea plaice for a fishing mortality of 0.5.

Fishing year	Age at midpoint (yr)[a]	F = yield to fishery	W_t = average weight (g)	N_t = fraction of recruits surviving to this age, $e^{-(F+M)_t}$	Product $(F) \times (N_t) \times (W_t)$
0–1	4.2	0.5	158	0.741	58.54
1–2	5.2	0.5	237	0.407	48.23
2–3	6.2	0.5	331	0.223	36.91
3–4	7.2	0.5	435	0.122	26.54
4–5	8.2	0.5	546	0.067	18.29
5–6	9.2	0.5	664	0.037	12.28
6–7	10.2	0.5	784	0.020	7.84
7–8	11.2	0.5	904	0.011	4.97
8–9	12.2	0.5	1024	0.006	3.07
9–10	13.2	0.5	1143	0.003	1.71

Total yield per recruit 218.38g

[a]The age at recruitment is 3.7 years.

NOTE: The average weight is obtained from the growth curve shown in Figure 5.

The fraction of recruits is calculated by applying a constant loss per year of $(F + M)$, which in this example is $(0.5 + 0.1)$.

The total yield per recruit is obtained from the formula $Y = \sum_{t_c}^{\infty} FN_tW_t$.

After World War II, fishing effort was reduced to the level of maximum yield predicted by Beverton and Holt (see Figure 5), and the landings from the fishery nearly tripled over the next 30 years (Rijnsdorp and Millner 1996). By the 1990s fishing effort increased again, the stock declined, and yield fell 35%, as the model illustrated in the figure would predict. In 2006 fishing mortality on plaice in the North Sea was 0.58, and the European Commission ruled in 2007 that the stock was being fished unsustainably. It introduced a management plan to reduce fishing mortality to 0.3, a level that would sustain high yields (see Figure 5).

This approach can identify the annual equilibrium yield of the fishery, but within it are several potential problems. For one thing, it assumes that a constant number of recruits enter the usable stock every year. But does any fishery in fact have a constant recruitment? A constant recruitment implies that the number of recruits does not depend on population size; to put it another way, it assumes that two adult fish could produce the same number of progeny as 10,000 adults. This is on the face of it quite impossible, and thus we are led to inquire into the relationship between population size (stock) and recruitment.

Stock and Recruitment in Fish Populations

One of the keys to understanding the impact of fishing on populations of commercially valuable fish is to try to determine the **stock-recruit relationship**, which is just another way of discussing the problem of population regulation. Fish populations, even when exploited, are still subject to population regulation. Recruitment in exploited populations is always measured as a rate, such as the number of young fish entering a fishery per year. Recruits are defined differently in each fishery because of size or age limits set by fishery regulations, and consequently the recruitment rate will be a combination of fecundity and early juvenile mortality.

Some component of the vital statistics—births, deaths, or dispersal—must be related to population density in order to prevent unlimited population growth. Population growth cannot be curtailed unless the net reproduction curve is depressed below 1.0 at high population densities. This can occur if adult mortality increases with density, but fishery ecologists think that natural mortality of adult fish is independent of density. Fecundity does decline at high population density in some fishes (Myers and Barrowman 1996), but most of the regulation in fish populations is believed to occur in the early life-cycle stages. One of the axioms of modern fisheries ecology is that the important density-

dependent processes in fish occur during the first few weeks or months of life (Myers 2002).

Two points on the stock-recruitment curve are fixed. Where there is no stock, there is no recruitment. The point at which stock equals recruitment is an equilibrium point. Given these two fixed points, there is much room for different relationships in the shape of recruitment curves. Two general shapes may occur (**Figure 6**). Beverton and Holt (1957) suggested a curve that rises to an asymptote at very high stock densities. Maximum recruitment in the Beverton-Holt model always occurs at maximum stock size. Ricker first suggested in 1958 that the recruitment curve may peak below equilibrium density, so that there would be a maximum in recruitment at intermediate stock sizes (Ricker 1975; Hilborn and Walters 1992).

The Beverton-Holt recruitment curve (see Figure 6) assumes that there is no falloff in recruitment at high population levels and gives rise to a stable population size. The Ricker recruitment curve is closely related to the discrete generation analogs of the logistic equation in which population growth may show large oscillations about the carrying capacity (Hilborn and Walters 1992). Species that are short-lived are more likely to show Ricker-type recruitment curves, whereas long-lived species may show Beverton-Holt–type recruitment.

Recruitment in fish populations is highly variable from one year to the next, and most stock-recruitment

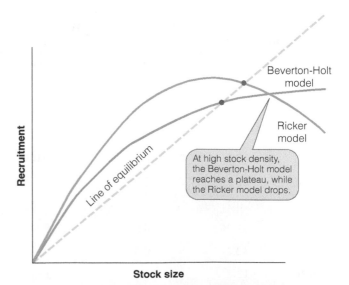

Figure 6 Two theoretical models of the relationship between stock and recruitment in an exploited fish population. The straight line represents all the conditions in which recruitment balances stock losses, and the point (dots) at which the recruitment curves cross this diagonal line represents an equilibrium point, which may be stable or unstable.

relations show great scatter. **Figure 7** illustrates this for sockeye salmon in the Wood River of Alaska. This variation is presumed to be caused by oceanographic effects on the survival of young fish. The variation seems to obscure any density effects and makes it difficult to fit the recruitment curves of Figure 6 to field data (Walters and Korman 1999). The variability in recruitment typically provides a coefficient of variation (standard deviation/mean recruitment) around 50% (Myers 2002). But some stocks have very high variability in recruitment. Salmonids in general show a coefficient of variation of about 65% but a few stocks have values over 200%. If the amount of recruitment is highly variable, a population being exploited may be susceptible to overfishing. The Peruvian anchovy (coefficient of variation = 69%) is a good example of this problem.

Variations in recruitment mask several general features of the stock-recruitment relationships illustrated in Figure 6, and some fishery managers have argued that recruitment is independent of stock size. Myers and Barrowman (1996) analyzed 364 time series of data like that shown in Figure 7 to ask if recruitment is related to stock size in general. To avoid variability problems they used ranking methods to analyze these relationships, and they found that in almost all cases the data trends were similar to those illustrated in Figure 6. Higher stock size produced higher recruitment, and lower stock size produced lower recruitment.

Their analysis showed clearly that fishery managers cannot assume that stock size is unimportant because fish lay so many eggs.

Why do some year-classes fail? This is one of the most important and difficult problems being addressed by fisheries ecologists. The critical period or **match/mismatch hypothesis** (Hjort 1914; Cushing 1990) postulates that early in the life of most fishes there is a short time period of maximum sensitivity to environmental factors. It is commonly assumed that oceanographic effects (particularly current patterns, winds, and water temperature) on food availability for newly hatched fry are critical in determining year-class size in fishes and that either temporal or spatial mismatching can produce recruitment failure (**Figure 8**).

One way to search for explanations of year-class failures in fish is to look for correlations between the abundance of food organisms such as plankton and the relative success of recruitment of young fish. Larval Atlantic cod (*Gadus morhua*) feed on copepods in the plankton, and Beaugrand et al. (2003) showed that cod recruitment in the North Sea could be related directly to the abundance of their food organisms (**Figure 9**). The changes in copepod abundance in turn are caused by changes in sea surface temperatures in the North Sea. Increasing sea temperature reduces the abundance of copepods, and at the same time higher temperatures increase the metabolic rate of the larval cod, and thus

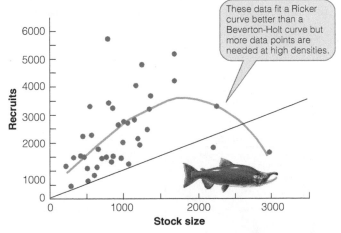

Figure 7 Stock-recruitment relationship for sockeye salmon (*Oncorhynchus nerka*) on the Wood River, Alaska, 1951–1995. Stock size is measured by the numbers of spawning salmon returning to the river in a particular year, and recruitment is the number of five-year-old fish returning to the river five years later plus the number of these caught in the fishery before they enter the river. The straight line represents equilibrium values at which stock = recruitment. (Data from R. A. Myers, Dalhousie University.)

These data fit a Ricker curve better than a Beverton-Holt curve but more data points are needed at high densities.

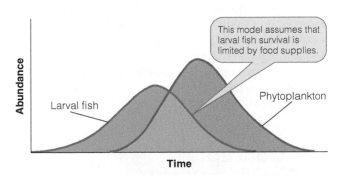

Figure 8 Schematic illustration of the critical period or match/mismatch hypothesis for recruitment in fish populations during one year. If fish spawning occurs too early (blue) there is inadequate food and the larval fish starve. If spawning occurs later (purple) and overlaps the spring phytoplankton bloom (red), there is adequate food, and larval fish survive well. The timing of fish spawning and phytoplankton increase determines year-class recruitment of juvenile fish. In a good year these two curves would overlap completely, and in a bad year they would hardly overlap at all.

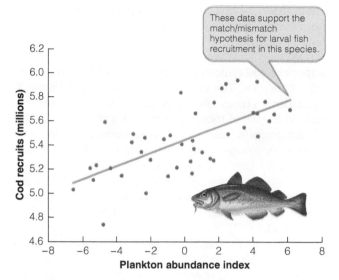

Figure 9 **Relationship between plankton food abundance and Atlantic cod recruitment as one-year-olds in the North Sea, 1959–1999.** The plankton population is an integrated measurement of the species eaten by larval cod and is standardized to be zero in an average year. (Data from Beaugrand et al. 2003.)

their energetic costs and food requirements. Recruitment is thus driven directly by the plankton community, as suggested by the match/mismatch hypothesis.

The details of how environmental factors affect recruitment are known for relatively few fishes, and slow progress is being made in evaluating whether the match/mismatch hypothesis is an adequate theory for many stocks. The key to testing this hypothesis lies in knowing the details of the spatial and temporal dynamics of the larval fish and their prey species, and then understanding the environmental factors that determine changes in dynamics. The challenge is important because if we can predict recruitment, we can adjust fishing pressure accordingly and avoid problems of overfishing.

The Concept of Optimum Yield

The concept of maximum sustained yield has been considered the "optimum" or "best" yield and has dominated fisheries management since the 1930s (Larkin 1977). In many situations, maximum yield is not a desirable goal. In sport fisheries, for example, the object is to maximize recreation, and the desirable fish are often the large ones. Hunters of large mammals may place more emphasis on the trophy status of the animals they harvest, and the harvesting of wildlife populations is often done without the goal of maximum sustained

yield. Forest harvesting in particular is difficult because it is often species-specific, and maximum yield from an area may include less desirable trees and also compromise biodiversity.

In any fishery or forest management area that must harvest several species at the same time, it is impossible to harvest at maximum sustained yield for all species. Multispecies fisheries are particularly difficult to manage effectively (Matsuda and Abrams 2006) and yet many tropical fisheries operate on complex multispecies assemblages. One species is often overharvested while another caught in the same nets is underharvested. Even within a single species, there are often subpopulations, or *stocks*, that have different resilience to harvesting. Harvesting of Pacific salmon operates on mixtures of stocks from different river systems and different spawning areas within one system. The result is that less productive salmon stocks are overfished, and even driven extinct, while more productive stocks are not fully utilized (Walters and Martell 2004).

In addition, any specification of optimum yield must include economic factors. The real yield from fisheries is not fish but dollars, and economists have long recognized that it is poor business to operate a fishery at maximum yield. H. Scott Gordon (1954) was one of the first to show that there is a level of harvesting associated with maximum sustained economic revenue, and that this usually occurs at a lower fishing intensity than the maximum sustained yield. What is optimal to an economist is not necessarily optimal to a biologist.

Figure 10 shows a simple economic model for a fishery. Total costs are assumed to be proportional to

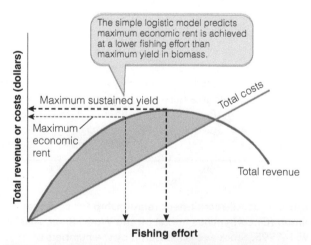

Figure 10 A simple economic model of a fishery in which costs are directly related to fishing effort, and revenue is directly related to yield. The shaded zone is the area in which revenue exceeds costs. Maximum economic rent is achieved when the difference between revenue and cost is maximal.

fishing effort. The revenue or benefit from fishing is assumed to be directly proportional to the yield. Thus the yield curve of Figure 1 is identical to the revenue curve of Figure 10 in this simple model. But the important point is that *maximum sustained yield*, the peak of the curve, is not at the same point as **maximum economic rent** (total revenue − total costs). The maximum economic profit will always occur at a lower fishing intensity than maximum yield. If this simple model prevailed, the economic management of fisheries would always be a safe biological management strategy. Alas, it is not always such. Scott Gordon (1954) showed that in an unmanaged fishery the only social equilibrium that will be reached occurs at the point where total costs equal total revenue, which is beyond the point of maximum sustained yield.

Clark (1990) has shown in an elegant analysis that under some situations it will pay fishermen to deplete the fishery to extinction, as might have happened to whales had the International Whaling Commission not intervened. The key economic idea in these cases is that of discounting future returns. If fishermen are given the choice of making $1000 today by overfishing, or $1500 in 10 years by delaying the harvest, most fishermen will take the money now and not wait. This type of exploitation makes perfect economic sense under our current economic theories, but it leads to overexploited populations and ecological disaster. Sustained yields can rarely be achieved without strong social or political controls on the allowable harvest.

Many but not all of the world's fisheries are overexploited beyond the limits of sustainability, and a historical perspective on fisheries management leads to pessimistic conclusions about the future unless management procedures are changed (Ludwig et al. 1993). The problem is that the concept of optimum yield is an equilibrium idea that works well when a harvestable population is stable over time. But if the harvestable resource fluctuates, a ratchet effect begins to operate (**Figure 11**). Estimates of maximum sustainable harvest rates are often too high, and if profit margins are good, additional investment in equipment is made, and the harvesting industry becomes increasingly susceptible to a sequence of poor years. Because of job losses, government will typically step in to subsidize the harvesting during the poor years, and this encourages even more overharvesting. The long-term result is a heavily subsidized industry that overharvests the resource until it completely collapses.

The problem with harvesting is typically that the management agencies are trying to maximize four objectives—biological, economic, social, and political. We have discussed mainly the biological objective, maximum sustained yield. Economic objectives are to maximize total economic gains from the fishery. Social objectives are to increase employment and income spread among the fish-

Figure 11 Ludwig's ratchet: For a fluctuating resource, continuing economic investment and ecological optimism fuel positive feedback that ratchets up the harvest rate to unsustainable levels and causes the eventual collapse of the fishery unless there are management agreements strong enough to be able to prevent this. (Modified from Ludwig et al. 1993.)

ermen, as well as to maintain traditional lifestyles. Political objectives are to minimize conflicts among the participants in the fishery (Hilborn 2007). Sometimes these objectives are in agreement but they are often in conflict. **Figure 12** shows in a schematic manner the relationships that flow from fishing effort and benefits.

The following three examples illustrate the difficulties of applying these harvesting theories and the concept of optimum yield to real-world fisheries.

Case Study: The King Crab Fishery of Bristol Bay, Alaska

The harvesting of king crabs in the North Pacific Ocean began commercially early in the twentieth century, when the Japanese began canning and exporting crabs to the United States. The Japanese gradually moved the crab fishery east and began harvesting in the eastern Bering Sea around 1930 using tangle nets dragged across the bottom. The king crab fishery was interrupted during World War II, and then both the Soviet Union and Japan took king crabs until the early 1970s, when the United States took over the king crab fishery. King crabs are now taken only in pot traps by U.S. fishermen.

The history of the king crab fishery in the eastern Bering Sea is shown in **Figure 13**. From initially low

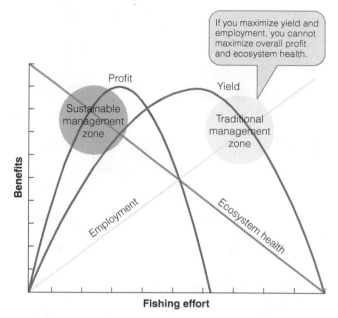

Figure 12 The relationship between fishing effort and the benefits in employment, yield, profit, and ecosystem preservation. Two zones are identified in which the objectives of management are not compatible. (Modified from Hilborn 2007.)

crab species. The snow crab fishery peaked in 1991 and has since collapsed to low numbers. Crab stocks are slowly rebuilding and the snow crab fishery in 2006 was about 10% of the peak catch of 1991.

The early life cycle of the king crab is complex. Eggs are brooded by females for 11 months, and individual females lay from tens of thousands to hundreds of thousands of eggs (Larkin et al. 1990). Larval king crabs go through four stages while swimming in coastal waters, and then transform into the adult form at a length of 2 mm. Mortality is very high in the larval stages. Growth occurs slowly (**Figure 14**), and individual females mature at about age 8 years and at a carapace length of 97 mm, while young males mature about age 9 to 10 at 105 mm (Loher et al. 2001). Young crabs aggregate in large pods of hundreds to thousands of individuals that move as a wave across the ocean floor. King crabs are predators and scavengers on a wide variety of invertebrates and fish. Their growth rate is affected by water temperature and can change with oceanographic conditions.

Why did the king crab fishery collapse in 1981? One assumption underlying the management of this fishery, which is unusual in several respects, has been the belief that by harvesting only males, the productivity of the population could be preserved. This assumption is true only if one mature male is able to service a large number of females. For behavioral reasons this assumption is now known to be faulty for king crabs. Fertilization occurs only after females have molted, and a male who has clasped a female may have to wait a few days for her to molt. After copulation the male may retain his clasp on the female. Females prefer to mate with larger males, and if a female cannot copulate within nine days of molting, the entire brood for that year is lost. Behavioral complications resulting from a male-only harvest may thus compromise the reproductive potential of the population, since the larger males are removed by the fishery.

catches the crab fishery reached a peak around 60 million pounds (9 million crabs) in 1964, and the catch was reduced over the next seven years while Japanese and Soviet boats were gradually eliminated from this fishery. The catch by U.S. fishermen grew rapidly during the 1970s to a peak catch of 130 million pounds (21 million crabs) in 1980. The king crab fishery collapsed in 1981–1982 and has recovered only very slightly since then. In 1994 and 1995 the major king crab fishery in Alaska was closed to all fishing, and fishermen have switched to snow crabs (*Chionoecetes opilio*) and other

Figure 13 Catch of a red king crab (*Paralithodes camtschatica*) from the Bristol Bay area of the Bering Sea, 1950–2005. The catch was taken mostly by Japanese and Soviet boats before 1969. The fishery collapsed in 1981, was closed in 1983 and again in 1994 and 1995, and has shown signs of recovery based on a strong 1990 year class coming into the fishery after 2002. (Data from Alaska Department of Fish and Game, Commercial Fisheries, 2007.)

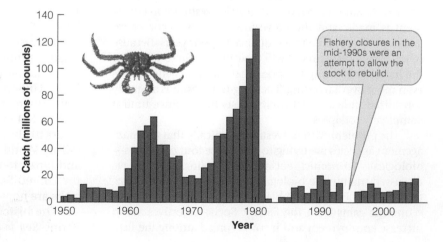

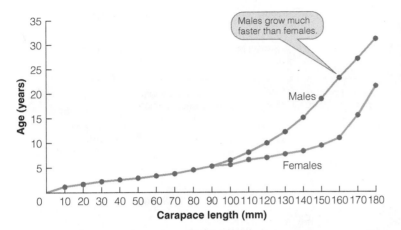

Figure 14 Average growth curves for male and female red king crabs (*Paralithodes camtschatica*). Sexual maturity is reached at age 5 years and 90 mm carapace length for females and 10 years and 120 mm for males. Only males over 136 mm in length are retained. (Data from Larkin et al. 1990.)

King crab growth and survival rates in the juvenile stages are also highly variable, perhaps due to changes in water temperature. During the 1970s the king crab fishery was relying on several very large cohorts from the 1960s, and these large year-classes of recruits provided a false sense of optimism in fishery managers and encouraged overinvestment in the fishery (see Figure 13).

Another factor affecting king crab abundance is the loss of immature crabs in pots (traps). For every legal-sized male crab (over 135 mm carapace length) taken, about seven immature males are caught and subsequently released. These immature crabs may be stressed or damaged during capture and may subsequently die. Other groundfish fisheries in the Bering Sea may capture king crabs incidentally or damage smaller crabs that are not retained in the nets. Finally, about 10% of the crab pots are lost each year due to storms and faulty ropes, and these lost pots continue to catch crabs, producing a hidden mortality on immature and mature crabs of both sexes.

In retrospect, the collapse of the king crab fishery was unavoidable because it built up too rapidly on the basis of a series of good year-classes, such that the maximum sustained yield was overestimated. The momentum associated with large capital investments in fishing boats makes it difficult to reduce rapidly the catch quotas when environmental conditions produce a series of poor year-classes. The compromise reached between socioeconomic realities and biological conservation measures almost always causes a further decline in stock abundance, and consequently an even longer time is required for recovery.

The **tragedy of the commons** was the term coined by Hardin (1968) to describe this type of exploitation. Whenever a resource is held in common by all the people, the best policy for each individual is to harvest as much of the resource as possible. There can be no incentive for individuals to stop harvesting at some optimum point because they can always make more money by overharvesting, and if you do not overharvest, your neighbor will. This tragedy of the overexploitation of common-property resources can be averted only by some form of management that restricts harvest, or by converting a common property resource to a private resource. Social control of harvesting is required for all large-scale fisheries, and for this reason good resource management is a creative mix of ecology, economics, and sociology. When applied properly, this mix can produce good fisheries management, as the next example shows.

Case Study: The Western Rock Lobster Fishery of Australia

Not all fisheries are basket cases and a contrast with the red king crab fishery is the highly successful western rock lobster fishery off Western Australia. It was awarded a Marine Stewardship Council certification in March 2000, the first fishery in the world to receive this award for sustainability.

Female western rock lobsters (*Panulirus cygnus*) reach sexual maturity at 6 to 7 years of age, and spawn in deep water in summer. The first larval stages are carried by oceanic currents for 9 to 11 months before the currents bring them onshore at almost 1 year of age in the last larval stage, about 25 mm long, called the puerulus stage. Juveniles feed and grow on shallow reefs for 4 to 5 years, and as they mature, they move to deeper water to spawn. Rock lobsters are caught near shore when they are 3 to 4 years old (about 500 g), before they have reached maturity (**Figure 15**).

The annual commercial catch of rock lobsters rose from very low numbers in 1944 to a peak around 11,000 tonnes by 1980, and has a commercial value of AU$200 million to AU$300 million (**Figure 16**). For the past 20 years the catch has fluctuated around this average. Lobsters are captured in pot traps, and part of the successful management of this fishery has

Oceanic phase   Nursery areas inshore

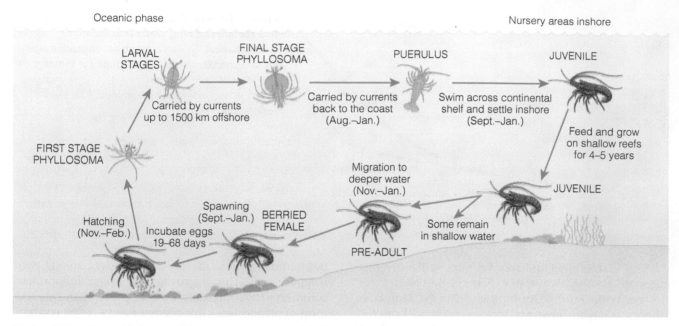

Figure 15 Life cycle of the western rock lobster off Western Australia. Eggs hatch offshore while juveniles feed and grow inshore. (Image courtesy of Bruce Phillips.)

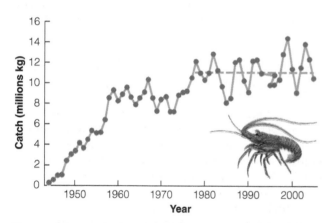

Figure 16 Trend in the catch of western rock lobsters, 1944–2005. Over the last 20 years the catch has been sustainable and has averaged about 11 million kg (red line), with fluctuations in the catch being caused by variation in survival and recruitment of juvenile lobsters. (Data from Caputi et al. 2003 and Australian Bureau of Agricultural and Resource Statistics.)

When the breeding stock declined in the early 1990s, stringent restrictions were applied to reverse this decline. These restrictions included a reduction of 18% in the number of pots to be used, and an increase in the minimum legal size of lobster that could be taken in the fishery. The western rock lobster fishery has the advantage of being able to monitor the abundance of the late larval stages each year. These small larvae settle inshore and must grow for 3 to 4 years before they reach catchable size for the fishery. By monitoring the abundance of these juveniles, managers can predict the allowable catch 4 years in the future, and set regulations accordingly. The allowable catch fluctuates from year to year because recruitment in the rock lobster is set by oceanographic conditions during larval life, and high water temperatures (>22°C) along with strong westerly winds increase larval growth rates and increase successful settlement on shallow reefs near shore. The fishing industry cooperates with management because with advance warning of good and poor years, they can adjust their effort and expectations of profit. Management decision rules have now been developed in consultation with the stakeholders in this fishery.

The success of this fishery is based on several factors. It is a limited-entry fishery in which each license holder is allowed to use a fixed number of pot traps, and this fishing effort can be adjusted annually to the recruitment success 3 to 4 years earlier. There is close governmental control of the industry, and because all the management is located in one state body, there are

centered on strong restrictions on fishing effort. The number of pot traps that can be used, and the number of days on which fishing is allowed are strongly controlled. The critical biological information that is needed for management is the size of the breeding stock, and this is estimated from the numbers of egg-carrying females captured per pot trap (**Figure 17**).

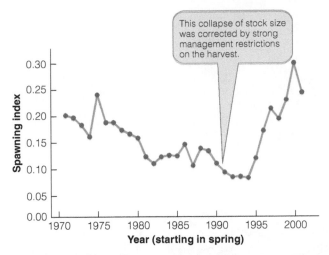

Figure 17 **Spawning stock index for the western rock lobster off Western Australia from 1971 to 2001.** This index is closely related to the number of reproductive females in the population. The drop in the spawning index in the early 1990s triggered management actions to reduce the catch and thereby reverse the population decline. The spawning stock index is a count of the number of eggs on females captured by commercial fishermen in one pot trap, relative to potential egg production per female. (Data from Caputi et al. 2003.)

no competing jurisdictions that have to be consulted and satisfied. All in all, it is a remarkable success story of a sustainable fishery.

Case Study: Antarctic Whaling

The exploitation of whale populations was the subject of vigorous and heated debate during the 1970s and 1980s. At the present time almost all commercial whaling has been stopped and most whales are protected. The large whales comprise ten species divided into two unequal groups. The sperm whale was the only toothed whale hunted commercially; the other nine species were all baleen whales, which have bony plates (baleen) in the roof of the mouth. Baleen whales are filter feeders whose principal food in the Antarctic is krill (shrimplike crustaceans) and other plankton.

The history of whaling is characterized by a progression from more valuable species to less attractive species as stocks of the original targets were reduced. Modern whaling dates from 1868, when a Norwegian, Svend Foyn, invented the harpoon gun and the explosive harpoon. In about 1905 whalers pushed south into the Antarctic and discovered large populations of blue whales and fin whales. Blue whales dominated the catches through the 1930s, but by 1955 few were being

taken (**Figure 18**). Attention was turned to the fin whale, originally the most abundant whale in the southern oceans. Fin whale numbers collapsed in the early 1960s. Sei whales, which were ignored as long as the bigger species were available, were not harvested until 1958. Sei whale catches were restricted after 1972 by the International Whaling Commission to prevent the collapse of these populations.

Harvesting models for whales have been developed extensively since 1961 (Baker and Clapham 2004). Simple logistic-type models have proved inadequate (**Figure 19**). Maximum sustained yield seems to occur at a density about 80% of equilibrium density, rather higher than the 50% predicted in the simple logistic model (Chapman 1981). Complications with these simple models are not difficult to find. Figure 19 assumes that all fin whales in the southern ocean belong to one population; it is now known that several subpopulations occur (Baker and Clapham 2004). Whales may interact, and most whale models are single-species models that do not recognize that many different species of whales and seals feed on krill in the Antarctic.

The current management of whales is directed to measuring the recovery rate of the depleted whale populations. Paradoxically, most of the data we have on whales came from whaling operations, and now that commercial whaling has stopped, additional research must be mounted to monitor how whale populations respond. Whale populations change slowly, and even 10 years is a short time to estimate accurately a population's response to protection from exploitation.

The principal food of the baleen whales is krill, a group of 85 species of shrimplike crustaceans that are on average about 6 cm long and weigh about 1–2 grams. Krill are so abundant in Antarctic waters that they were considered for potential harvest for many years before commercial harvesting began in the 1970s (**Figure 20**). Estimates of the sustainable harvest for krill are extremely large, nearly equal to the total production of all other fisheries on the planet (Ross and Quetin 1986). Commercial harvesting has been hampered by the remote location of the Antarctic and by processing problems once the krill are captured. Krill have powerful digestive enzymes that tend to spoil the catch by breaking down the edible tissues immediately after death. Krill also contain high amounts of fluoride, which must be removed before they can be used for human food. One of the emerging conservation problems of the southern oceans is to estimate the effect of krill harvesting on the many species of whales, seals, penguins, albatrosses, and fish that prey on krill (Reid et al. 2005). During the past 30 years krill populations in the South Atlantic region of the Antarctic have declined in direct relation to winter ice cover, which has been shrinking because of global warming (Atkinson et al.

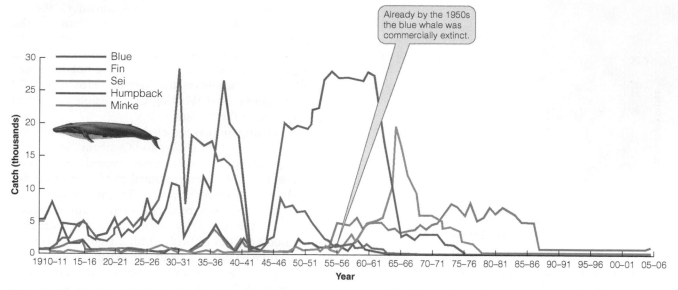

Figure 18 Catches of baleen whales in the Southern Hemisphere, 1904–2005. The usual lengths of whales in the commercial catches were: blue, 21–30 m; fin, 17–26 m; sei, 14–16 m; humpback, 11–15 m; and minke, 7–10 m. The blue whale is illustrated. As each species was overharvested, whalers moved to catching the next largest species. (Data from FAO Fishery Statistics and Allen 1980.)

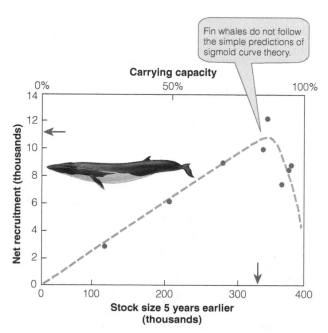

Figure 19 Sigmoid-curve-type model for Antarctic fin whales. Estimated stock size and yield at maximum sustained yield are indicated by arrows. The sigmoid curve model would predict maximum yield at 50% of carrying capacity (about 200,000 whales), but these data suggest maximum yield around 80%–85% of carrying capacity, indicated by the red arrow on the x-axis (approx. 330,000 whales).

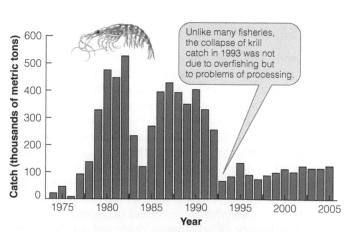

Figure 20 Krill commercial production in the Antarctic, 1974–2005. The krill fishery could become the largest fishery in the world, with an estimated potential harvest of 150 million tons a year. The biomass of krill in the Antarctic may be the largest of any animal species on the planet. (Data from FAO, Rome 2007.)

Principles of Effective Resource Management

Even though renewable resource management has historically failed, as many examples of fish harvesting will testify, we are currently committed to the general principle of sustainable use of resources. How can we do a better job in the future? According to Ludwig, Hilborn, and Walters (1993), five principles should underlie good resource management:

1. *Include humans as part of the system*. Human motivation, shortsightedness, and greed can underlie many of the problems of resource management. Instead of thinking of humans managing resources, we should think of resources managing human behavior, often with a short time frame.

2. *Act before scientific consensus is achieved*. For many management problems we do not need additional research to decide on management policies. Examples would include pollution impacts in the Great Lakes, tree harvesting on slopes subject to erosion, and harvesting of undersized fish. Calls for additional research on many topics are often just delaying tactics.

3. *Rely on scientists to recognize problems, but not to remedy them*. Good science is important for resource management, but it is not enough. The management of *human* activities is what is essential, and this is a sociological, psychological, and political problem.

4. *Distrust claims of sustainable resource use*. Because we have failed in the past to harvest sustainably, any new plan that represents itself as sustainable should be suspect and subject to detailed scrutiny. The linkage between basic research on fish populations and sustainable fisheries policies is a loose one, and good basic research does not automatically lead to better management.

5. *Confront uncertainty*. We often operate under the illusion that if we do enough research with enough funding we will be able to identify a solution to harvesting problems. But the large levels of natural variation found in most populations preclude any exact predictions about future dynamics. We need to favor management actions that are strong in the face of uncertainty and that are reversible if found to be damaging.

"Sustainable development" is the buzzword of the moment, and we must not pretend that scientific or technological advances alone will be sufficient to solve resource management problems. These are human problems that we have created many times in the past and under many types of political systems, and they will not necessarily be solved by more scientific data.

2004). Krill stocks in other parts of the Southern Ocean have been stable and some are increasing, but the major stocks (>50% of the total) off the Antarctic Peninsula have been in decline since the 1970s. The future development of a krill fishery in the Southern Ocean needs to be based on precautionary management principles.

Risk-Aversive Management Strategies

Because of the many failures of resource management in the past, resource managers have begun to search for management strategies that are designed to minimize risk. Two general approaches have been suggested. The first is to redirect management toward a harvest strategy that does not simply seek to achieve the maximum sustained yield (Hilborn and Walters 1992; Hilborn 2007).

In this approach we try to find a harvesting strategy that maximizes average yield over a long time period during which the population fluctuates naturally. At the same time that we try to maximize long-term yields, we need to minimize the risk of resource collapse or extinction. Two popular strategies for risk-aversive harvesting are (1) to impose a constant percentage harvest on a population, or (2) to harvest all individuals above a threshold population size, and no individuals below that threshold. In both cases there is a threshold or "escapement" level below which harvesting stops.[3] The problem with many fisheries is that this threshold is set too low to sustain the resource, and often the management authorities do not know very accurately where the population is with respect to the threshold. The problem with complete threshold harvesting is that in years

---

[3]Escapement is the unharvested portion of the population and may be defined as a percentage or as an absolute number.

What Are the Harvest Strategies for a Fishery?

The management of any harvested resource requires a clearly specified harvest strategy that is understood by biologists, managers, and harvesters alike. For a fishery, a harvest strategy is a plan stating how the catch will be adjusted from year to year depending on the size of the stock, the economic condition of the fishery, the condition of other stocks, and the uncertainty regarding our biological knowledge of the stock. Fisheries harvest strategies should be quantitative and explicit, not vague statements and wishes, and they need to be developed with the active participation of the fisherman and the industry that will be affected.

Three basic harvest strategies can be applied to a fishery (see **Figure 21**):

1. *Constant quota.* This strategy specifies a fixed catch or constant quota that does not change with stock size (blue area). If the stock is large, the catch will be a small fraction of the population; if the stock is small, the entire stock may be taken.

2. *Constant exploitation rate.* For this strategy the catch does not rise one-to-one with the stock, but instead at some lower rate (b). For example, the catch might rise 0.5 times the stock rise. The escapement thus increases as the stock grows larger. A variant of this strategy uses a threshold or lower limit point so that the fishery is closed at low stock abundance.

3. *Constant escapement.* This strategy implies that the catch will rise one-to-one with the stock size and that the stock size or escapement will not vary. This strategy has an implicit threshold so harvest begins only when the size of the stock exceeds this threshold (red arrow).

Different fisheries are managed with different harvest strategies. From a theoretical perspective, the optimal harvest strategy is usually a constant escapement strategy (Figure 21c) (Hilborn and Walters 1992). But there is a trade-off, because if we wish to maximize average yield to the fishery, we also maximize the variation in yields from year to year. Fishermen often prefer less variance in catch, so an optimal strategy from a human perspective might be to have a less-than-maximum yield with reduced variability from year to year.

Other types of harvesting strategies can be applied. In some cases it is more efficient to use a periodic harvest in which the fishery is not operated every year. This can be a useful strategy when it is more economically advanta-

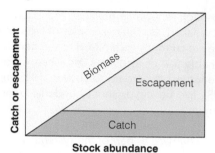

(a) Constant quota

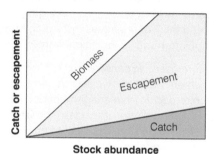

(b) Constant exploitation rate

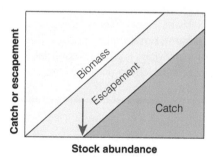

(c) Constant escapement

Figure 21 Three basic harvest strategies that can be applied to any exploited resource population. The red arrow indicates a threshold for the constant escapement model, and below this threshold no harvest is allowed. (Modified from Hilborn and Walters 1992.)

geous to take a large catch every few years than a smaller catch every year, or when older animals are much more valuable than younger animals. Some clam and abalone fisheries operate this way, as do most aquaculture industries. The most important message is the need to tailor the harvesting strategy to the particular resource being exploited by means of full consultation between industry workers and management biologists.

when the population is below the threshold there will be no harvest, with the attendant economic dislocations. For this reason Hilborn and Walters (1992) prefer a constant harvest-rate strategy.

The second general approach is to impose protected areas, or "no-take" zones, on the resource. This is a bet-hedging strategy in which we incur the cost of reducing the catch for the benefit of a reduced risk of catastrophic collapse of the fishery. This strategy has been discussed particularly for marine fishes (Roberts et al. 2005). The idea of a protected area in the aquatic realm is equivalent to the idea of national parks on land and is most useful for demersal fish that inhabit large areas of the ocean floor and are nonmigratory. The idea is simple: set aside a large enough "no-take" zone to ensure that the stock will remain at greater than 60% of carrying capacity over a given time horizon (for example, 20 years). Fishermen could harvest at a specified rate outside the "no-take" zone but would not be permitted to fish inside this protected area. The details of how to achieve these simple goals must be worked out for each resource, and the detailed trade-off of costs and benefits must be identified if this strategy is to obtain practical support among fishermen.

Marine protected areas or reserves are a recent concept in fisheries management and there is already enough data that has accumulated to suggest that they work very effectively to increase the biomass of aquatic organisms in adjacent areas. By measuring the density of marine organisms inside and outside of marine reserves of varying size, marine ecologists have shown that marine reserves do indeed work to increase local density of marine organisms (**Figure 22**). Density in-

side the reserves was up to three times that outside the reserve. Surprisingly, this percentage increase in fish and invertebrate density was only slightly affected by the area of the reserve. But clearly larger reserves are much better in a quantitative sense because increasing a fish population from 10 to 20 in a small reserve is less of an impact than increasing the population from 1000 to 2000 in a large reserve.

A protected area strategy for marine fishes is clearly an important means of preventing overfishing. Two things are essential for such a protected strategy—first, the area must indeed be protected from poaching, and second, long-term studies need to be implemented to study the population changes that may not be apparent in a short-term analysis of the success of marine reserves. The idea of marine reserves is an important new strategy for trying to prevent the kinds of disasters we have seen in the king crab fishery and in the history of whaling during the last century.

Harvesting and Natural Selection

Harvesting for sport has always concentrated on catching the largest fish or hunting for the largest grizzly bear. Wildlife and fishery ecologists have been concerned that harvesting can be genetically selective, and fishery scientists in particular have been apprehensive that sustained fisheries may cause undesirable evolutionary trends. Both sport and commercial fisheries typically select in favor of larger individuals, and one possibility is that this selection by the fishery removes the most fit individuals and leaves behind individuals with slow growth rates and a reduced size at sexual maturity. Fishery scientists have been reluctant to consider evolutionary changes in harvested stocks because there has been no hard proof of genetic changes in population productivity. Recent laboratory studies demonstrate that such genetic changes can occur rapidly, and inject a cautionary warning for fishery and wildlife management agencies.

The Atlantic silverside, *Menidia menidia*, is a common marine fish along the east coast of North America. Silverside have one generation per year, so they are suitable for laboratory selection experiments. To test for genetic changes associated with harvesting, fishery ecologists raised six experimental populations of silverside in large tanks and subjected them to three treatments: large fish harvest, small fish harvest, and random size harvest. In each case they removed 90% of the fish in an artificial fishery and studied the offspring of each group to measure the possible impact of Darwinian selection for size. **Figure 23** shows the results after four generations of laboratory selection. There was a

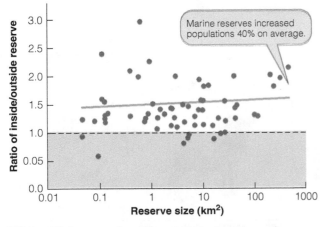

Figure 22 Impact of marine protected areas on the density of marine invertebrates and fishes. The ratio of density inside the reserve to that outside the reserve should be 1.0 if there is no effect. The zone below a ratio of 1.0 is shaded pink. Most reserves of all sizes showed large increases in density in the protected areas. Only eight reserves (red dots) failed to gain from protection. (Data from Halpern 2003.)

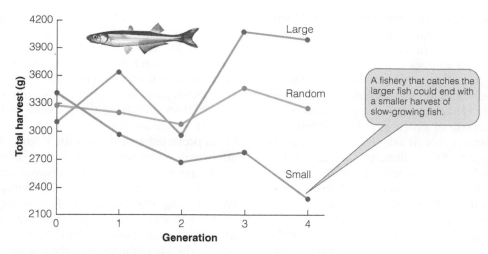

Figure 23 Size-selective fishing applied to Atlantic silverside for four generations in the laboratory. Small populations had 90% of the largest fish removed by the fishery each generation (= select for small fish and select against large fish), and large populations had 90% of the smallest fish removed. Random populations had random removal of 90% of the fish. The biomass harvested by generation 4 was 75% higher in the large-selected fish because they were growing faster. Growth rates are partly an inherited trait in many fish populations, and most marine fisheries operate to remove the largest fish, in effect selecting genetically for small or slow-growing individuals. (After Conover and Munch 2002.)

rapid divergence in size, with the large and small stocks differing almost twofold both in weight and yield to the laboratory fishery.

The importance of these considerations is that management plans for fisheries that aim at maximum sustained yield in the short run may in fact produce the opposite effect of reducing yields in the long run because of Darwinian selection for size. How could we reduce these unwanted side effects of fishing? One possible way is to establish marine reserves where there is no fishing and thus no selection for size. A second way would be to set a maximum limit on size, instead of the usual minimum size limit, so that large fish would be protected. Harvesting of fish, wildlife, and timber should adopt a Darwinian perspective on the possible consequences of natural selection operating through the harvesting process.

Hunting of trophy animals is another possible mechanism of harvesting selecting for traits that are in the long-term detrimental to the populations. Hunters are willing to pay large amounts of money to hunt trophy ungulates in many parts of the world. Bighorn rams (*Ovis canadensis*) in North America are hunted for the size of their horns, and this has resulted in one Alberta population showing a steady trend toward males of smaller horn size (**Figure 24**). But not all ungulate populations that are hunted show this trend. Loehr et al. (2007) show that a population of Dall sheep (*Ovis dalli*) in the Yukon show no systematic trend of horn size in hunted populations.

Dall sheep appear to live by the code of grow-fast-and-die-young in both hunted and unhunted populations. The key point is that selective harvesting may not have genetic consequences in all populations, and much depends on the level of harvest selection.

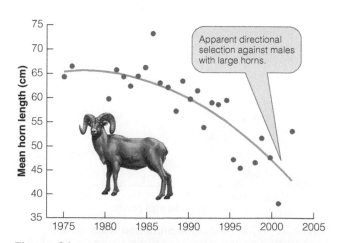

Figure 24 Horn length of four-year-old bighorn sheep at Ram Mountain Alberta, 1975–2005. A sample of 133 rams were measured in the field. This decline in horn size in males that mature at age 4 could be a consequence of hunting mortality selecting against fast-growing males with large horns. (Data from Coltman et al. 2003.)

Summary

To harvest a population in an optimal way, we must understand the factors that regulate the abundance of that population. That humans so frequently mismanage exploited populations like whales is partly a measure of our ignorance of population dynamics. When humans harvest a population, its abundance must decline, and the losses caused by harvesting are compensated for by increased growth, increased reproduction, or a reduced natural mortality. Harvested populations lose the older and larger individuals and often respond by a reduction in the age at sexual maturity. Species vary greatly in the amount of harvesting they can withstand.

Maximum sustained yield is often the goal of resource managers. Simple and complex models alike have been developed to estimate the maximum sustained yield for fisheries and forestry. Most models contain the hidden assumption that the environment remains constant, and for this reason they often fail in practice to prevent overexploitation and collapse. Economics and politics add further difficulties to achieving maximum sustained yield for valuable populations. Harvesting is subject to a ratchet effect in which the exploitation rate is pushed by ecological optimism and economic and social pressures toward overexploitation and collapse.

Management of forestry, fishery, and wildlife resources is at present based more on political and economic pressures than on scientific knowledge and forecasting. Because of the inherent ecological uncertainty in anticipating future changes in populations, resource management must adopt more risk-aversive strategies. The imposition of a protected-area strategy or "no-take" zones is one approach that may work for some fisheries.

Harvesting may be genetically selective on fast-growing individuals. Resource managers need to reconsider the common assumption that harvesting does not create undesirable long-term evolutionary paths that compromise sustainability. One of the great challenges of modern ecology is to help place resource management on a sustainable basis. We can all be very good at managing yesterday's populations. When will we be equally adept at managing tomorrow's?

Review Questions and Problems

1. Many Web sites that advise consumers on the fish species that they should eat (because they are harvested in a sustainable manner) recommend against buying orange roughy (*Hoplostethus atlanticus*). Hilborn et al. (2006) challenged this interpretation and showed that orange roughy were being harvested very near to maximum sustained yield and were not being overfished. Discuss this controversy and explain the principles you would use to define a stock that was overharvested.

2. Forest resources are another major natural resource subject to harvesting regimes. Are forest resources in your area being harvested in a sustainable manner? How could you determine if this were true or not? Which of the admonitions given in this chapter for fisheries would apply equally well to forest harvesting?

3. Suppose that in fact no relationship exists between stock and recruitment in the sockeye salmon (see Figure 7). Discuss the implications of this with respect to the various theories of population control.

4. Discuss the circumstances under which the constant quota model shown in Figure 21 could be a risky strategy for a fishery.

5. Fish communities consist of many different species, only a few of which are typically the focus of commercial or recreational fishing. One way to manage such fisheries is to use standard criteria for maximum sustained yield on the single species of interest, and to ignore the other species in the community. Discuss how this simplified approach might have disastrous consequences for the other species in the ecosystem. Walters et al. (2005) discuss this issue.

6. Examine the catch statistics for a fishery in your area or in an area of interest to you. Sources of data on the Web might be the *Fisheries Statistics of the United States*, *Fisheries Statistics of Canada*, or the *Food and Agricultural Organization*'s Web site. If the fishery you choose has been managed, is there any evidence of overfishing?

7. Black duck populations have been declining in North America since the mid-1950s. One hypothesis for this decline is that it is due to overhunting. Review the evidence for and against this hypothesis, and discuss the possible management actions that might help to stop this decline. Conroy et al. (2002) provide an overview of this management problem.

8. One of the assumptions of maximum sustained yield models is that birth, death, and growth responses to

population density are repeatable, such that a given population density will always be characterized by the same vital statistics. What mechanisms may make this assumption false?

9 The Peruvian anchovy fishery (Figure 4) is still among the largest fisheries in the world. What happens to this large biomass of fish once it is caught?

10 Ludwig and Walters (1985) showed in a computer simulation that the management of a hypothetical

fishery could be done better using simple yield models like the logistic equation than by using more realistic, detailed models like dynamic pool models. Discuss why this might be correct for a real fishery.

Overview Question

Suppose that you are in charge of a newly established fishery. Discuss the criteria you might use to detect when the population is being overexploited, and outline the relative merits of the different criteria.

Suggested Readings

- Caddy, J. F. 1999. Fisheries management in the twenty-first century: Will new paradigms apply? *Reviews in Fish Biology and Fisheries* 9:1–43.

- Coltman, D. W., P. O'Donoghue, J. T. Jorgenson, J. T. Hogg, C. Strobeck, and M. Festa-Bianchet. 2003. Undesirable evolutionary consequences of trophy hunting. *Nature* 426:655–658.

- Hilborn, R. 2007. Moving to sustainability by learning from successful fisheries. *Ambio* 36:1–9.

- Kuparinen, A., and J. Merila. 2007. Detecting and managing fisheries-induced evolution. *Trends in Ecology & Evolution* 22:652–659.

- Ludwig, D., R. Hilborn, and C. Walters. 1993. Uncertainty, resource exploitation, and conservation: Lessons from history. *Science* 260:17, 36.

- Morato, T., R. Watson, T. J. Pitcher, and D. Pauly. 2006. Fishing down the deep. *Fish & Fisheries* 7:24–34.

- Myers, R. A. 2002. Recruitment: Understanding density-dependence in fish populations. In *Handbook of Fish Biology and Fisheries*, ed. P. J. B. Hart and J. D. Reynolds, 123–148. Oxford: Blackwell Publishing.

- Pauly, D., R. Watson, and J. Alder. 2005. Global trends in world fisheries: impacts on marine ecosystems and food security. *Philosophical Transactions of the Royal Society of London, Series B* 360:5–12.

- Platt, T., C. Fuentes-Yaco, and K. T. Frank. 2003. Spring algal bloom and larval fish survival. *Nature* 423:398–399.

- Roberts, C. M., J. P. Hawkins, and F. Gell. 2005. The role of marine reserves in achieving sustainable fisheries. *Philosophical Transactions of the Royal Society of London, Series B* 360:123–132.

- Walters, C. J., and S. J. D. Martell. 2004. *Fisheries Ecology and Management*. Princeton, NJ: Princeton University Press.

Credits

Illustration and Table Credits

Photo Credits

Applied Problems II: Pest Control

Key Concepts

- Pest control is applied population ecology that asks what limits the average density of a pest species, and what we can do to change average density. Most pest control utilizes chemical poisons.

- Pest control is the reduction of damage caused by a particular pest below an economic threshold; consequently it involves information on the ecology of the pest and the economics of the damage.

- Classical biological control involves the introduction of nonnative predators, parasitoids, herbivores, or diseases to reduce the population density of the pest. In some cases it is spectacularly successful, in other cases it is less so.

- Genetic control involves either changing the genetic makeup of the host species to make it more resistant to pest attack, or changing the pest to make it sterile.

- Mating disruption uses the release of synthetic pheromones to prevent males from finding females for mating through natural pheromone attraction.

- Integrated pest management is the coordination of chemical control, biological control, and cultural control in an overall strategy to reduce pest numbers. Biotechnology is a key component of integrated control.

- Predators introduced for biological control may themselves become pests of native species, and careful evaluation must precede introductions of alien species.

From Chapter 16 of *Ecology: The Experimental Analysis of Distribution and Abundance*, Sixth Edition. Eugene Hecht.
Copyright © 2009 by Pearson Education, Inc. Published by Pearson Benjamin Cummings.

KEY TERMS

immunocontraception The use of genetic engineering to insert genes that stimulate the immune system of a vertebrate to reject sperm or eggs, thus causing infertility.

integrated pest management (IPM) The use of all techniques of control in an optimal mix to minimize pesticide use and maximize natural controls of pest numbers.

parasitoid An insect that completes larval development in another insect host.

pesticide Any chemical that kills a plant or animal pest.

push-pull strategies Management strategies that manipulate the behavior of insect pests to make the crop resource unattractive (push) and lure the pests toward an attractive source (pull) where the pests are destroyed.

resource concentration hypothesis The idea that agricultural pests are able to cause serious damage because crops are planted as monocultures at high densities.

sterile-insect technique The release of large numbers of sterilized males to mate with wild females and prevent the fertilization of eggs and production of viable young.

Some species interfere with human activities, in which case they are assigned the label "pests." Many of the most damaging pests we have are introduced species. The first response to pests is to control them, which in this context means to control damage and not necessarily to regulate the pest population around some reduced equilibrium density. One of the obvious ways of controlling damage is to reduce the average abundance of the pest species, but there are other ways of reducing damage by pests without affecting abundance (such as by the use of insect repellents).

A population is defined as being controlled when it is not causing excessive economic damage, and as being uncontrolled when it is. The boundary between these two states will depend on the particular pest. An insect that destroys 4%–5% of an apple crop may be insignificant biologically but may destroy the grower's profit margin. Conversely, forest insect pests may defoliate whole areas of forest without bankrupting the lumbering industry. To all questions of pest control, we must apply the concept of economic thresholds, including the cost of the damage caused by the pest, the costs of control measures, the profit to be gained with the crop, and interactions with other pests and their associated costs.

Pest control in most agricultural systems is achieved by the use of toxic chemicals, or **pesticides**. An esti-

mated 2.5 billion kilograms (nearly 5 billion pounds) of toxic chemicals are being used annually worldwide to control plant and animal pests (Pimentel et al. 1992). Despite the use of these pesticides, about 48% of the world's crops are lost to pests before and after harvesting, and despite increasing pesticide use in the last 60 years, crop losses have not been reduced (Oerke 2006). **Figure 1** gives an overview of the current global yields and the losses to pests that reduce the attainable yield to the actual yield. If pesticides were not used, the actual yield would be reduced by about 30% on average. Pesticide use has increased dramatically during the last 40 years, but in spite of this the losses of crops to pests has not declined (Oerke 2006). In essence there is a kind of Red Queen effect here in the coevolution of pesticide use and pest damage to crops.

Pesticides are essential to modern agriculture, but they have undesirable side effects and are thus only a short-term solution to the problem of pest control for several reasons. First, toxic chemicals have strong effects on many species other than pests. Rachel Carson gained fame as the first naturalist to point out to the public at large the ecological consequences of toxic chemicals. The well-known effects of DDT on bird populations,

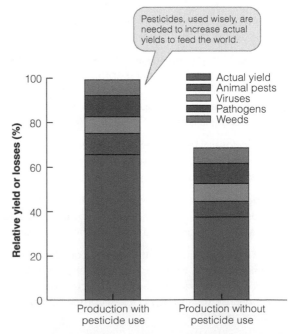

Figure 1 The average relative yield for all crops worldwide for 2005 and the estimated losses to pests. A relative yield of 100% indicates the attainable yield in field conditions in the absence of crop pests. If no pesticides were used in modern agriculture, the actual yield would be reduced about 30% below existing yields. These averages will vary with specific crops and local growing conditions. (Data from Oerke 2006.)

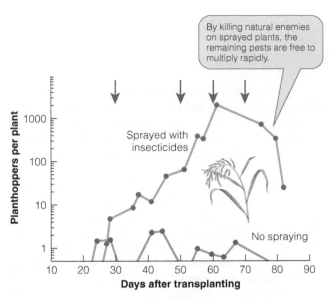

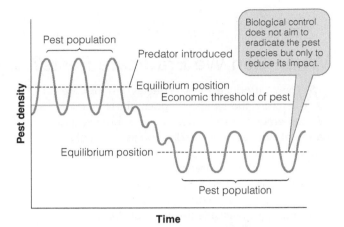

Figure 2 Rice crops in the Philippines contain more damaging insect pests on plots with insecticide spraying than on control plots in which no spraying was done. Broad-scale spraying of toxic chemicals (red arrows) kills many of the insect predators that limit the abundance of the critical rice pests, so that insect pest abundance and damage to the crop is worse after spraying. These data are counts of brown planthoppers (*Nilaparvata lugens*) on individual rice plants in unsprayed paddies (red) and sprayed paddies (blue). (After Way and Heong 1994.)

Figure 3 Classical biological control in which the average abundance of an insect pest is reduced after the introduction of a predator or disease agent. The economic threshold is determined by human activities, and its position is not changed by biological control programs. (After van den Bosch et al. 1982.)

which Carson highlighted in *Silent Spring* (1962), is a good example of how pesticides can degrade environmental quality. Second, many pest species are becoming genetically resistant to toxic chemicals that formerly killed them. Insects that attack cotton have evolved resistance to so many pesticides that it is no longer economically possible to grow cotton in parts of Central America, Mexico, and southern Texas. Third, and perhaps most surprisingly, the use of toxic chemicals in some situations can actually produce a pest problem where none previously existed. **Figure 2** illustrates how rice plants become infested by massive outbreaks of brown planthoppers when sprayed with pesticides. Toxic chemicals such as DDT or endrin destroy many insect parasitoids and predators that cause mortality in the pest species, and after treatment the few pest individuals that survive can multiply without limitation.

How can we achieve pest control without these problems? There are four primary strategies for dealing with pests:

1. **Natural control.** Pest populations are exposed to naturally occurring predators, parasites, diseases, competitors, and weather, and these factors can reduce their densities.

2. **Pesticide suppression.** Pest populations are treated with herbicides, fungicides, insecticides, or other chemical poisons to reduce their abundance.

3. **Cultural control.** Pests are reduced by agricultural manipulations involving crop rotation, strip cropping, burning of crop residues, staggered plantings, or other agricultural practices.

4. **Biological control.** Pests are reduced by introductions of predators, parasites, or diseases; by genetic manipulations of crops or pests; by sterilization of pests; or by mating disruption using pheromones.

Integrated control, or **integrated pest management (IPM)**, is the use of all four of these strategies; the goal is to reduce pest damage by minimizing pesticide use and maximizing natural control.

In this chapter we discuss the principles used in biological control and cultural control and relate them to ecological theory. Biological and cultural controls aim to reduce the average density of a pest population (**Figure 3**), and may be viewed as a practical application of the problem of what determines average abundance of organisms. The aim of pest management is not to eradicate the pest, which is usually impossible, but to reduce its effects to an acceptable level.

Examples of Biological Control

In this section we consider three successful examples of biological control from which general principles might be gained. The typical situation in which biological

When Can We Eradicate Pests?

When introduced pest species are discovered in a country, the typical cry goes out to eradicate them. We can eradicate some pest species, but we ought to be careful about declaring eradication as a goal of any control program. Eradication implies the removal of all individuals of a species from an area to which reintroduction will not occur. If a species has spread over a wide area, it is unlikely that eradication is going to be possible, no matter how much money is available (Myers et al. 2000).

Six Factors of a Successful Eradication Program

1. Sufficient resources to complete the project

2. Clear lines of authority for decision making during the eradication work

3. A target species that can be eradicated because it is easy to find and kill

4. Effective means to prevent reintroduction

5. Easy detection of the species when it is scarce

6. Plans for restoration management if the species has become dominant in the community (lest one pest be replaced by another bad pest)

At present, few countries except New Zealand and Australia have operational plans for dealing with new pests for which eradication is a possibility.

Rats introduced to islands have become serious pests of native wildlife species, especially seabirds, and some of the most successful eradication programs have been applied to rats on islands. Ninety New Zealand islands ranging in size from 1 to 11,300 ha have been cleared of Pacific rats, Norway rats, and black rats originally introduced by shipwrecks (Towns and Broome 2003). Poisoning with anticoagulant rodenticide baits has been the major technique used in these eradication programs, and poison baits were put out by hand or on larger islands by helicopter. The costs of these eradication programs for rats have declined to about $100 to $200 per ha in 2004, and for 10 islands an added conservation benefit has been the eradication of feral cats along with the rats.

Larger animals such as goats can be readily detected and eliminated on islands, but eradication becomes more expensive and difficult on larger islands with complex vegetation. Feral goats have been eliminated from 22 islands off New Zealand by shooting, but eradicating large animals on mainland areas is far more difficult because of immigration (Forsyth et al. 2003).

No one had expected that significant pests like rats could be eradicated on small or large islands just 10 years ago, and these success stories have encouraged scientists to work toward eradication of some pests, particularly in island situations. But while eradication is the holy grail of pest control, it will be possible only in a small number of situations, and we should not expect that pest control problems will be sorted out and solved so easily.

control is applied unfolds as follows: A pest, often an introduced species, is causing heavy damage. Efforts are then made to find predators and **parasitoids** from the pest's native habitat that can be introduced into its new habitat. If these introductions are successful, the pest population is reduced to a level at which no economic damage occurs. The first example of biological control we will consider is the classic case of the cottony-cushion scale.

Cottony-Cushion Scale
(*Icerya purchasi*)

One of the most striking and earliest successes of biological control concerned the cottony-cushion scale, a small coccid insect that sucks sap from leaves and twigs of citrus trees (**Figure 4**). This scale insect was first discovered in California in 1872, and by 1887 the whole citrus industry of southern California was threatened with destruction. Because of the size of the infested area, chemical control by cyanide and other sprays was a failure. In 1888 Albert Koebele of the Division of Entomology was sent to Australia by the U.S. government to represent the State Department at an international exposition in Melbourne. (All foreign travel for the Division of Entomology had been restricted to save money; this was the only subterfuge by which an entomologist could travel to Australia to search for parasites of the cottony-cushion scale, a native of Australia.) Koebele sent back to California two species of insects, a small dipteran parasite, *Cryptochaetum iceryae*, and a predaceous ladybird beetle called the vedalia beetle (*Rodolia cardinalis*) (**Figure 5**). The dipteran parasitoid was thought to be a potentially

Figure 4 Cottony-cushion scale (*Icerya purchasi*).

Figure 5 Vedalia beetle (*Rodolia cardinalis*).

important agent for control, but Koebele sent the ladybird beetles along as well, apparently without thinking that they could be very useful.

In late 1888 the first ladybird beetles were received in California, and by January 1889 a total of 129 individuals had been released near Los Angeles under an infested orange tree covered by a large tent. By April 1889 all the cottony-cushion scales on this tree had been destroyed; the tent was then opened. By June 1889 over 10,000 beetles had been sent to other citrus orchards from this first release point. By October 1889, scarcely

one year since *Rodolia* was found in Australia by Koebele, the cottony-cushion scale was virtually eliminated from large areas of citrus orchards in southern California. Within two years it was difficult to find a single individual of the scale *Icerya*, and this control continued so that the pest was effectively eliminated. The cost: about $1500; the saving: millions of dollars every year. The California legislature was impressed, and California became a center of activism in promoting the value of biological control (Caltagirone and Doutt 1989).

The cottony-cushion scale reappeared with the advent of DDT. Infestations of the scale that had not been seen in over 50 years were found after DDT had eliminated the vedalia beetle from some local areas. The vedalia beetle is currently the only significant natural enemy of cottony-cushion scale. Recent increased pesticide use against other citrus pests have caused high mortality in vedalia beetles (Grafton-Cardwell and Gu 2003). Under these circumstances, the beetle must be continually reintroduced.

Some of the host plants of the cottony-cushion scale are not suitable for the vedalia. For example, the scale infests scotch broom (*Cytisus scoparius*) and maples in central California, but the vedalia will not become established on these plants, for unknown reasons (Clausen 1956). Such host plants serve as a refuge or reservoir for the scale, which can then recolonize citrus trees.

The great success in controlling the cottony-cushion scale ushered in an era in which biological control was viewed as a panacea for all insect-pest problems. Large numbers of insects were collected from all over the world and released in North America without any testing or quarantine procedures. Eventually this dangerous policy was stopped—not, however, because of its dangers, but because of a sequence of repeated failures at control (Howarth 1991). Only approximately 1% of insects introduced for biological control have been shown to have negative ecological impacts, but little effort has gone into looking for these impacts (van Lenteren et al. 2006). Some biological control agents are considered to have driven native species to extinction.

Prickly Pear (*Opuntia* spp.)

Prickly pear is a cactus native to North and South America. Of the several hundred species of prickly pear, about 26 have been introduced into Australia as garden plants. One species, *Opuntia stricta*, became a serious weed in Australia.

In 1839 *O. stricta* was brought from the southern United States to Australia as a potted plant, and it was planted as a hedge plant in eastern Australia. It gradually got out of control and was recognized as a pest by 1880. By 1900 it occupied some 40,000 km² (15,600 mi²),

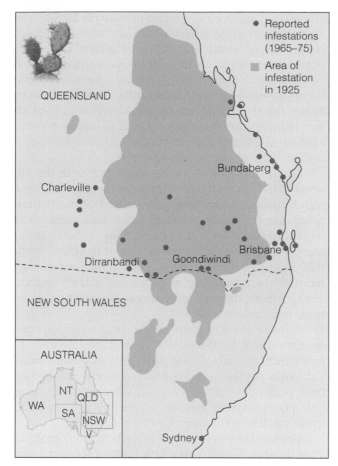

Figure 6 Distribution of prickly pear (*Opuntia*) in eastern Australia in 1925, at the peak of infestation, and areas of local infestation, 1965–1975. (After White 1981.)

and thereafter it spread rapidly in Queensland and New South Wales (**Figure 6**):

Area infested with Opuntia

	km²	mi²
1900	40,000	15,600
1920	235,000	90,600
1925	243,000	93,700

About half this area consisted of dense growth completely covering the ground (**Figure 7**), sometimes to a height of 1–2 meters and often at a density too high to walk through.

Prickly pear is propagated by seeds and by segments. The cactus pads, when detached from the parent plant by wind, animals, or people, can root and begin a new plant. Seeds are viable for at least 15 years. The problem of reducing this weed was largely one of cost. The grazing land it occupied in eastern Australia was worth only a few dollars an acre, and poisoning the cac-

(a)

(b)

Figure 7 Introducing biological control agents to the prickly pear. (a) A dense stand of prickly pear, October 1926, Chinchilla, Queensland, Australia. (b) The same stand is shown three years later, after attack by the moth (*Cactoblastis*). (After Dodd 1940; photographs courtesy of A. P. Dodd and Commonwealth Prickly Pear Board.)

tus cost about $25 to $100 an acre. Consequently, homesteads had to be abandoned to this invasion.

In 1912 two entomologists were sent from Australia to visit the native habitats of *Opuntia* and to learn of possible biological control agents that could be introduced. They sent back from Sri Lanka a mealybug, *Dactylopius indicus*, which was released, and in a few years it had destroyed a minor pest cactus, *Opuntia vulgaris*. But the major pest, *O. stricta*, continued to spread, and after World War I it was subjected to a more intensive effort of biological control. Investigations in the United States, Mexico, and Argentina resulted in 50 insect species being shipped to Australia for evaluation as possible control agents. Of these, only 12 species were released; three were of some help in controlling *O. stricta*, but only one, the moth *Cactoblastis cactorum*, was capable of controlling it.

Cactoblastis cactorum is a moth native to northern Argentina. Two generations occur each year. Females lay about 100 eggs on average, and the adults live for about two weeks. The larvae damage the cacti by burrowing into and feeding within the pads and by introducing bacterial and fungal infections with these activities. Two introductions of *Cactoblastis* were made. The first introduction, in 1914, failed (Osmond and Monro 1981). For the second introduction approximately 2750 eggs were shipped from Argentina in 1925, and two generations were raised in cages until March 1926, when 2 million eggs were set out at 19 localities in eastern Australia. The moth was immediately successful, and further efforts were expended from 1927 to 1930 in spreading eggs and pupae from one field area to another.

By 1928 it was obvious that *Cactoblastis* would control *O. stricta*, so further introductions were curtailed. *Cactoblastis* multiplied rapidly up to 1930, and between 1930 and 1931 the *Opuntia* stands were ravaged by an enormous *Cactoblastis* population. This collapse of the prickly pear population caused the moth population to fall steeply during 1932 and 1933, and the cactus then began to recover in some areas. Between 1935 and 1940 *Cactoblastis* recovered and completely controlled the cactus. After 1940, prickly pear survived only as a scattered plant in the community (Dodd 1959).

The present picture is that *Opuntia* exists in a stable metapopulation at low density maintained by *Cactoblastis* grazing. The eggs of *Cactoblastis* are not laid at random but are clumped on some plants; other plants escape infestation entirely (Myers et al. 1981). Plants heavily loaded with larvae are subsequently completely destroyed, and many *Cactoblastis* larvae thus starve and die. Larvae cannot move from one plant to another if cacti are 2 m or more apart. The clumped distribution of the eggs of *Cactoblastis* thus both destroys *Opuntia* plants and ensures that not all plants are killed; as a result, the metapopulation does not go to extinction, although local populations do disappear.

Most of the areas where prickly pear is now periodically considered a pest are outside the original area of dense cactus infestation, and plants in these areas seem to be partly resistant to *Cactoblastis* attack. Without *Cactoblastis*, prickly pear would make a rapid recovery.

Why was *Opuntia* such a successful plant in eastern Australia? Three important physiological properties of *Opuntia* determine its success (Osmond and Monro 1981). First, the tissues of this cactus are almost entirely photosynthetic. There is minimal investment in structural tissues, and the root system is shallow and small.

Second, *Opuntia* is capable of crassulacean acid metabolism (CAM), a process in which CO_2 fixation largely occurs at night, when minimal water vapor is lost to the atmosphere. Thus, photosynthesis can be done with minimal water loss. Third, CAM plants retain photosynthetically competent tissues throughout periods of stress. When the rains come, CAM plants can immediately begin to photosynthesize and grow. Because of this combination of characteristics, *Opuntia* proved a near perfect opportunist with superior competitive ability over the native plants that lacked CAM metabolism.

In an ironic flip side to the successful biological control of *Opuntia* in Australia, *Cactoblastis* is not welcome everywhere. In Florida and the Caribbean islands there are 99 species of *Opuntia* cacti that are native to these areas, where *Cactoblastis* has never occurred. They are now potentially threatened by an accidental introduction of *Cactoblastis* into Florida, and there are now widespread efforts to control this moth in these areas where it may itself be a pest species (Stiling et al. 2004; Pemberton and Liu 2007).

Floating Fern (*Salvinia molesta*)

The floating fern *Salvinia*, a plant native to South America, was introduced to Sri Lanka in 1939 through the Botany Department at the University of Colombo, and over the next 50 years it spread to Africa, India, Southeast Asia, and Australia. It is an important aquatic weed, forming mats up to 1 m thick and covering lakes, rivers, canals, and irrigation channels. Because *Salvinia* clogged the waterways, all water transport and fishing was disrupted.

Salvinia had been incorrectly identified until 1972, when it was recognized to be a new species (Room 1990). Because of this taxonomic uncertainty, ecologists were not able to look for specialized herbivores of this plant in its native habitat until the plant was found in southeastern Brazil in 1978. *Salvinia molesta* is unusual in being sterile, and all ramets appear to be genetically identical no matter where they occur in its geographical range. Plants of *Salvinia* are colonies of ramets held together by horizontal, branching rhizomes. The rate of growth of *Salvinia* on the water's surface is limited by temperature and the nitrogen concentration of the water.

Three insect species (a weevil *Cyrtobagous singularis*, a moth, and a grasshopper) found attacking *Salvinia auriculata* in Trinidad in the 1960s were introduced into Sri Lanka, India, Africa, and Fiji in the 1970s. None of these introductions had any effect on the weed. *Salvinia molesta* was located in Brazil, what was thought to be the same insect species were collected, and the weevil was released in north Queensland, Australia, in 1981. The weevil increased dramatically and destroyed the *Salvinia* within one year (**Figure 8**). The weevil was then discovered to be a new species, *Cyrtobagous salviniae*, and it proved

(a)

(b)

(c)

Figure 8 Biological control of the floating fern *Salvinia molesta.* (a) The water above Wappa Dam, Nambour, Queensland, completely covered by *Salvinia*, October 1982. (b) The same scene in September 1983 after the population explosion of the weevil *Cyrtobagous salviniae* and the subsequent crash of both *Salvinia* and the weevil. (c) The weevil *Cyrtobagous salviniae*.

highly successful at control in Australia, India, Sri Lanka, Botswana, and Namibia (Room 1990).

The success of the weevil in controlling *Salvinia* is partly explained by its tolerance of high population densities before it shows interference competition and then emigrates. The weevil reaches densities of 1000 adults per square meter, and by feeding on the buds as adults and on roots and rhizomes as larvae, the weevil either kills the plants or greatly reduces their size.

Salvinia molesta has become a significant problem in the 12 states of the southern United States from Florida to Texas and has colonized the Lower Colorado River. The weevil *Cyrtobagous salviniae* has been successful in controlling these populations once it has been introduced, and the biological control of floating fern appears to be a worldwide success story (Flores and Carlson 2006).

The biological control of floating fern highlights the need for proper taxonomy of both pest species and their potential biological control agents. Closely related species are not ecologically interchangeable, and the adaptations that determine success may be found in small differences.

Theory of Biological Control

Most biological control has operated empirically with a few rules of thumb, and this approach has achieved some spectacular successes. But if we are to avoid a case-by-case approach, we need to develop some general theory of biological control that could guide empirical work (Murdoch and Briggs 1996). Most of the theory that has developed comes from the Nicholson-Bailey model of predator-prey interactions. The premise of this approach is that successful biological control resulted from the predator imposing on the prey a low, stable equilibrium (see Figure 3). Theoretical evaluation of these predator-prey models by many ecologists (Hawkins and Cornell 1999) suggests an array of properties of successful biocontrol agents:

1. They are host specific.

2. They are synchronous with the pest.

3. They have a high intrinsic rate of increase (r).

4. They are able to survive when few prey are available.

5. They have great searching ability.

These properties are more typical of insect parasitoids than of predators in general. Most predators are considered by this theory to be poor candidates for biological control because they are generalists (not host-specific), they are rarely synchronized with the pest, they have relatively low *r* values, and they may feed on other, beneficial prey species.

Four issues are crucial for setting up a theory of biological control:

1. **Parasitoid and predator aggregation.** The classical theory of biological control is an equilibrium theory in which the main issue is how to produce stability in parasitoid-host interactions. Nicholson and Bailey (1935) produced a model of parasitoid-host interaction that was a discrete-generation model with one generation of host and parasitoid each year. The result of this simple model was instability. Part of this instability arises from the assumption in the original Nicholson-Bailey model that encounters between parasitoids and hosts are random. If we introduce nonrandom search and allow the parasitoids to spatially aggregate to high densities of hosts, the models produce a stable interaction (Murdoch et al. 2003). But this stability is achieved in the model only if parasitoids are not allowed to move from patch to patch within a generation. Parasitoids in the model are always revisiting patches of the host that have already been heavily attacked. If we relax this assumption and allow parasitoids to move among patches within a generation, these models do not induce stability in population dynamics (Murdoch and Stewart-Oaten 1989). The trade-off with stability is effectiveness: parasitoids that concentrate their attacks on high-density patches of prey are more likely to be effective biological control agents because they suppress the pest, but they do not induce stability in population dynamics. We need to look for stabilizing effects elsewhere.

2. **Metapopulations of hosts.** If a pest population is a metapopulation composed of many local populations, then even if great instability of predator-prey interactions occurred at the local level, great stability could occur at the regional or metapopulation level. This suggestion is analogous to the observations of Huffaker's mite predator-prey system in the laboratory. The question is whether or not parasitoid-host systems in biological control are in fact distributed as metapopulations. At present few data are available to decide this issue, and this remains an important theoretical idea to be tested.

3. **Refuges for the host populations.** If the host population has a refuge habitat in which the parasitoid or predator cannot reach them, there is a potential for a stable interaction of parasite and host similar to that Gause (1934) observed with *Didinium* and *Paramecium*. The question is whether or not successful biological control agents operate best in systems containing prey refuges.

4. **Density dependence in the parasitoid attack rate.** If parasitoids attack a pest species in a density-dependent manner such that the parasitoids interfere with one another when they are at high density, the host population will be stabilized. This mechanism is an attractive one for maintaining a stable predator-prey interaction, but it may achieve stability only at high pest densities, rather than at the low pest densities we desire for pest control. Consequently, we must first determine if density dependence occurs in parasitoid attack rates, and then determine how this can act to suppress the pest population to low density.

Successful control agents cause density-dependent losses in the host population, and this density dependence may be spatial or temporal or both. Spatial density dependence occurs when predators or parasitoids cause a higher fraction of losses in dense host patches than in sparse host patches. If predators can aggregate in patches of high host density, then, according to this theory, biological control of the pest is much more likely.

We can test these ideas about biological control by comparing their predictions with observations on case histories of successful biological control efforts. Murdoch et al. (1985) compiled data on several highly successful control efforts, and even though in most cases biological control was successful, the mechanisms of success were not those predicted by the classical theory of population regulation. Why might this be, and what biological traits are responsible for control in successful cases? We can answer this question best by considering one of the most carefully studied success stories of biological control—that of the California red scale.

The California red scale (*Aonidiella aurantii*) is a worldwide pest of citrus trees. It was imported accidentally from China about 1900 and until 1950 threatened the citrus industry of California. During this time about 50 natural enemy species were introduced. Of these, eight established, and full economic control was achieved in 1959 by the wasp parasitoid *Aphytis melinus*. The density of the red scale in southern California after *A. melinus* was introduced dropped to about 1/200 of the levels reached in the absence of natural enemies. Since 1959 the red scale is no longer a pest on citrus. It is present at very low numbers and relatively constant in density from year to year (Murdoch et al. 2006).

The red scale is a classic example of successful biological control and as such a useful vehicle for examining biological control theory. In a series of critical field experiments, Murdoch and his colleagues examined and rejected the standard theoretical explanations for success and stability of control for this system. The first class of explanations for success and for stability is that the relationship between the parasitoid and the scale is density dependent. Reeve and Murdoch (1986) showed that mortality caused by the wasp *Aphytis* was not density dependent in time (**Figure 9**), as classical theory would predict. Nor was there any evidence for density dependence in space, or aggregation of the parasitoids toward high densities of red scales. The second class of explanations for stability is the existence of a refuge for the prey. Reeve and Murdoch (1986) found that the red scale has a refuge in the interior branches of citrus trees and in this refuge the density of red scales is 100-fold higher than it is on exterior, exposed branches of the grapefruit trees. In this refuge the red scale is protected from parasitoids by the Argentine ant. The ants do not actively protect the red scales but seem to disturb *Aphytis* and prevent them from attacking the scales. To test this mechanism Murdoch et al. (1996) removed the refuge from some grapefruit trees and compared those trees with control trees with a normal refuge. There was no difference in population stability in treated and control trees, and so the existence of a refuge is not the explanation for the stability and success of this biological control. A third class of mechanisms for stability is a metapopulation structure with movements between local populations. But this hypothesis was rejected when the isolation of individual trees with cages to prevent parasitoid immigration and emigration produced no effect on the population dynamics (Murdoch et al. 1996). By a process of elimination, stability in this system must arise from the life history details of the interaction. By modeling the details of the life history of the parasitoid and the scale, Murdoch et al. (2006) showed that two main mechanisms explained the success of this biological control. First, the adult stage of the red scale is not vulnerable to parasitism because of its size, so there is a prey refuge in size. Second, the parasitoid has a faster generation time than the prey; while red scale has two generations per year in southern California, the parasitoid *Aphytis* has six generations per year.

To test these conclusions, Murdoch et al. (2005) carried out a critical field experiment. They caged individual trees in the field and generated a scale outbreak by adding scales to the trees and later by adding the parasitoid *Aphytis* (**Figure 10**). Because each tree was

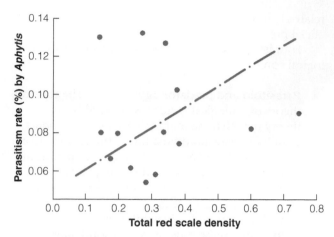

Figure 9 A test for temporal density-dependent predation by the wasp *Aphytis melinus* on California red scale over three years. The blue line shows the expected line if density dependence were occurring. The data do not fit the line, so there is no indication of temporal density dependence in this predator-prey system, despite its success as a biological control program. (From Reeve and Murdoch 1986.)

caged, there were no metapopulation effects or other immigration movements of the parasitoids. They followed the system for five generations of the red scale, which reached a low and stable equilibrium very rapidly within one generation of the scale. Because they had a detailed model of the biological interactions in this system, they could predict the outcome of the experiment in advance, and the model and the data form a perfect fit. The conclusion is that the local (within tree) interaction of the parasitoid and the pest scale species is sufficient both for control of the scale and for stability of the resulting low-density population.

This analysis of the successful biological control program for red scale shows that it is necessary to deconstruct the details of the species interactions to see how successful population control is produced. The best way to analyze these interactions is with simple models that explore the stability properties of particular biological interactions (Murdoch et al. 2003). Over 20 years of experimental and theoretical analysis were required to explain the characteristics of ths system that leads to the maintenance of a low equilibrium density of red scale.

Genetic Controls of Pests

Another alternative method of pest control is *genetic control*, a type of biological control that uses two strategies

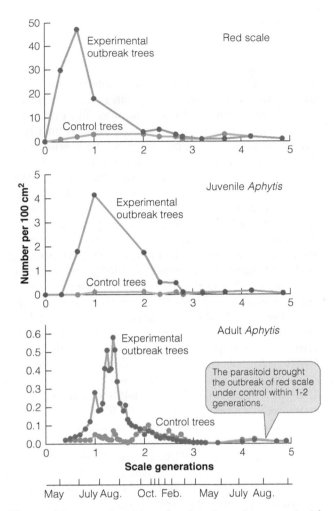

Figure 10 Mean densities of red scale and its parasitoid *Aphytis melinus* on four outbreak and 10 control trees over four scale generations occupying about 16 months. The scale development time is driven by temperature; thus the chronological time (shown at the bottom) does not map directly on generation time. Scales were added to the experimental trees during the first two months of the experiment. (From Murdoch et al. 2005.)

to reduce pest problems. Either crop plants can be genetically manipulated to increase their resistance to pests, or pests may be changed genetically so that they become sterile or less vigorous and thus decline in numbers.

The use of crop varieties resistant to attack by pests is one of the oldest and most useful techniques of pest control. In 1861 the grape *phylloxera*, an aphid that feeds on the roots of grape plants, was accidentally introduced into Europe from North America. The European grape (*Vitis vinifera*) was extremely susceptible to

the *phylloxera*, and the wine-making industry of France was on the brink of collapse by 1880. The American grape (*Vitis labrusca*) is resistant to phylloxera attack, so European grapevines were grafted onto American rootstocks to produce an artificial hybrid grape plant that was resistant to *phylloxera* attack (Granett et al. 2001).

Resistant varieties of many crop plants have been developed by selective breeding (Acquaah 2007). The method used is, in principle, very simple. Individual plants that are not being damaged are sought in an area where the pest species is common, and these plants are removed to the greenhouse for selective breeding. If resistance is inherited in the greenhouse lines, the new selected variety may be used for commercial production.

Selective breeding can be a two-edged sword, however, and must be used with care. For example, all species of cotton produce a plant chemical called *gossypol* (a sesquiterpene), which occurs in the green parts and seed of the cotton plant and is toxic when fed to chickens and pigs. To increase the value of cottonseed as an animal feed, plant breeders bred strains of cotton with low gossypol content and were able to reduce the concentration of gossypol to only one-fourth that of normal cotton. But breeding gossypol out of cotton deprived the plant of much of its resistance to insect pests and also made the cotton plant susceptible to a whole set of new pests (Klun 1974).

Resistant plants do not necessarily have chemical defenses. Morphological defenses, such as spines, prickles, hairs and tough leaves, can be highly effective (Myers and Bazely 1991). Soybeans are a major crop in the midwestern United States despite the presence of a serious potential pest, the potato leafhopper (*Empoasca fabae*). The potato leafhopper will not attack soybean varieties that have leaves covered with short hairs, whereas they attack and nearly destroy soybeans that have hairless, smooth leaves. The hairs deter insect movement and are highly effective as a defense mechanism.

Breeding resistant varieties of plants has been an important factor in limiting pest damage in many crops, but the rapid adaptability of plant pathogens has compromised much effort (Lucas 1998). For example, potato blight, a disease caused by the fungus *Phytophthora infestans*, first appeared in Europe in the 1840s, and it spread rapidly, wiping out potato crops and causing famine in Ireland. Plant geneticists have attempted to induce a high level of resistance to blight in cultivated potato plants by introducing into them single genes derived from closely related wild species. Four genes have been used in this way, but after the commercial introduction of each new gene for resistance, new races of the fungus appeared that could attack the "resistant" potatoes. Sexual recombination or asexual mutation of fungal pathogens results in rapid evolutionary changes in field populations, so that crop resistance breaks down over time.

One promising area of intense development currently is the production of resistant crop plants by means of genetic engineering. Genes that produce resistance in one species can be inserted into a crop plant to make the crop genetically resistant to specific pests. Alternatively, bacteria may be used as vehicles to carry biopesticide genes. *Bacillus thuringiensis* (*Bt*) is the main focus at present for developing insect-resistant crops (Ferry et al. 2006). This bacterium normally lives in the soil and carries a gene for a protein that is toxic to the larvae of butterflies and moths. By splicing this gene into the DNA of crop plants, genetic engineers can produce insect-resistant crops. Insects that feed on the plant and are thereby poisoned, and the plants are protected from damage.

Transgenic plants with *Bt* genes are considered by some to be the future in crop protection by creating plants that are genetically resistant to insect pests (Christou et al. 2006). One anticipated problem with this technology is that pest insects will become resistant to the biopesticide, just as they became resistant to chemical pesticides. More than a dozen species of moths have been selected for *Bt* resistance in the laboratory (Tabashnik et al. 1998) and two in the field, diamondback moth and cabbage loopers Tabashnik et al. 1990, Janmaat and Myers 2003). But no cases of resistance to *Bt* transgenic crops have yet been found in the field (Christou et al. 2006). This conflicts with the prediction of the evolution of rapid resistance to *Bt* crops. Tabashnik et al. (2006) used DNA screening to look for resistance alleles in the pink bollworm (*Pectinophora gossypiella*) in cotton fields in Arizona, Texas, and California, and found no resistance alleles in spite of 10 years of exposure of the bollworm to *Bt* cotton.

To avoid the development of resistance however, farmers have been required to plant 20% of cotton or corn crops in refuges of nontransgenic plants. Because genes for *Bt* resistance are generally recessive, homozygous resistant moths from fields of transgenic plants will likely mate with susceptible moths from the refuge and their offspring will be heterozygous and thus killed by the *Bt* toxin in the crop plants. But this strategy results in economic losses for the farmer. One approach to get around this problem is to "piggyback" two toxins into transgenic plants so that if one toxin does not kill the pest, the second one will (Christou et al. 2006). This strategy will work if the two toxins act independently in the pest insects. The success to date of transgenic *Bt* crops is encouraging, but past experience with insect pests should not lead us to underestimate their ability to develop resistance to both natural and artificial pesticides over the long term.

In addition to changing the genetic makeup of the plants, we can attempt to alter the genome of the pest species. The simplest genetic manipulation that can be carried out on a pest species is sterilization. Sterility can be produced in several ways, but the usual procedure is to sterilize large numbers of pest individuals by irradiation or by treatment with chemicals and then to release them into the wild, where they can mate with normal individuals. Because of these matings, the number of progeny produced in the next generation is greatly reduced, and control can be achieved. The **sterile-insect technique** cannot be used on all pest populations because it requires the rearing and sterilizing of large numbers of individuals. In addition immigration of fertile individuals to the control area must be negligible for this procedure to work. It was hoped that disease-carrying mosquitoes could be controlled using the sterile-insect technique, and many trials have been carried out (Benedict and Robinson 2003). Few have been very successful, and the reasons for failure are typically that the sterilized males are not vigorous, that the numbers of sterile individuals released were too small, and that immigration into the release area countered the sterility effect. One example of the successful use of the sterile-insect technique was the suppression of the mosquito *Culex pipiens quinquefasciatus* on a small island off Florida (Patterson et al. 1970). Between 8400 and 18,000 sterile males were released each day over a 10-week period during midsummer on a 0.3-km² island, and by the end of the experiment 95% of the eggs sampled on the island were sterile (**Table 1**). Thus the experiment was a success, but because the island was only 3 km from the mainland, recolonization by dispersing females occurred quickly once the experiment ended.

The sterile-mating technique of control may be rendered less effective if pest populations are genetically subdivided. One example of this difficulty is the screwworm control program in the southern United States (Richardson et al. 1982). Screwworms are larvae of several species of blowflies that lay their eggs on open wounds of warm-blooded animals. The larvae enter the wound and feed on the living tissue, possibly leading to the death of the host animal (often cattle, sheep, or deer) because of physical damage or secondary infections. A program to eradicate screwworms in the southern United States was begun in the 1950s. These programs were very effective until 1968, when a series of unexplained outbreaks began. Serious outbreaks occurred again in 1972, 1976, and 1978.

At least 11 chromosomal types of screwworms can be recognized (Richardson et al. 1982). Many of these types occur together geographically and could possibly be different species of screwworms. If there is a genetic mismatch between the sterile flies raised in the laboratory and the wild type causing an outbreak, then clearly the sterile-mating technique will not work because the flies will not mate.

Table 1 Sterile-male release experiment with the mosquito *Culex pipiens quinquefasciatus* on Seahorse Key, a small island off Florida.

Generation	Ratio of sterile to normal males	Eggs expected to be sterile (%)	Eggs actually sterile (%)	Reduction in eggs laid (%)
1	All normal	0	0	0
2 (begin releases)	3:1	75	62	36
3	4:1	80	85	34
4	12:1	92	82	79
5	100:1	99	84	96
6 (end)	100:1	99	95	96

NOTE: Each generation of mosquitoes took about two weeks during this summer period.

SOURCE: Data from Patterson et al. (1970).

Screwworms have now been successfully eradicated in the southern United States (by 1966), Mexico (by 1991), and Central America (by 2001) (Bowman 2006). An accidental introduction of the screwworm to Libya in 1988 was stopped by the importation of sterile males from Mexico, and by 1992 Libya was free of the flies (Vargas-Teran et al. 2005). Screwworm eradication has been one of the great success stories in the use of the sterile-insect technique, and because of its success has focused attention on the possibility of eradicating other animal diseases such as hydatid disease (Dyck et al. 2005).

Immunocontraception

A new method of biological control for vertebrates, **immunocontraception**, has emerged with recent developments in biotechnology. While much of biological control has aimed at ways of increasing the mortality rate in a pest population, an alternative approach is to reduce the fertility of the pest. This is the approach that has been used in the sterile-insect technique just discussed and underlies immunocontraception. Fertility could be reduced by the use of immunocontraceptive vaccines delivered in a bait, or by a virus or other contagious agent that spreads naturally through the target pest population (**Figure 11**). There are many different sperm and egg surface proteins that could be used for immunocontraception. One of the most commonly used set of proteins are the zona pellucida glycoproteins (ZPG), which facilitate sperm penetration of the egg. Antigens against zona pellucida proteins prevent sperm from attaching to the surface of the egg, thereby preventing fertilization.

Immunocontraception works. **Figure 12** illustrates the collapse and recovery of fertility of wild horses in Nevada that were injected once with zona pellucida protein in microcapsules designed to release the vaccine slowly (Turner et al. 2007). Wild horses on public lands in the western United States are difficult to manage because the mandate is to maintain a balance of wild horses, cattle, and other wildlife without a loss in range quality. This is ecologically impossible with increasing horse numbers, so the Bureau of Land Management, which is forbidden by law from culling horses, must either round them up and hold them or devise some method of reducing their population growth rate. Hence the interest in immunocontraceptive vaccines for wild horses. But the key point is to determine the impact of reduced fertility on population dynamics, and this is not simple.

The best illustration of the effect of sterilization of pests on population trends comes from long-term studies on the European rabbit in Australia. Female rabbits in

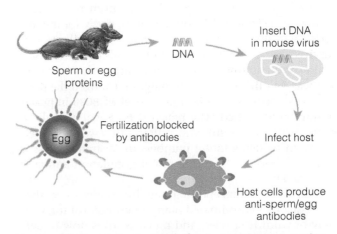

Figure 11 General procedure for immunocontraception in mammals, illustrated for the house mouse. The aim is to immunize the pest species against its own sperm or egg proteins, so that fertility is blocked. The same general procedure could be applied to any mammal or bird pest. (After Pech et al. 1997.)

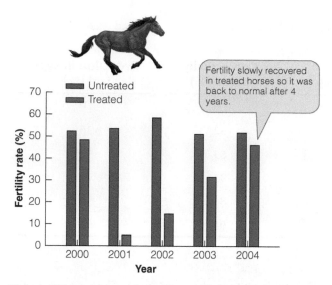

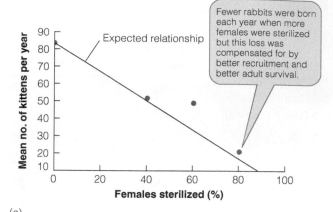

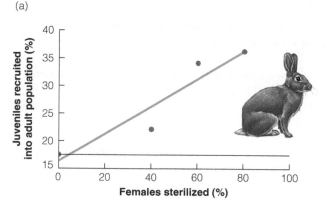

Figure 12 Fertility rates for free-roaming wild horses in Nevada. A total of 96 adult females were treated with a single dose of an immunocontraceptive vaccine in January 2000. Control sample size of untreated females varied from 43 to 69 in different years. The treatment could not work in 2000 because the females were already pregnant at the start of the study. (Data from Turner et al. 2007.)

wild populations were live trapped and surgically sterilized at two sites, one in Western Australia and one in southeastern Australia (Twigg and Williams 1999). Populations with 0%, 40%, 60%, and 80% sterility were then followed for four years. The reproductive output of rabbits decreased with increasing levels of sterility imposed (**Figure 13a**). The juveniles produced, however, survived much better in populations given the sterility treatments (Figure 13b), compensating partly for the imposed sterility. The adult rabbits that were sterilized also survived much better than fertile females (Figure 13c), again compensating for the sterility imposed. The net result was that this sterilization program changed the population density very little until a level of 80% imposed sterility was reached. The practical message from these experiments is that immunocontraception will not be effective in reducing rabbit numbers in Australia unless it can reach about 80% of the rabbit population annually (**Figure 14**).

Immunocontraception is an innovative idea that can assist in a coordinated plan of pest control for a variety of wildlife species, and much work is now underway in the United States, Australia, and New Zealand to determine its potential (Hardy et al. 2006). The greatest challege is to design mechanisms to introduce immunocontraception into wild populations cheaply and efficiently and this may well prevent this approach from being used effectively.

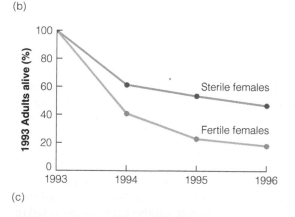

Figure 13 Sterility experiment on the European rabbit in Australia. (a) The reproductive output under the four levels of imposed female sterility. The number of rabbit kittens emerging from burrows is close to that expected by the percentage of sterility (blue line).
(b) Percentage of juvenile rabbits surviving to become adults in the four treatments. Juvenile survival compensated for the sterility by increasing above the expected (control) line shown in black. (c) Adult female survival during the experiment. Adult rabbits sterilized surgically in the first year of the study survived much better than females that bred, so there is a survival cost of reproduction in females that compensates for the imposed sterility. (Data from Twigg and Williams 1999.)

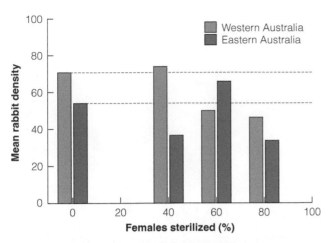

Figure 14 Density changes during the sterility experiment on the European rabbit in Australia from 1993 to 1996. The average number of rabbits over the entire experiment showed little change except at the most extreme level of 80% sterility, where rabbits declined in density 34%–37%. Control densities are shown by the horizontal dashed lines. (Data from Twigg and Williams 1999.)

Cultural Control of Pests

All of agriculture can be said to be an essay on cultural control of pests, as farmers have learned which agricultural practices reduce the damage caused by pests and the need for expensive chemical sprays. Because many of these methods of cultural control are discussed extensively in agricultural pest management books (Altieri and Nicholls 2004; Gurr et al. 2004), I will not discuss these methods extensively but provide only one example to illustrate what can be accomplished by a combination of ecological insight and agricultural manipulations. The guiding principle is to increase habitat heterogeneity and to move away from extensive monocultures of a single genetic strain of crop. Pest outbreaks in agricultural systems often can be anticipated from the **resource concentration hypothesis** (Root 1973). This hypothesis predicts that habitat patches with larger amounts of resources will have higher densities of organisms and is broadly supported in many animal groups (Connor et al. 2000). Thus, crop monocultures provide an ideal environment for pest species, and this helps to explain the origin of pest problems in agriculture. The converse prediction is that if we can decrease resources that a pest has available to use, we will decrease the problem of pest damage.

One illustration of how cultural controls can form part of a pest management program is found in the rice growing area of Yunnan Province of southwestern China (Zhu et al. 2000). Rice blast is a fungal disease that attacks many but not all varieties of rice. By interplanting a mixture of two varieties of rice—a traditional one

(a)

(b)

Figure 15 Cultural control of rice blast disease in rice crops in Yunnan Province, southwestern China. Two rice varieties are interplanted, and because they differ slightly in color (a), they give a striped appearance to the agricultural landscape. A close-up photo (b) shows the traditional taller rice variety separated by four rows of a high-yielding dwarf variety that is resistant to rice blast disease. The traditional variety of rice is preferred for its flavor, and provides higher income to the farmers, but is devastated by rice blast if grown as a monoculture. This intercropping experiment was developed by the International Rice Research Institute in the Philippines, and is now being applied to over 4 million hectares of southwestern China, one of the largest scale pest control experiments yet conducted. (Photos courtesy of International Rice Research Institute, Los Baños, Philippines.)

susceptible to rice blast and a new high-yielding variety selected to be resistant to rice blast (**Figure 15**)— Chinese farmers have been able to reduce the incidence of rice blast and increase rice yields by 10%–15%. This simple type of cultural control provides a way of making the agricultural landscape less of a monoculture in which pests thrive. Chinese farmers have developed a variety of interplanting methods to control pests and reduce pesticide use and have been important pioneers in the use of cultural control methods.

Integrated Control

Many important pests cannot be controlled by any one technique, so biologists concerned with pest management have been forced to take a wider view of pest problems. A unified approach, called integrated control or integrated pest management (IPM), uses biological, chemical, and cultural methods of control in an orderly sequence (**Figure 16**).

The objective of integrated pest management is to minimize economic, environmental, and health risks. Integrated control can be achieved only if the population ecology of the pest and its associated species, and the dynamics of the crop system, are known. Integrated control systems are ecologically sound because they rely on natural biological control as much as possible and resort to chemical treatments only when absolutely necessary. A considerable amount of information is needed to permit the effective use of an integrated control program. Density levels of the potential pest populations, stage of plant development, and weather data are often required to enable the pest manager to predict the future development of the crop and to judge the necessity for pesticide application.

One example of an integrated control program is that developed for rice by the International Rice Research Institute (Jahn et al. 2007). Rice is an important crop for 3.5 billion people on Earth, and many advances have occurred in pest control in rice by combining farmer knowl-

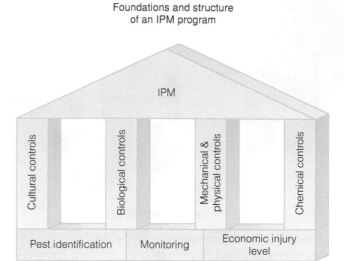

Foundations and structure of an IPM program

Figure 16 Foundations of an integrated pest management (IPM) system. No one type of pest control will be sufficient for many agricultural and forestry pests, and the various alternative methods of control need to be integrated ecologically. Detailed taxonomic identification of the pest is an important foundation, as is careful monitoring of the pest population and damage. (Diagram courtesy of U.S. Department of Agriculture.)

edge and practices with ecological insights of population dynamics. Much of this information is now available online from the International Rice Research Institute as Rice Doctor (http://www.knowledgebank.irri.org/ricedoctor/). The general breakdown of integrated pest management for rice is illustrated in **Figure 17**.

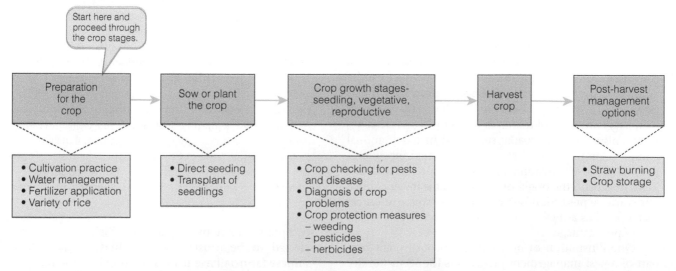

Figure 17 Key components of an integrated pest management system for rice ecosystems. Each of the actions indicated in the top row of boxes linked by blue arrows must be broken down into their components. Two examples of components are described in Table 2. (Courtesy of International Rice Research Institute.)

Table 2   Examples of integrated pest management (IPM) measures for two pests of rice crops in Asia.

	Green Leafhopper (*Nephotettix malayanus* and *N. virescens*)	Golden Apple Snail (*Pomacea canaliculata*)
The problem	The most common leafhopper in rice fields. They spread the virus disease tungro	An introduced pest that eats young rice plants and is particularly bad during crop establishment
Control measures		
Cultural and mechanical controls	1. If tungro virus is not present, do nothing as rice can tolerate large populations of these insects	1. Handpick snails and crush egg masses during the day when they are active
	2. Plant resistant varieties of rice	2. Place bamboo stakes in rice to attract adults for egg laying, and destroy the egg masses
	3. Synchronize planting across farms and plant early in the dry season	3. Use attractants like banana leaves to make hand picking easier
	4. Transplant older rice seedlings (>3 weeks)	4. Transplant older rice seedlings (>25 days)
	5. Do not use too much nitrogen fertilizer	5. Place toxic plant leaves in the fields (e.g., tobacco leaves)
	6. Control weeds on the field verges	6. Draining a field and after the crop is harvested tilling the drained field kills adult snails burrowed in the soil
	7. For upland rice intercrop rice with soybeans	7. Install screens on water inlets to rice fields
Biological control	1. Various parasitoids attack the eggs of green leafhoppers, and pathogens are also important in natural control	1. Red ants feed on snail eggs
	2. Reducing pesticide use allows natural enemies to control leafhoppers	2. Ducks and rats eat young snails. Ducks can be put in fields during final land preparation and after crops are well established
		3. Snails can be harvested, cooked, and eaten
Chemical control	1. Check 20 rice hills in a field. Control may be required if there are more than five green leafhoppers per hill	1. Chemical pesticides are rarely needed
	2. Choice of an insecticide depends on costs and the equipment available, the experience of the farmer, and the presence of fish. Expert advice from the local crop protection specialist is desirable	2. If chemical control is essential, check for products that have low toxicity to humans and to other organisms. Apply chemicals to low spots and water channels rather than to the entire rice field

The International Rice Research Institute (IRRI) has developed many similar protocols for IPM of other rice pests and diseases. (Data courtesy of IRRI.)

Table 2 presents the details of an integrated pest management program for two pests of rice crops in Asia. **Figure 18** illustrates these two species. There are many pests and pathogens of rice, and much of the effort of the International Rice Research Institute has been to develop these detailed protocols for preventing losses to rice crops. Similar details are now available for many agricultural crops around the world.

A novel approach for integrated pest management called **push-pull strategies** has evolved from behavioral ecology and holds promise to assist in the control of pests in agriculture and forestry (Cook et al. 2007). Push-pull

Figure 18 Two key animal pests of rice crops in Asia. (a) The green leafhopper *Nephotettix malayanus*. (b) The golden apple snail *Pomacea canaliculata*.

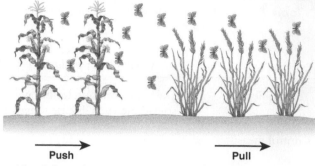

Push	Pull
- Visual distractions	- Visual stimulants
- Non-host volatiles	- Host volatiles
- Anti-aggregation pheromones	- Aggregation pheromones
- Alarm pheromones	- Sex pheromones
- Oviposition deterrents	- Oviposition stimulants
- Antifeedants	- Gustatory stimulants

Figure 19 Diagrammatic representation of the push-pull strategy as part of an integrated pest management system. A variety of chemicals that modify behavior is the key to these strategies, and have their origin in understanding the basic behavioral and chemical ecology of the pest species. The key is to push the pest species off crop plants like corn and to pull them toward non-crops or traps. (From Cook et al. 2007.)

strategies involve the behavioral manipulation of insect pests and their natural enemies via stimuli that make the resource unattractive (push) and luring the pests toward an attractive source (pull) from which the pests are subsequently destroyed. **Figure 19** illustrates these ideas schematically. A good illustration of the practical value of push-pull strategies comes from maize (corn) and sorghum crops grown by subsistence farmers in eastern and southern Africa. These crops are attacked by lepidopteran stem borers that cause 10%–50% yield loss. Pesticides are too expensive for these farmers to purchase. The solution to this pest problem was to use intercrops and trap crops in the fields. Stem borers are repelled (push) by molasses grass (*Melinis minutiflora*) and silverleaf desmodium (*Desmodium uncinatum*) planted between the crop rows (Khan et al. 2006). The borers are attracted to the trap plants Napier grass (*Pennisetum purpureum*) and Sudan grass (*Sorghum vulgare sudanense*) and concen-

trate there (pull). Stem borers oviposit heavily on Napier grass but few of the eggs survive. A variety of volatile chemicals, mainly sesquiterpenes, form the basis of this successful system for reducing pest damage without the use of expensive toxic pesticides.

Integrated control programs derive their validity from field studies and are thus empirical ecology in action. They have not been developed as theoretical strategies but as working programs, and they hold great promise for the future because they retain biological control as a core element of the integrated program (Barker 2003). They depend heavily on detailed ecological understanding of pest biology.

Generalizations about Biological Control

Why can we not control all pests by biological control? Biological control is something akin to a gambling system: it works sometimes. But how often? **Table 3** summarizes data from a global appraisal of the success rates of classic biological control against insect and arachnid pests. About one-third of the parasites and predators introduced get established more or less permanently after introduction (Hall and Ehler 1979; Mills 2006). If we define success in biological control according to economic benefits, only 17% of classic biological control attempts qualify as complete successes (Mills

Table 3 Summary of the success of biological control efforts against insect and arachnid pests throughout the world.

Category	No. of attempts	Established efforts		
		Established (%)	Partial or complete successes (%)	Complete successes (%)
Total*	2295	34	58	16
By order of insects introduced				
Homoptera	819	43	80	30
Diptera	258	37	31	0
Hymenoptera	105	34	56	0
Lepidoptera	628	27	48	6
Coleoptera	364	23	36	4
By demographic origin of pest				
Exotic pests	2163	34	60	17
Native pests	132	25	29	6
By geographic isolation				
Islands	827	40	60	14
Continents	1468	30	56	17
By habitat stability				
Unstable habitats (vegetable and field crops)	640	28	43	3
Intermediate (orchards)	916	32	72	30
Stable habitats (forests, rangelands)	535	36	47	8

*Not all minor orders of introduced insects are listed here.

SOURCE: Data compiled by Hall and Ehler (1979) and Hall et al. (1980).

2006). Why is this? What makes some biological control agents such as the vedalia work so well while others completely fail? A number of empirical generalizations have been suggested.

A series of life history characteristics are associated with successful biological control programs (Kimberling 2004). **Figure 20** shows the breakdown of life history traits that tend toward successful control. Predators and polyphagous species had the lowest proportions of successes. Species with female-biased sex ratios and multiple generations per year with respect to their hosts had the highest success rates. But still these trends are statistical in nature, and the practitioners of biological control would like to have a more predictive theory for individual cases. So far this has eluded researchers.

Other generalizations have been made. Most successful biological control programs have operated quickly. The rule of thumb is that three generations (or a maximum of three years) is the outside limit and that if definite control is not achieved in the vicinity of the colonization point within this time, the control agent will be a failure. This rule of thumb suggests that colonization projects should be discontinued after three years if no success is achieved and that prolonged efforts at establishment are wastes of money. Most examples of successful biological control to date support this rule, which suggests that major evolutionary changes in the host-parasite system seldom occur in introduced pests. If a parasitoid is not already adapted to control the host, it will not evolve quickly into a successful control agent.

A vital historical lesson is the frequency with which a species such as the vedalia was released more on faith than on any evidence that it could control the pest. Most biological control agents are only evaluated in retrospect, and biological control programs have been a gamble. An exception is the biological control of purple

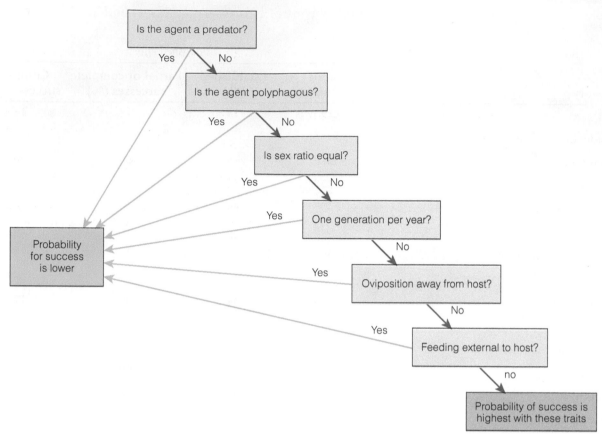

Figure 20 Life history traits of biological control agents that are significantly associated with success in control. (After Kimberling 2004.)

loosestrife, *Lythrum salicaria*, in North America. Blossey (1995) used a variety of approaches in prioritizing nine potential biological control agents for host plant specificity and release. In the end he used wide distribution, high impact on the plant and the feeding niches to prioritize the species. This led to the selection of a root boring beetle and two species of leaf feeding beetles for initial introduction to North America. These three species had been observed to reduce localized patches of high purple loosestrife density in Europe. Of these the root boring weevil was difficult to establish widely, but the two leaf feeding beetles, both in the genus *Galarucella*, were widely established and successful. In northern areas one of these beetles dominates and successfully controls loosestrife on its own.

Most successful biological control programs have resulted from a single species of parasitoid or predator, which raises a question: If one parasitoid species is good, are two species better? The argument about single versus multiple releases of biocontrol agents has raged for over 50 years. The argument has been that only one

species should be released at a time for pest control, because two parasitoids might interfere with each other when the pest is reduced to low numbers. This argument follows from the observation that native insect pests have numerous parasitoids, predators, and pathogens. The spruce budworm, for example, which is a serious forest pest, has over 35 species of parasitoids and many predators and pathogens. Is the spruce budworm a pest because it has many parasitoids? Or does it have many parasitoids because it is moderately abundant?

If competitive interactions occur between introduced parasitoids and predators, biological control is predicted to be more successful when fewer enemies are released. Denoth et al. (2002) looked at this question for both the biological control of weeds and of insects. For weed biological control programs the success of agent establishment was not influenced by the number of species introduced, but for insect biological control the rate of establishment on the introduced agents was higher for single versus multiple introductions. The success of introduced agents showed a slightly different pattern; success was higher for multiple species introductions in biological control of weeds but not for the biological control of insects. Interestingly, even though success was greater when multiple agents were introduced in weed control projects, a majority of successes were due to one species of biological control agent per program (**Figure 21**). The same was found for biological control of insects; usually one species is successful even though several species have been introduced. In a more targeted analysis of 10 studies where single-and multiple-control agents were released, Stiling and Cornelissen (2005) found that multiple-releases reduced pest abundance more than single-release efforts. Denoth et al. (2002) suggest that a relationship between biological control success and the number of agents released could result from the "lottery model"; with more introductions the probability of getting the "effective" agent will increase.

What can we conclude regarding the problem of natural regulation from these examples of biological control? What we do know is that there is an increasing skepticism about introducing more exotic species even for biological control. Thus, more emphasis should be placed on selecting effective agents; those that are capable of killing host plants or that are shown in the native habitat to respond to high host density. Environmental

costs such as the possibility of nontarget impacts or monetary costs associated with carrying out prerelease tests put increasing emphasis on increasing our understanding of why some biological control agents are effective and others not.

Integrated pest management, through the use of biological control with other types of control tactics, is rapidly becoming one of the most important practical applications of ecological theory to modern problems of food production. We are gradually replacing an outmoded version of attempted pest control using only toxic chemicals with a new view of crop management with minimal use of chemicals and less environmental disturbance. To achieve this goal, we need to know the population biology of both the crops and their associated pests. The challenge is great but the payoffs are vital.

Risks of Biological Control

Introducing a nonnative species into an ecological system to control a pest is not without some ecological risks. The danger is that the introduced species will attack nontarget organisms and cause more damage than it prevents. The clearest examples of this involve generalized predators released for biological control, such as the seven spotted lady bug, which is a very effective predator of aphids, but is reducing native lady bug species (Simberloff and Stiling 1996). The Indian mongoose (*Herpestes auropunctatus*) (**Figure 22**), which was introduced into Hawaii and many islands in the West Indies to control rats in sugarcane fields, has become an important predator of native birds on all these islands and is suspected of causing the extinction of some reptiles in the West Indies (Lever 1985).

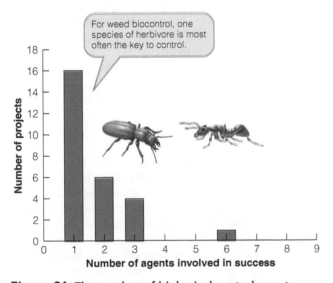

Figure 21 The number of biological control agents involved in the successful biological control of weeds for programs in which more than one species was released. In these 27 projects 153 species were released, yet only one species was the main control agent in most of the successful programs. Biological control of weeds does not seem to be a cumulative result from many introduced control agents but rather a result of one highly effective species. (Data from Denoth et al. 2002.)

Speech bubble in figure: "For weed biocontrol, one species of herbivore is most often the key to control."

Figure 22 Indian mongoose (*Herpestes auropunctatus*).

introduced snail *E. rosea* is now causing the extinction of many of the remaining native snails (Civeyrel and Simberloff 1996). The lesson learned from these mistakes is to very carefully evaluate the use of generalist predators as biological control agents.

Three aspects of biological control programs have come under scrutiny because of these potential problems. First, pest problems must be quantified before a biological control program is initiated. How much damage is being caused, and is the problem more aesthetic than economic? Some species that are thought of as pests do not actually cause economic or environmental damage (Hone 2007). Second, nontarget species must be tested more widely before a potential biocontrol agent can be considered for release. A broad array of nontarget species, not just other agricultural crops, must be considered for potential harm, because the risk of introductions typically involves conservation problems with native species. The potential of the biocontrol agent to attack new hosts also needs to be considered, for example, by determining the genetic basis of host preference. Third, more research is required after a biocontrol agent has been released. Studies of the demographic processes by which biological control is achieved are typically severely limited to control costs, so after the release we know only that it is a success or a failure but never why. As more and more pest species are spread around the world, the need for strict guidelines for the release of nonnative species becomes stronger, so that a clear estimate of the costs as well as the benefits can be evaluated.

Figure 23 Giant African Snail (*Achatina fulica*).

In other cases, the predatory snail *Euglandina rosea* has been introduced from Florida and Central America to many islands in the Pacific and Indian Oceans to control another introduced snail, the Giant African snail *Achatina fulica* (**Figure 23**), which can reach 15 cm in length and was originally introduced for food. It eats hundreds of plants and is considered an agricultural pest on many islands. Biological control has consisted of introductions of predatory snails that are not restricted in their feeding, so they have driven native snail species to low numbers or to extinction. Hawaii once had 931 species of land snails, but about 600 species have disappeared since European colonization (mostly because of deforestation for agriculture). The

Summary

Pests are species that interfere with human activities and hence need to be controlled. Most pest control in agricultural systems is achieved in a temporary manner using herbicides and pesticides, but these toxic chemicals affect other important species and become ineffective because pests develop genetic resistance to the chemicals. Biological control makes use of parasitoids, predators, and diseases to reduce the average abundance of a pest species.

There are many cases of major reductions in numbers of introduced pests by predators or by insect parasitoids that are introduced specifically for the purposes of control. Many other introductions have failed, leaving the pest to be controlled by chemical means. We cannot adequately explain most of the

successes, nor can we explain why failure is so common. The classical model of pest control through density-dependent predation does not seem to describe the situation of successful control programs in the field.

Genetic control of pests can be accomplished by producing resistant crop plants or by interfering with the fertility or longevity of the pest. Many techniques for the genetic control of pests have been proposed, but few have been used successfully in the field. Immunocontraception is a new method of pest control being developed for overabundant mammalian species. All forms of pest control raise ecological questions of how the pest may compensate for increased mortality or reduced fertility. Genetic engineering holds great promise for producing both new methods of reducing pest numbers and crops that are more resistant to insect pests.

Integrated pest management combines the best features of cultural, biological, and chemical control methods in an effort to minimize the environmental degradation that has been problematic with modern commercial agriculture. To achieve integrated control, we need to understand the population dynamics of the pest species; this is, at present, one of the greatest challenges in applied ecology. The application of behavioral ecology to the control of pests is a promising lead in integrated control programs.

Biological control programs entail the potential risk that the control agent will attack other native species in addition to the targeted pest. The cure must be better than the disease, and extensive testing must be carried out before and after any biological control program is activated.

Review Questions and Problems

1 Purple loosestrife (*Lythrum salicaria*) is a wetland plant introduced to North America from Europe in the early 1800s. It has been declared a severe environmental problem in Canada and the United States because it is believed to take over wetlands, displacing native vegetation and adversely affecting wildlife species. Discuss how you would test the hypothesis that purple loosestrife has detrimental effects on other species in wetlands so that biological control should be used. Compare your action plan with the data presented by Blossey et al. (2001), who decided that purple loosestrife does have adverse effects.

2 Why does *Bacillus thuringiensis* produce proteins that are toxic to insects? Review the biology of this bacterium and its geographic distribution, and discuss the evolution of its protein toxins. Lambert and Peferoen (1992) and Clark et al. (2005) provide references.

3 Figure 21 gives data to show that a majority of biological control successes are attributed to a single species even though several agents have been introduced for each program. Stiling and Cornelissen (2005) present an analysis that shows multiple-releases reduced pest abundance by 27% more than single releases. Discuss why these two sets of data appear to contradict one another.

4 Elton (1958) showed that introduced species often increase enormously and then subside to a more static, lower density level. How might this occur in a species that was not the subject of introductions for biological control? How could you distinguish this case from a decline that followed the introduction of some parasitoids for biological control?

5 Discuss the limitations of the push-pull strategy of integrated pest management. What pest problems is this strategy best suited for? Cook et al. (2007) provide references and information on push-pull methods and approaches.

6 Immunocontraception as a strategy of pest control will not work, according to the conclusions presented in Cooper and Larsen (2006). Discuss their pessimistic conclusions, compare them with the discussion in Hardy et al. (2006), and discuss how you might change a targeted immunocontraception program to take into account their critique.

7 Contact your local municipal authorities and find out how Norway rats are controlled in your area. Discuss any ecological problems you can see with their methods and approach. Buckle and Smith (1994), Colvin and Jackson (1999), and Ashley et al. (2003) provide background material on rat control methods.

8 Since the accidental introduction of the moth *Cactoblastis cactorum* into Florida via the Caribbean or South America, there has been great concern that this biocontrol agent could wipe out native species of the *Opuntia* cactus (Stiling 2002). Pemberton and Liu (2007) found that the impacts of *Cactoblastis* in the Caribbean were not catastrophic. Read their analysis and discuss why this biocontrol agent may not be as great a risk as originally thought.

9 For *r*-selected pests, Stenseth (1981) suggests that optimal control can be achieved by reducing reproduction rather than by increasing mortality. Discuss the ecological reasons behind this recommendation.

Overview Question

How do pests evolve resistance to chemicals used to control them? Will this problem arise with the most recent techniques that utilize genetic engineering and immunocontraception? How could you overcome the evolution of resistance in pest populations?

Suggested Readings

- Blossey, B. 1995. A comparison of various approaches for evaluating potential biological control agents using insects on *Lythrum salicaria*. *Biological Control* 5: 113–122.

- Bowman, D. D. 2006. Successful and currently ongoing parasite eradication programs. *Veterinary Parasitology* 139:293–307.

- Christou, P., T. Capell, A. Kohli, J. A. Gatehouse, and A. M. R. Gatehouse. 2006. Recent developments and future prospects in insect pest control in transgenic crops. *Trends in Plant Science* 11:302–308.

- Cook, S. M., Z. R. Khan, and J. A. Pickett. 2007. The use of push-pull strategies in integrated pest management. *Annual Review of Entomology* 52:375–400.

- Dyck, V. A., J. Hendrichs, and A. S. Robinson, eds. 2005. *Sterile Insect Technique: Principles and Practice in Area-wide Integrated Pest Management*. Dordrecht, Netherlands: Springer.

- Kimberling, D. N. 2004. Lessons from history: Predicting successes and risks of intentional introductions for arthropod biological control. *Biological Invasions* 6:301–318.

- Murdoch, W. W., S. L. Swarbrick, and C. J. Briggs. 2006. Biological control: Lessons from a study of California red scale. *Population Ecology* 48:297–305.

- Oerke, E.-C. 2006. Crop losses to pests. *Journal of Agricultural Science* 144:31–43.

- Simberloff, D., and P. Stiling. 1996. How risky is biological control? *Ecology* 77:1965–1974.

- Singleton, G. R., L. Hinds, H. Leirs, and Z. Zhang. 1999. *Ecologically-based Management of Rodent Pests*. Canberra, Australia: Australian Centre for International Agricultural Research.

- Van Lenteren, J. C., J. Bale, F. Bigler, H. M. T. Hokkanen, and A. J. M. Loomans. 2006. Assessing risks of releasing exotic biological control agents of arthropod pests. *Annual Review of Entomology* 51:609–634.

Credits

Illustration and Table Credits

Photo Credits

Applied Problems III: Conservation Biology

Key Concepts

- Conservation biology is the applied ecology of endangered species. It rests on two themes—the effects of small population size on fitness, and the causes for population decline and extinction.

- Small populations are subject to chance events associated with demography, environmental accidents, and genetic drift. All these events can contribute to an extinction vortex of positive feedbacks that result in declining fitness and finally extinction.

- Declining populations must be studied to diagnose the causes of the decline and to prescribe a remedy. Understanding the causes for a decline and the consequences of being a small population can together assist in developing action plans for conservation.

- Extinctions are increasing worldwide, primarily as a result of habitat loss and the introduction of nonnative species. Providing corridors that connect habitat fragments can facilitate movements, but this approach does not work for every species at risk.

- Reserves are part of an effective conservation strategy, but because none are large enough, they cannot conserve large vertebrates without explicit management of the surrounding landscape.

- The effects of increasing human populations and the continued loss of habitat for natural communities are the root causes behind the current conservation crisis.

From Chapter 17 of *Ecology: The Experimental Analysis of Distribution and Abundance*, Sixth Edition. Eugene Hecht.

KEY TERMS

coarse-grained habitat From a particular species' point of view a habitat is coarse grained if it spends its life in one fragment of habitat and cannot move easily to another patch.

declining-population paradigm The focus of this approach is on detecting, diagnosing, and halting a population decline by finding the causal factors affecting the population.

demographic stochasticity The random variation in birth and death rates that can lead by chance to extinction.

deterministic extinctions Losses of species due to the removal of an essential resource.

effective population size A population genetic concept of the number of breeding individuals in an idealized population that would maintain the existing genetic variability; it is typically much less than the observed population size.

environmental stochasticity Variation in population growth rates imposed by changes in weather and biotic factors, as well as natural catastrophes like floods and hurricanes.

fine-grained habitat From a particular species' point of view, a habitat is fine grained if it moves freely from one patch to another at no cost.

genetic stochasticity Any potential loss of genetic variation due to inbreeding or genetic drift (the nonrandom assortment of genes during reproduction).

minimum viable population (MVP) The size of a population in terms of breeding individuals that will ensure at some specified level of risk continued existence with ecological and genetic integrity.

small-population paradigm The focus of this approach is on rare species and on the population consequences of rareness, and the abilities of small populations to deal with rarity.

Conservation biology is the biology of population decline and scarcity and is a central focus of much public concern. Much of population ecology has been focused on abundant species and the factors preventing population growth. Many population biologists have pointed out that most species are rare, and rarity itself ought to be a focus for population research. Species that have become endangered or threatened are either rare or in sharp decline, and in this chapter we explore the causes

of decline and rarity of species and what we can do to alleviate the problems of threatened populations.

Conservation has become an important political issue during the last 20 years, and practical issues of conservation are continually in the newspapers and on television. The magnitude of the conservation issue can be illustrated most easily with data from well-known groups such as birds and mammals (**Figure 1**). Nearly one-quarter of all known mammals on Earth are classed as threatened species by the International Union for the Conservation of Nature. Many fewer species of reptiles, amphibians, and fishes are classed as threatened, but this reflects only the lack of study of these groups. For insects the problem is much worse: of the one million described species of insects, less than 0.1% have been evaluated for their conservation status.

Given the size of the problem of the numbers of threatened species, what theory and what understanding do ecologists bring to these practical problems?

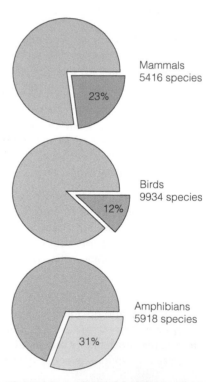

Figure 1 The total number of species and the percentage of species threatened with extinction in 2006 among the mammals, birds, and amphibians. For most groups only a small fraction of species have been evaluated for threatened status. Mammals, birds, and amphibians have been completely evaluated, but only 664 of the reptile species of the world have been evaluated. Thus, the fraction that is threatened is only poorly known. The species counts are the total number of described species in each group for the world. (Data from International Union for the Conservation of Nature Red List 2007.)

Conservation biology divides cleanly into two separate approaches that operate with different ideas and different goals. Caughley (1994) called these two paradigms the **small-population paradigm** and the **declining-population paradigm**. Both of these approaches are important for conservation of threatened or endangered species and need to be used together when possible. We examine the small-population paradigm first.

Small-Population Paradigm

This paradigm focuses on the population consequences of rareness and the abilities of small populations to deal with smallness as such. The ideal is a small island population, or a small group of endangered species of animals or plants in a zoo or a botanical garden, and the questions arising from this paradigm deal largely with population genetics and demographic models of extinction in small populations. The essence of the small-population paradigm is encapsulated in the extinction vortex (**Figure 2**). Small populations risk positive feedback loops of inbreeding depression, genetic drift, and chance demographic events that lead inexorably to extinction. An essential feature of the small-population paradigm is a set of strong theoretical predictions that follow from population genetics theory. The key element is the maintenance of genetic variability, which is essential for future evolution and thus for long-term persistence. The central issue is how a small population can maintain both ecological and genetic integrity, and this consideration has led to the concept of minimum viable populations.

Minimum Viable Populations

Rare species will still be able to sustain their numbers at some population density. This idea has been formalized as the concept of the **minimum viable population (MVP)**—that population size that will ensure at some acceptable level of risk that the population will persist for a specified time (Gilpin and Soulé 1986). The analysis of minimum viable populations involves the analysis of extinction. What factors cause a species to go extinct? Some extinctions are due to chance, and Shaffer (1981) recognized three kinds of variation that can contribute to population loss.

1. **Demographic stochasticity.** This source of variation reflects random variation in birth and death rates that can lead by chance to extinction. If only a few individuals make up the population, the fate of each individual can be critical to population survival. If a female produces only male offspring and then dies, the population goes extinct. These are examples of demographic stochasticity. In general, demographic variability is critical to extinction only when populations are less than about 30–50 individuals (Caughley 1994).

2. **Genetic stochasticity.** Because evolution cannot occur without genetic variability, any loss of genetic variation can be a cause of extinction. Many genetic studies have shown that individuals with more heterozygous loci are fitter than individuals with less genetic variation (Futuyma 2005). Genetic variability is lost by genetic drift, the nonrandom assortment of genes during reproduction, and by inbreeding. Both drift and inbreeding are minimized when populations become sufficiently large, so these phenomena are paradigms of the small-population problem.

3. **Environmental stochasticity and natural catastrophes.** These factors include variation in population growth rates imposed by changes in weather and biotic factors, as well as by fire, floods, hurricanes, and landslides, which can also be responsible for species declines. The key concept is how much variation the environment imposes on the rate of increase of the population. If the variance in the rate of population growth is greater than the population growth rate itself, environmental stochasticity can cause extinction (Lande 1993).

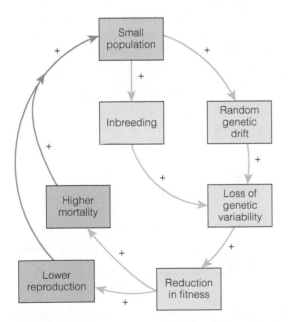

Figure 2 The extinction vortex of the small-population paradigm. Small populations, such as those on islands or in zoos, fall into a vortex of positive feedback loops in which small population size leads to inbreeding and genetic drift, resulting in loss of genetic variability. Because genetic variability is necessary for viability, fitness falls and the population size is reduced further.

Small populations may become small because of habitat changes, but much of the small-population paradigm focuses on small populations of rare species. Rare species are particularly important in conservation biology, but they are not always identical to what a conservation biologist is concerned with in the small-population paradigm. When a naturalist says that a bird or a plant is rare, he or she may mean several different things (Harper 1981; Rabinowitz 1981). The concept of rarity can be described best according to three characteristics: geographic range, habitat specificity, and local population size. If we divide each characteristic into two levels, we can classify rare species as in **Table 1**. Of the eight classes, only one describes "common" organisms, and thus "rare" may mean seven different things to an ecologist. If there are seven different classes of rare species, we must recognize different kinds of management to protect species that are threatened with extinction.

Classic rare species are often those with small geographic ranges and habitat specificity. Many plants of this type are restricted endemics, and are often endangered or threatened (Rabinowitz 1981). Other rare species have very large geographic ranges and occur widely in different habitats but are always at low density. These species are ecologically interesting but almost never appear on lists of endangered species. The important point is that not all rare species are concerns of conservation biology.

The general principle underlying the small-population paradigm is that the smaller the population, the greater the risk that chance events can lead to extinction. But how small is small, and how can we determine the minimum viable population for a particular species? The task is not easy (Beissinger and McCullough 2002). Genetics and demography provide two very different approaches for estimating minimum viable populations. Population genetics has provided a very simple rule for minimum viable populations: the 50/500 rule. Franklin (1980) pointed out that inbreeding could be kept to a low level with a minimum population of about 50 animals, and that this rule worked well for animal breeders working in agriculture. He suggested that this level was high enough to prevent inbreeding depression, one factor in the extinction vortex (see Figure 2). To prevent genetic drift a larger population is needed, and Franklin (1980) suggested that a minimum of 500 animals or plants would be sufficient to allow evolution to proceed unimpeded. Note that population geneticists count individuals in units of **effective population size**, and that these units are not the same as the individuals population ecologists count. Real populations would often have to be three to ten times larger than their effective population size counterpart. The 50/500 rule was proposed as a rule of thumb, and it should not be applied as a law for all species (Sanderson 2006). Because it is based purely on genetic concepts, it cannot be applied to animals and plants that are subject to varying levels of environmental variation and different breeding systems. Soulé and Mills (1992) have pointed out that simple rules for minimum viable populations need critical evaluation before we can use them in practical decisions about endangered species.

The demographic approach to setting minimum viable population size is much more data intensive. There are two approaches. First, if a population model like the theta-logistic can be fitted to a time series of population counts, the model can be run with a variety of initial population sizes, and if several thousand runs are carried out on the computer, it is possible to count how many simulated populations go to extinction (Brook et al.

Table 1 A classification of rare species based on three characteristics: geographic range, habitat specificity, and local population size.

	Geographic range			
	Large		Small	
	Habitat specificity			
Population size	Wide	Narrow	Wide	Narrow
Large, dominant somewhere	Locally abundant over a large range in several habitats	Locally abundant over a large range in a specific habitat	Locally abundant in several habitats but restricted geographically	Locally abundant in a specific habitat but restricted geographically
Small, nondominant	Constantly sparse over a large range and in several habitats	Constantly sparse in a specific habitat but over a large range	Constantly sparse and geographically restricted in several habitats	Constantly sparse and geographically restricted in a specific habitat

SOURCE: Modified after Rabinowitz (1981).

What Is Effective Population Size?

Geneticists and ecologists talk about population sizes in two quite different ways. To an ecologist a population consists of immature and mature plants and animals, some of which are breeding and some of which are not. A geneticist, by contrast, is concerned about the genetic population—those individuals that contribute genes to the next generation. If an individual does not breed or breeds unsuccessfully, it does not exist in a geneticist's count, even though it uses resources and is potential food for predators.

We begin with a definition from Sewall Wright (1931), "The effective size of a population is the size of an ideal population that would undergo the same amount of genetic drift as the population under consideration." *Genetic drift* is change in the genetic composition of a population arising as a consequence of sampling of gametes in a finite population. Drift is a sampling problem, and sampling problems are always more difficult in small populations. Genetic drift produces random changes in population composition, and a gradual increase in homozygosity, and thus a loss of genetic variation. For a given size of population, the rate of genetic drift will depend on the mating structure and the variation in number of successful offspring. In the ideal world of genetics, we can imagine a world in which each individual contributes gametes equally to a pool for the next generation—this is the *effective population size* or N_e. The important point for conservation is the relationship between actual population size and effective population size (Franklin 1980; Waples 2002).

1. *Variation in progeny number.* If the population consists of N individuals who have variance in progeny number σ^2, we obtain

$$N_e = \frac{4N}{2 + \sigma^2} \qquad (1)$$

If the mean family size is 2 and the variance is 4, the effective population size is 2/3 observed numbers.

2. *Unequal breeding numbers in the two sexes.* If the breeding system is a lek or harem system, as in fur seals or zebras, and only a few males do most of the breeding, the effective population size is given by

$$N_e = \frac{1}{\left(\dfrac{1}{4N_m} + \dfrac{1}{4N_f}\right)} \qquad (2)$$

where N_m = number of breeding males
N_f = number of breeding females

Thus if we have a population of 100 fur seals, with 95 breeding females and five harem bulls, you have an effective population size of 19 seals, not 100.

3. *Fluctuating population size.* If the population fluctuates in size from generation to generation, the effective population size is again reduced. Crow and Kimura (1970) show that for t generations:

$$\frac{1}{N_e} = \frac{1}{t}\left(\frac{1}{N_1} + \frac{1}{N_2} + \frac{1}{N_3} + \frac{1}{N_4} + \dots + \frac{1}{N_t}\right) \qquad (3)$$

The effective population size is strongly affected by the lowest observed population. For example, if we have a lion population that is stable for nine generations at 100 individuals but because of a severe drought drops to 10 individuals for one generation, the effective population size for this time period is 53 lions, not 100.

These are all simple illustrations of the general notion that effective population sizes are almost always much less than observed population sizes, often by margins of three or four times or even more. **Figure 3** illustrates this for populations of chinook salmon from Oregon.

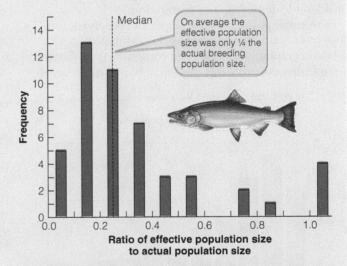

Figure 3 Estimates of the ratio of effective population size to total population size for stocks of chinook salmon from the Snake River in Oregon. Two genetic methods were used to estimate the effective number of breeders each year, and this was compared to a count of the number of spawners. Each sample represents one year from one population with an average genetic sample size of 76 fish. (Data from Waples 2002.)

2006). This has been done for 225 bird populations with the results shown in **Figure 4**. For these birds the average MVP is 722 breeding birds, not far above the recommended 500 from the genetics approach to MVP. Brook et al. (2006) ran data from 1198 species and obtained an average MVP of 1377 individuals (with 90% probability of surviving 100 years). There is however a great deal of scatter among the different species.

A second demographic approach uses a detailed demographic model such as VORTEX which demands demographic data on sex ratios, age at maturity of each sex, proportion of females breeding each season, survival rates of all age classes, amount of environmental variation affecting demographic rates, and the rate of population growth under optimal conditions (Lacy 2000). For example, Brito and Grelle (2006) estimated the minimum viable population size for a Brazilian endemic forest primate, the northern muriqui or wooly spider monkey *(Brachyteles hypoxanthus)* by the use of the VORTEX program. This primate is the largest South American primate, and is among the rarest mammals in the world. Brito and Grelle (2006) estimate an MVP of 40 individuals, but a population this small would lose genetic variability over the long term. Consequently they recommend a population of about 700 breeding individuals to maintain genetic variation for the long term, requiring about 11,570 ha of forest habitat. VORTEX is a general model that will estimate minimum viable population size for any species for which there is considerable demographic information.

The small-population paradigm in conservation biology has a strong theoretical base in population genetics and demography, and is useful to conservation biologists for exploring the problems that small populations face if they are to survive in the short term and in the long term. It does not often *solve* the problem of small populations, which is the focus of the next paradigm.

The Declining-Population Paradigm

The declining-population paradigm focuses on the ways of detecting, diagnosing, and halting a population decline. The problem is viewed in demographic terms—as a population in trouble—and for this reason this paradigm is action oriented. Some external agent must be identified as the cause of the decline, and research efforts focus on what can be done about it. Because it is action oriented, there is little theory in this paradigm. Research efforts are concentrated on each specific case study, and at least in the short term no great theoretical advances in understanding the causes of extinction will occur. But if done properly, the research gets the job done. This paradigm does not consider the current size of the population to be as important—it is the downward trend that is the main concern. In the best cases the declining-population approach can be combined with the small-population approach to solve conservation problems. An excellent example of this is the recovery of the prairie chicken in Illinois.

The prairie chicken (*Tympanuchus cupido*) was a common grouse from New England and Virginia, and west to Kansas, when Europeans first arrived in North America. The prairie chicken was distributed widely across the central plains of the United States, but it has been fragmented by agriculture into scattered populations in the central United States. In Illinois, prairie chickens numbered in the millions in the nineteenth century but declined to 25,000 birds by 1933 (Westemeier et al. 1998). By 1993 only 50 prairie chickens survived in Illinois, but large populations remained in Kansas, Minnesota, and Nebraska. **Figure 5** shows the decline of one population in Jasper County, central Illinois, from 1970 to 1997. The population decline was mirrored in a decline in the hatching rate of eggs, which was thought to be due to low levels of genetic diversity. In 1992 a translocation program was begun to move prairie chickens from the large populations in Kansas and Nebraska into Illinois. Over the next five years a total of 271 prairie chickens were translocated to Illinois. Egg viability quickly improved (see Figure 5), and the population rebounded. Reduced genetic variability in the declining Illinois population was verified by analyzing microsatellite loci isolated from feather roots of museum specimens and recent collections (Bouzat et al. 1998). **Table 2** shows that fewer alleles were found in the recent Illinois population compared with the number found either in other large

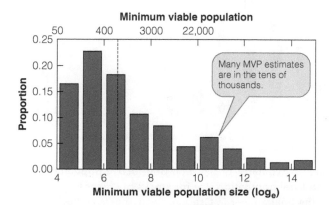

Figure 4 Frequency distribution of minimum viable population (MVP) size for 225 bird populations for which there are long-term population data. Note the MVP scale is a log axis; a few arithmetic equivalents are given at the top of the graph. The overall average for birds is 722, indicated by the dashed line. (Data from Brook et al. 2006.)

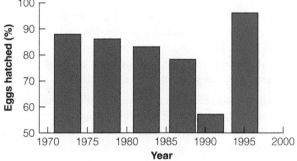

Figure 5 The decline of the prairie chicken (*Tympanuchus cupido*) in central Illinois, 1970–1997. The population collapse was mirrored in a reduction in fertility. In 1992 prairie chickens from Minnesota, Kansas, and Nebraska were translocated to increase genetic variability (blue arrow). The population rebounded strongly after this introduction. (Modified after Westemeier et al. 1998)

populations or in Illinois before the population decline of the past 50 years. These genetic data confirm that the collapse of the Illinois prairie chicken population followed the extinction vortex until it was rescued from imminent extinction in 1992 by translocating new genetic stock into the Illinois population (Bouzat et al. 1998a).

The prairie chicken rescue illustrates the key ideas of the declining-population paradigm: recognize the decline, assess the potential causes, and treat them experimentally to stop the decline. Other populations of the prairie chicken in the central United States are suffering the same problems of a decline in genetic variation due to small, isolated populations (Johnson et al. 2004), and management actions to reduce inbreeding are required to prevent further extinctions (Johnson and Dunn 2006).

Paul Ehrlich of Stanford University has been instrumental in putting declining populations and extinction on the worldwide agenda through his books and popular writings on the biodiversity crisis. He has highlighted the many ways in which humans are contributing to population losses. Ehrlich's main point is that many extinctions are not due to "chance" in the broad sense. Many population declines are completely determined by some inexorable change from which there is no escape without action. Shaffer (1981) called these **deterministic extinctions**. Deforestation is one such change; glaciation is another. If an area is deforested, all species that require trees are eliminated. Deterministic extinctions occur when some essential resource is removed or when something lethal is introduced into the environment. Loss of habitat leads to deterministic extinctions and is a major problem in almost every ecosystem on Earth (Ehrlich and Ehrlich 1981).

Why do deterministic extinctions occur? The four causes of extinction, called the "evil quartet" by Diamond (1989), are:

- Overkill
- Habitat destruction and fragmentation
- Introduced species
- Chains of extinction

We consider each of these in turn.

Table 2 Number of alleles per locus in the greater prairie chicken from a survey of the current Illinois population (1974–1993 birds) and the large populations from Kansas, Minnesota, and Nebraska.

	Illinois *Before 1950*	Illinois *After 1974*	Kansas	Minnesota	Nebraska
Mean number of alleles	5.12	3.67	5.83	5.33	5.83
Standard error	0.87	0.56	0.75	0.84	1.05
Sample size	15	32	37	38	20

A sample of museum specimens was used to obtain the pre-1950 Illinois data.

Six microsatellite loci were used from DNA isolated from feather roots.

SOURCE: From Bouzat et al. (1998b).

Diagnosing a Declining Population

Much of the practical conservation biology depends on the careful diagnosis of declining populations. Caughley (1994) laid out a series of logical steps to determine what is driving a species toward extinction.

1. *Confirm that the species is presently in decline, or that it was formerly more widely distributed or more abundant.* This will require some qualitative or quantitative assessment of population trends and distribution. A species in decline may be common or rare. Both types can become conservation problems if the decline continues.

2. *Study the species' natural history and collect all information on its ecology and status.* For many species a considerable amount of background knowledge, both formal and informal, exists. Information on related species may be useful here.

3. *List all the possible causes of the decline, if enough background information is available.* This is the method of multiple working hypotheses; cast a wide net to consider all possible causes. Remember that direct human actions may be an agent of decline, but do not restrict hypotheses to human causes.

4. *List the predictions of each hypothesis for the decline, and try to specify contrasting predictions from the different hypotheses.* Do not assume that the answer is already known by scientific or folk wisdom.

5. *Test the most likely hypothesis by experiment to confirm that this factor is indeed the cause of the decline.* Often factors are correlated with the decline but are not causing it. The best experiment involves removing the suspected agent of decline.

6. *Apply these findings to the management of the threatened species.* This will involve monitoring subsequent recovery until the problem of decline is resolved.

Applying this approach to an endangered species already in low abundance will be difficult, but there is no alternative. Several suspected agents of decline may have to be removed at once, and additional studies undertaken to identify exactly which one was most responsible. It is better to save the species than to achieve scientific purity.

Overkill

Overkill consists of fishing or hunting at a rate that exceeds a population's capacity to rebound. The species that are most susceptible to overkill are the large species with low intrinsic rates of natural increase (r)—elephants, whales, rhinoceros, and others that are considered valuable by humans. Species on islands are also vulnerable to extinction if the island is small. The great auk, a large flightless seabird (**Figure 6**), was hunted to extinction on islands in the Atlantic Ocean in the 1840s because of a demand for feathers, eggs, and meat (Montevecchi and Kirk 1996).

The decline of the African elephant is a classic example of the effect of hunting on a large mammal. The African elephant is the largest living terrestrial mammal, weighing up to 7500 kg. Sexual maturity is reached only after 10–11 years, and a single calf is born every three to nine years. The potential rate of increase was estimated by Sinclair (1997) to be about 6% per year, a low population growth rate. Between 1970 and 1989 half of Africa's elephants were killed for the ivory trade. This decline prompted CITES to ban all trade in ivory, and the response has been a dramatic increase in elephant numbers (Blake et al. 2007). From 2002 to 2006 elephant numbers in Africa increased on average 4% per year. Southern Africa holds about 58% of the continent's elephants, while east Africa holds 30%. The situation in Central and West Africa is less clear because of little data (Blanc et al. 2007). In some of the national parks in

Great Auk

Figure 6 Great Auk (*Pinguinis impennis*), extinct by 1844. These large flightless seabirds in the North Atlantic were easy prey for sailors needing food.

Females

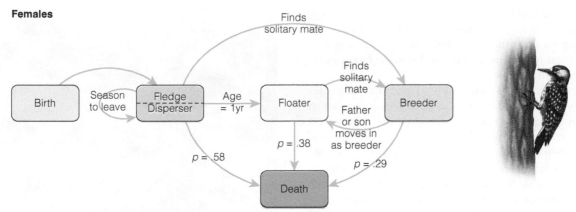

Figure 7 Annual transition probabilities of female red-cockaded woodpeckers in the sandhills of North Carolina. Females disperse to new territories, males must decide whether to remain on the same territory or disperse to another. Most males tend to remain as helpers on their natal territory. (From Letcher et al. 1998.)

Southern Africa elephants are considered overabundant and must be culled. The key to the extensive decline of elephants before 1990 was clearly poaching for ivory, and once this incentive was removed, populations have been recovering.

Overkill, or excessive human exploitation, will remain a problem for all animals and plants that are valuable or large.

Habitat Destruction and Fragmentation

The second factor in the "evil quartet" that promotes extinctions is habitat loss. Habitats may simply be destroyed to make way for housing developments or agricultural fields. Cases of habitat destruction appear to provide the simplest examples of the declining-population paradigm. An example can illustrate how subtle the effects of habitat destruction can be.

The red-cockaded woodpecker is an endangered species endemic to the southeastern United States. It was once abundant from New Jersey to Texas and inland to Missouri. It is now nearly extinct in the northern and inland parts of its geographic range. The red-cockaded woodpecker is adapted to pine savannas, but most of this woodland has been destroyed for agriculture and timber production. These birds feed on insects under pine bark and nest in cavities in old pine trees. Because most old pines have been cut down, the availability of nesting holes has become limiting (Walters 1991).

Designing a recovery program for the red-cockaded woodpecker has been complicated by the social organization of this species. They live in groups of a breeding pair and up to four helpers, nearly all males. Helpers do not breed but assist in incubation and feeding. Young birds have a choice of dispersing or staying to help in a breeding group. If they stay, they become breeders by inheriting breeding status upon the death of older

birds. Helpers may wait many years before they acquire breeding status. **Figure 7** and **Figure 8** show a schematic of the life history events of the female and male red-cockaded woodpecker respectively, along with the probabilities of moving between states.

From a conservation viewpoint, the problem is that red-cockaded woodpeckers compete for breeding vacancies in existing groups, instead of forming new groups that might occupy abandoned territories or start at a new site by excavating nesting cavities. The key problem is the excavation of new breeding cavities.

Males

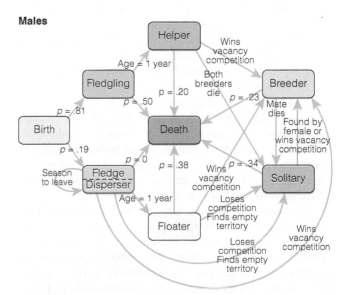

Figure 8 Annual transition probabilities of male red-cockaded woodpeckers in the sandhills of North Carolina. Whereas females disperse to new territories, males must decide whether to remain on the same territory or disperse to another. Most males tend to remain as helpers on their natal territory. (From Letcher et al. 1998.)

Because of the time (typically several years) and energy needed to excavate new cavities, birds are better off competing for existing territories than establishing new ones. Habitat loss appeared to be the main factor causing population decline.

To test this idea, Walters (1991) and his colleagues artificially constructed cavities in pine trees at 20 sites in North Carolina. The results were dramatic—18 of 20 sites were colonized by red-cockaded woodpeckers, and new breeding groups were formed only on areas containing artificial cavities. This experiment showed clearly that much suitable habitat is not occupied by this woodpecker because of a shortage of cavities. Management of this endangered species should not be directed toward reducing mortality of these birds but instead should focus on providing tree cavities suitable for nesting.

An additional complication of cavity-nesting species is competition for cavities. The endangered red-cockaded woodpecker population at the Savannah River Site in South Carolina was rescued from near extinction by a combination of adding artificial nest cavities and translocating birds from larger nearby populations (Franzreb 1997). To prevent competition for the artificial cavities, 2304 southern flying squirrels (*Glaucomys volans*) were removed between 1986 and 1995. The woodpecker population responded dramatically, increasing from four to 99 individuals in response to these management actions. Problems of inbreeding persist in small populations of the red-cockaded woodpecker because females typically disperse only short distances (Schiegg et al. 2006). Management actions to increase the size of local populations will significantly reduce the likelihood of extinction.

The rescue of the red-cockaded woodpecker is a good example of how successful conservation biology must depend on a detailed understanding of population dynamics and social organization, so that limiting factors can be identified and made more abundant. There are no general prescriptions for rescuing endangered species, and we must operate on a case-by-case approach. Detailed information on resource requirements, social organization, and dispersal powers are required before recovery plans can be specified for species suffering from habitat loss and fragmentation.

Humans have appropriated a large fraction of the land surface of the Earth for agriculture, and many plants and animals cannot survive in an agricultural landscape. Of the remaining areas, many have been fragmented, or broken up into small patches (**Figure 9**), a common situation in every country on Earth (Echeverria et al. 2006; Ewers et al. 2006). Habitat fragmentation has many components with varying effects on population dynamics (**Table 3**). The impact of fragmentation is species specific. A habitat is called **fine-grained** for a species if the patches are short distances apart and the species can move back and forth between patches with little cost.

Figure 9 Forest fragmentation in the Daintree region of North Queensland. Coastal tropical forest has been fragmented by agricultural fields into patches of various sizes and connectivity. (Photo courtesy of the Wet Tropics Management Authority, Australia.)

Conversely, a **coarse-grained habitat** for a species requires long-distance dispersal, and individuals in coarse-grained habitat typically live most or all their life in one patch. Species such as eagles that move over large areas may treat a fragmented habitat as continuous, whereas the exact same habitats may appear very coarse-grained to a plant with limited dispersal powers (Laurance et al. 2006). Scale is critical in fragmentation, and ecological scales are highly species specific.

Habitat fragmentation has the potential to reduce genetic variation in plants partly because of isolation of fragments that prevents gene flow and partly because of the small populations remaining in the fragments (Young et al. 1996). **Figure 10** illustrates how forest fragmentation in the white box woodlands of southeastern Australia have resulted in smaller populations of white box having less genetic variability. Many plant species show this pattern of reduced genetic diversity once habitats are fragmented into small pieces (Honnay and Jacquemyn 2007).

Fragmentation of habitats can be analyzed by considering the dynamics of populations subdivided into small patches. At one extreme, when patches are too small the species cannot survive. We can see this very clearly by looking at incidence functions, the occupancy rate of a species in habitats of differing size. **Figure 11** illustrates this concept with data on the great spotted woodpecker (*Dendrocopos major*) in England. The incidence function for 1991 was shifted toward larger patches because of severe winter weather, and the following year small woodlands were occupied again. Incidence functions for different species will vary, and are most useful for conservation planning because they indicate the minimum area of habitat required to support the species. In general there is a good relationship between the body size of animals and the area

Table 3 Changes associated with habitat fragmentation and their possible effects on population dynamics.

	Habitat change	Consequences for population dynamics
Population-level effects	Reduced connectivity, insularization, increased interfragment distance	Directly affects dispersal and reduces the immigration rate
	Reduced fragment size, reduced total area	Directly affects population size and increases the extinction rate
Landscape or community-level effects	Reduced interior-edge ratio	Indirectly affects mortality and production through increased pressure from predators, competitors, parasites, and disease
	Reduced habitat heterogeneity within fragments	
	Increased habitat heterogeneity in surrounding matrix	Indirectly affects population size through reduced carrying capacity within the fragment
	Loss of keystone species from the habitat	Indirectly affects mortality and production through increased carrying capacity of predators, competitors, etc. in the surrounding matrix
		Indirect effect through disruption of mutualistic guilds or food webs

SOURCE: From Rolstad (1991).

required for survival and reproduction (Biedermann 2003). Larger animals need a larger area of habitat.

Small patches are subject to chance extinction due to weather or disease more often than are large patches. In western Europe the European red squirrel (*Sciurus vulgaris*) occupies patches of forest interspersed in a mosaic of agricultural land (Celada et al. 1994). Home ranges of this squirrel range from 4.5 ha for females to 6.4 ha for males. Red squirrels in northern Italy were always present in woodlots larger than 7 ha and were never present in woodlots smaller than 2 ha (Celada et al. 1994). These small woodlots may be connected by fencerows or trees along roads, and what is crucial for all fragmented populations is how readily individuals can move between patches. The study of fragmented patches thus becomes a study of metapopulations; when subpopulations in patches become extinct, the patches can be recolonized by dispersing individuals.

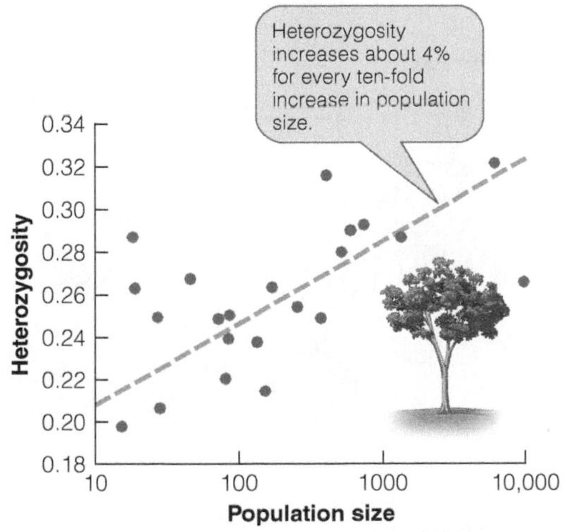

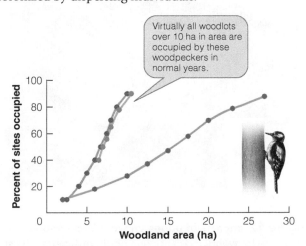

Figure 10 Relationship between reproductive population size in the tree *Eucalyptus albens* (white box) and heterozygosity. Population size is measured by the number of white box trees in fragments of differing area. Heterozygosity is measured by the proportion of loci at which an individual carries more than one allele. (Data from Prober and Brown 1994.)

Figure 11 Incidence functions for the great spotted woodpecker (*Dendrocopos major*) in English woodlands from 1990 to 1992. Severe winter weather in 1991 caused populations in smaller woodlots to disappear. In the following year, 1992, the smaller woodlands were reoccupied. Red, 1990; blue, 1991; green, 1992. (Data from Hinsley et al. 1996.)

Recolonization may not always occur in isolated patches. The Bogor Botanical Garden was established in 1817 on 86 ha in west Java. Until 1936 the Botanical Garden was connected with other forest areas to the east, but for the past 60 years it has been an isolated patch of forest with the nearest patch 5 km away (Diamond et al. 1987). Of the 62 bird species recorded as breeding in the Botanical Garden during 1932–1952, 20 species had disappeared by 1980–1985 and four more were close to extinction. The species that were lost were the less common species, and their low abundance combined with the lack of recolonization from surrounding areas has been the main cause of extinction (Diamond et al. 1987). The result is that much of the conservation value of the Botanical Garden for birds has been lost because it is too small by itself to support a secure population of many tropical forest birds.

In almost all cases habitat fragmentation leads to species loss. The prairies of North America are a good example. Prairie covered about 800,000 ha of southern Wisconsin when Europeans first arrived, and now prairie occupies less than 0.1% of its original area (Leach and Givnish 1996). Plant surveys of 54 Wisconsin prairie remnants studied between 1948 and 1954 were repeated in 1987–88. Between 8% and 60% of the plant species were lost during these four decades, at av-

erage rates between 0.5% and 1.0% per year. At this rate of extinction approximately half the plant species would disappear in 50 to 100 years. Losses were particularly high among the shorter plant species and the rare species. The control of fire in prairies seems to be the agent of decline for prairie plants, and controlled burns should be done to reverse these population declines (Leach and Givnish 1996).

One of the important consequences of fragmentation is that it increases the amount of edge in a habitat (see Table 3). If predators search habitat edges, higher predation rates might occur in smaller fragments because of the edge effect. Particular attention has been focused on roads, which provide edges as they cut through continuous habitats. The impact of oil and gas exploration on woodland caribou in Canada is a good example of the impact of development on animal populations. Woodland caribou have been declining in numbers during the last 100 years and are a threatened species in Alberta. Oil and gas exploration in northern Alberta has produced a series of roads, seismic lines, and oil and gas wells in a mosaic habitat of otherwise undisturbed coniferous forest (**Figure 12**). The

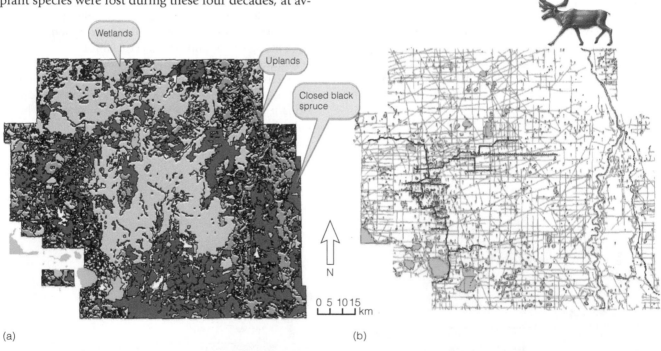

(a) (b)

Figure 12 Impacts of edges on woodland caribou. (a) Habitat mosaic of the 6000-km² study site for woodland caribou in northern Alberta. The study site is dominated by coniferous black spruce forest in wetlands (tan), closed black spruce wetlands (dark green), and uplands (orange) with aspen and white spruce. (b) Human development in this study area, showing wellsites (green triangles), roads (black), and seismic lines (blue). Only 1% of the land area is occupied by these human developments, yet because woodland caribou avoid these areas of human disturbance, up to half of the potential habitat is not used by the caribou. (Modified from Dyer et al. 2001.)

important point to note is that only 1% of the forest has been occupied by these gravel roads, seismic lines (5- to 8-m-wide cut lines for exploration), and well sites (1-ha gravel pads). No one expected that a 1% land use change would affect these caribou.

By putting satellite radio collars on 36 woodland caribou in this study area, Dyer et al. (2001) could follow the movements of each animal with several locations per day through an annual cycle. The question they asked was whether caribou used the areas adjacent to the roads, seismic lines, and well sites as often as they used areas away from disturbed sites. Caribou avoided human developments, staying up to 1000 m away from well sites and 250 m away from roads and seismic lines. Avoidance of disturbed sites was maximal in late winter, a time of food stress. Caribou reduced their use of 22%–48% of the entire study area due to these human disturbances on 1% of the landscape. Roads and seismic lines have also facilitated the travel of predators such as wolves in the region, and hunters have increased access with the road network. The net result of what would appear to be a minor loss of habitat translates into a major impact on woodland caribou and a declining population from these combined stresses.

One of the most critical variables in the dynamics of populations in fragmented habitats is migration between patches. At present we have few data on movements of animals and plants between patches. Much discussion in conservation agencies has focused on providing corridors between refuges so that species can disperse from one patch to the next. Corridors, if used, help to prevent inbreeding depression and allow recolonization (Simberloff and Cox 1987). But there are potential costs to corridors, because they may facilitate disease transmission, conduct fires, and expose individuals to increased predation risk (**Table 4**). The Florida panther (*Felis concolor*) has been reduced from approximately 1400 individuals to about 30 animals isolated in undeveloped areas of south Florida. By providing a corridor system between wildlife refuges, managers hope to increase the effective population size of panthers (Simberloff and Cox 1987; Kautz et al. 2006). But there are only limited data to determine how wide a corridor must be before large mammals like the panther will use them. Moreover, it may be difficult to stop poaching in corridors, which may be expensive to purchase and maintain. To reduce inbreeding in the small Florida panther population, eight female panthers were brought to Florida

Table 4 Potential advantages and disadvantages of conservation corridors.

Potential advantages	Potential disadvantages
1. Increase immigration rate to a reserve, which could: a. Increase or maintain species richness and diversity (as predicted by island biogeography theory). b. Increase population sizes of particular species and decrease probability of extinction (provide a "rescue effect") or permit reestablishment of extinct local populations. c. Prevent inbreeding depression and maintain genetic variation within populations.	1. Increase immigration rate to a reserve, which could: a. Facilitate the spread of epidemic diseases, insect pests, exotic species, weeds, and other undesirable species into reserves and across the landscape. b. Decrease the level of genetic variation among population or subpopulations, or disrupt local adaptations and coadapted gene complexes ("outbreeding depression").
2. Provide increased foraging area for wide-ranging species.	2. Facilitate spread of fire and other abiotic disturbances ("contagious catastrophes").
3. Provide predator-escape cover for movements between patches.	3. Increase exposure of wildlife to hunters, poachers, and other predators.
4. Provide a mix of habitats and successional stages accessible to species that require a variety of habitats for different activities or stages of their life cycles.	4. Riparian strips, often recommended as corridor sites, might not enhance dispersal or survival of upland species.
5. Provide alternative refuges from large disturbances (a "fire escape").	5. High cost, and conflicts with conventional land preservation strategy for preserving endangered species habitat (when inherent quality of corridor habitat is low).
6. Provide "greenbelts" to limit urban sprawl, abate pollution, provide recreational opportunities, and enhance scenery and land values.	

SOURCE: From Noss (1987).

from Texas in 1995. Kittens from these females survived much better than those from purebred Florida females, and the panther population has increased from about 30 individuals to about 87 in 2003. This genetic rescue has been a conservation success but has been highly controversial (Pimm et al. 2006). If sufficient habitat is not set aside in South Florida, the Florida panther will not be able to survive (Kautz et al. 2006).

Detailed studies of the movements of individuals between patches and along corridors are rare. As we have seen, corridors (which seems like a good idea to human observers) do not always fit the needs of the targeted organism. **Figure 13** illustrates this problem with corridors for grizzly bears in the Rocky Mountains of Alberta. Corridors for wildlife have been designated in the absence of any data and are sandwiched between human occupations and steep mountainous terrain (Figure 13a). By obtaining locations on radio-collared bears, Chetkiewicz et al. (2006) could describe

in detail the habitat characteristics used by grizzly bears in this region (Figure 13b). Given these data, corridors could be designed properly with the needs of the particular species in mind. Too often corridors are not wide enough to facilitate movements of large animals.

Corridors are not necessarily useful for the conservation of all species in all situations, and thus conservation recommendations will not be the same for all species affected by fragmentation (Beier and Noss 1998). Experimental manipulations of local populations could be used to test the general hypothesis that patches of remnant habitat connected to source areas by habitat corridors will be recolonized more readily than patches without corridors. More well-designed studies are needed to measure the effects of corridors on plants and animals (Chetkiewicz et al. 2006). Corridors can be an effective adjunct to conservation planning in fragmented landscapes, and it is prudent to retain landscape connectivity where possible.

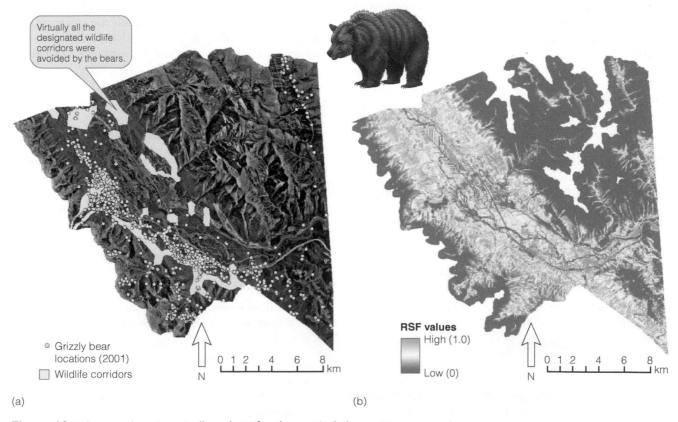

(a)   (b)

Figure 13 Telemetry locations (yellow dots) for three grizzly bears (*Ursus arctos*) during 2001 in the Bow Valley area of Alberta. (a) The corridors set aside for wildlife (green areas) were rarely used in this area of heavy human use, and by being poorly sited resulted in one human fatality in 2001. (b) By analyzing the habitats actually used by grizzlies, Chetkiewicz et al. (2006) could specify a resource selection function, indicating the areas bears preferred to use (green). By combining these data, wildlife corridors in this area could be redesigned to be more appropriate for these large mammals. (From Chetkiewicz et al. 2006.)

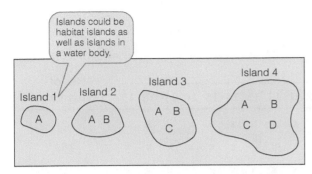

Figure 14 **Hypothetical island faunas forming a series of nested subsets.** A through D represent species occurring on the islands. Because all species in smaller faunas also occur in all larger faunas, the smaller faunas are subsets of the larger faunas and the fauna is completely nested. Thus if only the larger island can be conserved, no species are lost from the system. (From Cutler 1991.)

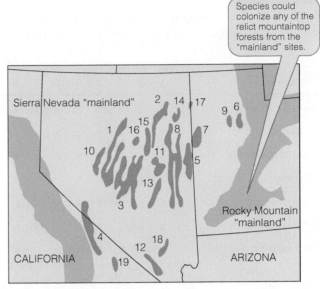

Figure 15 **Distribution of montane forests above 2300 m (7500 ft.) in the Great Basin region of western United States.** Stippled green areas indicate relict forests on isolated ranges: 1, Toiyabe–Shoshone; 2, Ruby; 3, Toquima–Monitor; 4, White–Inyo; 5, Snake; 6, Oquirrh; 7, Deep Creek; 8, Schell Creek–Egan; 9, Stanbury; 10, Desatoya; 11, White Pine; 12, Spring; 13, Grant–Quinn Canyon; 14, Spruce–South Pequop; 15, Diamond; 16, Robert Creek; 17, Pilot; 18, Sheep; 19, Panamint. Areas of lighter blue indicate "mainland" forests in the Sierra Nevada and Rocky Mountains. (Modified from Brown 1978.)

When landscapes are fragmented, species on the smaller patches may begin to go extinct. If extinction occurs at random among all the species, the extinction pattern of the remaining species will be random. But for many faunas extinction does not occur at random and the resulting patterns form nested subsets (Patterson 1987). **Figure 14** illustrates the concept of a nested subset. A nested subset can be considered as a time series—species D, which occurs only on island 4, is the first to go extinct if the island is devastated, followed by species C, which occurs on islands 3 and 4, and so on, in sequence (Cutler 1991). If this is correct, extinction is more predictable than random, and conservation biologists can focus on those species that need special protection because they occur only on larger areas.

The coniferous mountain forests in the Great Basin of the western United States (**Figure 15**) contain a good example of nested subsets. Brown (1978) tabulated the occurrence of 14 mammal species on these isolated areas, and Skaggs and Boecklen (1996) added some additional records (**Table 5**). These forest populations are relicts from the Ice Age when coniferous forests were more widespread; the forests now constitute patches or islands in a sea of relatively unsuitable desert habitat. If the mammal patterns were completely nested, there would be no holes or outliers in the table. For example, the chipmunk *Eutamias dorsalis* is "missing" from mountain range 2 (a hole), and the rabbit *Sylvilagus nutalli* is present in mountain range 19 (an outlier). The pattern shown in Table 5 is clearly nonrandom and is a good example of a nested subset.

Nested subsets may result from selective extinction or selective colonization. In the Great Basin mammals, selective extinction is usually given as the explanation of the nested structure, but data on potential colonization movements across the intervening barriers are lacking. Wright et al. (1998) found that about half of the 279 data sets like that shown in Table 5 were significantly nested, and they suggested that extinction is more often the process that leads to nested subsets. It is important to determine if nested subsets occur in fragmented habitats. Not all species are equally vulnerable to extinction, and it is important to direct conservation efforts toward the most vulnerable species (Donlan et al. 2005).

Impacts of Introduced Species

Introduced animals are responsible for about 40% of historic extinctions. Most of these data involve mammals and birds, for which we have more detailed information, and these are no doubt biased (Caughley and Gunn 1996). But no one doubts the adverse effects of introduced species. The Nile perch, which was introduced into Lake Victoria in the early 1980s, caused the extinction or near-extinction of over 200 endemic species of cichlid fish between 1984 and

Table 5 Species distribution matrix for boreal mammals of Great Basin mountain ranges. Mountain range numbers refer to those shown in Figure 15.

Species	1	2	3	4	5	6	7	8	9	10	11	12	13	14	15	16	17	18	19	No. of occurrences
Eutamias umbrinus	x	x	x	x	x	x	x	x	x	x	x	x	x	x	x	x		x		17
Neotoma cinerea	x	x	x	x	x	x	x	x	x	x	x	x		x	x	x	x	x	x	18
Eutamias dorsalis	x		x	x	x	x	x	x	x	x	x	x	x	x	x		x	x	x	17
Spermophilus lateralis	x	x	x	x	x		x	x		x	x	x	x	x	x	x	x			15
Microtus longicaudus	x	x	x	x	x	x	x	x	x	x	x			x		x				13
Sylvilagus nutalli	x	x	x	x	x		x	x		x	x	x		x	x	x			x	14
Marmota flaviventris	x	x	x	x	x	x	x	x	x	x	x				x	x				13
Sorex vagrans	x	x	x	x	x	x	x	x	x			x								10
Sorex palustris	x	x	x	x	x	x			x							x				8
Mustela erminea	x		x	x	x	x		x												6
Ochotona princeps	x	x	x	x					x											5
Zapus princeps	x	x				x			x							x				5
Spermophilus beldingi	x	x	x																	3
Lepus townsendi		x				x			x	x	x	x								6
No. of Species	13	12	11	11	10	10	9	8	10	9	8	7	5	4	6	8	3	3	3	

SOURCE: Data from Skaggs and Boecklen (1996).

1997 (Seehausen et al. 1997). The Nile perch is now subject to a large fishery, and its abundance has been greatly reduced during the 1990s. This reduction in predation pressure has resulted in a few of the endemic species recovering in Lake Victoria (Witte et al. 2000).

Nearly 50% of the mammal extinctions of the past 200 years occurred in Australia. Neither very small nor very large mammals have been affected in these recent losses. A critical weight range from 35 to 4200 g contains all the missing mammals (Burbidge and McKenzie 1989). Many causes can be suggested to explain these extinctions, from habitat clearing associated with agriculture, to changes in the fire regime, to introduced herbivores as competitors, to introduced predators. The main culprit seems to be introduced predators, particularly the red fox (Kinnear et al. 1998; Short 1998). The details of the loss of medium-sized marsupials in Australia is a mirror image of the spread of the red fox (Short 1998). If the red fox can be controlled, some of the threatened species, now confined to offshore fox-free islands, could be reintroduced to their former range (Richards and Short 2003).

There are many examples of introduced predators causing conservation problems. *Working with the Data: Recovery of Petrels after Eradication of Feral Cats on Marion Island, Indian Ocean* gives one illustration for introduced cats.

Introduced species are one of the most serious conservation problems today and the leading cause of animal extinctions (Clavero and Garcia-Berthou 2005). For birds the estimates are that introduced species are the cause of 50% of recent extinctions; for fish in North America, 67%; and for mammals, 48%. As global trade has increased, many inadvertent or deliberate introductions are occurring, often with little regard for their conservation consequences (Ruesink et al. 1995).

Chains of Extinctions

The last of the "evil quartet" causing extinctions is a set of secondary extinctions that follow from a primary

Recovery of Petrels after Eradication of Feral Cats on Marion Island, Indian Ocean

Example of the application of the declining-population paradigm described by Caughley and Gunn (1996) to solve a particular conservation problem.

PROBLEM

Marion Island (290 km²) in the southern part of the Indian Ocean had breeding populations of 12 petrel species, which breed in burrows. Five house cats were introduced in 1948 to control introduced house mice on the island, and the cat population increased at 23% per year to reach 3045 cats in 1977. The cats preyed on the adults, chicks, and eggs of eight species of burrowing petrels. As the petrel populations shrank, the cats shifted their attention to house mice. The great-winged petrel, *Pterodroma macroptera*, was especially vulnerable to the cats because it breeds in winter, has a long breeding season, and used larger burrows than other petrels.

DIAGNOSIS OF THE CAUSE OF THE DECLINE

In 1975 cats killed around 48,000 great-winged petrels, which became relatively rare compared to numbers on the neighboring (and cat-free) Prince Edward Island.

	Prince Edward Island	Marion Island
Cats	Present	Not Present
Nests with Chick (%)*	33% (n = 30)	1% (n = 109)

*Data from 1979

An introduced disease reduced cats, but then the survivors again increased in numbers. Outside a cat-proof exclosure, no petrel nests contained chicks, compared to 50% of nests inside the predator exclosure. On this evidence the factor driving the petrel decline was postulated to be cat predation.

RECOVERY TREATMENT

1977	Feline panleukopenia (FLP) introduced
1982	Cats reduced to 620; FLP antibodies subsequently decreased in the cats
1986–1990	952 cats removed by shooting and trapping
1990	Petrel survival increased from 100% chick mortality in 1979–1984 to 0% chick mortality in 1990
1991	Cat eradication believed complete
1992	Reports of increases in house mice abundance

extinction. If other species depend on a lost species for survival, these other species must also go extinct. Chains of extinctions require obligate specialist relationships that are more typical of tropical areas than of temperate or polar zones. One obvious chain of extinctions involves the loss of parasite species when their host goes extinct. This matter has received scant attention to date and there are few examples that are well documented.

The clearest examples of chains of extinctions involve large predators that disappeared when their prey went extinct. The extinct forest eagle of New Zealand (*Harpagornis moorei*) (**Figure 16**), which weighed 10–13 kg, and preyed on large ground birds, died out around AD 1400 when moas became extinct in New Zealand (Holdaway 1989). The decline of the black-footed ferret in North America was associated with the decline of its main food, prairie dogs, on the Great Plains (Caughley and Gunn 1996, p. 91). Currently the black-footed ferret is being reintroduced into areas where prairie dog colonies are safe (Biggins et al. 1998), but its future is not secure because it is highly susceptible to canine distemper, which is endemic in carnivores on the Great Plains.

Reserve Design and Reserve Selection

One way to conserve species in danger of extinction is to set up reserves or protected places. National parks in many countries have been viewed as protected areas for populations and communities. The selection and design of nature reserves is an important part of conservation biology, and much effort has gone into developing good methods of reserve selection and design. To begin

Approximately 7% of the world's land area is now set aside as some form of a reserve, and the goal of many governments is to protect about 12% of terrestrial habitats. If we are given the job of selecting and locating reserves, how should we proceed? One way is to identify "hotspots" that are particularly rich in species, and to locate reserves in these areas (Reid 1998). One problem with this approach is that areas that are hotspots for birds are typically not hotspots for butterflies, so we cannot choose reserves on the basis of only one taxonomic group and expect that it will protect other groups as well. Nevertheless, some small areas are much richer in species than others, and we should use this kind of information to help select reserves. Caughley and Gunn (1996, p. 321) have given the following overview of how to proceed in reserve selection:

Step 1. Decide on the objective of the reserve system clearly and unambiguously.

Step 2. Identify which areas of land are available for designation as reserves within the terms of the objectives decided in step 1.

Step 3. Survey each patch that might become a reserve and obtain a list of species present, and if possible, an estimate of abundance of each.

Step 4. Formulate a starting rule for selecting the first reserve, and how subsequent patches will be chosen in sequence.

Working with the Data: An Algorithm for Choosing Reserves for a Taxonomic Group illustrates one method of formulating objective rules for reserve selection (Margules and Pressey 2000). It is important to realize that many different ways of selecting reserves are possible, depending on the objectives. The preference criterion may be to preserve rare species, or to preserve sites with many different species, or to preserve the largest number of taxonomic units such as genera or families. Most reserve selection algorithms use presence/absence as the relevant criterion rather than species abundance because it is easier to determine presence/absence than it is to estimate abundances for many species.

To create a reserve system that is useful for conservation, it is necessary to know the ecological requirements of the species of concern. A special problem exists for species that use temporary habitats. Many butterflies use areas for egg laying and larval development that are temporary. If the protected area set aside in a reserve—for example, from a meadow to a forest—the butterfly loses its host plants (Warren 1994; Hanski et al. 1995). Butterflies are often distributed as metapopulations, and movement between suitable patches of habitat is critical to survival.

Figure 16 The now extinct forest eagle of New Zealand (*Harpagornis moorei*). This large eagle fed on moas, which were driven extinct after humans colonized New Zealand around AD 1400. (From Gill and Martinson 1991.)

we need to specify exactly what a reserve is intended to accomplish. Two quite divergent aims are often stated for reserves:

1. To conserve specific animal and plant communities that are subject to change because of fire, grazing, or predation. These reserves must be managed by intervention to set the permissible levels of fire, grazing, and predation.

2. To allow the system to exist in its natural state and to change as governed by undisturbed ecological processes, so that no attempt will be made to influence the resulting changes in populations and communities.

Often reserves such as national parks have both these aims, creating a recipe for conflict over what kinds of changes are acceptable and what kinds are unacceptable to the managers or to the general public (Caughley and Sinclair 1994).

An Algorithm for Choosing Reserves for a Taxonomic Group

Many methods exist for selecting reserves for conservation. Once the objective of the reserves is decided, we must specify objective rules for evaluating which areas are best to select. Nicholls and Margules (1993) have suggested the following method for selecting reserves for conservation.

Step 1. State the objective as clearly and specifically as possible; for example, "To create a reserve system that captures 10% of the range occupied within a region by each species in the genus *Eucalyptus.*"

Step 2. This is an optional step that allows the inclusion of some sites before the selection process begins. Examples might be reserves already set aside, national parks, or protected sites with known rare and endangered species.

Step 3. Select all sites that have a species that occurs in no other site.

Step 4. Find the next rarest species and select the sites that, when added to those already selected, will represent that species plus the greatest number of additional species at or above the required proportion (10%) of their area of distribution.

Step 5. If there is a choice, select the site that is closest in proximity to a site already selected.

Step 6. If there is still a choice, select the site that also contributes the largest number of as yet inadequately represented species.

Step 7. If there is still a choice, select the site that achieves the required level of representation of the rarest species remaining underrepresented.

Step 8. If there is still a choice, select the site that contributes the most to achieving the required level of representation of the rarest group of species remaining underrepresented.

Step 9. If there is still a choice, select the site that either contains the smallest percentage area needed to achieve the required level of representation of the species under consideration or that contributes the largest percentage of that species' range if no one site achieves adequate representation.

Step 10. If there is still a choice, select the smallest site.

Step 11. If there is still a choice, select the first suitable site on the list.

Step 12. Go to Step 4.

This objective method of site selection assumes a list of available sites and the species that occur in them.

SOURCE: Modified from Caughley and Gunn (1996).

One of the most significant contributions of conservation biology has been to show that viable populations of some species are large; thus, it may be impossible to maintain the required number of animals in parks or sanctuaries (Soulé 1987). **Figure 17** shows the situation for the grizzly bear in the area containing Yellowstone and Grand Teton National Park. If we draw a biotic boundary for a minimum viable population of 500 grizzly bears, the area needed to support this population is 122,330 sq. km, about 12 times the actual park area of 10,328 sq. km. Our existing parks are far too small to maintain large mammals and birds on the scale we now expect (Newmark 1985, 1995). Areas of private land outside of parks must also contribute to the preservation of diversity, and the integration of land use for agriculture and forestry with conservation is an important area of focus.

About 12%–13% of the Earth's land area is now protected (IUCN 2007). Most of the protected areas in the world are small (**Figure 18**), with 59% being smaller than 1000 ha in area and occupying only 0.2% of the total protected area. By contrast, the six largest protected areas (including Greenland National Park at 972,000 km²) constitute nearly 15% of the total protected area. Protected areas are not always protected from poaching and hunting, and setting aside land for conservation is an important first step but not the end point of conservation.

Examples of Conservation Problems

Two examples of conservation problems will illustrate the practical realities of applying conservation principles to endangered plants and animals. Many more examples are given in Caughley and Gunn (1996) and in Lindenmayer and Fischer (2006).

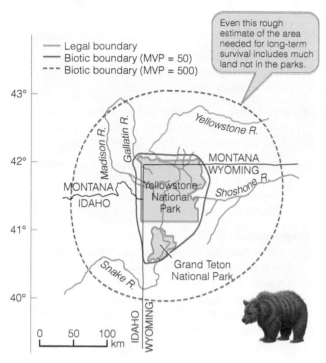

Figure 17 The legal and biotic boundaries for the grizzly bear (*Ursus arctos*) in the Yellowstone-Grand Teton National Park assemblage in western United States. The biotic boundaries are defined by the entire watershed for the parks and the area necessary to support a minimum viable population (MVP) of 50 bears for short-term survival, and 500 bears for long-term survival. (From Newmark 1985.)

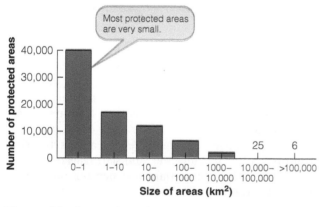

Figure 18 The number of protected areas in the world in 2003. There are few large areas (too small to show up on the scale so the number is given). Most protected areas are small, and the six largest protected areas include Greenland (972,000 km²), which has more ice and rock than biodiversity, and Ar-Rub'al-Khali in southwestern Saudi Arabia (640,000 km²), which is desert. The six largest areas comprise 13% of the total area of the globe that is protected in one form or another. (Data from IUCN 2007.)

The Northern Spotted Owl in the Pacific Northwest

The northern spotted owl (*Strix occidentalis caurina*) has been the focus of intense debate and confrontations over how the remaining old-growth forests of the western United States should be managed. The northern spotted owl is a territorial owl that lives in old-growth conifer forests. Each pair of owls utilizes about 250–1000 ha (1–4 sq. mi) of valuable old-growth forest, nesting in hollow trees and feeding on small mammals, birds, and insects. Heavy logging on private land in the past 40 years has destroyed most of the old-growth forest upon which these owls depend. Most of the remaining old growth is on lands managed by the U.S. Forest Service and the National Park Service. The northern spotted owl is not now an endangered species, and the total population in the Pacific Northwest is roughly 1200 pairs in 2007.

Old-growth forests are being rapidly reduced in the Pacific Northwest, as they are elsewhere on the globe. A large part of the controversy over the northern spotted owl concerns the questions of what type of habitat this owl requires, and how much its habitat can be fragmented by logging without causing a population decline. Northern spotted owls highly prefer old-growth forests for feeding and for roosting (Carey et al. 1992). In fragmented forests owls move more but still feed and roost only in old growth (**Figure 19**). The home range size of owls varies with the prey base. The most common prey in Washington and Oregon is the northern flying squirrel (*Glaucomys sabrinus*). In Washington State, owl home ranges include about 1700 ha of old-growth forests, but in Oregon ranges are less than half that size. These differences in home ranges are directly related to the prey base:

	Home range (ha)	Prey available (g/ha)
Washington	~ 1700	61
Oregon		
Douglas fir	813	244
Mixed conifer	454	338

Additional diet studies in northern California showed that the woodrat (*Neotoma fuscipes*) was a major prey item, and that owls preferred larger prey like woodrats (mean weight: 230 g) when they were available, with flying squirrels (110 g) a second choice (Ward et al. 1998).

Bart and Forsman (1992) surveyed 11,057 sq. km throughout the range of the northern spotted owl. They found no owls in forests that were only 50–80 years old and confirmed that owls occurred only where

Figure 19 Two examples of areas used by northern spotted owls in southwestern Oregon. (a) A lightly fragmented old-growth forest; (b) a heavily fragmented old-growth forest. The owls make very little use of young forest. (From Carey et al. 1992.)

old-growth stands were present. **Figure 20** shows that northern spotted owls were both more common and more successful reproductively in old-growth forests. Landscapes with less than 20% old-growth forest rarely supported an owl population. Spotted owls nest in trees that are much larger and older than the average tree in old-growth stands (LaHaye and Gutierrez 1999). In northern California more than 80% of their nest trees were older than 300 years old, and most were greater than 1.2 m in diameter.

One surprising result of studies on the northern spotted owl is that wilderness areas are not very suitable as habitat for the owls (Bart and Forsman 1992). Productivity within protected wilderness areas was only 30%–50% as much as that in old-growth forest outside these designated areas. Much of the wilderness areas and national parks in the Pacific Northwest are high-elevation areas that are less suitable for these owls. The surprising result is that currently protected stands of old-growth forest in parks and wilderness areas may be unable to sustain the northern spotted owl.

How much old-growth forest must be kept to preserve the northern spotted owl? The key parameters for making this estimate are the dispersal and colonization success of young owls and the survival and reproductive rates of territorial owls living in landscapes with variable amounts of old forest. Anthony et al. (2006) did an exhaustive analysis of the status of the northern spotted owl from 14 study areas in Washington, Oregon, and California. Fecundity showed no systematic time trend for these owls. Population growth rates of the northern spotted owl are most sensitive to the adult

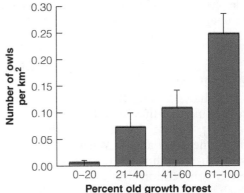

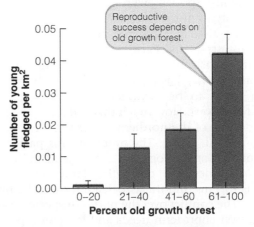

Figure 20 Density and reproductive success of northern spotted owls in relation to the amount of older forest on 145 forest areas in Washington, Oregon, and northern California. These owls do well only in areas with a large fraction of old-growth forest remaining. (From Bart and Forsman 1992.)

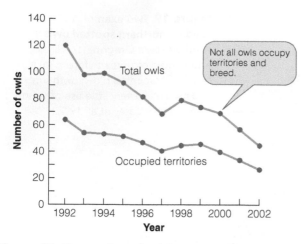

Figure 21 The number of northern spotted owls counted and the number of territories occupied in the Cle Elum Study Area of Okanogan-Wenatchee National Forest, Washington from 1992 to 2002. Since 1992 owl numbers at this site have been declining at 6% per year on average, implying that the current conservation plan is not working properly. (Data from Anthony et al. 2006.)

survival rate. For 12 of 13 study areas for which they had high-quality data, the rate of population change (λ) was less than 1.0, and the mean for all areas was 0.963, suggesting that they are all declining on average 3.7% per year. **Figure 21** illustrates the population trends from one study area in Washington. One possible cause of these declines in northern spotted owls is interference competition for territories with barred owls, which are larger and are moving their geographic range west into spotted owl range.

All analyses of the northern spotted owl concur in recognizing that a large part of the remaining old-growth forests in the Pacific Northwest must be preserved if we wish this species to persist. Since 2000 the U.S. Forest Service has cut back on the amount of old-growth forest to be protected from logging in the Pacific Northwest. How much this will affect the population decline of the northern spotted owl is not clear (Anthony et al. 2006). The problem thus passes from the conservation biologist to the general public as a matter of policy. The competing land use for these forests is logging and the associated jobs in the timber industry. The conflict over the northern spotted owl is a conflict over short-term needs and long-term goals. At the current rate of harvesting, most of the old-growth forests in the Pacific Northwest will be gone within 20 years, and at that time the problems of the timber industry will still be with us, but the northern spotted owl may not. The present conflict over land use in old-growth forests is but one example of a much broader

question: How can human populations and the Earth's biota coexist without serious disruptions? This is the central issue for conservation biology in the twenty-first century.

Leadbeater's Possum in Australia

The conservation of the endangered Australian marsupial Leadbeater's possum (*Gymnobelideus leadbeateri*) has been a contentious conservation issue in Australia with many similarities to the northern spotted owl problem. Leadbeater's possum has a highly restricted geographic distribution (60 km by 50 km) in old-growth eucalypt forests in the Central Highlands of Victoria in southern Australia. It is a small (130 g), nocturnal, arboreal marsupial that lives in colonies of up to 12 animals, with a life span of about five years (Lindenmayer 1996). The two key limiting resources for Leadbeater's possum are the availability of nest sites in large trees with hollows and the availability of food in dense understory stands of acacia trees. They are omnivores and feed on sap from acacia trees and a variety of arthropods. They live in tree hollows that occur mainly in mountain ash trees in excess of 150 years of age and in older trees damaged in wildfires (Lindenmayer and McCarthy 2006).

The conservation conflict over Leadbeater's possum is caused by clear-cut logging in the mountain ash forests with a rotation time of 80–120 years. Given this type of forest management, soon there will be no old trees with hollows available for Leadbeater's possum and other species of arboreal marsupials like the mountain brushtail possum (*Trichosurus cunninghami*). Both forest harvesting and fires are key disturbance factors in mountain ash forests. After fires, foresters typically undertake salvage logging of the trees damaged by fire, which effectively removes the trees that produced hollows for wildlife.

Which forest management options would be best for the conservation of Leadbeater's possum? Possingham et al. (2002) used a population viability model to assess the current status of the possum and to investigate the relative benefits of alternative management options. **Figure 22** shows a decision tree for alternative management strategies for Leadbeater's possum with the key parameter being the probability of surviving for 150 years under each strategy. Under the present forest management that includes salvage logging after a fire, the probability of extinction in 150 years is 100%. To change forest management, two general options are available. One option is to extend the rotation time from 80 years to 100 years or more. This would reduce the probability of extinction but at the cost of a complete cessation of logging for more than 100 years,

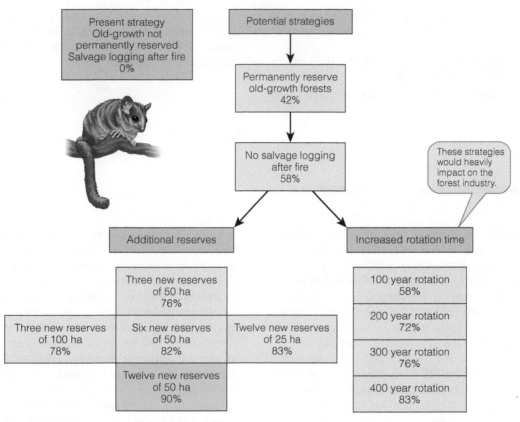

Figure 22 Flowchart showing the impact of different management options on the conservation of Leadbeater's possum in Australia. A population viability analysis for this possum provided estimates of the probability of the population surviving for 150 years (expressed as a percentage). (From Possingham et al. 2002.)

a solution not acceptable to the forest industry. The second option is to add additional reserves of old-growth forest, and this turns out to be the best option. Adding one or two large reserves would not be optimal because these could be burned by the same fire, and a network of smaller reserves is a better option (Figure 22). Setting up several 50- to 100-ha reserves in a forest block of 10,000 ha with adequate corridors between the reserves is an optimal strategy for reducing the chances of extinction (Possingham et al. 2002).

One suggestion for increasing populations of Leadbeater's possum is to add nest boxes to forest stands that are too young to have natural hollows (Lindenmayer et al. 2003; Harley 2006). The addition of nest boxes has been controversial because of the cost of the boxes and the fact that many other animals use the boxes in addition to Leadbeater's possum. The addition of nest boxes can, however, be a useful strategy in the short term to provide nest sites during the interval in which forest management practices such as salvage logging are being changed. Once enough trees with hol-

lows are available, nest boxes would no longer be needed and tree hollows would not be a limiting resource (Harley 2006).

The key finding of the extensive work on this arboreal marsupial is that it is possible to use population viability analysis in conjunction with decision theory to provide guidance to management agencies that are charged with conservation goals. The resulting management strategies will benefit not only Leadbeater's possum but all the associated animals in old-growth mountain ash forests.

Conclusion

In Part Three we have considered a complex set of ecological questions about the abundance of populations. We used population mathematics to illustrate how we can deal with populations in a precise, quantitative manner. Herein lies the strength and the weakness of population ecology, because to some degree we must abstract the population from the matrix of other species in the community in order to describe its dynamics.

For many populations, other species in the community are essential neighbors, and hence we need to broaden our frame of reference beyond the population level. Thus, we are led to consider the whole biological community and, in particular, to ask how distribution and abundance interact to structure the biological communities that cover the globe.

Summary

Conservation biology is focused on the ecology of rare and declining species. Two threads of conservation biology are a focus on small populations and the consequences of being small (the small-population paradigm), and a focus on declining populations (the declining-population paradigm). Small populations are subject to an array of uncertainties, from chance demographic events (having all male offspring) to chance environmental events (a flood), to chance genetic events (genetic drift). Even though not all small populations are conservation problems, being small increases the chances of extinction for many populations and can lead a species into an extinction vortex powered by positive feedback of chance processes. An elegant body of theory has given us a good description of the hazards of being a small population.

The declining-population paradigm focuses on identifying the ecological causes of decline and designing alleviation measures to stop the decline. It contains little ecological theory but is focused on individual action plans. Only by understanding the population biology of an endangered plant or animal can we provide a rescue plan for a declining population. In some cases, such as the African elephant, the causes of population decline are clear. In other cases we do not have the ecological understanding to recommend action, and we need to develop insights for action plans. The best conservation programs combine the small-population approach with the declining-population approach to solve problems with endangered species.

Extinction is the ultimate conservation focus, and four causes are prominent: excessive hunting or harvesting, habitat destruction and fragmentation, introduced species, and chains of extinctions. The major causes of recent extinctions are habitat destruction and introduced species. Habitat destruction leads to population reductions that may trigger the extinction vortex, so protecting habitat is a major goal for all conservation efforts. At present about 12%–13% of the world's land areas is protected, but most protected areas are small. Existing parks and reserves are seldom large enough to contain viable populations of larger vertebrates, and conservation efforts on private lands surrounding the reserves are essential to maintaining populations of flora and fauna.

Habitat fragmentation has been a side effect of agriculture and forestry and has many adverse effects on populations. Populations in isolated patches may go extinct, and unless recolonization occurs, a species may be lost. Corridors between reserves may assist dispersal between patches, but some potential problems, such as the spread of disease, can be aggravated by corridors. Maintaining connectedness of reserves has become an important goal of conservation biology.

The ecological challenge of conservation biology is to develop specific management plans for individual species, whereas the political challenge to the broader conservation movement is to protect large natural areas from destruction. Without parks and reserves there can be no conservation, but with them there is no guarantee of success unless conservation biology can solve the challenging ecological problems of threatened and endangered species.

Review Questions and Problems

1. Barro Colorado Island was formed 85 years ago in central Panama when Gatun Lake was created as part of the Panama Canal. Since that time 65 of 394 species of birds have disappeared from the island, 21 of them in the past 25 years. Discuss what mechanisms might cause extinctions of birds that can fly in an undisturbed area of tropical forest. Robinson (1999) discusses these changes.

2. Review the claim of Sagoff (2005) that introduced species are not a serious problem for conservation, and the rebuttal of this by Simberloff (2005) and by Clavero and Garcia-Berthou (2005).

3. When organisms of the same species are brought together to breed from divergent geographic areas, outbreeding depression may occur in which the fertility or viability of the offspring is impaired (Templeton 1986). This is one reason why some biologists were opposed to the addition of panthers from Texas to the Florida panther population in 1995. Discuss the reasons outbreeding depression occurs and its implications for the conservation of the Florida panther. Pimm et al. (2006) give an overview of the controversy.

4. One of the most extensive ecological experiments is being carried out by the Biological Dynamics of Forest Fragments Project in the Brazilian Amazon. One experiment has involved creating small isolated patches of rain forest and following the extinction and colonization of these patches over time. Ferraz et al. (2007) have reported on the impacts of fragmentation on Amazon birds. Discuss what predictions you would make for this experiment from island biogeography theory, and review the results to date.

5. Kirtland's warbler is an endangered species that breeds in northern Michigan jack pine forests. From the 1950s to the 1970s the population of this species declined, and it numbered about 200 individuals in 1971. It was stable in numbers from 1971 to 1986. The most important factor in the population decline seemed to be increasing parasitism of nests by brown-headed cowbirds. Cowbirds were removed from the breeding area of Kirtland's warbler starting in 1971, but no change occurred in warbler numbers by 1986, and speculation began that it was being limited on its wintering grounds in the Bahamas. The alternative hypothesis was that habitat was becoming unsuitable for these birds. Extensive habitat management was begun in 1987 and two large wildfires rejuvenated the jack pine stands in which it breeds. The warbler increased in numbers fourfold in the 1990s. What management plans would you now recommend for this endangered species? Probst et al. (2003) provide recent evidence on this endangered bird.

6. One of the possible reasons for the continuing decline of the northern spotted owl is interference competition with the larger barred owl (Anthony et al. 2006). Discuss how you would test this competition hypothesis, and what measurements you would require to do so.

7. Review the history of the successful rehabilitation of the endangered Lord Howe Island woodhen (*Tricholimnas sylvestris*) on Lord Howe Island in the Pacific (Caughley and Gunn 1996, pp. 75–81). Discuss the reasons for the success of this project and the general principles it illustrates for conservation problems.

8. One possible impact of invasive species is called "invasional meltdown" to describe situations in which invasive species facilitate more invasive species, thus accelerating the impact of invasives (Simberloff 2006). The best case so far described is the invasion of yellow crazy ants on Christmas Island (O'Dowd et al. 2003). Review this case and the message it provides for conservation biology.

9. Discuss the assumptions underlying the nested subset model of patch occupancy (see Figure 14). Explain what ecological processes could produce "holes" in the data matrix (see Table 5), and what processes could produce "outliers."

10. Amphibian populations have been declining in many parts of the world during the past 20 years (Stuart et al. 2003). Discuss the hypotheses proposed to explain these declines and suggest a research plan to rescue these populations. Davidson and Knapp (2007) discuss multiple causes for these declines, and Whitfield et al. (2007) provide a global overview of the problem.

Overview Question

Debate the following proposal: Resolved, that conservation biology is a crisis-oriented discipline and consequently should not be subject to the normal procedures of science for creating hypotheses and testing them experimentally.

Suggested Readings

- Beissinger, S. R., and D. R. McCullough, eds. 2002. *Population Viability Analysis*. Chicago: University of Chicago Press.

- Blackburn, T. M., P. Cassey, R. P. Duncan, K. L. Evans, and K. J. Gaston. 2004. Avian extinction and mammalian introductions on oceanic islands. *Science* 305:1955–1958.

- Caughley, G. 1994. Directions in conservation biology. *Journal of Animal Ecology* 63:215–244.

- Chetkiewicz, C.-L. B., C. C. St. Clair, and M. S. Boyce. 2006. Corridors for conservation: Integrating pattern and process. *Annual Review of Ecology, Evolution and Systematics* 37:317–342.

- Dobson, A. P., J. P. Rodriguez, W. M. Roberts, and D. S. Wilcove. 1997. Geographic distribution of endangered species in the United States. *Science* 275:550–553.

- Donlan, C. J., Knowlton, J., Doak, D. F., and Biavaschi, N. 2005. Nested communities, invasive species and Holocene extinctions: Evaluating the power of a potential conservation tool. *Oecologia* 145:475–485.

- Gaston, K. J., and R. A. Fuller. 2008. Commonness, population depletion and conservation biology. *Trends in Ecology & Evolution* 23:14–19.

- Lindenmayer, D., and J. Fischer. 2006. *Habitat Fragmentation and Landscape Change: An Ecological and Conservation Synthesis*. Washington, DC: Island Press.

- Pimm, S. L., L. Dollar, and O. L. Bass. 2006. The genetic rescue of the Florida panther. *Animal Conservation* 9:115–122.

- Sanderson, E. W. 2006. How many animals do we want to save? The many ways of setting population target levels for conservation. *BioScience* 56:911–922.

- Watson, J. E. M., R. J. Whittaker, and D. Freudenberger. 2005. Bird community responses to habitat fragmentation: How consistent are they across landscapes? *Journal of Biogeography* 32:1353–1370.

Credits

Illustration and Table Credits

Photo Credits

Community Structure in Space: Biodiversity

Key Concepts

- Biodiversity can be measured at the genetic level, at the species level, or at the ecosystem level. Species are usually the units of concern, and by counting all the species in an area we measure species richness as an index of biodiversity.

- There is a strong gradient in species diversity from the tropical regions toward the poles in most groups of plants and animals. This is one of the most striking patterns in community ecology.

- Six factors act jointly to enhance and maintain species richness in communities. The ambient energy hypothesis, including temperature, water, and solar energy, is the best predictor of large-scale patterns in biodiversity, while the evolutionary speed hypothesis helps to explain long-term trends over evolutionary time. Interspecific interactions such as predation and competition can help to explain local diversity patterns.

- Most species in a community are rare and only a few are common. This pattern of species abundances can be explained by two hypotheses. The sequential breakage hypothesis postulates competition and niche differences as the explanation, while the neutral theory of biodiversity explains this pattern by assuming all species are identical in their niche requirements and come and go at random.

- Local species richness tends to increase linearly with regional species richness, suggesting that local communities are never saturated with species.

- The answer to the general question, *What controls biodiversity?* depends on the species group and the scale of study. What is important at the local level will not necessarily be critical at the global level.

From Chapter 19 of *Ecology: The Experimental Analysis of Distribution and Abundance*, Sixth Edition. Eugene Hecht.

KEY TERMS

ambient energy hypothesis The idea that species diversity is governed by the amount of energy falling on an area.

biodiversity The number of species in a community or region, which may be weighted by their relative abundances; also used as an umbrella concept for total biological diversity including genetic diversity within a species, species diversity (as used here), and ecosystem diversity at the community or ecosystem level of organization.

endemic species Species that occur in one restricted area but in no other.

hotspots of biodiversity Areas of the Earth that contain many endemic species (typically 1500) and as such are of important conservation value.

intermediate disturbance hypothesis The idea that biodiversity will be maximal in habitats that are subject to disturbances at a moderate level, rather than at a low or high level.

keystone species Relatively rare species in a community whose removal causes a large shift in the structure of the community and the extinction of some species.

log-normal distribution The statistical distribution that has the shape of a normal, bell-shaped curve when the x-axis is expressed in a logarithmic scale rather than an arithmetic scale.

niche breadth A measurement of the range of resources utilized by a species.

niche overlap A measure of how much species overlap with one another in the use of resources.

umbrella species In conservation biology, species that serve as a proxy for entire communities and ecosystems, so that the entire system is conserved if they are conserved.

Ecological communities do not all contain the same number of species, and one of the currently active areas of research in community ecology is the study of species richness or **biodiversity**. Alfred Wallace (1878) recognized that animal life was on the whole more abundant and varied in the tropics than in other parts of the globe, and that the same applies to plants. Other patterns of variation have long been known on islands; small or remote islands have fewer species than large islands or those nearer continents (MacArthur and Wilson 1967). The regularity of these patterns for many taxonomic groups suggests that they have been produced in conformity with a set of basic principles rather than as accidents of history. How can we explain these trends in species diversity?

Biodiversity measurement is an important part of conservation biology, because we need an inventory of what is to be protected. Whereas conservation biologists often concern themselves with particular species, community ecologists tend to lump the species and condense information into counts of species. Often this is done within specific groups, such as the bird species or the tree species of an area. This community-based approach looks for large patterns in groups of species and tries to understand what has caused them. To do this we first need to know how to identify species of plants and animals, and then how to measure biodiversity.

Measurement of Biodiversity

The simplest measure of biodiversity is the *number of species*. In such a count we include only resident species, not accidental or temporary immigrants. It may not always be easy to decide which species are accidentals: Is a bottomland tree species growing on a ridgetop an accidental species or a resident one? The number of species is the first and oldest concept of species diversity and is called **species richness**.

A second concept of species diversity is that of **heterogeneity**. One problem with counting the number of species as a measure of diversity is that it treats rare species and common species equally. A community with two species might be divided in two extreme ways:

	Community 1	Community 2
Species A	99	50
Species B	1	50

The first community is very nearly a monoculture, and the second community intuitively seems to be more diverse than the first. We can combine the concepts of number of species and relative abundance into a single concept of heterogeneity: heterogeneity is higher in a community when there are more species and when the species are more nearly equally abundant.

Several measures of heterogeneity are in use (Krebs 1999; Magurran 2004), and the most popular has been borrowed from information theory. The main objective of information theory is to measure the amount of order (or disorder) contained in a system. We ask the question, How difficult would it be to predict correctly the species of the next individual collected? This is the same problem faced by communication engineers interested in predicting correctly the next letter in a message. This uncertainty can be quantified by a measure of information content, the Shannon-Wiener function; the

Biodiversity: A Brief History

Between 5 million and 30 million species of animals and plants live on Earth, and at present the best educated guess is about 15 million species. About 1.5 million of these are described by taxonomists, perhaps 10% of all life. This situation is a scandal that few nonbiologists seem to recognize. If only 10% of the companies being traded on Wall Street were known, or if the catalog of the Louvre Museum included only 10% of its paintings, right-thinking people would be outraged. Not so with biodiversity.

Taxonomists are the heroes of biodiversity, and without them working quietly in the background we would not know even the 10% we do, and our appreciation of community organization and dynamics would be much reduced. Fortunately a few taxonomists have risen to public recognition, including Edward O. Wilson of Harvard University. Wilson is an ant taxonomist by training and a naturalist by nature. While working on ant distributions on islands, he met and joined forces in 1961 with Robert MacArthur from the University of Pennsylvania to produce one of the most famous books on community ecology, *The Theory of Island Biogeography* (1967). Wilson has become a champion of biodiversity through his books on ants, and more recently through a series of popular books on biodiversity and its conservation. He is one of the few ecologists to have written an autobiography (*Naturalist*, Island Press, Washington DC, 1994).

Many other ecologists cooperated to bring biodiversity into the public eye at the close of the twentieth century. We have already met some of them, including Paul and Anne Ehrlich. But many who are less well known work hard to bring biodiversity to the fore in biological research agendas, and in the realm of political and social action. Given our ignorance of biodiversity, the exploding human population and its expanding effects on the globe is producing extinctions of species we will never have named or even described, a loss that we should not bequeath to our children and grandchildren.

---

details of how to calculate this measure of heterogeneity are provided.

To measure biodiversity we need a combination of two types of data: (1) the number of species in the community, and (2) the relative abundance of the species making up the community.

A difficult problem arises in trying to determine the number of species in a biological community: *species counts depend on sample size.* Adequate sampling can usually get around this difficulty, particularly with vertebrate species, but not always with insects and other arthropods, in which species counts cannot be complete.

Ecologists have adopted two different strategies to deal with the problem of measuring species richness for taxa, such as insects, that cannot be completely enumerated. First, a variety of statistical distributions can be fitted to data on the relative abundances of species. A second approach is to use species accumulation curves to quantify how species counts accumulate with larger and larger sample sizes.

One very characteristic feature of communities is that they contain comparatively few species that are common, and comparatively many species that are rare. Because it is relatively easy to sample any given area and count both the *number of species* on the area and the *number of individuals* in each of these species, a great deal of information of this type has accumulated and can be fitted to a variety of statistical distributions (Williams 1964). The first attempt to analyze these data was made by Fisher, Corbet, and Williams (1943) in one of the classic papers of community ecology.

In many faunal samples, the number of species represented by a single specimen is very large (the rare species), species represented by two specimens are less numerous, and so on, such that only a few species are represented by many specimens. When Fisher, Corbet, and Williams (1943) plotted the data, the result was a "hollow curve" (**Figure 1**) that could be described

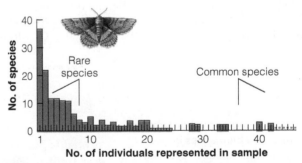

Figure 1 Relative abundance of Lepidoptera (butterflies and moths) captured in a light trap in Rothamsted, England, in 1935. A total of 6814 individuals of 197 species were caught (some of the abundant species are not shown). Thirty-seven species were represented in the catch by only a single specimen, and six common species constituted 50% of the catch. One very common species was represented by 1799 individuals in the catch. (Modified from Williams 1964.)

mathematically by a logarithmic series. The most significant ecological observation is that the largest number of species in a community fall into the "very rare" category.

Even though the logarithmic series implies that the greatest number of species have minimal abundance—that the number of species represented by a single specimen is always maximal—this is not the case in all communities and a search was made for a more general mathematical description. Preston (1948) suggested expressing the *x*-axis (number of individuals represented in the sample) on a geometric (logarithmic) scale rather than an arithmetic scale. When this conversion of scale is done and the species are combined into classes whose ranges of species abundances increase geometrically (for example, 1, 2–3, 4–7, 8–15, etc.), relative abundance data take the form of a bell-shaped, normal distribution, and because the *x*-axis is expressed on a geometric or logarithmic scale, this distribution is called **log-normal** (**Figure 2**). The essential point is that populations tend to increase geometrically rather than arithmetically, so the natural way to analyze abundances is as the *logarithm* of population density.

The log-normal distribution fits a variety of data from surprisingly diverse communities. **Figure 3** gives two more examples of relative abundance patterns in different communities. The log-normal distribution arises in all communities in which the total number of species is large, and the relative abundances of these species is determined by many factors operating independently. The log-normal distribution is thus the expected statistical distribution for many biological communities (May 1975). There is something very compelling about the log-normal distribution. The fact that moths in England, freshwater algae in Spain, snakes in Panama, and birds in New York all have a similar type of species abundance curve suggests regularity in community structure. The log-normal distribution can describe all the data that fit the logarithmic series and is a more general model of species abundance patterns in natural communities.

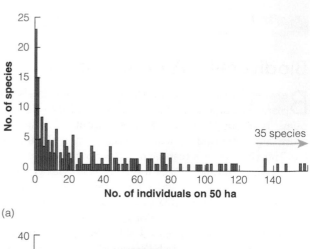

(a)

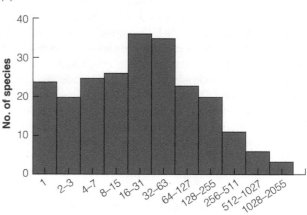

(b)

Figure 2 Relative abundance of tropical rainforest trees >10 cm diameter on a 50-ha plot on Barro Colorado Island, Panama, in 2005. A total of 229 tree species were present on this plot. Twenty-four species were represented in the census by only a single tree, and nine common species constituted 40% of the total tree count of 21,456 trees. One very common species was represented by 1909 individuals. (a) Presents the data on a linear scale, while (b) uses a logarithmic scale for the *x*-axis. (Data from Barro Colorado Island Web site http://ctfs.si.edu/datasets/bci, courtesy of Hubbell et al. 2005.)

Figure 3 Log-normal distribution of relative abundances in two diverse communities: (a) snake species in Panama and (b) British birds. Most species are intermediate in abundance in both these communities, and consequently the log-normal distribution fits the data better than the logarithmic series. (Data from Williams 1964.)

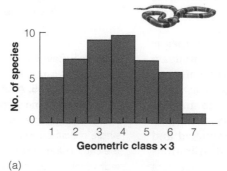

(a)

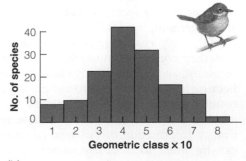

(b)

There has been much controversy over the log-normal distribution, because while it is a good statistical description of the data, it does not have built into it any clear ecological theory. This has stimulated two attempts to explain the patterns shown in Figures 1–3, the sequential niche breakage hypothesis and the neutral theory of biodiversity.

Sequential Niche Breakage Hypothesis

The log-normal distribution was recognized as an empirical regularity long before a theoretical justification was presented (Preston 1962). Sugihara (1980) has provided an explicit biological mechanism that leads to log-normal distribution. The essential hypothesis can be presented as follows. Assume that a community has a set of total niche requirements, which enables us to define a communal niche space. This niche space can be likened to a unit mass that is sequentially split up by the various component species such that each fragment denotes relative species abundance. Consider, for a simple example, Sugihara's model for the simple case of a three-species community. First, the total niche space is broken randomly to produce two fragments (**Figure 4**). The larger of the two fragments must range in size from 0.5 to 1.0 (of the original unit mass) and statistically will average 0.75 units. Next, one of these two fragments is chosen at random and broken to yield a third fragment. If the larger fragment is broken in the second step (breakage sequence A in Figure 4), we end up with three "species" with average relative abundances of 0.57, 0.28, and 0.15 (Sugihara 1980). If the smaller of the original two fragments is broken (breakage sequence B in Figure 4), we end up with

three "species" with average relative abundances of 0.75, 0.19, and 0.06. (These relative abundance estimates are averages that would apply to a large sample of three-species communities; any given community will vary because the breakage occurs at random.) Multispecies communities are more difficult to do these calculations for, but the principles remain the same. The important point is that the subdividing is done sequentially and not instantaneously, which corresponds with the biological assumption that the niche structure for communities is hierarchical. The niche space of a community has many dimensions and must not be thought of as a single resource axis.

The sequential breakage hypothesis predicts relative abundance patterns that are log-normal. Data from various communities fit this hypothesis very well, and the empirical findings of Preston (1962) can thus be biologically interpreted as a consequence of sequential niche subdivision.

The Neutral Theory of Biodiversity

A second major attempt to interpret species abundance patterns that appear to fit a log-normal type of curve was developed by Steve Hubbell in an influential book (Hubbell 2001). Hubbell challenged the classic niche-based view of ecological community structure. Hubbell's theory is a neutral theory because it is entirely based on chance. Its origins lay in long-term studies conducted by Hubbell and his colleagues of tropical rain forests and attempts to explain their structure (Condit et al. 2005). It is an analog to the neutral theory in population genetics described by Crow and Kimura (1970).

The neutral theory of biodiversity makes a series of simplifying assumptions. All species in an ecological community are assumed to be ecologically equivalent, so that no one species is competitively superior. Species arise at random when an individual mutates to become a new species, a process similar to mutation of alleles in genetics. As individuals die, they are replaced by the offspring of another individual, chosen at random, regardless of species. This assumption is equivalent to genetic drift in population genetics and produces a neutral drift of species abundances. A third assumption is that communities are saturated with species and in equilibrium.

The neutral theory of biodiversity has been widely criticized for its simplifying assumptions, which do not support the normal competition-dominated view of community structure (McGill 2003; Nee 2005). But it has strong adherents that marvel at its success even though it ignores differences among individual organisms and species (Alonso et al. 2006). In most of the tests to determine if the neutral model fits relative abundance patterns better than the log-normal model,

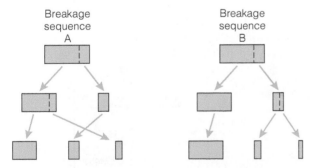

Figure 4 A hypothetical illustration of the sequential breakage hypothesis of Sugihara (1980). Two possible breakage sequences are illustrated for a hypothetical three-species community. If niches are subdivided in this manner in natural communities, the resulting abundance patterns for the species in the community will be log-normal. (From Sugihara 1980.)

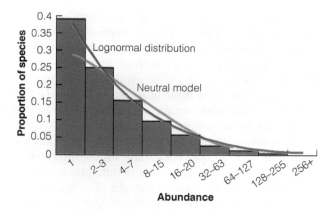

Figure 5 Coral species abundance patterns for the Indo-Pacific region and the fit of the two models to these relative abundance data. The neutral model of Hubbell (green line) does not fit the observed data as well as the log-normal model does (red line). (Data from Dornelas et al. 2006.)

the log-normal is slightly better. **Figure 5** illustrates this with relative abundance data from coral reefs in the Indo-Pacific region (Dornelas et al. 2006).

There is an advantage to using null models that assume very little about community dynamics to see how much one can predict with minimal assumptions. This approach to the relative abundance of species in a community forces us to think about the relative importance of different factors in generating the patterns we see in nature. Ecologists know that species have different life history properties (violating the first assumption of neutral

theory) and that competition is commonly observed among species in nature (violating another assumption of neutral theory), but the key question is how much difference these factors make in determining community structure. It is possible to relax the assumptions of neutral theory one by one to see how sensitive the neutral model is to each of its assumptions (Fuentes 2004; Zillio and Condit 2007). Recent investigations of the neutral model suggest that it is most sensitive to the speciation process in a community, and this is the key process generating species abundance patterns that are similar to log-normal curves (Nee 2005; Zillio and Condit 2007). Further analysis of this model is needed (Alonso et al. 2006).

Some Examples of Diversity Gradients

Tropical habitats support large numbers of species of plants and animals, and this diversity of life in the tropics contrasts starkly with the relatively impoverished faunas of temperate and polar areas. A few examples will illustrate this global gradient. A 50-ha plot of tropical rain forest in Malaysia contained 830 species of trees, and a 6.6-ha area in Sarawak contained 711 tree species (Whitmore 1998). A deciduous forest in Michigan contains 10 to 15 species on a plot of 2 hectares, and the whole of Europe north of the Alps has 50 tree species.

The 620 native tree species in North America north of Mexico are arrayed along a gradient that roughly follows latitude (**Figure 6**). More species in the United

Figure 6 Number of tree species in Canada and the United States. Contours connect points with the same number of species. (From Currie and Paquin 1987.)

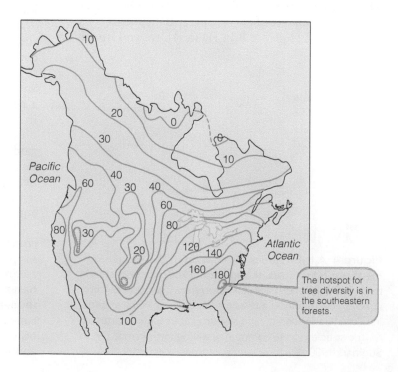

The hotspot for tree diversity is in the southeastern forests.

States occur in southeastern forests than in western forests, and minima occur in the rain shadows just east of the Rocky Mountains and the Sierra Nevada (Currie and Paquin 1987).

Ants are much more diverse in the tropics than in the high latitudes (Fischer 1960):

	No. of ant species
Brazil	222
Trinidad	134
Cuba	101
Utah	63
Iowa	73
Alaska	7
Arctic Alaska	3

There are 293 species of snakes in Mexico, 126 in the United States, and 22 in Canada. **Figure 7** shows the number of breeding land-bird species located at different latitudes, and illustrates the large increase in species richness in tropical regions.

Freshwater fishes are much more diverse in tropical rivers and lakes. Lakes Victoria, Tanganyika, and Malawi in east Africa each contain about 1450 species of freshwater fish. Over 1000 species of fishes have been found in the Amazon River in South America, and exploration is still incomplete in this region. By contrast, Central America has 456 fish species, and the Great Lakes of North America have 173 species (Rohde 1998). Lake Baikal in Asia has 39 fish species; Great Bear Lake in northwestern Canada has 14 species of fish.

Not all floras and faunas show a smooth trend of biodiversity with changing latitude. **Figure 8** shows the diversity of Alcid seabirds and of seals and sea lions in relation to latitude. Alcids occur only in the Northern Hemisphere and reach maximum species richness around 60°N. Seals and sea lions occur in both hemispheres and reach maximum diversity even closer to the poles (Proches 2001).

Species-diversity patterns of North American mammals, analyzed in detail by Simpson (1964), are a good example of a complex gradient. **Figure 9** shows that the number of land-mammal species increases from 15 in northern Canada to over 150 in Central America. Simpson recognized five notable features of this pattern:

- *North-south gradient.* The north-south gradient is not smooth. Some mammal groups—pocket gophers, shrews, and ungulates—are most diverse in the temperate zone and become less diverse toward the tropics. Bats contribute most of the high species richness for mammals in the tropics (Wilson 1974).

- *Topographic relief.* Areas like the Rocky Mountains or the Appalachians support a higher-than-average number of mammal species.

- *East-west trends.* Superimposed on the topographic variation is a general trend toward more species in the west than in the east. The topographically

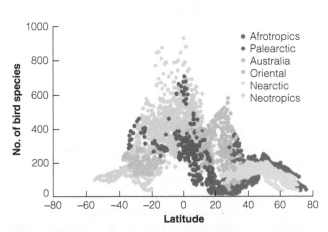

Figure 7 Bird species richness for the biogeographic regions of the Earth. The tropical bulge in bird diversity is not symmetrical about the equator. Biogeographic regions do not all follow exactly the same pattern. (Data courtesy of Bradford Hawkins, University of California, Irvine.)

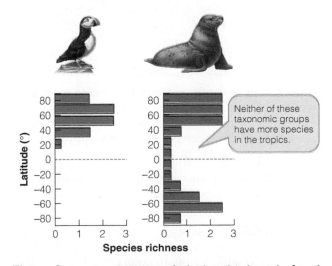

Figure 8 Species diversity of Alcid seabirds and of seals and sea lions (Pinnipedia) in relation to latitude. Southern latitudes are represented by a minus sign. Neither of these marine groups has a tropical-to-polar gradient in biodiversity. (Data from Proches 2001.)

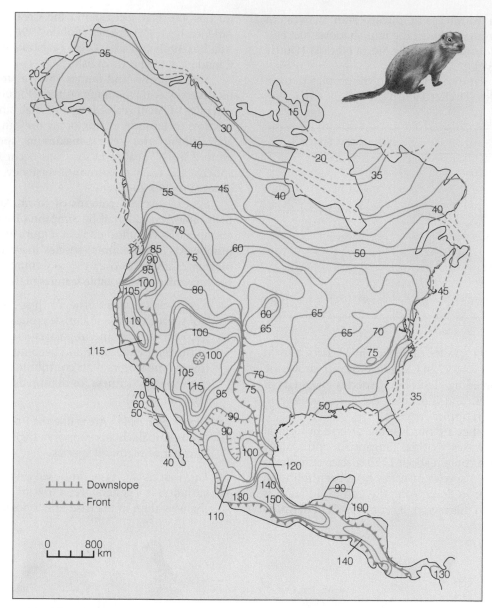

Figure 9 Species-density contours for existing mammals of continental North America.
The contour lines are isograms for numbers of continental (nonmarine and noninsular) species in 150-mi² (240-km²) quadrats. The "fronts" are lines of exceptionally rapid change that are multiples of the contour level for the given region. (After Simpson 1964.)

uniform Great Plains contain as many mammal species as the topographically diverse Appalachian Mountains.

- *Fronts of abrupt change.* Areas of rapid change in species diversity are often (but not always) associated with mountain ranges.

- *Peninsular "lows".* On peninsular areas such as Florida, Baja California, the Alaska Peninsula, and Nova Scotia, the number of mammal species is smaller than on adjacent continental areas.

The species-diversity gradient seen in North America is apparent in South America as well, even though the mammals of these two continents have distinct evolutionary histories (Kaufman and Willig 1998). This suggests that the ecological and evolutionary factors that produce the polar-equatorial gradients in species richness are general and not confined to any particular taxonomic lineage.

This brief look at some details of species-diversity gradients are a prelude to examining the causal factors that determine latitudinal gradients in species diversity. The overall pattern of increase in biodiversity from the poles toward the tropics is only a general trend, and the exceptions to this rule are useful because they permit us to untangle some of the ecological factors that influence biodiversity.

Hotspots of Biodiversity

The global pattern of tropical-to-polar gradients in biodiversity interact with evolutionary history to produce an unequal distribution of species around the Earth. Species arise primarily by geographic isolation, and the patterns of isolation that have arisen from continental drift have resulted in some areas being much more species-rich than others. Many, but not all, of these areas of high biodiversity occur in the tropics. Identification of these hotspots of biodiversity has become important in recent years because humans have cleared more and more areas for agriculture and forestry, thereby endangering many species. **Hotspots** are defined in several different ways, but in general the measure used to define a hotspot is the number of endemic species that it contains. **Endemic species** are those that occur in only one relatively small geographic area. The Hawaiian goose, for example, is an endemic bird found only on the islands of Hawaii and Maui.

There are 34 hotspots of biodiversity around the globe (**Figure 10**). Hotspots are defined as areas containing at least 1500 endemic plant species. One surprising feature of this map is that not all the hotspots are in tropical countries. Many hotspots are tropical, but the Cape Floristic Province of South Africa and New Zealand are two examples of temperate hotspots. Polar regions contain no hotspots. **Table 1** lists the size of the most important hotspots and the number of plant species and vertebrate species they contain. These data are incomplete and err in the direction of minimal species counts. For example, Brazil has the world's richest flora, and probably has at least 50,000 species of plants, but there is no up-to-date list of Brazil's plant species.

The 34 hotspots mapped in Figure 10 contain a minimum of 50% of the world's plant species and 77% of the world's vertebrate species, all within 2.3% of the land surface of the Earth. The implication of this concentration of biodiversity is that these hotspots should be a focus of our conservation efforts at the present time.

The hotspot concept has an underlying assumption that hotspots occur in the same geographic region for all the different plant and animal groups. This general idea is the concept of **umbrella species**—that one species or a group of species will serve as a guide to many other groups of species that are less well known or less studied. For example, butterflies could serve as umbrella species for the community of all insects and plants in a region. Consequently, before designing a

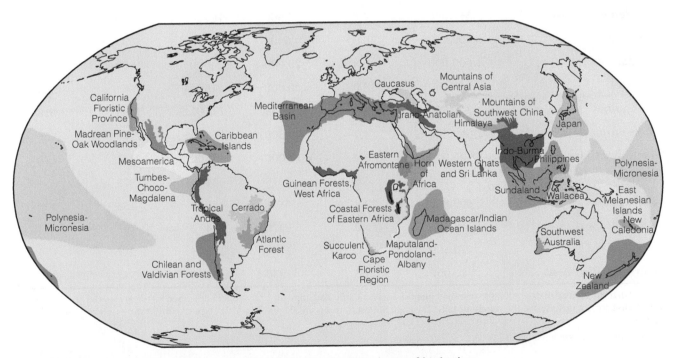

Figure 10 The 34 global hotspots of biodiversity, defined on the basis of high plant species richness. Table 1 lists the number of endemic plant species in the top 25 global hotspot areas. (From www.biodiversityhotspots.org.)

Table 1 Characteristics of 25 of the highest ranked biodiversity hotspots.

Hotspot	Original extent of vegetation (km²)	Percent remaining original vegetation	No. of plant species	No. of endemic plant species	No. of vertebrate species	No. of endemic vertebrate species
Tropical Andes	1,258,000	25.0	30,000	15,000	3389	1567
Mesoamerica	1,155,000	20.0	17,000	2941	2859	1159
Caribbean	263,500	11.3	13,000	6550	1518	779
Brazil's Atlantic Forest	1,227,600	7.5	20,000	8000	1361	567
Turnbes/Choco/Western Ecuador	260,600	24.2	11,000	2750	1625	418
Brazil's Cerrado	1,783,200	20.0	22,000	10,000	1268	117
Chile/Valdivian Forest	300,000	30.0	3892	1957	335	61
California	324,000	24.7	3488	2124	584	71
Madagascar	594,150	9.9	13,000	11,600	987	771
Eastern Afromontane and Coastal Forests of East Africa	30,000	6.7	11,598	4106	1019	121
Guinean West African Forests	1,265,000	10.0	9000	1800	1320	270
Cape Floristic Province	74,000	24.3	9000	6210	562	53
Succulent Karoo	112,000	26.8	6356	2439	472	45
Mediterranean Basin	2,362,000	4.7	22,500	11,700	770	235
Caucasus	500,000	10.0	6400	1600	632	59
Sundaland	1,600,000	7.8	25,000	15,000	1800	701
Wallacea	347,000	15.0	10,000	1500	1142	529
Philippines	300,800	3.0	9253	6091	1093	518
Indo-Burma and Himalaya	2,060,000	4.9	23,500	10,160	2185	528
Southwest China	800,000	8.0	12,000	3500	1141	178
Western Ghats/Sri Lanka	182,500	6.8	5916	3049	1073	355
SW Australia	309,850	10.8	5571	2948	456	100
New Caledonia	18,600	28.0	3270	2432	190	84
New Zealand	270,500	22.0	2300	1865	217	136
Polynesia/Micronesia	46,000	21.8	5330	3074	342	223

NOTE: There are approximately 300,000 described plant species on Earth, and approximately 28,595 described vertebrate species (excluding fish). Fishes are not included in the vertebrate tally. The eight hottest hotspots are shown in boldface type. Figure 10 shows a map of these regions.

(From www.biodiversityhotspots.org.)

conservation action plan, it is important to check how much overlap there is among hotspots for different taxonomic groups. Hotspots for one group of species, such as plants, might not coincide with hotspots for other groups, such as butterflies. For example, by mapping the geographic distribution of endangered plants, birds, fish, and molluscs in the United States, Dobson et al. (1997) found that the hotspots for one taxonomic group did not coincide with hotspots for other groups. This means that recovery plans for species will have to be area-specific, and for any specific group of species the hotspots for that group can be targeted for conservation.

The global distribution of hotspots provides another dimension to the overall trend of a drop in biodiversity as we move from the tropics toward the poles. These patterns raise the question of what environmental factors cause these large differences in species diversity.

Six Factors That Cause Diversity Gradients

Differences in species richness may be produced by six interrelated causal factors (**Table 2**). Many causes have interacted over evolutionary and ecological time to produce the assemblages we see today, so that no one cause will explain all the patterns we have described. For any particular diversity gradient we can ask which of these six factors are involved and which are most important.

Evolutionary Speed Hypothesis

The idea that history affects diversity via evolutionary speed, proposed chiefly by zoogeographers and paleontologists, has two main components. First, biotas in the warm, humid tropics are likely to evolve and diversify more rapidly than those in the temperate and polar regions (**Figure 11**) because of a constant, favorable environment and a relative freedom from climatic disasters like glaciation. The assumption here is that evolutionary rates depend on temperature. Second, biotic diversity is a product of evolution and therefore is dependent on the length of time through which the biota has developed in

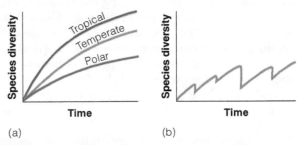

(a)   (b)

Figure 11 Evolutionary speed as a factor in biodiversity. (a) Hypothetical increase in species diversity with decreasing latitude in the absence of interruptions; (b) actual pattern of change in species diversity of a temperate or polar habitat subjected to glaciation and climatic variations. (After Fischer 1960.)

Table 2 Ecological and evolutionary factors that can have an influence on biodiversity.

Factor	Rationale
1. Evolutionary speed	More time and more rapid evolution permits the evolution of new species
2. Geographic area	Larger areas and physically or biologically complex habitats furnish more niches
3. Interspecific interactions	Competition affects niche partitioning and predation retards competitive exclusion
4. Ambient energy	Fewer species can tolerate climatically unfavorable conditions
5. Productivity	Richness is limited by the partitioning of production or energy among species
6. Disturbance	Moderate disturbance retards competitive exclusion

NOTE: More than one of these factors can operate in any particular ecological community or in any particular taxonomic group. Some factors operate at a local scale and others at regional scales.

(Modified after Currie 1991, and Willig et al. 2003).

an uninterrupted fashion (Fischer 1960; Turner 2004). Tropical biotas are examples of mature biotic evolution, whereas temperate and polar biotas are immature communities, continuously interrupted by glaciation and severe climate shifts. So, even if evolutionary rates are the same everywhere, more species will evolve in tropical communities. In short, all communities diversify over time, and older communities consequently have more species than younger ones.

The evolutionary speed hypothesis does not necessarily predict a smooth gradient from the tropics to the poles. The key point is that an ecological community must have a long, uninterrupted evolutionary history to achieve high species richness. Lake Baikal in the former Soviet Union is a particularly striking illustration of the role of time in generating species diversity. Situated in the temperate zone, Baikal is one of the oldest lakes in the world and contains a very diverse fauna (Kozhov 1963). For example, there are 580 species of benthic invertebrates in the deep waters of Lake Baikal. A lake of comparable area in glaciated northern Canada, Great Slave Lake, contains only four species in this same zone (Sanders 1968).

Some paleontological data support the assumption that species diversity increases over geological time. The number of species of terrestrial plants, as reflected in the fossil record, appears to have increased in two waves during the past 450 million years (**Figure 12**). No plateau in biodiversity has yet been reached for terrestrial plants (Knoll 1986). Detailed data from mammalian fossils permit a nearly complete analysis of the accumulation of species of mammals and changes in

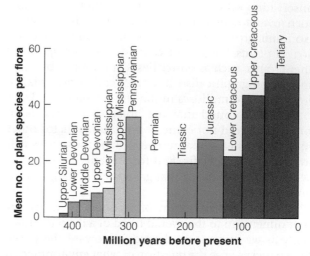

Figure 12 **Pattern of increase in the number of terrestrial plant species over evolutionary time; data are derived from fossils.** (Data from Nicklas et al. 1980.)

the rate of speciation over the last 150 million years (Bininda-Emonds et al. 2007). **Figure 13** shows the evolutionary history of the mammals and the variations in speciation rates, which are far from constant.

Note that the species diversity of a community is a function not only of the rate of addition of species through evolution but also of the rate of loss of species through extinction or emigration. Compared with polar communities, the tropics could have both a more rapid rate of evolution and a lower rate of extinction, and these two rates act together to determine species diver-

Figure 13 **The rate of mammalian evolution over the last 170 million years.** (a) The number of taxa for placental mammals (green), marsupial mammals (orange), and all mammals (blue). (b) The speciation rate (no. per million years) for all mammals (smoothed from the raw data). The Cretaceous-Tertiary (K/T) boundary is shown in red, indicating the time of the mass extinction of the dinosaurs. (From Bininda-Emonds et al. 2007.)

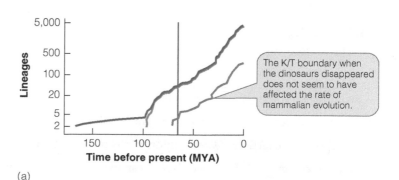

(a)

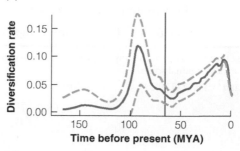

(b)

sity. There are only limited data at present to test the assumption that tropical organisms undergo more rapid evolution. Allen and Gillooly (2006) found that speciation rates were higher in ocean plankton species in areas of high species diversity, thus supporting the evolutionary speed hypothesis. But Bromham and Cardillo (2003) did not find any latitudinal difference in the rate of molecular evolution in birds, and their results cast doubt on one of the assumptions of the evolutionary speed hypothesis for birds. We need to also consider a second possibility that the rate of extinction is lower in the tropics. There are at present no data to test this assumption.

The evolutionary speed hypothesis suggests that species richness never reaches a limit but continues rising over time. We do not know if this is a correct interpretation of the fossil record (Gould 1981). While evolution brings new species into ecological communities, it does not by itself explain what maintains this increased diversity. Factors must operate in ecological time to maintain the biodiversity that evolution has produced in communities, and we need to analyze these factors to understand the best ways of protecting biodiversity in the future.

Geographic Area Hypothesis

This hypothesis begins with the assumption that larger areas support more species than smaller areas, which seems to be universally true. Given this assumption, the postulate is that the tropics support more species than the temperate zone because it has a larger land area. Larger areas contain more habitats and more individuals, reducing the risk of extinction. If there are more habitats in a region, we would expect there to be more species. There could be a general increase in the number of habitats per square km as one proceeds toward the tropics, and the more complex the plant and animal communities, the higher will be the species diversity.

But geographic area does not explain everything, as is evident if we examine the detailed distribution of

some species groups. Topographic relief—mountains and hills—may have a strong effect on species diversity in some groups of organisms. The highest diversities of mammals in the United States occur in the mountainous areas of the western states (see Figure 9). The explanation for this gradient seems simple: areas of high topographic relief contain many different habitats and hence more species. Also, mountainous areas produce more geographic isolation of populations and may thus promote speciation. But this conclusion does not fit all taxonomic groups. Neither the trees (see Figure 6) nor the land birds of North America show diversity patterns related to topographic relief.

MacArthur (1965) suggested that we should recognize two components in trying to analyze latitudinal gradients in species diversity: *within-habitat* diversity (also called α diversity), and *between-habitat* diversity (called β diversity) (**Figure 14**). We can use this distinction by inventing two simple schemes to explain high tropical diversity:

Hypothetical scheme A	Temperate location	Tropical location
No. of species per habitat	10	10
No. of different habitats	10	50

Hypothetical scheme B	Temperate location	Tropical location
No. of species per habitat	10	50
No. of different habitats	10	10

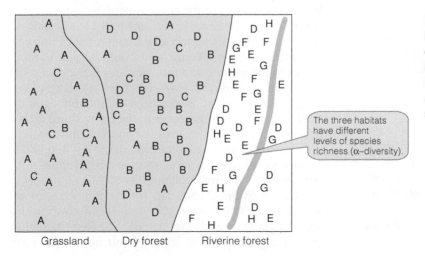

Grassland Dry forest Riverine forest

Figure 14 A schematic illustration of biodiversity in a region that includes three different habitats. Each of the three habitats has an α diversity. For example, the grassland has three species. The entire region including all three habitats has a β diversity of eight species.

The three habitats have different levels of species richness (α–diversity).

How Spatial Scale Affects Biodiversity Measurements

Counts of the numbers of species will always be affected by the spatial scale of the measurements. The larger the area sampled the greater the number of species that will be counted, and this makes it difficult to compare the data obtained from different studies. A simple example will illustrate this problem. Consider the herbaceous plant species in a region that has three different types of habitat—a grassland, a dry forest, and a wet riverine forest (Figure 14). The grassland (yellow) has three species in it, the dry forest (green) has four species, and the riverine forest (white) has five species. There are two simple ways to describe the diversity of this artificial landscape. If we stay within one community, we can measure *within-habitat diversity* (also called α diversity), but if we sample two of the communities together, we can measure *between-habitat diversity* (also called β diversity).

Habitats sampled	Total number of species	No. of species shared in the combined communities	No. of new species added by combining
Grassland	3	—	—
Dry forest	4	—	—
Riverine forest	5	—	—
Grassland + dry forest	4	3	1
Dry forest + riverine forest	8	1	4

For this hypothetical example, between-habitat diversity is low if we compare the grassland and the dry forest, since we add only one more species by combining these two communities. But if we consider the two forest types, between-habitat diversity is high, since four more species are added.

The key point is that the species diversity of a region is not simply the number of species in each community (α diversity) added together. The number of shared species can be used to measure a second component of biodiversity, between-habitat or β diversity.

In scheme A, between-habitat diversity or β diversity accounts for all the increase in diversity for tropical species; in scheme B, all the increase in tropical diversity is due to within-habitat diversity or α diversity.

Tropical-to-polar gradients in the oceans seem unlikely to be explained by geographic area. The oceans are not uniform water masses, yet they provide fewer opportunities for area effects. Benthic marine invertebrates become more diverse as one moves from shallow waters on the continental shelf to deeper waters at the edge of the shelf (Sanders 1968). There is no obvious change in the area of bottom sediments to explain this increase in biodiversity.

We conclude that geographic area does not explain very many of the observed tropical-to-polar diversity gradients (Rohde 1992). It cannot be a general explanation because many aquatic habitats such as shallow saltwater mudflats show these gradients in the absence of any change in spatial heterogeneity. In cases in which geographic area can be used to explain latitudinal gradients in species diversity, we must still identify the ecological "machinery" behind this effect.

Interspecific Interactions

Several hypotheses suggest that high tropical diversity is associated with greater interspecific competition and higher predation rates. How might interactions among species affect the latitudinal gradient in species diversity? Two general hypotheses have been developed with different explanations for the diversity gradient. The first suggests that competition is keener in the tropics, compared with more polar ecosystems, and that intense competition is an explanation for the diversity gradients we observe. The second hypothesis is that predation is stronger in the tropics, so that there is less competition and thus more species can coexist in tropical communities.

Many naturalists have argued that natural selection in the temperate and polar zones is controlled mainly by physical factors of the environment, whereas competition becomes a more important part of evolution in the tropics. For this reason, the argument goes, animals and plants are more restricted in their habitat requirements in the tropics, and this increases between-habitat (β) diversity. Animals may also have more restricted diets in each habitat, increasing within-habitat (α) diversity. Competition is keener in the tropics, and niches are smaller. Tropical species are more highly evolved and possess finer adaptations than do temperate species. Consequently, more species can occupy a given habitat in the tropics (Dobzhansky 1950).

Competition theory can be expanded in an attempt to explain species diversity in an equilibrium world (Chesson and Case 1986). The key prediction that emerges is that at least n limiting resources are needed for the coexistence of n species in a community. For plants there are at most four or five limiting resources (Tilman 1986), and thus competition by itself cannot explain the large number of plant species in natural communities. The conclusion is that this hypothesis is not a possible explanation for within-habitat diversity of plants.

For animal species, many more limiting resources potentially exist. The effect of competition on species richness can be made apparent by looking at the niche relations of the species in a community. Consider the simple case of one resource, such as soil water for plants or food-item size for animals. Two niche measurements are critical: **niche breadth** and **niche overlap (Figure 15a)**. We can recognize two extreme cases. If there is no niche overlap between species, then the wider the average niche breadth, the fewer the number of species in the community (Figure 15b). At the other extreme, if niche breadth is constant, then the smaller the niche overlap, the fewer the species in the community (Figure 15c). In this hypothetical analysis, tropical animal communities might have more species because tropical species have smaller niche breadths or greater niche overlaps. Both these arguments assume that Gause's hypothesis is true for natural communities.

To evaluate the competition factor, we must measure these niche parameters in a variety of tropical and temperate animal communities. The problems of measuring niche overlap and niche breadth are discussed in detail by Magurran (2004) and Krebs (1999). The basic problem is to decide which resource axes are relevant to any particular group of species; if the resource axes can be linearly ordered and measured, these niche parameters can be measured as indicated in Figure 15.

In relatively few cases have detailed measurements been made to test the schematic model of Figure 15. A good example is the work on Caribbean lizards summarized by Roughgarden (1986). Lizards of the genus *Anolis* are small, diurnal, insect-eating iguanid lizards that are a dominant component of the vertebrate community on islands in the Caribbean. Most species perch on tree trunks or bushes. They are sit-and-wait predators, and food size is a critical niche dimension. Roughgarden (1974) tested the prediction that niche breadth would decrease as more species occurred together on an island. **Figure 16** shows the results for two *Anolis* species. The results are consistent with the predictions from competition theory and support the suggestion that niche breadth is reduced in species-rich communities. Pacala and Roughgarden (1985) showed in enclosure experiments that *Anolis* species show strong effects of competition when their diets are similar; competition for food is a major factor determining the species diversity of these lizards.

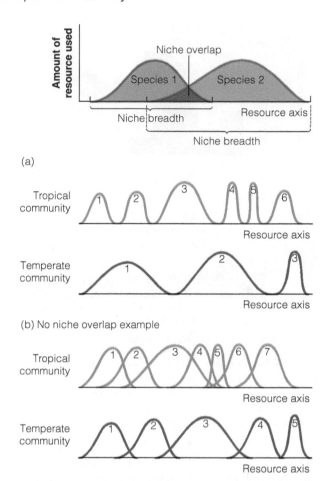

(a)

(b) No niche overlap example

(c) Constant niche breadth example

Figure 15 Diagram to illustrate two extreme hypothetical cases of how niche parameters may differ in tropical and temperate communities. (a) Both niche breadth and niche overlap are determined by competition within the communities. If there is no niche overlap (b), the number of species that can be in a community is determined by niche breadth. If there is constant niche breadth (c), the number of species is determined by the amount of niche overlap.

It is clear that competition does not play a large role in the maintenance of plant biodiversity (Austin 1990, 1999), but it is less clear how much it affects animal biodiversity in modern communities. The evolution of niche parameters due to competition could be a cause of greater tropical diversity, or it could be an effect of higher species numbers (Rohde 1992).

The polar opposite prediction about the role of competition in generating the polar gradients in biodiversity was presented by Paine (1966), who argued that predators and parasites are more abundant in the tropics than

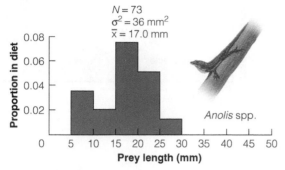

(a) *A. Cybotes* (Jarabacoa)

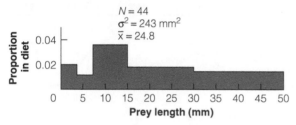

(b) *A. Marmoratus ferreus* (Marie Galante)

Figure 16 Niche breadth of two *Anolis* lizard species on islands in the Caribbean. (a) *Anolis cybotes* coexists with five other *Anolis* species on Jarabacoa and has a narrow niche. (b) *Anolis marmoratus* is the only species on Marie Galante. (After Roughgarden 1974.)

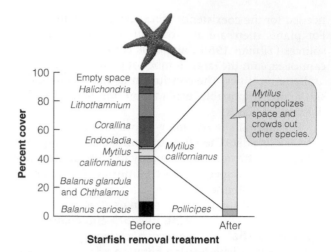

Figure 17 Keystone predator effect in the rocky intertidal zone. Paine (1974) removed predatory starfish (*Pisaster ochraceous*) from rocky intertidal sites in Washington State and observed a collapse of the community to a near-monoculture of California mussels (*Mytilus californianus*) over five years. (Data from Paine 1974.)

elsewhere, and they hold their prey populations to such low levels that competition among prey organisms is reduced. This reduced competition allows the addition of more prey species, which in turn support new predators. Thus, in contrast to the competition proposal, *less* competition should exist among prey animals in the tropics. Providing we can measure "intensity of competition," we can distinguish quite clearly between these two ideas.

Paine (1966, 1974) supported his ideas with some experimental manipulations of rocky intertidal invertebrates of the Washington coast (**Figure 17**). The food web of these intertidal areas on the Pacific coast is remarkably constant with about 15 species of herbivores and predators.

Paine removed the starfish *Pisaster* from a section of the shore and observed a *decrease* in diversity from a 15-species system to an 8-species system. A bivalve, *Mytilus*, tended to dominate the area, crowding out the other species. Four of the species that disappeared were not eaten by *Pisaster* but were affected by the increase in *Mytilus*. "Succession" in this instance is toward a simpler community. By continual predation, the starfish prevent the barnacles and bivalves from monopolizing space. Thus local species diversity in intertidal rocky zones appears to be directly related to predation intensity. Paine called the starfish a **keystone species** in this community.

The prediction from Paine's work that increased predation will lead to greater diversity of prey species

depends on the ability of one prey species to be competitively dominant. For the predation hypothesis to operate on a broad scale, the predators involved must be very efficient at regulating the abundance of their prey species. In terrestrial food webs, predators are usually specialized and in some cases do not seem to regulate prey abundance. Note that the predation hypothesis cannot be a sufficient explanation for tropical species diversity unless it can be applied to all trophic levels. If the species diversity of herbivores is determined by their predators, we are left with explaining the diversity of the primary producers. Keystone species, such as the starfish *Pisaster*, should be more common in tropical communities, but currently there is no evidence that this is correct (Power et al. 1996).

The effect of predation can be extended to the primary-producer level. Tropical lowland forests contain many species of trees and corresponding low densities of adult trees of each species. Most adult trees of a given species are also spread out in a regular pattern in the tropical forest, leading Janzen (1970) and Connell (1971) to suggest that these characteristics of tropical trees can be explained by the predation hypothesis, with the species that eat seeds or seedlings filling the role of the predators. The Janzen-Connell model for the maintenance of tropical tree species diversity, shown schematically in **Figure 18**, predicts that tree seedlings will do poorly if they are close to a large tree of the same species. To test this, Steve Hubbell and Robin Foster in 1981 established a 50-ha forest plot on Barro Colorado Island in Panama. This plot was censused in 1981–1983, 1985, and 1990–1991, and 244,000 stems of 303 species were measured and geographically mapped

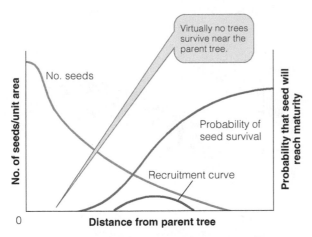

Figure 18 The Janzen-Connell model to account for high tropical-forest diversity. The amount of seed dispersed declines rapidly with distance from the parent tree, and the activity of host-specific seed and seedling herbivores and diseases is most evident near the parent tree. The product of these two factors determines a recruitment curve that peaks at the distance from the parent tree at which a new adult tree is likely to appear. (Modified after Janzen 1970.)

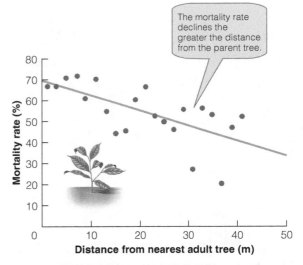

Figure 19 Mortality rate of 843 *Ocotea whitei* saplings on Barro Colorado Island in Panama in relation to distance of each sapling from the nearest adult tree of the same species. This mortality is caused primarily by a stem canker disease, and was measured over the interval from 1982 to 1991. The Janzen-Connell model predicts higher mortality nearer to conspecific adults, exactly as shown here. (Data from Gilbert et al. 1994.)

in 1990–91 (Condit 1995). Many of the species in this rain-forest plot show intraspecific effects of density on recruitment (Wills et al. 1997). Most of the negative effects occur within a species, so interspecific competition for resources does not seem important in maintaining diversity. Pests and pathogens may be particularly significant. **Figure 19** shows that the effect of stem canker disease in a laurel tree (*Ocotea whitei*) sapling is more severe near an adult tree, as the Janzen-Connell model would predict. Thus, each tree casts a "seed shadow," in which survival of its own kind is reduced. As one moves from the lowland tropical forests to temperate forests, the seed and seedling herbivores are hypothesized to be less efficient at preventing establishment of seedlings close to the parent tree. Data from tropical rain forests have supported the Janzen-Connell model in some but not all cases (Hyatt et al. 2003; Wyatt and Silman 2004; Wills et al. 2006).

The effects of predation and competition on species diversity may be complementary (Lubchenco 1986). Competition may be more important in maintaining high diversity among parasites and predators, whereas the process of predation and disease may be more important among herbivores and plants, respectively. Superimposed on these effects is another pattern: in complex communities with many species, predation may be the dominant interaction affecting diversity, whereas competition may be the dominant interaction in simple communities.

Ambient Energy Hypothesis

This hypothesis states that energy availability generates and maintains species richness gradients. Climate deter-

mines energy availability, and the key variables for terrestrial plants and animals are solar radiation, temperature, and water. Climates that are more stable and more favorable cause higher productivity, and all these factors work together to support more species. This idea was first suggested by Brown (1981), who called it the *species richness–energy model*.

The **ambient energy hypothesis** is the simplest and most elegant of the climatic explanations for the polar-tropical gradient in terrestrial biodiversity. A wealth of data have now been presented in support of this hypothesis. This simple hypothesis fits data for trees (Currie and Paquin 1987), British birds (Turner et al. 1988), South American raptorial birds (Bellocq and Gomez-Insausti 2005), and vertebrates (Currie 1991) and butterflies (Huntley et al. 2004) from North America and Europe. **Figure 20** illustrates the relationship between biodiversity and available energy for trees and vertebrates. Available energy can be measured by annual evapotranspiration, which measures the energy balance at a site and can be calculated from solar radiation and temperature. Vertebrate biodiversity in North America below about 45°–48°N (approximately the Canadian border) is not correlated with ambient energy but is more directly correlated with water availability. There seems to be a general threshold in midlatitudes of the Northern Hemisphere, where, in all terrestrial groups, water becomes the major factor predicting species richness (**Figure 21**).

How might we test the energy hypothesis? The energy hypothesis makes specific predictions about

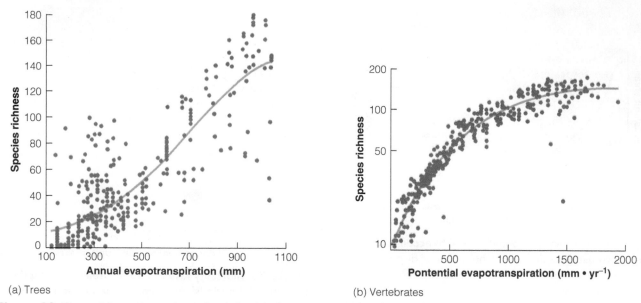

(a) Trees

(b) Vertebrates

Figure 20 The ambient energy hypothesis for biodiversity. Species richness of (a) trees and (b) vertebrates from North America are related to annual available energy at each site, as measured by evapotranspiration (which combines solar radiation and temperature). (From Currie 1991.)

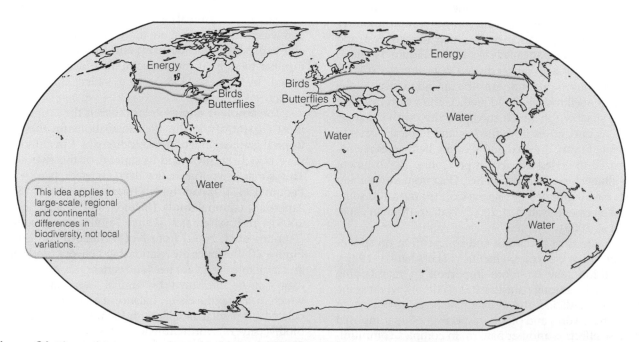

Figure 21 The ambient energy hypothesis to explain the polar-to-tropics gradient in species richness of terrestrial animals. Across the globe animal species richness is constrained by the interaction of energy and water. The bold red lines show the geographic position of thresholds—energy is the limiting component north of these lines for birds and butterflies, whereas south of these lines water is the key limiting component. Energy is less relevant in the Southern Hemisphere because landmasses do not extend so far south to extreme latitudes. (Modified from Hawkins et al. 2003.)

seasonal bird migrants, and this is one test of the idea. In temperate areas, energy levels in summer should control the diversity of summer birds, while energy levels in winter should control the numbers of resident birds in winter. In Britain, the biodiversity of summer birds is correlated with summer temperature and the diversity of winter birds is correlated with winter temperature at 75 localities, exactly as the energy hypothesis predicts (Turner et al. 1988).

A second test of the energy hypothesis can be made with coral reefs. Coral reefs are species-rich in tropical waters, and the number of taxa falls off rapidly as you move from warm tropical seas to cooler temperate waters (**Figure 22**). High species diversity in corals has usually been attributed to the historical factor associated with high rates of speciation in the Indo-Pacific region. The best predictors of coral species diversity are ocean temperature and coral biomass, so

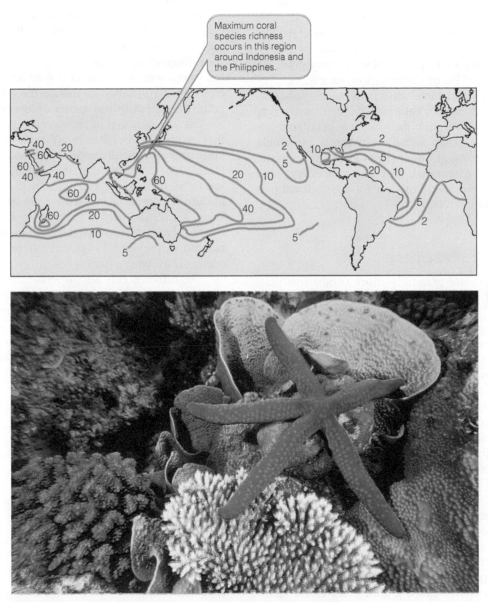

Figure 22 Biodiversity of corals in all the tropical and subtropical areas of the world. Contour lines connect areas with the same number of coral genera. The Indo-Pacific region is much richer in corals than is the eastern Pacific or the Atlantic region. Within these three regions, coral diversity can be predicted from the available energy, measured by ocean temperature. Coral photo from the Great Barrier Reef of Australia. (Modified after Fraser and Keddy 1996.)

that energy-rich areas had more coral genera and species (Hawkins et al. 2003). Historical evolutionary factors are responsible for the major differences between the number of coral species in the Atlantic and the Indo-Pacific regions, so that both history and available energy are important overall. Caribbean coral reefs are less than 10,000 years old and were affected by glaciation in the Northern Hemisphere, while Pacific coral reefs are up to 60 million years old and have been less affected by climatic oscillations. Caribbean reefs contain only 10%–20% of the number of coral species found on Pacific reefs.

Favorable climates on a broad geographic scale thus support high biodiversity. This idea explains a large fraction of the global tropical-polar diversity gradient, on average about 63% of the variation. Ambient energy theory works well for large-scale patterns in global diversity. It will not explain variations in species diversity on the scale of local habitats—why a grassland has more herb species than an adjacent forest. No single factor will explain all gradients in biodiversity from the local scale to the regional, but ambient energy and water provide a good explanation for the regional patterns (Hawkins et al. 2003).

Productivity Hypothesis

Tropical habitats are generally more productive than temperate and polar habitats, and this insight might lead one to suggest that productivity is a key to biodiversity and a good explanation of the tropical-polar trend in species richness. The productivity hypothesis in its pure form states that greater production results in greater diversity, everything else being equal. The data available do not support this idea. For example, Tilman (1986) describes several examples in which plant biodiversity is maximal in resource-poor habitats of low productivity. Two of the world's most diverse plant communities, the fynbos of South Africa and the heath scrublands of southeastern Australia (see Figure 10), both occur on nutrient-poor soils, and in both cases adjacent areas with better soils and more productive vegetation have fewer species. Productivity in plant communities seems to lead to reduced biodiversity on a local scale, exactly the opposite of what one might expect (Tilman 1986).

Productivity has been considered more important for animal communities on a global scale, but again the available data do not agree with this conclusion. Currie (1991) could find no relationship between productivity and vertebrate biodiversity in North America. Productivity by itself does not seem to be the key to understanding diversity gradients on a global scale.

Intermediate Disturbance Hypothesis

If ambient energy can explain much of the global pattern of species richness, local patterns on much smaller scales need to involve other mechanisms that affect diversity. Disturbance is one such local factor. If natural communities exist at equilibrium and the world is spatially uniform, then competitive exclusion ought to be the rule, and each community should come to be dominated by a few species—the best competitors (Crawley 1986). But if communities exist in a nonequilibrium state, competitive equilibrium is prevented. A whole range of factors can prevent equilibrium, including predation, herbivory, fluctuations in physical factors, and catastrophes such as fires, and we lump these together as "disturbance." When disturbances occur too often, species go extinct if they have low rates of increase. When disturbances are rare, the system goes to competitive equilibrium and species of low competitive ability are lost. The idea that in between is a level of disturbance that maximizes biodiversity (**Figure 23**) is called the **intermediate disturbance hypothesis** (Grime 1973; Horn 1975; Connell 1978). If population growth rates are low for all members of a community, the competitive equilibrium is approached so slowly that it is never reached. Thus, species diversity is maintained by periodic disturbance or by environmental fluctuations. If this model is in fact correct, the worst thing we can do to a community is to prevent disturbances such as fire.

Disturbance can also operate on a local scale to produce patches that undergo succession. Within each patch on a local area the species composition may be changing, but on a larger spatial scale the species composition may be constant and include both pioneer species and climax species (Connell 1987).

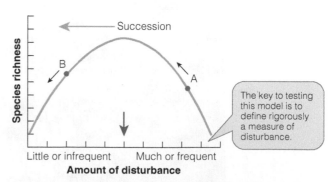

Figure 23 The intermediate disturbance hypothesis of species diversity. This model predicts that at some intermediate level of disturbance (red arrow) biodiversity will be maximum. Succession is assumed to be proceeding in the direction of the blue arrow. Note that if a community is at point A, a reduction in disturbances would result in an increase in species richness. At point B, the opposite effect would be observed, if this model is correct. (Modified from Huston 1979.)

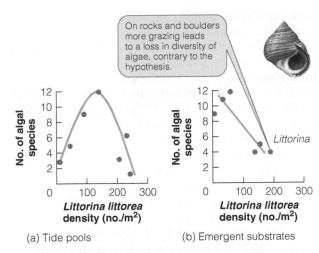

(a) Tide pools   (b) Emergent substrates

Figure 24 The effect of the disturbance of periwinkle snail (*Littorina littorea*) grazing on the diversity of algae in (a) high-tide pools and (b) on emergent rocks in the low intertidal zone in Massachusetts and Maine. The intermediate disturbance hypothesis applies only in the tide pools. (From Lubchenco 1978.)

Disturbance does not always produce maximum diversity at intermediate levels of disturbance, as predicted by the intermediate disturbance hypothesis (Wootton 1998; Li et al. 2004); some data are at variance with this prediction. Three examples illustrate some of the patterns. On rocky shores in Massachusetts the periwinkle snail (*Littorina littorea*) is the most common herbivore (Lubchenco 1978). In tide pools, moderate grazing by *Littorina* on the algae that are competitively dominant permits many competitively inferior algae to survive (**Figure 24a**), as predicted by the intermediate disturbance hypothesis. But on emergent rocks, the snails do not eat the perennial brown and red algae that are competitively superior, but instead feed on the competitively inferior algae. Consequently, on emergent rocks *Littorina* grazing reduces algal diversity (**Figure 24b**). The critical factors are the food preferences of the grazer and the competitive abilities of the plants.

Streams may have variable water flow and changeable water temperatures. Death and Winterbourn (1995) found that aquatic invertebrates did best when there was minimal disturbance, contrary to the predictions of the intermediate disturbance hypothesis (**Figure 25a**). Similarly, in prairie grassland fire is a major disturbance, and plant diversity declines with more fires (**Figure 25b**). The intermediate disturbance hypothesis does not apply to many grazing systems (Collins et al. 1995; Li et al. 2004).

The intermediate disturbance hypothesis is an attractive hypothesis for the maintenance of high species diversity in communities, but it does not apply to all communities, and further work is needed to delimit its

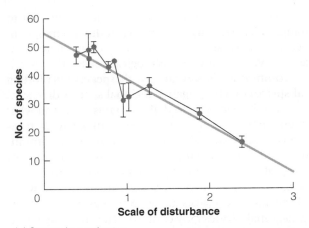

(a) Stream invertebrates

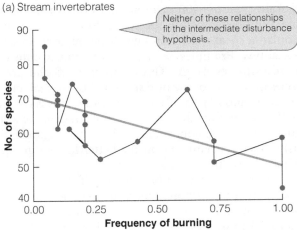

(b) Konza Prairie

Figure 25 The effect of disturbance on species richness in two contrasting communities. (a) Aquatic invertebrates in streams on the South Island of New Zealand. The scale of disturbance is a composite measure of variation in temperature, stream flow, and bottom stability. (b) Plant species richness in tallgrass prairie in Kansas. The frequency of burning is the probability of being burned each year between 1972 and 1990. Because species richness declines with the amount of disturbance, the intermediate disturbance hypothesis does not apply to either of these communities. ((a) From Death and Winterbourn 1995. (b) From Collins et al. 1995.)

range of application. In particular, land managers should not assume the validity of the intermediate disturbance hypothesis in making management plans for national parks or other protected areas.

Local and Regional Diversity

Biodiversity in local habitats could be limited by either evolutionary or ecological causes. The mixing of evolutionary processes on a long time scale and ecological processes on a short time scale has made it difficult to untangle the reasons for the latitudinal change in species diversity, as we have just seen. One way to separate out evolutionary and ecological causes is to see if each community is saturated with species by plotting local species diversity against regional species diversity.

To do this we need to define what is local and what is regional (Srivastava 1999). Local diversity is measured on a scale in which all the species in the community could interact with each other in some unit of ecological time, typically a generation. For example, fish species in a lake or stream, herb species in a meadow, or bird species in woodlots are all examples of local diversity. Regional diversity, by contrast, refers to a larger spatial scale, typically 100 or more times that of the local scale. Within the region, species could disperse to and colonize a local patch through dispersal over tens of generations. Examples of regional diversity would be the fish species of the Great Barrier Reef, the grass species of Britain, or the bird species of the boreal forest of Canada and Alaska. Regional species richness can be specified only if the flora and fauna of a region are well known. For this reason, studies of local and regional diversity have concentrated on the better known groups such as birds, butterflies, and trees.

Communities would be saturated if there were intense competition among the existing species such that no new species could fit into the suite of species. **Figure 26** illustrates the idea of testing for local community saturation; the key is that we expect a linear re-

lationship if communities are unsaturated, and a curvilinear relationship if communities are saturated. The best data for comparisons are from a single defined habitat sampled in several geographically distinct regions. **Figure 27** shows a comparison of local and regional diversity at the global scale for a whole range of taxa (Caley and Schluter 1997). In a broad, global sense, there is no evidence of local community saturation, which implies that biodiversity at the local level is not constrained by intensive competition, and that communities are not closed to new invaders.

On a smaller spatial scale, Pearson (1977) studied the bird communities of six undisturbed lowland tropical forest sites in Amazonia, Borneo, New Guinea, and West Africa. He censused local plots of about 15 ha, spending 200–700 hours on each plot to census the birds. He found that local and regional richness were linearly related, suggesting that bird communities in these tropical forests were not saturated with species.

The majority of studies to date have shown communities to be unsaturated. Srivastava (1999) summarized 36 studies, of which about two-thirds reported communities to be unsaturated. But several pitfalls exist in the analyses of local-regional diversity plots. Sample sizes must be large, or species will be missed (Caley and Schluter 1997). Determining the number of species in the regional pool must be done carefully, and the size of the regions should be equal to avoid bias. At present the assumption that ecological communities are saturated with species does not appear to be correct for many species groups (Ricklefs and Schluter 1993).

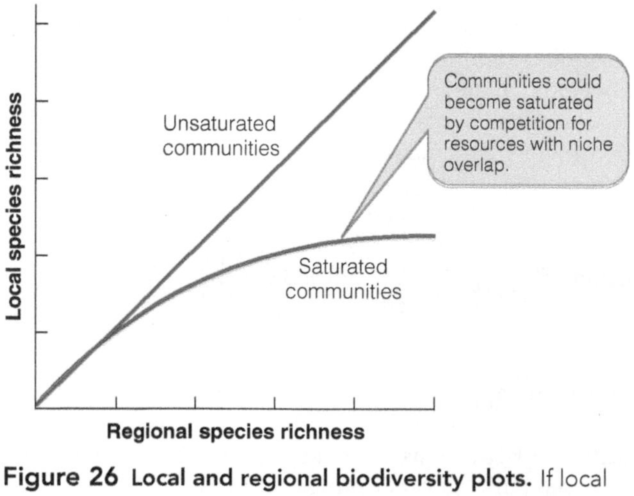

Figure 26 Local and regional biodiversity plots. If local communities are unsaturated, community diversity will continue increasing with regional diversity in a linear manner (blue line); richer regions will have richer local communities. By contrast, local communities that are saturated with species will reach an asymptote or maximum species richness (red line). Testing for saturation requires examining several communities in several different regions.

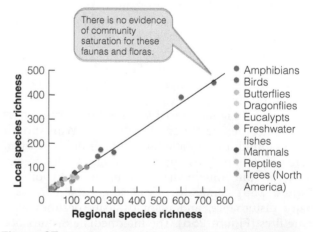

Figure 27 Local and regional species richness across continents for nine different taxonomic groups. No indication of saturation is apparent in this relationship, and remarkably all the different groups on different continents appear to follow the same linear regression. (Data from Caley and Schluter 1997.)

Summary

Biodiversity can be measured at the genetic level, at the species level, or at the ecosystem level. Species are most often the units of concern, and by counting all the species in an area we measure species richness as an index of biodiversity. Communities vary greatly in the numbers of species they contain, and in this chapter we ask *How does biodiversity vary over the globe?* and *What limits biodiversity?*

The first pattern we can recognize is that most communities consist of a few common species and many rare species. This pattern can be predicted by two contrasting hypotheses. The sequential breakage hypothesis assumes that niche dimensions are acquired sequentially by species and that competition is the fundamental process underlying this pattern of rare and common species. The neutral theory of biodiversity assumes all species are equivalent in their niche dimensions and that competition is not relevant to community patterns of diversity.

The second pattern we can recognize is that there are hotspots of biodiversity, regions with large numbers of rare species. Twenty-five hotspots that occur largely in tropical areas and in a few temperate regions like California contain nearly 50% of all species on only about 1%–2% of the Earth's land surface. Hotspots are thus areas of strong conservation interest.

The third pattern we can recognize is that tropical environments in general support more species in almost all taxonomic groups than do temperate and polar areas. The latitudinal gradient in biodiversity from the tropics to the polar regions is one of the most important patterns found in community ecology. The trees, birds, and mammals illustrate the complexity of these species-diversity gradients, which are not always gradual trends from the equator to the poles.

Combinations of six factors—evolutionary speed, geographic area, interspecific interactions, ambient energy, productivity, and disturbances—control biodiversity. On a regional scale, *ambient energy*, involving temperature, water, and solar energy, is the best predictor of species richness in terrestrial plants and animals. Evolutionary speed, which summarizes the rate of speciation and the evolutionary history of a region, is the most difficult factor to evaluate and is potentially important on a regional scale. On a local scale of a few tens of meters, disturbance, competition, and predation interact to affect species richness.

There is no simple, general answer to the question, *What controls biodiversity?* The answer depends on both the taxonomic group and the scale of analysis. Communities do not appear to be saturated with species, and we are led to consider the broader question of what controls community organization. To protect and preserve biodiversity we need to know what controls community organization.

Review Questions and Problems

1   The tree flora of Europe is less diverse compared with that of eastern North America or eastern Asia (Grubb 1987). Why should this be? Compare your explanations with those of Grubb (1987) and of Currie and Paquin (1987).

2   Hydrothermal vents are like geysers in the deep ocean on the seafloor. They continuously spew super-hot, mineral-rich water that helps support a diverse community of organisms. Vents are typically 2–100 km apart on the ocean floor, and in the Pacific exist for 10–100 years before they collapse. How could you test the intermediate disturbance hypothesis for these vents? Tsurumi (2003) discusses diversity at hydrothermal vents in the Pacific Ocean.

3   Marine algae along the west coast of North America do not increase in species richness toward the tropics but peak at about 70 species per 100 km of coastline around 40°N latitude (Gaines and Lubchenco 1982). Along the east coast of North America, species richness gradually increases as you move toward the tropics. Discuss why these patterns might hold.

4   In analyzing the role of fire as a disturbance in tallgrass prairie, Collins et al. (1995) found that the intermediate disturbance hypothesis was supported if, instead of plotting fire frequency as in Figure 25b, they plotted time since the last fire on the x-axis. Why should they get different results for these two plots of the same data?

5   Would you expect to have latitudinal gradients in the species richness of macroparasites of mammals and birds? What factors might control species richness in macro- and microparasites?

6   The merits of neutral theory in ecology are not clear to many ecologists yet they remain a powerful method to utilize in model building (Bell 2000; Alonso et al. 2006). Discuss the value of neutral models and in particular whether there is only one

type of neutral model for a particular question like global biodiversity abundance patterns.

7   The longest experiment in ecology is the Park Grass Experiment begun in 1856 at Rothamsted, England. A mowed pasture was divided into 20 plots, and a series of plots were fertilized annually with a variety of nutrients, including nitrogen. Discuss the predictions you would make regarding biodiversity on fertilized and unfertilized plots for this experiment, using the six factors discussed in this chapter. Silvertown et al. (2006) give the observed results.

8   In Antarctica, species richness in soft-bottom invertebrates (sponges, bryozoans, polychaetes, and amphipods) is higher than that of almost all other tropical- and temperate-zone soft-bottom communities (Clarke 1990). What observations or experiments would you perform to find out why this high biodiversity occurs in Antarctica?

9   Figure 20 shows that, on a global scale, species richness increases smoothly with solar energy and temperature. Why should this occur? Why is the available energy not monopolized by a few superspecies? Compare your ideas with those of Currie (1991, p. 46).

10   The evolutionary speed hypothesis has been tested with genetic marker data in only one group, birds (Bromham and Cardillo 2003). They found no evidence in favor of a higher rate of speciation in tropical areas for birds. Discuss how this finding would change the interpretation of the evolutionary speed hypothesis (e.g., Figure 11) if similar results are found in many other taxonomic groups. Are these data sufficient to reject the evolutionary speed hypothesis?

11   Does the Janzen-Connell hypothesis for diversity maintenance in tropical rain forests imply that mortality in small trees should be density dependent? Describe an experimental design (including the time frame required) that would allow you to test this hypothesis. Read Hyatt et al. (2003) and discuss whether their data are sufficient to reject the Janzen-Connell model for most forest systems.

Overview Question

To *preserve* biodiversity, how much do we need to understand about the factors that *control* biodiversity? Sketch the outlines of a management plan for preserving biodiversity in a large national park in a tropical rain forest.

Suggested Readings

- Bellocq, M. I., and R. Gomez-Insausti. 2005. Raptorial birds and environmental gradients in the southern Neotropics: A test of species-richness hypotheses. *Austral Ecology* 30:900–906.

- Butler, S. J., J. A. Vickery, and K. Norris. 2007. Farmland biodiversity and the footprint of agriculture. *Science* 315:381–384.

- Currie, D. J., et al. 2004. Predictions and tests of climate-based hypotheses of broad-scale variation in taxonomic richness. *Ecology Letters* 7:1121–1134.

- Hawkins, B. A. et al. 2003. Energy, water and broad-scale geographic patterns of species richness. *Ecology* 84:3105–3117.

- Hubbell, S. P. 2001. *The Unified Neutral Theory of Biodiversity and Biogeography*. Princeton, NJ: Princeton University Press.

- James, C. D., and R. Shine. 2000. Why are there so many coexisting species of lizards in Australian deserts? *Oecologia* 125:127–141.

- Li, J., W. A. Loneragan, J. A. Duggin, and C. D. Grant. 2004. Issues affecting the measurement of disturbance response patterns in herbaceous vegetation—A test of the intermediate disturbance hypothesis. *Plant Ecology* 172:11–26.

- Magurran, A. E. 2004. *Measuring Biological Diversity*. Oxford: Blackwell Publishing.

- Myers, N., R. A. Mittermeier, C. G. Mittermeier, G. A. B. da Fonseca, and J. Kent. 2000. Biodiversity hotspots for conservation priorities. *Nature* 403:853–858.

- Nee, S. 2005. The neutral theory of biodiversity: Do the numbers add up? *Functional Ecology* 19:173–176.

- Willig, M. R., D. M. Kaufman, and R. D. Stevens. 2003. Latitudinal gradients of biodiversity: pattern, process, scale, and synthesis. *Annual Review of Ecology, Evolution and Systematics* 34:273–309.

- Zillio, T., and R. Condit. 2007. The impact of neutrality, niche differentiation and species input on diversity and abundance distributions. *Oikos* 116:931–940.

Credits

Illustration and Table Credits

F1 C. B. Williams, *Patterns in the Balance of Nature*, Academic Press, 1964. F4 G. Sugihara, "Minimal community structure," *American Naturalist* 116:770–787, 1980. Copyright © 1980 The University of Chicago Press. F6 From D. J. Currie and V. Paquin, "Large-scale biogeographical patterns of species richness of trees," *Nature*, Vol. 329, p. 326. Copyright © 1987 Macmillan Magazines Ltd. Reprinted by permission. F9 From G. G. Simpson, "Species density of North American recent mammals," *Systematic Zoology* 13:57–73, 1964. F16 J. Roughgarden, "Niche width: Biogeographic patterns among *Anolis* lizard populations," *American Naturalist* 108:429–442, 1974. Copyright © 1974 The University of Chicago Press. F21 B. A. Hawkins et al., "Energy, water and broad-scale geographic patterns of species richness," *Ecology* 84:3105–3117, 2003. Copyright © 2003 Ecological Society of America. Used with permission. F22 L. H. Fraser and P. Keddy, "The role of experimental microcosms in ecological research," *Trends in Ecology & Evolution* 12(12):4, 1997. Copyright © 1997. Published by Elsevier Science Ltd. F24 J. Lubchenco, "Plant species diversity in a marine intertidal community," *American Naturalist* 112:23–39, 1978. Copyright © 1978 The University of Chicago Press. F25a R. G. Death and M. J. Winterbourn, "Diversity patterns in stream benthic invertebrate communities: The influence of habitat stability," *Ecology* 76:1446–1460. Copyright © 1995 Ecological Society of America. Used with permission. F25b S. L. Collins et al., "Experimental analysis of intermediate disturbance and initial floristic composition: Decoupling cause and effect," *Ecology* 76:486–492, 1995. Copyright © 1995 Ecological Society of America. Used with permission.

Photo Credits

Community Structure in Time: Succession

Key Concepts

- Succession is the process of directional change in communities over time.

- Succession proceeds through a series of seral stages toward a climax community that remains relatively stable on an ecological time scale. There are many different climax communities for a region, depending on soil type, water, grazing, and other environmental gradients.

- The key question is how species interact during succession. Existing species may facilitate, inhibit, or not influence invading species.

- Communities contain a mosaic of patches undergoing local dynamics, often in cycles. Patch or gap dynamics are controlled by the life history-traits of the species.

- Plant succession is largely controlled by competition among plants for light, and shade tolerance is a key parameter for succession models.

From Chapter 18 of *Ecology: The Experimental Analysis of Distribution and Abundance*, Sixth Edition. Eugene Hecht.

KEY TERMS

climatic climax The final, equilibrium vegetation for a site that is dictated by climate and toward which all successions are proceeding, according to Frederic Clements.

climax-pattern hypothesis The view that climax communities grade into one another and form a continuum of climax types that vary gradually along environmental gradients.

facilitation model The classic view that succession proceeds via one species helping the next species in the sequence to establish.

inhibition model Succession proceeds via one species trying to stop the next species in the sequence from establishing.

monoclimax hypothesis The classic view of Frederic Clements that all vegetation in a region converges ultimately to a single climax plant community.

polyclimax hypothesis The view of Whittaker that there are several different climax vegetation communities in a region governed by many environmental factors.

primary succession Succession occurring on a landscape that has no biological legacy.

secondary succession Succession occurring on a landscape that has a biological legacy in the form of seeds, roots, and some live plants.

succession The universal process of directional change in vegetation during ecological time.

tolerance model The view that plants in a successional sequence do not interact with one another in either a negative or a positive manner.

Neither individual organisms nor species populations exist by themselves in nature; they are always part of an assemblage of populations living together in the same area. When we previously discussed the interactions of two or more of these populations in predation and competition for food, the focus was on individual populations. Now we focus on the assemblage of populations in an area, the community. Most generally, a community is *any assemblage of populations of living organisms in a prescribed area or habitat*. So we can speak of the community of animals in a rotting log or the community of plants in the beech-maple deciduous forest or a community of plankton in part of the ocean. A community may be of any size and may be restricted to a taxonomic group like birds or include many different taxonomic groups.

Why do we need to be concerned about communities? The key to answering this question lies in the relationships between species in a community. There are two extreme views of the structure of any community. At one extreme, if the community is a complex ecological unit, a kind of superorganism, the populations should be bound together in a network and organized by obligate interrelations. This is the basic idea of the "web of life" and the general view that if you change one thing in a community, the change will reverberate throughout the whole community. The alternate extreme view is the individualistic view of a community as a group of populations that have few obligate relationships, so that each species operates under its own rules, and the community exists because a group of species have the same physical and chemical niche requirements. These extreme views highlight the first important question we must ask about any community: How strong are the connections between species—how strong is the web of life?

Much of what we discuss can be viewed as attempts to answer the question of how tightly linked species are in particular communities. There is no reason to assume that all communities will fit into one or the other of these extreme models. But there are important practical consequences that flow from these models if we are trying to manage a community (Walker 1992). If species are tightly linked in natural communities, losses of species may have cascading effects on other species. In these cases, conservation biology must be concerned about community dynamics rather than single-species dynamics. Within a community some species are "drivers" and others are "passengers," and the loss of some species is more critical than the loss of others.

These two extreme models of community interactions are similar to the historical conflict in plant ecology over the organismic and individualistic views of plant communities (Crawley 1997). Plant ecologists put the question this way: *Is the community an organized system of recurrent species, or a haphazard collection of populations with minimal integration?* The answer to this question has turned out to be somewhere in the middle rather than at the two extremes.

One of the most important features of biotic communities is change, and in this chapter we focus on the factors that cause community structure to change in ecological time. If we sat in a prairie for 10 years, or in a forest for 20 years, we would see the surrounding community change, whether slowly or dramatically. The consequences for land and water management and conservation can be severe. If we designate a prairie grassland as a protected area, for example, and keep grazing animals and fire off the site, the grassland will turn into shrubland and finally forest, and we will have lost the community we set out to protect.

Community change has important repercussions for conservation, and for all forms of water and land management. Two main types of changes occur in communities: directional changes and cyclic changes. Throughout this chapter we focus on two questions: (1) What factors cause community changes? and (2) How predictable are community changes?

When stripped of its vegetation by fire, flood, glaciation, or volcanic activity, the resulting area of bare ground does not long remain devoid of plants and animals. The area is rapidly colonized by a variety of species that subsequently modify one or more environmental factors. This modification of the environment may in turn allow additional species to become established. This development of the community by the action of vegetation on the environment leading to the establishment of new species is termed **succession**. Succession is the universal process of directional change in vegetation during ecological time. It can be recognized by a progressive change in the species composition of the community.

Most observed successions are called **secondary succession**, the recovery of disturbed sites. A few successions are called **primary succession** because they occur on a new sterile area, such as that uncovered by a retreating glacier or created by an erupting volcano. It is this latter situation that we explore in the next example.

Primary Succession on Mount St. Helens

Mount St. Helens in southwest Washington state erupted catastrophically on May 18, 1980. About 400 m was blown off the cone of this volcano, and the blast from the eruption devastated a wide arc extending some 18 km north of the crater (Dale et al. 2005). Three main areas were affected by the eruption: the blast zone, in which trees and vegetation were blown down but not eliminated; the pyroclastic flows (a hot mixture of volcanic gas, pumice, and ash) to the north of the crater; and the extensive mudflows and ash deposits away from the crater toward the south, east, and west (**Figure 1; Figure 2**). In addition, the eruption spewed tephra (ash) over thousands of square kilometers. The eruption produced a landscape with low nutrient availability, intense drought, and frequent surface movements and erosion, a great variety of conditions for vegetative recolonization (del Moral et al. 1995).

Primary succession following volcanic eruptions has been studied less than other forms of succession, and Mount St. Helens provided a good opportunity to study the mechanisms determining the rate of primary succession. Permanent plots have been established at several sites above treeline around the crater, and early

(a)

(b)

Figure 1 (a) The eruption of Mount St. Helens, Washington, May 18, 1980. (b) A view of the mountain the day before the eruption.

(a)

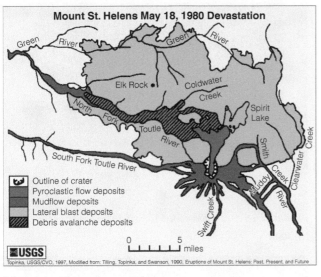

(b)

Figure 2 (a) The devastated area of Pine Creek Ridge, showing the scour from mudflow and ash, 4 months after the eruption (September 1980). Note the helicopter in this photo. (b) A map of the large area devastated by the eruption.

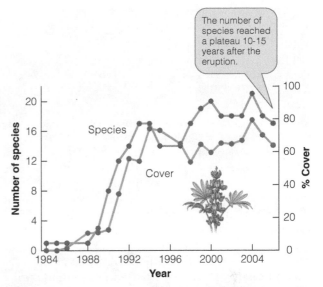

Figure 3 Number of species occurring in 1-m² quadrats in the Studebaker Ridge area on the south cone of Mount St. Helens, Washington. The total plant cover on this area has increased slowly because of the harsh conditions on these volcanic deposits. Lupine (*Lupinus lepidus*) is an early colonizer that facilitates further colonization by serving as a shelter or nurse plant. Photos of these plots are shown in Figure 4. (Data courtesy of Roger del Moral, 2007; photo courtesy of Thayne Tuason.)

succession has been described in a series of papers by Roger del Moral and his students (summarized in del Moral and Lacher 2005).

Colonization of habitats above treeline on Mount St. Helens has been slow. Vegetation on a small mudflow south of the crater at Studebaker Ridge had no surviving plants in 1980, and only one species by 1984. But eight species had colonized the area by 1990 and 20 species by 2000 (**Figure 3; Figure 4**). Only a few additional species invaded after 2000, and plant cover has increased slowly so that few of the species are common 28 years after the eruption.

Early primary succession on volcanic substrates rarely produces plant densities sufficient to inhibit the colonization of other species. Neither space nor light are limiting resources for plants in this environment. So-called nurse plants facilitate the establishment of other species. Lupines (*Lupinus lepidus*) have heavy seeds and are poorly dispersed, but they have become locally common on mudflows and pyroclastic surfaces (del Moral and Wood 1993). Before lupines get very common, other wind-dispersed plants such as *Aster ledophyllus* and *Epilobium angustifolium* become established in lupine clumps and survive better in the shelter of these nurse plants. Lupines die after four or five years, and because they fix nitrogen they contribute on a local scale to increased soil-nitrogen levels.

Chance events strongly affected primary succession on Mount St. Helens. Biological mechanisms are initially very weak in the severe environments produced by volcanic flows. The ability to become established in these severe environments is directly related to seed size (Wood and del Moral 1987), but dispersal ability is inversely related to seed size. Consequently, subalpine areas on Mount St. Helens received many wind-blown

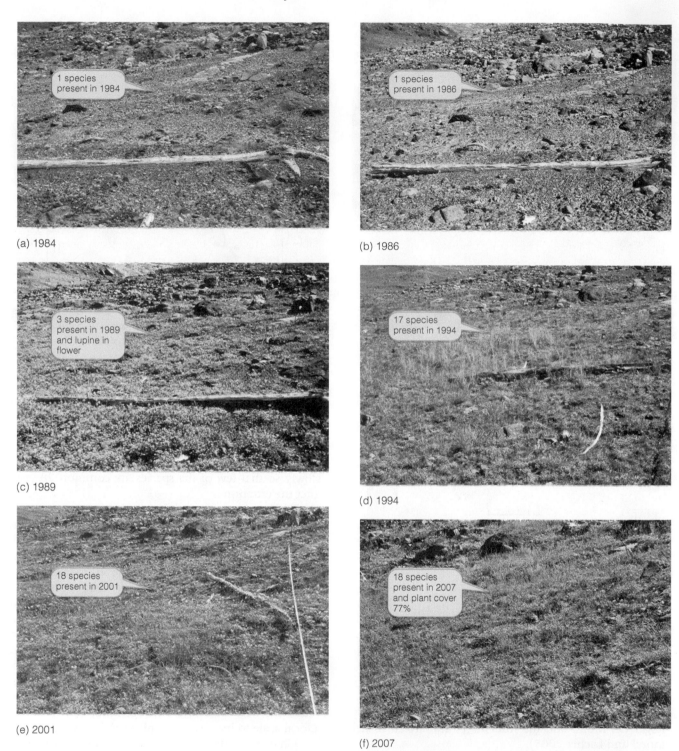

(a) 1984

(b) 1986

(c) 1989

(d) 1994

(e) 2001

(f) 2007

Figure 4 Primary succession on mudflow deposits on Studebaker Ridge on the south side of Mount St. Helens following the eruption of May 1980. Already in 1981 plants had colonized this devastated area, and plant cover has been slowly increasing as the site undergoes primary succession. Figure 3 shows data from these plots. (Photos courtesy of Roger del Moral, 2007.)

seeds, but almost none of these small seeds germinated and achieved colonization under the stressful conditions. When plants with large seeds colonize by chance, they become a focus of further community development. If by chance a single individual plant survives in the devastated landscape, it quickly becomes a locus for seed dispersal to adjacent areas, so positive feedback occurs during the early years of succession. Primary succession on Mount St. Helens has been very slow because of erosion, low-nutrient soils, chronic drought stress, and limited dispersal of larger seeds to areas distant from undisturbed vegetation.

Mount St. Helens provides a graphic example of plant succession after extreme disturbance. Measuring the speed of change on the mudflow areas allows us to estimate that it will require more than 100 years for the landscape on Mount St. Helens to return to a stable plant community. Understanding succession also requires understanding the mechanisms that drive changes in communities, and one focus has been on the effects that early successional species have on later successional species. Early species can help, hinder, or not affect the establishment of later species. Competition between individual plants for resources such as water, light, or nitrogen may drive succession. On Mount St. Helens we can see these processes occurring slowly, which enables us to understand them more easily. Understanding how naturally disturbed landscapes renew themselves can help us to understand how human-disturbed landscapes might respond. We now turn to these broader issues, and to the theory of succession and the mechanisms involved.

Concepts of Succession

The concept of succession was largely developed by the botanists J. E. B. Warming (1896) in Denmark and Henry C. Cowles (1899) in the United States. Warming's book greatly influenced Cowles, who studied the stages of sand-dune development at the southern edge of Lake Michigan. Henry Cowles was one of the most influential plant ecologists in the United States in the early years of the twentieth century, and the students he taught at the University of Chicago became a who's who of American plant ecology.

Successional studies pioneered by Warming and Cowles have led to four major hypotheses of succession. The first is the classical theory of succession, which was called relay floristics by Egler (1954) because it postulates an orderly hierarchical system of change in the community (**Figure 5a**). The classical ideas of succession were elaborated in great detail by

F. E. Clements (1916, 1936), who developed a complete theory of plant succession and community development called the **monoclimax hypothesis**. The biotic community, according to Clements, is a highly integrated superorganism that develops through a process of succession to a single end point in any given area—the **climatic climax**. The development of the community is gradual and progressive, from simple pioneer communities to the ultimate or climax stage. This succession is due to biotic reactions only; the plants and animals of the pioneer stages alter the environment such that it favors a new set of species, and this cycle recurs until the climax is reached. Development through succession in a community is therefore analogous to development in an individual organism, according to Clements' view (Phillips 1934–1935). Thus, reverse succession (retrogression) is not possible unless some disturbance such as fire, grazing, or erosion intervenes. Secondary succession differs from primary succession in having a seed bank from plants that occur later in succession, so that late succession species are already present in the early stages of secondary succession (see a_1 and a_2 of Figure 5).

The key assumption of the classical theory of succession is that species replace one another because at each stage they modify the environment to make it less suitable for them and more suitable for others. Thus species replacement is orderly and predictable and provides directionality for succession. These characteristics led Connell and Slatyer (1977) to call this the **facilitation model** of succession. The early species in succession facilitate the arrival of the later species.

The climax community in any region is determined by climate, in Clements' view. Other communities may result from particular soil types, fire, or grazing, but these *subclimaxes* are understandable only with reference to the end point of the climatic climax. Therefore, the natural classification of communities must be based on the climatic climax, which represents the state of equilibrium for the area.

A second major hypothesis of succession was proposed by Egler (1954), who called it initial floristic composition. In this view, succession is very heterogeneous because the development at any one site depends on who gets there first. Species replacement is not necessarily orderly because each species excludes or suppresses any new colonists (Figure 5b and 5c). Thus, succession becomes more individualistic and less predictable because communities are not always converging toward the climatic climax. Egler's hypothesis of initial floristics actually contained two ideas. Part of his hypothesis was called the **inhibition model**

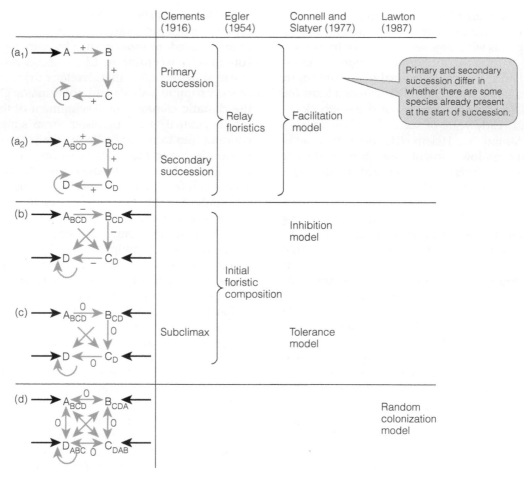

	Clements (1916)	Egler (1954)	Connell and Slatyer (1977)	Lawton (1987)

Primary and secondary succession differ in whether there are some species already present at the start of succession.

Figure 5 Four models of succession (a–d) proposed by different authors. The letters a–d represent hypothetical vegetation types or dominant species; subscript letters indicate species that are present as minor components or as propagules. Light blue arrows represent species or vegetation sequences in time; bold, black arrows represent alternative starting points for succession after disturbance. Curved arrows indicate that the species replaces itself. +, facilitation; −, inhibition; 0, no effect. The relay floristics model has two patterns (a_1, a_2), depending on whether primary or secondary succession is occurring. (Modified after Noble 1981.)

of succession by Connell and Slatyer (1977): the species present early in succession inhibit the establishment of the later species (see Figure 5b). No species in this model is competitively superior to another; whoever colonizes the site first holds it against all comers until it dies. Succession in this model proceeds from short-lived species to long-lived species and is not an orderly replacement because "who gets there first" is a matter of chance. Wilson et al. (1992) called this model the preemptive initial floristics model to emphasize that the first species at a site preempt the course of succession.

Egler's initial floristics model can also describe the third major model of succession proposed by Connell and Slatyer (1977), who called it the **tolerance model**.

In the tolerance model, the presence of early successional species is not essential—any species can start the succession (see Figure 5c). Some species are competitively superior, however, and they eventually come to predominate in the climax community. Species are replaced by other species that are more tolerant of limiting resources. Succession proceeds either by the invasion of later species or by a thinning out of the initial colonists, depending on the starting conditions. The tolerance model includes Egler's emphasis on the initial floristics as a major influence on how succession proceeds.

A fourth model of succession was proposed by Lawton (1987) to provide a null model with no ecological interactions. The random colonization model suggests that succession involves only the chance survival of different species and the random colonization by new species. There is no facilitation and no interspecific competition (Figure 5d) and succession can move in any direction.

The first three hypotheses of succession agree in predicting that many of the pioneer species in a succession will appear first because these species have evolved colonizing characteristics, such as rapid growth, abundant seed production, and high dispersal powers (**Table 1**).

The critical feature of the life-history traits listed in Table 1 is that there is a trade-off, or inverse correlation, between traits that promote success in early succession and traits that are advantageous in late succession (Huston and Smith 1987).

The critical distinction among the four hypotheses of succession is in the mechanisms that determine subsequent establishment. In the classical facilitation model, species replacement is *facilitated* by the previous stages. In the inhibition model, species replacement is *inhibited* by the present residents until they are damaged or killed. In

Table 1 Physiological and life history characteristics of early- and late-successional plants.

Characteristic	Early succession	Late succession
Photosynthesis		
Light-saturation intensity	high	low
Light-compensation point	high	low
Efficiency at low light	low	high
Photosynthetic rate	high	low
Respiration rate	high	low
Water-use efficiency		
Transpiration rate	high	low
Mesophyll resistance	low	high
Seeds		
Number	many	few
Size	small	large
Dispersal distance	large	small
Dispersal mechanism	wind, birds, bats	gravity, mammals
Viability	long	short
Induced dormancy	common	uncommon?
Resource-acquisition rate	high	low?
Recovery from nutrient stress	fast	slow
Root-to-shoot ratio	low	high
Mature size	small	large
Structural strength	low	high
Growth rate	rapid	slow
Maximum life span	short	long

SOURCE: From Huston and Smith (1987).

What Is the Gaia Hypothesis?

The superorganismic view of communities that was advanced by F. E. Clements and A. G. Tansley over 70 years ago is similar to the Gaia Hypothesis proposed by James Lovelock during the past 20 years. (Gaia is the name the Greeks gave to their Goddess of the Earth.) The Gaia Hypothesis arose from Lovelock's observations of the atmospheres on Mars and other planets. His answer to the question, Why is the Earth so different and so suitable for life? is that the Earth's atmosphere and organisms are tightly coupled in a feedback loop, such that organisms control the makeup of the atmosphere and keep it at or near the chemical composition that favors life. Thus, the living biota and the physical atmosphere act in a feedback system to control oxygen and carbon dioxide levels on Earth.

Two crucial aspects of the Gaia Hypothesis require investigation. First, we need to ascertain if in fact there are feedback loops between the biota and the Earth's atmosphere that act to control changes in oxygen and carbon dioxide levels. One suggested mechanism of climate control is cloud production due to dimethylsulphide (DMS) production by marine phytoplankton. Plankton produce DMS, which aids in cloud formation, which in turn affects global climate by stabilizing temperature (Lovelock 1988).

Second, if the mechanisms of the Gaia Hypothesis do exist, we need to find out how such a system could have evolved. Most evolutionary ecologists are skeptical of the Gaia Hypothesis because natural selection maximizes fitness at the level of the individual not at the level of groups of species. Traits that operate for a good of the species or for the good of the whole biota are believed to have evolved by individual selection, not by group selection, which is typically very weak compared to individual selection. The Gaia Hypothesis postulates group selection to the extreme, such that evolution selects for systems that operate for the good of all living things. The challenge is to derive individual-based selective advantages for mechanisms that act to control climate (Kirchner 2002; Wilkinson 2003).

A second criticism of the Gaia Hypothesis is that if such mechanisms do exist, they have not been very effective in the past in controlling climatic changes. Large changes in the concentrations of carbon dioxide over the past million years (Steffan et al. 2005) do not suggest any effective stabilizing mechanisms that the Gaia Hypothesis postulates. But the control of the Earth's atmosphere postulated by the Gaia Hypothesis could allow large swings in gas composition within some limits that are never exceeded. From a human perspective, even if the Gaia Hypothesis is correct, we must not neglect the causes of climate change in the naive hope that Gaia will rescue us from our folly of increasing emissions of greenhouse gases.

---

the tolerance and random colonization models, species replacement is *not affected* by the present residents.

The utility of these four models of succession is that they immediately suggest experimental manipulations to test them. Removing or excluding early colonizers, transplanting seeds or seedlings of late-succession species into earlier stages, and other experiments can shed light on the mechanisms involved in succession.

To explain a successional sequence we must add to these idealized models of succession some additional information on seed availability, insect and mammal herbivory, mycorrhizal fungi, and plant pathogens (Walker and Chapin 1987). **Figure 6** summarizes the relative importance of a variety of factors in succession. The primary processes underlying successional changes could be competition or mutualistic interactions between plants, but these plant-plant interactions are affected by animal grazing and diseases, and by seed dispersal and storage. The resulting successional sequences are thus complex and do not proceed in a single direction to a fixed end point (Walker and del Moral 2003).

A Simple Mathematical Model of Succession

We can construct a simple mathematical model of succession by assuming that succession is a replacement process (Horn 1981). For each plant, we ask a simple question: What is the probability that this plant will be replaced in a given time by another plant of the same species or of another species? A matrix of replacement probabilities can be constructed in forests by counting the number of saplings of various species growing under the canopy of the mature trees. For example, of a total of 837 saplings growing beneath large gray birch trees, Horn (1975a) found that (to mention but a few species) zero were gray birch saplings, 142 were red maple saplings, and 25 were beech saplings. Thus, the probability that a gray birch will be replaced by another gray birch (self-replacement) is $0/837 = 0$, by a red maple is $142/837 = 0.17$, and by a beech is $25/837 = 0.03$. These probabilities are entered as percentages in **Table 2**.

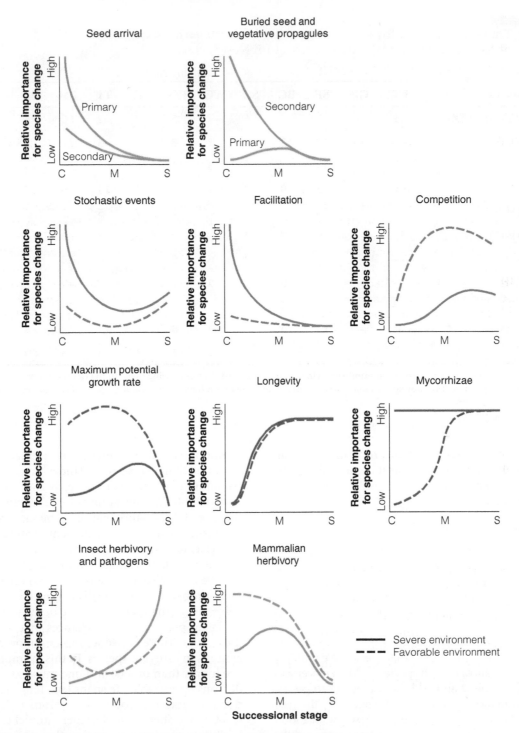

Figure 6 Influence of type of succession and environmental severity upon major successional processes that determine change in species composition during colonization (C), maturation (M), or senescence (S) stages of succession. (Modified after Walker and Chapin 1987.)

Table 2 Transition matrix for saplings growing beneath various species of trees at Institute Woods in Princeton, New Jersey.

Canopy species	BTA	GB	SF	BG	SG	WO	RO	HI	TT	RM	BE	Total no.
Big-toothed aspen (BTA)	**3**	5	9	6	6	—	2	4	2	60	3	104
Gray birch (GB)	—	—	47	12	8	2	8	—	3	17	3	837
Sassafras (SF)	3	1	**10**	3	6	3	10	12	—	37	15	68
Blackgum (BG)	1	1	3	**20**	9	1	7	6	10	25	17	80
Sweetgum (SG)	—	—	16	—	**31**	—	7	7	5	27	7	662
White oak (WO)	—	—	6	7	4	**10**	7	3	14	32	17	71
Red oak (RO)	—	—	2	11	7	6	**8**	8	8	33	17	266
Hickories (HI)	—	—	1	3	1	3	13	**4**	9	49	17	223
Tulip tree (TT)	—	—	2	4	4	—	11	7	**9**	29	34	81
Red maple (RM)	—	—	13	10	9	2	8	19	3	**13**	23	489
Beech (BE)	—	—	—	2	1	1	1	1	8	6	**80**	405

Values are expressed as percentages of the total number of saplings (final column) found growing beneath the canopy species listed. Entries are interpreted as the percentages of canopy species that will be replaced one generation hence by the sapling species listed across the top; the percentages of "self-replacements" are in boldface.

SOURCE: After Cody (1975).

We can use these replacement probabilities to calculate what will happen to this forest community in the future. For example, working down the columns of the table, we can calculate

$$\left\{\begin{array}{c}\text{Proportion of the next generation}\\\text{that will be white oak}\end{array}\right\}$$

$$\begin{array}{l}= 0.02 \text{ gray birch} + 0.03 \text{ sassafras} +\\0.01 \text{ blackgum} + 0.10 \text{ white oak} +\\0.06 \text{ red oak} + 0.03 \text{ hickory} +\\0.02 \text{ red maple} + 0.01 \text{ beech}\end{array} \quad (1)$$

where all the species on the right side of this equation refer to the current abundance of that species. We multiply the current abundances of each species by the replacement probabilities in Table 2 and add them together to get the predicted abundance of each canopy species in the next generation. We can cycle through these calculations, which are more tedious than difficult, and predict the abundances of all species any number of generations into the future (**Table 3**). After several generations, the abundances of all species settle down to a stationary distribution that will not change over time. In this example, the stationary distribution predicted will contain 50% beech, 16% red maple, 7% tulip tree, and so on.

To use this simple model of succession, we must make a number of assumptions. The most critical assumption is that the table of replacement probabilities does not change over time. Under this assumption, the community will approach a steady state that is independent of both the community's initial composition and of the type of disturbance that starts the succession going. In this form, the model is a statement of any type of succession (facilitation, inhibition, or tolerance) that predicts a regular, repeatable change culminating in a stable climax.

We must also assume that we can calculate replacement probabilities in a realistic way. For forests, this involves assuming that abundance in the sapling stage is a sufficient predictor of the chances that trees will reach the canopy. There is also a problem of overlapping generations in different species. In this example, some trees live longer than others, and one must correct these predictions for variable life spans (see Table 3). This correction is tedious but not difficult (Horn 1975b).

Can we alter the replacement model to describe the inhibition model of succession that does not culminate in a stable, fixed climax? If the table of replacement probabilities depends on the present composition of the forest, predictions from the model change dramatically. Assume, for example, that the recruitment of young plants depends on the density of trees of their own species. The transition probabilities for any one tree are not constant under this assumption but change depending on who neighbors are. Succession in this model

Table 3 Theoretical approach of the Institute Woods in Princeton, New Jersey, to a stationary distribution ("climax forest").

	Species[a] (%)										
	BTA	GB	SF	BG	SG	WO	RO	HI	TT	RM	BE
Theoretical generation											
0	—	49	2	7	18	—	3	—	—	20	1
1	—	—	29	10	12	2	8	6	4	19	10
2	1	1	8	6	9	2	8	10	5	27	23
3	—	—	7	6	8	2	7	9	6	22	33
4	—	—	6	6	7	2	6	8	6	20	39
5	—	—	5	5	6	2	6	7	7	19	43
...	...	...	...	...	...	...	...	...	...	...	...
...	...	...	...	...	...	...	...	...	...	...	...
...	...	...	...	...	...	...	...	...	...	...	...
Stationary distribution (%)	—	—	4	5	5	2	5	6	7	16	50
Longevity (yr)	80	50	100	150	200	300	200	250	200	150	300
Age-corrected stationary distribution (%)	—	—	2	3	4	2	4	6	6	10	63

[a]Abbreviations refer to the names of tree species listed in Table 2

The starting point is an observed 25-year-old forest stand dominated by gray birch. The theoretical predictions are obtained by multiplying the starting tree composition (generation 0) by the transition matrix of probabilities in Table 2

SOURCE: After Cody (1975).

does not converge to one point; instead, alternative communities could be produced, depending on the accidents of history, as suggested by the inhibition model.

The replacement model of succession is useful because it focuses our attention on the local regeneration of species, and how species replace themselves and other species as disturbances open up new areas for succession. It is not, however, a mechanistic model of succession, and to understand why succession is occurring we need to focus on the mechanisms controlling succession. One possible mechanism driving succession is competition for limiting resources (Tilman 1985).

One way to model succession mechanistically is to use an individual-based plant model that explicitly incorporates light as the limiting resource (Huston and Smith 1987). Each individual plant is given species-specific traits of maximum size and age, maximum growth rate, and tolerance to shading. Most of these models have been used for trees to model forest succession, but they could be applied to other kinds of plants as well. The key variable in these models is light availability, and each individual plant is analyzed to see how much shading is produced by its neighbors. If light is limited, growth rates and survival rates are reduced accordingly. This type of simple mechanistic model can produce successional sequences of tree species that resemble natural succession (Shugart 1984). **Figure 7** illustrates two scenarios with species of different life-history traits. In every case the species that is most shade tolerant and grows to the largest size wins out in succession. The seral stages vary greatly depending on which trees are present.

The additional effects of competition for soil nitrogen can be added to these simple models, such that both light and nitrogen become limiting resources (Shugart 1984; Tilman 1985). Mechanisms of nutrient limitation make these models more realistic, but also more complex. The most successful models of succession are for forests. Because of their economic importance a wealth of details are known about the life-history traits of individual tree species.

How well do natural communities fit the four hypotheses of succession? Does succession in a region converge to a single end point, or are there multiple stable states? For Mount St. Helens we have seen that the facilitation model of succession is a good description of the

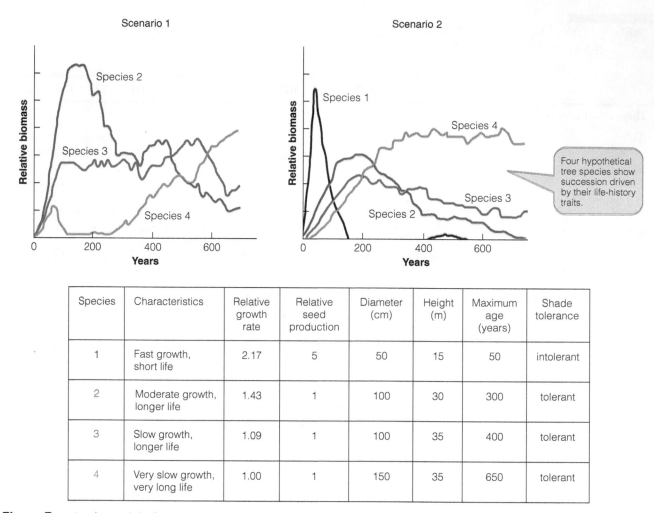

Figure 7 A simple model of succession for trees. Species biomass dynamics and community biomass for hypothetical successional sequences with three and four idealized species in which competition for light is the driving variable. In scenario 1, all three species are shade-tolerant and thus have late-successional characteristics but differ sufficiently in relative competitive abilities (growth rate) to produce a "typical" successional replacement. In scenario 2, an early-successional species with a rapid growth rate and shade-intolerance is added to the three species in scenario 1. (Modified from Huston and Smith 1987.)

early stages of succession. As vegetation cover on the mudflows becomes more complete, inhibition may begin to operate. In the next section we look at some additional examples of succession to evaluate these models.

Case Studies of Succession

Although numerous studies of succession have been conducted, in few of them can succession be related to a time scale. Here we examine three investigations of succession in which the time scale is known: a volcanic island in Iceland, sand dunes near Lake Michigan, and the succession of insects on carrion.

Surtsey, Iceland: Primary Succession on a Volcanic Island

The general principles of primary succession can be seen most clearly on barren surfaces that are free of a biological legacy. One good example of this has been the new island of Surtsey which rose out of an underwater volcanic eruption in the sea off the south coast of Iceland on November 14, 1963, and continued to grow until June 5, 1967. It reached 2.8 km² at its maximum and has since been steadily eroded by ocean waves so that now it is less than half the original size (**Figure 8**).

Figure 8 The volcanic island of Surtsey off the south coast of Iceland. The island is about 1.3 km in diameter. The soil of the island consists of volcanic tephra (volcanic materials of all sizes), the rocks are from lava flows, and the geological material is alkali olivine basalt.

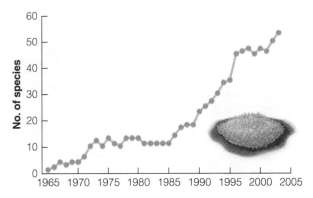

Figure 9 Number of higher plants present on the volcanic island of Surtsey, Iceland, after the eruption, which extended from 1963–1967. The sea sandwort (*Honkenya peploides*) is the most common higher plant growing on Surtsey. Mosses and lichens followed much the same time sequence of colonization. (Data from Friäriksson 1987 and the Surtsey Research Society.)

The first higher plants colonized Surtsey in 1965, and three species were found on the island in the following two years. But none survived overwinter, as they were buried by the volcanic sand or washed away by storm action. In 1968 sea sandwort became the first plant to establish on Surtsey, and colonization of other species of higher plants continued (**Figure 9**). From 1975 to 1985 there was a lull and few species were added, possibly because of difficult nutrient conditions in the volcanic soil. In the late 1970s gulls began to colonize the lava beds on the southern part of Surtsey and the gull population increased rapidly after 1986. Two factors then accelerated plant colonization. First, gulls brought seeds of various plants to the island, and second, the gulls transferred nutrients from the sea to the land in their droppings, enriching the soil nutrients, particularly nitrogen. The first shrubs appeared on Surtsey in 1995, possibly wind-dispersed as seeds.

The lower plants first colonized the steam vent areas where the rocks were kept damp. Two mosses were found in 1967 and six more species in 1968. At this time nitrogen-fixing cyanophyta also appeared around steam vents. Other areas of the island were slow to be colonized by mosses and lichens. All these plants disperse by spores carried by the wind, and they are common on other lava-covered areas of Iceland. The increasing abundance of gulls also aided the growth of mosses and lichens. By 2003 there were 53 species of moss on Surtsey and at least 45 species of lichens.

In 2007, 40 years after the eruptions stopped, Surtsey is a well-vegetated volcanic island that will continue to increase in species numbers as the soil improves. It appears to be a good example of the facilitation model of succession with the twist that it was the pioneer plants as well as the birds that helped the later plants establish.

Lake Michigan Sand-Dune Succession

Henry Cowles (1899) worked on the sand-dune vegetation of Lake Michigan and made a classic contribution to our understanding of plant succession. The sand dunes around the southern edge of Lake Michigan have been a model system for examining the theories of succession in a dynamic landscape. Cowles did not have access to radiocarbon dating methods and Lichter (1998, 2000) reexamined the successional stages in dune systems at the northern part of Lake Michigan in relation to an absolute time scale determined by radiocarbon dates.

During and after the retreat of the glaciers from the Great Lakes area, the resulting fall in lake level left many distinct "raised beaches" and their associated dune ridges. These systems, which run roughly parallel to the eastern shoreline of Lake Michigan, consist of 72 dune ridges formed over the last 2375 years above the present lake level (**Figure 10**).

The dunes, like glacial moraines, offer a near-ideal system for studying plant succession because many of the complicating variables are absent. The initial substrate for all the area is dune sand, the climate for the whole area is similar, the relief is similar, and the available flora

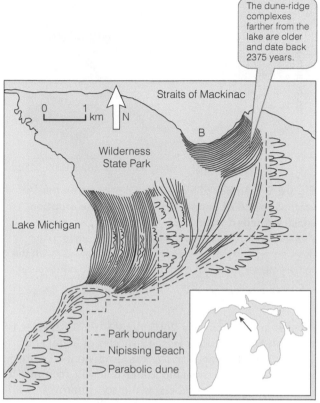

> The dune-ridge complexes farther from the lake are older and date back 2375 years.

Straits of Mackinac

0   1
└──┴──┘ km   N

Wilderness
State Park

B

Lake Michigan

A

--- Park boundary
--- Nipissing Beach
⊃ Parabolic dune

Figure 10 The dune-ridge complexes of Wilderness State Park in northern Lake Michigan in the central United States. Seventy-two dune ridges can be dated in this area, studied by Lichter (1998, 2000). (From Lichter 1998.)

and fauna are the same. The only disturbances during succession were some selective logging of large trees during the late nineteenth century and occasional fires. So the differences between the dunes should be due only to *time*, to *the biological processes of succession*, and to *chance events* associated with dispersal and colonization.

Two processes produce bare sand surfaces ready for colonization. One is the slow process of a fall in lake level; the other is a rapid process, the movement of sand to form a new dune ridge from the strong winds that come off the lake during storms. This wind erosion creates a moving dune that is gradually stabilized (only by vegetation) after migrating inland.

The bare sand surface is colonized first by dune-building grasses, of which the most important is marram grass (*Ammophila breviligulata*) **(Figure 11a)**. Marram grass propagates by rhizome migration, only rarely by seed. It spreads very quickly and can stabilize a bare area in six years. After the sand is stabilized, marram grass declines in vigor and dies out. The reason for this is not known, but the result is that this grass is not found in stable dune areas after about 20 years.

(a)

(b)

(c)

Figure 11 Plant succession on the sand dunes of northern Lake Michigan. (a) Marram grass (*Ammophila breviligulata*) growing on a 25-year-old dune at Wilderness State Park, Michigan. (b) A 150-year-old dune ridge with a few scattered pine trees, willow shrubs, and juniper bushes in the center and left side of the photo. (c) A 400-year-old dune ridge colonized by red pine and white pine with an understory of bracken fern. (Photos courtesy of John Lichter, Bowdoin College.)

Two shrubs are important in dune stabilization: willows (*Salix* spp.) and sand cherry (*Prunus pumila*). As dunes age, little bluestem bunchgrass (*Schizachyrium scoparius*), bearberry (*Arctostaphylos uva-ursi*), and juniper (*Juniperus communis*) come in to stabilize the sand surfaces (**Figure 11b**). The first trees to appear in the older dunes are pines (**Figure 11c**), and a mixed pine forest appears after about 200 years (**Figure 12**).

Cowles believed that this succession from pines to hardwoods might be part of the succession sequence that would proceed to a white oak–red oak–hickory forest and finally after a long time to the "climatic climax," a beech-maple forest. The general belief has been that soil development is the key driver of succession on the dunes of Lake Michigan, the classic facilitation view of succession proposed by Clements (1916) and supported by soil data from the dune sequence. Both nitrogen and phosphorus may limit plant growth during primary succession on the dunes. Nitrogen and phosphorus both increase dramatically during the first 500 years of dune succession (Lichter 1998), consistent with the classical hypothesis of succession in which species composition and abundance approach an equilibrium controlled by soil resource supply. Late-successional species can colonize only when the soil nutrients are available.

The alternative hypothesis is that late-successional species are constrained by colonization events, which are limited by chance arrival of seeds, mortality due to water shortage, and seed and seedling predation by herbivores. Lichter (2000) tested this alternative hypothesis by adding seeds and transplanting seedlings to early-successional dunes. He added late-successional pine and oak seeds and seedlings to experimental plots that were watered and fertilized and followed their survival in comparison to control plots without water or fertilizer.

The first question is whether seeds of pines and oaks can emerge and survive in young dunes. Seed addition experiments showed that pine and oak seeds emerged at the same rate as the seeds of open dune species (Lichter 2000). Thus there is no soil nutrient impediment keeping pines and oaks from early-succession sites, and the limitations must be found elsewhere. The seeds of pines and oaks in the young dunes were eaten by small rodents much more than were the seeds of the pioneer species (**Figure 13**). Herbivory of seeds is one important limiting factor for tree colonization of the early stages of succession. The second limiting factor is water. Lichter (2000) found that when water was added to seeded plots, in wire cages that excluded seed predators, nearly 40% of seeds germinated, while only about 3% germinated when no water was added. The conclusion was that pines and oaks do not colonize the early stages of plant succession not because of low soil nutrients but rather because they are constrained by limited seed dispersal

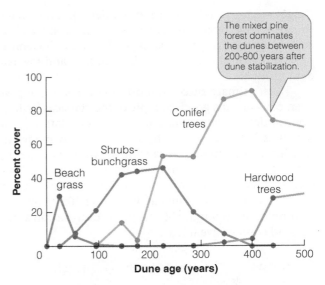

Figure 12 Changes in percentage of cover for the dominant plant species groups in the Lake Michigan sand dunes of Figure 11 during the first 500 years of primary succession. The original dune-building grasses are replaced by evergreen shrubs and bunchgrass, which are in turn replaced by a mixed pine-oak forest. (Data from Lichter 1998, Table 1.)

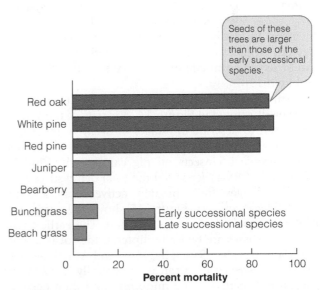

Figure 13 Overwinter losses of experimentally added seeds to young sand dune systems in northern Lake Michigan. Losses of seeds are largely due to seed predators, particularly small rodents, so that even if tree seeds reach young dunes, they are likely to be removed by herbivores before they get established. Green, early successional species; red, late successional species. (Data from Table 1 in Lichter 2000.)

(large seeds), seed predation (rodents), and water shortage. Tree establishment in sand dune succession around Lake Michigan thus depends on a coincidence of seed dispersal events, favorable rainfall, and low rodent numbers.

Thus, primary succession on dunes has in the past been considered a good example of the classical facilitation model of succession, but these recent data suggest that it is better described by the inhibition model in which the resident species causing the inhibition are rodents that occupy the early-successional stages. In later stages of succession, competition between species for light becomes a dominant factor (Lichter 2000). Succession proceeds as a trade-off between colonization abilities and competitive traits.

Succession of Insects on Carrion

Carcasses of animals are broken down by scavengers and by a variety of insects that feed on the decaying flesh. A succession of insects occupy decaying carcasses, and already in 1888 scientists developed the idea of using seral stages of invading insects to identify the time of death for medical-legal cases. This potential use has stimulated a great deal of work on the successional sequences shown on carrion (Schoenly 1992). There are many advantages for the use of carcasses to investigate processes of succession. The carcass microhabitat is small and has clear boundaries. In many ecosystems only about 30 species occur on small carcasses, and they can be easily counted. Because carcasses can be put out in large numbers, good experimental designs can be used to evaluate successional patterns (Schoenly and Reid 1987).

Arthropod succession on a carcass produces data of the type illustrated in **Figure 14**. A variety of terms have been used to describe the stages of decay (seral stages) of a carcass. For example, in an analysis of the succession patterns of insects on pig carcasses in Virginia, Tabor et al. (2004) separated the total decay succession into four stages: fresh, bloated, active decay, and advanced decay and dry. **Figure 15** shows the succession of decomposers during two summers. The earliest visitors to carcasses are typically dipteran flies of the families Calliphoridae, Sarcophagidae, and Muscidae. The number of species on a carcass is typically low early in succession, increases to a midpoint, and then declines toward the final stages of succession. The key questions asked by forensic investigators relate to the predictability of the successional sequence on a carcass and the species that might be good indicators of the time since death. Species that appear, disappear, and then reappear are not very useful in determining the time since death (Anderson 2001).

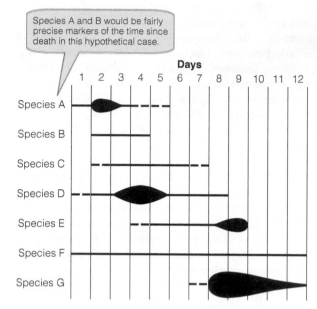

Figure 14 A hypothetical diagram of arthropod succession on a carcass. A horizontal line of varying thickness indicates populations of differing sizes of a given species as the succession proceeds. Dashed lines indicate only a single individual. (Modified from Schoenly 1992.)

One way to identify how samples differ from day to day during a period of succession on carrion is to use a similarity index (Krebs 1999, p. 376). The Jaccard similarity coefficient is a simple one that measures the similarity between two community samples based on the number of species that occur in each sample. It is defined as

$$S_j = \frac{a}{a + b + c}$$

where: S_j = Jaccard's similarity coefficient
  a = number of species in sample A and sample B (joint occurrences)
  b = number of species in sample B but not in sample A
  c = number of species in sample A but not in sample B

Jaccard's coefficient ranges from 0 for no similarity to 1.0 for maximum similarity between the community samples. If we calculate Jaccard's coefficient for a sequences of samples in a carrion succession (day 1 versus day 2, day 2 versus day 3, etc.), we obtain the type of data shown in **Figure 16**. The patterns of succession on these pig carcasses were similar in the two years of the study and could be used to estimate the time since death for these pigs, which are most similar in decomposition to humans and therefore useful for legal forensic analyses.

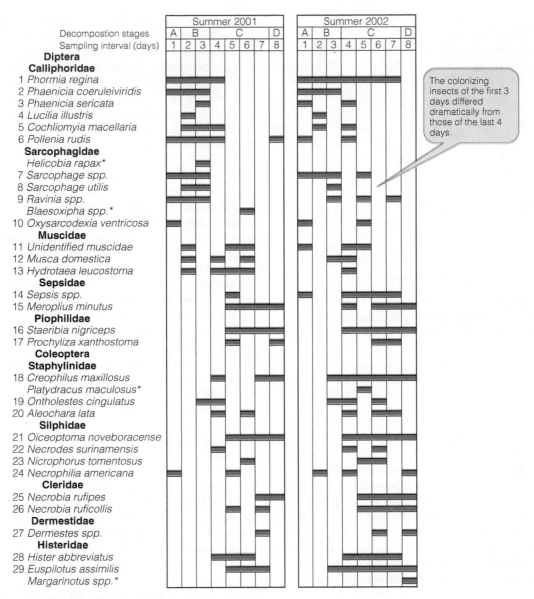

Figure 15 Succession diagrams for insect taxa on pig carcasses put out in summer and sampled over eight days. The stages of decomposition indicated at the top of the diagram are (A) fresh, (B) bloat, (C) active decay, and (D) advanced decay and dry. (Data from Tabor et al. 2004.)

The model of succession most appropriate to the decomposition of carcasses is not clear because the interactions between the taxa colonizing carcasses have not been analyzed. It seems unlikely that the inhibition model will operate during carcass decomposition, and the most likely model is the tolerance model for this system. The system is ideal for the kinds of experiments that could distinguish among the models of succession, but since the focus has been on forensic studies, this information on species interactions is not yet available.

The three examples of succession that we have just discussed do not always fit any single model of succession, as some replacements are facilitated whereas others are inhibited or tolerated. Walker and del Moral (2003) and Finegan (1996) have reviewed succession in forested regions and concluded that most forest succession does not conform to the classical facilitation model of succession originally proposed by Frederic Clements. If the classical model is not always correct, we must reconsider the nature of the climax state, the "end point" of succession.

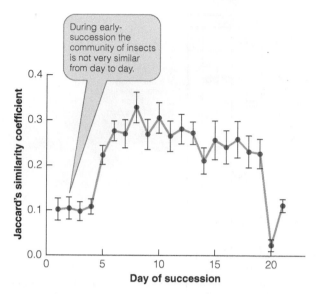

Figure 16 **Jaccard's measure of taxonomic similarity between day-to-day samples during the decomposition of a pig carcass in Virginia in the summer of 2001.** The estimation of the time since death would be most precise during the first five days after death. Pigs are used as analogs for humans in obtaining forensic samples to establish a baseline in each region. (From Tabor et al. 2004.)

The Climax State

In the examples of plant succession described in the preceding section, the vegetation developed to a certain stage of equilibrium. This final stage of succession is called the climax state, defined as *the final or stable community in a successional series. It is self-perpetuating and in equilibrium with the physical and biotic environment.* There are three schools of thought about the climax state: the monoclimax school, the polyclimax school, and the climax-pattern view.

The *monoclimax hypothesis* was an invention of the American F. E. Clements (1916, 1936). According to the monoclimax theory, every region has only one climax community toward which all communities are developing. Clements' fundamental assumption was that given time and freedom from interference, a climax vegetation of the same general type will be produced and stabilized, irrespective of earlier site conditions. Climate, Clements believed, was the determining factor for vegetation, and the climax of any area was solely a function of its climate.

However, it was clear in the field that certain areas had communities that were not climax communities but appeared to be stable. For example, tongues of tallgrass prairie extended into Indiana from the west, and isolated stands of hemlock occurred in what is supposed to be deciduous forest. In other words, we observe communities in nature that are nonclimax according to

Clements but apparently in equilibrium nevertheless. These communities are determined by topographic, edaphic (soil), or biotic factors.

The **polyclimax hypothesis** arose as the obvious reaction to Clements' monolithic system. Tansley (1939) was one of the early proponents of the polyclimax idea—that many different climax communities may be recognized in a given area, including climaxes controlled by soil moisture, soil nutrients, activity of animals, and other factors. Daubenmire (1966) also suggested that several stable communities may be found in a given area such that no single climax existed for a region.

The real difference between these two schools of thought lies in the time frame for measuring relative stability. Given enough time, say the monoclimax proponents, a single climax community would develop, eventually overcoming the edaphic climaxes. The question is, should we consider time on a geological scale, or on an ecological scale? If we view the problem on a geological time scale, we would classify communities such as the coniferous forest as a seral stage to the establishment of deciduous forest. The important point is that climate fluctuates and is never constant. We see this vividly in the Pleistocene glaciations, and more recently in the advances and retreats of mountain glaciers in the past 1000 years. As a result, *equilibrium can never be reached because the vegetation is subject not to a constant climate but to a variable one.* Climate varies on an ecological time scale as well. In a sense, then, succession is continuous because we have a variable vegetation interacting with a variable climate.

Whittaker (1953) proposed a variation of the polyclimax idea, the **climax-pattern hypothesis**. He emphasized that a natural community is adapted to the whole pattern of environmental factors in which it exists—climate, soil, fire, biotic factors, and wind. Whereas the monoclimax theory allows for only one climatic climax in a region and the polyclimax theory allows for several climaxes, the climax-pattern hypothesis allows for a continuum of climax types that varies gradually along environmental gradients and is not neatly divisible into discrete climax types. Thus, the climax-pattern hypothesis is an extension of the continuum idea and the approach of gradient analysis to vegetation (Whittaker 1953). The climax is recognized as a steady-state community in which its constituent populations are in dynamic balance with environmental gradients. We do not speak of a climatic climax, but of prevailing climaxes that are the end result of climate, soil, topography, and biotic factors, as well as fire, wind, salt spray, and other influences, including chance. The utility of the climax as an operational concept is that similar sites in a region should produce similar climax stands. This stand-to-stand regularity should allow prediction for new sites of known environment, and we can say that in

100 years a particular site should develop, for example, to a stand of sugar maple and beech of specified density.

A good example of how biotic factors may affect plant succession is found in the coastal dunes of Queensland in eastern Australia (Ramsey and Wilson 1997). Coastal foredune vegetation in southern Queensland is dominated by the perennial grass sand spinifex (*Spinifex sericeus*). Spinifex occupies the same dune stabilization niche as marram grass in the dunes of Lake Michigan. Spinifex stabilizes the dune sand and facilitates the establishment of other herb and grass species. Grazing pressure on the foredune vegetation can have a detrimental effect on plant succession. On South Stradbroke Island cattle had historically degraded the dunes, and they were removed in the early 1970s. The Department of Environment and Conservation then became concerned that grazing by macropods, primarily the agile wallaby (*Macropus agilis*), were replacing the cattle and preventing vegetation recovery through succession. Ramsey and Wilson (1997) constructed nine exclosures (10 m × 5 m) in January 1992 to stop wallaby grazing, and followed the development of succession over the next two years. They found that macropod grazing inhibited succession in the foredunes (**Figure 17**). Spinifex itself was grazed by macropods, but only minor effects could be found and grazed plants responded strongly with regrowth. The major impacts of grazing was on the herb and grass colonizers of the foredunes. In particular the nitrogen-fixing species *Vigna marina* (yellow beach bean) was removed by grazing, thus preventing the accumulation of nitrogen in the poor sandy soils of the foredunes. In this way, macropod grazing greatly delayed the facilitation that occurred between these herbs and grasses and the later colonists of the seral stages. Herbivore grazing acted as an inhibitor of this successional sequence.

A similar pattern of grazing effects occurs in the vegetation of the uplands of northwest Scotland, further complicated by the addition of fires in the plant communities (Miles 1987). Sheep grazing selects against trees and favors grassland. Under low grazing pressure and no fire, Scot's pine and birch woodlands develop. Under high grazing pressure from sheep and frequent fires, only grassland is able to survive. When grazing is moderate and fires are occasional, both heather moor (*Calluna vulgaris*) and bracken fern (*Pteridium aquilinum*) communities are favored. There is no one climatic climax on these Scottish uplands, and the plant communities are a mosaic, changing as sheep grazing and fire frequency change. These Scottish plant communities are best described by Whittaker's climax-pattern hypothesis.

How, then, can we recognize climax communities? The operational criterion is the attainment of a steady state over time. Because the time scale involved is very long, observations are lacking for most presumed successional sequences. We assume, for example, that we can

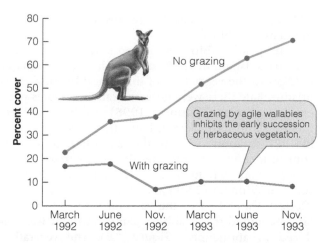

Figure 17 Response of herbaceous vegetation in herbivore exclosures during early succession on the sand dunes of South Stradbroke Island, Queensland, Australia. For two years exclosure fences prevented agile wallabies from grazing on herbaceous plants in 10-m × 5-m plots. The dunes are dominated by *Spinifex* vegetation. Grazing by macropod herbivores acts to inhibit succession in these dunes. (Data from Ramsey and Wilson 1997.)

determine the time course of succession for a spatial study of younger and older dune systems around Lake Michigan (see Figure 12), but this translation of space and time may not be valid. In forests, we can use the understory of young trees to look for changes in species composition, because the large trees must reproduce themselves on a one-for-one basis if steady state has been achieved. Forest changes may be very slow. Lertzman (1992) studied a subalpine forest stand at 1100 meters elevation near Vancouver, Canada. This site was undisturbed by fire for almost 2000 years and still was not in equilibrium: the dominant hemlocks were not replacing themselves while amabilis fir seedlings were invading the understory in large numbers. Climax vegetation on this site was not reached after two millennia.

Some communities may appear to be stable in time and yet may not be in equilibrium with climatic and soil factors. Some of the most striking demonstrations of this have come from changes in European rabbit populations. A striking example of this occurred after the outbreak of the disease myxomatosis in the European rabbit in Britain. Before 1954, rabbits were common in many grassland areas. Myxomatosis devastated the rabbit population in 1954, and the consequent release of grazing pressure caused dramatic changes in grassland communities (Thomas 1960, 1963). The most obvious change was an increase in the abundance of flowers. Species that had not been seen for many years suddenly appeared in large numbers. There was also an increase in woody plants, including tree seedlings that were commonly grazed by rabbits. No one anticipated these effects following the removal of the rabbits.

European rabbits were introduced to subantarctic Macquarie Island in 1878, and they have caused major changes to the vegetation of this island because in its pristine state there were no vertebrate herbivores present (Copson and Whinam 1998). Rabbits on Macquarie Island have been reduced from about 150,000 in 1977 to about 5000 in the late 1990s by various control measures, and the results of this major herbivore population reduction have been reported by Copson and Whinam (2001) and by Kirkpatrick and Scott (2002). About half of the vascular plants on this island benefited from rabbit grazing, and half were reduced in abundance (**Figure 18**). The vegetation changes on Macquarie Island after rabbit reduction have not yet reached an equilibrium or climax stage, since they are being affected both by climatic warming and by a reduction in rabbit grazing (Kirkpatrick and Scott 2002).

We conclude from this discussion that climax vegetation is an abstract ideal that is, in fact, seldom reached, owing to the continuous fluctuations of climate. The climate of an area has clear overall control of the vegetation, but within each of the broad climatic zones are many modifications caused by soil, topography, and animals that lead to many climax situations. The rate of change in a community is rapid in early succession but becomes very slow as it nears the potential climax community. But the climax community, like Nirvana, may never quite be attained.

Figure 18 A rabbit exclosure on Macquarie Island, two years after setup. The status of the large herb *Stilbocarpa polaris* changed from rare to dominant within two years, while grazing by the introduced European rabbit (*Oryctolagus cuniculus*) prevented its recovery outside the wire exclosure.

Patch Dynamics

We have seen that communities are dynamic and changing continually. In 1947 A. S. Watt, a British plant ecologist, first called attention to cyclic events that occur repeatedly in communities occupying spatially small patches. A plant community over a region may be moving slowly toward a climax state, while on a local scale the internal dynamics of the community are producing more-rapid cyclic changes in patches or gaps in the community. The study of gap dynamics has shed interesting light on the overall processes of community change. Here we discuss two examples of patch dynamics.

Watt (1947) studied several examples of cyclic changes in British vegetation. One of these was the *Calluna* heath that covers large areas in Scotland and has made heather almost synonymous with Scotland. The dominant shrub in this community is heather (*Calluna*), which loses its vigor as it ages and is invaded by lichens of the genus *Cladonia*. The lichen mat in time dies back, leaving bare ground. This bare area is invaded by bearberry (*Arctostaphylos*), which in turn is invaded by *Calluna*.

Heather (*Calluna*) is the dominant plant, and *Arctostaphylos* and *Cladonia* are allowed to occupy the area that is temporarily vacated by *Calluna*.

The cycle of change can be divided into four phases (**Figure 19**):

- *Pioneer.* Establishment and early growth in *Calluna*; open patches, with many plant species (years 6 to 10).

- *Building.* Maximum cover of *Calluna* with vigorous flowering; few associated plants (years 7 to 15).

- *Mature.* Gap begins in *Calluna* canopy and more species invade the area (years 14 to 25).

- *Degenerate.* Central branches of *Calluna* die; lichens and bryophytes become very common (years 20 to 30).

Barclay-Estrup and Gimingham (1969) describe this sequence in detail from maps of permanent quadrats in Scotland. The life history of the dominant plant *Calluna* controls the sequence.

A second example of patch dynamics that Watt studied involved cyclic changes associated with microtopography in a grassland in England: the hummock-and-hollow cycle. The vegetation of the grassland Watt studied was very patchy, and he could recognize four stages (**Figure 20**). The whole scheme centers around the grass *Festuca ovina*. Once the seedlings of this grass become established in the bare soil of the hollow stage, the plant builds a "tussock" by trapping windborne soil particles and by its own growth. The vigor of this grass declines with age; it begins to degenerate in the mature

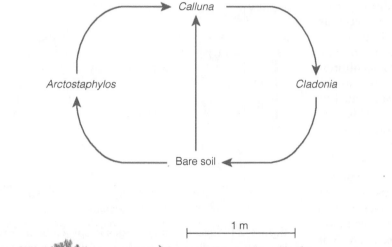

Figure 19 Four phases of the *Calluna* cycle in Britain, and a profile of the phases viewed from ground level. Like many perennial plants, heather loses vigor with age. *Calluna vulgaris* is a dominant plant on many of the moors of Scotland. (After Watt 1955.)

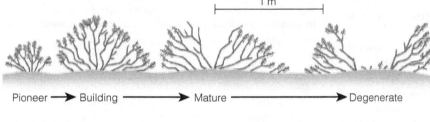

phase and is invaded by lichens in the early degenerate phase. These lichens use up the organic matter and in turn die, and the hummock is eroded down to base level, only to begin the process again. At any given time, all four stages can be found in a *Festuca* grassland. Seedlings cannot usually get established except in the hollow and building phases. Lichens dominate the degenerating phase, when they can use the organic matter that has accumulated. Bryophytes seem to suffer competition from fescue and cannot get established except in the degenerate or hollow phase.

Watt divided these cycles of change into an *upgrade* series and a *downgrade* series and pointed out that the total productivity of the series increases to the mature phase of the cycle and then decreases. What initiates the downgrade phase? A possible explanation is that there seems to be a general relationship between age and performance (vigor) in most perennial plants, and consequently between age and competitive ability (Kershaw and Looney 1985). Several studies on the relation of leaf diameter to age also support this idea.

For this reason, a stable community will be in a constant state of phasic fluctuation, one species becoming locally more abundant as another species reaches its degenerate phase. All these dynamic interrelationships in natural communities tend to operate on small spatial scales and may not be apparent without detailed measurement.

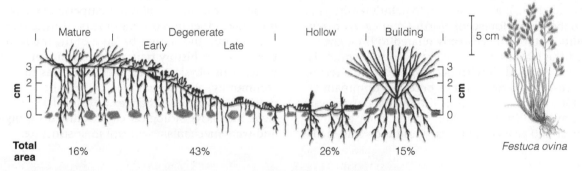

Figure 20 Phases of the hummock-and-hollow cycle, showing change in flora and habitat and indicating the "fossil" shoot bases and detached roots of *Festuca ovina* in the soil. The whole cycle centers around the grass *Festuca ovina*. (After Watt 1947b.)

Patch dynamics will result in monocultures if one species is able to replace itself and all other species as well. But in many old-growth forests there is a mixture of tree species, and this presents the question of why these old-growth stands do not become monocultures. As early as 1905, French foresters suggested that in virgin forests, individual trees tended to be succeeded in time by those of another species (Fox 1977). This phenomenon, called reciprocal replacement, occurs at the individual tree level, and could be one explanation of why old-growth forests have a mixture of tree species rather than a single dominant. If, for example, seedlings of species A were found predominately under large species B trees, and seedlings of species B were found predominately under large species A trees, we would have reciprocal replacement at the individual tree level.

American beech and sugar maple are codominant trees in old-growth forests in southern Michigan, and reciprocal replacement of individual trees has been suggested as one possible mechanism of codominance. Poulson and Platt (1996) found that reciprocal replacement did not occur at the individual tree level, but that codominance was caused by differing light intensities in regenerating gaps once older trees die. Sugar maples grow quickly upward in gaps in the forest, whereas beech seedlings spread laterally and capture light in flecks in the understory. Thus, beech does better in the understory of these forests and will decrease in relative abundance when tree-fall gaps become more frequent. Stable coexistence depends on a mixture of gaps of different sizes being created over periods of hundreds of years.

The boreal forests of northern Finland are dominated by two tree species, Norway spruce (*Picea abies*) and birch (*Betula pubescens*). In the absence of fires, particularly on north-facing slopes, these two tree species maintain mixed stands that regenerate via a cyclic patch dynamic (Doležal et al. 2006). This reciprocal replacement is driven by soil nitrogen changes (Pastor et al. 1987), and the same patch dynamics are found in the boreal forests of North America. Nitrogen availability in the soil is directly related to litter chemistry, and the leaf litter produced by Norway spruce is low in nitrogen and slow to decompose. In this reciprocal succession Norway spruce becomes dominant at ages of 100–140 years, when its litter begins to depress nitrogen availability in the soil. Spruce then declines from about 140 years to 260 years, and the stand is in-

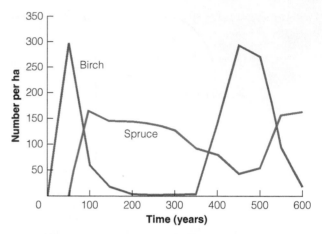

Figure 21 **Successional changes in spruce-birch forests in Ontario predicted on the basis of nitrogen availability in the litter.** The poor litter quality of white and black spruce trees leads to a decline in spruce tree density. High nitrogen in leaf litter early in succession from birch trees is taken over by spruce, which self-destructs over the next 200+ years, opening up the forest and speeding decomposition. Birch, which is shade-intolerant, colonizes at this time and starts another cyclical replacement series. These data are based on a computer simulation model. (Modified from Pastor et al. 1987.)

vaded by birch trees. Birch produce leaf litter that decomposes rapidly, and nitrogen in the soil becomes available again, encouraging a reinvasion of Norway spruce seedlings. This pattern of reciprocal replacement also occurs in the boreal forests of eastern North America (**Figure 21**).

Gap dynamics play an important role in the regeneration of forest and other communities in which space is a key resource. Space in forests (and in many other plant communities) is equivalent to light and soil nutrients, and competition for space is competition for light and nutrients. Light and soil nutrients are not the only resources for which plants compete, and water may modify competition among plants. Thus, communities are dynamic and change because of the interactions between the life history traits of the dominant species and patterns of physical disturbances such as fire. A stable community at a regional level may be a mosaic of patches undergoing changes at a local level. We consider next how biological communities are organized, and what mechanisms control their structure.

Summary

Communities change over time, and the study of succession has been an important focus of community ecologists for a century. Succession is the process of directional change in communities. Most of the work on succession has been done on plants, but the same principles apply to animal communities.

Four conceptual models of succession have been proposed to explain directional vegetation changes. All models agree that pioneer species in a succession are usually fugitive or opportunistic species with high dispersal rates and rapid growth. How are these pioneer species replaced? The classical model states that species replacements in later stages of succession are *facilitated* by organisms present in earlier stages. The inhibition model, at the other extreme, suggests that species replacements are *inhibited* by earlier colonizers and that successional sequences are controlled by "who gets there first." The tolerance model suggests that species replacements are not affected by earlier colonizers, and that later species in succession are those able to *tolerate* lower levels of resources than earlier species. The *random colonization* model is a null model that suggests that species replacements occur completely randomly, with no interspecific interactions. No single model explains an entire successional sequence. We now view succession as a dynamic process resulting from a balance between the colonizing ability of some species and the competitive ability of others. Succession does not always involve progressive changes from simple to complex communities.

Succession proceeds through a series of seral stages from the pioneer stage to the climax stage. The *monoclimax hypothesis* suggested that a single predictable end point existed for whole regions and that given time, all communities would converge to the climatic climax. This hypothesis has been superseded by the *climax pattern hypothesis*, which suggests a continuity of different climaxes varying along environmental gradients controlled by soil moisture, nutrients, herbivores, fires, or other factors.

A stable community contains patches repeatedly undergoing cyclic changes that are part of the internal dynamics of the community. The life cycle of the dominant organisms dictates the cyclic changes, many of which are caused by the decline in vigor of perennial plants with age. In many forests, tree-fall gaps create a mosaic of patches undergoing cyclic changes within a relatively stable climax community.

Communities are not stable for long periods in nature because of disturbances—short-term changes in climate, fires, windstorms, diseases, or other environmental factors. For most communities we can observe changes over time, but we need to determine the mechanisms that cause the changes. Unless we understand the mechanisms behind succession, we will be unable to suggest manipulations to alleviate undesirable trends caused by human activities.

Review Questions and Problems

1   Discuss the inhibition, facilitation, and tolerance models of succession with respect to the following simple experiment. In this hypothetical succession, species A normally precedes species B. Two treatments are applied: (1) All of species A are removed from a series of replicate plots, and (2) a portion of species B equal to the biomass of A is removed from another set of plots. Growth in biomass is then measured over several years. Interpret all the possible outcomes of this experiment. Compare your analysis with that of Botkin (1981).

2   Discuss how you would decide if a given successional sequence is a primary or a secondary succession, and how this distinction would affect your evaluation of the seral stages.

3   Pham et al. (2004) observed the following transition probabilities for a forest in northeastern Quebec:

Species in canopy	Species in saplings		
	Balsam fir	Paper birch	Black spruce
Balsam fir	0.25	0.00	0.74
Paper birch	0.28	0.02	0.71
Black spruce	0.20	0.01	0.79

Calculate the changes over five generations in the composition of a forest containing these three species, starting from equal numbers of each species. What is the climax forest in this area?

4 Herbivores may increase or decrease the rate of succession. Which of these impacts would you predict for early- and late-successional species in a successional sequence that is driven by the inhibition model of succession? Would these same predictions apply to a sequence driven by the facilitation model? Walker and del Moral (2003, p. 220) discuss these predictions.

5 Discuss the application of the concept of succession to marine communities.

6 Relate the adaptive strategies of species in early and late stages of succession to what you know about the ideas of *r* and *K* selection.

7 In discussing forest succession as a plant-by-plant replacement process, Horn (1975b, p. 210) states: "Copious self-replacement does not guarantee a species' abundance or even its persistence in late stages of succession." How can this be true?

8 If successional changes in a landscape are partly a result of the regional climate, what utility will this concept have in an era of rapid climate change, such as we are now undergoing? How predictable will future successional sequences be, and will the past be a good guide to the future?

9 In the primeval forest landscape, where were the plants that are abundant today in old fields? Discuss the evolution of colonizing ability in plants that evolved in temporary forest openings, and those that evolved in persistent open, marginal habitats. Compare your conclusions with those of Marks (1983).

10 Discuss how much the current state of a plant community such as a forest depends on history. Does the simple model of succession presented in this chapter have any history in it?

11 How can species that facilitate other species in a successional sequence evolve? For example, why should species that fix nitrogen from the air leak this nutrient into the soil to assist their competitors who will replace them in the successional sequence? Is this an example of altruistic behavior?

Overview Question

Is it possible to construct a theory of succession in plant communities solely on the mechanism of competition between species? What would be missing from such a theory?

Suggested Readings

• Anderson, G. S. 2001. Insect succession on carrion and its relationship to determining time since death. In *Forensic Entomology: The Utility of Arthropods in Legal Investigations*, (ed. J. H. Byrd and J. L. Castner), 143–175. Boca Raton, FL: CRC Press.

• Anderson, K. J. 2007. Temporal patterns in rates of community change during succession. *American Naturalist* 169:780–793.

• Connell, J. H., and R. O. Slatyer. 1977. Mechanisms of succession in natural communities and their role in community stability and organization. *American Naturalist* 111:1119–1144.

• del Moral, R. 2007. Limits to convergence of vegetation during early primary succession. *Journal of Vegetation Science* 18:479–488.

• Henry, H., and L. Aarssen. 1997. On the relationship between shade tolerance and shade avoidance strategies in woodland plants. *Oikos* 80:575–582.

• Johnson, E. A., and K. Miyanishi. 2008. Testing the assumptions of chronosequences in succession. *Ecology Letters* 11:419–431.

• Kirkpatrick, J. B., and J. J. Scott. 2002. Change in undisturbed vegetation on the coastal slopes of subantarctic Macquarie Island, 1980–1995. *Arctic, Antarctic, and Alpine Research* 34:300–307.

• Lichter, J. 2000. Colonization constraints during primary succession on coastal Lake Michigan sand dunes. *Journal of Ecology* 88:825–839.

• Torti, S. D., P. D. Coley, and T. A. Kursar. 2001. Causes and consequences of monodominance in tropical lowland forests. *American Naturalist* 157:141–153.

• Turner, M. G., W. L. Baker, C. J. Peterson, and R. K. Peet. 1998. Factors influencing succession: Lessons from large, infrequent natural disturbances. *Ecosystems* 1:511–523.

• Walker, L. R., and R. del Moral. 2003. *Primary Succession and Ecosystem Rehabilitation*. Cambridge: Cambridge University Press.

Credits

Illustration and Table Credits

T1 M. Huston and T. Smith, "Plant succession: Life history and competition," *American Naturalist* 130:168–198, 1987. Copyright © 1987 The University of Chicago Press. F6 From L. R. Walker and F. S. Chapin, III, "Interactions among processes controlling successional change," *Oikos*, Vol. 50, pp. 131–135. Copyright © 1987 Munksgaard International Publishers Ltd., Copenhagen, Denmark. Reprinted by permission. T2 Reprinted by permission of the publisher from *Ecology and Evolution of Communities*, edited by Martha L. Cody and Jared M. Diamond, pp. 199–200. Cambridge, Mass.: The Belknap Press of Harvard University Press. Copyright © 1975 by the President and Fellows of Harvard College. T3 Reprinted by permission of the publisher from *Ecology and Evolution of Communities*, edited by Martha L. Cody and Jared M. Diamond, pp. 199–200. Cambridge, Mass.: The Belknap Press of Harvard University Press. Copyright © 1975 by the President and Fellows of Harvard College. F10 J. Lichter, "Primary succession and forest development on coastal Lake Michigan sand dunes," *Ecological Monographs* 68:487–510, 1998. Copyright © 1998 Ecological Society of America. Used with permission. F14 K. Schoenly, "A statistical analysis of successional patterns in carrion-arthropod assemblages," *Journal of Forensic Sciences* 37:1489–1513, 1992. Copyright © 1992. Reprinted by permission of Blackwell Publishing. F15 From K. L. Tabor et al., "Analysis of the successional patterns of insects on carrion in southwest Virginia," *Journal of Medical Entomology* 41:785–795, 2004.

Photo Credits

Unless otherwise indicated, photos provided by the author.

CO Andrew Darrington/ Alamy. 1a Harry Glicken/USGS. 1b USGS. 2a Austin Post/USGS. 2b Tom Casadevall/USGS. 4a-f Roger del Moral, University of Washington. 8 Arctic Images/ Alamy. 11a-c John Lichter, Bowdoin College. 18 Ecological Society of America.

Community Dynamics I: Predation and Competition in Equilibrial Communities

Key Concepts

- Communities can be organized by competition, predation, or mutualism. Almost all discussion centers on competition and predation, and mutualism has been largely ignored in studies of community integration.

- Two general models of community organization exist. The *equilibrium model* focuses on community stability and biotic coupling, the *nonequilibrium model* on stochastic effects and species independence.

- Food webs describe who eats whom in a community. Within food webs, food chains are short, predator-prey ratios relatively constant, and the number of linkages each species has rises with species richness of the community.

- Not all species in a food web are of equal importance. *Keystone species* are low in abundance, but their removal causes high community impact. *Dominant species* are high in abundance and help determine community structure.

- Stability of community composition and dynamics is produced by species diversity in many communities. The hypothesis that *diversity promotes stability* is a key argument in conservation programs.

KEY TERMS

apex predator In a food chain the highest trophic level. Apex predators do not have other predators feeding on them within the food web.

equilibrium model of community organization The global view that ecological communities are relatively constant in composition and are resilient to disturbances.

food chain The transfer of energy and materials from plants to herbivores to carnivores.

global stability Occurs when a community can recover from any disturbance, large or small, and go back to its initial configuration of species composition and abundances.

guild A group of species that exploit a common resource base in a similar fashion.

keystone species Rare species of low abundance in a community but whose removal has drastic effects on many other species in the community.

local stability Occurs when communities recover from only small disturbances and return to their former configuration of species composition and abundances.

neighborhood stability Also called local stability, the ability of a community to return to its former configuration after a small disturbance.

nonequilibrium model of community organization The global view that ecological communities are not constant in their composition because they are always recovering from biotic and abiotic disturbances, never reaching an equilibrium.

trophic levels The source of energy for organisms divided into primary producers, herbivores, carnivores, and higher carnivores.

Communities can be organized by four biological processes—competition, predation, herbivory, and mutualism. Competition among plants, herbivores, and carnivores might control the diversity and abundance of species in a community. Predation and herbivory might organize the community according to "who-eats-whom," such that the framework of community organization is set by the animals. Mutualism, an important process that links species, might serve to increase community organization by linking species to the benefit of all. Physical processes set limits to these four biological processes, and variation in temperature, salinity, and other physical factors have potential implications for the species in a community. To study community orga-

nization, we need to look at the component species and the processes that interconnect them.

To speak of community organization implies that there is some regularity in the biomass or the numbers of species that make up the community. Naturalists looking for particular birds, butterflies, or flowers have an implicit model of community organization in mind. Conservation biologists have an implied model of community organization when they discuss the preservation of the Florida Everglades or other natural landscapes. Natural communities could be very loosely organized, or be very tightly organized. How can we determine this for any particular community?

Communities contain so many different species that we cannot study each species separately. One way to reduce the complexity of communities is to measure the biodiversity of the community. If we measure the species richness of a community, we implicitly assume that each species is equal to every other species in the community. A second way to simplify the analysis is to define feeding roles in the community and to group species according to their roles. We can group species into **trophic levels** (such as herbivores), or at a finer level into **feeding guilds**. A third way is to look at particular types of species and to ask, *Are all species of equal importance in a community?* This question is purposely vague in that we must define *importance,* and we can do this in several ways (see *Working with the Data: Measuring Community Importance*). We could consider a species important if its removal changes the diversity or abundance of other species in the community. Such species are **keystone species**. Alternatively, we could determine which species are most common in the community—the **dominant species**. Dominant species could be major players in defining the organization of the community. In this chapter we discuss each of these approaches to understanding community organization.

We begin this analysis with the classical assumption that communities are in equilibrium. Communities are in equilibrium when species abundances remain constant over time, when nature is in a "state of balance." In most cases the **equilibrium model** refers to a *stable equilibrium* (**Figure 1a**). In different habitats the equilibrium point may differ, such that there is spatial variation in species numbers, but the key point is that at each spatial location the community is in equilibrium and remains constant (Chesson and Case 1986). This equilibrium will usually be *locally stable* within a specified environmental range. In some cases the equilibrium can be *globally stable*, such that over all environmental conditions the system will return to the equilibrium point following any disturbance (**Figure 1b**).

Measuring Community Importance

Even if all species in a community are not equally important, we need to develop a measure of community importance. Mills et al. (1993) were the first to define community importance values for particular species:

$$CI_x = \begin{bmatrix} \text{Percentage of species lost} \\ \text{from a community upon} \\ \text{removal of species } x \end{bmatrix} \quad (1)$$

Thus, if you remove the starfish *Pisaster* from a rocky intertidal area, and nine of 23 invertebrate species are lost from that area, the community importance value of *Pisaster* would be 39%. By contrast, if a redundant species were removed, nothing would happen and the community importance value for that species would be 0%.

Power et al. (1996) recognized that not all of the effects of a keystone species would show up as species losses, so they devised the following metric of community importance to make it more general:

$$CI_x = \frac{(t_N - t_D)/t_N}{p_x} \quad (2)$$

where CI_x = community importance of species x
t_N = quantitative measure of community trait in intact community
t_D = quantitative measure of community trait after species x is removed
p_x = proportional abundance of species x before removal

Any community trait can be used—species richness, productivity, or the abundance of indicator species. For example, Fagan and Hurd (1994) studied the effects of praying mantids on the numbers of other arthropods and found that the arthropod community without mantids had 316 individuals in 4 m², whereas plots with mantids had 194 individuals. Mantids averaged 14 individuals, or 7.2% of the arthropods. Consequently, for these mantids:

$$CI_x = \frac{(t_N - t_p)/t_N}{p_x} = \frac{(194 - 316)/194}{0.072} = -8.73$$

Negative values indicate that species x reduces the community measure when it is present.

Community importance measurements are similar to Paine's measurement of interaction strength (I_x) in communities (Paine 1992):

$$I_x = \frac{(t_N - t_D)/t_D}{n_x} \quad (3)$$

where terms are as previously defined and n_x = number of individuals of species x in unmanipulated plots. For the mantid data just given, the interaction strength is

$$I_x = \frac{(t_N - t_D)/t_D}{n_x} = \frac{(194 - 316)/316}{14} = -0.028$$

The interaction strength is a per capita estimate of effects that measures how much a single individual of species x changes the community. Negative values of interaction strength indicate that species x reduces the abundance or other trait of the community being analyzed, in this case by about 3% per mantid individual. Community importance values and interaction strengths are two similar ways of measuring effects of species, and because they are highly correlated measures, either one may be used to quantify effects of species removals on community structure.

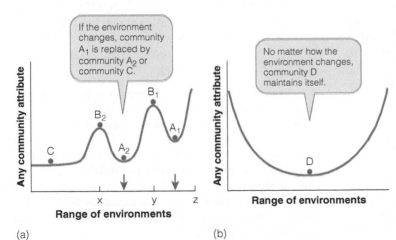

(a)

(b)

Figure 1 Schematic view of types of community equilibrium points. (a) Local stability: two locally stable equilibrium points A₁ and A₂ are shown. B₁ and B₂ are unstable equilibrium points, and C is a neutrally stable equilibrium point. The community at A₁ is locally stable between the environmental range from y to z, and A₂ is locally stable between x and y. (b) Global stability: a globally stable community at D will come to the same point no matter what the environmental change. Most real-world communities are only locally stable; only a few are globally stable. (Modified from DeAngelis and Waterhouse 1987.)

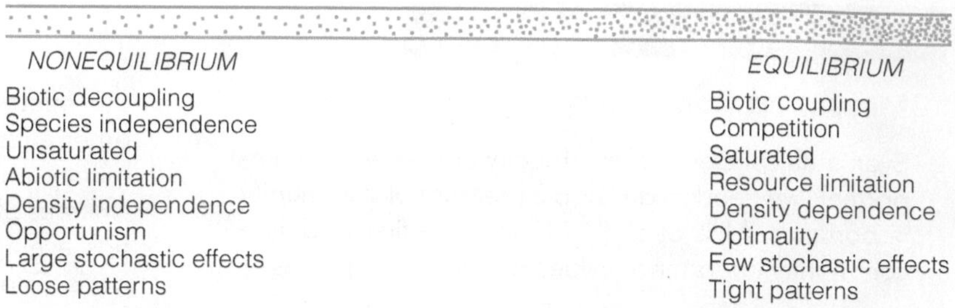

Figure 2 Natural communities may be arrayed along a continuum of states from nonequilibrium to equilibrium. At either extreme, several attributes of community organization and dynamics can be anticipated. In this chapter we discuss equilibrium models. (From Wiens 1984.)

NONEQUILIBRIUM	EQUILIBRIUM
Biotic decoupling	Biotic coupling
Species independence	Competition
Unsaturated	Saturated
Abiotic limitation	Resource limitation
Density independence	Density dependence
Opportunism	Optimality
Large stochastic effects	Few stochastic effects
Loose patterns	Tight patterns

The classical equilibrium assumption of community ecology is an abstraction and will not be found in its pure state in natural communities. Real communities will be spread along a continuum from equilibrium to nonequilibrium (**Figure 2**). Equilibrium communities purportedly show stability, and stability can be measured in several different ways (Pimm 1991). The mathematician's idea of **local stability** (points A_1 and A_2 in Figure 1a) is the simplest meaning. Stability can be measured by the *time* it takes for a community to recover from disturbance; accordingly, stable communities recover quickly from disturbances. Stability can also be measured as the *variability* of a community over time, so that if the populations that make up the community fluctuate in size dramatically from year to year, the community would be considered unstable. (This is the most common meaning ecologists attach to the word *stability*.) Stability can also be measured as the *persistence* of a community over time.

An ideal equilibrium community would score high on all these measures of stability. Such a community would have many biotic interactions involving competition and predation, and these processes would operate in a density-dependent manner to regulate population size. Equilibrium communities would also be saturated with species, such that species invasions would be rare. Weather catastrophes would rarely occur, and the community would form a tightly coupled biotic unit, an interlocking web of life.

By contrast, ideal nonequilibrium communities would score low on all these measures of stability. Species would operate individualistically, and density-dependent population regulation would be difficult to find. Climatic catastrophes would occur frequently, and species would come and go regularly, such that the composition of the community would be highly variable.

The three major equilibrium theories of community organization are the classical competition theory, the competition-predation theory, and the competition-spatial patchiness theory:

1. **Classical competition theory.** Hutchinson (1959) argued that competition is the major biological process controlling community structure. The subsequent development of this theory is reviewed by Armstrong and McGehee (1980). The essential assumptions of this theory are the following:

   a. Population growth rates can be described with deterministic equations, and environmental fluctuations can be ignored.

   b. The environment is spatially homogeneous, and migration is unimportant.

   c. Competition between species is the only significant biological interaction.

   d. The coexistence of competing species requires a stable equilibrium point for population densities.

   This theory predicts that *n* limiting resources are required for the coexistence of *n* species. Moreover, there will be a limiting similarity of species such that species differ in their use of the available resources (Chesson and Case 1986).

2. **Competition-predation theory.** The classical theory was clearly deficient in allowing only competition to operate. Adding predation to assumption (c) of the classical theory produces a new equilibrium model that will allow *n* species to coexist on fewer than *n* resources (Levin 1970).

3. **Competition-spatial patchiness theory.** Another equilibrium model that was modified from the classical theory allowed the environment to be subdivided into patches, such that different species would be favored in different patches (Levin 1974). Each patch has its own distinct stable equilibrium, and the resulting model is similar to a metapopulation model.

In real communities, competition for nutrients or food and competition for space both occur (Yodzis 1986). By adding predation and spatial patchiness we can construct more realistic models of community organization, and the equilibrium model of community organization includes competition, predation, and spatial patchiness. Despite the fact that mutualism between species could also help to structure ecological communities, almost no attention has been paid to it as a factor in community organization.

Given this background, let us consider the ways in which we can classify species in communities; then we can see how much of this structure we can explain by classical equilibrium theories involving competition, predation, and spatial patchiness.

Food Chains and Trophic Levels

One component of community organization is "who-eats-whom." The transfer of food energy from its source in plants through herbivores to carnivores is referred to as the **food chain**. Elton (1927), one of the first to apply this idea to ecology and to analyze its conse-

quences, pointed out the great importance of food to organisms, and he recognized that the length of food chains was limited to four or five links. Thus, we may have a pine tree–aphids–spiders–warblers–hawks food chain. Elton recognized as well that these food chains were not isolated units but were linked into **food webs**. Let us look at a few examples of food chains.

The Antarctic pelagic food chain is a good example of a food chain found in seasonally productive oceans. Phytoplankton are fed on by the dominant herbivores, euphausids (krill), and copepods. These zooplankton species are fed on by an array of carnivores, including fish, penguins, seals, and baleen whales (**Figure 3**).

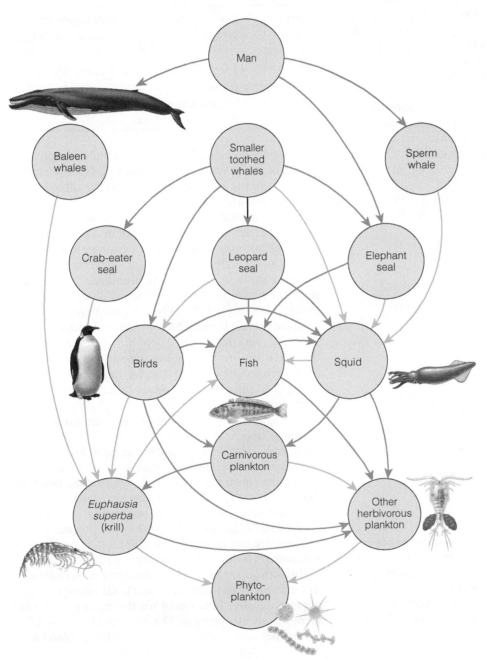

Figure 3 A simplified version of the Antarctic marine food chain. Blue arrows indicate the major trophic interactions before whaling began. (Modified after Knox 1970.)

Squid, which are carnivores that feed on fish as well as zooplankton, are another important component of this food chain because seals and the toothed whales feed on them in turn. During the whaling years, humans became the top predator of this food chain. Having reduced the whales to low numbers, humans are now harvesting krill.

Partial food webs are commonly used when data on all trophic levels are not available. A good example is the partial food web for rocky intertidal communities on the east coast of the United States (**Figure 4**). Birds can be important links between marine and terrestrial ecosystems along coasts. Many birds exploit intertidal species during low tide. In New England and throughout the North Atlantic great black-backed gulls (*Larus marinus*) and herring gulls (*Larus argentatus*) have dramatically increased in numbers (Ellis et al. 2007). The Jonah crab (*Cancer borealis*) is a major component of the diet of these gulls, and gull predation could reduce the abundance of these crabs. When gulls were prevented from feeding in three 50-m sections of the rocky intertidal zone on the coast of Maine, *Cancer borealis* increased dramatically in density and reduced the density of the intermediate predators *Carcinus maenas* and *Nucella lapillus*, as would be predicted from the food web (see Figure 4). Gull predation on the major intertidal predator *Cancer borealis* indirectly benefits the herbivores in this system, the snail

Littorina littorea and the mussel *Mytilus edulis* (Ellis et al. 2007).

In many cases ecologists simplify food webs, typically by taking two approaches. First, some taxonomic groups are lumped together. Often all the vertebrate species are identified individually, but plants or invertebrates are lumped together. The Antarctic marine food web in Figure 3 shows this approach. Second, only a part of the whole food web is isolated for analysis to keep things relatively simple, as we saw in the rocky intertidal food web of Figure 4.

Within food webs we can recognize several different trophic levels:

Producers	= Green plants	= First trophic level
Primary consumers	= Herbivores	= Second trophic level
Secondary consumers	= Carnivores, insect parasitoids	= Third trophic level
Tertiary consumers	= Higher carnivores, insect hyperparasites	= Fourth trophic level

For long food chains, fifth and even higher trophic levels are possible. When tertiary consumers are present in a food web, the secondary consumers are often labeled as **mesopredators**.

The classification of organisms by trophic levels is one of *function* and not of species as such, because a given species may occupy more than one trophic level. For example, male horseflies feed on nectar and plant juices, whereas the females are blood-sucking ectoparasites. The partial food web in Figure 4 illustrates two important concepts in food webs. Animals may feed on two or more trophic levels. The crab *Cancer borealis* feeds on prey that function at two trophic levels, and consequently it can be classified as functioning at a trophic level intermediate between 3 and 4. Animals that feed on both plants and other animals are called **omnivores**, and humans are the classic omnivore. Thompson et al. (2007) have shown that omnivory is common above the herbivore trophic level, and trophic levels could be more graphically described as trophic tangles for secondary consumers. Second, in food webs that contain secondary and higher consumers there is always an **apex predator** that itself has no predators. Grizzly bears in North America would be an example of an apex predator, and in the partial web of Figure 4 gulls are the apex predators.

The size of organisms has a great effect on the organization of food chains, as Elton (1927) recognized. Animals of successive trophic levels in a food chain tend to be larger (except for parasites). Of course, definite upper and lower limits exist for the size of food a carnivorous animal can eat. The size and structure of an animal put some limits on the size of food it can ingest.

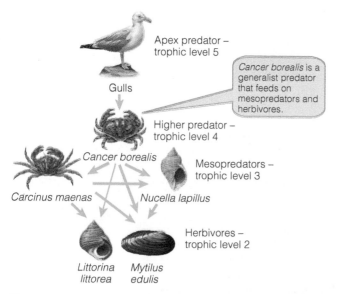

Figure 4 illustration labels:
Apex predator – trophic level 5
Gulls

Cancer borealis is a generalist predator that feeds on mesopredators and herbivores.

Higher predator – trophic level 4
Cancer borealis

Mesopredators – trophic level 3

Carcinus maenas   *Nucella lapillus*

Herbivores – trophic level 2

Littorina littorea   *Mytilus edulis*

Figure 4 Partial food web for the moderately exposed rocky shores in the Gulf of Maine. The thick lines indicate strong interactions in this community. Predation by great black-backed and herring gulls prevents the Jonah crab *Cancer borealis* (a generalist predator) from establishing large populations in the lower intertidal zone. (From Ellis et al. 2007.)

Table 1	Definitions of food web terminology.

Top predators: species eaten by nothing else in the food web (also called apex predators)

Basal species: species that feed on nothing within the web (usually plants)

Intermediate species: species that have both predators and prey within the web

Trophic species: groups of organisms that have identical sets of predators and prey

Cycles within a food web: species A eats species B and species B eats A

Cannibalism: a cycle in which a species feeds upon itself

Interactions: any feeding relationship (line with an arrow in a food web diagram)

Possible interactions: among s species in a food web, there can be s^2 possible interactions, including cannibalism

Connectance: number of actual interactions in a food web divided by the number of possible interactions

Linkage density: average number of links or interactions per species in the web

Omnivores: species that feed on more than one trophic level

Compartments: groups of species with strong linkages among group members but weak linkages to other groups of species

Figure 5 illustrates these definitions.

SOURCE: Modified from Cohen (1978) and Pimm (1982).

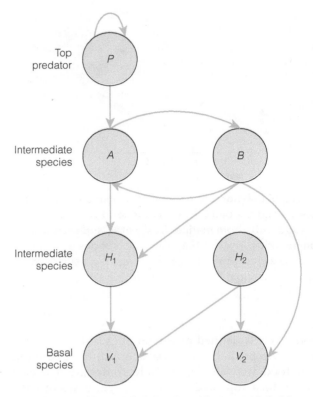

Figure 5 Hypothetical food web to illustrate definitions of food web properties defined in Table 1. Arrows point toward the species that is being eaten. Species A and B illustrate a cycle in a food web. Top predator P illustrates cannibalism. There are 10 interactions in this hypothetical web of seven species, including cannibalism. There are 7^2 or 49 possible interactions, so connectance is 10/49 or 0.20. The linkage density is 10/7 or 1.43. Species A and B are both omnivores.

Except in a few cases, large carnivores cannot live on very small food items because they cannot catch enough of them in a given time to meet their metabolic needs. The one obvious exception to this general size rule is the omnivore *Homo sapiens*, and part of the reason for our biological success is that we can prey upon almost any level of the food chain and can eat any size of prey.

Food webs can form a useful starting point for the theoretical analysis of community organization (Pimm et al. 1991). **Table 1** provides definitions for terms used in food web theory, and **Figure 5** illustrates these properties schematically. More than 400 food webs have now been described, and attempts have been made to draw generalizations about food web structure (Pimm et al. 1991; Vander Zanden and Fetzer 2007).

One question we can ask about food webs is whether there are limits to their complexity. As more and more species are involved in a food web, the number of possible linkages increases, and the empirical question is whether or not the number of *actual* linkages also goes up or remains constant. **Figure 6** shows that the number of links per species does indeed increase as food webs with greater diversity are considered (Dunne et al. 2004). The result of this increasing number of links is that each species is on average connected to more and more species as the species richness of the food web increases. This means that species-rich, complex food webs will be progressively more difficult to untangle in order to understand community organization. Dunne et al. (2004) suggested that marine food webs may be slightly more highly connected than freshwater and terrestrial webs with about 12 links per species compared with an average value of 8 for other food webs. But there is high variation among food webs, and generalizations always have exceptions.

A second generalization about food webs is that food chains are short (Elton 1927). (The length of a

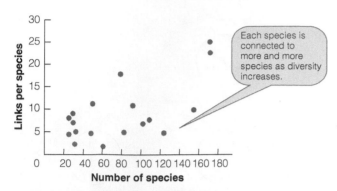

Figure 6 Relationship between the number of links per species and the biodiversity of the food web for 18 food webs from marine, freshwater, and terrestrial communities. The number of links per species tends to increase with species richness, but there is considerable scatter. (Data from Dunne et al. 2004.)

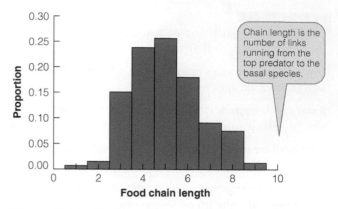

Figure 7 Distribution of food chain lengths in the Ythan Estuary of northeast Scotland. A total of 95 species in this community have been studied in detail to construct a complex food web from which 5518 food chain lengths were counted. The most common chain length was five. (From Hall and Raffaelli 1997.)

food chain is defined as the number of links running from a top predator to a basal species.) If we count for each food web all the possible routes from a basal species to a top predator, we can get a set of chain lengths and identify the maximum chain length for that food web. Hall and Raffaelli (1997) have done this for the Ythan Estuary in northeast Scotland, where they have detailed studies of 95 species in this community.[1] **Figure 7** shows the distribution of all the 5518 possible chains in this community. The most common food chain length is five links, and the range was one to nine links. There is some suggestion that the length of food chains increases as food webs contain more species. But there is a limit on food chain length; few chains exceed eight or nine links, and the mean chain length rarely exceeds five links.

There are several hypotheses for why food chains should be relatively short like this (Hall and Raffaelli 1997). The **energetic hypothesis**, the most popular explanation for food chain length, suggests that the length of food chains is limited by the inefficiency of energy transfer along the chain (Pimm 1991). This classical hypothesis for food chain length was articulated by Charles Elton in 1924. If this idea is correct, food chains should be longer in habitats of higher productivity, a clear prediction that can be tested.

The **dynamic stability hypothesis** explains short food chains by the fact that because longer food chains are not stable, fluctuations at lower levels are magnified at higher levels causing top predators to go extinct.

Moreover, in a variable environment top predators must be able to recover from catastrophes, and the longer the food chain, the slower the recovery rate from catastrophes for top predators. If catastrophes occur too often, the top predators will again go extinct. This hypothesis predicts shorter food chains in unpredictable environments, a prediction that again can be tested when detailed data are available.

These two hypotheses to explain food chain length are difficult to test in large communities but can be tested in simple communities that have only a few species. Jenkins et al. (1992) used organisms that inhabit natural, water-filled tree holes in the subtropical rain forest of Queensland to test these ideas in a small system. They simulated tree holes using 1-liter plastic containers to which they added leaf litter in various amounts. By reducing leaf litter input to 1/10 and 1/100 the natural rate over one year, Jenkins et al. (1992) found that both the number of species supported and the number of trophic links were reduced as leaf-litter input was reduced, results that support the prediction of the energetic hypothesis (**Figure 8**) that reduced energy input will result in reduced food chain lengths.

A third generalization about food webs is that the proportions of species that are top predators, intermediate species, and basal species are nearly constant, regardless of the size of the food web. **Figure 9** shows this relationship for predator-prey ratios. There is an approximately constant ratio of two to three prey species for every predator species in food webs, regardless of the total number of species in the web (Martinez 1991;

---

[1]The food web for this community is too complex to illustrate.

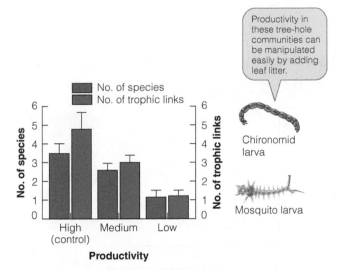

Figure 8 Experimental test of the energetic hypothesis for the restriction on food chain length. Tree-hole communities in Queensland were simulated using litter input at three levels: high litter input = natural (control) rate of litter fall, medium = 1/10 natural rate, and low = 1/100 natural rate. Reducing energy input reduced food chain length, in agreement with the hypothesis. The tree-hole community consists of microbes that break down leaf litter, mosquito larvae that feed on these microbes, predatory midges (chironomids), and other insects that feed directly on leaf litter. (From Jenkins et al. 1992.)

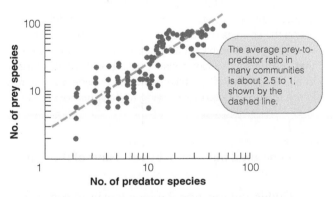

Figure 9 The relationship between the number of predator species and the number of prey species in 92 freshwater invertebrate food webs. The prey-predator ratio varied between 2:1 and 3.5:1 in these communities. The dashed line shows the expected average ratio of 2.5 prey per predator. (Data from Jeffries and Lawton 1984.)

Raffaelli 2000). In the Ythan Estuary food web analyzed in Figure 7, the ratio of intermediate to top species is 2.4 (Hall and Raffaelli 1997).

A fourth generalization is that omnivory—feeding on more than one trophic level—seems to be common in food webs. Aquatic communities often have fishes that eat their way up the food chain as they grow in size. Also, the detritus that sustains some organisms originates in several trophic levels (Pimm et al. 1991). Omnivory is difficult to estimate accurately unless the taxonomic breakdown of species in the food web is comprehensive (Woodward et al. 2005). The general conclusion from more detailed food web studies is that omnivory is more common than originally thought (Thompson et al. 2007).

Many of the early generalizations about food webs have now been revised as more-detailed data on food web structure becomes available. It is clear that real food webs are often very complex, and that their properties are not simple. The importance of understanding food webs was emphasized by Pimm (1991) because the structure of food webs has implications for community persistence. Some food webs can support additional species without suffering any losses, whereas other food webs are unstable and thus subject to species losses. If we can better understand the structure of food webs, we can design better management strategies for conservation. The addition of new tools such as stable isotopes for uncovering feeding relationships in communities has greatly added to our understanding of who eats whom in ecological communities.

Functional Roles and Guilds in a Community

Trophic levels provide a good description of a community, but by themselves they are not sufficient for defining community organization. A more refined approach is to use food webs to subdivide each trophic level into **guilds**, which are groups of species exploiting a common resource base in a similar fashion (Root 1967). For example, hummingbirds and other tropical nectar-feeding birds form a guild exploiting a set of flowering plants (Feinsinger 1976). We expect competitive interactions to be potentially strong between the members of a guild. By grouping species into guilds, we may also identify the basic **functional roles** played in the community. Functional roles can be defined broadly, such as producers and decomposers, but they are more usefully defined as groups of species such as nitrogen-fixers or leaf-chewers in order to combine a number of species under one umbrella term.

Use of Stable Isotopes to Analyze Food Chains

Stable isotopes are alternate forms of elements. For example, nitrogen in the air contains two isotopes, ^{14}N and ^{15}N, only a small fraction (0.4%) of which is ^{14}N. Carbon contains two isotopes, ^{13}C and ^{12}C, and only about 1% of carbon in nature is ^{13}C. The ratio of the isotopes in any material is expressed as δ values, which are parts-per-thousand differences from a standard substance. For nitrogen, we have

$$\delta^{15}N = \left[\left(\frac{^{15}N/^{14}N_{sample}}{^{15}N/^{14}N_{air}} \right) - 1 \right] 1000 \qquad (4)$$

Nitrogen in the air is the standard for nitrogen analysis, and carbon in a particular limestone is the standard for carbon isotopes. The $\delta^{15}N$ values are in parts per thousand, and positive values indicate that the sample material is richer in the heavy isotope than the standard is. So far, this is interesting chemistry, but what does it have to do with ecology?

Different organisms take up nitrogen and carbon in ways that discriminate among these isotopes, so that the isotopic composition of plants and animals varies. On average, $\delta^{15}N$ increases 3.4‰ in animals relative to their diet. The result is that animals in different trophic levels have different isotopic signatures. For example, **Figure 10** shows for Lake Ontario the pattern of isotopic signatures for the pelagic food chain (Cabana and Rasmussen 1994).

The crustacean *Mysis relicta* eats zooplankton and has a $\delta^{15}N$ ratio of about 9‰, alewife and smelt feed on *Mysis* and have a $\delta^{15}N$ of 13‰–14‰, and lake trout that feed on alewife and smelt have a $\delta^{15}N$ of 16‰. The key point is that if we have a species of unknown position in the food chain, we can determine its trophic level by measuring its $\delta^{15}N$ ratio.

Marine and terrestrial plants differ in their $\delta^{15}N$ values, and this difference can be used to see if coastal animals utilize marine base foods more than terrestrial foods. Anderson and Polis (1998) found that coastal spiders on islands in the Gulf of California had $\delta^{15}N$ values of 20‰,

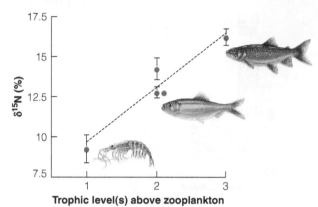

Figure 10 **Changes in the isotopic ratio of nitrogen in aquatic organisms in relation to their trophic status.** The differential between ^{15}N and ^{14}N is measured as indicated in Equation (4). On average this ratio increases about 3‰–4‰ for animals at each higher trophic level. Isotopic differentials are becoming increasingly useful to determine diets and can even be used to determine the diet of fossil animals (Clementz et al. 2003; Bocherens et al. 2004).)

compared with inland spiders with values of 12‰, showing that marine-based resources were sustaining these coastal spider populations.

One of the most important advantages of the stable isotope method of diet analysis is that it integrates the diet of the animals over a long time period, as opposed to measuring diet from stomach samples, which identifies the contents of the last meal only. In communities that process detritus, the use of stable isotopes can be helpful in separating the sources of the organic material in the system. For wide-ranging seabirds, stable isotopes can distinguish birds that feed inshore from those that feed offshore (Hobson et al. 1994; Kelly 2000). Stable isotopes provide another tool in the toolbox of ecologists trying to decipher food webs.

There are four advantages to using guilds and functional roles in the analysis of community organization (Wilson 1999):

- Guilds and functional roles focus attention on all competing species living in the same community, regardless of their taxonomic relationship.

- The use of the term *guild* or *functional role* clarifies the concept of niche: groups of species having similar ecological roles can be members of the

same guild but cannot be occupants of the same niche.

- Guilds and functional roles allow us to compare communities by concentrating on specific groups of taxa. We need not study the entire community but can concentrate on a manageable unit.

- Guilds and functional roles might represent the basic building blocks of communities and thus aid our analysis of community organization.

A community can be viewed as a complex assembly of component guilds or functional roles, each containing one or more species. Guilds may interact with one another within the community and thus provide the organization we observe. No one has yet been able to analyze all the guilds in a community, and at present we can deal only with a few guilds that make up part of an entire community. Two examples of the organization of guilds illustrate how this concept can be applied to communities.

Root (1973) grew collards (*Brassica oleracea* var. *acephala*) in two experimental habitats: in pure stands and in single rows bounded on each side by meadow vegetation. Three herbivore guilds were associated with collard stands (**Figure 11**). Pit feeders are insects that rasp small pits into leaf surfaces; they comprise 18 species, of which two chrysomelid beetles were

abundant. Strip feeders, insects that chew holes in the leaves, included 17 species, of which only one was abundant. Sap feeders suck the juices of the collard plants and included 59 species, many of them aphids. The pit feeders usually formed the most important herbivore guild, particularly in the pure collard stands.

The species composition of the three herbivore guilds changed from year to year, and these changes were most striking among the sap feeders. The cabbage aphid (*Brevicoryne brassicae*) was the most abundant aphid in 1966 and 1968 but was absent entirely in 1967, when other aphids increased in abundance. The implication is that within some guilds, species can replace one another and perform the same functional role.

The nectar-eating birds of successional montane forests in Costa Rica form a guild clearly organized around competition for food (Feinsinger 1976). This

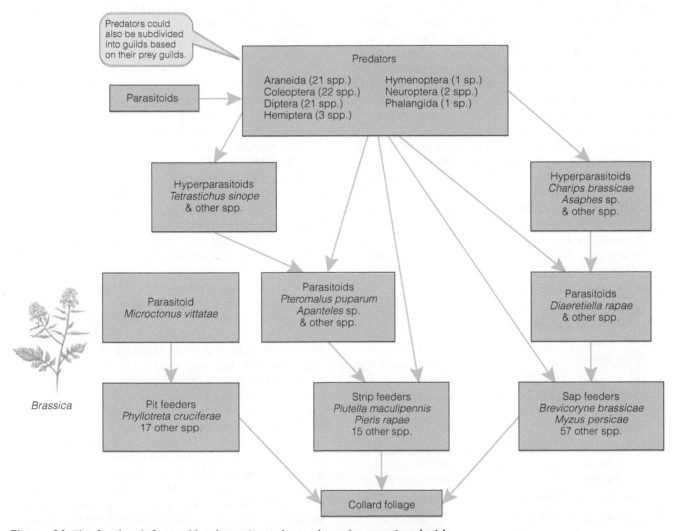

Figure 11 The food web formed by the major arthropod species associated with collard stands at Ithaca, New York. The herbivores, parasitoids, and hyperparasitoids are divided into guilds. (After Root 1973.)

guild of hummingbirds is organized around the dominant species, *Amazilia saucerottei*, the blue-vented hummingbird. *Amazilia* specializes on plants that produce large quantities of nectar and sets up individual feeding territories that each bird defends against other hummingbirds. *Amazilia* is aggressively dominant over most other hummingbird species. A second common species, *Chlorostilbon canivetii*, is excluded from this rich flower resource by aggressive *Amazilia* individuals and exhibits "trap-line" feeding, following a regular route between scattered flowers. *Chlorostilbon* spends much time in flight and only rarely defends any flowers. Two other hummingbirds complete the core group of this guild. *Philodice bryantae* sets up feeding territories in flower-rich areas but defends these territories only against other *Philodice*. This species seldom elicits attack behavior from the dominant *Amazilia*, partly because *Philodice* looks more like a bee than a bird. The final member of the core species in this guild is *Colibri thalassinus*, which is highly migratory and moves in to exploit seasonal flowering. Ten other hummingbird species forage in the study area, and most of these species were important in adjacent communities. The foraging of all these species is affected by the territorial behavior of the dominant *Amazilia*, and the high diversity of this bird guild is related to the highly migratory strategy of many of these hummingbird species.

Most of the hummingbird species in the guild that Feinsinger (1976) analyzed were general nectar feeders, and hence functional equivalents. If one of these species were removed from the community, we would predict that the other hummingbird species would take its place, and the community would be little changed.

The guild or role concept of community organization is not yet fully developed (Simberloff and Dayan 1991). We can recognize four hypotheses that require testing in natural communities:

- Many species form interchangeable members of a guild from the point of view of the rest of the community. These species are functional equivalents.

- The number of functional roles within a community is small in relation to the number of species and might be constant among different communities.

- There may be a limit on the number of species that can simultaneously fill a given functional role. A community always has a set of roles, but the guilds may be packed with different numbers of species.

- Species within guilds fluctuate in abundance in such a way that the total biomass or density of the guild remains stable.

At present we can define roles or guilds only crudely via the analysis of food webs. There is a need to define objectively the criteria used to assign species to guilds in natural communities (Simberloff and Dayan 1991). The utility of the guild concept is that by reducing the number of components in a community, it should help us to study how communities are organized. It also emphasizes that ecological units are not taxonomic units. Ants, rodents, and birds can all eat seeds in desert habitats, and thus they form a single guild of great taxonomic diversity (Brown and Davidson 1979).

Keystone Species

A role may be occupied by a single species, and the presence of that role may be critical to the community. Such important species are called *keystone species* because their activities determine community structure. Bob Paine was the first ecologist to recognize keystone species in his research on the rocky intertidal zone (Paine 1969). Keystone species typically are not the most common species in a community, and their effects are much larger than would be predicted from their relative abundance. One way to recognize keystone species is through removal experiments.

The starfish *Pisaster ochraceous* is a keystone species in rocky intertidal communities of western North America (Paine 1974). When *Pisaster* was removed manually from intertidal areas, the mussel *Mytilus californianus* was able to monopolize space and exclude other invertebrates and algae from attachment sites. *M. californianus* is an abundant species that is able to compete for space effectively in the intertidal zone. Predation by *Pisaster* removes this competitive edge and allows other species to use the space vacated by *Mytilus*. *Pisaster* is not able to eliminate mussels because *Mytilus* can grow too large to be eaten by starfish. Size-limited predation provides a refuge for the prey species, and these large mussels are able to produce large numbers of fertilized eggs (Paine 1974).

Sea otters are a keystone predator in the North Pacific. Once extremely abundant, they were reduced by the fur trade to near extinction by 1900. Once they were protected by international treaty, sea otters began to increase and by 1970 had recovered in most areas to near maximum densities (Estes and Duggins 1995). Sea otters feed on sea urchins, which in turn feed largely on macroalgae (kelp). Early natural history observations showed that in areas where sea otters were abundant, sea urchins were rare and kelp forests were well developed. Similarly, where sea otters were rare, sea urchins were common and kelp was nonexistent. Sea otters are thus a good example of a keystone species in a marine subtidal community. Since about 1990 sea otters have declined precipitously in large areas of western Alaska (**Figure 12**), often at rates of 25% per year. The loss of this keystone species has allowed sea urchins to increase and has resulted in the destruction of kelp forests. Killer whales are the suspected cause of the sea otter decline (Williams et al. 2004). Killer whales have begun to attack sea otters in the last 15–20 years because their prey base (seals, sea lions) has declined along with the fishes that constitute the seals' prey base. Fish have probably declined from human overharvesting in the North Pacific, illustrating that the interactions in food webs can propagate from top predators to basal species in unexpected ways.

A third example of a keystone species is the African elephant (Laws 1970), a relatively unspecialized herbi-

vore that relies on a diet of browse supplemented by grass. By their feeding activities, elephants (**Figure 13**) destroy shrubs and small trees and push woodland habitats toward open grassland. Elephants feeding on the bark can destroy even large mature trees. As more grasses invade the woodland habitats, the frequency of fires increases, which accelerates the conversion of woods to grassland. This change works to the elephants' disadvantage, however, because grass alone is not a sufficient diet for elephants, and they begin to starve as woody species are eliminated. Other ungulates that graze the grasses are favored by the elephants' activities.

The critical effect of keystone predators is that they can reverse the outcome of competitive interactions. The impact of keystone predators is clearly evident in aquatic communities. Amphibians are a major component of temporary ponds. In the coastal plain of North Carolina, a single pond can support five species of salamanders and 16 species of frogs and toads (Fauth and Resetarits 1991). Salamanders are the major predators in these temporary ponds, and the broken-striped newt *Notophthalmus viridescens* acts as a keystone predator. By selectively preying on the dominant competitors *Rana utricularia* and *Bufo americanus*, it allows less competitive frogs such as the cricket frog (*Hyla crucifer*) to survive.

Kangaroo rats in the Chihuahuan Desert form a keystone guild (Brown and Heske 1990). Kangaroo rats (*Dipodomys* spp.) prefer to eat large seeds. When they were excluded by experimental fences from areas of desert shrubland, large-seeded winter annuals increased greatly in abundance, raising the vegetative cover and reducing the ability of ground-feeding birds to feed on seeds. After 12 years grasses increased, and the area became desert grassland. Grassland species of rodents, such as the cotton rat, which were previously absent, began colonizing the habitat from which kangaroo rats

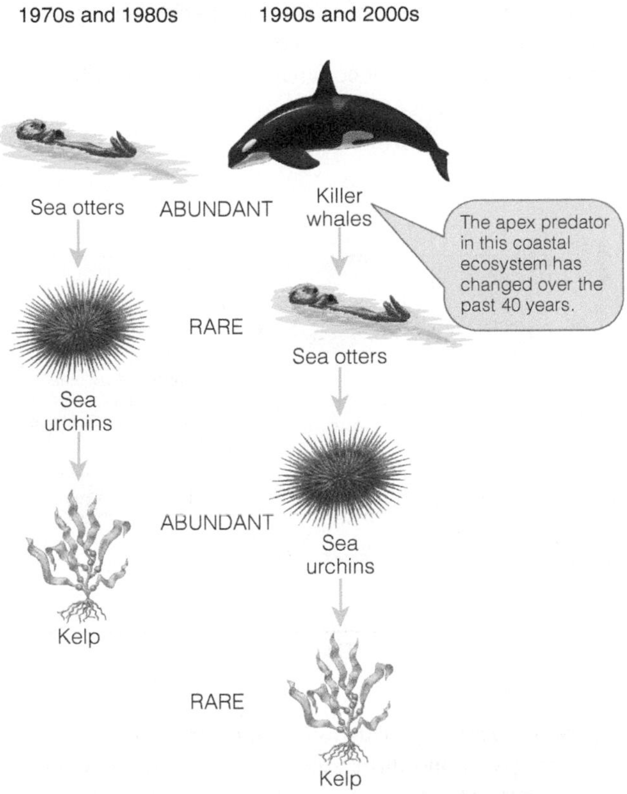

Figure 12 Sea otters as keystone predators in the North Pacific. The food chain of this nearshore ecosystem in the Aleutian Islands has changed over the past 40 years, as illustrated in this schematic diagram. The food chain on the left shows how the kelp forest ecosystem was organized in the 1970s and 1980s before the sea otter's decline, and the one on the right shows how this ecosystem changed with the addition of killer whales as an apex predator. (Modified after Estes et al. 1998.)

Figure 13 African elephants browsing in open woodland in the Serengeti area of East Africa.

were excluded. Fifteen species of rodents live in the Chihuahuan Desert, but only the three species of kangaroo rats seem to play this keystone role.

Keystone species may be rare in natural communities, or they may be common but unrecognized. At present, few terrestrial communities are believed to be organized by keystone species, but in aquatic communities keystone species may be more common. There seems to be no simple way of recognizing keystone species in food webs without doing detailed studies (Power et al. 1996). The important message is that some species of low abundance can have strong effects on community structure, so land managers and conservationists must be concerned with both common and uncommon species in communities. We cannot determine which species might be keystone species unless we gain a detailed understanding of food web structure by analyzing it experimentally.

Dominant Species

Dominant species in a community may exert powerful control over the occurrence of other species, and the concept of dominance has long been engrained in community ecology. Dominant species are recognized by their numerical abundance or biomass and are usually defined separately for each trophic level. For example, the sugar maple is the dominant plant species in part of the climax forest in eastern North America; its abundance determines in part the physical conditions of the forest community. Dominance usually means numerical superiority, and keystone species are not usually the dominant species in a community (**Figure 14**).

Dominant species are usually assumed to achieve their dominance by competitive exclusion. Buss (1980) identified several possible configurations of competing species (**Figure 15**). The simplest case, **transitive competition**, occurs when a linear hierarchy exists (species *A* outcompetes *B* and *B* outcompetes *C*). In this case, competitive exclusion can occur. A more complex case of **intransitive competition** occurs when a circular network exists in which no one species can be called dominant. In circular networks of spatial competition, species *A* outcompetes species *B*, *B* outcompetes *C*, but *C* in turn is able to outcompete *A*. This type of competitive interaction is called *intransitive* because it has no end point. Competitive exclusion does not occur in circular networks, and if intransitive competition is the rule in natural communities, species diversity need not decline because of competitive exclusion (Buss and Jackson 1979).

Competitive dominance is not the only explanation for a species becoming dominant. Predation may

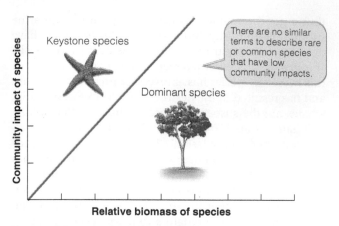

Figure 14 Schematic illustration of the difference between dominant species and keystone species. Species whose total impact is exactly proportional to their abundance would fall on the red line. Both the dominant and keystone species in a community are assumed to have a high community impact, but keystone species have low biomass. Trees, giant kelp, and corals are examples of dominant species in their communities. (Modified after Power et al. 1996.)

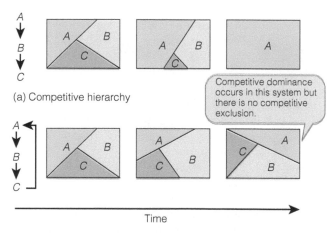

Figure 15 Schematic illustration of competitive relationships among three species competing for space on a simple surface. Each rectangle represents a plot of ground or a rock surface in the intertidal zone. (a) The conventional competitive hierarchy showing transitive competition, in which species A outcompetes species B, which in turn outcompetes species C. As time progresses, competitive exclusion will occur. (b) A competitive network showing intransitive competition, in which species A outcompetes B and B outcompetes C, but C is able to outcompete A. This system changes in time but does not move toward competitive exclusion. (After Buss and Jackson 1979.)

override competition in some communities. Australian mangrove forests show a complex zonation across the intertidal zone (Smith 1987). This zonation has usually been explained either by mechanisms of physiological tolerance to seawater inundation or by tidal sorting of seeds by size. Smith (1987) found that seed predation by small grapsid crabs was very high, and dominance in four of five mangrove species was correlated with the amount of seed predation.

Because humans commonly target large mammal predators, there is a long history of change within the dominance structure of predator communities. The dingo (*Canis lupus dingo*) has been reduced in large areas of Australia since the 1800s because they prey on sheep. The result has been a strong increase in red foxes and cats, to the detriment of small herbivores (Glen et al. 2007, Johnson et al. 2007).

Dominance in plant communities can be changed by the addition of nutrients. Humans are adding nitrogen to plant communities via the burning of fossil fuels and the production of nitrogen fertilizers. Due to industrialization, there has been a tenfold increase in atmospheric deposition of nitrogen within the last 200 years (Gilliam 2006). The consequences for plant communities that are nitrogen limited are dramatic. **Figure 16** illustrates the change in dominance of two common species in the boreal forests of Sweden in response to 20 years of nitrogen addition. Dominance in the herbaceous layer of these boreal forests is completely changed by the addition of nitrogen from the atmosphere. The dominance structure of these forest communities is highly responsive to extrinsic inputs of nutrients (Strengbom et al. 2001).

Communities that develop under similar ecological conditions in a given geographic region are expected to be dominated by the same species. For example, a deciduous forest in Ohio is expected to be dominated by beech and sugar maple, and botanists would be surprised if a rare species such as black walnut or white ash became dominant. Ecologists have long wondered if the same pattern occurs in aquatic communities, particularly those in the open ocean. The central gyre of the North Pacific Ocean has a rich diversity of phytoplankton, zooplankton, and fish species (McGowan and Walker 1979). Because the ocean mixes and the gyre is so large, the dominant species might be expected to vary from place to place within this large area, but this does not seem to occur. McGowan and Walker (1985) found that about 30 species of copepods were abundant, and that the dominance structure of this community remained the same in samples collected up to 16 years apart. The same constancy of community structure was evident in the phytoplankton and fish communities of the central gyre. These oceanic

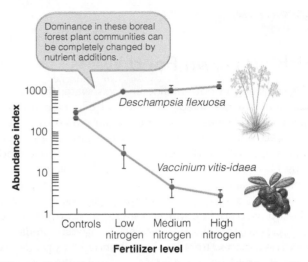

Figure 16 Changes in species dominance in the herb layer of Swedish boreal forest plots to which nitrogen fertilizer was added from 1971 to 1990.
Low nitrogen was 34 kg N/ha/year, medium nitrogen was 62 kg N/ha/year, and high nitrogen was 108 kg N/ha/year. *Deschampsia flexuosa* is called wavy hair grass, and *Vaccinium vitis-idaea* is called huckleberry or cowberry or lingon berry. These experiments were carried out to mimic the impacts of nitrogen deposition from the air from pollution. (Data from Strengbom et al. 2001.)

communities appear to be as constant in their dominant species as are temperate-zone forests.

The human-induced removal of a dominant species in a community has occurred frequently, but unfortunately few of these removals have been studied in detail. The American chestnut was a dominant tree in the eastern deciduous forests of North America before 1910, making up more than 40% of overstory trees. This species has now been eliminated as a canopy tree by chestnut blight. The effects of this removal have been negligible, as far as anyone can tell, and various oaks, hickories, beech, and red maple have replaced the chestnut (Keever 1953). Of the 56 species of Lepidoptera that fed on the American chestnut, seven species went extinct, but the other 49 species apparently did not rely only on the chestnut for food and still survive (Pimm 1991).

Dominance has been studied in freshwater communities in considerable detail. The zooplankton community of many temperate-zone lakes is dominated by large-sized species when fish are absent and by small-sized species when fish are present. Brooks and Dodson (1965) observed this change in Crystal Lake, Connecticut, after the introduction of a herring-like fish, the alewife (*Alosa pseudoharengus*)

Fishing Down Food Webs

Overfishing in both marine and freshwater fisheries is having a predictably devastating effect on aquatic food webs. Fisheries do not operate at random within the food web. The valuable fishes humans prefer to eat, such as tuna and cod, tend to be top predators in marine food webs. Pauly et al. (1998) asked whether there were any systematic global trends in overfishing. To do this they had to assign each species in the commercial catch a trophic value from 2.0 (for herbivorous fishes like anchovies) to 5.0 (for top predators like killer whales). Secondary predators would be assigned trophic value 3.0, and tertiary predators like cod would be assigned 4.0. To make these assignments they had to know at least approximately the diet of each of the main species in the fishery. In general, the higher the trophic level, the larger the size of the fish.

Many of the world's fisheries are overexploited, and overall catches have been declining since 1989. To their surprise, Pauly et al. also found a global pattern of collapse in average trophic level for both marine and freshwater fisheries, as shown in **Figure 17**.

This change in trophic status reflects a gradual global shift from catching long-lived predatory fish to catching short-lived plankton-feeding fish and invertebrates. Clearly this pattern of decline in the trophic level of the catch is not sustainable in the future unless we wish to dine on zooplankton.

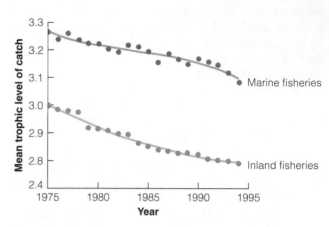

Figure 17 Global trends toward fisheries being depleted of apex predators like tuna so that catches are increasingly being dominated by lower trophic level species. (From Pauly et al. 1998.)

Fishing down food webs can be explained in part by overfishing at lower trophic levels as well as overfishing of apex predators (Essington et al. 2006). But the trends toward overfishing at all trophic levels has now been extended to fishing for deeper water species, which have little resilience to overfishing (Moroto et al. 2006).

(**Figure 18**). They proposed the *size-efficiency hypothesis* as a wide-ranging explanation of the observed shift in dominance in the zooplankton community. The size-efficiency hypothesis is based on two assumptions: (1) that planktonic herbivores (zooplankton) all compete for small algal cells (1–15 μm) in the open water, and (2) that larger zooplankton feed more efficiently on small algae than do smaller zooplankton, and that large animals are able to eat larger algal particles that small zooplankton cannot eat.

Given these two assumptions, Brooks and Dodson (1965) made three predictions:

- When predation on zooplankton is of low intensity or absent, the small zooplankton herbivores will be completely eliminated by large forms (dominance of large cladocera and calanoid copepods).

- When predation is of high intensity, predators will eliminate the large zooplankton and allow the small zooplankton (rotifers, small cladocera, and small copepods) to become dominant.

- When predation is of moderate intensity, predators will reduce the abundance of the large zooplankton such that the small zooplankton species are not eliminated by competition.

Thus, competition forces communities toward larger-bodied zooplankton, whereas fish predation forces them toward smaller-bodied species. These three predictions of the size-efficiency hypothesis are consistent with the keystone-species idea discussed in the preceding section.

The second and third predictions of the size-efficiency hypothesis have been tested in several lakes, and the predictions seem to describe adequately the zooplankton distributions in many lakes (Kerfoot 1987). Fish predation does seem to fall more heavily on the larger zooplankton species, but invertebrate predators in the plankton seem to prey more heavily on the smaller zooplankton species. Large zooplankton may predominate in lakes with no fish either because they are superior competitors (as the size-efficiency hypothesis predicts) or because small zooplankton are selectively removed by invertebrate predators.

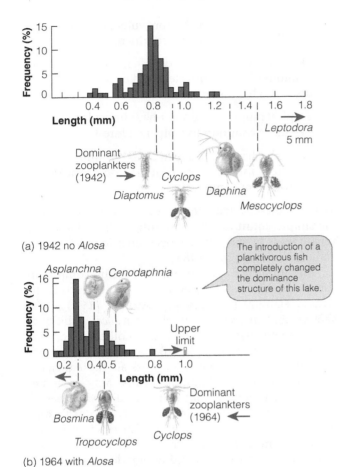

(a) 1942 no *Alosa*

The introduction of a planktivorous fish completely changed the dominance structure of this lake.

(b) 1964 with *Alosa*

Figure 18 The composition of the crustacean zooplankton of Crystal Lake, Connecticut. (a) Before the introduction (1942) and (b) after the introduction (1964) of the alewife (*Alosa*), a plankton-feeding fish. *Daphnia* is a large cladoceran; *Ceriodaphnia* and *Bosmina* are small cladocerans. *Mesocyclops* is a large copepod; *Tropocyclops* is a small copepod. Some of the larger zooplankton species are not represented in the 1942 histogram because they were not abundant. (After Brooks and Dodson 1965.)

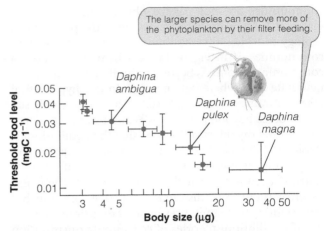

The larger species can remove more of the phytoplankton by their filter feeding.

Figure 19 A test of the central assumption of the size-efficiency hypothesis that large zooplankton are competitively superior to small zooplankton. Eight *Daphnia* species of different sizes were raised in the laboratory on a constant food source. Larger *Daphnia* can maintain weight on a lower food concentration and are thus competitively dominant. (From Gliwicz 1990.)

The second assumption of the size-efficiency hypothesis is that large-sized zooplankton are superior competitors for food resources. In a laboratory microcosm, Gliwicz (1990) fed eight species of *Daphnia* on constant food levels and measured the concentration of algae they needed to maintain body weight. **Figure 19** shows that large copepods can subsist on much lower algal concentrations than small copepods, as assumed by the size-efficiency hypothesis.

The importance of fish predation in structuring zooplankton communities is now well established (Kerfoot 1987), but the competitive nature of feeding relationships among the zooplankton may not always favor large species because of fluctuating food conditions in lakes and ponds. In some ponds and lakes small zooplankton predominate in the absence of fish predators, and this is likely due to predation by invertebrate predators.

Dominance is an important, although poorly understood component of community organization. Dominant species may be the focal point of interactions that structure much of the species makeup of a community. Moreover, the characteristics of dominant species may affect the stability of the community as well.

Community Stability

Stability is a dynamic concept that refers to the ability of a system to recover from disturbances. If a brick is raised slightly from the floor and then released, it will fall back to its original position. This is the physicists' concept of **neighborhood stability** or local stability, in which the system responds to small, temporary disturbances by returning to its original position. Thus, for example, a rabbit population may show neighborhood stability to moderate hunting pressure if it returns to its normal density after hunting is prohibited.

Physicists discuss stability in terms of small perturbations, but ecological systems are subject to large disturbances. To deal with these, we must consider a second type of stability, **global stability**. A system that has local stability shows global stability only if the system returns to the same point after large disturbances. That brick, for example, shows both local and global stability because if we raise it either 10 mm or 10 meters from the floor and release it, it will fall back to the floor. Ecological communities are not passive objects like bricks, and global stability is probably rare. One of the problems of ecology is to identify the limits of stability for various communities. If in Figure 1a we schematically move a community at point A_2 beyond point B_2, it will move to a new state somewhere near point C. In this hypothetical case community A_2 is locally stable, but not globally stable, if forced beyond point B_2. All equilibrium theories of community organization assume that the equilibrium is stable, and the usual assumption is that the equilibrium is globally stable. Note that the shape of stability "basins" need not be circular in cross section. There may be great stability to disturbances in one direction but little stability to disturbances in other directions.

If equilibrium theories of community organization are correct, there are four important consequences for our understanding of community dynamics (Chesson and Case 1986):

- *Community conservation.* An equilibrium community will show no tendency to lose species over time. Global stability implies that in the absence of external perturbations no losses of species will ever occur.

- *Community recovery.* An equilibrium community can recover from events that drive any of its constituent species to low density.

- *Community composition.* An equilibrium community can be built up by immigration of species from outside the system. Combinations of species that can coexist will increase to their equilibrium values.

- *Independence of history.* Because of global stability, past events have no effect on community structure, and within a broad range the order of arrival of member species is irrelevant to the final community composition.

Equilibrium communities are stable in the sense of persistence, but they may not be stable in the sense of being resilient to disturbances, particularly disturbances caused by humans.

One of the classic tenets of community ecology and a hallowed tenet of conservation biology has been that biodiversity *promotes* stability. Elton (1958) suggested several lines of circumstantial evidence that support this conclusion:

- Mathematical models of simple systems show how difficult it is to achieve numerical stability.

- Gause's laboratory experiments on protozoa confirm the difficulty of achieving numerical stability in simple systems.

- Small islands are much more vulnerable to invading species than are continents.

- Outbreaks of pests are most often found in simple communities on cultivated land or on land disturbed by humans.

- Tropical rain forests do not have insect irruptions like those common to temperate forests.

- Pesticides have caused irruptions by eliminating predators and parasites from the insect component of crop plant communities.

Many of these statements are only partly correct, and the simple, intuitive, and appealing notion that biodiversity leads to stability has been questioned as a general conclusion (Pimm 1984).

The intuitive argument that increasing community complexity in the food web automatically leads to increased stability was attacked by May (1973), who showed that increasing complexity *reduces* stability in general mathematical models. In hypothetical communities in which the trophic links are assembled at random, the more diverse communities are more unstable than the simple communities. Thus, May cautioned community ecologists that if species diversity does indeed result in stability in the real world, it is not an automatic mathematical consequence of species interactions. Natural communities are products of evolution, and evolution may have produced nonrandom assemblages of interacting species in which diversity and stability are related. These theoretical conclusions were questioned by Haydon (1994) who showed that with slightly less restrictive assumptions, stability could potentially increase with diversity, as Elton (1958) had suggested.

Even though theoretical ecologists had produced mathematical models showing that greater species diversity does not always lead to greater stability, field ecologists followed Elton in believing that complex communities were indeed more stable than simple ones. As more data on the composition and complexity of food webs became available during the 1990s, empirical ecologists became more and more convinced that complexity does indeed favor stability. McCann et al. (1998) have constructed a set of theoretical models that predict that complex food webs will be more stable. The key to this stability is the interaction strength between species in the food web. If there are many weak links between species in a food web, complex communities are more stable than simple ones. The conclusions reached by McCann et al.'s model are exactly the reverse of those reached by May's (1973) model. The reasons for these differences are that McCann et al. used nonlinear models for interactions between species and assumed realistic optimal foraging constraints for feeding interactions. The result is that we now have both theoretical and empirical agreement that greater diversity

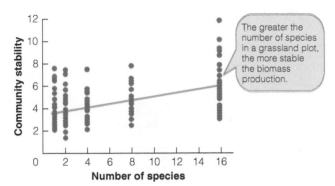

Figure 20 Temporal stability of experimental grassland plots in Minnesota that were seeded with 1, 2, 4, 8, and 16 perennial grassland species and followed for 10 years. Stability increases with diversity as predicted by Charles Elton. Stability was measured by the average yield over 10 years of each plot in late summer divided by its standard deviation in yield over the same time period. (Data from Tilman et al. 2006.)

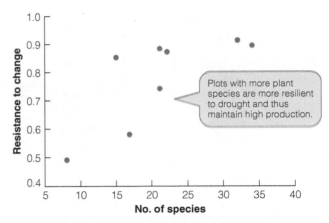

Figure 21 The relationship between stability and diversity for grasslands in Yellowstone National Park. Eight plots of 0.5 m² were measured in 1988, a year of severe summer drought, and 1989, a year of normal rainfall. Resistance is a measure of change in species abundances (high resistance = little change, low resistance = much change). Biomass was measured on grazed plots at the peak of the summer growth season. (Data from Frank and McNaughton 1991.)

contributes to greater stability in ecological communities (McCann 2000).

Are there any data from field experiments that quantify the diversity-stability relationship? The few detailed studies conducted on the relationship between stability and diversity suggest that in many systems greater diversity does lead to greater stability (Johnson et al. 1996). All the studies to date are on plant communities.

In 1994 David Tilman and his colleagues began a long-term study of 168 plots in a Minnesota grassland to identify the relationships between species diversity in plants and community functions. They seeded these plots with 1, 2, 4, 8, or 16 perennial grassland species and measured the plant biomass on each plot at the peak of 10 growing seasons from 1996 to 2005. With these manipulations Tilman et al. (2006) directly assessed the diversity-stability hypothesis. **Figure 20** shows that plots with greater plant diversity showed more stability, measured by less fluctuation in yield, than plots with low diversity, in keeping with the expectations of the diversity-stability hypothesis of Elton.

Stability is usually measured as variability in numbers or biomass, but it can also be measured as *resistance* to change. More-stable communities will change less when external stress is imposed on them. For temperate plant communities, drought is a major stress, and plant ecologists have used droughts to test the diversity-stability hypothesis. In Yellowstone National Park a severe drought in 1988 allowed Frank and McNaughton (1991) to study the effects of diversity on stability in grassland communities. **Figure 21** shows that species-rich communities had higher resistance to change, as predicted by the diversity-stability hypothesis.

The conjecture of Charles Elton in 1958 that species diversity indeed imparts stability to ecological communities has been supported for many, but not all, ecological systems (Ives and Carpenter 2007). The practical application of this idea to our human-degraded landscapes is the focus of the applied area of restoration ecology. What progress have we made in restoring damaged ecological systems?

Restoration Ecology

Communities recover from disturbances through a whole series of biological restoration mechanisms. Succession is a major pathway in restoration ecology, which aims to harness natural processes to restore systems adversely affected by humans. The key starting principle in restoration ecology is that the spatial scale of the impact and the recovery time are related, such that the larger the scale of the disturbance, the longer the time frame for restoration (**Figure 22**). There appears to be no difference in this relationship between man-made and natural disturbances (Dobson et al. 1997a). If we can identify the processes that limit the speed of recovery, we can alter this curve to reduce the effects of human disturbances.

The first principle of restoration ecology is that environmental damage is not irreversible. This optimistic principle must be tempered by the second principle of restoration ecology—that communities are not infinitely resilient to damage. Figure 1a illustrates these

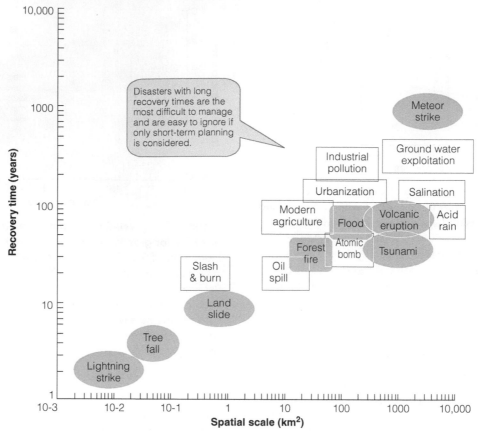

Figure 22 The relationship between the spatial scale of natural and artificial disasters and the approximate expected time to recovery. Natural disasters are depicted as orange ellipses, and human-caused changes are represented by black rectangles. The aim of restoration ecology is to reduce the recovery time by manipulating ecological factors restricting the time sequence of recovery. (From Dobson et al. 1997.)

ideas schematically—there is only a finite range of environments over which a community is locally stable. To illustrate the first principle of restoration ecology, let us consider one example of lake restoration that has been successful. Aquatic communities have been disturbed by pollution of human origin for many years, and the stability of aquatic systems under pollution stress is a critical focus of restoration ecology today. Much experimental work during the last 30 years has accompanied many large-scale uncontrolled experiments involving the diversion of nutrients into lakes near cities (Jeppesen et al. 2007).

The broad picture is that lakes can exist in several configurations depending on nutrients. **Figure 23** shows schematically the range of conditions for shallow lakes, from lakes dominated by large aquatic plants (macrophytes) to those dominated by phytoplankton. The mechanisms underlying these alternative states in lakes are primarily nutrient related. The amount of nitrogen in

lake water is a strong predictor of whether submerged macrophytes are present in lakes (**Figure 24**). At the other extreme, the amount of phytoplankton in the water is directly related to the phosphorus content of the lake (**Figure 25**).

To manage these shallow lakes, it would appear that one need only manipulate the nutrient regime of the lake. The usual problem is that nutrients from sewage and excess use of fertilizers has moved the lake into the "green water" zone with excess phytoplankton. The strategy to restore these lakes thus involves ways of reducing nutrient inputs. One of the success stories of this approach has been Lake Arres in Denmark (Jeppesen et al. 2007) and another has been Lake Washington in the United States (Edmondson 1991).

Lake Washington is a large, formerly unproductive lake in Seattle, Washington. In the early phases of city's development, Lake Washington was used for raw sewage disposal, but this practice was stopped between

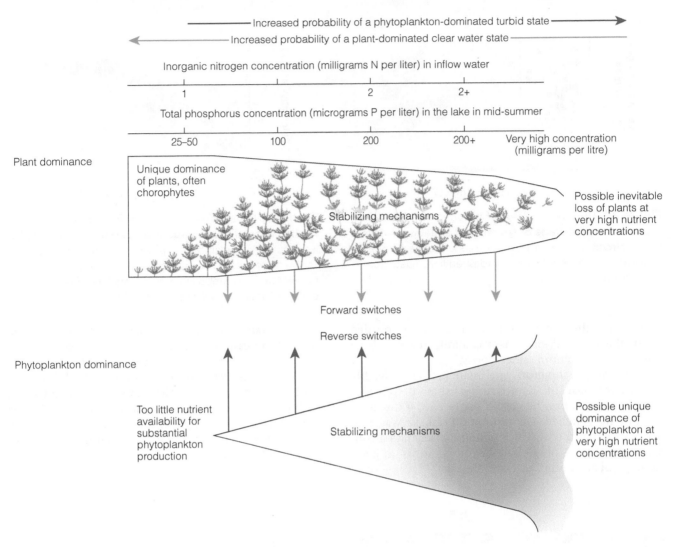

Figure 23 The alternative states for shallow lakes in the temperate zone. Low phosphorus and low nitrogen levels favor aquatic macrophytes, while high phosphorus and nitrogen favor the growth of phytoplankton, which shade out the aquatic macrophytes. (From Moss 2002.)

1926 and 1936. However, with additional population pressure, a number of sewage-treatment plants built between 1941 and 1959 began discharging treated sewage into the lake in increasing amounts. By 1955 it was clear that the sewage was destroying the clear-water lake, and a plan to divert sewage from the lake was voted into action. More and more sewage was diverted to the ocean from 1963 through 1968, and almost all was diverted from March 1967 onward. Thus, the recent history of Lake Washington consists of two pulses of nutrient additions followed by complete diversion.

What happened to the organisms in Lake Washington during this time? Some information can be obtained by looking at the sediments in the bottom of the

lake. After sewage had been added to the lake, the sedimentation rate rose to about 3 mm per year. The organic content of this sediment progressively increased since the early 1900s, which suggests an accelerated rate of primary production. The recent lake sediments also contain a greater amount of phosphorus. Edmondson (1991) has recorded the changes in Lake Washington in detail since the diversion of sewage began in 1963. There was a rapid drop in phosphorus in the surface waters within five years and a closely associated drop in the standing crop of phytoplankton. Nitrogen content of the water has dropped very little, which suggests that phosphorus is a limiting nutrient to phytoplankton growth. The water of the lake has become noticeably

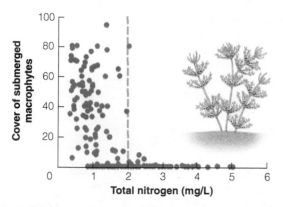

Figure 24 The average level of nitrogen in 44 Danish lakes and the coverage of submerged aquatic plants in late summer. Submerged plants disappear once nitrogen levels are above 2 mg N per liter (dashed line). (Data from Gonzalez-Sagrario et al. 2005.)

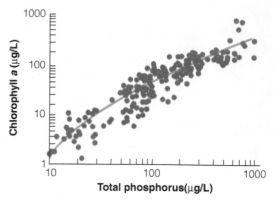

Figure 25 Summer mean chlorophyll *a* concentration in relation to total phosphorus in Danish lakes. Chlorophyll a is a measure of the amount of phytoplankton in the water. Ten lakes were sampled over 15 years or more to obtain these data. Each symbol represents data taken from one lake per year. (Data from Jeppesen et al. 2007.)

clearer since the sewage diversion. The phosphorus tied up in the lake sediments is apparently released back into the water column rather slowly.

The Lake Washington experiment is of considerable interest because it demonstrates that detrimental changes in lakes may be *stopped and reversed* if the input of nutrients is halted. The restoration of Lake Washington shows that this aquatic community displays a considerable degree of global stability and is a good example of an equilibrium community.

Restoration ecology can depend on the natural time scale of succession, as occurred in Lake Washington, or it can speed processes of recovery by adding nutrients when they are deficient, by seeding areas that have a shortage of available colonists, and by adding microbes to break down organic compounds such as oil. This area of applied ecology will assume more importance in the years to come as we try to speed the recovery of degraded landscapes.

Summary

Communities can be organized by competition, predation, and mutualism working within a framework set by the physical components of the environment. Of these three processes, the greatest emphasis has been placed on the roles of competition and predation in organizing communities, and the role of mutualism is largely unstudied.

Two broad views of communities are postulated as explanations of community organization. The classical model is that communities are in equilibrium, and their species composition and relative abundances are controlled by biotic interactions. According to this equilibrium model, interspecific competition, predation, and spatial heterogeneity are the major processes controlling organization. The nonequilibrium model does not assume stable equilibria but holds instead that communities are always recovering from disturbances.

Species in a community can be organized into food webs based on "who-eats-whom." Trophic levels may be recognized in all communities from the level of producers (green plants) to the higher carnivores and hyperparasites. Within a trophic level, we can recognize *guilds* of species exploiting a common resource base. Guilds may serve to pinpoint the functional roles species play in a community, and species within guilds may be interchangeable in some communities.

Food webs can be analyzed theoretically and empirically. Patterns of food chain length and complexity can be described, but all of them may vary with the complexity of the food web. The analysis of complex tropical food webs containing more than 100 species is a difficult task and is only just beginning.

Keystone species single-handedly determine community structure and can be recognized by removal experiments. Dominant species are the species of highest abundance or biomass in a community. Dominance is often achieved by competitive superiority, and some dominant species can be removed from the community and be replaced by subdominants with little effect on community organization. In aquatic communities, dominance in zooplankton herbivores may be determined by competition when fish predators are absent, and by predation when fish are present.

The characteristics of dominant species may affect community stability, with respect to the system's ability to return to its original configuration after disturbance. The ecological generalization that *diversity promotes*

stability is supported by field data and by theoretical analyses, but we do not know if it is true for all communities. The attributes of individual species and compartments in food webs may be significant in determining community stability.

Restoration ecology strives to apply ecological knowledge of community dynamics to restore damaged landscapes. Succession can heal damaged landscapes but may take more time than we might like. We can try to speed recovery by knowing how succession operates in a community and what limits its rate of progress. Communities can recover from disasters, but there is a limit to their resilience.

Review Questions and Problems

1 Elton (1958, p. 147) claims that natural habitats on small islands are much more vulnerable to invading species than natural habitats on continents. Find evidence that is relevant to this assertion, and evaluate its importance for the question of community stability. Gimeno et al. (2006) discuss one test case for plants.

2 Coyotes are commonly stated to be important predators of sage grouse in western United States. But coyotes also prey on foxes, badgers, and ground squirrels, all of which also prey on sage grouse nests or chicks. How could you decide if coyote control would be a benefit or a detriment to the conservation of sage grouse? Mezquida et al. (2006) discuss this problem.

3 How could you determine if an ecological community was an equilibrium or a nonequilibrium system, if you were placed on a new continent to study a community for which you had no background data? Discuss your research plan and the time scale needed to answer this question.

4 Compare and contrast the following statements of an evolutionist and an ecologist about species diversity and the stability of biological communities:

a. Simpson (1969, p. 175) states: "If indeed the earth's ecosystems are tending toward long-range stabilization or static equilibrium, three billion years has been too short a time to reach that condition."

b. Recher (1969, p. 79) states: "The avifaunas of forest and scrub habitats in the temperate zone of Australia and North America have reached equilibrium and are probably saturated."

5 How is it possible for stable isotope ratios to change between trophic levels? List several possible physiological mechanisms that might cause such

changes. Are there any population mechanisms for achieving these changes? Would you expect differences in isotope ratios if you measured different parts of an animal or plant? Kelly (2000) discusses the use of stable isotopes and their limitations for studying diets.

6 What is the length of time a community must be studied before all components of its food web are identified? Discuss the implications of constructing a time-specific food web versus a cumulative food web over a long period. Compare your analysis with that of Schoenly and Cohen (1991).

7 Compare the definitions of a *trophic species* and a *guild*. How does the aggregation of species into trophic species affect the analysis of a food web? Conversely, how would poor taxonomic resolution within species groups affect estimates of connectance in food webs—for example, if species are grouped into categories such as "insects"? Hall and Raffaelli (1993) discuss these problems of aggregation and taxonomic resolution in food web analysis.

8 Keystone species can be discovered by species removal experiments. Is there any other way to identify potential keystone species in a community, or must we always proceed by trial and error? Libralato et al. (2006) discuss this question.

9 Bracken fern (*Pteridium aquilinum*) often invades lowland heaths in Britain and develops a dense, uniform stand with a very depauperate flora and fauna. To reverse this habitat deterioration, various chemical and physical control methods were carried out over 18 years, but the objective of this restoration scheme (to restore heather heathland) was not achieved. Read Marrs et al. (1998) and discuss the reasons for the failure of this restoration program.

10   A shift in the shallow lake community from the dominance of rooted aquatic plants to the dominance of phytoplankton is partly driven by nutrient additions to the lake water but this is not the complete story (Moss 2007). What are the switches that are included in Figure 23 and how do they relate to the scheme illustrated in Figure 1?

Overview Question

You are asked to determine the food web for a lake that is proposed to be part of a national park and to assess the web's stability. Discuss how you would construct this food web and measure its stability, and the operational decisions you would have to make about what to include in the web.

Suggested Readings

- Blankenship, L. E., and L. A. Levin. 2007. Extreme food webs: Foraging strategies and diets of scavenging amphipods from the ocean's deepest 5 kilometers. *Limnology & Oceanography* 52:1685–1697.

- DeMaster, D. P., A. W. Trites, P. Clapham, S. Mizroch, P. Wade, R. J. Small, and J. V. Hoef. 2006. The sequential megafaunal collapse hypothesis: testing with existing data. *Progress in Oceanography* 68:329–342.

- Dunne, J. A., R. J. Williams, and N. D. Martinez. 2004. Network structure and robustness of marine food webs. *Marine Ecology Progress Series* 273:291–302.

- Ellis, J. C., M. J. Shulman, M. Wood, J. D. Witman, and S. Lozyniak. 2007. Regulation of intertidal food webs by avian predators on New England rocky shores. *Ecology* 88:853–863.

- Estes, J. A., M. T. Tinker, T. M. Williams, and D. F. Doak. 1998. Killer whale predation on sea otters linking oceanic and nearshore ecosystems. *Science* 282:473–476.

- Jeppesen, E., M. Søndergaard, M. Meerhoff, T. Lauridsen, and J. Jensen. 2007. Shallow lake restoration by nutrient loading reduction—some recent findings and challenges ahead. *Hydrobiologia* 584:239–252.

- Kelly, C. K., M. G. Bowler, O. Pybus, and P. H. Harvey. 2008. Phylogeny, niches, and relative abundance in natural communities. *Ecology* 89:962–970.

- Power, M. E., D. Tilman, J. A. Estes, B. A. Menge, W. J. Bond, L. S. Mills, G. Daily, J. C. Castilla, J. Lubchenco, and R. T. Paine. 1996. Challenges in the quest for keystones. *BioScience* 46:609–620.

- Reisewitz, S., J. Estes, and C. Simenstad. 2006. Indirect food web interactions: Sea otters and kelp forest fishes in the Aleutian archipelago. *Oecologia* 146:623–631.

- Thompson, R. M., M. Hemberg, B. M. Starzomski, and J. B. Shurin. 2007. Trophic levels and trophic tangles: The prevalence of omnivory in real food webs. *Ecology* 88:612–617.

- Tilman, D., P. B. Reich, and J. M. H. Knops. 2006. Biodiversity and ecosystem stability in a decade-long grassland experiment. *Nature* 441:629–632.

- Woodward, G., D. C. Speirs, and A. G. Hildrew. 2005. Quantification and resolution of a complex, size-structured food web. *Advances in Ecological Research* 36:85–135.

Credits

Illustration and Table Credits

Photo Credits

Community Dynamics II: Disturbance and Nonequilibrium Communities

Key Concepts

- Communities are not in equilibrium if their recovery times exceed the frequency of disturbance.

- Patchiness in communities can result from disturbances caused by physical or biotic factors.

- Two extreme alternative models of community organization are the *top-down model* in which predators drive all the lower trophic levels, and the *bottom-up model* in which nutrients drive all the higher trophic levels.

- Many intermediate models are possible between these two extremes. Reciprocal interactions between trophic levels and omnivory can complicate predictions based on food web manipulations.

- The general rule that larger areas contain more species is one of the oldest generalizations in community ecology and is codified in the *species-area curve*.

- Island communities can be equilibrium systems in which immigration balances extinction, but this is rarely the case. Some islands never reach their expected species equilibrium because they are limited by colonization and human disturbance.

- Some communities may exist in multiple stable states in which disturbances move them from one state to another. Whether a community has several stable states or is a nonequilibrium system is a critical distinction for conservation and management strategies.

From Chapter 21 of *Ecology: The Experimental Analysis of Distribution and Abundance*, Sixth Edition. Eugene Hecht.
Copyright © 2009 by Pearson Education, Inc. Published by Pearson Benjamin Cummings. All rights reserved.

KEY TERMS

biomanipulation The management practice of using a trophic cascade to restore lakes to a clear water condition by removing herbivorous or planktivorous fishes or by adding piscivorous (predatory) fishes to a lake.

bottom-up model The idea that community organization is set by the effects of plants on herbivores and herbivores on carnivores in the food chain.

disturbance Any discrete event that disrupts an ecological community.

lottery competition A type of interference competition in which an individual's chances of winning or losing are determined by who gets access to the resource first.

nonequilibrium model of community organization The global view that ecological communities are not constant in their composition because they are always recovering from biotic and abiotic disturbances, never reaching an equilibrium.

patch Any discrete area, regardless of size.

species-area curve A plot of the area of an island or habitat on the x-axis and the number of species in that island or habitat on the y-axis, typically done as a log-log plot and typically restricted to one taxonomic group such as plants or reptiles.

supply-side ecology The view that population dynamics are driven by immigration of seeds or juveniles from sources extrinsic to the local population, so there is no local control of recruitment processes.

top-down model The idea that community organization is set by the effects of carnivores on herbivores and herbivores on plants in the food chain.

trophic cascade model The idea that a strict top-down model applies to community organization so that impacts flow down the food chain as a series of + and – impacts on successive trophic levels.

The equilibrium model of community organization has been the classical model of community organization for the past 60 years. Like many classical models in ecology, more and more exceptions have been found to its predictions. The equilibrium model is focused on stability, and in some communities it is a good description of community organization. But in many other communities of a few hectares in area, *change* seems to be the rule rather than stability, causing many ecologists to search for a broader model for communities. To replace the classi-

cal equilibrium model, ecologists are now forging a new model of community organization called the **nonequilibrium model**, which focuses on the small spatial scale and emphasizes two central ideas—the concepts of **patches** and **disturbance** (DeAngelis and Waterhouse 1987). In this chapter we explore some of the new concepts the nonequilibrium model has stimulated.

Patches and Disturbance

Patchiness, a term that refers to the spatial scale of a system, has been recognized as a factor in how a system is described. The patchiness of different communities varies, and conclusions that apply to one spatial scale will not necessarily apply to others. We can recognize five spatial scales at which ecologists work (Wiens et al. 1986):

- Space occupied by one plant or sessile animal, or the home range of an individual animal

- Local patch, occupied by many individual plants or animals

- Region, occupied by many local patches or by local populations linked by dispersal

- Closed system (if it exists), or a region large enough to be closed to immigration or emigration for the species under study

- Biogeographical scale, including zones of different climate and different communities

At sufficiently small spatial scales, all ecological systems are short-lived and can never be at equilibrium. Understanding community dynamics at small spatial scales of a few hectares and aggregating the resulting dynamics into a regional scale is an important approach to predicting large-scale community dynamics (Chesson and Case 1986).

Most field studies of communities and virtually all experimental manipulations of communities are conducted at the local patch scale. No single general definition of a "patch" can cover all ecological communities, but in general a patch will cover a few square meters to a few hectares. Note that patches need not be completely spatially discrete, nor do they need to be completely homogeneous (Pickett and White 1985).

A disturbance is any discrete event that disrupts community structure and changes available resources, substrate availability, or the physical environment. Note that disturbances can be destructive events like fires or an environmental fluctuation like a severe frost. The notion of "normal" is excluded from ecologists' view of disturbance in communities, an important change of view that has implications for conservation and management. We cannot assume under the nonequilibrium model that communities in the "good old days" were "normal," and that the job of conservationists or land managers is to get back to what the community was like in the "good old days." For some communities disturbances are frequent, but in others disturbances are rare.

Disturbances can affect ecological communities in many different ways depending on their strength and frequency of occurrence. In general, ecologists have considered communities subject to disturbances to be recovering slowly back to the original community through a process of succession (**Figure 1**). But if sev-

eral disturbances hit a community at the same time or in rapid succession, the community may not be able to recover, and will be pushed into an altered state. These impacts of disturbances can be particularly severe on communities already stressed by human impacts of pollution or climate change.

Disturbances can be measured in a variety of ways (**Table 1**), most of which provide either a spatial or temporal perspective. Disturbances may also be classified as *exogenous* (arising from outside the community, like fire) or *endogenous* (resulting from biological interactions, like predation). These two classes are the extremes of a continuum of types of disturbances, and

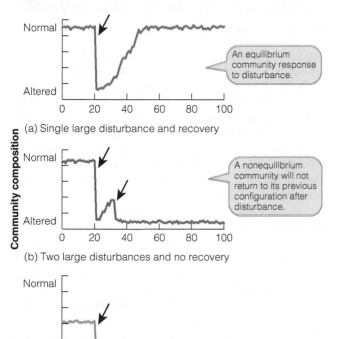

(a) Single large disturbance and recovery

An equilibrium community response to disturbance.

(b) Two large disturbances and no recovery

A nonequilibrium community will not return to its previous configuration after disturbance.

(c) Human-altered with added disturbance

Figure 1 Schematic illustration of the effects of disturbances (arrows) on ecological communities. In (a), a community is subjected to a single large disturbance, such as a fire at time 20, and then recovers through a process of succession to its original state. In (b), a community is subjected to two disturbances in sequence at time 20 and time 35, and the combined effects lead to a change in community composition and no recovery. In (c), a community altered by human activities, such as farming or forestry, is then subjected to a disturbance by fire or flooding, and the combination of stresses changes community composition and prevents it from recovering in a short time. "Normal" is used here as a shorthand for "previous community state." (Modified from Paine et al. 1998.)

Table 1 Definitions of measures of disturbance.

Measure	Definition
Distribution	Spatial distribution, including relationship to geographic, topographic, environmental, and community gradients
Frequency	Mean number of events per time period
Return interval, or turnover time	The inverse of frequency: mean time between disturbances
Rotation period	Mean time needed to disturb an area equivalent to the study area (the study area must be defined)
Predictability	An inverse function of variance in the return interval
Area or size	Area disturbed. This can be expressed as area per event, area per time period, area per event per time period, or total area per disturbance type per time period
Magnitude	
Intensity	Physical force of the event per area per time (e.g., wind speed for hurricanes)
Severity	Effect on the community (e.g., basal area removed)
Synergism	Effects on the occurrence of other disturbances (e.g., drought increases fire intensity, or insect damage increases susceptibility to windstorm)

SOURCE: Modified from Pickett and White (1985).

many communities are affected by a combination of endogenous and exogenous disturbances. The challenge to ecologists is to incorporate the various measures of disturbance in the field in their efforts to understand how disturbances affect particular communities.

The Role of Disturbance in Communities

In this section we examine the effects of disturbance in two communities: coral reef communities and rocky intertidal communities.

Coral Reef Communities

Coral reefs have existed in tropical oceans for at least 60 million years, and this long history is postulated to have produced the great diversity of organisms that are present on reefs today. Coral reefs have long been viewed as the classical equilibrium community living in tropical waters because on a geological time scale they have been stable for 200,000 years (Pandolfi et al. 2006). But on an ecological time scale this stability view has been challenged by long-term studies of coral reefs that show continuous change. We consider here two aspects of coral reefs: the corals and the reef fishes.

Coral reefs are subject to a variety of physical disturbances associated with tropical storms. At the Heron Island reef at the southern edge of the Great Barrier Reef off Queensland, Connell et al. (1997) followed changes in coral cover over a 30-year period using permanently marked quadrats. They measured the amount of coral cover to estimate abundance because corals are modular organisms and colonies vary greatly in size, and they measured larval recruitment of corals by sequential photographs of the permanent quadrats.

Violent storms (cyclones or hurricanes) were the main source of disturbance to Heron Island reefs, and the amount of damage caused by cyclones was strongly affected by the position of the coral colonies on the reef (**Figure 2**). Five cyclones passed near Heron Island during the 30 years of study from 1963 to 1992. Of the four study areas shown in **Figure 3**, only the protected area of the inner flat was relatively unaffected by cyclones. Virtually every cyclone caused a reduction in coral cover in the exposed pools. The 1972 cyclone completely removed coral cover on the exposed crest, the most severe disturbance observed. Recovery on the exposed crest was slow for the next 25 years. Gradual declines in coral cover on the protected sites was caused by increasing exposure to air as the corals grew upward over the 30 years of study.

Recruitment rates of corals were highly variable, and this is typical of many marine invertebrates whose

Figure 2 Heron Island, Australia.

larvae drift before settling. **Figure 4** illustrates the differences in recruitment rates among years and among sites on the Heron Island reef. There were no particularly good or bad years for coral recruitment in the sense that the whole reef varied in unison. The variability in recruitment was partly associated with how much free space was available in different areas. Coral larvae cannot attach to other living coral or macroalgae and need free space to settle.

The picture that emerges from this work on the Great Barrier Reef is of a coral community that changes continually because of exogenous disturbance caused by tropical cyclones and the internal processes of growth and recruitment. The coral community is not in equilibrium at the spatial scale of the reef because the frequency of disturbance exceeds the rate of recovery. Coral reefs are a good example of a nonequilibrium community. The same patterns of change in coral cover and in recruitment have been described for Hawaiian coral reefs, and seem to be a general pattern for many coral reefs (Coles and Brown 2007).

Coral reef fishes are one of the primary drawing cards for ecotourism to coral reef areas. The diversity of coral reef fishes is astonishing. For example, at One Tree Reef, a small island at the southern edge of the Great Barrier Reef, Talbot and colleagues (1978) recorded nearly 800 species of fish. At the northern edge of the Great Barrier Reef, over 1500 species of fish have been recorded. What determines the community structure of a coral reef fish community? Are these fish communities stable in time and space?

There are two extreme schools of thought about what controls the organization of coral reef fish communities. The first view suggests that coral reef fish communities are equilibrium systems in which fish populations are controlled by density-dependent processes. Within this equilibrium viewpoint are two hypotheses concerning the mechanism by which equilibrium is achieved. The **niche-**

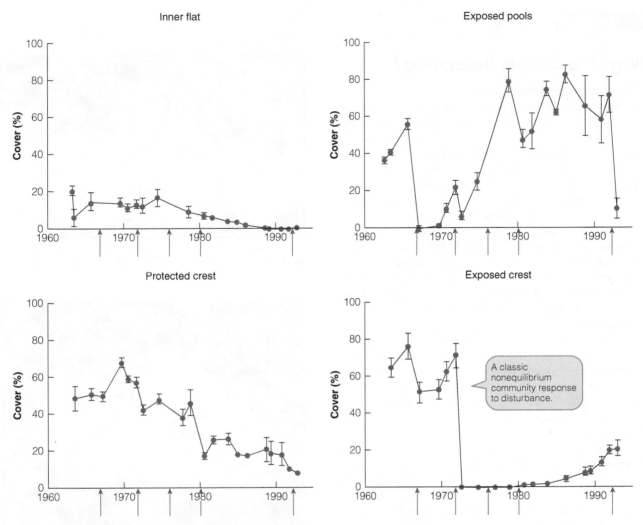

Figure 3 Percentage cover of corals in four areas of the coral reefs surrounding Heron Island at the southern edge of the Great Barrier Reef in Australia. Years with tropical cyclones are indicated by red arrows. Cover was measured on permanent quadrats in these shallow-water sites from 1963 to 1992. Damage from cyclones was highly variable depending on how much the site was protected by the island. (From Connell et al. 1997.)

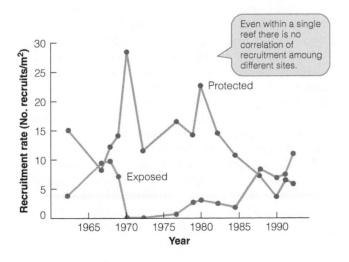

Figure 4 Coral recruitment rate, measured as number of recruits per m² per year, on permanent plots on protected and exposed crests of the Heron Island reef off Queensland, 1963–1992. Because recruitment is highly variable among sites and among years and does not correlate among the different sites, there were no generally "good" or "bad" recruitment years. (After Connell et al. 1997.)

Why Are Corals Bleaching?

Corals are animals that contain within their cells symbiotic algae that both contribute to coral growth (by conducting photosynthesis) and impart color to the corals. When corals lose their symbiotic algae, they lose their color—a process called bleaching—and often the corals die.

Coral reef bleaching has increased dramatically in many tropical areas around the globe over the past 25 years. Widespread bleaching can cause the death of whole coral reefs. The primary cause of coral bleaching is elevated sea surface temperatures. Many reef-building corals live very close to their upper lethal temperatures, and small increases of 0.5°C–1.5°C over a few weeks, or large increases of 3°C–4°C over several days, can kill corals (Lesser 2007). **Figure 5** shows the temperature profile of the Caribbean Basin during the severe coral bleaching event of 2005.

What might cause increased surface temperatures? Water temperatures in the equatorial Pacific Ocean follow a roughly cyclical pattern of warm and cool phases; broad-scale warming occurs every three to seven years, a phenomenon called El Niño. It has been suggested that if we know that El Niño conditions are due, perhaps we can predict episodes of coral bleaching. But part of the difficulty in accepting a simple temperature model of coral reef bleaching has been the observation that not all corals bleach in a given area, and that areas outside the Pacific (such as the Caribbean) are also affected. One possible explanation is that these large-scale oceanographic events have worldwide climatic repercussions and thus are not confined to the traditional El Niño regions of the Pacific and Indian Oceans.

If the temperature explanation for bleaching is correct, why should bleaching have increased dramatically in the past 25 years? Three factors may be involved. Global warming is a key suspect. In addition, degradation of coral reefs from pollution may have reduced the general resilience of reefs to damage, including bleaching. Finally, increased ultraviolet radiation could be combining with increased temperatures to induce bleaching. El Niño years are often associated with clear skies and calm seas, which increase the penetration of UV radiation to greater ocean depths.

The ecological consequences of the destruction of richly diversified coral reefs are large, and this problem deserves global attention.

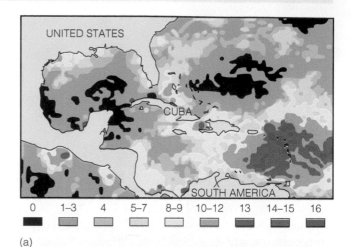

(a)

(b)

Figure 5 (a) **High temperatures in the Caribbean Basin as measured by the NOAA Coral Reef Watch during the coral bleaching event of 2005.** The units at the bottom of the image are degree-heating-weeks (DHW) color-coded for 12 weeks before October 28, 2005 in the Caribbean Basin with the highest thermal stress ever recorded. DHW values >4 indicate high sea surface temperatures under which coral bleaching is expected, whereas DHW values >8 indicate extremely high sea surface temperatures under which mass bleaching and mortality are expected. (From Lesser 2007.) (b) Partial bleaching of a *Pocillopora* coral in the Indian Ocean in 1998. When only moderately stressed some coral colonies become only partially bleached by expelling their photosynthetic zooxanthella algae that provide the color for the coral polyps.

diversification hypothesis suggests that coral reef fish communities are equilibrium competitive systems in which each species has evolved a very specific niche with respect to food and microhabitat. According to this view, current competition among species is strong and maintains niche differences, species segregation, and high diversity (Anderson et al. 1981a); alternatively, density-dependent predation on adults could control populations of reef fish and maintain equilibrium communities. The second school of thought champions the **variable recruitment hypothesis**, which suggests that coral reefs are nonequilibrium systems in which larval recruitment is as unpredictable as a lottery. Competition among species is present, but the winner in competition cannot be predicted, and the local community present on a reef is a random sample from a common larval pool (Sale 1977). Mortality after larval settlement is density independent, and consequently local populations fluctuate under the control of unpredictable recruitment. How can we evaluate these equilibrium and nonequilibrium hypotheses of community organization for coral reef fishes?

The first question we may ask is, How specialized are reef fishes? Reef fish exhibit both food and habitat specialization, but often several species are found within one restricted niche, and many generalist feeders are present (Sale 1977, 2002). **Table 2** gives a sample of data on the feeding specializations of butterfly fishes. Herbivorous fishes are more generalized feeders than predatory fishes. But even among the specialist feeders, it is common to find two or three species with identical specializations, as seen in Table 2. Thus, feeding niches are not organized as tightly as the niche-diversification hypothesis would predict, and competition for food does not appear to be an important process of importance in these fish communities.

Habitat specialization could be another way that reef fishes have evolved niche differences. Adult reef fishes are very sedentary and thus could have very narrow habitat requirements, but this does not seem to be the case. Habitat partitioning does occur to the extent that few species range over all regions of the reef. Species tend to occur in broadly defined habitats, such as "reef flat" or "surge channel," but when microhabitats are assigned more carefully, extensive overlap of species is observed. Thus, we do not find a high degree of habitat specialization among coral reef fishes (Sale 2002).

Natural history information thus does not tend to support the niche-diversification hypothesis. How can we test these hypotheses experimentally? If the stochastic recruitment hypothesis is correct, then reef fish communities ought to be unstable in species composition and highly variable from reef to reef. Also, the species structure at a given site should not recover following artificial removals or additions of species, and we should be able to predict population size on a reef from the number of recruits that arrive.

To test the first prediction, Talbot et al. (1978) put out standard cement building blocks to create artificial reefs of constant size and shape. Forty-two fish species colonized these artificial reefs, averaging 17 species per reef. The similarity among replicate reefs was only 32% which means that even though these reefs were set out in the same lagoon at the same time within a few meters of one another, only about 32% of the fishes colonizing them were of the same species. A survey of natural coral isolates of about the same size as the artificial reefs (0.6 m^3) showed only a 37% similarity (Talbot et al. 1978). Moreover, as one followed the artificial reefs over time, very high turnover occurred from month to month. Of the species on a reef one month, 20%–40% would have disappeared by the next month and been replaced by a species that was not previously present. Clearly, reef fish communities are very unstable and highly variable from one small reef to the next.

Table 2 Feeding specializations among 20 species of butterfly fishes near Lizard Island, Great Barrier Reef, Australia.

Hard coral	Soft coral and some hard coral	Noncoralline invertebrates	Generalists[a]
Chaetodon aureofasciatus	*Chaetodon kleinii*[b]	*Chaetodon auriga*	*Chaetodon citrinellus*
C. baronessa	*C. lineolatus*	*C. auriga*	*C. ephippium*
C. ornatissimus	*C. melannotus*	*Chelmon rostratus*	*C. ulietensis*
C. plebeius	*C. unimaculatus*	*Forcipiger* spp.	*C. vagabundus*
C. rainfordi	*C. speculum*	*C. trifascialis*	*C. trifasciatus*

[b]The most common food items for this species were polychaetes and crustaceans.

[a]Includes plankton in the diet.

SOURCE: After Anderson et al. (1984).

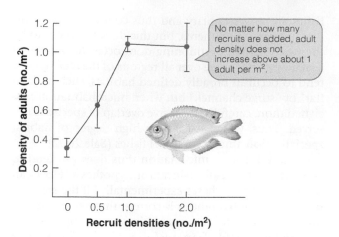

Figure 6 Test of the variable recruitment hypothesis for coral reef fish. Four different recruitment levels were experimentally provided for the damselfish *Pomacentrus amboinensis* for three years. Adult densities increased directly with recruitment, as predicted by the variable recruitment hypothesis, up to 1 recruit/m^2, but above 1 recruit/m^2 density was limited in a density-dependent manner. (From Jones 1990.)

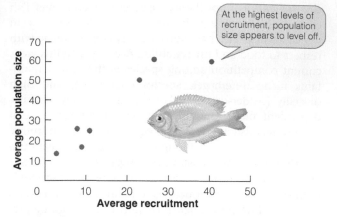

Figure 7 Survey of the recruitment of the damselfish *Pomacentrus moluccensis* on the southern Great Barrier Reef. Ten patch reefs in seven lagoons were surveyed for nine years. Population size units are number of fish per 100 m^2. Measuring recruitment allows predictions of subsequent population densities, as suggested by the variable recruitment hypothesis, but there is an upper limit to density at high recruitment levels. (From Doherty and Fowler 1994.)

The variable recruitment hypothesis assumes the absence of any resource limitations on populations and the lack of competitive effects; population size is set by how many recruits arrive at a reef. One prediction from this hypothesis is that if recruits are experimentally added to a coral reef population, adult numbers will rise. Jones (1990) transplanted juveniles of the damselfish *Pomacentrus amboinensis* for three years to natural patches of reefs of approximately 8 m^2 at the southern edge of the Great Barrier Reef in Australia. **Figure 6** shows that adult densities increased as more recruits were added, but only to a ceiling. At high recruitment levels, density-dependent interactions between adults and potential recruits put a ceiling on numbers, contrary to the predictions of the variable recruitment hypothesis. The important point illustrated by Figure 6 is the existence of both a range of recruitment that fits the variable recruitment hypothesis, and a threshold above which this hypothesis does not hold.

Another way to test the variable recruitment hypothesis is to survey a variety of reefs and measure recruitment rates. Doherty and Fowler (1994) did this for nine years on reefs at seven scattered islands in the southern part of the Great Barrier Reef. They found that average recruitment of the damselfish *Pomacentrus moluccensis* over the nine years was highly correlated with the average population size in each lagoon (**Figure 7**). These results agree with the predictions of the variable recruitment hypothesis. The mortality rate of damselfish after recruitment appeared to be constant rather than density dependent, suggesting a nonequilibrium model.

Discussions among coral reef ecologists about the importance of prerecruitment and postrecruitment processes in limiting adult fish densities have remained controversial because some results favor the variable recruitment hypothesis and others do not (Hixon and Webster 2002). Both recruitment and postrecruitment processes affect the abundance of coral reef fishes. Moreover, the results from these studies may depend on the spatial scale at which the fish populations were observed. Most studies have been done on patch reefs that were small enough to be censused by one or two divers. Forrester et al. (2002) found by the use of simulation models that small-scale data could be used to simulate large-scale population dynamics, but they pointed out that more data are needed at larger scales.

Coral reef fishes are similar to many marine organisms in having a life history that includes a planktonic larval phase that is transported by ocean currents. This type of life cycle implies that local reproduction is not linked to local recruitment, in complete contrast to the life cycle of birds and mammals. For these marine organisms the population or community can never be a closed system, and the physical factors controlling recruitment may control the system. This has been called **supply-side ecology** by Roughgarden et al. (1987). The ideas of supply-side ecology have a long history (Underwood and Denley 1984). The structure of an ecological community driven by supply-side ecology cannot be understood solely as a result of competition, predation, and disturbance, but only by identifying what controls variable recruitment, which keeps populations under the carrying capacity (Underwood and Fairweather 1989).

How is the coexistence of so many species permitted if the variable recruitment hypothesis is correct? Sale (1982) called the reef fish community a **lottery competition**. Individuals compete for access to units of resources (for these fish, space) without which they cannot join the breeding population. A lottery competition is a type of interference competition in which an individual's chances of winning or losing are determined by who gets access to the resource first. Lottery competitive systems are very unstable but can persist if there is high environmental variability in birth rates (Chesson 1986). Because recruitment of reef fishes depends on larvae settling from the plankton, high variability is the rule, and vacated space is allocated at random to the first recruit to arrive from the larval pool (Sale 1982). The lottery competition model has three important requirements if it is to explain the coexistence of many species: (1) environmental variation must be such that it permits each species to have high recruitment rates at low population densities, (2) generations must overlap, and (3) adult death rates should be unaffected by competition (Chesson and Warner 1981).

Thus, the high diversity of coral reef fish communities is not achieved by precise niche diversification in an equilibrium community, but rather by highly variable larval recruitment resulting in a competitive lottery for vacant living spaces in which the first to arrive wins. Reef fish communities are thus not in equilibrium, but instead continually fluctuate in local species composition while retaining high regional diversity.

Rocky Intertidal Communities

The rocky intertidal zone is a tension zone between land and sea featuring disturbances in the form of waves and storms. Space is the key limiting resource in the intertidal zone, and competition for space has been a key component of many studies of this community (Dayton 1971; Sousa 1985). The key concepts of keystone species and the intermediate disturbance hypothesis have their origins in the rocky intertidal zone. Two examples illustrate the ways in which rocky intertidal communities can be organized.

Communities of seaweeds on the rocky coasts of New England are jointly affected by predation and physical disturbances (Lubchenco 1986). Seaweeds in New England are of two groups: ephemeral seaweeds live for weeks or months, grow rapidly, and are eaten rapidly by herbivores such as limpets; perennial seaweeds can live for many years, grow more slowly, and except in their juvenile stages are relatively inedible. Seaweeds compete for space and for light, but the primary resource in the rocky intertidal zone is space. Using wire-mesh cages to exclude herbivores,

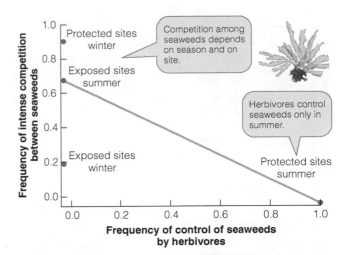

Figure 8 Effects of herbivores on the frequency of competition among ephemeral seaweeds (including *Ulva* sp.) in the rocky intertidal zone of New England. Season and wave exposure are indicated for each point. The effects of herbivores depend on the season and the physical environment. (From Lubchenco 1986.)

Lubchenco (1986) found no simple answer to the question of what controlled seaweed abundance (**Figure 8**). On protected areas in summer, limpet grazing reduced the abundance of ephemeral algae so much that competition for space among the algae was eliminated. On wave-exposed sites, limpets cannot easily live and algae are washed away by wave action, so the amount of competition among seaweeds is reduced. In this system herbivores set the stage for competitive interactions, and the exact dynamics of a small patch of rock depends on the physical environment (wave action) and the number of herbivores present.

Coralline algae are encrusting algae in the rocky intertidal zone that compete with each other via overgrowth. In the absence of herbivores like chitons and limpets on a smooth surface, Paine (1984) found a clear dominance of competitive interactions (**Figure 9**)—that is, competition for space was *transitive*. But in natural communities with and without grazers, three of the coralline algae showed *intransitive* competition, winning some encounters and losing others. Grazers thus act to slow down the rate of succession of coralline algae by inducing competitive uncertainty, and this acts to promote species diversity. In the absence of disturbance (grazing), a single competitive dominant would monopolize space in the rocky intertidal (Paine 1984). This algal community is not an equilibrium assemblage under natural conditions because grazing changes the system from a transitive competitive network with a definite stable outcome to an intransitive network with no stable equilibrium.

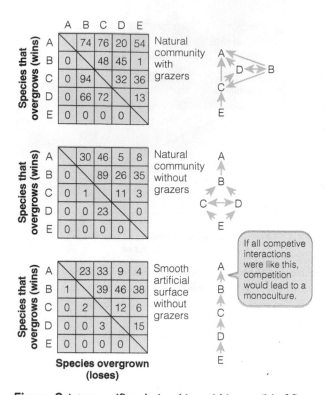

Figure 9 Interspecific relationships within a guild of five coralline algal species under three different conditions: (top) with grazers, on a natural surface; (center) without grazers, on a natural surface; (bottom) without grazers, on a smooth artificial surface. Letters refer to species: A, *Pseudolithophyllum lichenare*; B, *Lithothamnium phymatodeum*; C, *Pseudolithophyllum whidbeyense*; D, *Lithophyllum impressum*; E, *Bossiella* sp. Numbers in the array are observed overgrowths when two guild members come into contact. The diagrammed competitive interactions indicate the change of position induced under the various conditions. Arrows point toward competitive winners. Two-headed arrows indicate that no significant bias exists in the interaction's direction. (From Paine 1984.)

Theoretical Nonequilibrium Models

These two examples of community dynamics illustrate some of the ideas that have been central to the development of nonequilibrium models of communities. Communities can be positioned along a gradient from stable, biotically interactive communities that are equilibrium centered, to unstable, interactive communities in which biotic interactions do not lead to a stable equilibrium, to weakly interactive communities in which physical factors such as temperature, salinity, or fire prevent any stable equilibrium. Models have been developed all along this gradient to describe the ecological complexities of these systems. Chesson and Case (1986) recognize four types of nonequilibrium models of communities:

1. **Fluctuating environment models.** The simplest deviation from the classical equilibrium model of a community is a model with temporal variability. Competition is viewed in these models as the major biological interaction, but the environment changes seasonally or irregularly and the competitive rankings of species also fluctuate such that no one species can win out. These models may include movements between patches, each of which may have a different environment. Fluctuating environment models are similar to equilibrium models in that they produce stable communities, but they differ in emphasizing the dynamics of dispersal among patches and variable life history traits.

2. **Density-independent models.** These models assume that population densities change, but the classical models, fluctuations, are often density independent. Density-vague dynamics predominates in these theoretical communities, and populations are typically at levels at which competition for resources is rare (Strong 1986). Spatial patchiness is added to some of these models as another feature promoting fluctuations in the community.

3. **Directional changing environment models.** Variable environment models usually consider environments to fluctuate about some mean value that remains constant over time. When the mean itself changes, the result heavily depends on the amount of fluctuation and the speed of the community in reacting to change. The current concern over global climate change makes these models very significant for the future. Unlike many community models, these models cannot ignore history, and the response of a community to, for example, changing climate, depends on its past history. Modern communities, these theories argue, cannot be understood only by looking at present environmental conditions. Life history characteristics and dispersal abilities strongly affect the ability of a species to respond to environmental changes.

4. **Slow competitive displacement models.** If competitive abilities of species are nearly equal, the process of competitive exclusion will be very slow, and random variation in success will obscure any obvious displacements. Hubbell and Foster (1986) argue that the tropical rain forest has many species that are ecologically identical, and that community

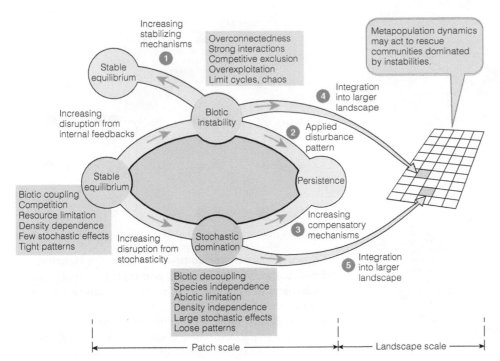

Figure 10 A schematic diagram showing five general mechanisms that may explain why communities tend to persist despite the prevalence of biotic instabilities and environmental fluctuations. (From DeAngelis and Waterhouse 1987.)

composition is the net consequence of a slow random-walk of tree species densities. Competition in these models occurs all the time, but because all species are identical in competitive abilities, there is no time trend or succession. Under these models, community structure is strongly affected by chance and by history, and changes occur only on a geological time scale.

The purpose behind all of these models is to understand what enables a community to persist over time. These four models utilize five general types of mechanisms to explain why communities tend to persist (**Figure 10**). The first mechanism can be characterized as mechanisms that stabilize interactions and that are usually additions to classical equilibrium models of community organization. If predator functional responses are sigmoid, or if consumers are self-regulated, the stability of the community could be increased (DeAngelis and Waterhouse 1987). The second mechanism that promotes community persistence is (paradoxically) disturbances that create nonequilibrium landscapes and prevent competitive exclusion. The third mechanism involves compensatory changes in reproduction, survival, or movements when populations reach low densities. Such changes could favor rare species over common ones. The fourth and fifth mechanisms that promote community persistence both involve spatial patchiness. If local populations are connected into metapopulations at the landscape level

(see Figure 10), the fact that each local patch is unstable may not matter, because species can recolonize by dispersal between patches. Stability in metapopulations is seen at the landscape level, not at the local population level. Species may go extinct in local patches, but as long as local patches are out of phase with one another, the species will persist in the landscape.

To translate these ideas on nonequilibrium community dynamics into the real world, ecologists have developed a series of conceptual models that can be tested in field studies.

Conceptual Models of Community Organization

A series of models have been proposed by field ecologists to try to capture the interrelations between physical factors and biological interactions in organizing communities. All of these models recognize that many different kinds of ecological communities exist and that the important processes will not be the same in all ecological systems.

The most comprehensive model of community organization was proposed by Menge and Sutherland (1987). They recognized three ecological processes as the main determinants of community organization—physical disturbance, predation, and competition—and they included variable recruitment as a part of the model (**Figure 11**). A central assumption of this model is that food web

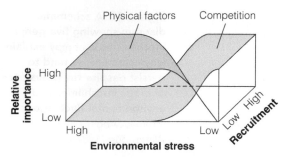

(a) Top level (carnivores)

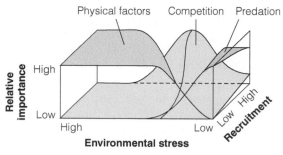

(b) Intermediate level (herbivores)

> In high stress environments physical factors always predominate.

> Predation on plants includes herbivory.

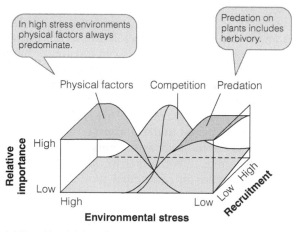

(c) Basal level (plants)

Figure 11 The Menge-Sutherland model, in which three factors—interspecies competition, predation, and physical factors—drive community organization. The relative importance of these factors changes with trophic level, harshness of the environment, and level of recruitment. (From Menge and Sutherland 1987.)

> One test of these two models would be to see how common competition is between plants in benign environments.

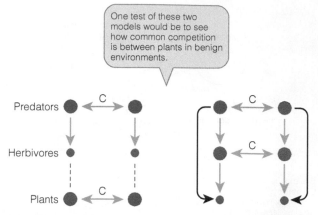

(a) Hairston-Smith-Slobodkin model   (b) Menge-Sutherland model

Figure 12 Schematic comparison of (a) the Hairston-Smith-Slobodkin (HSS) model and (b) the Menge-Sutherland (MS) model of community organization for benign environments. Size of the circles indicates the relative abundance of the trophic levels. Vertical arrows indicate feeding relationships, whereas horizontal (C) arrows indicate competition. Dashed lines indicate weak interactions. In a world driven by the HSS model, herbivores should be rare and plants abundant. In a world driven by the MS model, herbivores should be abundant and plants relatively rare. (Modified after Pimm 1991.)

complexity decreases with increasing environmental stress. This model makes three predictions for communities that have high recruitment. First, in stressful environments herbivores have little effect because they are rare or absent, and plants are regulated directly by environmental stress. Neither predation nor competition is significant. An example of such a community is the arctic tundra or a desert. Sec-

ond, in moderately stressful environments, consumers are ineffective at controlling plants, and plants attain high densities. Competition among plants is the dominant biological interaction in these communities. Third, in benign environments, consumers control plant numbers, and plant competition is rare. Predation is the dominant biological interaction under these benign conditions.

The Menge-Sutherland model is a general model of community organization that incorporates some ideas from a model proposed by Hairston, Smith, and Slobodkin (1960). Hairston et al. predicted that for terrestrial communities in benign environments, predators must limit herbivores, which are then unable to limit plants. Consequently, plants compete for nutrients and light, but herbivores do not compete. (They restricted their model to herbivores that feed on green plants and excluded seed and fruit eaters.) This model predicts that whereas herbivore removals will have little effect on plants, predator removals will strongly affect herbivore numbers. This model has also been called the "green-world hypothesis."

The Hairston-Smith-Slobodkin (HSS) model makes the same predictions as the Menge-Sutherland (MS) model for predator removal experiments, but not for herbivore removals. **Figure 12** compares these two

Table 3 The percentage of field experiments on predation showing large significant effects as a function of trophic level and system.

| | Species removed | | |
System	Herbivore	Primary carnivore	Secondary carnivore
Intertidal	84 (120)	70 (67)	—
Other marine	95 (82)	68 (57)	—
Lakes	—	73 (22)	75 (95)
Rivers and streams	—	61 (106)	55 (29)
Terrestrial	74 (112)	61 (36)	—

"Large effects" is defined as twofold or greater changes. Numbers in parentheses are sample sizes (= number of studies).
SOURCE: From Sih et al. (1985).

models and shows that the Menge-Sutherland model assumes omnivory to be a common feature of the food web (Pimm 1991). We can test these models by measuring the frequency of competitive effects in different communities in benign environments. Hairston et al. (1960) predict intense competition among plants, whereas Menge and Sutherland predict little competition among plants. Similarly, herbivores in benign environments are expected to compete in the MS model but rarely compete in the HSS model (see Figure 12).

Predator removal experiments provide one way to test these two models. Herbivore removals should to have no effects on plants, if the HSS model is correct (see Figure 12). Sih et al. (1985) surveyed removal experiments; the results are given in **Table 3**. The vast majority of herbivore removal experiments both in marine and terrestrial systems had large effects on the plants. This evidence favors the MS model and is contrary to the HS model.

Freshwater ecologists have proposed several models for community organization in freshwater ecosystems that parallel the MS and the HSS models. The key to these conceptual models is to consider the interactions between any two adjacent trophic levels. For example, for vegetation (V) and herbivores (H), three possible relationships are possible:

$$V \rightarrow H \qquad V \leftarrow H \qquad V \leftrightarrow H$$

$\rightarrow$ means that an increase in vegetation will cause an increase in the numbers or biomass of herbivores, but not vice versa. Similarly, $\leftarrow$ means that an increase in herbivore numbers will cause an effect on vegetation (a decrease), but not vice versa. A $\leftrightarrow$ means that an increase in vegetation will cause an increase in herbivore numbers and an increase in herbivore numbers will decrease the vegetation (a reciprocal interaction).

Given these simple interactions, we can define two polar views of community organization: the **bottom-up model** and the **top-down model**. The bottom-up model postulates $V \rightarrow H$ linkages, which means that nutrients control community organization because nutrients control plant numbers, which in turn control herbivore numbers, which in turn control predator numbers. The simplified bottom-up model is thus $N \rightarrow V \rightarrow H \rightarrow P$. By contrast, the top-down model postulates that predation controls community organization, because predators control herbivores, which in turn control plants, which in turn control nutrient levels. The simplified top-down model is thus $N \leftarrow V \leftarrow H \leftarrow P$. The top-down model has been called the **trophic cascade model** by Carpenter et al. (1985). The cascade model, which is similar to the HSS model (see Figure 12a), predicts for strong interactions among species a series of $+/-$ effects across all the trophic levels. Thus predators will strongly depress herbivore numbers, and depressed herbivore numbers will have only a minor effect on plant abundance, so the abundant plants can strongly depress nutrients. The trophic cascade model predicts for freshwater systems with four trophic levels that removing the top (secondary) carnivores will increase the abundance of primary carnivores, decrease herbivores, and increase phytoplankton.[1] The effects of any manipulation will thus be propagated down or up the trophic structure as a series of positive or negative effects.

The top-down and the bottom-up models are clearly not the only models that can be postulated for food chains. Sinclair et al. (2000) derived 27 different models from various combinations of $\rightarrow$, $\leftarrow$, and $\leftrightarrow$ arrows. For example, a *pure reciprocal model* would postulate two-way effect at all trophic levels: $N \leftrightarrow V \leftrightarrow H \leftrightarrow P$. The important point is that we can start with two simple models, but we must realize that many intermediate types of models are in fact possible for communities, and that it is unlikely that all communities will fit only one model.

These models immediately suggest experimental manipulations of communities to search for trophic-level effects. Many experiments have now tested these models in freshwater, marine, and terrestrial ecosystems.

---

[1] In aquatic systems, secondary carnivores are piscivorous (fish-eating) fish, and primary carnivores are planktivores (zooplankton-eating fish).

Biomanipulation of Lakes

Many freshwater lakes have been degraded by pollution, and one of the major thrusts of applied aquatic ecology has been to devise methods to aid lake recovery from pollution. The trophic cascade model of lake communities immediately suggests ways of improving water quality. In lakes with four trophic levels, adding top predators should improve water quality by reducing algal populations. In lakes with three trophic levels, removing fish should improve water quality. This tool for lake restoration, called **biomanipulation**, is illustrated in **Figure 13**. Many attempts at lake restoration using biomanipulation have been made, but mixed results have been obtained, possibly related to variation in the depth and size of the lake (Olin et al. 2006).

Horppilla et al. (1998) described one of the largest food web manipulation trials yet carried out. Lake Vesijärvi in southern Finland is a large lake (110 km²) with a mean depth of only 6 m. It was heavily polluted with city sewage and industrial wastewater until 1976; once these inputs were stopped, its water quality began to recover. But by 1986 massive blooms of blue-green algae began to appear, and these algal blooms coincided with a very dense population of roach, a planktivorous cyprinid fish that had built up during the years of nutrient input. To reverse these changes, from 1989 to 1993 some 1018 tons of fish were removed from Lake Vesijärvi, reducing roach to about 20% of their former abundance. At the same time the lake was stocked with pikeperch, a predatory fish that feeds on roach, adding a fourth trophic level to the lake.

Biomanipulation was a success in Lake Vesijärvi; the water became clear and blue-green algal blooms stopped in 1989. The lake remained clear for seven years after fish removal had stopped in 1993 but then began to deteriorate after 2000 as fish populations increased again and more people moved into the drainage basin of the lake. In

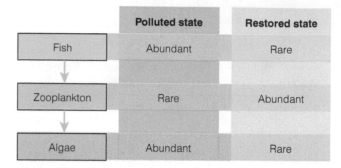

Figure 13 The simplified trophic cascade model that underlies the concept of biomanipulation of lakes to restore them to a relatively clear-water condition from a state of high turbidity and pollution.

2002 a new program of biomanipulation was begun along with attempts to control nutrient inputs from new human settlements in the lake basin in order to maintain the clear water status of the lake.

The mechanism for successful treatment from these biomanipulations was not as suggested in the trophic cascade shown in Figure 13 because zooplankton density in the lake did not change over the period of lake manipulation, and the same zooplankton species were present. The reduction of algal blooms was achieved because nutrient-rich excretion by roach was greatly reduced, and it was this source of nutrients that was stimulating the excessive algal growth in the lake. An additional nutrient pathway from fish excretion directly to the phytoplankton could be an important additional mechanism for achieving of lake restoration. Lake Vesijärvi may be an example of a lake with two alternate stable states defined by nutrient transfer from fish directly to algae.

Figure 14 illustrates the overall impacts of bottom-up manipulations of ecosystems with fertilizers, and the top-down effects from predator removal studies (Borer et al. 2006).

Trophic cascades can also occur in terrestrial communities. The most striking cascades are associated with specialist large predators with simple food webs. In Zion National Park, increasing human usage has caused cougars (mountain lions, *Puma concolor*), the apex preda-

tor, to avoid large areas of the Park, thus producing a trophic cascade illustrated in **Figure 15**. Ripple and Beschta (2006) showed that reduced cougar densities led to higher mule deer densities, which increased browsing density on riparian cottonwood trees, and the lack of cottonwood tree recruitment reduced soil stability and caused the stream banks to erode during high river flows. The erosion of stream banks in turn has led to a 75% reduction of streamside plants like rushes and cattails and

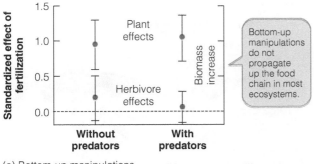

(a) Bottom-up manipulations

Bottom-up manipulations do not propagate up the food chain in most ecosystems.

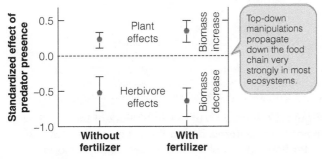

(b) Top-down manipulations

Top-down manipulations propagate down the food chain very strongly in most ecosystems.

Figure 14 Summary of the effect size for 121 community studies that manipulated nutrients (a) by adding fertilizer to test for bottom-up effects, and (b) by removing predators to test for top-down effects. Bottom-up manipulations have large effects on plant biomass but near zero effect on herbivore abundance. By contrast, top-down manipulations show strong negative effects on herbivores when predators are present, and at the same time moderate positive effects on plant biomass. (Data from Borer et al. 2006.)

a hundredfold drop in the abundance of frogs and toads along the streams (**Figure 16**). The surprising point about the Zion trophic cascade is the extensive changes it has caused throughout the food web, all stemming from the reduction of the apex predator, cougars.

Aquatic food webs in rivers provide a good model system for testing predictions of community organization models. **Figure 17** shows the food web of boulder-bedrock parts of the Eel River in northern California. Four trophic levels occur in this community, and by constructing cages in the river, Power (1990) was able to measure the effects of removing fish on community dynamics. **Table 4** summarizes the results of these manipulations and how they relate to the predictions of the two major models of community organization. The observations fit the trophic cascade or HSS model rather than the Menge-Sutherland model

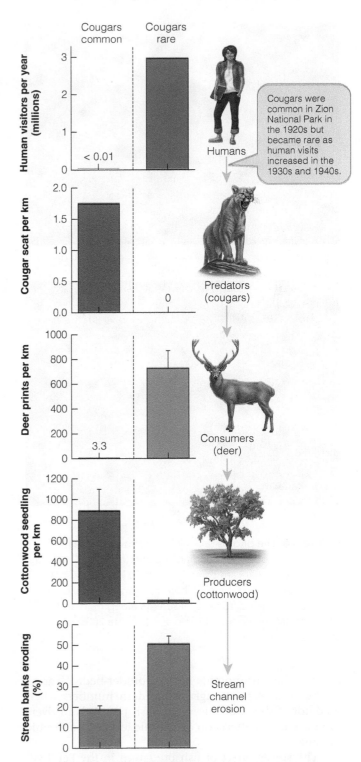

Cougars were common in Zion National Park in the 1920s but became rare as human visits increased in the 1930s and 1940s.

Figure 15 Trophic cascade in Zion National Park in Utah during the last century. The inverse patterns of abundance from apex predators to producers has resulted in destabilization of the stream banks and a loss of biodiversity in the streamside vegetation.

(a)

(b)

Figure 16 The photos taken in 2005 show the stream channel and floodplain conditions along (a) North Creek, an area where cougars are common, and (b) North Fork of the Virgin River in Zion Canyon, where cougars are rare. Census data were obtained from visual surveys along trails and transects. (Modified from Ripple and Beschta 2006; photos courtesy of Bill Ripple.)

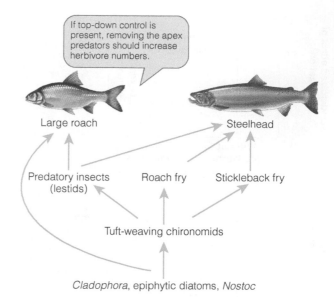

Cladophora, epiphytic diatoms, *Nostoc*

Figure 17 Food web of the South Fork of the Eel River in northern California during the summer period of low water flow. Four trophic levels are present. By removing the top predators it is possible to test the trophic cascade model of community organization. (From Power 1990.)

because the chironomids in the boulder-bedrock areas of the river were strongly reduced in numbers when predatory fishes were excluded. Fish removals in rivers can have major effects on all trophic levels via a trophic cascade.

The strong effect of fish predation in the Eel River was limited to boulder-bedrock substrates (Power 1992). In gravel beds of the river, fish predation had very little effect on algae or invertebrate abundance. Fish predation is relatively inefficient in gravel bars because invertebrate prey are relatively scarce and can hide in the spaces within the gravel.

The results of species removals are complex because the interactions among species are complex and may be habitat specific, as we have just seen. Food webs can also be affected directly by disturbances. Riverine food webs in northern California are of two types. In rivers regulated by dams, large caddisfly larvae become abundant because they have gravel cases that make them invulnerable to fish predators, and they graze algae to low levels. In unregulated rivers, floods move rocks, killing many caddisflies and, by reducing their numbers, allowing algae to increase (**Figure 18**). The food webs of dammed rivers thus change dramatically from their original composition, with implications for fish predators such as juvenile salmon (Wootton et al. 1996). The effects of disturbances in changing the dynamics of food webs can parallel those observed in species removal experiments. With frequent disturbances the community structure will differ dramatically from that expected under disturbance-free equilibrium conditions.

The Special Case of Island Species

Islands can be viewed as special kinds of traps that catch species that are able to disperse there and colonize successfully. Since Darwin's visit to the Galápagos

Table 4 Changes in the abundances of the lower three trophic levels when top predators are removed from the community.

| | Predicted changes in abundance | | |
Trophic level	Menge-Sutherland model	Trophic cascade model	Observations from Eel River, California
Plants (algae)	Increase	Increase	Increased about threefold for *Cladophora* and 120-fold for *Nostoc*
Herbivores (chironomids)	Increase	Decrease	Decreased about 80%
Primary carnivores (insects, fish fry)	Increase	Increase	Increased about tenfold

All changes are related to the intact, four-trophic-level food web illustrated in Figure 17.

SOURCE: Data from Power (1990).

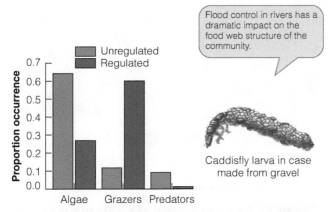

Caddisfly larva in case made from gravel

Flood control in rivers has a dramatic impact on the food web structure of the community.

Figure 18 Food web structure of northern California rivers, illustrating a trophic cascade driven by flooding. In regulated (dammed) rivers in which no flooding occurs, algae are relatively scarce, grazers (caddisflies) are abundant, and predators are scarce during the summer. Disturbance by flooding (scouring) reduces the caddisflies (which allows other insect grazers to increase) and releases the algae from grazing. Flooding produces effects similar to those observed in a grazer-removal experiment. (Data from Wootton et al. 1996.)

Islands, biologists have been using islands as microcosms for studying evolutionary and ecological problems. Because they are bounded, islands are useful for analyzing community structure, and for determining the role of disturbances in affecting communities.

One of the oldest generalizations of ecology is that the number of species on an island is related to the size of the island. Alexander von Humboldt wrote in 1807 that larger areas harbor more species than smaller ones. This phenomenon can be seen most easily in a group of islands like the Galápagos (**Figure 19**). The relationship between species and area can be described by a simple equation called the **species-area curve**:

$$S = c A^z \quad (1)$$

or, taking logarithms,

$$\log S = (\log c) + z(\log A) \quad (2)$$

where
S = Number of species
c = A constant measuring the number of species per unit area
A = Area of island (in square units)
z = A constant measuring the slope[2] of the line relating to S and A

For the Galápagos land plants shown in Figure 19 we obtain

$$S = 28.6 \, A^{0.32}$$

For these plants, the slope (z) of the species-area curve is 0.32, and the number of plant species on 1 km^2 of island (c) is predicted to be 28.6 on average.

Rosenzweig (1995) has championed the species-area relationship as a fundamental ecological law, claiming that the species-area curve is a useful descriptive model for both plants and animals. **Figure 20** illustrates this basic relationship between species and area for the amphibian and reptile fauna of the West Indies, where the relationship is

$$S = 3.3A^{0.30}$$

---

[2]Many species-area curves are reported in English units rather than metric units. Because the scales are logarithmic, the z value (slope) is independent of scale and thus does not depend on whether English or metric units are used. The c value, however, is completely dependent on the units used to measure area.

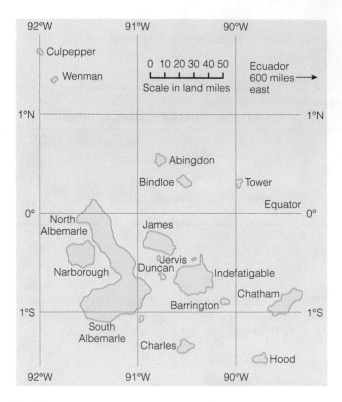

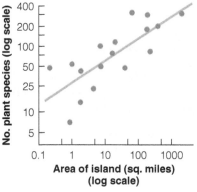

Figure 19 Number of land-plant species on the Galápagos Islands in relation to the area of the island. The islands range in area from 0.2 to 2249 mi² (0.5–5850 km²) and contain from seven to 325 plant species. (After Preston 1962.)

Preston (1962) noted that the slope of the species-area curve (z) tended to be about 0.3 for a variety of island situations, from beetles in the West Indies and ants in Melanesia to vertebrates on islands in Lake Michigan and land plants on the Galápagos. This raises an interesting question: What is the species-area curve for continental areas? Is its slope the same as that for islands—and is z some sort of ecological constant?

The number of species increases with area on continental areas as well as on islands, but there is much variation in the slope of species-area relationships. Species-area curves from continental samples typically

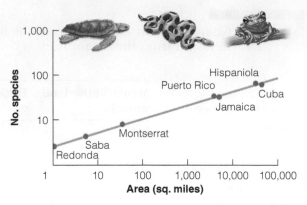

Figure 20 Species-area curve for the amphibians and reptiles of the West Indies. The slope of the species-area curve (z) is 0.30. Note that the axes are on a log scale. (After MacArthur and Wilson 1967.)

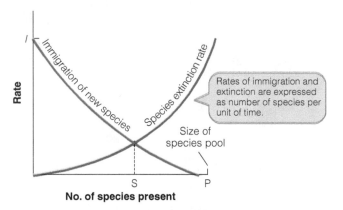

Figure 21 Equilibrium model of a biota of a single island. The equilibrial number of species (S) is reached at the point at which the curve of immigration of new species to the island intersects the curve of extinction of species on the island. (After MacArthur and Wilson 1967.)

show a lower slope (z) than are found in samples from islands. Why might this be, and what are the ecological mechanisms that determine the number of species on islands or parts of continents?

The number of species living on any plot, whether an island or an area on the mainland, is a balance between immigration and extinction. If the immigration of new species exceeds the extinction of old species already present, the plot or island will gain species over time. Thus, we can treat the problem of species diversity on islands by an extension of the approach used in population dynamics, in which changes in population size were produced by the balance between immigration and births on the one hand, and emigration and deaths on the other hand. MacArthur and Wilson (1967) developed this theoretical approach in detail, and **Figure 21** shows the simplest model from their theory of island biogeography.

The immigration rate, expressed as the number of new species per unit time, falls continuously because as more species become established on the island, most of the immigrants will be from species already present there. The upper limit of the immigration curve is the total fauna for the region. The extinction rate (the number of species disappearing per unit time) rises because the chances of extinction depend on the number of species already present. The point at which the immigration curve crosses the extinction curve is by definition the equilibrium point for the number of species on the island.

The shapes of the immigration and extinction curves are critical for making any predictions about island situations (see *Working with the Data: Measuring Immigration and Extinction Rates*). Assume for the moment that distance will affect the immigration curve only; near islands will receive more dispersing organisms per unit time than will distant islands. Assume also that extinction rates will differ between small islands and large islands such that the chances of becoming extinct are greater on small islands. **Figure 22** illustrates these assumptions and shows why more-distant islands should have fewer species than nearer islands (if island size is constant), and why small islands should have fewer species than large islands (if distance from the source area is constant).

We need long-term studies to obtain observational data to test the MacArthur-Wilson model, and so far most of the data available come from bird studies. Britain is surrounded by many islands of various sizes for which bird census data are available for 50 years or more (Russell et al. 1995). **Figure 23** shows the observed immigration and extinction curves for two

British islands. A central assumption of the MacArthur-Wilson model is that the immigration and extinction functions are concave, and while this seems to be correct for many islands, there are exceptions to the rule, and much variation occurs from year to year.

The relationship of species richness to island area is strong but it has always been questioned because there is

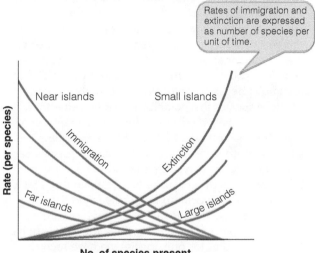

Figure 22 Equilibrium model of biotas of several islands of various sizes and distances from the principal source area. An increase in distance (near to far) lowers the immigration curve; an increase in island area (small to large) lowers the extinction curve. An equilibrium of species richness occurs at each intersection point of the immigration and extinction curves. (After MacArthur and Wilson 1967.)

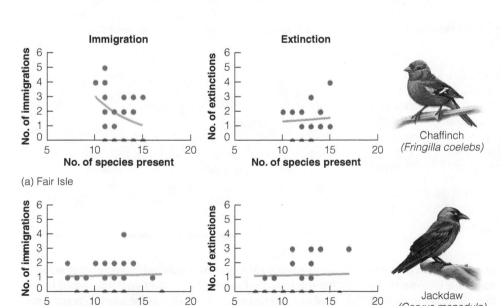

(a) Fair Isle

(b) Skokholm

Chaffinch
(*Fringilla coelebs*)

Jackdaw
(*Corvus monedula*)

Figure 23 Island biogeography theory applied to British birds on two islands: (a) Fair Isle and (b) Skokholm. Immigration curves are represented in blue; extinction curves, in red. Consecutive annual censuses established gains and losses for each island. The MacArthur-Wilson model assumes concave curves for these relationships. Extinction curves rise with species richness as predicted, but immigration curves do not always fall as predicted; note the considerable noise in the data. (Data from Manne et al. 1998.)

Measuring Immigration and Extinction Rates

Immigration and extinction rates are key to all of the predictions of the MacArthur-Wilson theory of island biogeography, but these rates are not easy to measure. The most common approach is to count all the species on an island each year during the breeding season. If a species was present last year and is absent this year, it is counted as an extinction, and conversely if a species that is present this year was absent last year, it is counted as an immigration. But this simple arithmetic belies some ecological complexity. For example, a species that colonized the island after the last census and then died out before the current census would not be counted. More problems arise if the census cannot be done every year, because species can come and go, or even come and go and come back again, between the censuses. In measuring these rates there can be no substitute for detailed field data.

If we conduct an annual census, we can estimate immigration and extinction rates as follows. Assume for simplicity that for one species we have records of presence (P) and absence (A) in a series of 30 years, as follows:

PPPPPPAAPPPAPPAAAAAPPPAPAPPPAA

We define:

k = number of transitions from absence to presence = 5

l = number of transitions from absence to absence = 6

m = number of transitions from presence to absence = 6

n = number of transitions from presence to presence = 12

Given these raw data, there are two rates that follow directly from the observed transitions. The first, the immigration rate λ, is the probability that a species not present in the community will enter it in a given time interval (usually a year). Immigration rate is estimated as

$$\lambda = \frac{k}{k + l} \qquad (3)$$

The second rate, the probability that a species becomes absent after being present the previous year (δ), is estimated as

$$\delta = \frac{m}{m + n} \qquad (4)$$

From these rates we can estimate that for a very long series of observations, the population will be absent from the island for the fraction $\delta/(\lambda + \delta)$ of the total number of censuses.

To estimate the extinction rate we note that a species may go extinct and recolonize in between the census times, such that it appears from the records that nothing has changed. To take this into account we note that

$$\begin{pmatrix} \text{Probability} \\ \text{of} \\ \text{becoming absent} \end{pmatrix} = \begin{pmatrix} \text{Probability} \\ \text{of} \\ \text{extinction} \end{pmatrix} \begin{pmatrix} \text{Probability} \\ \text{of not} \\ \text{recolonizing} \end{pmatrix}$$

$$\delta = \mu(1 - \lambda) \text{ or} \qquad (5)$$

$$\mu = \frac{\delta}{1 - \lambda}$$

where μ is the extinction rate. For these hypothetical data, $\lambda = 5/11$ or 0.45, and $\delta = 6/18$ or 0.33, so the extinction rate μ is 0.33/0.55 or 0.60. For this hypothetical species on this island, clearly the extinction rate is on average greater than the immigration rate. We expect that in the long run the species will be absent from the island for $\delta/(\lambda + \delta)$ or 42% of the censuses.

If we assume there is a common immigration rate and extinction rate for all species, these transitions can simply be added to obtain estimates for the island (see Clark and Rosenzweig 1994). These annual rates are all probabilities and thus range from 0 to 1.0. We can use these rates to answer questions about the likelihood of extinction of individual species or of species groups on islands, or questions about whether the immigration rate varies with island size.

If a census is not conducted each year, so that only sporadic records of presence and absence are available, it is more difficult to estimate rates of immigration and extinction because much can happen between census periods. Clark and Rosenzweig (1994) discuss this problem of estimation as well.

---

no clear mechanistic explanation of what area represents. Wright (1983) suggested that instead of area, one might use available energy to predict the number of species. *Species-energy theory* has been applied to the global patterns of species richness discussed earlier (Hawkins et al. 2003). Wright (1983) showed that species richness in birds on islands could be predicted by the energy arriving on the islands (**Figure 24**). Net energy, measured by actual evapotranspiration (AET), can be used to predict primary production on each island.

The species-energy theory for island biodiversity can be combined with the species-area relationships

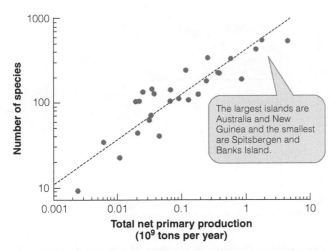

Figure 24 Number of breeding land and freshwater bird species on 28 islands around the world in relation to the annual net primary production on the island. For these data energy, estimated by net primary production, is a better predictor of the number of bird species than is the island area. (Data from Wright 1983.)

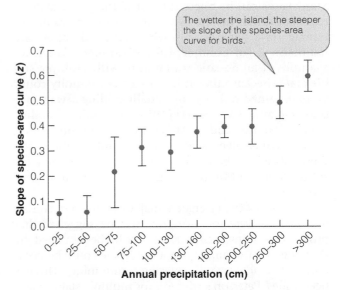

Figure 25 Species-area curves for birds on islands show an increased slope (z) for islands that have higher rainfall. The slope is thus not a universal constant in species-area relationships, and a comprehensive model must include precipitation as well as island area. The same effect occurs with island temperature. (Modified from Kalmar and Currie 2006.)

in the MacArthur-Wilson theory to provide a more comprehensive model of island biogeography (Kalmar and Currie 2006). The slope of the species-area curve changes systematically with rainfall (**Figure 25**), and a comprehensive model that includes island area, precipitation, distance to a continent, and temperature can explain statistically 88% of the variation in bird species richness in 346 islands scattered around the globe (Kalmar and Currie 2006).

The MacArthur-Wilson theory stimulated much work on island faunas, and the species-energy theory has extended this theory to link with theories explaining tropical-polar gradients in biodiversity. By concentrating on predictions of the *number* of species on islands, these models have ignored the more difficult questions of *which* particular species will occur where, which is often the question conservationists ask. Habitat heterogeneity is a major determinant of the species-area curve, and detailed studies of habitats are needed to further our understanding of islands (Rosenzweig 1995). Individual species differ greatly in their abilities to occupy islands, and by understanding the population and community dynamics of individual species we can improve our understanding of island faunas and floras.

Multiple Stable States in Communities

If communities are not equilibrium assemblages of species, their composition will change over time and we will not observe a single stable configuration. But if

natural communities can exist in multiple stable states, changes in community composition that appear to be nonequilibrial may instead be the result of two or more alternative states for the same community (Sutherland 1990). What evidence is required to demonstrate that natural communities have multiple stable states? Connell and Sousa (1983) defined the following criteria that must be met if at a given time a community exhibits multiple stable states:

- The community must show an equilibrium point at which it remains, or to which it returns if perturbed by a disturbance.

- If perturbed sufficiently, the community will move to a second equilibrium point, *at which it will remain after the disturbance has disappeared.*

- When multiple stable states are believed to exist, the abiotic environments of the community must be similar for the various sites.

- The community on both sites that are postulated to be alternate stable states must persist for more than one generation of the dominant species.

A community may show multiple stable states on a single site over time or simultaneously at two or more sites. By applying these criteria, Connell and Sousa (1983) questioned many of the examples of communities purported to be multiple stable states.

Many cases can be rejected because the physical environment differs on the two sites. In other cases the alternate state persists only when artificial inputs are maintained. For example, Lake Washington is not an example of an aquatic community with multiple stable states because the enriched lake community could be maintained only by continually adding sewage nutrients. Connell and Sousa (1983) concluded that they could find no studies of natural communities that showed conclusive evidence for multiple stable states. They issued a challenge to ecologists to search for solid evidence of multiple stable states in natural communities.

Peterson (1984) argued that Connell and Sousa's four criteria were too stringent, and he suggested that proof of multiple stable states could be obtained by showing experimentally that a single site could be occupied by two or more self-replacing communities. This has been called Peterson's criterion for multiple stable states and is the minimum requirement for the demonstration of multiple stable states (Sousa and Connell 1985). **Figure 26** illustrates these ideas schematically.

Marine communities could show multiple stable states but few analyses have been able to demonstrate Peterson's criterion for particular systems (Petraitis and Dudgeon 2004). One possible ecosystem showing alternate stable states is the rocky intertidal zone in the Gulf of Maine. Two communities on sheltered shores may be alternate stable states—mussel beds and rockweed stands. Rockweeds are thought to be poorer competitors for space in the rocky intertidal zone compared with mussels, and can dominate only where mussels are controlled by predators. To test for alternate states, Petraitis and Dudgeon (2005) cleared rockweeds (*Ascophyllum nodosum*) from 12 sites ranging from 1 to 8 m in diameter to mimic the impact of ice scours, and

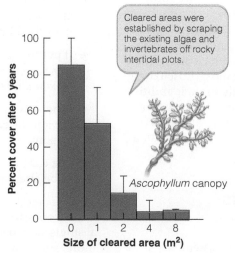

Figure 26 Schematic illustration of expected experimental outcomes for the rocky intertidal zone in the Gulf of Maine in which mussel beds and seaweeds may possibly exist as alternate stable states. At the start of the experiment all plots are dominated by seaweeds. If there is a single stable state, after the perturbation (red arrow) the community will return to its original configuration of seaweed dominance. If there are two stable states some plots will flip over the threshold and move into an alternative state dominated by mussels. Succession in this case has two divergent pathways. (Modified from Petraitis and Dudgeon 2004.)

then followed the succession for six years to see if they could obtain results similar to those diagrammed in Figure 26. After six years the largest cleared sites were dominated by the alga *Fucus vesiculosus* and the barnacle *Semibalanus balanoides*, and there was a clear successional divergence between small and large cleared areas (**Figure 27**). But after eight years there was no sign

Figure 27 Percentage cover in 2004 by fucoid algae after removal of the dominant alga *Ascophyllum nodosum* in 1996 from 60 intertidal plots in the Gulf of Maine. In the smaller sites (a) *Ascophyllum* recovers dominance, but on the larger areas (b) *Fucus* dominates, and this could possibly lead to an alternate stable state if succession continues in divergent directions. (Data from Petraitis and Methratta 2006.)

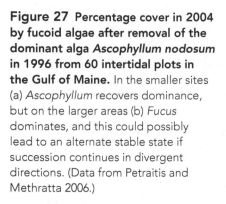

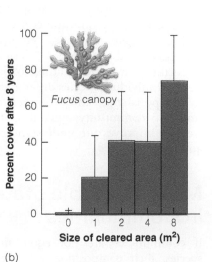

(a)

(b)

of a convergence to an alternate stable state dominated by mussels on any of the sites, and Bertness et al. (2002) questioned whether mussel beds and rockweed stands were alternate stable states and suggested that they were two communities that developed in different environmental conditions. What is clear is that this question will be answered only after experimental studies that are much longer than eight years because succession moves very slowly in the rocky intertidal zone on temperate coastlines.

Woodlands and grasslands of savanna regions around the world have been considered as alternate stable states maintained by fire and herbivory (van Langevelde et al. 2003). The key feedback loop is between grass biomass and fire intensity. The greater the grass growth, the more severe the fire intensity. The second feedback is between fire intensity and woody shrub and tree growth. The more severe the fire intensity, the greater the mortality of shrub and tree seedlings. Humans enter this set of feedbacks in several ways (**Figure 28**). Grazing by cattle removes grasses and herbs that form the fuel load, so that fire intensity

changes. Fire suppression for the protection of property also reduces the frequency of fires, but fires can also be started to move the vegetation in a desired direction (Noble 1997). Shrubs and trees can be removed mechanically to stimulate grass growth for cattle (**Figure 29**). All of these interventions push succession in different directions (Westoby et al. 1989).

White-tailed deer may be creating alternate stable states of woody plant communities in the eastern United States (Stromayer and Warren 1997). Between 1890 and 1920 much of the hardwood forests in Pennsylvania were clear-cut. These stands contained valuable trees such as white ash, sugar maple, red maple, and black cherry. Deer populations increased rapidly in the regenerating stands, due to increased browse. At the same time, predators such as wolves were removed from

(a)

(b)

Figure 29 Two alternate stable states in savanna grassland-woodlands of eastern Australia. (a) Grassland with scattered woody plants. (b) Dense shrub cover with little grass. Pastoralists wish to have their grazing lands in the grassland state, but grazing removes fuel loads and pushes the community toward dense woodland, often referred to as "woody weeds." (Noble 1997; Hill et al. 2005.)

Alternate stable state | Transient states | Alternate stable state

| STATE I Grassland with scattered woody shrubs and trees | →1→ | STATE II Grassland with many shrub seedlings | →2→ | STATE III Dense shrub cover with little grassland |

5 | | 3 4

STATE IV Fire with many shrub seedlings and resprouted trees

Figure 28 Two alternate stable states (dark green) for savannas and the transitions (light green) between them. Grasslands and woodlands are connected by two transient states that can move in one direction or another depending on disturbances. For semiarid rangelands of eastern Australia, Transition 1 involves two or more good rainfall years, and this transition could be reversed by fire that kills the shrub seedlings in transient state II. Transition 2 in the absence of fire over 10–20 years is inevitable (in the absence of small browsing marsupials, see Noble et al. 2007). Transition 3 occurs if high rainfall provides a fuel load of ephemeral plants, but transient state IV can move back via Transition 4 to dense shrubs over 5–15 years if there are no fires. Transition 5 back to grassland occurs if enough rainfall is repeated so that fire can propagate to suppress the shrub seedlings. (Modified after Westoby et al. 1989.)

the system, so deer numbers skyrocketed such that deer are now considered "overabundant." Deer browsing has been shown to reduce hardwood regeneration, particularly of valuable timber species. With sufficient browsing the seed bank of these hardwoods becomes exhausted within three or four years, and no regeneration is possible. Ferns and grasses invade the forest floor and completely suppress regeneration of desirable hardwoods (Tilghman 1989). The result is a community of trees dominated by black cherry and containing other trees less preferred by foresters and less browsed by deer, an alternate tree community that may be stable on the time frame of 300 years or more.

Multiple stable states may occur in other communities affected by humans and may be confused with nonequilibrium communities. In some of these cases the community may revert to its original configuration once human disturbance is removed, but in others the community may become locked into a changed configuration even after the disturbance is stopped. For the purposes of conservation and land management, it is important to determine which model of community organization applies to natural communities. We cannot assume that all communities subjected to human disturbance will return to their original configuration once the disturbance is ameliorated.

Summary

Many community ecologists question the existence of a single equilibrium state for biological communities. Patchiness is an inherent property of natural communities, and the disturbances that lead to patchiness have been the main focus of nonequilibrium theories of community organization. Nonequilibrium communities exist when disturbance intervals are shorter than recovery times, so the community never reaches equilibrium. Disturbances may include fires or weather events, as well as grazing, land clearing, predation, disease, or other biotic events.

Coral reefs were once thought to be classic examples of stable, equilibrium communities, but careful long-term studies have shown that reefs vary dramatically due to disturbances caused by cyclones and oceanographic changes resulting from weather fluctuations. Coral reef fish communities may be driven by the lottery of variable recruitment, which can cause irregular population fluctuations.

The two extreme conceptual models of community organization are the *top-down model* in which changes in food webs are driven from above by predators, and the *bottom-up model* in which nutrients and plants control the food web. Many intermediate models between these two extremes can be used to describe particular communities.

Studies in the rocky intertidal zone and in aquatic systems have stimulated several models of community organization. The Menge-Sutherland model includes the roles of environmental harshness and variable recruitment in its prediction of when community interactions will be dominated by competition, predation, or physical factors. The trophic cascade (or Hairston-Smith-Slobodkin model) is a top-down model that emphasizes the alternation of positive and negative effects in food webs. When top predators are removed, the effects cascade as alternating positive and negative effects down the trophic ladder. Trophic cascades are common in aquatic systems but also occur in many terrestrial communities. But not all systems follow trophic cascades, and cascades often attenuate as they move down the food chain. Some populations and communities are driven bottom-up by nutrients, and the lower trophic levels may be affected from below by plant dynamics and remain unaffected by changes in predator abundance.

Island communities are a special case in which species makeup is driven by the interaction between colonization and extinction. The species-area curve, which describes how biodiversity increases with island size, is one of the grand generalizations of community ecology and can be combined with species-energy theory to obtain a detailed description of the species richness of island faunas and floras. Not all islands are equilibrium systems in which immigration and extinction are balanced, however, and historical effects are an important component of many island faunas and floras.

Some communities may exist in multiple stable states, and these communities may be confused with nonequilibrium assemblages. If a community is perturbed sufficiently, it may change to a new configuration at which it will remain even when the disturbance is stopped. Considerable controversy exists concerning how common multiple stable states are in natural ecosystems, and the question is important for conservation and land management.

Review Questions and Problems

1 The species-area curve rises continually as area is increased, implying that there is no limit to the number of species in any community. Is this a correct interpretation? What hypotheses can you suggest to explain why the number of species rises as area increases?

2 The trophic cascade in Zion National Park illustrated in Figure 15 rests on a comparison of two geographically separate areas of the park because of human use patterns. Discuss whether this comparison of two separate areas at the same time can be used to validate a system of multiple stable states according to the criteria given from Connell and Sousa (1983).

3 Mammals on mountaintops in the Great Basin of western North America have been cited as a model case at variance with the MacArthur-Wilson theory of island biogeography. Brown (1971) postulated that mammals in the Great Basin were remnant populations subject only to extinction (no immigration is possible). Lawlor (1998) rejects this explanation. Review Lawlor's data and the changing interpretations of these mountaintop communities in relation to equilibrium and nonequilibrium theories of community dynamics.

4 Can nonequilibrium models of community organization be stable? Read Chesson and Case (1986) and DeAngelis and Waterhouse (1987) and discuss the relationship between stability and equilibrium/nonequilibrium concepts.

5 Islands in the Bahamas have a relatively simple food chain in which *Anolis* lizards are the apex predator feeding on spiders and insect herbivores. Hurricanes hit the Bahamas in 2000 and again in 2002, drastically reducing the abundance of lizards and spiders (that also prey on insect herbivores). Discuss what you would predict for this ecosystem after these disturbances if it is controlled top-down or if it is controlled bottom-up. Spiller and Schoener (2007) give the details of what actually happened after the hurricanes.

6 In western North American grasslands, bison (*Bison bison*) and prairie dogs (*Cynomys ludovicianus*) are considered keystone herbivores. What changes would you predict in the plant community of these grasslands if you set up an experiment in which bison were excluded from some plots, prairie dogs from other plots, and both species from a third set of plots. Fahnestock and Detling (2002) did this experiment for three years and got no vegetation changes in any of the plots. Is this sufficient information to reject the idea that these herbivores are keystone species?

7 Freshwater lakes have been suggested to have two alternate stable states, one of clear water dominated by macrophytes and one of murky water with high phytoplankton levels. If this is correct, would you predict that measurement of the clarity of lake water for many lakes would be bimodal with a peak of lakes at the clear end of the spectrum and another peak of lakes at the murky end, with few lakes in between? Why might this prediction be incorrect? Peckham et al. (2006) did this analysis for Wisconsin lakes and discuss the results obtained from satellite measurement of lake transparency.

8 Trophic cascades are weaker in terrestrial systems compared with aquatic ones (Hall et al. 2007). One suggested explanation for this difference is that differences in body size between plants and their herbivores could be responsible (Shurin et al. 2006). Discuss the relative size of plants and their herbivores in aquatic ecosystems and in terrestrial ecosystems, and speculate how these differences might affect trophic cascades.

9 Review the argument between Hairston (1991) and Sih (1991) over the interpretation of field data for testing the predictions of the Hairston et al. (1960) model of community organization. Discuss the problem of testing hypotheses about community organization.

10 Analyze the semiarid woodland multiple stable state model of Figure 28 using the criteria for the existence of multiple stable states given by Connell and Sousa (1983). Would this example be acceptable to Connell and Sousa? Would the white-tailed deer example satisfy their criteria?

11 The immigration and extinction curves in the MacArthur-Wilson theory of island biogeography are concave upwards (see Figure 21). What difference would it make if these curves were straight lines?

12 What role does history play in community organization as defined by the equilibrium model and the nonequilibrium model? Do we need to know anything about the history of a community to predict its future course? Tanner et al. (1996) discuss this issue for coral reefs.

Overview Question

List some of the possible manipulative experiments that could be applied to a community to test the Menge-Sutherland and the trophic cascade models of community organization. List the predictions each model would make for each possible manipulation. Are some manipulations more instructive than others?

Suggested Readings

- Berumen, M., and M. Pratchett. 2006. Recovery without resilience: Persistent disturbance and long-term shifts in the structure of fish and coral communities at Tiahura Reef, Moorea. *Coral Reefs* 25:647–653.

- Briske, D. D., S. D. Fuhlendorf, and F. E. Smeins. 2006. A unified framework for assessment and application of ecological thresholds. *Rangeland Ecology and Management* 59:225–236.

- Coles, S., and E. Brown. 2007. Twenty-five years of change in coral coverage on a hurricane impacted reef in Hawai'i: The importance of recruitment. *Coral Reefs* 26:705–717.

- Gillson, L. 2004. Testing non-equilibrium theories in savannas: 1400 years of vegetation change in Tsavo National Park, Kenya. *Ecological Complexity* 1:281–298.

- Hebblewhite, M., C. A. White, C. G. Nietvelt, J. A. McKenzie, T. E. Hurd, J. M. Fryxell, S. E. Bayley, and P. C. Paquet. 2005. Human activity mediates a trophic cascade caused by wolves. *Ecology* 86:2135–2144.

- Kalmar, A., and D. J. Currie. 2006. A global model of island biogeography. *Global Ecology & Biogeography* 15:72–81.

- Lesser, M. P. 2007. Coral reef bleaching and global climate change: Can corals survive the next century? *Proceedings of the National Academy of Sciences of the USA* 104:5259–5260.

- Noble, J. C., D. S. Hik, and A. R. E. Sinclair. 2007. Landscape ecology of the burrowing bettong: Fire and marsupial biocontrol of shrubs in semi-arid Australia. *Rangeland Journal* 29:107–119.

- Olin, M., M. Rask, J. Ruuhijarvi, J. Keskitalo, J. Horppila, P. Tallberg, T. Taponen, A. Lehtovaara, and I. Sammalkorpi. 2006. Effects of biomanipulation on fish and plankton communities in ten eutrophic lakes of southern Finland. *Hydrobiologia* 553:67–88.

- Petraitis, P. S., and S. R. Dudgeon. 2004. Detection of alternative stable states in marine communities. *Journal of Experimental Marine Biology and Ecology* 300:343–371.

- Ripple, W. J., and R. L. Beschta. 2007. Restoring Yellowstone's aspen with wolves. *Biological Conservation* 138:514–519.

- Scheffer, M., and S. R. Carpenter. 2003. Catastrophic regime shifts in ecosystems: Linking theory to observation. *Trends in Ecology & Evolution* 18:648–656.

- van der Wal, R. 2006. Do herbivores cause habitat degradation or vegetation state transition? Evidence from the tundra. *Oikos* 114:177–186.

Credits

Illustration and Table Credits

Photo Credits

Ecosystem Metabolism I: Primary Production

Key Concepts

- Communities process solar energy through green plants, and the resulting energy fixed via photosynthesis sustains all the trophic levels in the food web.

- Less than 1% of solar energy reaching the Earth is captured by plants.

- Primary production varies globally, and total production is nearly equally distributed between the land and the oceans, even though the oceans occupy more than twice as much surface area.

- Primary production in aquatic environments is limited by nutrients—primarily by nitrogen, iron, and silica in the open ocean, and by light, temperature, phosphorus, and nitrogen in freshwater lakes. Solar radiation does not limit oceanic productivity.

- Temperature and moisture are the master limiting factors on land because they determine the length of the growing season.

- On land, solar radiation, temperature, moisture, and nutrients limit primary production. Nitrogen and phosphorus are the master limiting nutrients, but trace metals can also be critical in some soils.

- Satellite imagery now makes it possible to study large-scale changes in primary production on land and in the ocean in real time.

From Chapter 22 of *Ecology: The Experimental Analysis of Distribution and Abundance*, Sixth Edition. Eugene Hecht. Copyright © 2009 by Pearson Education, Inc. Published by Pearson Benjamin Cummings. All rights reserved.

KEY TERMS

autotrophs Organisms that obtain their energy from the sun and materials from nonliving sources.

compensation point For plants the equilibrium point at which photosynthesis equals respiration.

eutrophication The process by which lakes are changed from clear water lakes dominated by green algae into murky lakes dominated by blue-green algae, typically caused by nutrient runoffs from cities or agriculture.

gross primary production The energy or carbon fixed via photosynthesis per unit time.

harvest method The measurement of primary production by clipping the vegetation at two successive times.

heterotrophs Organisms that pick up energy and materials by eating living matter.

net primary production The energy (or carbon) fixed in photosynthesis minus the energy (or carbon) lost via respiration per unit time.

photosynthetically active radiation (PAR) That part of the solar radiation spectrum in the range 0.4 to 0.7 μm that can be used for photosynthesis by green leaves.

productivity A general term that covers all processes involved in ecological production studies—carbon fixation, consumption, rejection, leakage, and respiration.

Redfield ratio The observed 16:1 atomic ratio of nitrogen to phosphorus found in organisms in the open ocean by A. C. Redfield in 1934—$C_{106}N_{16}P_1$.

We can take two broad approaches to the study of communities and ecosystems of plants and animals. First, we can treat species as biological entities, with all the specific adaptations and interrelationships they show; this can be considered a population-ecological approach to community and ecosystem dynamics. Second, we can move beyond the details of particular species and concentrate on the physics of ecosystems as energy machines and nutrient processors. Exactly how plants and animals process energy and materials has important implications for humans.

The metabolism of ecosystems is most easily understood as the sum of the metabolism of individual animals and plants. Individual organisms require a continual input of new energy to balance losses resulting from metabolism, growth, and reproduction. Individuals can be viewed as complex machines that process energy

and materials. Organisms pick up energy and materials in two main ways. **Autotrophs** pick up energy from the sun and materials from nonliving sources. Green plants are autotrophs. **Heterotrophs** pick up energy and materials by eating living matter. Herbivores are heterotrophs that live by eating plants, and carnivores are heterotrophs that live by eating other heterotrophs. Communities are mixtures of autotrophs and heterotrophs. Energy and materials enter a biological community, are used by the individuals, and are transformed into biological structure only to be ultimately released again into the environment. The **ecosystem** level of integration includes both the organisms and their abiotic environment and is a comprehensive level at which to consider the movement of energy and materials. (We could also discuss the flow of matter and energy at the individual level or at the population level.) The basic unit of metabolism is always the individual organism, even when individuals are assembled into communities and ecosystems.

The first step in the study of ecosystem metabolism is to identify the food web of the community. Once we know the food web, we must decide how we can judge the significance of the different species to community metabolism. Even though some 5000 species of animals live on the 5 km² of Wytham Woods in Britain (Elton 1966), we feel intuitively that many of these 5000 species are not significant, and that many or most of them could be removed without affecting the metabolism of the woodland.

Three measurements might be used to define relative importance in an ecosystem:

1. **Biomass.** We could use the weight or standing crop of each species as a measure of importance. This is useful in some circumstances, such as the timber industry, but it cannot be used as a general measure. In a dynamic situation in which *yield* is important, we need to know how rapidly a community produces new biomass. When metabolic rates and reproductive rates are high, production may be very rapid, even from a small standing crop. **Figure 1** illustrates the idea that yield need not be related to biomass.

2. **Flow of chemical materials.** We can view an ecosystem as a superorganism taking in food materials, using them, and passing them out. Note that all chemical materials can be recycled many times through the community. A molecule of phosphorus may be taken up by a plant root, used in a leaf, eaten by a grasshopper that dies,

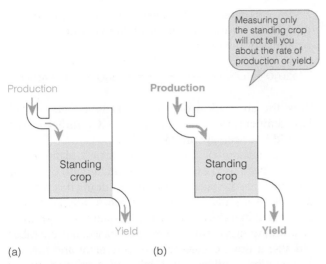

Figure 1 **Hypothetical illustration of two equilibrium communities in which input equals output: (a) low input, low output, slow turnover; (b) high input, high output, rapid turnover.** Standing crop is not related to production or yield because turnover time is not a constant for all systems. Production (input) must equal yield (output) in equilibrium communities, but many communities are not always in equilibrium, so standing crop may rise and fall.

and released by bacterial decomposition to reenter the soil.

3. **Flow of energy.** We can view the ecosystem as an energy transformer that takes solar energy, fixes some of it in photosynthesis, and transfers this energy from green plants through herbivores to carnivores. Note that most energy flows through an ecosystem only once and is not recycled; instead it is transformed to heat and ultimately lost from the system. Only the continual input of new solar energy keeps the ecosystem operating. Again we may draw the analogy between an ecosystem and an organism that processes food energy.

To study the dynamics of ecosystem metabolism, we must decide what to use as the *base variable*. Most ecologists have decided to use either carbon or energy. Elements other than carbon are often tied up in biological peculiarities of organisms. For example, vertebrates and molluscs contain much more calcium than do most freshwater invertebrates because of the presence of bone or shell. Some marine invertebrates concentrate certain chemical elements. Even within an individual there are variations. Calcium in the teeth and bones of a mammal may be stable for long periods, whereas calcium in the blood serum may turn over rapidly because of ingestion

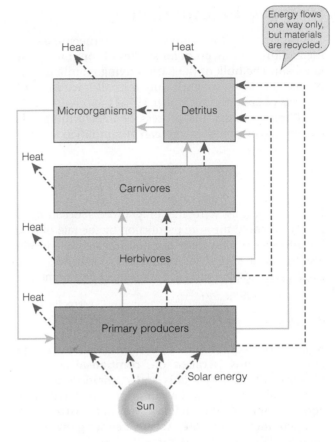

Figure 2 **General representation of energy flows (red) and material cycles (blue) in the biosphere.** Energy flows included are solar radiation, chemical energy transfers (in the ecological food web), and radiation of heat into space. Materials flow through the trophic levels to detritus and eventually back to the primary producers. (From DeAngelis 1992.)

and excretion. As a result, the description of the calcium flow through an ecosystem is very difficult. Because most energy is not recycled, it is easier to measure than are chemical materials. **Figure 2** illustrates the flows of energy and materials through the food chain.

Energy or carbon content is just another way of describing an individual, a population, or a community, and the convenience and precision of these measures should not blind us to their limitations as a way of describing organisms. The great strength (and weakness) of the energetics approach is that it can allow us to add together different species in a community. It reduces the fundamental diversity of a biological community to a single unit—either the joule (for energy) or the gram (for carbon).

Primary Production

The process of photosynthesis is the cornerstone of all life and the starting point for studies of community metabolism. The bulk of the Earth's living mantle is green plants (99.9% by weight); only a small fraction of life consists of animals (Whittaker 1975).

Photosynthesis, the process of transforming solar energy into chemical energy, can be simplified as

$$12H_2O + 6CO_2 + \text{solar energy} \xrightarrow[\text{enzymes}]{\text{chlorophyll} +} C_6H_{12}O_6 + 6O_2 + 6H_2O$$
$$\text{(from air)} \qquad\qquad\qquad\qquad \text{(carbohydrate)} \quad \text{(to air)}$$

If photosynthesis were the only process occurring in plants, we could measure production by the rate of accumulation of carbohydrate, but unfortunately plants also respire, using energy for maintenance activities. In an overall view, respiration is the opposite of photosynthesis:

$$C_6H_{12}O_6 + O_2 \xrightarrow[\text{enzymes}]{\text{metabolic}} CO_2 + H_2O + \begin{array}{c}\text{energy for work} \\ \text{and maintenance}\end{array}$$
$$\text{(carbohydrate)} \quad \text{(from air)} \qquad\qquad \text{(to air)}$$

At metabolic equilibrium, photosynthesis equals respiration, and this is called the **compensation point**. Measures of photosynthesis and respiration are rates; they are always expressed as amount of material or energy per unit of time. If plants always existed at the compensation point, they would neither grow nor reproduce. We define two terms:

Gross primary production	= energy (or carbon) fixed via photosynthesis per unit time
Net primary production	= energy (or carbon) fixed in photosynthesis − energy (or carbon) lost via respiration per unit time

Production is always measured as a rate per unit of time. How can we measure these two aspects of primary production in natural systems?

For terrestrial plants, the direct way is to measure the change in CO_2 or O_2 concentrations in the air around plants. Most studies measure CO_2 uptake by an enclosed branch or a whole plant. During daylight conditions, CO_2 uptake rate is a measure of net production because both photosynthesis and respiration are operating simultaneously. At night only respiration occurs, and the rate at which CO_2 is released can be used to estimate the respiration component. Photosynthesis and respiration are both affected by temperature; photosynthesis is also affected by light intensity. The daily changes in leaf temperature and light intensity determine the daily net production for an individual plant.

We can determine the energetic equivalents of photosynthesis measurements from the chemical thermodynamics of the reaction:

$$12H_2O + 6CO_2 + 2966kJ \xrightarrow[\text{energy}]{\text{solar}} C_6H_{12}O_6 + 6O_2 + 6H_2O$$

Thus, the absorption of 6 moles (134.4 liters at standard temperature and pressure) of CO_2 indicates that 2966 kJ has been absorbed.[1]

The measure of gas exchange around plants in the field has been used extensively to estimate photosynthetic rates now that the instrumentation for doing this is readily available. A slightly different approach to measuring CO_2 uptake is to introduce radioactive ^{14}C–labeled CO_2 into the air surrounding a plant (covered by a transparent chamber) and after a time to harvest the whole plant and measure the quantity of radioactive ^{14}C taken up by photosynthesis.

The simplest method of measuring primary production is the **harvest method**. The amount of plant material produced in a unit of time can be determined from the difference between the amount present at the two times:

$$\Delta B = B_2 - B_1 \qquad (1)$$

where ΔB = biomass change in the community between time *1* (t_1) and time *2* (t_2)

B_1 = biomass at t_1

B_2 = biomass at t_2

Two possible losses must be recognized:

L = biomass losses by death of plants or plant parts

G = biomass losses to consumer organisms

If we know these values, we can determine net primary production:

$$\text{Net primary production} = NPP = \Delta B + L + G \qquad (2)$$

This may apply to the whole plant, or it may be specified as *aerial* production or *root* production.

The net primary production in biomass may then be converted to energy by measuring the caloric equivalent of the material in a bomb calorimeter. This measurement should be done for each particular species studied as well as for each season of the year. Golley

---

[1] Energy units have been reported in many forms in the literature, often in calories, and can be standardized to *joules* with the following conversion factors: 1 joule (J) = 0.2390 gram calorie (cal) = 0.000239 kilocalorie (kcal); conversely, 1 gram calorie = 4.184 joules and 1 kilocalorie = 4184 joules or 4.184 kilojoules (kJ).

(1961) showed that different parts of plants have different energy contents:

	Mean of 57 plant species	
	(kcal/g dry wt)	**(kJ/g dry wt)**
Leaves	4.229	17.694
Roots	4.720	19.748
Seeds	5.065	21.192

Vegetation collected in different seasons also varied in energy content.

The harvesting technique of estimating production is used in a variety of situations. Foresters have used a modified version of it for timber estimation, and agricultural research workers use it to determine crop yield. The application of harvesting techniques to natural vegetation involves some specialized techniques that we will not describe here; Moore and Chapman (1986) and Pieper (1988) give details of techniques.

In aquatic systems, primary production can be measured in the same general way as in terrestrial systems. Gas-exchange techniques can be applied to water volumes, and usually oxygen release instead of carbon dioxide uptake is measured. This procedure is usually repeated with a dark bottle (respiration only) and a light bottle (photosynthesis and respiration), so that both gross and net production can be measured. Vollenweider (1974) discusses details of techniques for measuring production in aquatic habitats.

How does primary production vary over the different types of vegetation on the Earth? This is the first general question we can ask about community metabolism. **Table 1** gives some average values for global net primary production in biomass for different vegetation types, and **Figure 3** illustrates the yearly primary production for ocean and land areas of the globe as estimated from satellite data. In general, primary production is highest in the tropical rain forest and decreases progressively toward the poles. **Productivity** of the open ocean is very low, approximately the same as that of the arctic tundra. But because oceans occupy about 71% of the total surface of the Earth, total oceanic primary production adds up to about 46% of the overall production of the globe. Grassland and tundra areas are less productive than forests of equivalent latitude.

The efficiency of plants in converting solar energy into primary production can be broken down into a set of seven factors that begin with the solar constant and are modified by latitude, cloudiness, dust and water in

Table 1 Net primary production for land and ocean estimated from satellite data, as illustrated in Figure 3.

Vegetation type	Annual net primary production
Ocean	48.5
Land	
Tropical rainforests	15.9
Broadleaf deciduous forests	4.4
Broadleaf and needleleaf forests	3.0
Needleleaf evergreen forests	4.5
Needleleaf deciduous forests	1.6
Savannas	12.4
Perennial grasslands	2.8
Broadleaf shrubs	0.2
Tundra	2.3
Desert	1.6
Cultivated areas	8.0
Total for land vegetation	56.7
Total for globe	105.2

All values of net primary production are in petagrams of carbon (1 petagram = 10^{15} grams = 10^9 metric tons).

SOURCE: Field et al. (1998) and Ito and Oikawa (2004) estimates.

the atmosphere, leaf arrangement and leaf area, and the concentration of CO_2 (Monteith 1972). How efficient is the vegetation of different communities as an energy converter? We can determine the overall efficiency of utilization of sunlight by the following ratio:

$$\text{Efficiency of gross primary production} = \frac{\text{energy fixed by gross primary production}}{\text{energy in incident sunlight}} \quad (3)$$

All the estimates of the overall efficiency of primary production report values that range from 0.004% to 0.2%, indicating that only a tiny fraction of solar energy is utilized by plants. The process of photosynthesis in green leaves uses radiant energy in the wavelengths from 0.4 to 0.7 μm, and this is referred to as **photosynthetically active radiation (PAR)**. For ecological purposes it is

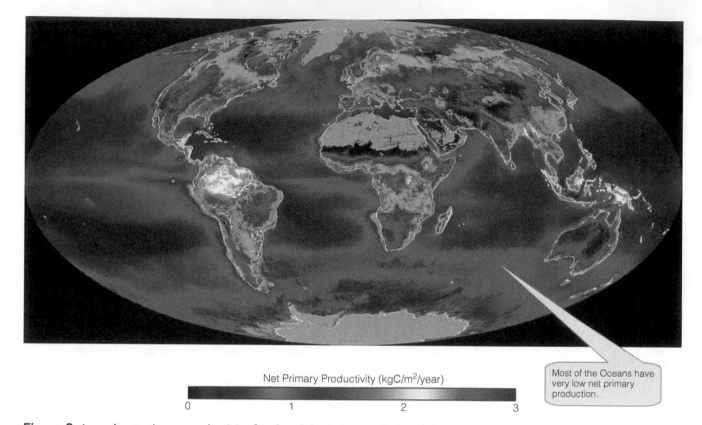

Net Primary Productivity (kgC/m²/year)

0   1   2   3

Most of the Oceans have very low net primary production.

Figure 3 **Annual net primary productivity for the globe in 2002 calculated from satellite data gathered by the Moderate Resolution Imaging Spectroradiometer (MODIS).** The yellow and red areas show the highest rates of net production, and the green, blue, and purple areas show progressively lower productivity. Gray areas indicate no net primary production. Of total global primary production, the ocean contributes 46% and the land 54%. Plant matter is about 50% carbon, so these carbon data can be readily converted to vegetation biomass by multiplying by 2. (From NASA, Earth Observatory 2007.)

more useful to express the use of light in different plant communities as *light use* **efficiency** (LUE), defined in the following equation (Bradford et al. 2005):

$$NPP = APAR \times LUE \qquad (4)$$

where NPP = net primary production (gC per m² per unit time)

$APAR$ = absorbed photosynthetically active radiation (MJ per m² per unit time)

LUE = light use efficiency (gC per MJ)

APAR is highly variable because it is strongly affected by the limiting resources for photosynthesis, such as water supplies and soil nutrients for land plants. **Figure 4** shows that light use efficiency varies dramatically in terrestrial ecosystems in different parts of the globe. Ruimy et al. (1999) estimated that global light use efficiency averaged 0.43 gC per MJ of light energy.

How much of the energy plants fix by photosynthesis is subsequently lost via respiration? A great deal of energy is lost in converting solar radiation to gross primary production. Net primary production, which is what interests us, must therefore be even less efficient. Over a wide range of measurements at the whole plant level, the ratio of autotroph respiration to gross photosynthesis ranges from 0.35 to 0.6, with a mean value around 0.53 (Gifford 2003). Thus, approximately half of the energy fixed in photosynthesis is used in respiration. Forests may be less efficient than herbaceous communities, with 50%–75% of the gross primary production lost to respiration in forests (Kira 1975). Forests have larger amounts of stems, branches, and roots to support than do herbs, and thus less energy is lost to respiration in herbaceous and crop communities (40%–45%). The result of these losses is that for a broad range of terrestrial communities, less than 1% of the sun's energy is converted into net primary production during the growing season.

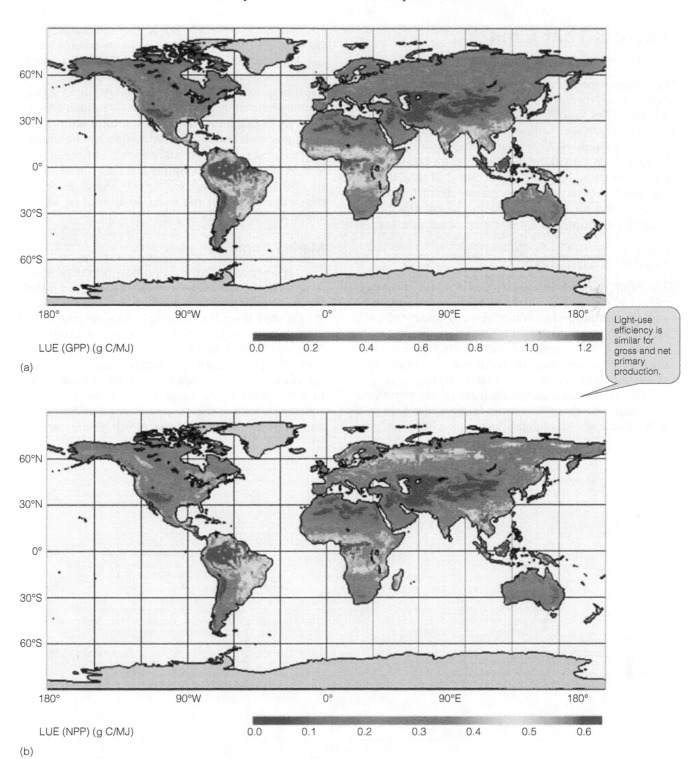

Light-use efficiency is similar for gross and net primary production.

LUE (GPP) (g C/MJ)

0.0   0.2   0.4   0.6   0.8   1.0   1.2

(a)

LUE (NPP) (g C/MJ)

0.0   0.1   0.2   0.3   0.4   0.5   0.6

(b)

Figure 4 **Estimated annual mean light use efficiency for (a) gross primary production and (b) net primary production.** Because of seasonal temperature changes, much of the incident light cannot be used for photosynthesis in the temperate and polar regions. Water availability and low plant biomass also limit light use efficiency in areas like the Sahara and much of Australia. (From Ito and Oikawa 2004.)

Factors That Limit Primary Productivity

One important question about primary production is, *What controls the rate of primary production in natural communities?* Put another way, what factors could we change to increase the rate of primary production for a given community? Note that this question could be broken down into many questions of the same type for each population of plants. The control of primary production has been studied in greater detail for aquatic systems than for terrestrial systems, so we first look at some details of production in aquatic communities.

Aquatic Communities

The depth to which *light* penetrates in a lake or ocean is critical in defining the zone of primary production in aquatic communities. Water absorbs solar radiation very readily. More than half of the solar radiation is absorbed in the first meter of water, including almost all the infrared energy. Even in "clear" water, only 5%–10% of the radiation may reach a depth of 20 m.

Figure 5 illustrates the decrease in photosynthesis with depth in three California lakes. Clear Lake is a **eutrophic** lake with high production and little light pen-

etration. Castle Lake is a lake of intermediate productivity, in which the zone of photosynthesis extends below a depth of 20 m. Lake Tahoe is an alpine **oligotrophic** lake of remarkably clear water in which the zone of photosynthesis extends to a depth of 100 m, although there is little photosynthesis at any depth (Goldman 1988).

Very high light levels can also inhibit photosynthesis of green plants, and this inhibition can be found in tropical and subtropical surface waters throughout the year. When surface radiation is excessive, the maximum in primary production will occur several meters beneath the surface of the water, as illustrated for Lake Tahoe and Castle Lake in Figure 5.

Marine Communities

Light is a necessary factor for primary production in the ocean but paradoxically it is not usually the limiting factor in oceanic primary production (Platt et al. 1992). If light were the primary variable limiting primary production, we would expect a gradient of productivity from the poles toward the equator. Figure 3, which shows the global distribution of primary production in the oceans, indicates that no such latitudinal gradient of production exists. Large parts of the tropics and subtropics, such as the Sargasso Sea, the Indian Ocean, and the Central Gyre of the North Pacific, are very

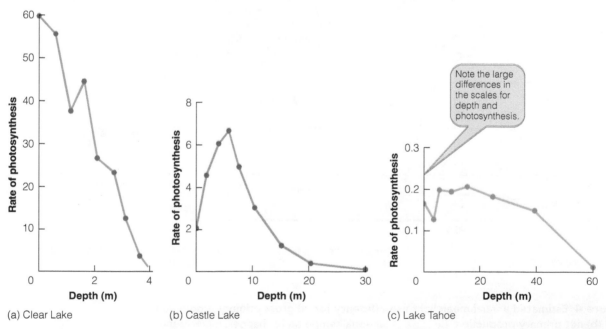

(a) Clear Lake   (b) Castle Lake   (c) Lake Tahoe

Figure 5 Change in photosynthesis at varying depths in three California lakes during the summer: (a) Clear Lake, (b) Castle Lake, and (c) Lake Tahoe. Clear Lake is a eutrophic lake with green water, whereas Lake Tahoe is one of the clearest lakes in the world. Rate of photosynthesis is measured in grams of carbon fixed per m² per day. (Data from Goldman 1968.)

unproductive even though they have abundant solar radiation. In contrast, the North Atlantic, the Gulf of Alaska, and the Southern Ocean off New Zealand are quite productive. The most productive areas are coastal areas off the western side of Africa and North and South America (Falkowski et al. 1998).

Why are tropical oceans unproductive when the light regime is good all year? *Nutrients* appear to be the primary limitation on primary production in the ocean through their effects on the biomass of chlorophyll in the phytoplankton. Two elements, nitrogen and phosphorus, often limit primary production in the oceans. One of the striking generalities of many parts of the oceans is the very low concentrations of nitrogen and phosphorus in the surface layers where phytoplankton live (see **Figure 6**), whereas the deep water contains much higher concentrations of nutrients.

Nitrogen may be a limiting factor for phytoplankton in many parts of the ocean (Elser et al. 2007). A good illustration of this is shown by pollution (nutrient runoff) from duck farms along the bays of Long Island, New York, which adds both nitrogen and phosphorus to the coastal water. Unlike phosphorus, the nitrogen added is immediately taken up by algae, and no trace of nitrogen can be measured in the coastal waters (**Figure 7**). That

nitrogen was limiting was confirmed by nutrient-addition experiments (Figure 7c). The addition of nitrogen (in the form of ammonium) caused a heavy algal growth in bay water, but the addition of phosphate did not induce algal growth. This work has some obvious practical conclusions: if nitrogen is the factor currently limiting phytoplankton production, the elimination of phosphates from sewage entering the ocean will not help the problem of coastal pollution.

The discovery that nitrogen limits primary production in many parts of the ocean was completely unexpected because nitrogen is abundant in the air and can be converted into a usable form by nitrogen-fixing cyanobacteria (Falkowski 1997). The expectation had been that phosphorus must be limiting productivity in the ocean because phosphorus does not occur in the air. But this has turned out to be completely wrong, and nitrogen seems to be the major nutrient limiting oceanic primary production. But this conclusion raises other questions because several large regions of the oceans contain high amounts of nitrate and few phytoplankton. For example, the surface waters of the equatorial Pacific have both high nitrate and high phosphate concentrations but low algal biomass (Behrenfeld et al. 1996). One explanation of these regions is that they are communities

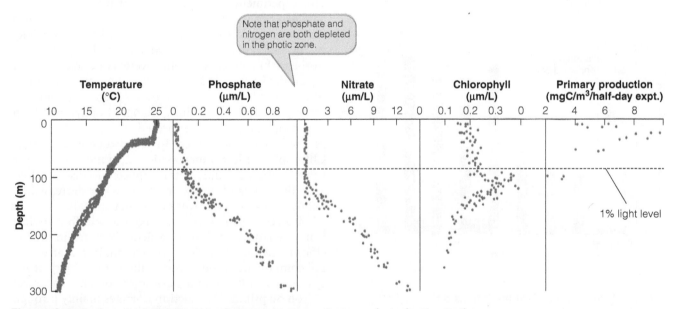

Figure 6 Typical vertical distribution of temperature, nutrients, and production in the upper layer of the North Pacific Central Gyre during summer. The curves are composites of several vertical profiles made over a two-day period at a single location in the North Pacific Ocean (28° N, 155° W). The dashed line illustrates the depth to which only 1% of surface light penetrates (the traditional definition of the lower limit of the euphotic zone). Note that nitrate has been depleted to undetectable levels above the 1% light level, and that most of the primary production (measured from ^{14}C uptake) takes place above the depth where nitrate can be detected with conventional techniques. (From Hayward 1991.)

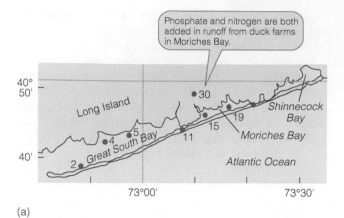

(a)

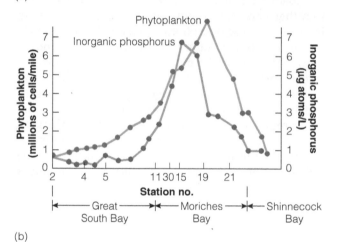

(b)

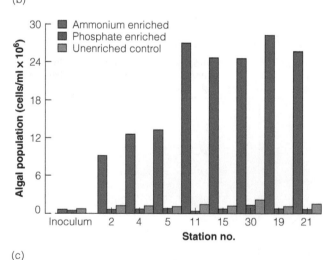

(c)

Figure 7 Effects of nutrient limitations on phytoplankton production in coastal waters of Long Island. (a) Coast of Long Island, New York; (b) abundance of phytoplankton and distribution of phosphorus arising from duck farms; (c) nutrient-enrichment experiments with alga *Nannochloris atomus* in water from the bays. Phosphorus is superabundant, and nitrogen (added in ammonium) seems to limit algal growth. (After Ryther and Dunstan 1971.)

dominated by top-down processes in which herbivores control plant biomass. Alternatively, these could be bottom-up communities limited by some nutrient other than nitrogen or phosphorus.

The Sargasso Sea is an area of very low productivity in the subtropical part of the Atlantic Ocean. The seawater there is among the most transparent in the world, and the surface waters are very low in nutrients. Nitrogen and phosphorus, however, do not seem to be limiting primary production; instead, iron seems to be critical (Menzel and Ryther 1961b), as was shown by a series of nutrient-enrichment experiments in which surface water from the Sargasso Sea was placed in bottles and enriched with various nutrients for three days. The results were as follows:

Nutrients added to experimental culture	Relative uptake of ^{14}C by cultures
None (controls)	1.00
N + P only	1.10
N + P + metals (excluding iron)	1.08
N + P + metals (including iron)	12.90
N + P + iron	12.00

This experimental demonstration of iron limitation in Sargasso Sea water stimulated the hypothesis that iron limitation could be responsible for the low productivity of the equatorial Pacific. Two large-scale open ocean iron-enrichment experiments were conducted in the equatorial Pacific in 1993 and 1995. Low concentrations of dissolved iron were spread over 72 km^2 of ocean over seven days, and the resulting changes in phytoplankton and zooplankton were measured. A massive bloom of phytoplankton developed in the iron-enriched area. Chlorophyll *a* levels increased to 27 times the starting value. At the same time, nitrate uptake increased 14-fold such that levels of nitrate in the seawater decreased by about 35% within one week (Coale et al. 2004).

Iron comes to the oceans largely as windblown dust from the land, and dust is particularly scarce in the Pacific Ocean and in the Southern Ocean. Iron is an essential component of the photosynthetic machinery of the cyanobacteria that fix nitrogen in the oceans. The effect of iron on primary production operates mainly through its role in nitrogen fixation, resulting in the following sequence of potential limitations in iron-poor parts of the ocean (Falkowski et al. 1998):

iron→cyanobacteria→nitrogen fixation→phytoplankton

In most of the open oceans, light is always available for photosynthesis, but nitrogen is not.

To quantify the relative effects of different limiting nutrients in the oceans, Downing et al. (1999) analyzed 303 controlled nutrient-addition experiments carried out over the past 30 years. They found that nitrogen addition stimulated phytoplankton growth most strongly, followed closely by iron addition (**Figure 8**). Doubling times for phytoplankton populations averaged 3.3 days for nitrogen addition and 4.1 days for iron addition. In general these results are consistent with the conclusion that nitrogen and iron are key limiting resources in the oceans, and that silica may limit diatom production when diatoms are a dominant component of the phytoplankton.

The role of iron limitation in the Southern Ocean was the subject of one of the largest experiments in oceanic fertilization yet carried out. It arose from the statement by the oceanographer John Martin who made the challenge at a scientific meeting that "Give me a half tanker of iron, and I will give you an ice age." Martin argued that iron in the Southern Ocean would stimulate phytoplankton to take up CO_2, thus cooling the climate (Martin et al. 1990). If this was correct, we could reverse the current global warming by fertilizing the Southern Ocean with iron. Three large-scale experiments were carried out to test this idea. The Southern Ocean Iron Experiment (SOFeX) was the largest of these and was carried out in January and February 2002. It involved the spreading of 631 kg of iron over 225 km^2 with two repeated treatments over the following four weeks in one area, and a replicate experiment with slightly less iron with three repeated treatments in a more southerly location 1400 km south (Coale et al. 2004). **Figure 9** shows

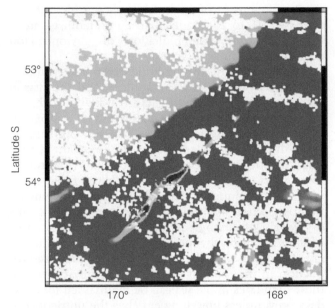

(a) **North Patch**

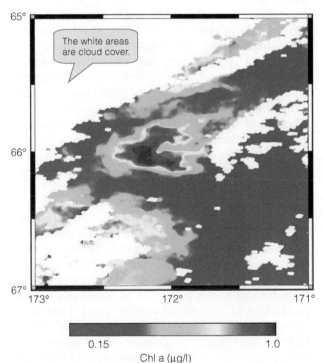

(b) **South Patch**

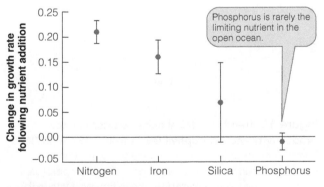

Figure 8 Effects of nutrient addition on marine phytoplankton growth rates in 303 experiments. Excess nutrients were added to a large water sample and followed for two to seven days. Doubling times for nitrogen addition were 3.3 days, for iron 4.1 days, and for silica 10 days. Silica limitation occurs when diatoms are the dominant species in the phytoplankton. The black line marks the line of zero effect. (After Downing et al. 1999.)

Figure 9 Satellite images of the Southern Ocean Iron Experiment (SOFeX) in 2002. (a) The North Patch experiment on day 28 after the start on 10 January 2002, showing an elongate band of chlorophyll extending over 250 km. The elongate band was produced by local currents that moved the added iron from a 15 km × 15 km square into a 7-km-wide band. (b) The South Patch experiment on day 20 showing a circular area of high primary production. In both cases chlorophyll concentrations increased 10- to 20-fold in the experimental areas. (From Coale et al. 2004.)

the satellite images of the enriched areas. Chlorophyll *a* increased 10- to 20-fold inside the iron-fertilized areas, thus confirming iron limitation of phytoplankton growth in the Southern Ocean.

Compared with the land, the ocean is very unproductive; as can be seen in Figure 3, and the reason seems to be that fewer nutrients are available. Rich, fertile soil contains 5% organic matter and up to 0.5% nitrogen. One square meter of surface soil can support 50 kg dry weight of plant matter. In the ocean, by contrast, the richest water contains 0.00005% nitrogen, four orders of magnitude less than that of fertile farmland soil. A column of rich seawater one square meter in cross section could support no more than 5 grams dry weight of phytoplankton (Ryther 1963). In terms of standing crops, the sea is a desert compared with the land. And although the maximal rate of primary production in the sea may be the same as that on land, these high rates in the sea can be maintained for a few days only, unless upwelling enriches the nutrient content of the water.

Areas of upwelling in the ocean are exceptions to the general rule of nutrient limitation. The largest area of upwelling occurs in the Antarctic Ocean (see Figure 3), where cold, nutrient-rich, deep water comes to the surface along a broad zone near the Antarctic continent. Other areas of upwelling occur off the coasts of Peru and California, as well as in many coastal areas where a combination of wind and currents moves the surface water away and allows the cold, deep water to move up to the surface. In these areas of upwelling, fishing is especially good, and in general a superabundance of nitrogen and phosphorus is available to the phytoplankton.

One of the most exciting recent developments in marine ecology is the ability to estimate primary production from satellite remote sensing data (Falkowski et al. 1998; Behrenfeld et al. 2006). Chlorophyll concentration in the surface water can be estimated by spectral reflectance using blue/green ratios. **Figure 10** illustrates how satellite information can be used to track changes in primary production in the world's oceans. From 1997 to 1999 chlorophyll biomass increased dramatically, but from 1999 to 2005 there was a slower overall decrease in primary production. Increasing temperature in the oceans is highly correlated with decreasing primary production (Behrenfeld et al. 2006), a worrying trend in an era of climatic warming.

Total primary production in the ocean is thus rarely limited by light but instead by the shortage of nutrients, particularly nitrogen and iron, which are critical for plant growth. Diatom growth may be limited by silica in areas of low silica content. Limitation of primary production by phosphorus alone is rare in oceanic ecosystems.

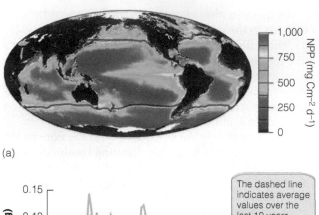

(a)

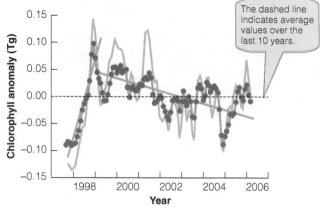

> The dashed line indicates average values over the last 10 years.

(b)

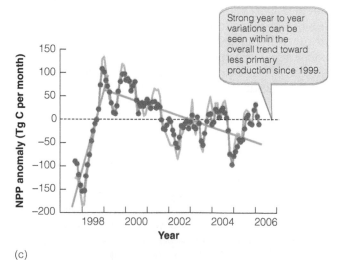

> Strong year to year variations can be seen within the overall trend toward less primary production since 1999.

(c)

Figure 10 Trends in global ocean phytoplankton productivity and chlorophyll levels from 1997 to 2006. The dark contour lines in (a) indicate the permanently stratified ocean waters with average surface temperatures over 15°C that were analyzed for these figures. Data in (b) and (c) are expressed as anomalies, which are deviations from the overall mean value for this time period. Values above the zero line indicate periods of higher than average primary production. The thick blue lines indicate overall trends. (From Behrenfeld et al. 2006.)

Freshwater Communities

In freshwater communities the same conclusions do not seem to hold. Solar radiation limits primary production on a day-to-day basis in lakes, and within a given lake one can predict the daily primary productivity from the solar radiation (Horne and Goldman 1994). Temperature is closely linked with light intensity in aquatic systems and is difficult to evaluate as a separate factor. Nutrient limitations operate in freshwater lakes, and the great variety of lakes are associated with a great variety of potential limiting nutrients.

For growth, plants require nitrogen, calcium, phosphorus, potassium, sulfur, chlorine, sodium, magnesium, iron, manganese, copper, iodine, cobalt, zinc, boron, vanadium, and molybdenum. These nutrients do not all act independently, which has made the identification of causal influences very difficult (Wetzel 2001). The conclusion of early work—that nitrogen and phosphorus were the major limiting factors in freshwater lakes—was a practical one based on the fertilization of small farm ponds to increase fish production.

During the 1970s the problem of what controls primary production in freshwater lakes became acute because of increasing pollution. Nutrients added to lakes directly in sewage or indirectly as runoff had increased algal concentrations and had shifted many lakes from phytoplankton communities dominated by diatoms or green algae to those dominated by blue-green algae. This process is called **eutrophication**. Before we can control eutrophication in lakes, we must decide which nutrients need to be controlled. Three major nutrients were considered: nitrogen, phosphorus, and carbon. Phosphorus is now believed to be the limiting nutrient for phytoplankton production in the majority of lakes (Edmondson 1991).

A series of elegant whole-lake nutrient-addition experiments conducted in the Experimental Lakes area of northwestern Ontario by David Schindler and his Winnipeg-based research group pinpointed the role of phosphorus in temperate-lake eutrophication (Schindler 2006). In one experiment, Lake 227 was fertilized for five years with phosphate and nitrate, and phytoplankton levels increased 50–100 times over those of control lakes. To separate the effects of phosphate and nitrate, Lake 226 was split in half with a curtain and fertilized with carbon and nitrogen in one half and with phosphorus, carbon, and nitrogen in the other (**Figure 11**). Within two months a highly visible algal bloom had developed in the basin to which phosphorus was added (Schindler 1977). All this experimental evidence is consistent with the hypothesis that phosphorus is the master limiting nutrient for phytoplankton in freshwater lakes.

Figure 11 Lake 226 in the Experimental Lakes area of northwestern Ontario, showing the role of phosphorus in eutrophication. The far basin, fertilized with phosphorus, nitrogen, and carbon, is covered with an algal bloom of the blue-green alga *Anabaena spiroides*. The near basin, fertilized with nitrogen and carbon, showed no changes in algal abundance. Photo taken September 4, 1973. (Photo courtesy of D. W. Schindler.)

When phosphorus is added to a lake, algae may show signs of nitrogen limitation, but long-term processes cause these deficiencies to be corrected (Schindler 1990; Smith 2006). Physical factors such as water turbulence and gas exchange seem to regulate CO_2 availability, so it rarely becomes limiting for algae. Nitrogen can be fixed by blue-green algae. Nitrogen deficiency in freshwater lakes frequently selects for cyanobacteria that can fix nitrogen, and in this way lakes are often restored to conditions of phosphorus limitation. The net result is that the standing crop of phytoplankton in lakes is highly correlated with the total amount of phosphorus in the water (**Figure 12**).

The practical significance of these and other experiments is the advisability of controlling phosphorus input to lakes and rivers as a simple means of checking eutrophication (Likens 1972; Schindler 2006). The amount of phosphorus that a lake can withstand can be calculated so that planners can determine the effects of human developments on a lake (Dillon and Rigler 1975).

Part of the difficulty of studying nutrient limitations of phytoplankton production is that nutrients may occur in several chemical states in aquatic systems. In some conditions, nutrients are present but not available to the organisms because they are bound up in organic complexes in the water or mud (Wetzel 2001). This has been shown dramatically in acid-bog lakes, which contain large amounts of phosphorus in forms not available to the phytoplankton. Waters (1957)

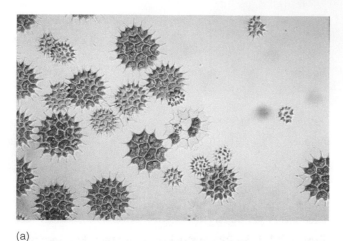

(a)

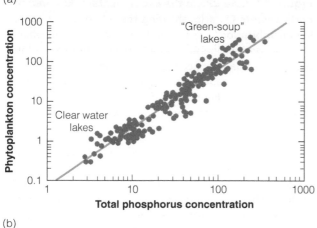

(b)

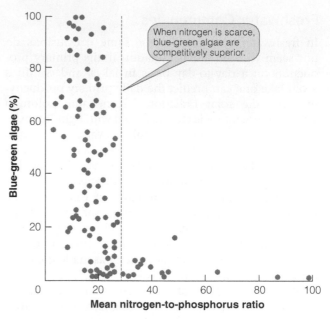

Figure 13 Relation between proportion of blue-green algae in the phytoplankton and nitrogen-to-phosphorus ratio in 17 lakes around the world. Each point represents data from one growing season. Blue-green algae are dominant when the nitrogen-to-phosphorus ratio is less than 25–30 (measured by moles), indicated by the vertical dashed line. (After Smith 1982.)

Figure 12 Phosphorus limitation of primary production in 170 freshwater lakes. The relationship between phosphorus concentration (g/L) and summer phytoplankton standing crop (measured by chlorophyll as g/L). Phosphorus concentration varies 100-fold in these lakes and algal concentration in the water varies 1000-fold. Phosphorus levels limit primary production in many freshwater ecosystems. The algae pictured are *Pediastrum boryanum*, a colonial green alga. (Data from Ahlgren et al. 1988.)

showed that fertilizing acid-bog lakes in Michigan with lime ($CaCO_3$) increased the pH, allowed phosphorus to be released from sediments, and greatly increased phytoplankton abundance.

One of the changes that often accompany eutrophication in lakes is that blue-green algae tend to replace green algae (see Figure 11). Blue-green algae are "nuisance algae" because they become extremely abundant when nutrients are plentiful and form floating scums on highly eutrophic lakes. Blue-green algae become dominant in the phytoplankton for several reasons. They are not heavily grazed by zooplankton or fish, which prefer other algae. Zooplankton often cannot

manipulate the large colonies and filaments of blue-green algae. Some species of blue-green algae also produce secondary chemicals that are toxic to zooplankton (DeMott and Moxter 1991). Blue-green algae are also poorly digested by many herbivores, so they are low-quality food for them. Finally, many blue-green algae can fix atmospheric nitrogen, putting them at an advantage when nitrogen is relatively scarce (Smith 2006). In eutrophication, more and more phosphorus is continually loaded into a lake, so the nitrogen-to-phosphorus ratio falls and nitrogen can become a limiting factor (**Figure 13**) (Smith 1982). The critical nitrogen-to-phosphorus ratio is slightly above the Redfield ratio of 16:1. The phytoplankton community in many temperate freshwater lakes therefore may have two broad configurations at which it can exist, one with low nutrient levels (organized by predation and dominated by green algae) and one with high nutrient levels (organized by competition and dominated by blue-green algae).

Estuaries are often heavily polluted with nutrients from sewage and industrial wastes. Because they form an interface between saltwater, in which nitrogen is often limiting to phytoplankton, and freshwater, in which phosphorus is typically limiting, estuaries are

Nutrient Ratios and Phytoplankton

Chemistry is an important aspect of ecosystem science, and one example of its utility involves the nutrient ratios of primary producers. In 1934 the oceanographer A. C. Redfield discovered that samples of organisms from the open ocean consistently exhibited the atomic ratio that mirrored that of the major dissolved nutrients in the deep ocean, $C_{106}N_{16}P_1$, which is now referred to as the **Redfield ratio** in his honor. In contrast to the constant Redfield ratio observed in some areas of the open ocean, the composition of phytoplankton from both marine and freshwater is highly variable (Arrigo 2005). The ratio of C:N:P in phytoplankton varies with the ratio present in the water and with the pH of the water (Sterner and Hessen 1994). In a series of 51 lakes surveyed, the C:N ratio varied from 4 to 20 (**Figure 14**), and the C:P ratio from 100 to 550, so that Redfield proportions are the exception rather than the rule in freshwater lakes as well as in the ocean.

In general, much more carbon is present in freshwater phytoplankton relative to nitrogen and phosphorus. Why should this variation matter? Algae with high C:P ratios are poor-quality food for herbivores such as zooplankton but are good-quality food for microbes. The C:N:P ratio could affect the structure of the food web. Phytoplankton with low N:P ratios are adapted to exponential growth, while those with high N:P ratios are *K*-strategists adapted to survive when resources are low. Different zooplankton species have differ-

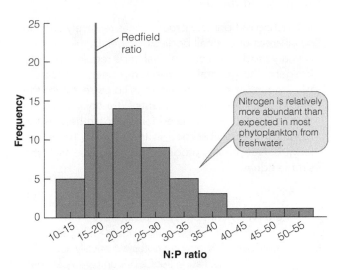

Figure 14 The nitrogen to phosphorus ratio observed in organisms in the deep ocean by A. C. Redfield is 16 (blue line). In the phytoplankton of freshwater lakes the N:P ratio varies more widely than a constant Redfield ratio would suggest. (Data from Heckey et al. 1993.)

ent C:N:P ratios, and consequently survive better feeding on different algal species. In general, there is much variation in C:N:P ratios in plants, less variation in bacteria, even less in zooplankton, and still less in fish species (Sterner et al. 1998).

complex gradients of nutrient limitation in which added phosphorus and nitrogen from pollution can strongly affect primary production throughout the estuary (Doering et al. 1995).

To summarize, the major controlling factors for primary production in freshwater communities are light (and temperature), phosphorus, and silicon (for diatoms), and occasional controlling factors include nitrogen and iron.

Terrestrial Communities

In terrestrial habitats, temperature ranges are much greater than in aquatic habitats, and the great variation in temperature from coastal to alpine or continental areas makes it possible to uncouple the solar radiation-temperature variable, which is so closely linked in aquatic systems. The large seasonal changes in radiation and temperature are reflected in the global pat-

terns of primary production. Using satellite imagery, we can identify continental and global patterns of terrestrial productivity. Satellites, such as the NOAA[2] meteorological satellite operated by the United States, have onboard sensors that record spectral reflectance in the visible and infrared portions of the electromagnetic spectrum.

As green plants photosynthesize, they display a unique spectral reflectance pattern in the visible (0.4–0.7 µm) and the near-infrared (0.725–1.1 µm) wavelengths (Goward et al. 1985). Vegetation indices that discriminate living vegetation from the surrounding rock, soil, or water have been developed by combining these spectral bands (see *Working with the Data: Estimating Primary Production from Satellite Data*). One

---

[2]National Oceanic and Atmospheric Administration of the United States.

WORKING WITH THE DATA

Estimating Primary Production from Satellite Data

Estimating net primary production is relatively easy on small areas or in small bodies of water, but estimating primary production on a global basis requires satellite imagery. The general approach has been to measure the amount of solar radiation and to correct it for the efficiency of light use by plants. The solar radiation that can potentially be used for photosynthesis is the radiation that is absorbed in the 400–700 nm wavelengths. Net primary production is calculated from the simple equation:

$$NPP = (APAR)(\varepsilon) \qquad (5)$$

where NPP = net primary productivity
        (g carbon per unit area per year)
    $APAR$ = absorbed photosynthetically active
        radiation (joules per unit area per year)
    ε = average light utilization efficiency
        (in g carbon per joule)

Satellites can measure radiation (both visible and invisible), and from measures of radiation we can derive estimates of $APAR$. For the oceans, $APAR$ can be correlated to measurements of surface chlorophyll. For land areas, satellites can measure greenness and obtain from it a vegetation index ($NDVI$) that corresponds to the amount of chlorophyll in land plants:

$$NDVI = \frac{NIR - RED}{NIR + RED} \qquad (6)$$

where $NDVI$ = normalized difference vegetation
        index
    NIR = near-infrared reflectance
        (0.725–1.1 µm)
    RED = red reflectance (0.6–0.7 µm)

The $NDVI$ is closely related to $APAR$ and can be used to estimate $APAR$.

The difficult parameter to be estimated from Equation (5) is ε, light use efficiency, and this estimate cannot be made from satellites but must be calculated from field measurements. This is relatively tedious, particularly on land, and much of the uncertainty in estimating net primary production comes from uncertainty of the exact value of ε. For the ocean, ε is estimated as a function of sea-surface temperature. For terrestrial systems, ε depends on ecosystem type (forest, grassland, tundra) and stresses from unfavorable levels of temperature, water, and nutrients (Field et al. 1998). There is now broad agreement among different models that estimate net primary productivity for terrestrial systems (Cramer et al. 1999; Ito and Oikawa 2004).

of the most common spectral vegetation indices, the normalized difference vegetation index (*NDVI*), is a ratio of near-infrared and visible red spectral bands. This index is closely correlated with primary productivity (Graetz et al. 1992). The NASA Earth Observing System Terra satellite with the Moderate Resolution Imaging Spectroradiometer (MODIS) sensor and the SeaWiFS (Sea Viewing Wide Field-of-view Sensor) on the SeaStar spacecraft are especially useful for monitoring global vegetation because they have global coverage at a resolution of 1.1 km at least once per day in daylight hours (Signorini et al. 1999). Global carbon models are now able to estimate primary production with high precision by calibrating the models to the observed satellite data. **Figure 15** illustrates the estimates of primary production for terrestrial ecosystems that can be achieved with current mechanistic models of primary production (Ito and Oikawa 2004; Running et al. 2004).

What limits primary production in terrestrial communities and produces the patterns shown in Figure 14? Rosenzweig (1968) quantified the conventional view that temperature and moisture were the master limiting factors for primary production in terrestrial ecosystems. He showed that actual evapotranspiration could predict the aboveground production of terrestrial communities with good accuracy. Actual evapotranspiration—the amount of water pumped into the atmosphere by evaporation from the ground and via transpiration from vegetation—is a measure of solar radiation, temperature, and rainfall. **Figure 16** illustrates these primary limiting factors to terrestrial productivity. Later models to predict primary production of terrestrial communities have become more complex as they include more biochemical details regarding respiration and photosynthesis as well as changes to atmospheric CO_2 levels in their predictions (Adams et al. 2004).

Primary production data for different types of plant communities have been summarized by Zheng et al. (2003). **Figure 17** shows the range of primary production for 12 growth forms of plants from around the world. In the temperate zone deciduous broadleaf forests are among the most productive

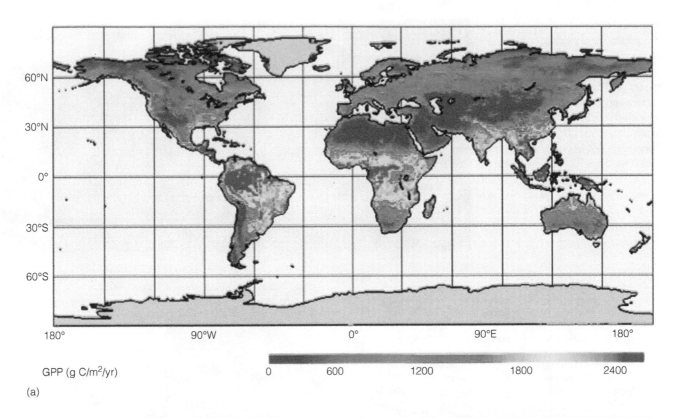

GPP (g C/m²/yr)

0   600   1200   1800   2400

(a)

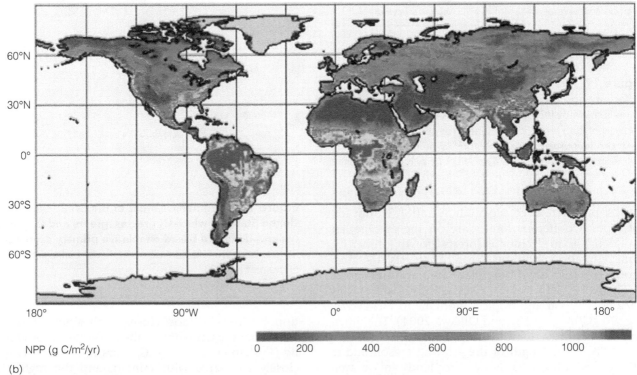

NPP (g C/m²/yr)

0   200   400   600   800   1000

(b)

Figure 15 Terrestrial primary productivity estimated from the Sim-CYCLE model. (a) Gross primary production of carbon (g/m²/y). (b) Net primary production. Global carbon models are now able to replicate observed patterns of primary production estimated from satellite images such as that shown in Figure 3. (From Ito and Oikawa 2004.)

Figure 16 Net primary production of terrestrial vegetation depends on solar radiation, temperature, and moisture. These primary limiting factors vary over the Earth and are modified locally by soil nutrient availability. (From Boisvenue and Running 2006).

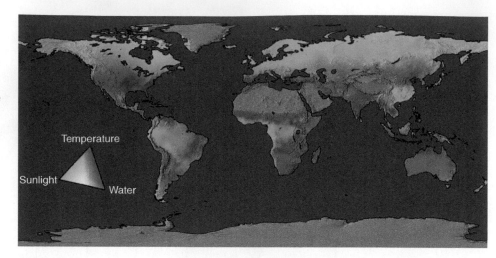

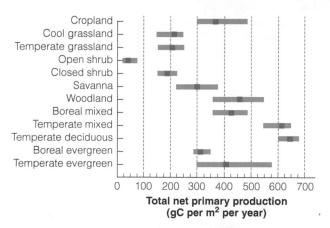

Figure 17 Total net primary production ranges for different vegetation types around the globe. The median value is given by the red square, and data represent a high and low of 25th and 75th percentile ranges, respectively. Total production includes both aboveground and belowground components. (Data from Zheng et al. 2003.)

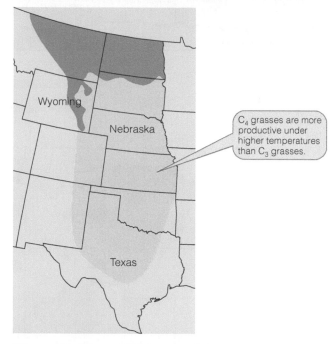

C$_4$ grasses are more productive under higher temperatures than C$_3$ grasses.

Figure 18 Geographical areas of the Great Plains of the United States in which C$_3$ grasses (green) and C$_4$ grasses (yellow) dominate based on relative primary production. C$_4$ grasses dominate on 74% of the Great Plains. (After Epstein et al. 1997.)

stands, but there is considerable variation within forest types. Coniferous forests are on the average less productive than deciduous forests growing under the same climatic conditions. Tropical forests are more productive than any of the plant communities shown in Figure 17, with tropical evergreen forests averaging 1072 gC/m^2/year and tropical montane forests averaging 910 gC/m^2/year (Ito and Oikawa 2004). The differences in productivity among forests are due to variation in the length of the growing season and to differences in leaf-area index. Croplands are on average less productive than forests.

Primary production in grasslands is strongly affected by the relative amounts of C$_3$ and C$_4$ grasses. In the Great Plains of the United States, C$_3$ grass produc-

tion is correlated most closely with temperature, and the higher the temperature the lower the C$_3$ productivity (Epstein et al. 1997). C$_4$ grass production is most closely correlated with rainfall, and the higher the rainfall the higher the production. C$_4$ grasses dominate about 74% of the Great Plains, and C$_3$ grasses are dominant only in the northern, cooler parts of this rea (**Figure 18**). Within the primary

Why Does Primary Production Decline with Age in Trees?

Forest managers have long known that tree growth and wood production decline with tree age. Net primary production in forests reaches a peak early in succession and then gradually declines by as much as 76% from that peak (Gower et al. 1996). For example, in Russia primary production of 140-year-old Norway spruce trees declines 58% from the peak reached at about 70 years of age (**Figure 19**).

Why should this occur? Three hypotheses have been advanced to explain age-related forest decline. The classical explanation is that a change occurs in the balance of photosynthesis and respiration. As trees grow larger with age, they have more tissues that respire and lose energy and proportionally less leaf area to photosynthesize. But this explanation is not supported by recent measurements showing that respiratory losses do not increase very much with tree age, because most of the sapwood uses little energy. The second hypothesis is nutrient limitation by nitrogen as the forest ages. Nitrogen is commonly found to be the limiting factor to tree growth, and as forests age, more woody litter accumulates on the soil surface. Woody litter decomposes very slowly compared to fine litter from leaves, so nitrogen becomes locked up in woody debris on the forest floor. The third and newest explanation is that as trees grow larger, water transport to the leaves becomes limited because of increased hydraulic resistance associated with the greater distance between the roots and the stomata of the leaves. Trees reduce stomatal conductance to conserve water in their tissues, and because photosynthesis is tightly coupled with stomatal conductance, production declines. This hypothesis is consistent with the observation that leaf stomata close earlier in the day in older trees compared with young trees.

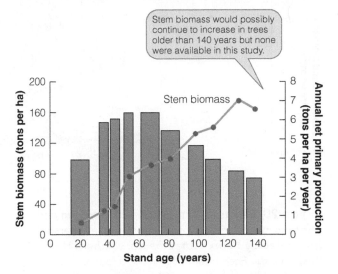

Figure 19 **Net primary production of Norway spruce trees in Russia in relation to the age of the trees in the forest stand.** Net production (green) increases only to about age 70 years, while stem biomass (red) continues to grow until at least 130 years of age. (Data from Gower et al. 1996.)

Current forest growth models suggest that nutrient limitation (hypothesis 2) and water flow limitations (hypothesis 3) are of nearly equal importance in reducing net primary production as trees age. An increase in respiration seems to contribute little to decreasing production (Gower et al. 1996). Knowing what limits primary production in forests is critical for understanding the effects of climate change on ecosystems.

control exerted by temperature and moisture, secondary limitations are imposed by the type of soil and its water holding capacity, and by nutrient availability (Sala et al. 1988).

Nutrient-addition experiments on local sites can be used to determine how much primary production can be limited by nutrients. Cargill and Jefferies (1984) added nitrate and phosphate to salt-marsh sedges and grasses in the subarctic zone to test for nutrient limitation. **Figure 20** shows that in the absence of grazing, the addition of nitrate doubled primary production of the sedges and grasses, and the joint addition of phosphate and nitrate quadrupled production.

In this marsh, as in many terrestrial communities, nitrogen is the major nutrient limiting productivity, and when nitrogen is suitably increased, phosphorus becomes limiting.

In unexploited virgin grassland or forest, all nutrients that the plants take up from the soil and hold in various plant parts are ultimately returned to the soil as litter that decomposes. The net flow of nutrients must be stabilized (such that input equals output), or the site would deteriorate over time. But in a harvested community, the situation is fundamentally different because nutrients are being continuously removed from the site. This makes it necessary to study

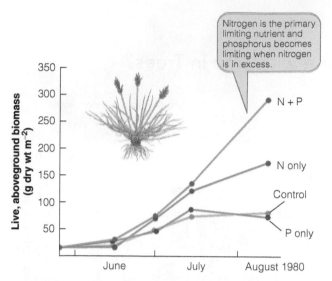

Figure 20 Effects of fertilization with nitrogen and phosphorus on primary production in a salt marsh dominated by *Carex subspathacea*, southern Hudson Bay, Canada. Each treatment included four replicates. (From Cargill and Jefferies 1984.)

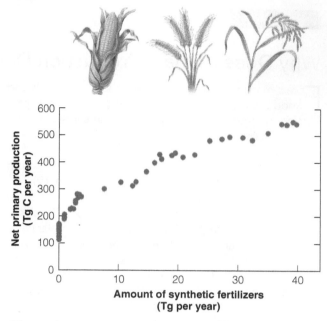

Figure 21 Net primary production of all crops in China from 1950 to 1999 in relation to the amount of synthetic fertilizer applied. Synthetic fertilizers include N, P_2O_5, and K_2O. Net primary production includes aboveground and belowground components of production. (Data from Huang et al. 2007.)

the nutrient demands of crops so that the nutrient capital of soils will not be progressively exhausted. Crops show clearly how plant production can be increased with added nutrients in fertilizer. Net primary production of crops in China has increased about fivefold from 1950 to 1999 and this increase has been associated with the use of more and more fertilizer (**Figure 21**).

For all terrestrial ecosystems that have been studied to date, both nitrogen and phosphorus are limiting nutrients (**Figure 22**). Elser et al. (2007) have suggested that there is a common thread of nitrogen and phosphorus limitation across all marine, freshwater, and terrestrial ecosystems, and that the combination of nitrogen and phosphorus produces a synergistic effect in which the sum is often greater than the separate parts.

Terrestrial communities, especially forests, have large nutrient stores tied up in the standing crop of plants. In this way they differ from marine and freshwater communities. This concentration of nutrients in the standing vegetation has important implications for nutrient cycles in forest communities. If the community is stable, the input of nutrients should equal the output, and a considerable amount of research effort is now being directed at studying nutrient cycles in terrestrial communities.

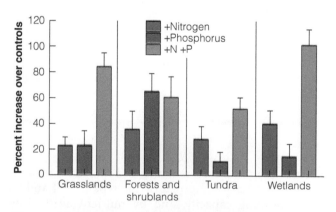

Figure 22 Response of aboveground biomass of terrestrial plants to a single addition of fertilizer as nitrogen, phosphorus, and combined nitrogen + phosphorus in four habitat types of terrestrial ecosystems. Results from 173 experiments are summarized and only those experiments in which N and P were manipulated were included in the analysis. N is represented by blue; P, by red; and N and P combined, by green. (Data from Elser et al. 2007.)

Plant Diversity and Productivity

One reason to conserve biodiversity might be that more-diverse sites are more productive. The hypothesis that diversity enhances productivity is an old one, and Charles Darwin suggested that it was true for terrestrial plants (Jolliffe 1997). The idea behind it is that different plant species have different resource needs such that the niches of several species would be complementary (Tilman et al. 2001). Under this hypothesis, resources would be more completely utilized when species richness is greater. But more species means more competition among plants, and it is possible that excessive competition could reduce productivity rather than permit it to increase. The net result of these opposing forces suggests a possible parabolic relationship between productivity and species diversity (**Figure 23**).

Over about half of the range of productivity we might expect a negative relationship between productivity and species richness. This has been observed many times in both plant and animal communities (Huston 1994). One example illustrates the point. The longest running experiment on productivity and diversity is the Park Grass Experiments at Rothamsted Experimental Station in England (Silvertown et al. 2006). These experiments measured the effects of liming and fertilizing pastures on hay production. From the earliest measurements in 1856 to 1978, a negative relationship between pasture productivity and biodiversity has always existed. On one pasture fertilized with nitrogen each year since 1856, species numbers declined over time, as follows:

Year	Number of plant species
1856	49
1862	28
1872	16
1903	10
1919	8
1949	3

In spite of the loss in biodiversity, productivity in the fertilized plots remained high. The Park Grass Experiments illustrate the right side of the parabolic curve shown in Figure 23.

In plant communities in low-nutrient soils or in severe physical environments, we expect a positive relationship between productivity and diversity. Tilman

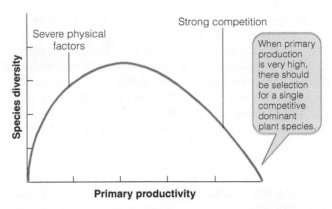

Figure 23 The expected relationship between primary productivity and biodiversity in plant communities. In areas of low productivity, we would expect productivity to rise with diversity, but in moderate to highly productive communities, productivity declines.

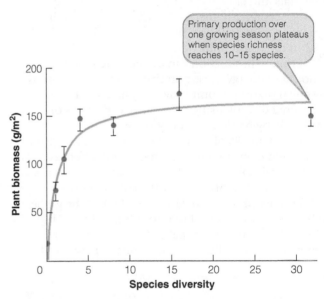

Figure 24 Relationship of species diversity to plant productivity in Minnesota grasslands. Varying numbers of species were seeded into 13 m × 13 m plots, and plants were harvested after completing growth over one summer. Only aboveground biomass was measured. Some plant biomass occurred even in plots with no species seeded because weeds invaded the plots before the final biomass clipping. (After Tilman et al. 1996.)

et al. (1997) have shown this in Minnesota grasslands on low-nitrogen soils. They seeded 289 plots with 1, 2, 4, 8, 16, or 32 perennial species and observed a positive relationship between plant productivity and species diversity (**Figure 24**). It is clear from these experimental studies that in this prairie community, productivity has reached a plateau after about 10 species. The highest productivity in these experiments came from plots that had the most functional groups of plants—C_3 grasses, C_4 grasses, legumes, forbs, and woody plants. Productivity is driven by a few species of dominants that make up a large fraction of the plant biomass (Grime 1997). The key to ecosystem productivity is in the biological characteristics of the dominant plants in a community.

From a global perspective, primary production is largely driven by the physical environment in the form of light, temperature, rainfall, and nutrient availability. Plants have adapted to these environmental constraints to produce through photosynthesis the materials that drive all the subsequent biota, including ourselves.

Summary

A community can be viewed as a complex machine that processes energy and materials. To study ecosystem metabolism, we must identify the food web of the community and then trace the flows of chemical materials and energy through the food web. Many ecologists prefer to measure energy use in studying community metabolism because energy is not recycled within the community.

Primary production can be measured by the amount of energy or carbon fixed via photosynthesis by green plants per unit time. CO_2 uptake can be measured directly using radioactive carbon-14 or indirectly by harvesting new growth.

Less than 1% of solar energy is captured by green plants and converted into primary production. Forests are relatively efficient, and aquatic communities are relatively inefficient, at capturing solar energy.

Primary production varies greatly over the globe; it is highest in the tropical rain forests and lowest in arctic, alpine, and desert habitats. Global primary production is distributed nearly equally between the oceans and the land. The sea is less productive than the land per unit of area (except for coastal areas and upwelling zones) because of limitations imposed by nitrogen, phosphorus, and iron. In freshwater lakes and streams, light, temperature, and nutrients restrict primary production, and phosphorus is the major limiting nutrient in many lakes.

Terrestrial primary productivity can be predicted from the length of the growing season, temperature, and rainfall. Nutrient limitations further restrict productivity levels set by these climatic factors, and the stimulation of plant growth achieved by fertilizing forests and crops indicates the importance of studying nutrient cycling in biological communities.

Remote sensing with satellites has provided new methods for measuring the spatial and temporal variability of primary production in the oceans and on land on large spatial scales. Identifying the factors that limit primary production is important for understanding how climatic changes will affect both natural and agricultural communities.

Review Questions and Problems

1   "Red tides" are spectacular dinoflagellate blooms that occur in the sea and often lead to mass mortality of marine fishes and invertebrates. Human deaths from eating shellfish poisoned with red tide algae is a worldwide problem. Review the evidence available about the origin of red tides, and discuss the implications for general ideas about what controls primary production in the sea. Landsberg (2002), Kubanek et al. (2005), and Wong et al. (2007) discuss this problem.

2   Iron fertilization of the Southern Ocean could occur either by dust blowing in winds off the continents or by upwelling of nutrient-rich deep water. Discuss in general how you might decide between these two mechanisms of iron transport. Is the chemical form of iron compounds important in stimulating phytoplankton growth? Read Meskhidze et al. (2007) for a discussion of this problem.

3   In the Great Plains grasslands of the United States, Epstein et al. (1997) showed that primary production of C_3 grasses could be predicted from mean annual temperature, with minimal contribution from mean annual precipitation. Discuss why precipitation and soil nutrients do not appear to be relevant variables for C_3 grass production in this ecosystem.

4   In discussing the effect of light on primary productivity in the ocean, Nielsen and Jensen (1957, p. 108) state:

It is thus quite likely that a permanent reduction of the light intensity at the surface (to, e.g., 50 percent of its normal value without the other factors being affected—a rather improbable condition in Nature) in the long run would have very little influence on the organic productivity as measured per surface area.

How could this possibly be true?

5   Even though the concentration of inorganic phosphate in the water of the North Atlantic Ocean is only about 50% of that found in the other oceans, the North Atlantic is more productive than most of the other oceans. How can one reconcile these observations if nutrients limit primary productivity in the oceans?

6   The discovery that iron was a primary factor limiting primary production in large areas of the ocean caused the oceanographer John Martin to say in the late 1980s that "Give me a half tanker of iron, and I will give you an ice age." List the causal links that could make this prediction come true, and read the evaluation by Buesseler et al. (2004) that suggests this prediction could not possibly be correct.

7   Is it possible to have more than one limiting factor for primary production at any given time? How can we interpret the synergy between two nutrients like nitrogen and phosphorus as shown for example in Figure 22 if we accept Liebig's law of the minimum?

8   Tilman et al. (1982, p. 367) state:

We suggest that the spatial and temporal heterogeneity of pelagic environments will prevent us from meaningfully addressing questions on short time scales or small spatial scales.

Discuss the general issue of whether there are some questions in community ecology that we cannot answer because of scale.

9   North American grasslands are similar in structure to South African grasslands but the grass species differ because of their divergent evolutionary history. Both areas have dominant C_4 grasses with less abundant C_3 forbs and woody plants. But South Africa has greater climatic variability and poorer soils than North America. Would you expect the two areas to show the same relationships between rainfall and net primary production? Knapp et al. (2006) present an analysis of these questions.

10   Many studies of nutrient limitation in freshwater lakes and in the ocean use small water bottles as experimental units to which nutrients of various types are added. Other aquatic ecologists use mesocosms of plastic that hold several cubic meters of water for their experiments. Discuss why these small-scale experiments might give less reliable results than whole-lake manipulations. Schindler (1998) and Howarth and Marino (2006) discuss the scale problem for freshwater lakes and coastal marine ecosystems.

11   Photosynthetic organisms produce about 300×10^{15} g of oxygen per year (Holland 1995). If this oxygen accumulated, the oxygen content of the atmosphere would double every 2000 years. Why does this not happen? Is the global system regulated? If so, how is this regulation accomplished?

Overview Question

What limits primary production in agricultural systems? List the differences in the controls of primary production in natural plant communities and agricultural crops, and discuss the implications for sustainable agriculture.

Suggested Readings

- Arrigo, K. R. 2005. Marine microorganisms and global nutrient cycles. *Nature* 437:349–355.

- Boisvenue, C., and S. W. Running. 2006. Impacts of climate change on natural forest productivity—evidence since the middle of the 20th century. *Global Change Biology* 12:862–882.

- Boyd, P. W., T. Jickells, C. S. Law, S. Blain, E. A. Boyle, K. O. Buesseler, K. H. Coale, et al. 2007. Mesoscale iron enrichment experiments 1993–2005: Synthesis and future directions. *Science* 315:612–617.

- Elser, J. J., M. E. S. Bracken, E. E. Cleland, D. S. Gruner, W. S. Harpole, H. Hillebrand, J. T. Ngai, et al. 2007. Global analysis of nitrogen and phosphorus limitation of primary producers in freshwater, marine and terrestrial ecosystems. *Ecology Letters* 10: 1135–1142.

- Running, S. W., R. R. Nemani, F. A. Heinsch, M. Zhao, M. Reeves, and H. Hashimoto. 2004. A continuous satellite-derived measure of global terrestrial primary production. *BioScience* 54:547–560.

- Schindler, D. W. 2006. Recent advances in the understanding and management of eutrophication. *Limnology & Oceanography* 51:356–363.

- Silvertown, J., P. Poulton, E. Johnston, G. Edwards, M. Heard, and P. M. Biss. 2006. The Park Grass Experiment 1856–2006: Its contribution to ecology. *Journal of Ecology* 94:801–814.

- Smith, V. H., S. B. Joye, and R. W. Howarth. 2006. Eutrophication of freshwater and marine ecosystems. *Limnology & Oceanography* 51:351–355.

- Turkington, R., E. John, S. Watson, and P. Seccomb-Hett. 2002. The effects of fertilization and herbivory on the herbaceous vegetation of the boreal forest in north-western Canada: A 10-year study. *Journal of Ecology* 90:325–337.

- Wolff, E. W., H. Fischer, F. Fundel, U. Ruth, B. Twarloh, G. C. Littot, R. Mulvaney, et al. 2006. Southern Ocean sea-ice extent, productivity and iron flux over the past eight glacial cycles. *Nature* 440:491–496.

Credits

Illustration and Table Credits

F2 D. L. DeAngelis, *Dynamics of Nutrient Cycling and Food Webs*, p. 416. Copyright © 1992 with kind permission from Springer Science and Business Media. F3 NASA, Earth Observatory. F6 Thomas L. Hayward, "Primary production in the North Pacific Central Gyre: A controversy with important implications," *Trends in Ecology & Evolution* 6(9):4, 1991. Copyright © 1991. Published by Elsevier Science, Ltd. F7 Reprinted with permission from J. H. Ryther and W. M. Dunstan, "Nitrogen, phosphorus, and eutrophication in the coastal marine environment," *Science*, Vol. 171, pp. 1008–1009, 1971. Copyright © 1971 American Association for the Advancement of Science. Used with permission. F8 From J. A. Downing et al., "Meta-analysis of marine nutrient-enrichment experiments" *Ecology* 80:1157–1167, 1999. Copyright © 1999. Reprinted by permission of The Ecological Society of America. F9 K. H. Coale et al., "Southern Ocean Iron Enrichment Experiment," *Science* 304:408–414, 2004. Copyright © 2004 American Association for the Advancement of Science. Used with permission. F15 A. Ito and T. Oikawa, "Global mapping of terrestrial primary productivity," in M. Shiyomi, editor, *Global Environmental Change in the Ocean and on Land*, pp. 343–358, 2004, p. 349. Copyright © 2004.

Reprinted by permission of Terra Scientific Publishing Company. F16 C. Boisvenue and S. W. Running, "Impacts of climate change on natural forest productivity," *Global Change Biology* 12:862–882, 2006. Copyright © 2006. Reprinted by permission of Blackwell Publishing. 17 D. Zheng, S. Prince, and R. Wright. 2003. Terrestrial net primary production estimates for 0.5° grid cells from field observations—a contribution to global biogeochemical modeling. *Global Change Biology* 9:46–64. F20 S. M. Cargill and R. L. Jefferies, "Nutrient limitation of primary production in a sub-arctic salt marsh," *Journal of Applied Ecology* 21:657–668, 1984. Copyright © 1984. Reprinted by permission of Blackwell Publishing. 24 David Tilman et al., "Productivity and sustainability influenced by biodiversity in grassland ecosystems," *Nature* 379:718–720, 1996. Copyright © 1996 Nature Publishing Group.

Photo Credits

Unless otherwise indicated, photos provided by the author.

CO. Polka Dot Images/Jupiter Images. 11 D. W. Schindler. 12a Y. Tsukii, Hosei University, Japan.

Ecosystem Metabolism II: Secondary Production

Key Concepts

- Energy fixed by green plants flows to either herbivores or detritus, or is lost in respiration.

- In forest ecosystems, most primary production goes directly into detritus, and only 3%–4% goes into the herbivore food web. Aquatic ecosystems are more productive, and typically 20% or more of primary production is consumed by herbivores.

- Homeotherms use more than 98% of their ingested energy to maintain body temperature, and are much less efficient energy users than are poikilotherms such as insects.

- Secondary production is limited by primary production and the second law of thermodynamics, which states that no energy transfer is completely efficient.

- About 5%–20% of the energy passes from one trophic level to the next; the remainder is lost to respiration or goes to detritus.

- Most of the animals humans consider important are a trivial part of the energetics of the ecosystem. Plants and detritus are the main players in all ecosystems.

- Metabolic theory can derive many ecological patterns from individual organisms to ecosystems by tracing their origin to the relationship of basal metabolic rate to body size and temperature.

From Chapter 23 of *Ecology: The Experimental Analysis of Distribution and Abundance*, Sixth Edition. Eugene Hecht.
Copyright © 2009 by Pearson Education, Inc. Published by Pearson Benjamin Cummings. All rights reserved.

KEY TERMS

basal metabolic rate The amount of energy expended by an animal while at rest in a neutral temperate environment, in the post-absorptive (fasting) state; the minimum rate of metabolism.

detritus The plant production not consumed by herbivores.

Eltonian pyramid Abundance or biomass of successive trophic levels of an ecosystem, illustrating the impact of energy flows through successive trophic transfers.

field metabolic rate The amount of energy used per unit of time by an organism under normal conditions of life in a natural ecosystem.

green world hypothesis The proposed explanation for the simple observation that the world is green, that herbivores are held in check by their predators, parasites, and diseases, although other explanations have been suggested.

gross productivity The assimilation rate of an animal, which includes all the digested energy less the urinary waste.

metabolic theory of ecology An attempt to derive patterns of individual performance, population, and ecosystem dynamics from the fundamental observation that the metabolic rate of individuals is related to body size and temperature.

trophic efficiency Net production at one trophic level as a fraction of net production of the next lower trophic level.

We have seen that primary production by green plants drives the entire biosphere. Primary production in agriculture is a particularly important part of global ecosystem metabolism, and we now ask how the energy fixed by green plants is dissipated by all the other parts of the food chain in both natural and agricultural ecosystems.

Measurement of Secondary Production

The biomass of plants that accumulates in an ecosystem as a result of photosynthesis can eventually move to one of two fates: consumption by herbivores or degradation by detritus feeders. The fate of the energy and materials captured in primary productivity can be illustrated most simply by looking at the metabolism of an individual herbivore.

The partitioning of food materials and energy for an individual animal can be seen as a series of dichotomies. With respect to energy, we have the energetic approach, championed by Eugene Odum (see Researcher Profiles), working at the University of Georgia, and his brother Howard Odum, which has formed the basis of major insights into how ecosystems work in both natural and agricultural systems. **Figure 1** illustrates how energy from any trophic level is partitioned into components that can be measured. This scheme could be presented for carbon intake or any essential nutrient, and again we have the choice of using chemical materials or energy to study the system.

Let us look at this scheme illustrated in Figure 1 in detail. Every animal will remove some energy or material from a lower trophic level for its food. Some of this energy will not be used, as for example, when a beaver fells a whole tree and eats only some of the bark. In this case, most of the energy removed from the plant trophic level is not used by the beaver but is left to decompose. Of the material consumed, some energy passes through the digestive tract (is egested) and is lost in the feces. Of the remaining digested energy, some is lost as urinary output, and the rest is available for assimilation or metabolic energy. Assimilated energy can be subdivided into two general pathways, maintenance and production. All animals must expend energy in the process of respiration just to subsist. Production occurs by using assimilated energy for growth and for reproduction.

How can we measure the components of secondary productivity in an animal community? Several techniques are available (Petrusewicz and MacFadyen 1970), but the general procedure is as follows. Each species of animal is considered separately. To determine the gross energy intake of the population, we must know the feeding rate of individuals. This can be measured by confining a herbivorous animal to a feeding plot and measuring herbage biomass before and after feeding. In some predators, such as birds of prey, the number of food items being consumed can be counted by direct observation. Indirect techniques such as weight of stomach contents can also be used but require knowledge of the rates of digestion and feeding.

Assimilated or metabolizable energy can be measured very simply in the laboratory where gross intake can be regulated and feces and urine can be collected, but in the field it is extremely difficult to estimate

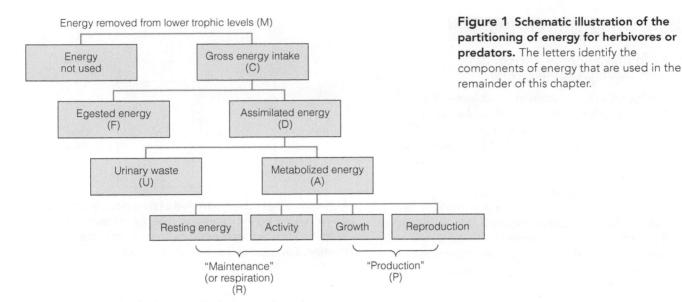

Energy removed from lower trophic levels (M)

Figure 1 Schematic illustration of the partitioning of energy for herbivores or predators. The letters identify the components of energy that are used in the remainder of this chapter.

assimilation directly. The usual approach is to measure it indirectly by use of the relation

$$\text{Assimilation rate} = \text{Respiration rate} + \text{net productivity} \quad (1)$$

If we can measure the rates of respiration and production, we can get assimilation rate by addition. Note that all these are rates (per unit time), and that assimilation rate is also called **gross productivity**.

Respiration can be measured very easily in laboratory situations by confining an animal to a small cage and measuring oxygen consumption, CO_2 output, and heat production directly. There is a minimum rate of metabolism, the **basal metabolic rate**, which increases with body size in warm-blooded animals. The relationship between basal metabolic rate and body size is a key relationship recognized many years ago by physiologists (White and Seymour 2005). **Figure 2** illustrates this fundamental relationship for 619 species of mammals. Considerable controversy has raged over the slope of the relationship between basal metabolic rate and body mass. The slope of ¾ has been suggested as a universal constant for all living organisms from bacteria to elephants (West and Brown 2005). But it is not constant in all taxonomic groups (Glazier 2005). For example, within the mammals, bats of a given body size have lower basal metabolic rates than rodents, and hedgehogs show a slope of 0.5 in the regression of basal metabolic rate versus body mass (Duncan et al. 2007). Basal metabolism is measured under conditions in which the animal is at rest and has no food in its stomach, at a temperature at which the animal is not required to expend energy for extra heat production or cooling. Mea-

sured in such an abstract way, basal metabolism is not closely related to respiration losses in field situations, in which activity is necessary, temperature varies, and digestion is occurring.

Energy metabolism in the field can be measured directly by means of doubly labeled water (Nagy 2005). This method involves injecting wild animals with water in the form $^3H_2O^{18}$ and then measuring the loss rates of the hydrogen (tritium) and oxygen isotopes. The hydrogen isotope relates to water loss, whereas the oxygen isotope is lost in both CO_2 and water. The difference between these isotope loss rates represents CO_2 loss alone,

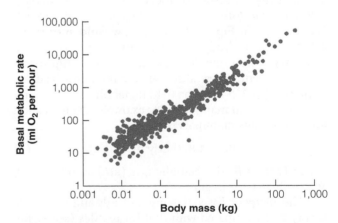

Figure 2 The classic mouse-to-elephant curve of the relationship of basal metabolic rates in mammals to body mass. Data from 619 species of mammals are included, and the slope of the log-log regression is 0.686. (Data from White and Seymour 2003.)

Estimating Energy Expenditure with Doubly Labeled Water

Estimating energy expenditure in free-living animals has always been a Holy Grail for ecologists interested in bioenergetics. In 1966 N. Lifson and R. McClintock suggested that turnover rates of body water labeled with radioactive isotopes could be used to measure energy balance in animals. They showed that one could do this with doubly labeled water, $^2H_2^{18}O$ or $^3H_2^{18}O$. The method operates schematically in an individual animal as shown in **Figure 3**. Tritium (^3H) leaves the body only in water, whereas ^{18}O leaves the body both in water and in carbon dioxide in respiration. We can get the amount of energy utilized by the difference:

$$^{18}O \text{ elimination} - {}^3H_2 \text{ elimination} = CO_2 \text{ production}$$

In addition to measuring energy utilization, the doubly labeled water method can measure water turnover rates.

In practice this method is applied by injecting a known amount of the two isotopes, waiting 12–24 hours to allow the isotopes to mix throughout the body, and then taking a blood sample to define the start of the experiment. The animal is released and then recaptured after a specified time period, and a second blood sample is taken to define the end of the

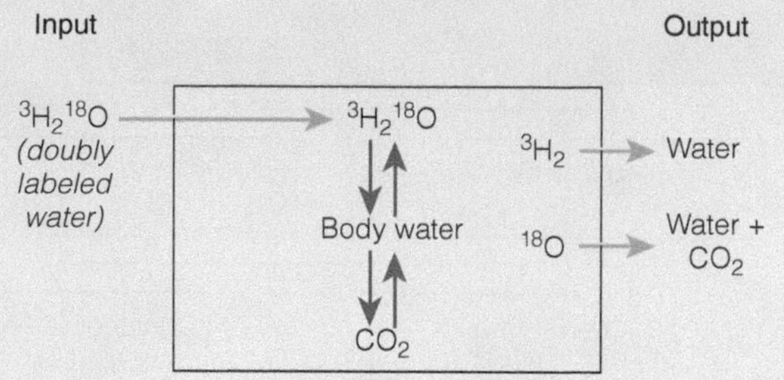

Figure 3 Schematic illustration of how doubly labeled water is used in metabolism so that, by measuring the output of the two isotopes, one can estimate the energy utilization of the individual animal.

experiment. The isotope concentrations in the blood samples are measured on a liquid scintillation counter for tritium and proton activation analysis, or by mass spectrometry for ^{18}O.

The length of time over which a doubly labeled water experiment can be conducted depends on the turnover of the two isotopes. For studies of mammals and birds, the optimal metabolic interval is 7–21 days, but for reptiles and amphibians with lower metabolic rates, 1–2 months can be used for the sampling interval. By reinjecting individuals, field metabolic rates can be determined for long time periods (Nagy 1989).

and is a field measure of metabolic rate (see *Working with the Data: Estimating Energy Expenditure with Doubly Labeled Water*). **Figure 4** shows how **field metabolic rates** vary with body size in mammals and birds. No single regression line describes this relationship for all groups of vertebrates, although for medium-sized endotherms (250–550 g) the field metabolic rates are similar for birds and mammals. For example, the regression for herbivorous mammals is

$$FMR = 4.82(\text{body mass}^{0.734}) \tag{2}$$

where FMR = field metabolic rate (kJ/day/individual), and body mass is in grams.

The energetic costs of living for birds and mammals is very high compared with that for reptiles (see Figure 4). A 250-g mammal or bird will use about 320 kJ of energy daily, whereas a 250-g iguanid lizard will use about 19 kJ per day, a 17-fold difference in energy use. Respiration in cold-blooded organisms is highly dependent on temperature, as well as body size.

Net production in a population can be measured by the growth of individuals and the reproduction of new animals. We have discussed techniques for measur-

ing changes in numbers, and growth can be measured by weighing individuals at successive times. The only admonition we must make here is that sampling must be frequent enough so that individuals are not born and then die in the interval between samples. Net production is usually measured as biomass and converted to energy measures by determination of the caloric value of a unit of weight of the species.

Figure 5 gives a schematic representation of how production can be determined from information on population changes. In this hypothetical population, production is the sum of growth and natality additions:

$$\begin{aligned} \text{Production} &= \text{growth} + \text{natality} \\ &= 20 + 10 + 10 + 10 + 10 \\ &\quad + 30 - 10 - 10 \\ &= 70 \text{ units of biomass} \end{aligned} \tag{3}$$

Note that we can calculate production in a second way:

$$\begin{aligned} \text{Production} &= \frac{\text{net change}}{\text{in biomass}} + \frac{\text{losses by}}{\text{mortality}} \\ &= 30 + 40 = 70 \text{ units of biomass} \end{aligned} \tag{4}$$

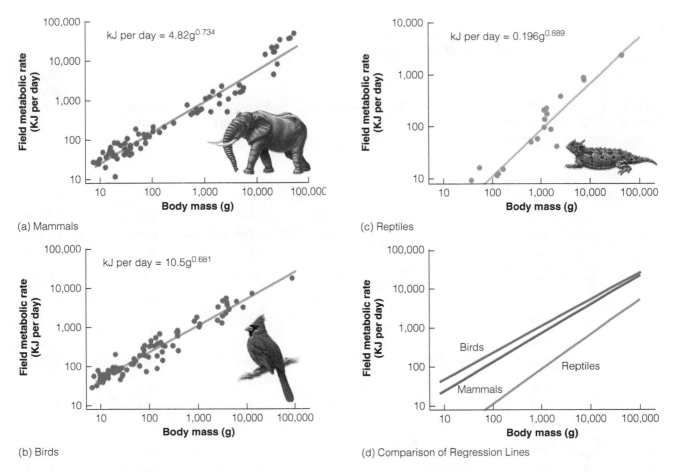

Figure 4 Field metabolic rate in relation to body size for (a) mammals, (b) birds, and (c) reptiles. In (d) the regression lines for birds and mammals are superimposed and contrasted with that for reptiles. (Modified from Nagy 2005.)

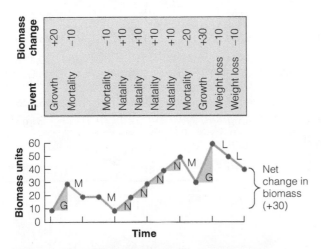

Figure 5 **Changes in biomass of a hypothetical population to illustrate the factors that contribute to secondary production.** The net change in biomass over a given time period is the outcome of gains from net growth and reproduction and losses from death and emigration.

Losses caused by mortality (including harvesting) or emigration are a part of production and should not be ignored. We can see this very clearly by looking at a population that is stable over time (net change in biomass is zero): a ranch that has the same biomass of steers this year as last year does not necessarily have zero production over the year.

Now we will look at an actual example of the calculations of the secondary productivity in an African elephant population. Petrides and Swank (1966) estimated the energy relations of the elephant herds in Queen Elizabeth National Park (now Ruwenzori National Park) in Uganda. To do these calculations, we must assume a stable population of elephants with a stationary age distribution. For convenience we will also assume that no births or deaths occur during the study interval. The population was counted and the age structure estimated to construct a life table. The maximum age was estimated at 67 years, and the survivorship schedule is given in **Table 1**. Weight growth was estimated from some records of zoo

Table 1 Elephant life table and production data, Queen Elizabeth (now Ruwenzori) National Park, Uganda, November 1956 to June 1957.

Age, x (years)	No. alive at beginning of age x	Mortality rate per year	Median no. alive	Weight average (kg)	Weight increment (kg)	Population weight increment[a] (kg)
1	1000	0.03	850.0	91	91	77,350
2	700	0.02	630.0	205	114	71,820
3	560	0.10	532.0	318	114	60,648
4	504	0.10	478.5	455	136	60,076
5	453	0.10	430.5	614	159	68,450
6	408	0.10	387.5	795	182	70,525
7	367	0.10	348.5	1000	205	71,443
8	330	0.10	313.5	1205	205	64,268
9	297	0.10	282.0	1409	205	57,810
10	267	0.05	260.5	1614	205	53,403
11	254	0.05	247.5	1818	205	50,738
12	241	0.05	235.0	2023	205	48,175
13	229	0.05	223.5	2205	182	40,677
14	218	0.05	212.5	2386	182	38,675
15	207	0.02	205.0	2591	205	42,025
16	203	0.02	201.0	2795	205	41,205
17	199	0.02	197.0	3000	205	40,385
18	195	0.02	193.0	3182	182	35,126
19	191	0.02	189.0	3386	205	38,745
20	187	0.02	185.0	3591	205	37,925
21	183	0.02	181.0	3750	159	28,779
22	179	0.02	177.0	3864	114	20,178
23	175	0.02	173.0	3955	91	15,743
24	171	0.02	169.5	4045	91	15,425
25	168	0.02	166.5	4091	45	7493
26–67	3107[b]	0.02–0.20	3026.5[b]	4091	0	0
Total	**10,933**		**10,487.5**	**2291**[c]	**4091**	**1,162,084**

[a]Median no. alive times weight increment.

[b]Sum of the numbers of animals in each year-class for ages 26 to 67.

[c]Average body weight.

SOURCE: After Petrides and Swank (1966).

animals and a limited amount of field data on weights. The average weight at each age and the average increase in weight from one year to the next are also given in Table 1. If the age structure is stationary, then

$$\text{Growth in biomass} = \sum_{\text{all ages}} \begin{pmatrix} \text{median no. alive} \\ \text{during age} \\ x \text{ to } x+1 \end{pmatrix} \times \begin{pmatrix} \text{average weight} \\ \text{growth for ages} \\ x \text{ to } x+1 \end{pmatrix} \quad (5)$$

This is calculated in the last column of Table 1. The caloric value of elephants is 6.276 kJ/g of live weight. We determine growth as follows:

From the bottom line in Table 1, 1000 elephants lived 10,487.5 elephant-years and produced 1,162,084 kg of growth.

$$\text{Average growth (in weight)/elephant/yr} = \frac{1,162,084}{10,487.5} = 110.8 \text{ kg}$$

$$\text{Average growth (in energy)/elephant/yr} = 110,800 \text{ g} \times 6.276 \text{ kg/g}$$
$$= 695,381 \text{ kJ}$$

The population density of elephants was 2.077 elephants/km^2 or 0.000002077 elephant/m^2. Thus

$$\text{Growth} = 695,381 \text{ kJ} \times 0.000002077$$
$$= 1.44 \text{ kJ/m}^2/\text{yr}$$

A large amount of the food consumed by elephants passes through as feces. From studies on captive elephants, an average 2273-kg (5000-lb) elephant would consume 23.59 kg dry weight of forage per day and produce from this 13.25 kg dry weight of feces. The food plants are worth approximately 16.736 kJ/g dry weight, so we can calculate

$$\text{Average consumption} = 23,590 \times 16.736$$
$$= 394,802 \text{ kJ}$$
$$\text{Average fecal production} = 13,250 \times 16.736$$
$$= 221,752 \text{ kJ}$$

Counting these for a whole year and multiplying by the number of elephants per square meter, we obtain:

$$\text{Food consumed} = 394,802 \text{ kJ/day/elephant} \times 365 \text{ days}$$
$$\times 0.000002077 \text{ elephant/m}^2$$
$$= 299 \text{ kJ/m}^2/\text{yr}$$
$$\text{Feces produced} = 221,752 \text{ kJ/day/elephant} \times 365 \text{ days}$$
$$\times 0.000002077 \text{ elephant/m}^2$$
$$= 168.1 \text{ kJ/m}^2/\text{yr}$$

We know that

$$\frac{\text{Food energy}}{\text{consumed}} = \text{feces} + \text{growth} + \text{maintenance}$$
$$299 = 168.1 + 1.44 + \text{maintenance}$$

Maintenance must be about 130 kJ/m^2/yr if we ignore the losses due to urine production and the production due to newborn animals.

We can also estimate maintenance from Equation for the field metabolic rate of a standard 2273-kg (5000-lb) elephant:

$$FMR = 4.82M^{0.734}$$
$$= 4.82(2,273,000^{0.734})$$
$$= 223,250 \text{ kJ/day/elephant}$$
$$\text{Estimated maintenance} = 223,250 \text{ kJ/day} \times 365 \text{ days}$$
$$\times 0.000002077 \text{ elephant/m}^2$$
$$= 169 \text{ kJ/m}^2/\text{yr}$$

This is slightly greater than the maintenance estimate of 130 kJ/m^2/yr previously obtained.

Finally, we can determine the standing crop of elephants in energetic terms:

$$\text{Standing crop} = 0.000002077 \text{ elephant/m}^2$$
$$\times 2,273,000 \text{ g} \times 6.3 \text{ kJ/g}$$
$$= 30 \text{ kJ/m}^2$$

A rough estimate of primary productivity by the harvest method produced an estimate of net primary productivity of 3125 kJ/m^2/yr for the foraging area of the elephants.

We can summarize these estimates for energy dynamics of the African elephant population of Queen Elizabeth Park, as follows:

	Energy (kJ/m^2/yr)
Net primary production	3125[a]
Secondary production	
Food consumed	299
Fecal energy lost	168
Maintenance metabolism	130
Growth	1.44
Standing crop of elephants	30

[a]Probably a low estimate.

Clearly, over 99% of the energy intake of these elephants is used in maintenance or lost in fecal production.

The details of estimating secondary production obviously vary from species to species, and the number of assumptions that must be made depend on how well the species is studied. The procedure is to repeat these calculations for all dominant species in a community and by addition to obtain the secondary production of the community. This procedure is more tedious than conceptually difficult, and we can now consider the results of this kind of analysis.

Problems in Estimating Secondary Production

In principle, ecologists can apply the kind of techniques just described for elephants to all the major consumer species in a community. In practice these calculations have led ecologists into three conceptual problems (Cousins 1987).

The first difficulty is that the individuals of a particular species do not fit clearly into discrete trophic levels. Plants are usually easily assigned to the producer level, trophic level 1. But the next trophic level includes animals that eat other animals as well as plants (**Figure 6**). House mice (*Mus musculus*) are thought of as herbivores, yet they consume substantial amounts of insects at some seasons of the year. The red fox (*Vulpes fulva*) eats herbivores such as rabbits, but they can also eat plant material, detritus, and other carnivores. In general, the higher up the food chain, the less clear it is how to categorize a species.

A second problem in estimating secondary production is what to do with **detritus**. The normal procedure has been to include detritus and dung in the first trophic level—to treat detritus as plant material—but this is not correct. Detritus from herbivores should be kept separate from plant detritus. Moreover, typically a complex food web exists within detritus itself, such that it is difficult to assign to the producer-herbivore-carnivore type of trophic organization.

The third major difficulty in estimating secondary production is the practical one of sampling a complex community adequately and in allowing for nonequilibrium conditions in natural ecosystems. Aquatic ecologists have been particularly aware of these problems (Kokkinn and Davis 1986), and the rapid changes that occur in plankton populations and benthic invertebrates may render accurate estimates of production impossible on a finite financial budget.

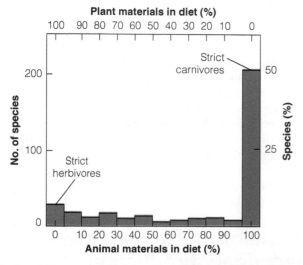

Figure 6 The composition of the diets of 430 species of North American birds. While 50% of birds are strictly carnivores and 8% are strict herbivores, over 40% are omnivores that feed on more than one trophic level. (Modified from Peters 1977.)

Given these problems, ecologists have moved away from trying to estimate secondary production for whole trophic levels, and have begun to analyze parts of food webs taxonomically (Cousins 1987). We can analyze single species such as the elephant in detail to estimate their effects on the community. We now consider how efficient different species are in their use of energy.

Ecological Efficiencies

If we view animals as energy transformers, we can ask questions about their relative efficiencies. A large number of ecological efficiencies can be defined, and here we are concerned with efficiency *within* a species on a particular trophic level. One useful measure of efficiency at the level of a single species is defined as follows:

$$\text{Production efficiency} = \frac{\substack{\text{net productivity} \\ \text{of species } n}}{\substack{\text{assimilation} \\ \text{of species } n}} \quad (6)$$

Data on production efficiency for individual species are readily obtained, and Humphreys (1979, 1984) has summarized 235 energy budgets measured in natural populations. The first question we can ask about these energy budgets is whether different taxonomic groups have different efficiencies. Based on efficiency, homeotherms separate into four groups: insectivores, birds, small mammals, and other mammals; poikilotherms separate into three groups: fish and social insects, nonsocial insects, and other invertebrates. Within each of these groups, as respiration goes up 1 kJ, production likewise goes up by 1 kJ. This means that production efficiencies $P/(R + P)$ are the same for all sizes of animals within the seven groups. Analyzing average production efficiencies illustrates an important generalization in secondary production. For mammals and birds in general, respiration seems to utilize 97%–99% of the energy assimilated, and consequently only 1%–3% of the energy goes to net production in these groups. For insects the loss is less; approximately 59%–90% of the energy assimilated is used for respiration. This difference between insects and mammals is a reflection of the cost of homeothermy. There appears to be no variation in production efficiency between animals in different habitats. Aquatic and terrestrial poikilotherms seem to have equal production efficiencies (Humphreys 1979) (**Table 2**).

Another measure of efficiency that we can use to describe ecosystems involves transfers between trophic levels and is called **trophic efficiency**:

$$\text{Trophic efficiency} = \frac{\substack{\text{net production at} \\ \text{trophic level } i + 1}}{\substack{\text{net production at} \\ \text{trophic level } i}} \quad (7)$$

Thermodynamics and Ecology

The energetic approach to ecosystem science has a strong foundation in thermodynamics and in physics in general. Whatever happens in an ecosystem can be described as a transfer of energy from one place to another, or as a transformation of energy between different forms such as chemical energy to heat. In 1964 Eugene Odum dubbed energetics the "new ecology," and his brother Howard T. Odum has been a champion of the idea that energy is the central object of study in ecosystem ecology (Odum 1983; Patten 1993). The essential features of Howard Odum's approach to ecosystem analysis can be stated as five conjectures (Månsson and McGlade 1993):

Conjecture 1. All significant aspects of ecosystems can be captured by the single concept, energy.

Conjecture 2. The formalism of energy circuit language is sufficient for a holistic approach to ecosystem analysis. Odum invented energy circuit language as a substitute for linear mathematical models. A few examples of this symbolic language are shown in **Figure 7**:

Conjecture 3. Ecosystems evolve such that power is maximized. This idea, the *maximum power principle*, is a new principle for the evolution of ecosystems.

Conjecture 4. Hierarchical structures can be deduced from the flow of energy in ecosystems.

Conjecture 5. Ecological succession is due to the maximum power principle and culminates in stable systems with maximum biomass and maximum gross production.

Even though these conjectures represented a bold attempt to bring thermodynamic theory into ecology, they have been superceded by new ecological concepts, and few ecologists now believe that energy can play such a central role in ecosystem analysis. Energy circuit language does not easily permit nonlinear systems analysis, which is now a common approach in defining system dynamics. There does not seem to be any evolutionary theory to support the maximum power principle as a guiding principle in ecosystem evolution, and ecological succession does not lead to maximum biomass and stable equilibrial systems. The complexities of interactions between species cannot be reduced to the single currency of energy.

The scope of energetics in the analysis of ecosystems has now become more limited than Howard Odum first proposed. Energetics has made an important contribution to ecological understanding, particularly with respect to the role of humans in ecosystems. Progress in science often involves the rejection of bold conjectures, and we gain deeper insights into how the world works via the process of conjecture and refutation.

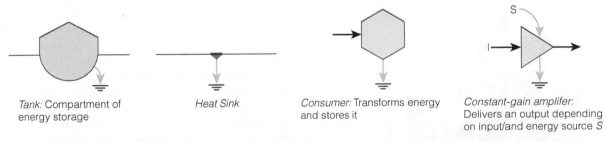

Tank: Compartment of energy storage

Heat Sink

Consumer: Transforms energy and stores it

Constant-gain amplifer: Delivers an output depending on input/and energy source *S*

Figure 7 **Four examples of energetics symbolism derived from electrical engineering and applied by Howard Odum to decipher energy flows in ecosystems.**

This measure of the efficiency of energy transfer gives the fraction of production passing from one trophic level to the next (Pauly and Christensen 1995). The energy not transferred is lost in respiration or to detritus. For aquatic ecosystems, trophic efficiencies vary from 2% to 24% and average 10.1% (**Figure 8**). If we assume a trophic efficiency of 10%, we can calculate how much primary production is required to support a particular fishery.

Consider the case of tuna caught in the open oceans. Tuna are top predators operating at trophic level 4, and in 1990 2,975,000 tons of tuna were taken, or 0.1 g carbon per m^2 of open ocean per year. To support this yield of tuna to the fishery, and assuming equilibrium conditions and trophic efficiency of 10%, we can calculate the production values of the other trophic levels in grams of carbon per m^2 as illustrated in **Figure 9**.

Table 2 Average production efficiencies.

Group	Production efficiency (%)	No. of studies
Insectivores	0.86	6
Birds	1.29	9
Small mammals	1.51	8
Other mammals	3.14	56
Fish and social insects	9.77	22
Other invertebrates (excluding insects)	25.0	73
Herbivores	20.8	15
Carnivores	27.6	11
Detritivores	36.2	23
Nonsocial insects	40.7	61
Herbivores	38.8	49
Detritivores	47.0	6
Carnivores	55.6	5

NOTE: Data are from 235 natural populations. A breakdown into trophic groups is presented for two of the groups for which adequate data are available.

SOURCE: After Humphreys (1979).

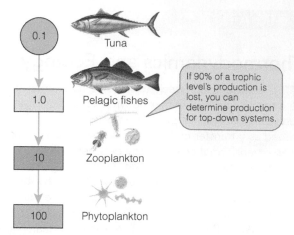

Figure 9 Hypothetical example illustrating the estimation of primary production for a simple food web with constant 10% trophic efficiencies. For every 100 g of tuna, there must be 100 kg of phytoplankton primary production utilized. Calculations of this type can be useful in determining the impact of fishing on aquatic food chains.

Note that these values are not standing crops but are production or yield values. For example, to provide 0.1 g C of tuna per m² we need to have 1 g C per m² of pelagic fishes to be eaten by the tuna, and 10 g C per m² of zooplankton to be eaten by the pelagic fishes, and finally 100 g C per m² of phytoplankton. Note that we do not know the standing crop in each box of this trophic chain, but instead only the production that comes out and moves up the chain. But if we know the net primary production of the plants, we can calculate what fraction of this production the tuna fishery is taking.

Using this approach, Pauly and Christensen (1995) aggregated all the data for the fisheries of the world and showed that on average 8% of global aquatic primary production was being used to produce the global fisheries catch. But this average masked high variation among different fisheries (**Table 3**). In continental shelf and upwelling ecosystems, fisheries harvest one-fourth to one-third of the net primary production, a very high fraction that leaves little margin for maintaining ecosystem integrity and a sustainable fishery (Pauly et al. 1998).

This analysis for aquatic ecosystems raises the question of whether terrestrial and aquatic ecosystems operate in the same way. Many terrestrial systems are dominated by decomposers, and most of the energy in the system flows through the decomposer link in the food web. **Figure 10** illustrates this for a temperate deciduous forest. For a typical deciduous forest about 96% of the net primary production moves directly into dead organic matter and thence to the decomposers. This loss is greatly reduced at higher trophic levels, such

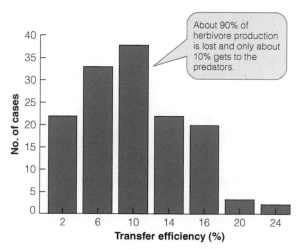

Figure 8 Frequency distribution of energy transfer efficiencies (Equation 7) for 48 trophic models of freshwater and marine aquatic ecosystems. The 140 estimates express for herbivores to carnivores the fraction of production passing from one trophic level to the next. The mean transfer efficiency is 10.1%. (From Pauly and Christensen 1995.)

Table 3 Global estimates of net primary production and the total catch to world fisheries (including discarded catches), and the calculated percent of primary production required to support the observed fishery catches.

Ecosystem type	Area (10^6 km^2)	Net primary production (g C m^{-2}yr^{-1})	Fishery catch[a] (g C m^{-2}yr^{-1})	Primary production required (%)
Open ocean	332.0	103	0.012	1.8
Upwellings	0.8	973	25.560	25.1
Tropical shelves	8.6	310	2.871	24.2
Temperate shelves	18.4	310	2.306	35.3
Coastal/reef systems	2.0	890	10.510	8.3
Rivers and lakes	2.0	290	4.300	23.6
Weighted means		126	0.330	8.0

Data from 1988–1991 were used in these estimates.

[a]Includes an estimated 25% discards that are not counted in official fishery catch statistics.

SOURCE: From Pauly and Christensen (1995).

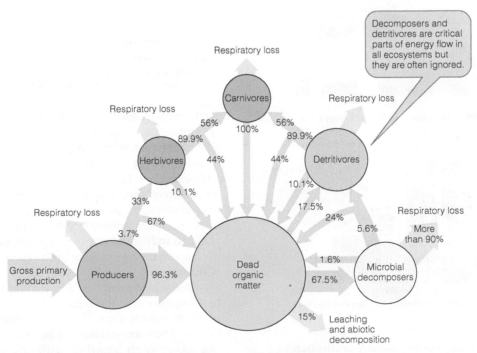

Figure 10 Estimates of energy flow through a temperate deciduous forest ecosystem. The fraction of energy going to the next trophic level (gross energy intake) is subdivided into two arrows representing egested energy and assimilated energy. The vast majority of the net primary production in temperate deciduous forests goes directly into detritus and then to decomposers, and about two-thirds of the organic matter is in the detritus pool. (From Hairston and Hairston 1993.)

that most of the production of herbivores is taken by carnivores and only 10% flows directly into the decomposer food chain (Hairston and Hairston 1993).

The amount of herbivory varies in different ecosystems. Herbivores in aquatic ecosystems consume a higher fraction of the primary production than they do in terrestrial ecosystems (**Figure 11**). Zooplankton in aquatic food webs consume an average of 79% of the net primary production of phytoplankton, whereas only 18% of the terrestrial primary production is eaten

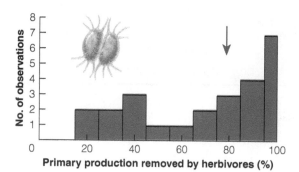

(a) Aquatic algae

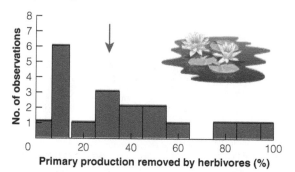

(b) Aquatic macrophytes

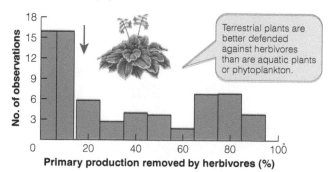

(c) Terrestrial plants

Figure 11 **Percentage of net primary production removed by herbivores in ecosystems dominated by (a) algae (phytoplankton), (b) rooted aquatic plants, and (c) terrestrial plants.** Red arrows indicate average values. Herbivores have a significantly greater effect on phytoplankton than on aquatic plants or terrestrial plants. (From Cyr and Pace 1993.)

(Cyr and Pace 1993). Thus, we can distinguish ecosystems dominated by grazers from those dominated by decomposers. Whittaker (1975) gives the following average values:

	Net primary production going to animal consumption (%)
Tropical rain forest	7
Temperate deciduous forest	5
Grassland	10
Open ocean	40
Oceanic upwelling zones	35

Thus, in forest ecosystems, almost all of the primary production goes into the decomposer food chain.

How do these differing consumption rates affect the standing crop of plants in different ecosystems? Lodge et al. (1998) summarized the reduction in standing crop of vascular plants in terrestrial, marine, and freshwater communities, citing the following averages:

	Reduction in standing crop of vascular plants from herbivores
Terrestrial ecosystems	26%
Marine ecosystems	65%
Freshwater ecosystems	31%

Herbivores reduce standing crops of plants in all systems by a significant amount. Thus, by excluding herbivores one would expect between a twofold and threefold larger effect in marine ecosystems compared with terrestrial ones.

Although much of the work on secondary production has centered on energy flow, an increasing amount of research on nutrient cycles is being done, because work on individual populations has suggested that nutrients, not energy, may be limiting animal populations (Cousins 1987).

One consequence of low ecological efficiencies is that organisms at the base of the food web are much more abundant than those at higher trophic levels. Charles Elton recognized this in 1927, and when he put this observation together with the observation that predators are usually larger than the prey they consume, the result was a pyramid of numbers or of biomass that has been called an **Eltonian pyramid** in his honor. **Figure 12** illustrates a pyramid of numbers of invertebrate individuals on a tropical forest floor in Panama.

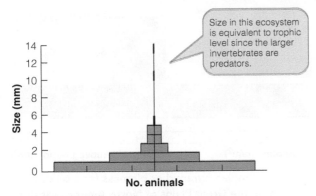

Figure 12 Pyramids of numbers of forest floor invertebrates in a tropical forest in Panama. Very few invertebrates in this ecosystem are over 5 mm in length. Each tick mark on the x-axis represents 500 individuals. (After Cousins 1985.)

Note that pyramids can be constructed on the basis of numbers, biomass, or energy of standing crop. They illustrate graphically the rapid loss of numbers and biomass as one moves from smaller to larger animals in an ecosystem, a biological illustration of the second law of thermodynamics and the constraints of foraging (Cousins 1985).

What Limits Secondary Production?

This is one of the critical questions we need to answer. As a first approximation, we could state that secondary production is limited by primary production and by the second law of thermodynamics, which states that no process of energy conversion is 100% efficient. **Figure 13** illustrates these broad patterns for a large array of 69 terrestrial communities from tundra to tropical forests. Herbivore biomass and consumption rise rapidly with increasing primary productivity, and secondary production increases with primary production in a 1:1 ratio (McNaughton et al. 1989). There is considerable scatter in the relationships shown in Figure 13, and, as one would predict from Figure 11, forest communities tend to fall below the regression line, whereas grassland communities fall above the line. To understand what limits secondary production in individual communities, we need to study the details of energy and nutrient flows. From 1965 to 1980 the International Biological Program (IBP) conducted out a series of studies of production in specific types of plant communities. Let us look at one example of these studies—grassland ecosystems—and then consider secondary production in the Antarctic.

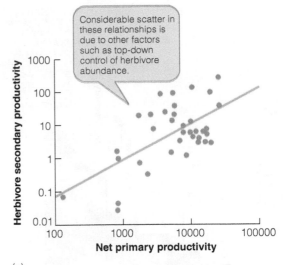

(a)

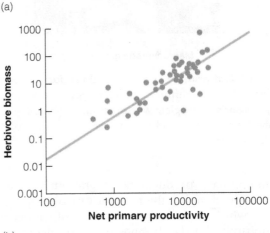

(b)

Figure 13 Relationships (a) between net aboveground primary production and net secondary productivity, and (b) between net primary production and herbivore biomass. Secondary production increases with primary production in a 1:1 ratio. Biomass measured as kJ/m², all others as kJ/m²/yr. Data from 69 studies from arctic tundra to tropical forests. (From McNaughton et al. 1989.)

Grassland Ecosystems

Grassland is the potential natural vegetation on 25% of the Earth's land surface. Grasslands occur in a great diversity of climates and are defined by having a period of the year when soil water availability falls below the requirement for forest. In 1968 the IBP began a series of studies on grassland sites throughout the world (Coupland 1979). The grassland biome study of the North American International Biological Program was a major attempt to analyze how natural grasslands work (French 1979).

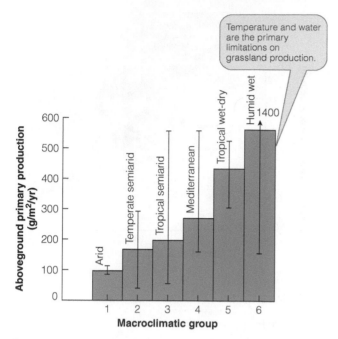

Figure 14 **Net primary aboveground production for grasslands from six different climatic types ranging from arid grasslands to tropical wet grasslands.** Vertical lines indicate the range of values within each climatic group. (After Lauenroth 1979.)

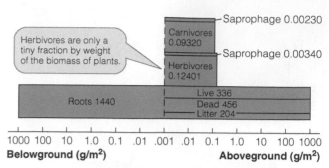

Figure 15 **Eltonian pyramid as biomass for a tallgrass prairie site in the Great Plains of North America for the height of the growing season in mid-July.** The base of the pyramid represents biomass (g dry wt/m²) of producers; the middle level, herbivores; and the top level, carnivores. Aboveground (right) and belowground (left) biomass are separated by the vertical dashed line. In this grassland, herbivores are only 0.04% of the aboveground biomass of living plants. Much of the plant biomass is belowground. (After French et al. 1979.)

The primary productivity of grassland increases with precipitation, but the relevant variable is periodic drought, which arises as a combination of temperature and moisture. Grasslands range from arid desert grasslands with nearly continuous drought to tropical humid grasslands with nearly no drought. The average primary production of these six climatic types ranges from 100 g to 600 g dry weight/m²/yr (**Figure 14**). Within these ranges, North American grasslands tend to fall at the lower end (Lauenroth 1979):

	Aboveground net primary production (g dry wt/m²/yr)
Tallgrass prairie	500
Annual grassland	400
Mixed grassland	300
Shortgrass prairie	200
Bunchgrass and desert grassland	100

Eltonian pyramids were calculated to determine the distribution of biomass among producers, consumers, and decomposers. **Figure 15** illustrates the trophic pyramid for a tallgrass prairie site. Little variation between years occurred in these pyramids, and the most striking finding was that belowground biomass in these grasslands greatly exceeds that aboveground.

Herbivores and carnivores were grouped into taxonomic groups in order to summarize consumption and production figures for these ecosystems. **Figure 16** shows the energy flow through tallgrass and shortgrass prairie sites. Several results stand out in these data. The most conspicuous animals are the least important energetically. Birds and mammals contribute almost nothing to production and very little to consumption (Scott et al. 1979). Birds consume 0.05% of the aboveground primary production in tallgrass prairie sites, and mammals consume 2.5%. The aboveground insects consume a greater amount, but the most important consumers are the soil animals—especially nematodes, which consume more than half of all plant tissue consumed. Nematodes are a major factor controlling total grassland primary production.

Only a small fraction of the primary production in grasslands is consumed by animals. **Table 4** shows that, aboveground, only 2%–7% of primary production is eaten by herbivores, but belowground, 7%–26% is eaten. In contrast, virtually all of the secondary production by herbivores seems to be eaten by carnivores in grasslands. Predators may control consumer populations, at least aboveground. Both production and consumption percentages increase in moving from shortgrass to tallgrass prairie.

The hypothesis that emerges from this analysis of grassland ecosystems is that grassland plants may be limited by nematode consumption of roots, by soil water, and by competition for nutrients and light,

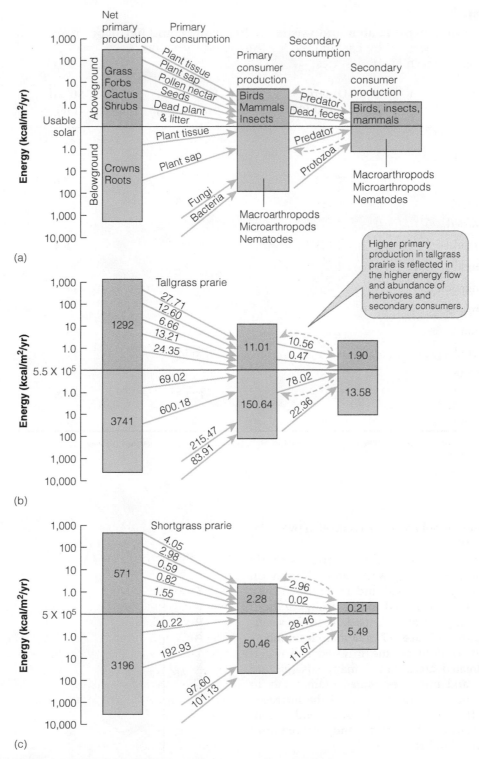

Figure 16 Primary and secondary production and consumption for a tallgrass prairie and a shortgrass prairie in the United States. All values are expressed as energy (kcal/m²/yr). (a) Key to the components of energy flow measured. (b) Tallgrass data. (c) Shortgrass prairie. Much of the energy flow in these systems occurs belowground. (After Scott et al. 1979.)

Table 4 Plant production eaten and wasted by herbivores, and herbivore production eaten and wasted by carnivores, for four grassland sites in the western United States.

	Desert grassland (%)	Shortgrass prairie (%)	Mixed (%)	Tallgrass prairie (%)
Plant Production				
Eaten by herbivores				
Aboveground	4.3	1.7	3.6	6.5
Belowground	—	7.3	26.4	17.9
Eaten and wasted				
Aboveground	6.3	3.4	8.4	10.2
Belowground	—	13.0	41.1	28.9
Herbivore Production				
Eaten by carnivores				
Aboveground	110.9[a]	119.7[a]	51.1	85.4
Belowground	—	50.9	76.1	47.5
Eaten and wasted				
Aboveground	120.5[a]	124.5[a]	60.5	97.2
Belowground	—	61.0	91.3	57.0

[a]Because consumption cannot exceed 100%, these values are too high, possibly because consumption was slightly overestimated or production underestimated.

SOURCE: Scott et al. (1979).

whereas consumer populations are controlled by predators (Scott et al. 1979).

This hypothesis was tested on a shortgrass prairie by experimentally providing water and nitrogen on 1-ha plots for six years (Dodd and Lauenroth 1979). Primary production increased dramatically in both treatments involving irrigation, especially in the water-plus-nitrogen plots (**Figure 17**). Nematode numbers increased about fourfold on the irrigated and water-plus-nitrogen treated areas. Small mammals (ground squirrels, voles, and mice) responded dramatically to the water-plus-nitrogen plots because of the increased herbage cover. These experimental results tend to confirm the hypothesis that both water and nitrogen limit production in grassland ecosystems.

A Krill-Centered Food Web in the Southern Ocean

The Southern Ocean around Antarctica has a relatively simple food chain and relatively constant environmental

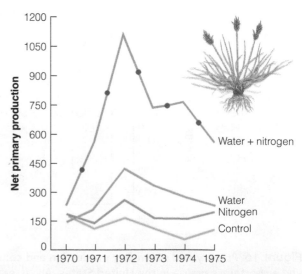

Figure 17 Net primary production (g/m²/y) over six years on plots of shortgrass prairie in north-central Colorado subjected to nitrogen fertilization, irrigation, and a combination of irrigation and fertilization. (After Dodd and Lauenroth 1979.)

Why Is the World Green?

The world is green, according to the **green world hypothesis**, because herbivores are held in check by their predators, parasites, and diseases such that they cannot consume all the plant biomass. On land about 83 × 10^10 metric tons of carbon is tied up in plant biomass, and about 5 × 10^10 metric tons of plant matter are produced each year (Polis 1999). Only 7%–18% of this production on land is consumed by herbivores, so it is quite correct to say that grazing by herbivores plays a comparatively minor role for land plants on a global scale. But this conclusion must be tempered by the observation that because on occasion herbivores such as the gypsy moth do indeed destroy their plant resources, it is possible for the world not to be green. How can we reconcile these observations?

There are at least six reasons why the world is green:

1. *Plants are not passive agents waiting to be eaten.* Plants contain much woody lignin as well as many secondary compounds that inhibit herbivores. Not all that is green is edible.

2. *Nutrients limit herbivores, not energy.* Nutrients such as nitrogen are critical for animals and are often in short supply in plant materials (White 1993). Even in a world full of green energy, many herbivores cannot obtain sufficient nutrients to grow and reproduce.

3. *Abiotic factors limit herbivores.* Seasonal changes in temperature, precipitation, and other climatic factors depress herbivore numbers.

4. *Spatial and temporal heterogeneity reduce the availability of plants.* The world is not uniform; herbivores must search for food plants and cannot always locate them efficiently.

5. *Herbivores limit their own numbers.* Self-regulation through intraspecific competition—territoriality, cannibalism, or other forms of interference competition—can limit the numbers of some herbivores.

6. *Enemies limit herbivore numbers.* This limitation is the primary one suggested by the *green world hypothesis.* Enemies are effective in some communities in which predators, parasites, and diseases limit herbivores and prevent them from consuming all green plants. But not all herbivores are so limited, and the previous five mechanisms also act on predators to limit their numbers. On a global scale enemies may not be the most important limitation on herbivores.

All six of these mechanisms act in varying ways to limit herbivores, and the key issue for any particular plant community is the relative importance of each of these factors. No one mechanism by itself explains why the world is green.

conditions, and consequently of all the ocean ecosystems it may be the easiest to analyze (Murphy et al. 2007). The Scotia Sea is a part of the Southern Ocean to the south of South America and north of the Antarctic Peninsula. The Antarctic Circumpolar Current moves through the Drake Passage off South America and into the Scotia Sea (**Figure 18**). The keystone species in the Scotia Sea is the krill (*Euphausia superba*) and the food web for the predators in this Antarctic ecosystem is shown in **Figure 19**. Seabirds and seals are the major predators of krill in the Scotia Sea; fish and squid could also be important predators of krill but no quantitative data are available to estimate their offtake.

Variations in winter sea ice distribution and sea surface temperature strongly affect the production of this oceanic ecosystem. **Figure 20** gives the estimated transfer efficiencies of the krill-based food web. Krill abundance changes from year to year, and when krill are scarce the food web changes to a domination by copepods and amphipods (Siegel 2005). However, this alternative food pathway cannot support the same level of predator demand as the krill pathway, and predators can thus suffer food shortage in years when krill are scarce. Long-term monitoring data on the breeding success and population sizes of upper trophic level predators such as macaroni penguins and crabeater seals have highlighted the fact that, when krill are scarce, predators suffer reduced breeding performance. Low years for krill also affect the distribution and feeding of blue whales and fin whales. The overharvesting of whales during the last century and the present-day rapid changes in climate are key changes to Antarctic ecosystems whose impacts on secondary production we do not yet fully understand.

Figure 18 Map of the Drake Passage off South America and the Scotia Sea. The Antarctic Circumpolar Current moves from west to east through the Drake Passage and into the Scotia Sea, where it is strongly affected by submarine ridges and the geometry of the Antarctic Peninsula.

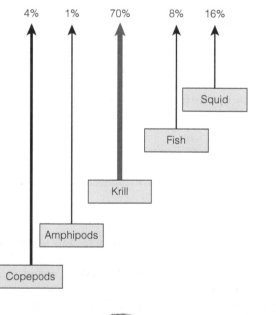

(a)

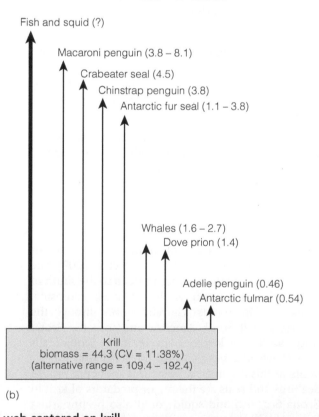

(b)

Figure 19 Predator links in the Antarctic Scotia Sea food web centered on krill.
(a) Proportional consumption of different groups of prey by the major predators in the ecosystem. (b) Estimates of the annual consumption of krill biomass (in units of 10^6 tonnes per year [1 tonne = 1000 kg]) by the main krill predators. Estimates are based mainly on summer studies. Estimates of krill consumption by fish and squid is unknown and could be very large. The dove prion is a common petrel (seabird) of the Southern Ocean. (From Murphy et al. 2007.)

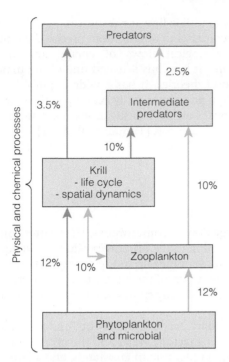

Figure 20 Estimated trophic transfer efficiencies in the Scotia Sea krill-based food web (green) with transfer efficiencies for alternative routes (blue) through other zooplankton species and intermediate predators. The final transfer of energy to predators is less efficient than the expected value of 10%. (From Murphy et al. 2007.)

Metabolic Theory of Ecology

Metabolism is a key process in all organisms, and could serve as a basis for linking the biology of individuals to the ecology of populations, communities, and ecosystems (Brown et al. 2004). According to this theory, the major limitations at all levels of biological organization are temperature and body size. We have already seen the starting point of the metabolic theory in Figure 2, which illustrates that basal metabolism is directly related to body size. We have also seen a second relationship that is part of the metabolic theory which shows that population density declines with body size in birds and mammals. The value of the metabolic theory is that it makes a range of predictions that can be tested with observed data.

One prediction of the metabolic theory is that the maximum rate of population increase will be related to body size with the slope $-\frac{1}{4}$. **Figure 21** shows this relationship for 294 mammal species (Duncan et al. 2007). The critics of metabolic theory praise these general predictions but call attention to the scatter of data around these regressions. While mammals in general may fit the predictions, different taxonomic groups within the mammals do not. **Figure 22** compares the

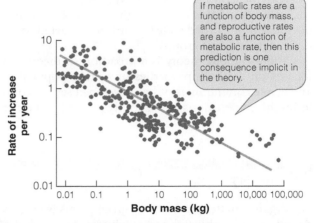

Figure 21 Observed relationship between body mass and instantaneous rate of increase per year (r_m) for 294 species of mammals. The metabolic theory of ecology predicts a slope of –0.25 for this relationship and the observed slope is –0.251. (Data from Duncan et al. 2007.)

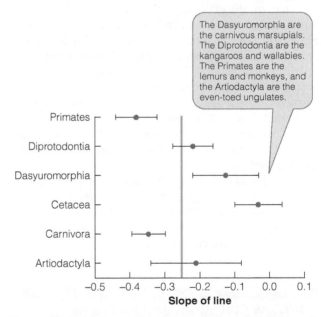

Figure 22 Variation in the slopes of the relationship of rate of population increase (r_m) and body mass. The expected value under the metabolic theory is –0.25, as shown by the vertical blue line. Most of the mammalian orders (with data from more than 20 species) do not fit the prediction. (Data from Duncan et al. 2007.)

slopes of the regressions of population growth rate on body mass for six different taxonomic orders of mammals. Only two of the six orders of mammals fit the expected slope of –0.25. The carnivores and the primates have a steeper negative slope than expected, whereas the cetaceans (whales and dolphins) have a shallower slope. It is clear that the metabolic theory is operative in the broad scale but fails in the details of individual taxonomic groups.

The metabolic theory is a broad brush theory that focuses on the bulk properties of average populations and stable communities. It has developed several counterintuitive predictions of the ability of species of differ-

ing sizes to utilize limiting resources, and this has stimulated ecologists to obtain more precise data on fundamental ecological flows of energy and matter in ecosystems. It has thus satisfied one of the major criteria of good theories: it has a wide applicability across population and community ecology with many precise predictions. Further investigations are now needed to test these predictions (Tilman et al. 2004).

Summary

The organic matter produced by green plants is used by a food web of herbivores and carnivores. A great deal of the matter and energy that animals eat is lost in feces and urine, or is used for maintenance metabolism. In warm-blooded vertebrates, often 98% of energy intake is used for maintenance. Invertebrates and fish use less energy for maintenance, typically 60%–90% of the energy taken in.

Data on the efficiency of secondary production of individual species is relatively easy to obtain, but the aggregation of this information into trophic levels is problematic. Many species of animals do not feed on only one trophic level. Typical "herbivores" may obtain part of their energy from other animals, and typical "carnivores" may feed on plants, herbivores, and other carnivores. The trophic level concept can be mapped directly onto species only if fractional trophic levels are defined. Omnivory becomes more common at higher trophic levels.

Considerable energy is lost at each step of the food chain, and thus for a given biomass of green plants, only a much smaller biomass of animals can be supported. Many of the animal species that humans consider important constitute a small component of

the energy flow in communities. Herbivores consume a higher fraction of the primary production in aquatic ecosystems than they do in forest or grassland ecosystems. Much of the energy flow in terrestrial systems goes directly from plants to the decomposer food chain.

Secondary production may be limited by a variety of interacting factors. Water and nitrogen limit secondary production in grasslands, and a large fraction of secondary production occurs belowground. Levels of production in oceanic ecosystems vary in response to nutrient inputs and temperature variations. Until we understand the environmental and biotic factors that limit primary and secondary production, we cannot predict the effects of environmental changes on a community.

Metabolic theory can be extended from individual organisms up to ecosystems, and a variety of predictions flow from the basic observation that metabolic rates are related to body size and to temperature. This theory can explain many observed ecological patterns but is not able to explain the details of individual taxonomic groups in which factors other than body size and temperature are key limitations.

Review Questions and Problems

1 In assessing the metabolic theory of ecology, Sterner (2004) notes that it assumes that limiting resources are simple and consistent across all plant and animal groups. But he notes that the ratios of carbon, nitrogen, and phosphorus as well as other elements vary widely in different organisms. Read Sterner (2004) and discuss whether there could be a universal currency for ecological systems so that we could ignore the chemical peculiarities of individual species.

2 Does any increase in primary production lead to an increase in herbivore grazing pressure, thus maintaining a low standing crop of plants? Discuss

what ecological processes might prevent this from happening. Van de Koppel et al. (1996) discuss this question and provide data from a salt marsh grazed by hares, rabbits, and geese.

3 In discussing the reality of trophic levels, Murdoch (1966a, p. 219) states:

Unlike populations, trophic levels are ill-defined and have no distinguishable lateral limits; in addition, tens of thousands of insect species, for example, live in more than one trophic level either simultaneously or at different stages of their life histories. Thus trophic levels exist only as abstractions, and unlike populations they have no empirically measurable properties or parameters.

Discuss.

4　Suggest two possible impact pathways for the Scotia Shelf food web shown in Figure 20 if blue whales regain their former abundance in the Antarctic.

5　How would it be possible to have an inverted Eltonian pyramid of numbers in which, for example, the standing crop of large animals is larger than the standing crop of smaller animals? In what types of communities could this occur? Do Eltonian pyramids apply to both animals and plants? Del Giorgio et al. (1999) discuss these issues.

6　How does the answer to the question *What limits secondary production?* differ from the answer to the question of whether trophic structure is controlled top-down or bottom-up in communities?

7　Would you expect that the relationship of metabolic rates to body size would also apply to bacteria and other prokaryotes? Would this imply a universal constant of metabolism for all living things? Makarieva et al. (2005) attempt to answer this question.

8　The basis for estimating secondary production is the estimation of population size, biomass, and growth, and the accuracy of any estimate of production depends on the accuracy of these three measurements. Read Morgan (1980) and then discuss the relative difficulty of measuring these

three variables in freshwater ecosystems for zooplankton, benthic invertebrates, and fish.

9　How would you expect trophic level biomass to change as the primary productivity of the community increases? Use your knowledge about hypotheses of community organization and discuss the assumptions underlying your predictions. Compare your expectations with those of Power (1992).

10　Lodge et al. (1998) found that in freshwater ecosystems nonvascular plant biomass was reduced nearly 60% by herbivores, whereas vascular plant biomass was reduced only 30% on average. Discuss two reasons why this might occur.

11　Could herbivores remove a high fraction of the net primary production in an ecosystem without depressing the standing crop of plants? How might this happen?

12　Population density (no. of individuals per m^2) of all organisms in all ecosystems falls with increasing body size, so that larger animals are less common. But for species of equal body size, aquatic organisms are 10–20 times more abundant in lakes than terrestrial organisms on land. Suggest two reasons why this might be. Cyr et al. (1997) discuss this issue.

Overview Question

Debate the following resolution: Resolved, that the ecological efficiencies of agricultural crops are higher than those of natural communities, and secondary production from agricultural systems is a more efficient use of solar energy.

Suggested Readings

- Brown, J. H., J. F. Gillooly, A. P. Allen, V. M. Savage, and G. B. West. 2004. Toward a metabolic theory of ecology. *Ecology* 85:1771–1789.

- Cohen, J. E. 1994. Marine and continental food webs: Three paradoxes? *Philosophical Transactions of the Royal Society of London, Series B*, 343:57–69.

- Cyr, H., and M. L. Pace. 1993. Magnitude and patterns of herbivory in aquatic and terrestrial ecosystems. *Nature* 361:148–150.

- Duncan, R., D. Forsyth, and J. Hone. 2007. Testing the metabolic theory of ecology: Allometric scaling exponents in mammals. *Ecology* 88:324–333.

- Glazier, D. S. 2005. Beyond the "3/4-power law": Variation in the intra- and interspecific scaling of metabolic rate in animals. *Biological Reviews* 80:611–662.

- Hairston, N. G., Jr., and N. G. Hairston, Sr. 1993. Cause-effect relationships in energy flow, trophic structure, and interspecific interactions. *American Naturalist* 142:379–411.

- Hall, S. R., J. B. Shurin, S. Diehl, and R. M. Nisbet. 2007. Food quality, nutrient limitation of secondary production, and the strength of trophic cascades. *Oikos* 116:1128–1143.

- Long, Z. T., and P. J. Morin. 2005. Effects of organism size and community composition on ecosystem functioning. *Ecology Letters* 8:1271–1282.

- McNaughton, S. J., M. Oesterheld, D. A. Frank, and K. J. Williams. 1989. Ecosystem-level patterns of primary productivity and herbivory in terrestrial habitats. *Nature* 341:142–144.

- Murphy, E. J., J. L. Watkins, P. N. Trathan, K. Reid, M. P. Meredith, S. E. Thorpe, N. M. Johnston, et al. 2007. Spatial and temporal operation of the Scotia Sea ecosystem: A review of large-scale links in a krill centred food web. *Philosophical Transactions of the Royal Society B: Biological Sciences* 362:113–148.

- Nagy, K. A. 2005. Field metabolic rate and body size. *Journal of Experimental Biology* 208:1621–1625.

Credits

Illustration and Table Credits

T2 From W. F. Humphreys, "Production and respiration in animal populations," *Journal of Animal Ecology* 48:427–453, 1979. F8 D. Pauly and V. Christensen, "Primary production required to sustain global fisheries," *Nature* 374:255–257, 1995. Copyright © 1995 Nature Publishing Group. T3 D. Pauly and V. Christensen, "Primary production required to sustain global fisheries," *Nature* 374:255–257, 1995. Copyright © 1995 Nature Publishing Group. F11 H. Cyr and M. L. Pace, "Magnitude and patterns of herbivory in aquatic and terrestrial ecosystems," *Nature* 361:148–150, 1993. Copyright © 1993 Nature Publishing Group. F10 N. G. J. Hairston and N. G. S. Hairston, "Cause-effect relationships in energy flow, trophic structure, and interspecific interactions," *American Naturalist* 142:379–411, 1993. Copyright © 1993. Reprinted by permission of the University of Chicago Press. F13 S. J. McNaughton et al., "Ecosystem-level patterns of primary productivity and herbivory in terrestrial habitats," *Nature*, Vol. 341, p. 143. Copyright © 1989 Macmillan Magazines Ltd. Reprinted by permission. F14 From W. K. Lauenroth, "Grassland primary production: North American grasslands in perspective" from *Perspectives in Grassland Ecology*, ed., N. R. French, p. 18. Copyright © 1979 Springer-Verlag. Reprinted by permission. F15 From N. R. French et al., "Grassland biomass trophic pyramids," from *Perspectives in Grassland Ecology*, pp. 59–87. Copyright © 1979 Springer-Verlag. Reprinted by permission. F16 From J. A. Scott et al., "Patterns of consumption in grasslands" in *Perspectives in Grassland Ecology*, ed., N. R. French, pp. 89–105. Copyright © 1979 Springer-Verlag. Reprinted by permission. T4 J. A. Scott et al., "Patterns of consumption in grasslands," pp. 89–105 in N. R. French, ed., *Perspective in Grassland Ecology*. Copyright © 1979. Reprinted by permission. F19 E. J. Murphy et al., "Spatial and temporal operation of the Scotia Sea ecosystem," *Philosophical Transactions of the Royal Society B: Biological Sciences* 362:113–148, 2007. Copyright © 2007. Reprinted by permission of The Royal Society of London.

Photo Credits

Ecosystem Metabolism III: Nutrient Cycles

Key Concepts

- Nutrients cycle and recycle in ecosystems. Humans are drastically modifying these nutrient cycles.

- Global nutrient cycles such as the nitrogen cycle include a gaseous phase. Local nutrient cycles contain nonvolatile elements such as phosphorus.

- To achieve sustainable harvesting of ecosystems such as forests, nutrient input must equal output.

- Nutrient use efficiency is higher in low-nutrient systems, and plants have evolved mechanisms to recycle limiting nutrients efficiently.

- Acid rain is one consequence of human alteration of the sulfur and nitrogen cycles. Reducing sulfur emissions is necessary for ecosystem recovery, but the time scale of recovery is not yet clear.

- Human additions of nitrogen to the nitrogen cycle have enriched water and land areas, and are reducing nitrogen limitation of plant growth in terrestrial ecosystems.

From Chapter 24 of *Ecology: The Experimental Analysis of Distribution and Abundance*, Sixth Edition. Eugene Hecht.

KEY TERMS

bioelements The chemical elements that move through living organisms.

biogeochemical cycles The movement of chemical elements around an ecosystem via physical and biological processes.

compartment Any component of study for an analysis of nutrient cycling, such as a lake, a species of plant, or a functional group of nitrogen fixers, measured by its standing crop or amount of nutrient.

critical load The amount of a nutrient such as nitrogen that can be absorbed by an ecosystem without damaging its integrity.

eutrophic soils Soils with high nutrient levels, mostly recent and often volcanic in origin.

flux rate The rate of flow of nutrients or biomass from one compartment to another.

global nutrient cycles Nutrient cycles that operate at very large scales over much of the Earth because the nutrients are volatile, such as oxygen.

local nutrient cycles Nutrient cycles that are confined to small regions because the elements are nonvolatile, such as phosphorus.

oligotrophic soils Soils of very low nutrient levels that are common in tropical areas and regions with geologically old, highly eroded soils with most of the nutrients in the litter layer.

residence time The time a nutrient spends in a given compartment of an ecosystem; equivalent to turnover time.

Living organisms are composed of chemical elements, and one way to describe an ecosystem is to follow the transfer of chemical elements between the living and the nonliving worlds. Interest in the nutrient content of plants and animals has been an important focus in agriculture for over 100 years. Nutrients often set some limitation on the primary or secondary productivity of a population or a community, and nutrient additions as fertilizer have become increasingly common in agriculture and forestry. In this chapter, we consider how nutrients cycle and recycle in natural systems and in the process link together the living and the dead material in the ecosystem.

Nutrient Pools and Exchanges

Nutrients can be used as an organizing focus in ecosystem studies. We can view the biological community as a complex processor in which individuals move nutrients from one site to another within the ecosystem. These biological exchanges of nutrients interact with physical exchanges, and for this reason nutrient cycles are also called **biogeochemical cycles**. Chemical elements that cycle through living organisms are called **bioelements**. **Figure 1** illustrates the general pattern of bioelement or nutrient cycles on a global scale. Nutrient cycles are closed on a global scale but are open on a local scale. The individual atoms that make up the cycle are indestructible and can be recycled in plants and animals. Ecologists are interested in understanding and measuring *global nutrient cycles* because human activities are altering these cycles, with possible effects on global climate. An analysis of nutrient cycling thus ends with an assessment of human impacts on nutrient cycles and their consequences for animals and plants.

Global nutrient cycles represent the summation of local events occurring in different biotic communities, and to make progress in understanding global nutrient cycles we must begin at the level of the local community. A simple example of a nutrient cycle in a lake is depicted in **Figure 2**. All nutrients reside in **compartments**, which represent a defined space in nature. Compartments can be defined very broadly or very specifically. Figure 2 includes all of the plants in the ecosystem as one compartment, but we could recognize each species of plant as a separate compartment, or even the leaves and the stem of a single plant as separate compartments. A compartment contains a certain quantity, or **pool**, of nutrients in the standing crop. In the simple lake ecosystem shown in Figure 2, the phosphorus dissolved in the water is one pool, and the phosphorus contained in the bodies of herbivores is another pool.

Compartments exchange nutrients, and thus we must measure the uptake and outflow of nutrients for each compartment. The rate of movement of nutrients between two compartments is called the **flux rate** and is measured as the quantity of nutrient passing from one pool to another per unit of time. The flux rates and pool sizes together define the nutrient cycle within any particular ecosystem. Ecosystems are not isolated from one another, and nutrients come into an ecosystem through meteorological, geological, or biological transport mechanisms and leave an ecosystem via the same routes. Meteorological inputs include dissolved matter in rain and snow, atmospheric gases, and dust blown by the wind; geological inputs include weathering

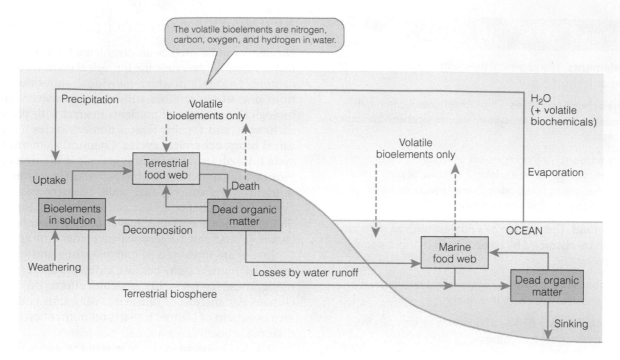

Figure 1 General schematic of nutrient cycling on a global scale. Movement of nonvolatile elements, such as phosphorus, is largely one way, toward ocean sediments. (From DeAngelis 1992.)

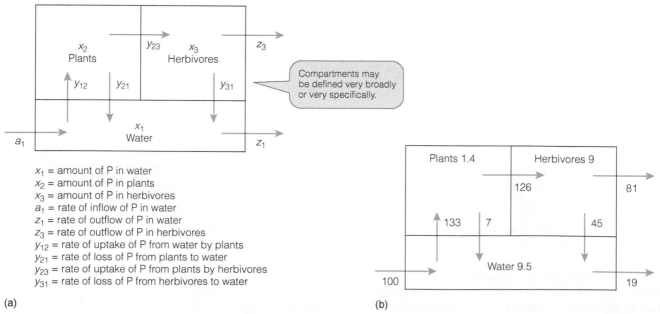

x_1 = amount of P in water
x_2 = amount of P in plants
x_3 = amount of P in herbivores
a_1 = rate of inflow of P in water
z_1 = rate of outflow of P in water
z_3 = rate of outflow of P in herbivores
y_{12} = rate of uptake of P from water by plants
y_{21} = rate of loss of P from plants to water
y_{23} = rate of uptake of P from plants by herbivores
y_{31} = rate of loss of P from herbivores to water

(a)

(b)

Figure 2 Hypothetical nutrient cycle for phosphorus in a simple lake ecosystem composed of three compartments: plants, herbivores, and water. (a) Definition of compartments and flux rates (inflow or outflow). Compartments are standing crops or amounts. (b) Hypothetical distribution (mg) and flux rates (mg/day) of phosphorus after equilibration to a constant input rate of 100 mg/day. (After Smith 1970.)

and elements transported by surface and subsurface drainage; and biological inputs include movements of animals between ecosystems.

Nutrient Cycles in Freshwater Ecosystems

Many freshwater ecosystems are limited in productivity by phosphorus. The nutrient cycle of phosphorus in natural lakes and reservoirs is strongly affected by subsidies of phosphorus coming in from the surrounding landscape and by regeneration processes within the lake or reservoir (Vannie et al. 2005). Phosphorus as well as other nutrients tend to accumulate in the sediment of lakes such that continual nutrient inputs are required to maintain high productivity. A considerable amount of

study of phosphorus cycling has been carried out in reservoirs because they are highly subsidized ecosystems, affected by human activities. Most reservoirs drain large watersheds and receive large inputs of nutrients and detritus, particularly in agricultural regions.

In eastern North America the fish community of most reservoirs are dominated by gizzard shad (*Dorosoma cepedianum*), an omnivorous fish that consumes detritus from the lake bottom as well as zooplankton. Gizzard shad form a link between watersheds and the pelagic grazing food chain in reservoirs (**Figure 3**). Phosphorus in particular is often locked up in the sediments of lakes, and by feeding on detritus, gizzard shad bring phosphorus as well as nitrogen back into the water column by excretion. This action forms a positive feedback loop because juvenile gizzard shad feed on zooplankton, and the more productive the reservoir the

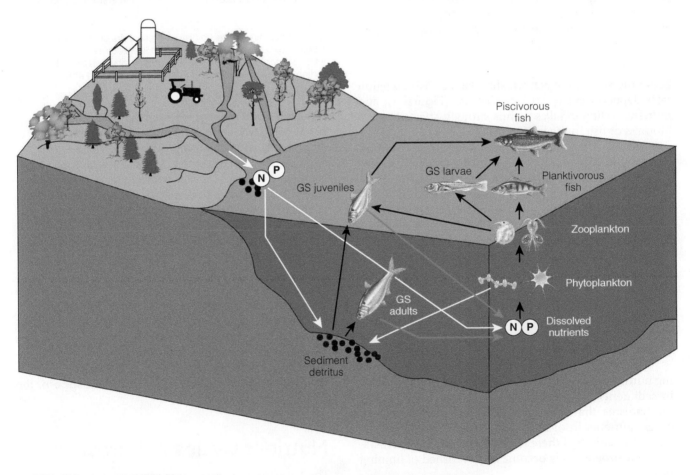

Figure 3 Reservoirs in eastern North America are often dominated by gizzard shad.
The watershed brings detritus and nutrients into the reservoir, and gizzard shad feed on the detritus and excrete nitrogen and phosphorus (green arrows) that stimulate the phytoplankton. Gizzard shad juveniles feed on zooplankton and reduce zooplankton resources for other fish. N = nitrogen, P = phosphorus, GS = gizzard shad. (From Vanni et al. 2005.)

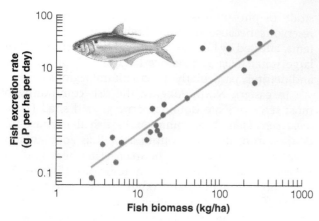

Figure 4 Phosphorus regeneration rates by fish in lakes and reservoirs as a function of the standing crop of fish in the lake. If fish abundance is high, this can feed a positive feedback to higher primary and secondary productivity in lakes because of the recycling of phosphorus. The fish illustrated is a gizzard shad (*Dorosoma cepedianum*). (From Griffiths 2006.)

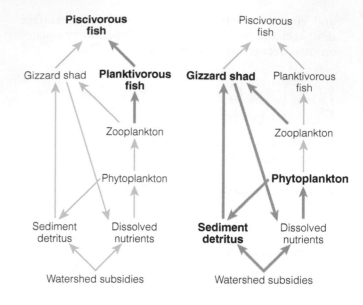

(a) Forested watersheds

Reservoirs characterized by
• Low sediment input
• Low phytoplankton biomass
• High sport-fish abundance

(b) Agricultural watersheds

Reservoirs characterized by
• High sediment input
• High phytoplankton biomass
• Low sport-fish abundance

Figure 5 Food webs of reservoirs in which the surrounding landscape is (a) primarily forest or (b) primarily agricultural land. The nutrient and energy flows are indicated by the thickness of the arrows, and the larger print indicates the dominant species. These are not alternate stable states because they depend on the continued input of different levels of nutrients and detritus from the watersheds. (After Vanni et al. 2005.)

better the survival of gizzard shad larvae. Fish excretion rates depend directly on fish biomass (**Figure 4**), and nutrient cycling in lakes is thus strongly affected by the biomass of fish in the lake or reservoir.

Because of the impacts of gizzard shad on the phosphorus and nitrogen cycles, reservoirs can exist in two different states that depend on the surrounding landscape (**Figure 5**). If a reservoir is surrounded by forest, there is relatively little input of nutrients and detritus. This results in a low abundance of gizzard shad and a low abundance of phytoplankton because of nutrient limitation. Low shad abundance releases plankton-feeding fish species, which reach higher abundance. By contrast, in an agricultural landscape, high inputs of nutrients and detritus stimulates phytoplankton growth and provides more food for gizzard shad, which reach high densities and subsequently depress the abundance of other fish in the reservoir.

The key problem in aquatic ecosystems is that limiting nutrients such as phosphorus can become locked up in sediments, thus restricting primary production. Any mechanisms that take up and release phosphorus or other nutrients from sediments can restore an ecosystem. Alternatively if there is no regeneration, the ecosystem may progressively become impoverished of limiting nutrients.

Nutrient cycles may be subdivided into two broad types. The phosphorus cycle just described in reservoirs is an example of a sedimentary or **local nutrient cycle**, which operates within an ecosystem. Local cycles involve the less-mobile elements (nonvolatile elements of

Figure 1) that have no mechanism for long-distance transfer. By contrast, the gaseous cycles of nitrogen, carbon, oxygen, and water (volatile elements) are called **global nutrient cycles** because they involve exchanges between the atmosphere and the ecosystem. Global nutrient cycles link together all of the world's living organisms in one giant ecosystem called the **biosphere**, the entire Earth ecosystem.

Nutrient Cycles in Forests

The harvesting of forest trees removes nutrients from a forest site, and this continued nutrient removal could result in a long-term decline in forest productivity unless nutrients are somehow returned to the system. Because of the economic importance of forest productivity, an increasing amount of work is being directed toward the

How Does Phosphorus Get to Hawaii?

Soils develop from rock weathering, and because phosphorus does not occur as a gas, once a rock is laid down, it contains all the phosphorus that the subsequent soil will ever have. As rock weathers, some phosphorus is lost to insoluble forms, and so soils should continually lose this critical element needed for plant growth, unless there is some outside source of input. On isolated oceanic islands such as the Hawaiian Islands, we would expect older soils to have less and less phosphorus. Because this chain of islands was formed over a time span of 5 million years, it presents ecosystem ecologists with a near-perfect laboratory to analyze nutrient cycling and nutrient limitation (Vitousek 2004). **Figure 6** illustrates the ages of the soils on different islands in the Hawaiian group.

The guiding model for plant growth is that nitrogen should be the limiting major nutrient on newly formed soils (which have relatively large amounts of phosphorus), and old soils should show phosphorus limitation (because as time passes nitrogen is fixed by organisms). Vitousek (2004) used the transect shown in **Figure 7** to test this model. By carrying out fertilizer trials with a tree that was common on all the islands, Vitousek (2004) showed that plant growth on very young soils was limited primarily by nitrogen and secondarily by phosphorus (because rock weathering was slow), but on older soils, nitrogen limitation disappeared and phosphorus was limiting, just as the guiding model had predicted (see Figure 7).

But rock weathering turned out not to be the only source of inputs of phosphorus to the Hawaiian system. Dust from Asia has accumulated in ocean sediments and could be measured in ocean cores to show the long-term pattern of dust deposition over the last million years in the Pacific Ocean (**Figure 8**). In Hawaiian sites that are more than 150,000 years old, dust from Asia contributes 80% or more of the phosphorus that is available in these old soils (Chadwick et al. 1999). Nevertheless, the rate of phosphorus input in dust from Asia is relatively small, so that plants on these older soils are still phosphorus limited.

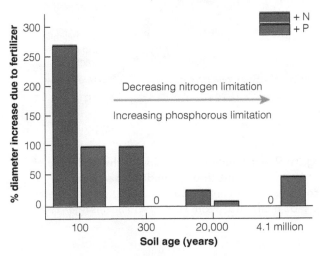

Figure 7 Nutrients limiting plant growth across the soil gradient mapped in Figure 6. Young soils are nitrogen deficient but this limiting factor is gradually removed by the colonization of nitrogen-fixing species of algae and higher plants. In very old soils, phosphorus is lost by erosion and becomes the limiting nutrient. (Data from Vitousek 2004.)

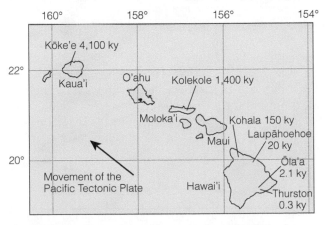

Figure 6 Transect across the Hawaiian Islands in the Pacific Ocean. Because these are volcanic islands at the edge of the Pacific Plate, the age of the soils on the different islands varies from 300 years on new volcanic soils in the southeast to over 4 million years in the northwest. Ages of soils are given in thousands of years. (Modified from Vitousek 2004.)

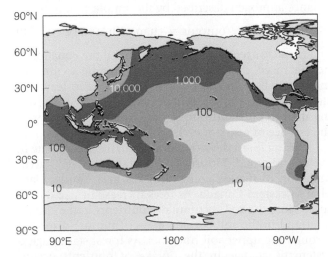

Figure 8 Inputs of dust from Asia to the Pacific Ocean over the last million years. The units of dust deposition are mg per square meter per year. A relatively low amount of dust reaches the Hawaiian Islands. (From Nakai et al. 1993.)

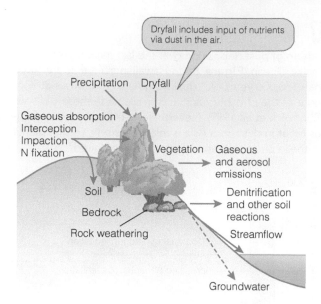

Figure 9 A schematic illustration of the pathways of nutrient movement through undisturbed forest ecosystems. To quantify nutrient cycling, all these pathways must be measured. (From Waring and Schlesinger 1985.)

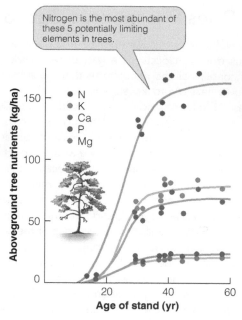

Figure 10 Accumulation of nutrients during the postfire development of jack pine (*Pinus banksiana*) stands in New Brunswick, Canada. More nitrogen accumulates in trees than do other nutrients, and nitrogen is often the limiting factor for forest tree growth. (From MacLean and Wein 1977.)

analysis of nutrient cycles in forests. **Figure 9** depicts the factors that must be quantified in order to describe the nutrient cycle in a forest. Some examples of nutrient cycles in forest stands will illustrate these concepts.

Nutrient budgets for forest ecosystems attempt to balance the inputs and outputs of nutrients in the system under study. The key to nutrient budgets is the mass balance approach described by the simple equation

$$\text{Inputs} - \text{outputs} = \text{change in storage} \qquad (1)$$

The term *change in storage* implies that if the equation is not balanced, then nutrients must either be accumulating somewhere in the ecosystem, such as in leaves or stems, or declining somewhere, such as in the soil. Tracing these inputs and outputs through the ecosystem is the essence of nutrient cycling.

During forest development, nutrients accumulate in leaves and wood. **Figure 10** illustrates the rapid accumulation of five different nutrients in a stand of jack pine in eastern Canada. As trees increase in size during succession, the soil accumulates nutrients in the surface litter and in the soil organic matter (humus) dispersed through the upper soil horizons. As forest stands age, a systematic change in the uptake of nutrients occurs. **Figure 11** illustrates changes in nitrogen cycling in a spruce forest in Russia. In this forest the spruce canopy becomes more open after 70 years, and understory vegetation increases in volume and importance in nutrient

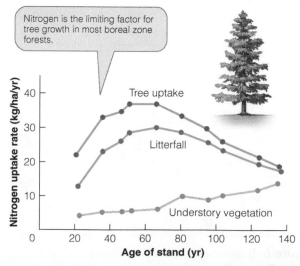

Figure 11 Uptake and cycling of nitrogen in spruce (*Picea abies*) from the ages of 22 years to 138 years. The rate of nitrogen uptake is maximal about 60 years of stand age, when tree growth is maximal. (After Kazimirov and Morozova 1973.)

Table 1 Aboveground accumulation of organic matter and nitrogen in trees for various forest regions.

Forest region	No. of sites	Organic matter (kg/ha)			Nitrogen (kg/ha)		
		In trees	Total	Aboveground (%)	In trees	Total	Aboveground (%)
Boreal coniferous	3	51,000	226,000	19	116	3250	4
Boreal deciduous	1	97,000	491,000	20	221	3780	6
Temperate coniferous	13	307,000	618,000	54	479	7300	7
Temperate deciduous	14	152,000	389,000	40	442	5619	8
Mediterranean	1	269,000	326,000	83	745	1025	73
Average	—	208,000	468,000	45	429	5893	7

SOURCE: From Cole and Rapp (1981).

cycling. Not all forest successions produce the same pattern of changes in nutrient cycling, but in general nutrient cycling varies with forest age. In old growth forests, little accumulation of nutrients occurs, and ecosystem inputs and outputs should be balanced.

The cycling of different nutrients in forest ecosystems is highly variable (Johnson et al. 2000). Nutrients that are in short supply are recycled far more efficiently than those present in excess of requirements. In most forest sites, nitrogen is a major limiting factor and is present at deficiency levels. **Table 1** lists the organic matter and nitrogen amounts present in the aboveground component of 32 forests from five different climatic zones. In the boreal forests of Alaska, only about 20% of the organic matter in the trees is present aboveground. Low decomposition rates in these cold Alaskan forests cause most of the nitrogen and organic matter to be tied up in the soil. Coniferous forests have the largest forest floor accumulation of organic matter of all forests, on average about four times the biomass of tropical forests (Waring and Schlesinger 1985).

Nutrient cycles operate more quickly in warmer forests than in colder ones. If we assume as an approximation that forest soil nutrients are in equilibrium for the short time they can be studied (three to ten years), then for each nutrient we can calculate average turnover time, the time an average atom will remain in the soil before it is recycled into the trees or shrubs (see *Working with the Data: Estimating Turnover Time for Nutrients*). **Table 2** gives the mean turnover times for five elements. All northern forests have very slow turnover of nutrients. On average, boreal conifer forests

Table 2 Mean turnover time for five mineral elements in the forest floor for five forest regions.

Forest region	No. of sites	Mean turnover time (yr)					
		Organic matter	N	K	Ca	Mg	P
Boreal coniferous	3	353	230.0	94.0	149.0	455.0	324.0
Boreal deciduous	1	26	27.1	10.0	13.8	14.2	15.2
Temperate coniferous	13	17	17.9	2.2	5.9	12.9	15.3
Temperate deciduous	14	4	5.5	1.3	3.0	3.4	5.8
Mediterranean	1	3	3.6	0.2	3.8	2.2	0.9
All stands	32	12	34.1	13.0	21.8	61.4	46.0

A steady-state condition is assumed.

SOURCE: From Cole and Rapp (1981).

Estimating Turnover Time for Nutrients

How long does a molecule of phosphorus stay in a lake? This simple question is difficult to answer because ecosystems have many compartments with different turnover times (*residence times*). Estimating turnover time for nutrients in compartments can be achieved most easily using radioactive tracers. It is useful to consider a very simple situation to clarify what ecologists mean by turnover time (DeAngelis 1992).

Consider a single compartment or pool with a constant inflow and outflow (an equilibrial system) as shown in **Figure 12**.

This compartment could represent an entire lake, or an individual organism in a lake. The rate of flow (q) is expressed as liters per minute (or some similar rate); concentrations are expressed as grams of nutrient per liter (or some similar measure). Given rapid mixing in the compartment, the outflow concentration will equal that inside the compartment. If the system starts with an inflow concentration of C_1, then it will eventually equilibrate with a concentration in the compartment of C_1.

Given that this is a steady state system, we can ask, What is the expected time of residence of a molecule entering the pool? This quantity, the **residence time**, is given by the simple equation

$$T_{res} = \frac{C_1 V}{C_1 q} = \frac{V}{q} \qquad (2)$$

where T_{res} = mean residence time = mean turnover time
V = volume
q = rate of flow
C_1 = inflow concentration of nutrient

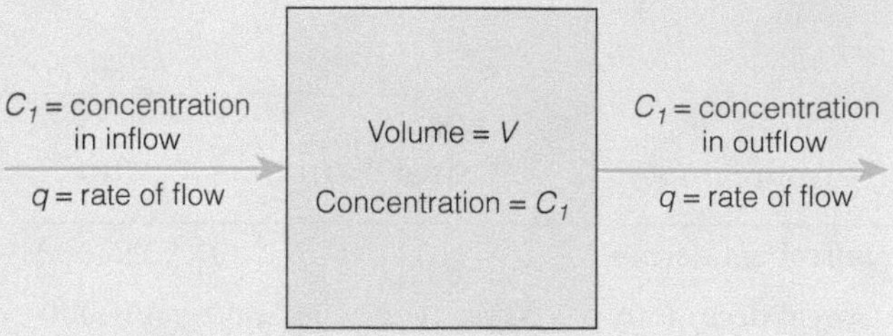

Figure 12 A schematic illustration of a single compartment in an ecosystem model indicating the six quantities that must be measured to estimate turnover time for a nutrient.

The mean residence time can be thought of as a measure of the rate of flushing. In particular, if the input of nutrients is stopped such that $C_1 = 0$, the concentration of nutrients in the compartment will decline from that moment according to the equation

$$C_t = C_0 e^{-(q/V)t} \qquad (3)$$

where C_t = concentration of nutrient at time t
C_0 = concentration of nutrient in the compartment at the instant input ceases
t = time after nutrient input cases
V = volume
q = rate of flow

This simple model of exponential decay is useful for obtaining an estimate of recovery time for a compartment that has been subjected to pollution input. If the system is not in equilibrium, turnover time is more difficult to estimate. For lakes, turnover time of nutrients in sediments differs greatly from that of nutrients in the water column. For any ecosystem it is crucial to define compartments carefully.

---

retain nitrogen 100 times longer than a Mediterranean evergreen oak forest (Cole and Rapp 1981). Deciduous forests turn over nutrients more rapidly than coniferous forests. Coniferous forests use nutrients more efficiently because they retain their needles and do not need to replace all their foliage each year.

Nutrients are lost from forest ecosystems in several ways. Streams transport both dissolved and particulate matter, and measurements of stream water chemistry can provide a good way to monitor overall forest function. Anaerobic soil bacteria produce methane and hydrogen sulfide gases. Plants release hydrocarbons such as terpenes from their leaves, and these compounds may add to atmospheric haze in summer. Both ammonia and hydrogen sulfide can be released from plant leaves (Waring

and Schlesinger 1985). During forest fires, nutrients are released in both gases and in particles. Finally, forest harvesting removes nutrients in wood from the ecosystem.

Nitrogen cycling has received much attention in forestry studies because it is unique among the nutrients needed for trees: it has no soil mineral source and thus must be derived ultimately from nitrogen in the atmosphere. Almost all of the nitrogen in the soil is tied up in organic compounds (detritus) or in soil bacteria and fungi. The key questions in forest development are where the nitrogen is stored in the ecosystem and what compartments are turning over quickly. **Figure 13** illustrates the nutrient pools of nitrogen in forests of temperate and tropical ecosystems. In warmer regions there is relatively little nitrogen tied up in litter on the

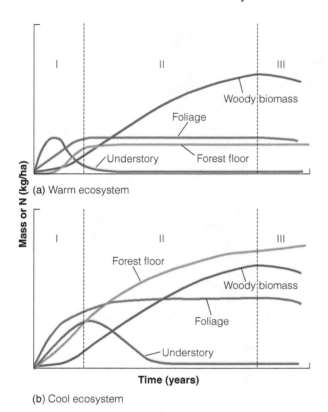

(a) Warm ecosystem

(b) Cool ecosystem

Figure 13 Schematic representation of changes in nitrogen pools during the development of a forest stand in (a) a warm ecosystem in which the forest floor litter compartment reaches a steady state, and (b) in a cool or cold ecosystem in which the litter continues to accumulate. Three phases can be recognized. In Phase I the understory has much of the nitrogen and plays a major role in cycling. In Phase II the forest canopy closes and the understory pool declines. In Phase III senescence of the forest stand occurs and tree mortality may ensue. (From Johnson 2006.)

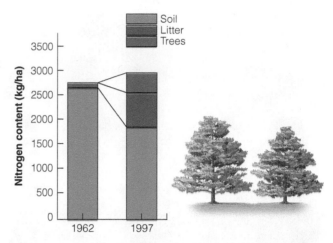

Figure 14 Changes in the nitrogen content of a loblolly pine forest in South Carolina over 35 years. This forestry plantation began in abandoned farmland so there was no vegetation at the start of the study. As the forest developed, more and more of the nitrogen accumulated in the litter and in the vegetation, and on average the site accumulated nitrogen at a rate of 6 kg N/ha/year. This is approximately the expected amount from atmospheric deposition per year in this area of eastern North America. (Data from Binkley et al. 2000.)

forest floor, and much of the nitrogen is in the woody biomass. But in colder climates the litter on the forest floor decomposes more slowly and is the major compartment for nitrogen in the system, typically exceeding that of the woody biomass.

There have been relatively few measurements of changes in the total nitrogen content of a forest to determine if nitrogen is being progressively lost after forest harvesting (Binkley et al. 2000). Some forests such as a loblolly pine forest in South Carolina accumulate nitrogen as the trees grow (**Figure 14**). The problem then comes when the trees are harvested. Sustainable forest management demands that nutrient budgets in a forested ecosystem must be balanced over the long term. Akselsson et al. (2007) calculated nutrient budgets for all of the regions in Sweden and found that the prospective nitrogen budget of Swedish forests depended greatly on how trees were harvested. If whole

trees were removed from the forest, it would result in losses of nitrogen and base cations (K, Ca, Mg) in large parts of Sweden (**Figure 15**). If only the tree stems were removed, so the bark and branches remained on site, there was considerably less ecosystem stress. The balancing act is to replace the nutrient losses from forest operations by artificial fertilizers while not adding so much nitrogen that runoff causes freshwater and marine pollution problems (Akselsson et al. 2007).

There is considerable controversy over whether or not harvesting of trees induces a long-term decline in forest productivity (Morris and Miller 1994). Unless the nutrients removed in the harvest of trees are replaced by natural sources or by fertilization, productivity may decline. To date, little evidence indicates that this is happening, but most experiments are of short duration, and the key is the long-term response of the ecosystem (**Figure 16**). Even though many management practices in forestry, such as slash removal or burning, can alter nutrient cycling, at present there is not enough evidence to decide whether there is or is not a decline in long-term forest productivity in temperate forests utilized for tree harvest.

Research on nutrient cycling in forests has shown the need for guidelines specifying sound management procedures in forestry. For example, bark is relatively rich in nutrients, and hence timber operations ought to be designed to strip the bark from the trees in the field,

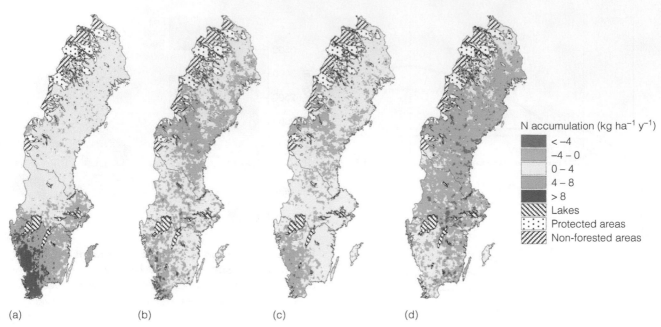

Figure 15 Nitrogen accumulation in Swedish forests according to four possible future scenarios. (a) Nitrogen aerial deposition levels of 1998 and harvesting of only the tree stem. (b) Nitrogen deposition of 1998 and harvesting of the entire tree. (c) Decreased aerial deposition of nitrogen by 2010 and tree stem harvesting. (d) Decreased aerial deposition of nitrogen by 2010 and whole tree harvesting. Southern Swedish forests have no need for nitrogen fertilization, but central and northern Swedish forests require some fertilization to replace losses due to forest harvesting. (From Akselsson et al. 2007.)

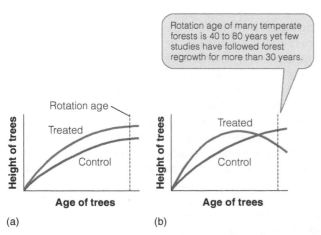

Figure 16 Two potential patterns of forest tree growth in relation to management treatments such as fertilization or burning. (a) Long-term improvement in forest productivity. This is the model assumed by foresters. (b) Short-term improvement with long-term detrimental effect. Short-term studies may confuse (a) with (b). (Modified after Morris and Miller 1994.)

not at some distant processing plant. The conservation of nutrients in forest ecosystems can be done intelligently only when we understand how nutrient cycles operate in these systems.

Efficiency of Nutrient Use

Large areas of the Northern Hemisphere have been glaciated, and the soils, derived from till in which the bedrock has been pulverized, are very fertile, with a high availability of nutrients. Areas of volcanic activity can also have rich soils. But in much of the world, soils are very old, highly weathered, and basically infertile. For example, the continents derived from Gondwanaland—Australia, South America, and India—have large areas covered by very old, poor soils. The vegetation supported on these soils has adapted remarkably well to efficient nutrient use by recycling within the plant and by leaf fall and reabsorption.

Australian soils are typical of very old, highly weathered soils and contain almost no phosphorus, and eucalyptus trees native to them are adapted to grow on soils of low phosphorus content (Attiwill 1981). **Table 3** gives the biomass and nutrient

Table 3 Aboveground living biomass and its nutrient content of temperate forest and Australian eucalyptus forest ecosystems.

Forest	Location	Age (yr)	Biomass (t/ha)	N	P	K	Na	Mg	Ca
Coniferous									
1. *Abies amabilis—Tsuga mertensiana*	Washington	175	469	372	67	980	—	160	1046
2. *Abies balsamea—Picea rubens*	Quebec	various	132	387	52	159	—	36	413
3. *Abies mayriana*	Japan	?	129	527	63	278	—	415	515
4. *Cryptomeria japonica*	Japan	?	114	386	36	244	—	118	314
5. *Larix leprolepis*	Japan	?	100	230	28	212	—	40	107
6. *Picea abies*	England	47	140	331	37	161	5	39	212
7. *Picea abies*	Sweden	55	308	770	87	437	38	69	459
8. *Picea glauca*	Minnesota	40	151	383	58	230	—	41	720
9. *Picea mariana*	Quebec	65	107	167	42	84	—	27	276
10. *Pinus banksiana*	Ontario	30	81	171	15	85	—	19	114
11. *Pinus banksiana*	Minnesota	40	150	276	27	106	—	40	226
12. *Pinus maricata*	California	88 (mean)	409	510	418	321	83	140	589
13. *Pinus radiata*	New Zealand	35	305	319	40	324	—	—	187
14. *Pinus resinosa*	Minnesota	40	204	373	46	205	—	62	335
15. *Pinus sylvestris*	Southern Finland	28	21	82	9	40	—	—	50
		45	79	194	19	98	—	—	115
		47	45	136	14	58	—	—	63
16. *Pinus sylvestris*	England	47	164	364	34	314	8	43	204
17. *Pseudotsuga menziesii*	Washington	36	173	326	67	227	—	—	342
18. *Pseudotsuga menziesii*	Oregon	450	540	371	49	265	—	—	664
19. *Sequoia sempervirens*	California (slopes)	?	1155	1474	616	941	102	232	1068
20. *Sequoia sempervirens*	California (flats)	?	3190	3846	1695	2412	271	569	2574
Hardwood									
1. *Alnus rubra*	Washington	34	185	516	40	224	—	114	334
2. *Betula verrucosa*	England	22	63	264	33	76	3	36	327
3. *Betula verrucosa—Betula pubescens*	Southern Finland	40	91	232	24	136	—	—	185
4. *Betula platyphylla*	Japan	?	114	265	20	84	—	95	401
5. Evergreen broadleaf forest	Japan	?	114	329	52	249	—	94	259

(continued)

Table 3 Aboveground living biomass and its nutrient content of temperate forest and Australian eucalyptus forest ecosystems. *(continued)*

Forest	Location	Age (yr)	Biomass (t/ha)	Nutrient content (kg/ha)					
				N	P	K	Na	Mg	Ca
6. *Fagus sylvatica*	England	39	134	285	38	187	4	42	151
7. *Fagus sylvatica*	Sweden	a. 90	314	800	53	458	24	118	980
		b. 90	324	1050	84	452	32	105	602
		c. 100	226	640	65	318	17	85	478
8. *Populus tremuloides*	Minnesota	40	170	383	48	297	—	62	881
9. *Quercus alba*	Missouri	35–92	100	204	20	115	—	35	601
10. *Quercus alba— Quercus rubra— Acer saccharum— Carya glabra*	Illinois	150	190	478	29	310	—	105	1603
11. *Quercus robur*	England	47	130	393	35	246	5	45	257
12. *Quercus robur— Carpinus betulus— Fagus sylvatica*	Belgium	30–75	121	406	32	245	—	81	868
13. *Quercus robur— Fraxinus excelsior*	Belgium	115–160	328	947	63	493	—	126	1338
14. *Quercus stellata— Quercus marilandica*	Oklahoma	various	195	902	75	1093	—	230	3895
Eucalyptus									
1. *E. regnans*	Australia	38	654	399	38	1389	138	192	849
2. *E. regnans*	Australia	27	831	—	17	—	—	—	—
3. *E. oblique—E. dives*	Australia	38	373	426	17	111	103	71	264
4. *E. oblique*	Australia	51	316	—	31	256	—	204	336
5. *E. sieberi*	Australia	27	929	—	14	—	—	—	—
6. Mixed dry sclerophyll	Australia	?	176	395	—	—	—	—	—
7. *E. signata—E. umbra*	Australia	?	104	456	18	192	169	77	344
8. *E. diversicolor*	Australia	37	263	473	27	296	82	211	1133
9. *E. diversicolor— E. calophylla*	Australia	?	305	449	31	424	125	344	1266

SOURCE: After Feller (1980)

content of temperate forest ecosystems growing on good soils and from Australian sites on poor soils. There is no suggestion from Table 3 that eucalyptus trees growing on poor soils contain fewer nutrients than other tree species from other sites, with the single exception of phosphorus. Eucalyptus trees have only 20%–50% the amount of phosphorus in their tissues compared to trees from Northern Hemisphere forests.

One might expect plants growing in nutrient-poor soils to contain fewer nutrients than plants in fertile soils. In fact, the opposite is true (Chapin 1980). Plants from infertile habitats consistently have higher nutrient concentrations than plants from fertile habitats when grown under the same controlled conditions. Plants from nutrient-poor habitats may achieve this nutrient-rich status by being more efficient than plants from nutrient-rich habitats. Peter Vitousek in 1982 proposed a measure of nutrient use efficiency that captures the different abilities of plants to take up a critical nutrient such as nitrogen and use it for the production of biomass. Vitousek's original formulation of nutrient use efficiency (*NUE*) was generalized by Berendse and Aerts (1987) as

$$NUE = (A)\left(\frac{1}{L}\right) \tag{4}$$

where *NUE* = nutrient use efficiency
      A = nutrient productivity (dry matter production per unit nutrient in the plant)
      L = nutrient requirement per unit of plant biomass

If we express the nutrient requirement on a relative basis (for example, as the amount of nitrogen uptake needed to maintain each unit of nitrogen in a plant for a given time period), and if we assume a steady state in the plant community, we can estimate the mean residence time for the nutrient as

$$MRT = \frac{1}{L_n} \tag{5}$$

where *MRT* = mean residence time for a unit of nutrient in plant tissue
     L_n = relative nutrient requirement for maintenance

Figure 17 illustrates the concept of nutrient use efficiency. An abundance of data is available from forests to investigate nutrient use efficiency. We can use litterfall in a forest as a measure of aboveground primary production, and the amount of nutrients in litterfall as a measure of nutrient uptake.

Vitousek (1982, 1984) showed that nutrient use efficiency increases as nutrients become scarcer in tropical forests. **Figure 18** plots nitrogen availability

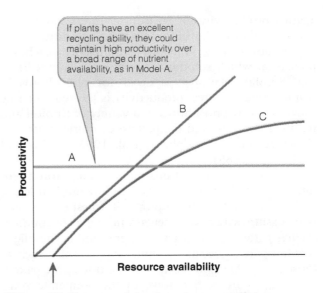

Figure 17 Three models of nutrient use efficiency (NUE). Model A: Productivity is not related to the amount of nutrient available. This model implies higher nutrient use efficiency at lower resource levels. Model B: Productivity increases linearly with more nutrients. This model assumes constant nutrient use efficiency at all resource levels. Model C: Productivity increases to an upper limit as resources increase, and there is a minimum level of nutrient needed before there is any production (red arrow). This model implies that nutrient use efficiency is higher at lower resource levels. (After Pastor and Bridgham 1999.)

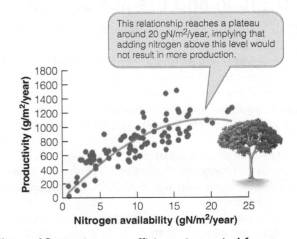

Figure 18 Nutrient use efficiency in tropical forests. The relationship between the amount of nitrogen available (indexed by nitrogen in the litter) and the productivity of the site (measured by the amount of litterfall) follows a curve of diminishing returns. Model C in Figure 17 is supported by these data, which show that tropical forests growing on low-nutrient sites use nitrogen more efficiently. (Data from Vitousek 1984.)

against productivity for 90 tropical forest stands, and shows the law of diminishing returns—as nitrogen becomes abundantly available, productivity reaches a plateau at which some other resource becomes limiting. Similar relationships occur for phosphorus in tropical forests. Forest productivity is limited either by nitrogen or by phosphorus in a variety of tropical and temperate forests, and the pattern of nutrient use efficiency is similar (Bridgham et al. 1995; Waring and Schlesinger 1985).

One consequence of this finding is that forest productivity may be high on soils with low nutrient levels. A classic example is the tropical rain forest of the Amazon Basin, which represents one type of nutrient cycling pattern (Jordan and Herrera 1981). The **oligotrophic** pattern occurs on nutrient-poor soils, such as those in the Amazon Basin, and the **eutrophic** pattern occurs on nutrient-rich soils. In the temperate zone, where most forest research has been done, forests are usually of the eutrophic type on rich soils. Oligotrophic ecosystems on poor soils constitute a much greater proportion of the tropics. But exceptions occur, and not all tropical forests are oligotrophic; nor are all temperate forests eutrophic. Some striking differences exist between these forest types. Oligotrophic systems have a large biomass in the humus layer of the soil, and this layer of fine roots and humus is critical for nutrient cycling and nutrient conservation in these systems.

Productivity and nutrient cycling do not differ greatly in oligotrophic and eutrophic forests, as long as these ecosystems are not disturbed (Jordan and Herrera 1981). But when the forest is cleared for agriculture, the nutrient-poor systems quickly lose their productive potential, whereas the nutrient-rich systems do not. Once the humus and root layer on top of the mineral soil is disturbed in oligotrophic systems, the mechanism of efficient nutrient recycling is lost, and nutrients are leached out of the system. Oligotrophic ecosystems cannot be used for crop production unless critical nutrients are supplied in fertilizers (Sanchez et al. 1982).

Acid Rain: The Sulfur Cycle

One human activity—combustion of fossil fuels—has altered the sulfur cycle more than any of the other nutrient cycles. Even though human-produced emissions of carbon dioxide and nitrogen are only about 5%–10% of the level of natural emissions, for sulfur humans produce about 160% of the level of natural emissions (Likens et al. 1996). One clear manifestation of this alteration of the sulfur cycle is the widespread problem of

ESSAY

Acid Rain and the Sudbury Experience

The recovery of the ecosystems in the vicinity of the nickel and copper smelter at Sudbury, Ontario, is an encouraging sign that ecosystems damaged by air pollution can restore themselves. From 1900 to 1970 the Sudbury smelters spewed over 100 million tons of sulfur dioxide into the atmosphere, as well as thousands of tons of toxic trace metals such as lead and cadmium. At its peak, Sudbury alone accounted for 4% of total global emissions, an amount equal to the current emissions of the entire United Kingdom (Schindler 1997). Within 30 km of the smelter, vegetation was destroyed and thousands of lakes were acidified.

One way to combat acidification of lakes is to add lime (calcium carbonate) to raise their pH. The Swedes have been particularly active in liming lakes affected by acid rain. But because no one would pay for the enormous cost to lime the lakes around Sudbury, they constitute a natural experiment. Since 1980 the Sudbury smelters have emitted less than 10% of their original air pollution, and sources of sulfur dioxide in eastern Canada has been reduced by more than half. During the past 20 years the ecosystem around Sudbury has been recovering via natural processes. Trees and shrubs have recolonized, and the pH in most lakes has risen. Lake trout have colonized many lakes, but not all of them.

The idea that acidified lakes could recover only in geological time has been shown to be false, and there is room for some optimism. But acid rain continues to fall, even if in reduced amounts, and until we have further limited sulfur dioxide and nitrous oxides emissions, lakes will not be able to recover fully (Doka et al. 2003). The longer we wait to implement strict air pollution controls, the more damage will accumulate. Acid rain leaches cations such as calcium out of soils and thus reduces the natural ability of the soil to absorb acidity.

The important messages that Sudbury provides are that ecosystems can recover from disturbances, although it may take longer than a few years, and that we should not assume that we can achieve a technological fix for ecological damage. Adding lime to acidified lakes may solve some problems, but it creates another whole set in its wake (Steinberg and Wright 1994).

acid rain in Europe and North America. Acid precipitation is defined as rain or snow that has a pH lower than 5.6. Low pH values are caused by strong acids (sulfuric acid, nitric acid) that originate as products of the combustion of fossil fuels.

Acid rain emerged as a major environmental problem in the 1960s, when widespread damage to forests and lakes in Europe and eastern North America became apparent. It was one of the first widespread environmental problems recognized because oxides of sulfur and nitrogen can be carried hundreds of kilometers and then deposited in rain and snow. Lakes in eastern Canada were dying because of air pollution produced in the midwestern United States. Lakes in southern Norway were losing fish because of acid rain from England. Already by 1980, annual pH values of precipitation over large areas of western Europe and eastern North America averaged between 4.0 and 4.5, and individual storms produced acid rain of pH 2 to 3. **Figure 19** illustrates that the situation has gradually improved in North America, so that acid rainfall in the eastern states averaged about 4.6 in 2006.

Sulfur released into the atmosphere is quickly oxidized to sulfate (SO_4) and is redeposited rapidly on land or in the oceans (Schlesinger 1997). **Figure 20** illustrates the sources and sinks of the global sulfur cycle. Short-term events like volcanic eruptions contribute to the global sulfur cycle and make it difficult to estimate the equilibrium state of the atmosphere. Human-caused emissions are the largest component of additional sulfur to the atmosphere. Ore smelters and

electrical generating plants have increased emissions during the past 100 years. To offset local pollution problems, smelters and generating plants have built taller stacks, which reduce pollution at ground level. Tall stacks (over 300 m), now the standard, have exported the pollution problem downwind. Ice cores from Greenland show large increases in SO_4 deposition from the atmosphere in the past 50 years (Mayewski et al. 1986; Duan et al. 2007).

A net transport of SO_4 occurs from the land to the oceans. The ocean is also a large source of aerosols that contain SO_4. Dimethylsulfide $[(CH_3)_2S]$ is the major gas emitted by marine phytoplankton, and it is quickly oxidized to SO_4 and then redeposited in the ocean. Sulfate is abundant in ocean waters (12×10^{20} g of elemental S), and the mean residence time for a sulfur molecule in the sea is over 3 million years (Schlesinger 1997).

The United States and most developing countries have reduced sulfur dioxide emissions during the past 40 years (**Figure 21**). In the United States, sulfur dioxide emissions have been declining on average 2.4% per year since 1970, and in Britain the decline has been 4.5% per year over the same time period. Reduced emissions should reduce surface deposition of acid rain, but the effects of acid rain on the environment do not disappear immediately as sulfur dioxide emissions decline (Likens et al. 1996). A key question remains: *Will forest and aquatic ecosystems recover from the effects of acid rain, and if so, at what rate?* At Hubbard Brook the effect of acid rain has been to leach calcium from the soil to such an extent that available calcium, not nitrogen, now

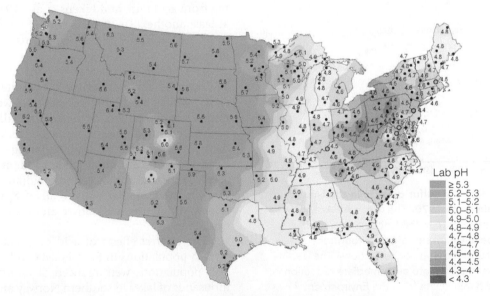

Figure 19 Distribution of acid precipitation in the United States in 2006. The lower the pH, the more acid the precipitation. Acid rain is a serious pollution problem in the eastern states. (Data from the National Atmospheric Deposition Program, 2007.)

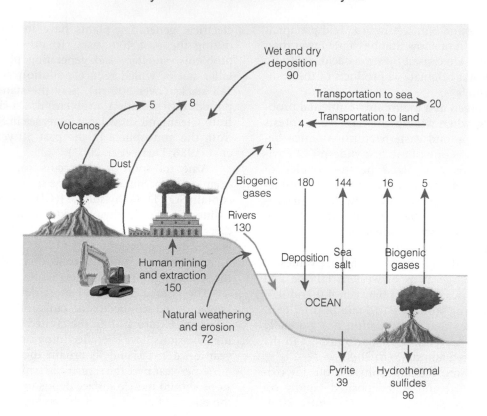

Figure 20 The global sulfur cycle. Burning of fossil fuels is the major component of atmospheric input of sulfur, which leads to acid rain. Humans have affected the sulfur cycle more than any other nutrient cycle. All values are 10^{12} g S/yr. (Modified from Schlesinger 1997.)

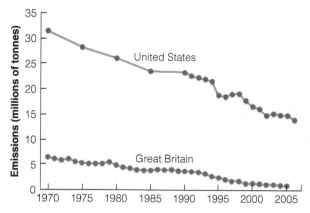

Figure 21 Emissions of sulfur dioxide in the United States and Britain since 1970. The Clean Air Act of 1970 required emissions to drop in the United States, and in Britain the European Union set targets through the Gothenburg Protocol for emission reductions. The decline in SO_2 emissions translates into an immediate reduction in the amount of acid rain. (Data from the Environmental Protection Agency of the United States and the Department for the Environment of the UK, 2007.)

appears to limit forest growth. Stream water chemistry at forest sites in New Hampshire is only slowly recovering from acid rain, and Likens et al. (1996) predict that at least another 10–20 years will be needed for streams to recover, even if sulfur dioxide emissions continue to decrease.

Freshwater ecosystems are particularly sensitive to acid rain. In areas underlain by granite and granitoid rocks, which are highly resistant to weathering, acid rain is not neutralized in the soil, so lakes and streams become acidified. Lakes in these bedrock areas typically contain soft water of low buffering capacity. Thus, bedrock can be an initial guide to identifying sensitive areas. The Precambrian Fennoscandian Shield in Scandinavia, the Canadian Shield, all of New England, the Rocky Mountains, and other areas are thus potential trouble spots.

The clearest effects of acid precipitation have been on fish populations in Scandinavia and eastern Canada. Fish populations were reduced or eliminated in many thousands of lakes in southern Norway and Sweden once the pH in these waters fell below pH 5 (Likens et al. 1979; Doka et al. 2003). In Canada, lakes containing

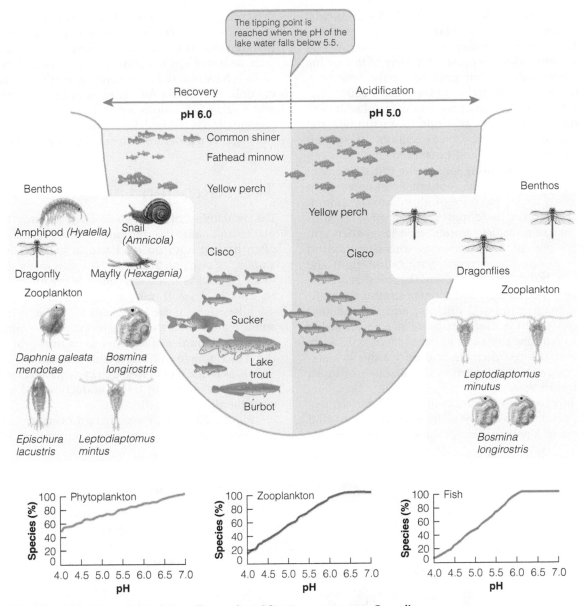

Figure 22 Schematic illustration of the effects of acidification on eastern Canadian lakes dominated by lake trout. Once the input of acid rain is curtailed, these lakes recover slowly, but full reversibility of the degradation has not yet been seen. (From Gunn and Mills 1998.)

lake trout have been the principal focus of research on the effect of acid rain. Lake trout disappear in lakes once the pH falls below 5.4, and the cause is reproductive failure because newly hatched trout die (Gunn and Mills 1998). Lake trout, which are a keystone predator in many Canadian lakes, disappear slowly in lakes of low pH. Adult trout do not seem to be affected by low pH, nor is there any food shortage at low pH. The effect is on mortality of the small juveniles, and this causes a slow decline in the trout population over 10–20 years. Once

lake trout are gone, acid-tolerant fish such as yellow perch and cisco become more abundant, and the food web shifts dramatically (**Figure 22**).

The effect of acid precipitation on terrestrial ecosystems is more complex. A good example of this complexity is found in forest declines in Europe (Tomlinson 2003). Forest declines have been particularly severe in central Europe, where needle yellowing and needle loss in Norway spruce (*Picea abies*) have brought public attention to the problem. Large-scale damage to spruce trees first became apparent in the late 1970s in Bavaria, Germany, and by the early 1980s about 25% of all European forests were classified as moderately or severely damaged from unknown causes. A variety of interacting factors associated with air pollution have now been shown to cause forest declines (Tomlinson 2003). The main effects of air pollution are on the forest soils. Soil acidification results from acid rain due to nitrate and sulfate deposition. Acid soil water reduces the amount of calcium, magnesium, and potassium that roots can absorb. Interactions among ions in the soil are particularly complex. For example, root uptake of magnesium is suppressed in the presence of aluminum or ammonium ions. Spruce seedlings growing in acid soils develop magnesium deficiencies, especially when ammonium is present. Spruce trees are thus stimulated to increased growth by nitrogen fertilization from the polluted air, but the acidification of the soil that accompanies the added nitrates and sulfates reduces the availability of magnesium and calcium, and the resultant deficiencies of these nutrients cause needle yellowing and loss. Plant diseases seem to have played only a secondary role in forest declines, and once a tree is weakened by nutrient imbalances, fungal diseases or insect attacks may increase.

The human-caused changes in the sulfur cycle thus have the potential to change nutrient cycling in natural ecosystems in a great variety of ways we cannot yet understand, much less predict. We cannot continue this aerial bombardment of ecosystems in the naive belief that nutrient cycles have infinite resilience to human inputs. Recent efforts to curtail sulfate emissions from fossil fuels have reduced the emissions of SO_4, and we must continue to press for further reductions because the current levels of emissions are still too large. Once acidic precipitation is reduced, both forest and lake ecosystems can begin to recover from the damage inflicted on them.

The Nitrogen Cycle

The availability of nitrogen is often limiting for both plants and animals, and net primary production is often limited both on land and in the oceans by the amount of nitrogen available. Nitrogen gas is abundant in air (it is 78% nitrogen), but few organisms can use N_2 directly. A small number of bacteria and algae can use nitrogen from the air directly and fix it as nitrate or ammonia. Many of these organisms work symbiotically in the root nodules of legumes to fix nitrogen, and this is a major source of natural nitrogen fixation. Human additions to the global nitrogen cycle have become substantial, particularly with the production and use of nitrogen fertilizers for agriculture.

Figure 23 shows the global nitrogen cycle. Human activities add about the same amount of nitrogen

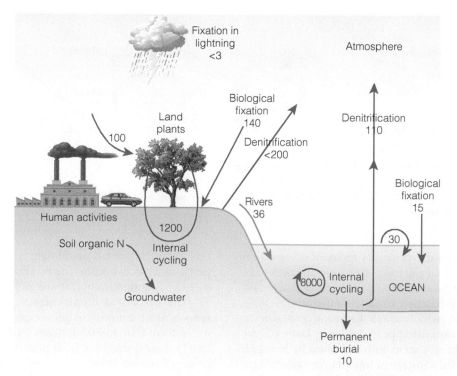

Figure 23 The global nitrogen cycle. Humans have had a strong effect on the nitrogen cycle. All fluxes are in units of 10^{15} g N/year. (From Schlesinger 1997.)

to the biosphere each year as do natural processes, but this human addition is not spread evenly over the globe. The effects of human additions of nitrogen have shown up particularly in changes in the composition of the atmosphere. Nitrogen-based trace gases—nitrous oxide, nitric oxide, and ammonia—have major ecosystem effects. Nitrous oxide is chemically unreactive and persistent in the atmosphere; it traps heat and thus acts as a greenhouse gas that changes climate. Nitrous oxide is increasing in the atmosphere at 0.25% per year (Vitousek et al. 1997). Nitric oxide, by contrast, is highly reactive and contributes significantly to acid rain and smog. Nitric oxide can be converted to nitric acid in the atmosphere, and in western United States acid rain is based more on nitric acid than on sulfuric acid. In the presence of sunlight, nitric oxide and oxygen react with hydrocarbons from auto exhaust to form ozone, the most dangerous component of smog in cities and industrial areas. Nitric oxide is produced by burning fossil fuels and wood. Ammonia neutralizes acids and thus acts to reduce acid rain. Most ammonia is released from organic fertilizers and domestic animal wastes.

The result of human activities on the nitrogen cycle has been an increased deposition of nitrogen on land and in the oceans. Because nitrogen additions are typically coupled with phosphorus additions, the result is eutrophication of freshwater lakes and rivers and coastal marine areas (Carpenter et al. 1998). Phosphorus additions to freshwater typically increase primary production, whereas nitrogen addition to estuaries increases primary production in marine environments. The adverse effects of eutrophication on aquatic systems are listed in **Table 4**.

Nitrates in rivers are rising everywhere in proportion to the human population along the rivers and the use of excess fertilizers in agricultural areas. One good example of this problem is the Mississippi River in central United States. The Mississippi River drains nearly one-third of North America, and changes in water quality in the river over the last 50 years have triggered drastic ecosystem impacts in the northern part of the Gulf of Mexico. The problem is nitrogen in the water, and the principal cause is a dramatic increase in fertilizer nitrogen input into the Mississippi River drainage basin between the 1950s and 1980s (**Figure 24**). Since 1980, the Mississippi River has discharged, on average, about 1.6 million metric tons of total nitrogen to the Gulf each year (Rabalais et al. 2002). The most significant trend in nutrient loads has been in nitrate, which has almost tripled from 0.33 million metric tons per year during 1955–1970 to 0.95 million metric tons per year during 1980–1996. Other nutrients such as phosphorus have not increased, and have possibly decreased over the last 50 years. About 90% of the nitrate in the river comes

Table 4 Adverse effects of nutrient additions of nitrogen and phosphorus on freshwater and coastal marine ecosystems.

Increased biomass of phytoplankton
Shifts in phytoplankton communities to bloom-forming species that may be toxic
Increase in blooms of gelatinous zooplankton in marine ecosystems
Increased biomass of benthic algae
Changes in macrophytic species composition and biomass
Death of coral reefs
Decreases in water transparency
Taste, odor, and water treatment problems for domestic water supplies
Oxygen depletion
Increased frequency of fish kills
Loss of desirable fish species
Reductions in harvestable fish and shellfish
Decrease in aesthetic value of bodies of water

SOURCE: From Carpenter et al. (1998).

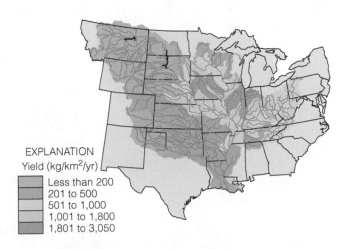

EXPLANATION
Yield (kg/km²/yr)
- Less than 200
- 201 to 500
- 501 to 1,000
- 1,001 to 1,800
- 1,801 to 3,050

Figure 24 Nitrogen inputs into the Mississippi River drainage system from different parts of the drainage basin. The largest inputs are from the corn belt of Indiana, Illinois, and Iowa. (From U.S. Geological Survey, Hypoxia Task Force, http://toxics.usgs.gov/hypoxia/task_force_workgroup.html)

from excess fertilizer draining off agricultural land and drainage from feedlots for cattle.

Nitrogen does not seem to be a primary limiting factor for primary production in river systems, and the ecological damage starts when these waters reach the coastal zone in the Gulf of Mexico. Coastal waters around the world are suffering from pollution—nutrients draining from the land and stimulating algal growth in the sea. In coastal waters off Louisiana, the excess nitrogen stimulates algal growth and associated zooplankton growth. Fecal pellets from zooplankton and dead algal cells sink to the bottom, and as this organic matter decomposes, the bacteria use all the oxygen in the bottom layer of water. Stratification of fresh and saline waters prevents oxygen replenishment that would normally occur by the mixing of oxygen-rich surface water with oxygen-depleted bottom water. At dissolved oxygen levels of less than 2 mg/L all animals either leave or die. This shortage of oxygen in the bottom layer of coastal waters is called **hypoxia**, and these zones are called "dead zones."

The Mississippi River outflow produces each summer a hypoxic zone in the northern Gulf of Mexico along the Louisiana-Texas coast that varies in size up to 20,000 sq. kilometers (**Figure 25**). The hypoxic zone is most pronounced from June to August but can begin as early as April and last until October, when storms and winds mix up the surface and bottom water. Spawning grounds of fish and migratory routes of commercially harvested fish species are affected by the hypoxia zones. To reduce hypoxia in the Gulf of Mexico the most effective actions would be to reduce the amount of fertilizer usage, to keep the nitrogen in the agricultural fields with alternative cropping systems, and to increase the area of wetlands, which pick up nitrogen from the river water. The important message is that alleviating the problem of hypoxia in the Gulf of Mexico requires an ecosystem

Figure 25 Size of the hypoxia zone (red) off the Mississippi River mouth in midsummer 1986, 1990, and 1996. The dots are the sampling stations in the Gulf of Mexico. Nitrogen-rich water coming down the Mississippi River increases primary production in coastal waters so much that decomposition causes all the oxygen to be depleted in summer in the red zones. (Modified from Rabalais et al. 2002.)

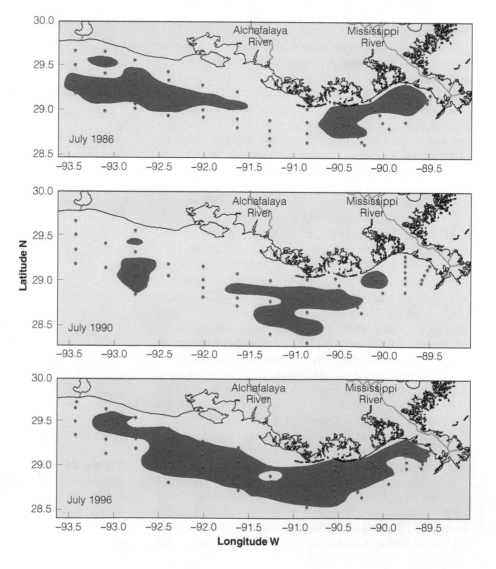

approach to the whole catchment of the Mississippi River. An ecological understanding is needed of how the whole catchment works and how nutrients applied in fertilizer to grow corn in Iowa can impact fish populations thousands of kilometers away in the Gulf of Mexico.

By contrast, the addition of nitrogen to terrestrial ecosystems can have positive effects. Nitrogen deposition on land can relieve the nitrogen limitation of primary production that is common in many terrestrial ecosystems. Swedish forests, all of which are nitrogen limited, have averaged 30% greater growth rates in the 1990s compared with the 1950s (Binkley and Högberg 1997). The important concept here is that of the **critical load**—the amount of nitrogen that can be input and absorbed by the plants without damaging ecosystem integrity. When the vegetation can no longer respond to further additions of nitrogen (see Figure 18), the ecosystem reaches a state of nitrogen saturation, and all new nitrogen moves into groundwater or streamflow or back into the atmosphere. Nitrate is highly water soluble in soils, and excess nitrate carries away with it positively charged ions of calcium, magnesium, and potassium. Excess nitrate can thus result in calcium, magnesium, or potassium limiting plant growth, and this is why most commercial garden fertilizers contain more than just nitrogen.

Increasing nitrogen in terrestrial ecosystems can have undesirable effects on biodiversity. In most cases, adding nitrogen to a plant community reduces the biodiversity of the community (Huston 1994). **Figure 26** illustrates the effect of 12 years of experimental nitrogen fertilization of grasslands in Minnesota (Wedin and Tilman 1996). Species that are nitrogen responsive, often grasses, can take over plant communities enriched in nitrogen. The Netherlands has the highest rates of nitrogen deposition in the world, largely due to intensive

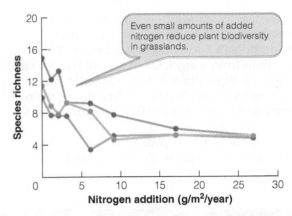

Figure 26 Vegetation responses to 12 years of nitrogen fertilization in Minnesota grasslands. Three fields were used, and six replicates were used for each level of nitrogen addition. Biodiversity declines dramatically as more nitrogen is added to these grasslands. (From Wedin and Tilman 1996.)

livestock operations, a consequence of which has been a conversion of species-rich heathland to species-poor grasslands and forest. The mix of plant and animal species adapted to sandy, infertile soils is being lost because of nitrogen enrichment (Vitousek et al. 1997).

The nitrogen cycle, like the sulfur cycle, has been heavily affected by human activities during the past 50 years. It is urgent that national and international efforts be directed to reversing these changes and moderating the adverse effects on ecosystems. The most obvious direct effect on humans resulting from these changes in nutrient cycling are now manifest in global climate change.

Summary

Nutrients cycle and recycle in ecosystems, and tracing nutrient cycles is another way of studying fundamental ecosystem processes. Human activities are changing cycles on a global basis, with consequences for biodiversity, ecosystem function, and climate change. Nutrients reside in compartments and are transferred between compartments by physical or biological processes. Compartments can be defined in any operational way to include one or more species or physical spaces in the ecosystem. Nutrient cycles may be local or global. Global cycles, such as the nitrogen cycle, include a gaseous phase that is transported in the atmosphere. Local cycles include less mobile elements such as phosphorus.

Nutrient cycles in forests have been studied because of nutrient losses associated with logging. If the inputs of nutrients do not equal the outflow for any ecosystem, it will deteriorate over the long run. Logging can result in high nutrient losses even if soil erosion is absent. An undisturbed forest site recycles nutrients efficiently. Nutrient use efficiency is important to ecosystem functioning because plants in poorer soils are more efficient in their nutrient use, such that a greater mass of plant tissue is produced per unit of nutrient.

The sulfur cycle is an example of a global nutrient cycle that is strongly affected by human activities. The burning of fossil fuels adds a large amount of SO_2 to the atmosphere and results in acid rain. Acid rain in combination with other airborne pollutants has caused forest declines in Europe and has eliminated fish populations from many lakes in eastern Canada and Scandinavia. Regulations in North America and Europe have caused SO_2 emissions to decline during the past 40 years, and ecosystems can slowly recover once the inputs are stopped.

The nitrogen cycle is critical because primary production in many terrestrial ecosystems and in coastal waters is limited by nitrogen. Nitrogen emissions by human activity have doubled the input of nitrogen into the air and waters of the globe. Smog in cities and more acid rain are two effects of this added nitrogen. Nitrogen and phosphorus leach from agricultural fertilizer and are a major cause of algal pollution in lakes and rivers. Unless we can curb these emissions of critical nutrients, global ecosystems will continue to be degraded.

Review Questions and Problems

1   In slash-and-burn agriculture, which is common in many tropical countries, forests are cut and burned, and crops are planted in the cutover areas. Yields are usually good in the first year but decrease quickly thereafter. Why should this be? Compare your ideas with those of Tiessen et al. (1994), and evaluate the sustainability of slash-and-burn agriculture.

2   Increased nitrogen deposition from the atmosphere may lead to nitrogen saturation of forests. Design a long-term experiment that would address the question of how much nitrogen loading a forest could sustain. Compare your design with that used by the NITREX project in Europe (Gundersen et al. 1998).

3   Discuss the relative merits of making a compartment model of a nutrient cycle very coarse (with only a few compartments) versus making it very fine (with many compartments).

4   Shallow lakes may have two alternate stable states depending on nutrient influx, one dominated by phytoplankton in turbid water and another dominated by macrophytes in clear water. Discuss how you might determine the critical nutrient loading that would trigger a transition between these states. Would you expect the transition from phytoplankton to macrophytes to occur at the same nutrient loading as the opposite transition? Janse (1997) discusses this problem.

5   Schultz (1969, p. 92) states:

The idea of one cause–one effect is left over from the nineteenth century when physics dominated science. The whole notion of causality is under question in the ecosystem framework. Does it make sense to say that high primary production causes a rich organic soil and a rich organic soil causes high production? This kind of reasoning leads up a blind alley.

Discuss.

6   Pacific salmon grow to adult size in the ocean and move into freshwater streams and lakes to spawn and die. In the process they transport nutrients from ocean ecosystems to freshwater ecosystems. These returning salmon are eaten by bears and other predators, bringing some of these nutrients into the terrestrial ecosystem. Discuss how you might measure the impacts of this nutrient transport system both for aquatic and for terrestrial ecosystems within the geographic range of Pacific salmon. Helfield and Naiman (2006) discuss this issue.

7   Land use changes in Brazil are frequently in the news because of the conversion of tropical forest to pasture. About 70% of the forest clearing in Brazil is for cattle pastures. Discuss the nutrient balance issues that arise from this type of land conversion from forest to pasture, and whether there could be no overall net loss of soil carbon and nitrogen in this conversion. Cerri et al. (2004) discuss this issue and provide data.

8   Soils in Australia contain very low amounts of phosphorus, from 10%–50% the amount in North American soils (Keith 1997). Would you predict that eucalypts growing in Australian soils would be phosphorus limited? What adaptations might plants evolve to achieve high nutrient use efficiency when growing on soils of low nutrient content?

9   A key question in restoration ecology is how long it will take for an ecosystem to recover from some disturbance caused by humans. Discuss how we might find out what the time frame is for ecological recovery from acid rain.

Overview Question

One suggested ecological response to help restore acidified lakes is to add lime. Explain the mechanisms behind this recommendation, and discuss whether any possible negative effects might result from this amelioration program. How could you determine if this restoration program was effective?

Suggested Readings

- Akselsson, C., O. Westling, H. Sverdrup, and P. Gundersen. 2007. Nutrient and carbon budgets in forest soils as decision support in sustainable forest management. *Forest Ecology and Management* 238:167–174.

- Arrigo, K. R. 2005. Marine microorganisms and global nutrient cycles. *Nature* 437:349–355.

- Galloway, J. N., A. R. Townsend, J. W. Erisman, M. Bekunda, Z. Cai, J. R. Freney, L. A. Martinelli, S. P. Seitzinger, and M. A. Sutton. 2008. Transformation of the nitrogen cycle: recent trends, questions, and potential solutions. *Science* 320:889–892.

- Helfield, J., and R. Naiman. 2006. Keystone interactions: Salmon and bear in riparian forests of Alaska. *Ecosystems* 9:167–180.

- Hooper, D. U., and P. M. Vitousek. 1998. Effects of plant composition and diversity on nutrient cycling. *Ecological Monographs* 68:121–149.

- Johnson, D. W. 2006. Progressive N limitation in forests: Review and implications for long-term responses to elevated CO_2. *Ecology* 87:64–75.

- McKenzie, V. J., and A. R. Townsend. 2007. Parasitic and infectious disease responses to changing global nutrient cycles. *EcoHealth* 4:384–396.

- Newman, E. I. 1997. Phosphorus balance of contrasting farming systems, past and present. Can food production be sustainable? *Journal of Applied Ecology* 34:1334–1347.

- Rabalais, N. N., R. E. Turner, and D. Scavia. 2002. Beyond science into policy: Gulf of Mexico hypoxia and the Mississippi River. *BioScience* 52:129–142.

- Schlesinger, W. H. 1997. *Biogeochemistry: An Analysis of Global Change*, 2nd ed. San Diego: Academic Press.

- Tomlinson, G. H. 2003. Acidic deposition, nutrient leaching and forest growth. *Biogeochemistry* 65:51–81.

- Vanni, M. J. 2002. Nutrient cycling by animals in freshwater ecosystems. *Annual Review of Ecology & Systematics* 33:341–370.

- Vitousek, P. M. 2004. *Nutrient Cycling and Limitation: Hawai'i as a Model System*. Princeton, NJ: Princeton University Press.

Credits

Illustration and Table Credits

Photo Credits

Ecosystem Dynamics under Changing Climates

Key Concepts

- Global warming has been underway since the 1970s and the average global temperature has increased about 0.5°C during that time.

- The cause of global warming is increased greenhouse gas emissions from the burning of fossil fuels and land clearing. Increasing CO_2 is one of the major problems and is now above the levels recorded in ice cores covering the last 650,000 years. Atmospheric CO_2 is increasing at 0.5% per year.

- Global warming and atmospheric CO_2 enrichment increase plant primary production, but carbon sequestration by plants is limited by other nutrients such as nitrogen.

- Increased primary production from CO_2 enrichment may not result in additional carbon storage in soils. There is a limit to how much CO_2 global ecosystems can absorb.

- Global warming is speeding spring flowering and breeding events and allowing organisms to move their geographic ranges toward the poles.

- Disease vectors and pathogens can increase emergent disease problems for plants and animals as the climate warms and they extend their ranges.

From Chapter 25 of *Ecology: The Experimental Analysis of Distribution and Abundance*, Sixth Edition. Eugene Hecht.
Copyright © 2009 by Pearson Education, Inc. Published by Pearson Benjamin Cummings. All rights reserved.

KEY TERMS

ENSO El Niño-Southern Oscillation, the coupled ocean-atmosphere change in the tropical Eastern Pacific Ocean in which warmer surface waters move east toward South America, causing major climatic changes in Pacific Rim countries.

FACE Free Air Carbon Enrichment experiments done in open environments to study the effect of increased CO_2 on plants and animals in natural settings.

greenhouse effect The process in which the *emission* of *infrared* (long-wave) radiation by the *atmosphere* warms a *planet's* surface.

greenhouse gases Gases in the atmosphere, such as carbon dioxide, that contribute to the *greenhouse effect*.

trace gases Refer to gases in the atmosphere that occur in small concentrations (<1% by volume) but contribute to the greenhouse effect, for example methane which occurs at 2 parts per million in air and has 72 times the impact of the same mass of carbon dioxide.

Climate is a variable that influences all population and community processes because the distribution and abundance of all organisms is affected by climate. Few people need to be convinced now that climates are changing around the Earth, and in this chapter we will review the evidence for climate change and the best estimates of the future course of climate under several possible scenarios. During your lifetime these impacts of climate change will lead to the most significant problems facing the globe, and our present preoccupation with the stock market and political chicanery will be viewed as the latest example of Nero fiddling while Rome is burning.

We have discussed multiple examples of how differences in temperature, rainfall, or solar radiation has affected particular species or ecosystems, and here we draw these threads together to try to estimate the impacts of changing climate. Translating science into policy is an art, and ecological scientists are not yet very good at it. Our first objective must be to get the science right. The central mandate of applied ecology is to assess the biological consequences of climate change scientifically and to suggest possible ways of ameliorating them.

In this chapter we first examine the evidence for a changing climate, and then explore the carbon cycle, the primary nutrient cycle involved in climate change. Then we summarize some of the changes that have already occurred in populations and communities, and consider the potential changes that may occur with continuing climate shifts, including the problem of invasive pathogens.

Climate Change in the 21st Century

Climate change is in the news every day, and reports from the Intergovernmental Panel on Climate Change (IPCC) are a focus of attention. Their latest report in 2007 paints a general picture of rapid climate change and discusses the requirements for adaptation that are necessary for human societies to adjust to these changes. We will consider only a small part of the data they have amassed on changes in temperature and precipitation over the Earth. Climate change involves a complex set of changes because not all areas are warming when global temperatures are increasing, and not all areas are having greater precipitation when on average global precipitation is increasing. The devil of climate change is in the details of how climate change can simultaneously lead to droughts in Australia and excessive rainfall in Europe.

Global temperatures have been rising somewhat irregularly since 1900 and overall global temperatures have risen about 1°C in the last 100 years (**Figure 1**). The past 1000 years in the Northern Hemisphere has been cooler than the 1961–1990 average that we tend to think of as "average." If we look in detail at the different parts of the Earth, we can see that the climatic warming during the last 30 years has not been evenly spread (**Figure 2**). The strongest temperature increases have occurred in the northwestern part of North America. The next largest shifts have been in parts of Eurasia, Africa, western Canada, and Greenland. Much of the oceans have warmed slightly (0.2°C to 1°C), along with South America and Australia. On average the warming trend for the last 50 years has been twice that of the last 100 years (Solomon et al. 2007). There is no question that the Earth is warming.

The next question is what are the causes of the recent warming, and part of the answer to this question lies in the **greenhouse effect**. The greenhouse effect is the process in which the *emission* of *infrared* radiation by the *atmosphere* warms a *planet's* surface. The name comes from an incorrect *analogy* with the warming of air inside a *greenhouse* compared to the air outside the greenhouse (LeTreut et al. 2007). The *Earth's* average surface temperature is much warmer than it would be without the

Variations of the Earth's surface temperature for...

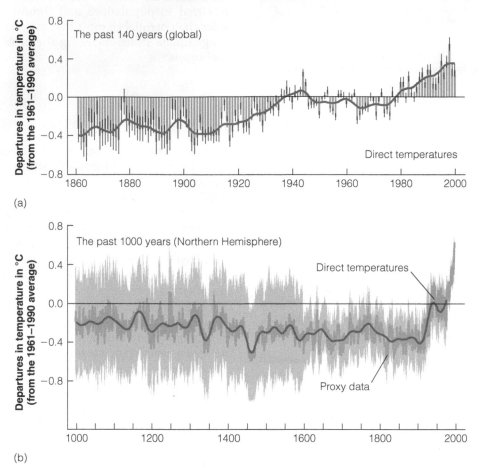

(a)

(b)

Figure 1 Changes in average annual temperatures for the Earth. (a) Global temperatures since 1860 when direct temperature measurements became widely available. (b) Northern Hemisphere temperature variation since the year 1000, including both direct temperature measurements since 1860 and a combination of proxy data (blue bars) for earlier years. All temperatures are expressed as deviations from the 1961–1990 average. (From Solomon et al. 2007.)

Figure 2 Changes in average annual temperature for different regions of the Earth over the period 1970 to 2004. Only the blue areas have cooled and most of the globe including the oceans have warmed over this time span. (From Parry et al. 2007.)

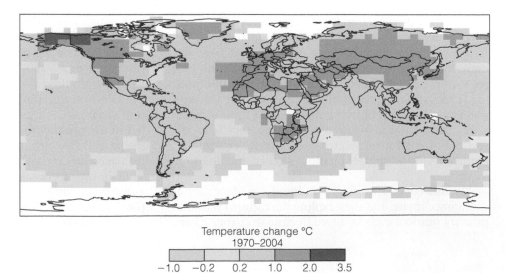

Temperature change °C
1970–2004

−1.0   −0.2   0.2   1.0   2.0   3.5

greenhouse effect (**Figure 3**). Incoming solar radiation averaged over the Earth represents 342 watts per m² per year.

Greenhouse gases represent one of the components that affect global warming. The major greenhouse gases are *water vapor*, which causes about 36%–70% of the greenhouse effect on Earth (*not including clouds*); *carbon dioxide*, which causes 9%–26% of the greenhouse effect; *methane*, which causes 4%–9%, and *ozone*, which causes 3%–7%. Other greenhouse gases include *nitrous oxide, sulfur hexafluoride, hydrofluorocarbons, perfluorocarbons,* and *chlorofluorocarbons* (Solomon et al. 2007). The major atmospheric constituents (nitrogen and *oxygen*) are not greenhouse gases. The majority of greenhouse gases are natural in origin but some are increased by human activities.

The contribution of each of the major components of recent global warming has been estimated by the Intergovernmental Panel on Climate Change. **Figure 4** gives the relative contributions of each factor and illustrates that CO_2 is the most important greenhouse gas. Humans contribute to many aerosols (dust, sulfate, nitrate, soot) that act to cool the Earth. Ozone-forming chemicals such as carbon monoxide act to warm the Earth, and the solar con-

stant has increased very slightly since 1750 (Solomon et al. 2007).

While trends in temperature are somewhat variable, the overall warming trend is clear for virtually all parts of the Earth. By contrast, precipitation trends are much more variable and no overall patterns are clear. Over the last 100 years there is no overall statistical trend in recorded precipitation on land (Trenberth et al. 2007). The ecological impacts of variation in precipitation are short-term physical events such as the El Niño-Southern Oscillation. The high variability in precipitation and the lack of measurements in many areas of the Earth make it difficult to obtain an overall picture of how precipitation is changing.

Aquatic ecosystems in temperate and polar regions are affected both by precipitation and by the melting of snow and ice that has accumulated over many years (Wrona et al. 2006). One of the features of global warming has been a shrinking of the glaciers and snowfields around the world. This melting will affect both the total flow of rivers and their seasonal variation (Zhang et al. 2001).

We now turn to explore in more detail the carbon cycle, the most critical nutrient cycle involved in changing climate.

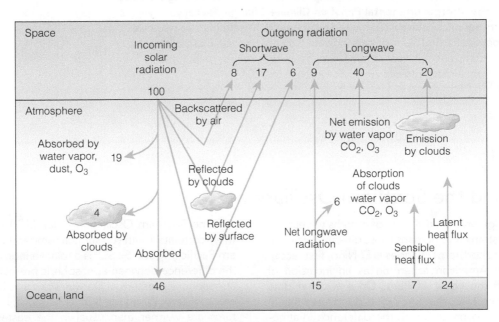

Figure 3 The greenhouse effect of CO_2 and other trace gases. The sun's radiation (expressed here as 100%) is dominated by short wavelengths, which are reflected or absorbed at or near the Earth's surface. The absorbed radiation is reradiated at longer wavelengths that can be absorbed by atmospheric gases, including CO_2. Higher concentrations of these gases in the atmosphere reduce the net emission of longwave radiation into space, warming the Earth. (U.S. Department of Energy.)

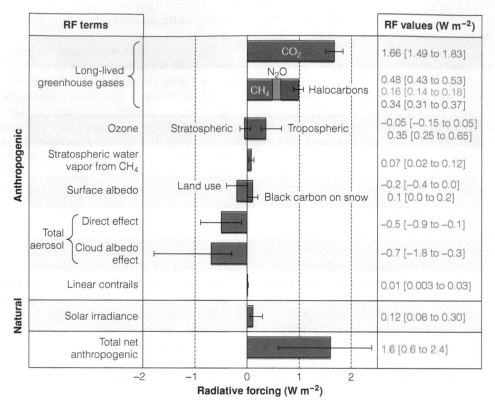

Figure 4 **Average global estimates of the components of radiative forcing for 2005 as estimated by the Intergovernmental Panel on Climate Change.** Radiative forcing is measured in watts per m²; positive values in red increase global warming and negative values in blue cool the Earth. The error bars indicate the uncertainties in estimating the exact size of each component. Linear contrails are the white ice vapor trails made by the exhaust of water from aircraft engines. (From Solomon et al. 2007.)

ESSAY

El Niño and the Southern Oscillation

Climate change on the time scale of months to years affects ecosystems and humans most directly. The most famous of these short-term changes is El Niño, first recognized by South American fishermen as an incursion of warm water off the coast of Peru. Once weather data began to be assembled, meteorologists discovered that El Niño events were correlated with the difference in atmospheric pressure between Tahiti and Darwin, Australia. The Southern Oscillation is the name given to this seesaw of change of atmospheric pressure between these two stations. Once it was realized that these oceanographic and atmospheric processes were coupled, the joint name El Niño–Southern Oscillation (ENSO) was coined.

The Southern Oscillation Index (SOI) is highly negatively correlated with sea surface temperatures in the eastern Pacific (**Figure 5**). SOI is a relative index, measured by the difference between atmospheric pressure at Tahiti and atmospheric pressure at Darwin, scaled to a long-term average of zero. When the SOI[1] is negative, ocean temperatures are warmer than usual in the eastern Pacific and colder than usual in the western Pacific and Indian Oceans.

---

[1]Many SOI indices have been devised and here I use one that correlates positively with sea surface temperatures to reduce confusion.

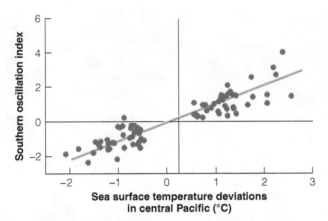

Figure 5 Relationship between sea surface temperatures in the central Pacific Ocean and one scaling of the Southern Oscillation Index, a common measure of the strength of El Niño events. The SOI Index is a relative measure of the atmospheric pressure difference (Tahiti – Darwin). Positive deviations indicate El Niño conditions (warm waters in the central Pacific). (Data from Columbia University Climate/ENSO Web site: http://iri.columbia.edu/climate/ENSO/background/basics.html.)

Data on the SOI date back to the start of the twentieth century (**Figure 6**), and more recently a network of fixed buoys has been deployed across the central Pacific to measure sea surface temperature in real time.

The effects of ENSO events are dramatic. When an El Niño event occurs, weather changes are triggered on a global scale, as the maps in **Figure 7** show. Mild winters in the northeastern United States and western Canada are one correlate of El Niño, and severe droughts in Australia, India, Indonesia, Brazil, and Central America are typical of El Niño years. The ENSO cycle has an average period of four years but varies from two to seven years in length. There is much variation in the strength of the Southern Oscillation for reasons that are not yet clear.

The effects of El Niño on global ecosystems are varied and significant. Warm surface temperatures lead to coral bleaching and the destruction of coral reefs. The pelagic upwelling ecosystem off Peru collapses because the warm water displaces the nutrient-rich cold water. Fisheries such as the Peruvian anchovy collapse, and seabirds, which also depend on these fish, suffer high mortality in El Niño events. Pacific salmon production in the North Pacific is linked to similar changes in oceanographic events (Mantua et al. 1997), and tree recruitment in the forests of South Dakota vary with El Niño events (Brown 2006). It is likely that many ecological changes are driven by large-scale weather changes caused by these oceanographic shifts, and the linkages between large-scale weather changes and ecosystem dynamics are a critical focus of current research.

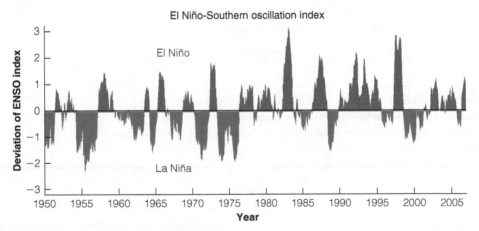

Figure 6 The El Niño–Southern Oscillation index since 1950. The long-term average index is set to zero. Positive deviations indicate warmer waters in the central Pacific (El Niño), and negative deviations indicate cool waters (La Niña). (Data from U.S. Department of Commerce, National Oceanographic and Atmospheric Administration, www.cdc.noaa.gov/ENSO/enso.climate.html.)

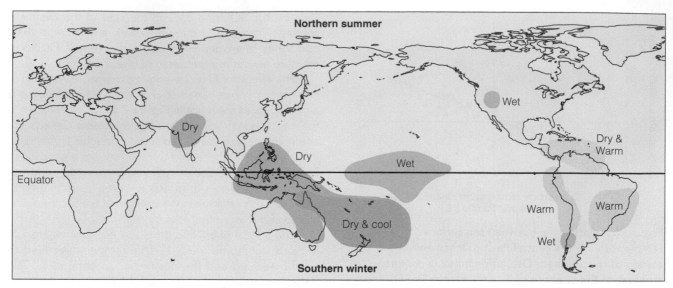

(a)

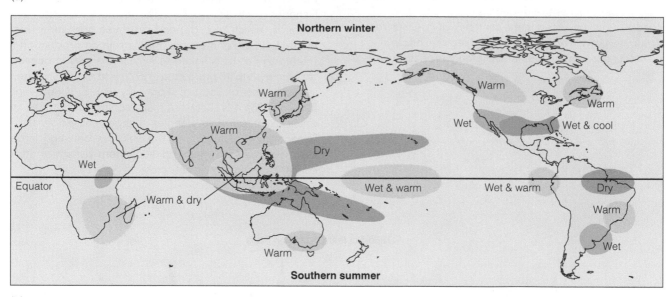

(b)

Figure 7 Expected ENSO climatic events during (a) the northern summer
(= southern winter) and (b) the northern winter (= southern summer). Depending on
the strength of the ENSO event, these climatic events may be more or less severe. (Maps
courtesy of NOAA.)

The Carbon Cycle

To understand the climate changes we have just described, we must understand the effects humans are having on the global carbon cycle. Ecosystems are critically
involved in the carbon cycle because plants and animals
are primarily composed of carbon, and the global carbon cycle is partly a reflection of primary and secondary

production. The fixation of carbon by plants via photosynthesis over geological time accounts for the oxygen
in the Earth's atmosphere. Humans have affected the
global carbon cycle nearly as much as they have the sulfur cycle, and intense public interest now focuses on the
resulting greenhouse effect and climate change.

Figure 8 shows the global carbon cycle. The carbon
cycle is mostly the carbon dioxide cycle. The largest

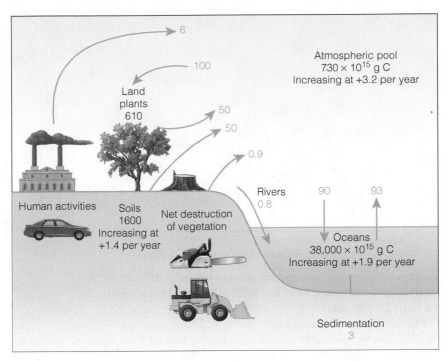

Figure 8 The present-day global carbon cycle. All pools are expressed in units of 10^{15} g C (black) and all annual fluxes in units of 10^{15} g C/yr (blue). (Modified after Schlesinger 1997 with data from the Australian Bureau of Meteorology, 2007.)

pool of carbon is dissolved inorganic carbon in the ocean, which contains about 56 times as much carbon as the atmosphere (Schlesinger 1997). The atmospheric pool of carbon is slightly larger than the total carbon bound up in vegetation. The largest fluxes of the global carbon cycle are between the atmosphere and land vegetation, and between the atmosphere and the oceans. These two fluxes are approximately equal, and the mean residence time of a molecule of carbon in the atmosphere is about five years.

The amount of CO_2 in the atmosphere has not been constant. From bubbles of gas trapped in ice cores in Antarctica we know that atmospheric CO_2 was about 270–280 ppm from AD 900 to AD 1750 (Barnola et al. 1995). The Vostok ice core, which was collected by the Soviet Antarctic Expedition, spans 420,000 years and shows that during the last ice age (20,000 to 50,000 years ago) CO_2 levels were 180–200 ppm, much lower than our current levels (**Figure 9**). Since 1750 and the start of the Industrial Revolution, CO_2 levels in the atmosphere have risen rapidly and continuously.

Detailed CO_2 measurements have been made since 1958 at Mauna Loa in Hawaii. **Figure 10** shows the rising CO_2 levels in the atmosphere since the 1950s. Superimposed on these long-term trends is a seasonal trend in CO_2 concentration. The atmospheric oscillations of CO_2 are the result of the seasonal uptake of CO_2 by plants via photosynthesis and seasonal differ-

ences in fossil fuel use and CO_2 exchange with the oceans. The majority of terrestrial vegetation occurs in environments with seasonal growth cycles, and plants fix CO_2 so that atmospheric CO_2 levels decline in summer (Figure 10). The atmospheric oscillations of CO_2 are superimposed on a long-term increase of 0.4% (1.5 ppm) per year from 1960 to 1995, and an increasing trend of 0.5% (1.9 ppm) per year from 1995 to 2005. Most of this increase comes from the burning of fossil fuels. If all the CO_2 released from fossil fuels accumulated in the atmosphere, CO_2 would be increasing about 0.7% per year. But only about 56% of the CO_2 released from fossil fuel is accumulating in the atmosphere (Solomon et al. 2007). What happens to the remainder?

We cannot yet balance the global carbon budget to answer this question, and this is one of the most vexing problems in ecosystem ecology today. **Figure 11** shows our current understanding of the sizes of major sources and sinks for carbon. Oceanographers believe that about 24% of the CO_2 from fossil fuels enters the ocean each year and about 30% enters terrestrial ecosystems (Canadell et al. 2007). CO_2 is exchanged only at the ocean's surface, and much of the carbon in the oceans is in the deeper waters. Exchange in the oceans between surface waters and deep waters occurs only very slowly. Turnover of carbon for the entire ocean occurs about every 350 years (Schlesinger 1997).

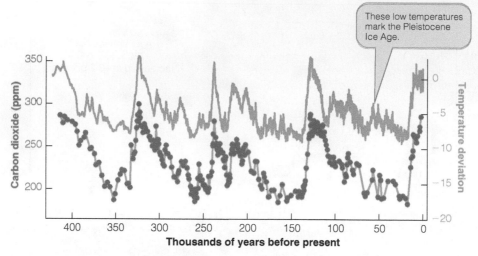

Figure 9 **Long-term variations in atmospheric carbon dioxide concentration (blue) and global temperature (red) as determined from the 3623-m-long Vostok ice core, Antarctica, covering the last 420,000 years.** Carbon dioxide levels can be measured from air trapped in the ice as it is formed. Temperature is measured as the deviation from present-day temperatures (°C). The temperature changes show the last four ice ages, each about 100,000 years long. There is a high correlation between CO_2 levels and global temperature. (Modified after Petit et al. 1999.)

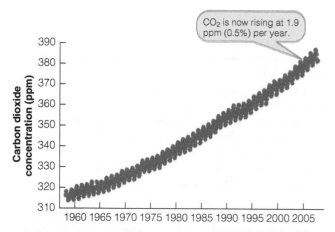

Figure 10 **The concentration of atmospheric CO_2 at Mauna Loa Observatory in Hawaii since 1958.** The annual oscillation reflects the seasonal cycles of photosynthesis and respiration by land biota in the Northern Hemisphere, while the overall increase is largely due to the burning of fossil fuels. (Data from Keeling and Whorf (2004) and NOAA 2007 www.esrl.noaa.gov/gmd/ccgg/trends/.)

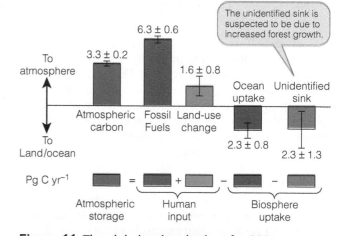

Figure 11 **The global carbon budget for 2006.** Atmospheric carbon is increasing because of fossil fuel combustion and land use changes. These additions to the atmosphere are partly counteracted by oceanic uptake and increased forest growth on land. A petagram is 10^{15} grams. (From Canadell et al. 2007 and the Global Carbon Project www.globalcarbonproject.com.)

Another source of atmospheric carbon dioxide is the destruction of terrestrial vegetation, often by clearing and burning for agriculture, especially in the tropics (Raupach et al. 2007). Destructive land use is offset by the increased growth of natural forests. Shifting cultivation is a dominant form of land use in tropical countries, and about 75% of all land-use changes fall under

this heading. Shifting cultivation is less destructive of forest because after one to three years the farmers move on and abandon the fields to secondary succession. Such temporary clearing of land contributes less CO_2 to the atmosphere than does permanent conversion of forest land to pastures. In 2006 the best estimate was that tropical areas were a net source of 1.5×10^{15} g carbon because of deforestation, about 16% of total carbon emissions (Canadell et al. 2007). At the same time the regrowth of forests in the temperate zones captured about 2.8×10^{15} g of carbon.

In trying to balance the global carbon budget, it is important to remember that we are focusing on the annual movements of carbon, not on the amount stored in the various reservoirs. The ocean contains the largest pool of carbon, but most of this carbon turns over very slowly. Desert soil carbonates contain more carbon than all terrestrial plants, but virtually no exchange of carbon occurs between the atmosphere and desert soils.

Recent work has focused on terrestrial vegetation as a sink for CO_2 (Canadell et al. 2007). If biomass is increasing in terrestrial vegetation because of a "fertilization" of vegetation by CO_2, and if this increase is rapid enough, we might have located the "unknown sink" of the global carbon cycle. What is the evidence that rising CO_2 levels stimulate plant growth? And how does changing climate affect animals and plants in natural ecosystems?

Climate Change Effects on the Biosphere

Increasing temperatures and changing precipitation are two major components of climate change but we need to consider also the direct effects of rising CO_2 levels on individuals and ecosystems.

Individual Plant Responses to CO_2

The effects of rising CO_2 and temperature have been analyzed in great detail for individual species of plants because this work can be done relatively easily in greenhouses or in open-top chambers in the field. The simple view is that more CO_2 will mean greater plant growth. When other resources are available in adequate amounts, additional CO_2 can increase growth of C_3 plants over a wide range of CO_2 levels, as this simple view would expect. The photosynthetic machinery of C_3 plants saturates at CO_2 levels of 1000 ppm, far above the current level of CO_2 in the atmosphere (Körner 2006). By contrast, C_4 plants do not exhibit increased photosynthetic rates at higher CO_2 levels (Bazzaz 1996). **Figure 12** illustrates these differences for two species of annual plants. Growth enhancement in C_3 plants is not always a simple phenomenon. Some plants acclimate to high CO_2 levels and show a decline in photosynthetic rate with time. Other resources for growth—water, light, and nutrients—must be present in excess to observe the effect of elevated CO_2 levels.

The expectation that increases in CO_2 would translate into higher rates of plant growth and thus carbon

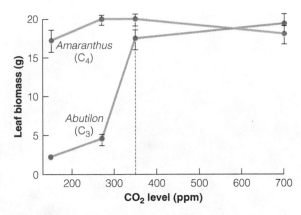

Figure 12 The response of annual C_3 and C_4 plants to four carbon dioxide levels in a greenhouse. The CO_2 levels were chosen to simulate ambient level during the last ice age (150 ppm), ambient level before the Industrial Revolution (270 ppm), current level (350 ppm, vertical black line), and a predicted future level (700 ppm). Whereas the C_4 species *Amaranthus retroflexus* showed no response to changes in CO_2 levels, the C_3 species *Abutilon theophrasti* showed a strong CO_2 response, increasing 22% in biomass from current to future doubled CO_2 levels. (Data from Dippery et al. 1995.)

sequestration have been tempered by long-term field experiments that indicate much less plant growth stimulation in the field than one observes in laboratory greenhouse chambers (Long et al. 2006). The technique that has been adopted around the world is to release additional CO_2 from a series of towers onto a study site, and then monitor the outcome within the FACE (Free Air Carbon Enrichment) ring (**Figure 13**). C_3 crops like wheat and soybeans grown in greenhouses under increased CO_2 typically increased yields by 20%–30%, but in open field studies yields increased only 12%–14% (Schimel 2006). C_4 crops like corn (maize) in greenhouse experiments increased yield 20%–30% but in open fields by 0%. The results obtained from FACE experiments permit us to obtain a better picture of how plants will respond to increased CO_2 in natural ecosystems.

In most plants stomatal conductance declines as CO_2 levels rise. Because stomata are closed more often at high CO_2 levels, water loss through transpiration is reduced, so the overall result is improved water use efficiency (Bazzaz 1996). C_4 crop plants may show improved growth under elevated CO_2 levels because of increased water use efficiency, even when photosynthetic rates are unaffected by CO_2 levels (Rogers et al. 1983).

The effect of CO_2 on plant reproduction has been studied in many herbaceous species. In general, elevated CO_2 levels increase reproductive output. Flower number, fruit number, and seed production all increase

(a)

(b)

Figure 13 The Aspen FACE experiment, located near Rhinelander, Wisconsin, consists of twelve 30-m rings in which the concentrations of carbon dioxide can be controlled. The design provides the ability to assess the effects of increased CO_2 on many plant attributes, including growth, leaf development, root characteristics, and soil carbon, as well as insect responses to plant changes. In these experiments carbon dioxide is released from the towers under computer control to increase atmospheric CO_2 within the plots to a specified level, typically twice the ambient level. (Photos courtesy of the U.S. Forest Service and the University of Wisconsin.)

in herbs exposed to high CO_2 levels (Ward and Strain 1999). The increase in reproductive output could be due in part to a simple increase in size of individual plants, rather than an increased allocation of resources to reproductive output.

The key to understanding CO_2 effects in long-lived plants such as trees is to determine how to extrapolate plant responses measured in the greenhouse on young plants to responses in older, larger plants in natural communities (Bazzaz et al. 1996). Yellow birch trees grown at high density show less enhancement from high CO_2 levels compared with trees grown in individual pots with no root competition (Wayne and Bazzaz 1995). The effect of competition for resources is even more complex in communities of many plant species, and CO_2 is only one factor that may be limiting plant growth.

Plant Community Responses to CO_2

To predict the global effects of CO_2 increases, ecologists must take these laboratory studies out into natural communities and expand short-term experiments into longer time frames. This has been done for more and more ecosystems using facilities such as the FACE site shown in Figure 13, and in reviewing these findings we can begin to appreciate the challenges being set by climate change.

Arctic ecosystems may be the most sensitive ecosystems to climate change. Not all plants in high latitudes respond the same to environmental changes. Increases in ultraviolet light radiation and increased CO_2 do not appear to affect arctic plants (Dormann and Woodin 2002). But changes in nitrogen additions via aerosols and increasing temperatures have a strong effect on some plant functional types (**Figure 14**). Arctic plant communities have several characteristics that make them susceptible to global warming. The active soil layer is shallow because of underlying permafrost, so that most root systems occur in the top 10–15 cm of the soil. Permafrost layers in the soil also contain large amounts of frozen organic matter that is not available to decomposers. Rising global CO_2 levels will lead to increased temperatures and increased evaporation on arctic tundra.

Billings et al. (1983) postulated that rising global temperatures could change arctic tundra communities from a sink for CO_2 to a source of CO_2. To test this idea they took cores of arctic tundra that were 8 cm in diameter into a greenhouse and measured net exchange of CO_2 from these cores at two levels of CO_2 and two levels of water table. **Figure 15** shows the results of these experiments. Under present conditions in which the water table is at the surface, CO_2 enrichment has a minor effect on the carbon cycle in tundra vegetation, indicating that in this ecosystem CO_2 is not an important limiting factor. By contrast, water table levels strongly affect the carbon cycle of tundra. As the water table falls in the tundra, decomposition of organic matter increases, and this is further exacerbated if temperature rises as well (Billings et al. 1983, 1984). The net result of this work is that a doubling of CO_2 and the

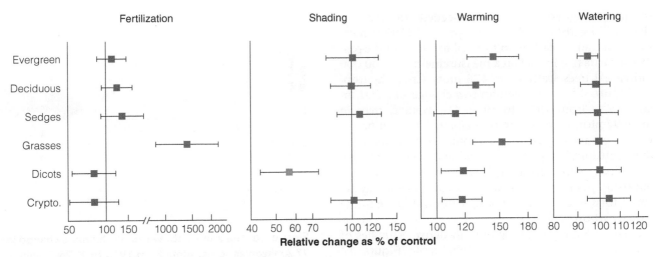

Figure 14 Biomass response of six functional plant types to fertilization with nitrogen, shading, warming in summer by greenhouses, and adding water. A total of 36 experiments are included in this analysis. Red points indicate changes not significantly different from control plots. Grasses respond strongly to both fertilization and warming, and all plant types, especially shrubs and grasses, respond to warming. Cryptograms includes mosses and liverworts. (From Dormann and Woodin 2002.)

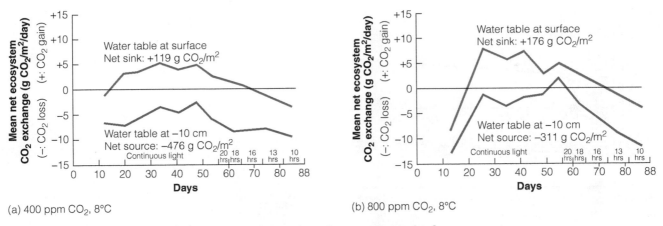

(a) 400 ppm CO_2, 8°C

(b) 800 ppm CO_2, 8°C

Figure 15 Net gain or loss of carbon dioxide from cores of wet arctic tundra from Barrow, Alaska, grown in a greenhouse through a simulated arctic summer at two levels of CO_2 enrichment: (a) current CO_2 levels, and (b) elevated CO_2 levels. The water table is presently at the surface in this tundra community (blue line), but is predicted to drop with global warming (red line). If this happens, the tundra will no longer absorb CO_2 from the atmosphere but will become a source of further additions to increasing atmospheric CO_2. (From Billings et al. 1983.)

associated climatic warming could convert the wet tundra ecosystem of northern Alaska from a CO_2 sink to a CO_2 source. Over the past 40 years in Alaska in moist and wet tundra areas, the carbon balance has fluctuated so that the tundra was a sink for atmospheric carbon in the 1960s and 1970s but then in the 1980s and 1990s became a source for carbon, adding to the increases shown in Figure 10 (Callaghan et al. 2004). Plant growth in tundra communities is limited more by nitro-

gen availability, and this limitation prevents rising CO_2 levels from enhancing the growth of individual plants. Shrubs are expanding in tundra ecosystems in Alaska and other parts of the arctic (Tape et al. 2006), and the consequences of this increase in shrub abundance for carbon sequestration are not yet known.

The ecosystem consequences of CO_2 enrichment for forest communities can now be analyzed with satellite imagery coupled with data from intensive field studies. Long-term plots are needed to measure changes in carbon storage. Phillips et al. (1998) measured the gain in carbon in tropical forests over the past 25–40 years from basal area measurements on 600,000 individual trees scattered in 478 plots across the tropics. **Figure 16** shows the biomass change in Amazonian forests from 1975 to 1996. On average, trees in these 97 plots increased in biomass by 1 ton per ha per year. Biomass in trees is accumulating particularly in the neotropical forests of Central and South America. This biomass increase is equivalent to fixing 0.62 tons of carbon per ha per year, and this carbon sink may account for some of the "missing" carbon dioxide in the global carbon cycle.

Satellite images can be used to derive estimates of net primary production on a global scale. **Figure 17** shows that primary production has increased on average about 6% over 18 years, and the largest increase in primary production occurred in the Amazon rain forest. Not all of this increase in primary production can be attributed to the rise in CO_2 levels over this time period. Decreased cloud cover in the Amazon region permitted

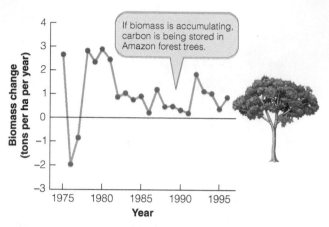

Figure 16 The annual aboveground biomass change in 97 Amazonian forest plots from 1975 to 1996. Points below the horizontal black line indicate biomass loss. All trees on these plots were measured each year. Biomass increased in almost every year, and thus carbon was being taken up from the atmosphere, and the Amazon forests were a carbon sink. Growth accumulation in tropical forests is probably one part of the "missing" carbon sink. (From Phillips et al. 1998.)

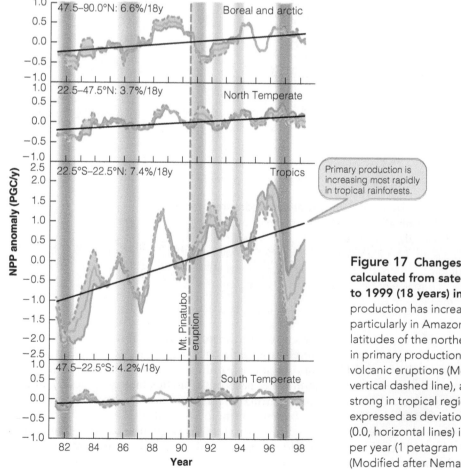

Figure 17 Changes in global net primary production calculated from satellite data for the period from 1982 to 1999 (18 years) in relation to latitude. Primary production has increased most strongly in tropical areas, particularly in Amazonia, and also strongly in the higher latitudes of the northern hemisphere. Year-to-year variation in primary production follows El Niño events (gray bars) and volcanic eruptions (Mount Pinatubo eruption in mid-1991, vertical dashed line), and the El Niño impact is particularly strong in tropical regions. Values of primary production are expressed as deviations from the long-term average value (0.0, horizontal lines) in units of Pg (petagrams) of carbon per year (1 petagram = 1×10^{15} g = 1 billion metric tons). (Modified after Nemani et al. 2003.)

an increase in solar radiation, stimulating tree growth. Increases in net primary production as shown in Figure 17 do not automatically translate into a net uptake of atmospheric CO_2 because decomposers and the resulting soil respiration might increase more than net primary production.

Few studies with time frames of 10–20 years have yet been done to measure the entire ecosystem responses to CO_2 enrichment. Grime (1997) has questioned the simple extrapolation of short-term greenhouse experiments to long-term ecosystem effects. Much of the initial research on the effects of CO_2 enrichment has been conducted on crop plants growing on rich soils. Grime has argued that a different model of CO_2 effects is needed for natural vegetation growing on infertile soils in which resources such as water, nitrogen, and phosphorus are often limiting (Diaz et al. 1993; Grime 1997). The effect of increasing CO_2 levels could be minimal if other resources determine plant community composition. A few experiments in natural ecosystems already suggest that little or no response to CO_2 enrichment occurs. Hungate et al. (2006) found that a seven-year enrichment of CO_2 in a scrub oak woodland in Florida resulted in increased growth for four years but then a declining uptake due to limitation from nitrogen availability. Oechel et al. (1994, 2000) showed that the Alaskan tundra ecosystem did not respond to enriched CO_2 after three years of additions, and tundra ecosystems were a net source of CO_2 because even though they took up CO_2 in summer, they released more CO_2 during winter. Temperature had a larger effect on the CO_2 dynamics of these tundra ecosystems. The simple idea that more CO_2 means increased plant production is not correct in natural ecosystems. De Graaff et al. (2006) pointed out that even though there is a large response of plant biomass to enhanced CO_2 levels, soil carbon enrichment was completely absent unless nitrogen was added to the ecosystem (**Figure 18**).

Much scientific effort is now being expended to define the limits of how ecosystems respond to carbon enrichment, and to define more clearly and more locally the sources and sinks for CO_2. All this research is important because of the implications of the carbon cycle for climate change. The simple assumption that with more CO_2 plants will grow more and serve as a sink for increasing human emissions has proven to be overly optimistic (Albani et al. 2006).

Animal Community Responses to Changing Climate

Since animals depend directly on plants, we need first to determine the effects of climate change on plant communities and then on the animals that depend on

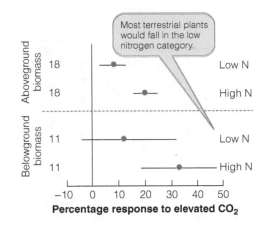

(a)

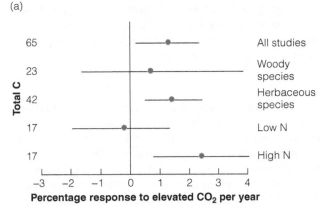

(b)

Figure 18 Synthesis of data from 117 studies of the effects of adding CO_2 to field studies using open-top chambers and FACE experiments. (a) Average percentage response of aboveground and belowground plant biomass production to elevated CO_2 in low and high nitrogen fertilizer experiments. (b) Annual response of soil carbon storage to elevated CO_2. Under low nitrogen conditions virtually no carbon is sequestered in the soil. Lines indicate 95% confidence limits. Numbers indicate number of studies for each category. (From De Graaff et al. 2006.)

the plants. What do we know about the effects of climate change on animals?

There is no evidence that animals will be affected *directly* by changing CO_2 levels in the twenty-first century, but there is concern that herbivorous animal populations and communities may be affected indirectly by changes in their food plants. For nine years Stiling and Cornelissen (2007) counted the numbers of leaf miners on the leaves of myrtle oak (*Quercus myrtifolia*) on CO_2 enriched and control plots in Florida. **Figure 19** shows that leaf mining insects were less common on leaves of plants that had enriched CO_2 in all years of the study. Nitrogen content of plant leaves tends to fall with increased atmospheric CO_2 levels, and this may result in a reduction of larval insect growth in nitrogen-limited

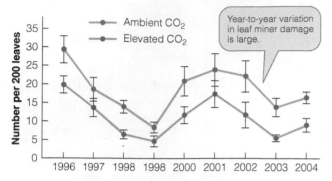

Figure 19 Average number of leaf miners on leaves of myrtle oak in Florida. CO_2 was added to experimental field plots in order to double ambient levels. In every year there were fewer leaf miners of three different species on the leaves of oaks that were exposed to higher CO_2 levels. (Data from Stiling and Cornelissen 2007.)

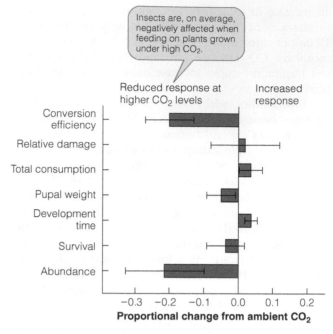

Figure 20 Insect responses to elevated CO_2 concentrations. These are average tendencies and particular taxonomic or functional groups of insects vary in their responses. Conversion efficiency is a measure of the biomass gained by the insects divided by the food ingested. The broad picture is that insects do not do as well when CO_2 is increased. (Data from Stiling and Cornelissen 2007.)

insect species (**Figure 20**). Stiling and Cornelissen (2007) report decreases of 16% in nitrogen concentration in plants grown with enhanced CO_2 levels and a simultaneous increase in tannins (+30%) and other phenolic secondary compounds. Insect herbivores are often limited by plant nitrogen in natural communities (White 1993), and reduced nitrogen reduces insect numbers on CO_2-enriched vegetation. To compensate for low nitrogen levels, insect herbivores feeding on plants grown under high CO_2 levels may increase their feeding rate, and the conversion efficiency of plant matter to insect biomass falls. Plant damage could increase even if insect numbers fall as CO_2 levels rise in the future.

Climatic warming could also disrupt the timing of hatching in insects (van Asch and Visser 2007). Insects that feed on newly emerging foliage are very sensitive to the age of the foliage. Maximum overlap between bud burst and larval emergence results in good survival and growth of the insects. In Scotland the emergence of the winter moth is strongly affected by spring temperatures, and consequently these larvae will emerge earlier if climatic warming occurs. By contrast, the vegetative buds of Sitka spruce, their food plant, are not greatly affected by spring temperatures and will open only slightly earlier when the climate warms. The net result will be a mismatch between the moth and its food plant that will reduce moth survival and growth (Dewar and Watt 1992). These short-term effects could be alleviated by natural selection in the longer term, but the disruption of life cycles in insect herbivores and their predators could be a major effect of global warming (Mondor et al. 2004).

Coral reefs have provided a graphic indicator of climate change because of coral bleaching events. Corals also provide us with a historical record of climate change so that the climatic impacts of the present can be viewed in the light of the past.

Biotic Invasions and Climate Change

One of the most difficult effects of climate change to alleviate is the resulting changes in species distributions. If species ranges are currently controlled by climate, a shift in climate implies a shift in geographic range. Because shifts in range have occurred many times in the history of the Earth, at first glance it might appear that this problem will take care of itself. Two factors argue against complacency. First, the speed of climate change is now many times greater than it has ever been in the past, raising the critical question: How fast can species move? Second, human changes in land use have disrupted many possible corridors of movement for both plants and animals, such that dispersal may no longer be possible.

The fingerprint of climate change is seen most clearly in the movement of geographic ranges toward the poles and the advancement of spring events of

On Corals and Climate Change

Because the tropical oceans leave a climate record in corals, one of the main techniques used for reconstructing climatic changes over the past several centuries involves coral reefs. Most reef corals live at depths of less than 20 m and grow continuously at rates of 6–20 mm per year. Many coral species produce annual density bands similar to tree rings that can be seen in x-rays or under ultraviolet light, as **Figure 21** shows.

Many coral records have absolute annual chronologies extending back at least to the fifteenth century. Fossil corals going back several thousand years can be dated with carbon-14 (Gagan et al. 1998). Given that we can detect these annual rings, what can we infer from them?

The skeletons of reef-building corals carry a diverse suite of isotopic and chemical indicators that track water temperature and salinity. For example, the ratio of strontium to calcium traces sea surface temperatures at a time scale of three weeks (**Figure 22**).

The ecological significance of these methods is that they permit us to reconstruct El Niño–Southern Oscillation events back in time before meteorological measurements were available, enabling us to assess their range of variability in the past. In collaboration with data from tree rings, ice cores, peat bogs, and other biophysical archives, we can begin to construct a picture of how climate has varied globally and locally, and to measure how the Earth's biota has responded to climatic fluctuations (Gagan et al. 2000).

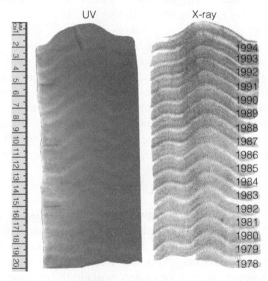

Figure 21 Cross sections of coral skeletons viewed under ultraviolet light and with x-rays. Annual rings in the coral skeletons are similar to tree rings and can be used to date different parts of coral reefs. (Photo courtesy of M. Gagan, Australian National University.)

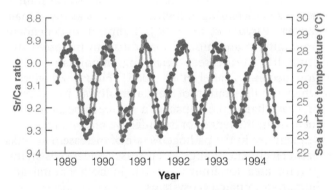

Figure 22 Correlation of strontium/calcium ratios (red) in coral skeletons to the sea surface temperature (blue) in West Java, Indonesia, from 1989 to 1994. The isotopic chemistry of coral skeletons can be calibrated in the present and then used to infer sea surface temperature changes thousands of years ago. (Data courtesy of M. Gagan, Australian National University.)

breeding or flowering (Parmesan and Yohe 2003). Overall ranges have been moving toward the poles at a rate of 6.1 km per decade, and spring phonological events have moved ahead on average 2.3 days. Root et al. (2003) reported an average change in spring events of about five days for a set of 145 studies spanning on average 35 years. **Figure 23** shows the frequency distribution of these values. Some spring breeding seasons have moved as much as 24 days in 10 years. Réale et al. (2003) reported the case of a particu-

larly well-studied mammal, the red squirrel (*Tamiasciurus hudsonius*), which changed its spring birth date by 18 days in 10 years.

The consequences of these range shifts are biotic invasions or species moving into new geographic ranges. None of this would be thought very significant except for the fact that some of the invasions are of disease-causing organisms (Harvell et al. 2002). The most obvious cases involve mosquito vectors. In Hawaii avian malaria cannot infect native birds at high elevations,

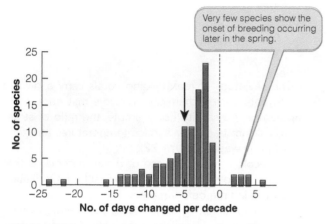

Figure 23 **In plants and animals that have shown a change in life history events, the average change observed has been about five days earlier over a time span of 10 years (arrow).** A total of 143 studies of 694 species were analyzed. (Data from Root et al. 2003.)

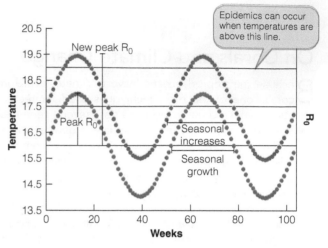

Figure 24 **The influence of an average 1.5°C rise in temperature on the reproductive rate R_0 of a hypothetical pathogen.** The blue curve represents the average weekly temperature before climate change, and the red curve illustrates temperatures after a climatic warming of 1.5°C. The lower horizontal line corresponds to the temperature level below which the pathogen cannot increase, and we assume that the disease becomes severe when temperatures are above the middle line and become epidemic when temperatures are above the upper line. The diagram illustrates that increasing temperatures lead to a higher rate of pathogen population growth (R_0) as well as a longer season of susceptibility to the disease. (From Harvell et al. 2002.)

but as global warming occurs, all the native bird fauna is at risk (Kilpatrick 2006). In East Africa human malaria is increasing at higher elevations as the climate warms (Pascual et al. 2006). Temperature-dependent immunity in amphibians may cause an increased susceptibility to disease, accelerating amphibian population declines (Raffel et al. 2006). **Figure 24** gives a schematic illustration of how a small rise in temperature can change the impact of a pathogen because of an increase in the length of the infective season and a consequent rise in the pathogen's rate of increase. The links between climate change and disease incidence is an important area for future research in both natural and human-dominated ecosystems.

As plant communities are disrupted by climatic warming during the next 100 years, the animal communities on which they depend will also be disrupted. At present few data are available from which to estimate these potential effects on animals, and the monitoring of climatic effects on plants and animals alike are part of the research agenda for all ecologists in the coming years.

Summary

Climatic warming is the most serious problem facing humanity during this century. The evidence for global warming is now very strong with the availability of remote sensing and increasing sophistication of the analysis of paleoclimates from proxies like coral isotopes, ice cores, and tree rings.

Climatic warming is not evenly spread over the Earth, and a few areas are cooling while others warm rapidly. The principal factor responsible for recent changes in global temperatures and precipitation are human emissions of greenhouse gases from the burning of fossil fuels. Natural swings of climate have always occurred, but none has been so rapid as the changes of the last 40 years. Carbon dioxide is a key greenhouse gas, and it has now reached levels in the atmosphere that have not occurred during the last 650,000 years.

The carbon cycle is key to understanding climate change. Rising levels of atmospheric CO_2 have occurred for 200 years due to fossil fuel burning and the destruction of native vegetation. The ocean takes up about 24% of the current CO_2 emissions, and land ecosystems about 30%. The global carbon budget does not balance; a large sink now missing from the overall equation may be located in growing trees in temperate and tropical forests.

Climate change is one crucial result of the altered carbon cycle. Greenhouse gases such as CO_2, methane, and nitrous oxide trap heat at the Earth's surface and increase global temperatures. Individual plants typically grow more when CO_2 levels are elevated in a greenhouse, but in natural communities other factors such as nitrogen or water may limit productivity, and respiration may use up much of the carbon fixed in photosynthesis. Some increased plant growth will probably accompany climatic warming, but the ecosystem consequences for plants are far from clear. Arctic ecosystems are particularly susceptible to the effects of climatic warming. Animal communities will be affected indirectly through changes in the chemistry of their food plants.

Climatic warming in the twenty-first century is having dramatic effects on the distribution of native animals and plants. Disease organisms can be favored by warmer temperatures, and the biotic invasions that allow disease vectors such as mosquitoes to move poleward will bring many more animals and humans into contact with newly emerging diseases.

Review Questions and Problems

1   Discuss the implications of adopting a "top-down" versus a "bottom-up" view of community organization for evaluating the effects of climatic warming on aquatic and terrestrial ecosystems.

2   Low-growing tundra vegetation is being slowly replaced by faster growing shrubs like willows as global warming continues (Sturm et al. 2005). What effect might this change in vegetation have on the CO_2 dynamics of tundra ecosystems? Discuss what you would need to measure to test your ideas. Mack et al. (2004) provide some data on this issue.

3   Discuss the implications of the response of C_3 and C_4 plants to changing CO_2 levels (see Figure 12) with respect to changes in community composition from the last ice age into the future. Ehleringer et al. (1997) review this question.

4   Pastor and Post (1988, p. 55) state that "the carbon and nitrogen cycles are strongly and reciprocally linked." Review the nitrogen cycle and discuss these linkages.

5   Red squirrels in the southwestern Yukon of Canada have advanced their season of birth in spring by more than two weeks over 10 years, associated with climatic warming. Red squirrels breed in mid-December in the middle of winter and have a fixed gestation period. Discuss how this change in birth season might happen and what the mechanism might be for detecting a change in climate. Réale et al. (2003) discuss this issue.

6   Discuss why experiments on the impact of increased CO_2 levels on crop growth should give different answers when conducted in the greenhouse and when conducted in open fields with FACE technology. Long et al. (2006) discuss this problem.

7   Norby et al. (1992) grew yellow poplar (*Liriodendron tulipifera*) trees for three years at three levels of CO_2 enrichment. Photosynthetic rate nearly doubled at high CO_2 levels, but no difference in aboveground biomass of trees grown at normal or elevated CO_2 levels was found after three years. How is this possible?

8   Rising CO_2 levels cause the surface waters of the ocean to become slightly more acidic because CO_2 dissolves in seawater as carbonic acid. The entire phytoplankton biomass of the oceans turns over every 2–6 days, taking up carbon dioxide and releasing it to decomposers and grazers. Carbonate skeletons protect many species of marine phytoplankton. What impact might increasing acidity have on ocean ecosystems? Ruttimann (2006) and Dybas (2006) explore this question.

9   In agricultural landscapes, farmers have the choice of managing roadside verges by mowing, burning, grazing, or doing nothing to them. Discuss the implications of these four treatments for the global carbon cycle.

10   Would you expect that different trophic levels would be affected in different ways by climate change? Or would many communities simply shift geographically and remain functionally intact? Voigt et al. (2003) discuss this question and provide data from two grassland communities.

Overview Question

One suggested ecological response to help arrest increasing atmospheric carbon dioxide levels is to plant trees. Explain the mechanisms behind this recommendation, and discuss how you could calculate how many trees would need to be planted to achieve this policy goal.

Suggested Readings

- Canadell, J. G., D. E. Pitaki, and L. F. Pitelka, eds. 2007. *Terrestrial Ecosystems in a Changing World*. Berlin: Springer.

- De Graaff, M.-A., K.-J. van Groenigen, J. Six, B. Hungate, and C. van Kessel. 2006. Interactions between plant growth and soil nutrient cycling under elevated CO_2: A meta-analysis. *Global Change Biology* 12:2077–2091.

- Harvell, C. D., C. E. Mitchell, J. R. Ward, S. Altizer, A. P. Dobson, R. S. Ostfeld, and M. D. Samuel. 2002. Climate warming and disease risks for terrestrial and marine biota. *Science* 296:2158–2162.

- Hoegh-Guldberg, O., et al. 2007. Coral reefs under rapid climate change and ocean acidification. *Science* **318**:1737–1742.

- Körner, C. 2006. Plant CO_2 responses: An issue of definition, time and resource supply. *New Phytologist* 172:393–411.

- Long, S. P., E. A. Ainsworth, A. D. B. Leakey, J. Nosberger, and D. R. Ort. 2006. Food for thought: Lower-than-expected crop yield stimulation with rising CO_2 concentrations. *Science* 312:1918–1921.

- Root, T. L., D. Liverman, and C. Newman. 2007. Managing biodiversity in the light of climate change: Current biological effects and future impacts. In *Key Topics in Conservation Biology*, eds. D. W. Macdonald and K. Service, 85–104. Oxford: Blackwell Publishing.

- Solomon, S., D. Qin, M. Manning, Z. Chen, M. Marquis, K. B. Averyt, M. Tignor, and H. L. Miller, eds. 2007. *Climate Change 2007: The Physical Science Contribution of Working Group I to the Fourth Assessment Report of the Intergovernmental Panel on Climate Change*. Cambridge: Cambridge University Press.

- Stiling, P., and T. Cornelissen. 2007. How does elevated carbon dioxide (CO_2) affect plant-herbivore interactions? A field experiment and meta-analysis of CO_2-mediated changes on plant chemistry and herbivore performance. *Global Change Biology* 13: 1823–1842.

- Ward, N. L., and G. J. Masters. 2007. Linking climate change and species invasion: An illustration using insect herbivores. *Global Change Biology* 13:1605–1615.

- Wrona, F. J., T. D. Prowse, J. D. Reist, J. E. Hobbie, L. M. J. Levesque, and W. F. Vincent. 2006. Climate change effects on aquatic biota, ecosystem structure and function. *Ambio* 35:359–369.

Credits

Illustration and Table Credits

Photo Credits

Ecosystem Health and Human Impacts

Key Concepts

- About 6.4 billion people now inhabit the Earth, and population growth is the root cause of all our environmental problems.

- The carrying capacity of the Earth is difficult to estimate, but the best guess is that we are already above a sustainable level of world population.

- Invasive species are a major problem that threaten the integrity of both agricultural and natural ecosystems. Not all invasive species cause ecological harm, and we must distinguish benign invaders from invaders that could become serious pests.

- Ecosystem services such as pollination are provided by the Earth's ecosystems for free, and are not valued in our current economic systems. Without them, human societies would collapse.

- We cannot at present construct an artificial ecosystem that can sustain human life indefinitely, no matter how much money we invest in it.

- Ecological indicators of ecosystem health are needed to chart progress toward sustainability.

From Chapter 26 of *Ecology: The Experimental Analysis of Distribution and Abundance*, Sixth Edition. Eugene Hecht.

KEY TERMS

demographic transition The change in human populations from the zero-population-growth state of high birth and high death rates to that of low birth and low death rates.

ecological footprint The total land and water area that is appropriated by a nation or a city to produce all the resources it consumes and to absorb all the waste it generates.

ecosystem services All the processes through which natural ecosystems and the biodiversity they contain help sustain human life on the Earth.

evapotranspiration The movement of water back into the atmosphere via the transpiration of plants and the physical evaporation of water from the soil and from water bodies.

sustainability The characteristic of a process that can be maintained at a certain level indefinitely. The original definition of the Bruntland Commission of 1987 defined sustainable development as development that meets the needs of the present without compromising the ability of future generations to meet their own needs.

Human impacts on the planet's ecosystems are significant and growing. Before we can alleviate human impacts we need to have a clear idea of the magnitude and direction of change of both positive and negative alterations of the Earth's ecosystems.

We have touched on many of the human impacts, from overfishing to pest control and the conservation of endangered species. This chapter brings these problems into central focus and asks both *What are the problems?* and *What can we do about them?* The role of ecologists in these matters is much like that of a medical doctor. As scientists our job is to diagnose problems and suggest cures. But just as a patient can ignore a doctor's advice to stop smoking, for example, the public can ignore all the advice and recommendations of ecologists to ameliorate problems. Media discussions over the magnitude of climate change are a current example. Translating science into public awareness is essential if the problems of which ecologists are all too aware are to stimulate public action. Our first objective as scientists in all these matters must be to get the science right.

In this chapter we first examine the human population problem, the root cause of all the adverse human impacts on ecosystem health, and the potential carrying capacity of the Earth for humans. Then we consider three critical aspects of ecosystem health: the challenge of invasive species, the measurement of ecosystem services, and the concept of sustainability.

Human Population Growth

If growth is good, as some business people seem to believe, then the human beings on the planet ought to be nearing perfection. Global human population has been growing throughout most of recorded history, and probably for more than 2 million years before that. In 2008 an estimated 6.64 billion people inhabit the Earth, and the human population is growing at 1.2% per year, or 211,000 people per day. **Figure 1** shows the growth of the human population over the past 500 years. The increase appears exponential, but is even more rapid than exponential (Cohen 1995). Because no population can go on increasing without limit, Figure 1 immediately raises two questions: (1) What is happening now? and (2) Can we estimate the long-term carrying capacity of the Earth for humans?

Current Patterns of Population Growth

The human population can exist in one of two stable configurations that lead to population stability:

$$\frac{\text{Zero population}}{\text{growth}} = \frac{\text{high birth}}{\text{rates}} - \frac{\text{high death}}{\text{rates}} \quad (1)$$

or

$$\frac{\text{Zero population}}{\text{growth}} = \frac{\text{low birth}}{\text{rates}} - \frac{\text{low death}}{\text{rates}}$$

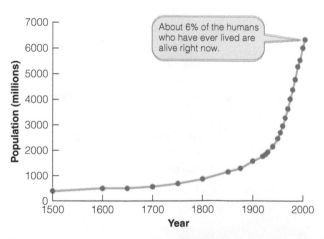

Figure 1 Human population growth during the last 500 years. The population increase appears approximately like geometric growth. Population growth accelerated dramatically after about 1900. Note that the impact of the millions killed in World War I, the Spanish Influenza Pandemic of 1918, and World War II cannot even be seen on this graph. (Data from Cohen 1995, Appendix 2 and Population Reference Bureau 2007.)

WORKING WITH THE DATA

How Large Is a Billion Anyway?

We are the supreme counters in the universe, and we learn in school how to use decimals and exponents and how to manipulate them correctly. But at some point our numerical system loses meaning, and such is the case of a billion. There are now about 6 billion people on the Earth, the U.S. debt in 2007 has been increasing at $1.52 billion per day, and in 2006 the U.S. government had to borrow $248 billion to balance its budget. How can we get a grasp on how big a billion is? Try these two simple games:

Question: How long would it take to spend $1 billion if you were able to spend $10,000 a day, every day of the year, on anything you wished?

Answer: To spend $1 billion you would need 274 years with no days off.

Question: How far would you walk if you took a billion steps?

Answer: If we assume each of your steps is 61 cm (about 2 feet), you would walk 610,000 kilometers (379,000 miles). This would mean you could walk to the moon and almost all of the way back, or if you are more conservative you could walk around the equator about 15 times. Remember to take some water with you.

The point is that while we can talk about and manipulate large numbers like a billion, they exceed our ability to comprehend how large they really are. So when we note that the human population increased 1 billion from 1995 to 2006, the size of this increase is nearly impossible to relate to our normal existence.

The movement between these two states has been called the **demographic transition**. **Figure 2** illustrates the demographic transition for Sweden and for Mexico. The demographic transition is a descriptive theory rather than a law of human population growth (Gelbard et al. 1999). After 1950, mortality rates declined rapidly in all countries, but birth rates have declined in a more variable manner (**Figure 3**). Fertility decline has been most dramatic in China. In 1970 the average Chinese woman could expect to have 5.9 children; by 2006 the expected family size was 0.84 children. In India, fertility rates have fallen more slowly from 5.4 in 1970 to 3.1 in 2006. In much of Africa the transition to lower fertility is just beginning and average family size remains at 5 children in 2006 (Population Reference Bureau 2007).

One consequence of variable fertility rates on the growth of the world's population is that the current rapid rise in human population is composed of two quite different elements. In the developed nations, populations are near to equilibrium, with net reproductive rates near replacement (total fertility rate = 2.1 or R_0 = 1.0). In many developed countries such as Canada and the United Kingdom, net reproductive rates are in fact below 1, and these populations will decline in the long term if there is no immigration and if the net reproductive rate does not change. Most developed countries, however, are still increasing in population in the absence of immigration because the age structure is not in equilibrium such that births exceed deaths (**Table 1**). This increase is due to population momentum and will stop in about 30 years. About 80% of the world's people now live in the less developed countries, and most of the population growth is occurring in these nations.

The projected human population of the globe depends on assumptions about future changes in fertility and mortality. The United Nations in 2006 projected a population in 2050 of 9.2 billion with a possible range from 7.8 billion to 10.8 billion people. No matter what projection is used, without some catastrophe at least

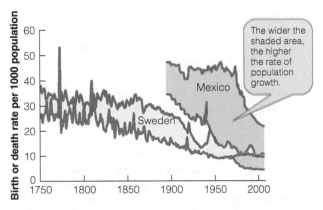

Figure 2 Demographic transition in the human populations of Sweden and Mexico, 1750 to 2006. The transition took 150 years in Sweden but about half that time in Mexico. When the birth rate exceeds the death rate, the population grows (shaded areas), if emigration is low. In 2006 Mexico had a crude birth rate of 22 and a crude death rate of 5. In Sweden at the same time the comparable rates were 11 and 10. (Data from the Population Reference Bureau 2007.)

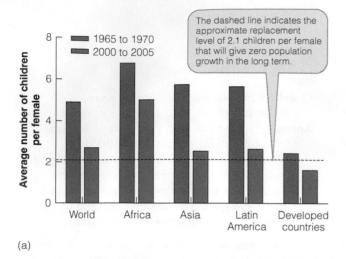

(a)

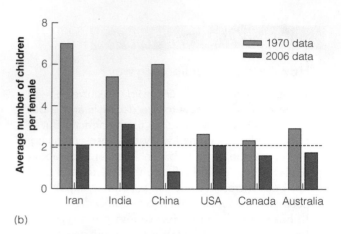

(b)

Figure 3 Changes in human birth rates from 1965–1970 to 2000–2006 (a) in select regions of the world and (b) in selected countries. The transition from high birth rates to low birth rates has been particularly rapid in Asia and Latin America. (Data from Population Reference Bureau 2007.)

ESSAY

The Demographic Transition: An Evolutionary Dilemma

The demographic transition involves two puzzling aspects of human fertility. First, a large decrease occurs in the number of children that parents produce, despite the fact that resources are increasing. This change was shown very well in Europe during the nineteenth century. Second, rich families reduce their fertility rate earlier than does the rest of the population, and often they have fewer than the average number of children. Thus, a negative correlation exists between wealth and fertility (Mulder 1998). If we assume that humans should follow the principles of natural selection that apply to other animals, these observations create a puzzle. When birds and other animals have more resources, they have more offspring. Why should parents with access to plentiful resources choose to have low fertility rates?

Three hypotheses have been suggested to explain this puzzle of low fertility:

1. *Lowered rates are optimal because of the competitive environment in which offspring are raised.* This idea is the classic view of evolutionary anthropologists, who suggest that high levels of parental investment are critical to a child's success and are costly to the parents. Thus, maximizing fitness is achieved by fewer offspring with higher parental investment, and there is a trade-off between offspring quality and offspring number.

2. *Lowered rates are a consequence of Darwinian selection, but on nongenetic mechanisms of inheritance.* Cultural selection through imitation drives the demographic transition in fertility because people see that successful people have fewer children. This hypothesis is attractive because it postulates that *ideas,* rather than economic resources, can drive fertility rates, but it suffers some serious flaws. It begs the question of why rich and successful people should reduce their fertility in the first place.

3. *Low fertility is maladaptive, a by-product of rapid environmental change.* This hypothesis suggests that lowered fertility is indeed an inappropriate evolutionary response not favored by natural selection. The availability of contraceptives is often cited as one explanation of why this maladaptive situation has prevailed in human societies, but it may be difficult to apply this idea to events in the nineteenth century. This explanation is intriguing but remains vague and untestable until the precise environmental changes can be identified.

There is currently much interest in applying the principles of evolutionary ecology to human behavior. However, a shortage of data in human societies at the individual level makes it difficult to find out, for example, whether individuals who have fewer offspring in fact have more grandchildren. At the present time the best guess is that the demographic transition can be explained by the first model, in which people can have more descendents in the long term by pursuing wealth at the cost of immediate reproductive success—so long as wealth is inherited (Penn 1999).

Table 1 Human population of the different continents and some selected countries as of July 2007.

Country	Population in July 2007 (millions)	Rate of increase (%/yr)	Doubling time (years)	Projected population 2025	Projected population 2050
World	6555	1.2	58	7940	9243
More developed	1216	0.1	693	1255	1261
Less developed	5339	1.5	47	6685	7982
Africa	924	2.3	30	1355	1994
North America	332	0.6	116	387	462
Canada	32.6	0.3	231	38	42
United States	299.1	0.6	116	349	420
Central America	149	1.9	37	187	214
South America	378	1.4	50	465	528
Asia	3968	1.2	58	4739	5277
Saudi Arabia	24	2.7	26	36	47
Yemen	22	3.2	22	39	68
Bangladesh	147	1.9	37	190	231
India	1122	1.7	41	1363	1628
Pakistan	166	2.4	29	229	295
China	1311	0.6	116	1476	1437
Japan	128	0.0	—	121	101
Europe	732	−0.1	—	717	665
Norway	5	0.3	231	5	6
Sweden	9	0.1	693	10	11
France	61	0.4	174	63	64
Germany	82	−0.2	—	82	75
Poland	38	0.0	—	37	32
Russia	142	−0.6	—	130	110
Italy	59	0.0	—	59	56
Spain	45	0.2	347	46	44

Population projections for 2025 and 2050 are based on United Nations estimates of projected future trends in reproduction and mortality. The rate of increase is the natural rate from births and deaths and excludes immigration and emigration. Doubling times assume the rate of increase does not change in the future.

SOURCE: The Knowers Ark.

What Is Population Momentum?

One of the more elusive ideas in human demography is that of population momentum. The central idea is that population growth continues after fertility falls to replacement levels due to time lags in the adjustment of the age structure. The transition from the stable to the stationary age distribution is not instantaneous and population momentum is the result of this time lag.

Population momentum is relatively easy to calculate. Goldstein (2002) showed that population momentum is driven by the speed at which the demographic transition occurs. This is indexed by a parameter m, which is 0 when the demographic transition is instantaneous and is 1 when the demographic transition is very gradual, as in Sweden. For a relatively fast transition (e.g., China) $m = 0.5$, and most countries will have m values between 0.5 and 1.0. The critical message is that the population increase caused by momentum critically depends on how fast the transition occurs (m value).

To calculate the factor by which the population will grow due to population momentum, Goldstein (2002) gives this equation:

$$PMF = (CBR/1000)(LE)(R_0^{[m-0.5]}) \qquad (2)$$

where PMF = population momentum factor of increase

$\quad CBR$ = crude birth rate per 1000

$\quad\quad LE$ = life expectancy in years

$\quad\quad R_0$ = net reproductive rate

$\quad\quad m$ = factor measuring speed of the demographic transition (0 to 1)

All of these variables are available for each country in the World Population Data Sheet of the Population Reference Bureau with the exception of the net reproductive rate and m. For simplicity one can use m estimates between 0.6 and 0.7 for most less developed countries and m values of 0.9 for most developed countries. The net reproductive rate for human populations is not often given in human population tables. The approximation of the net reproductive rate is usually given as

$$R_0 = e^{ur} \qquad (3)$$

where R_0 = net reproductive rate

$\quad\quad e$ = 2.71828

$\quad\quad u$ = mean age at childbearing in the population

$\quad\quad r$ = instantaneous population growth rate per year

Tables usually give the human population growth rate as percent per year and these percentages can be converted to instantaneous rates by the conversion

$$r = \log_e\left[1 + \left(\frac{GR}{100}\right)\right] \qquad (4)$$

where r = instantaneous population growth rate per year

$\quad GR$ = percent growth rate per year

As an example consider Pakistan in 2006 in which the population is 166 million, the crude birth rate is 33 per 1000, the average age at childbearing is 23 years, the population is growing at 2.4% per year, and life expectancy for women is 63 years. The instantaneous rate of population growth is given by

$$r = \log_e\left[1 + \left(\frac{GR}{100}\right)\right] = \log_e\left[1 + \left(\frac{2.4}{100}\right)\right] = 0.0237$$

The net reproductive rate is thus calculated as

$$R_0 = e^{ur} = 2.71828^{(23.2*0.0237)} = 1.73 \text{ per generation}$$

Finally we can calculate the factor by which the population will increase from equation (2) as

$$PMF = (CBR/1000)(LE)(R_0^{[m-0.5]})$$
$$= (33/1000)(63)(1.73^{[0.7-0.5]}) = 2.32$$

Given that Pakistan has 166 million people in 2006, once it completes the demographic transition it will have (2.32*166) or 385 million people.

How much of these expected population increases are due to population momentum and how much is due to excess births is shown in **Figure 4**.

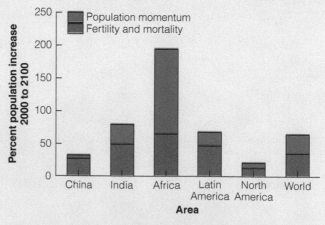

Figure 4 The estimated contribution of population momentum to the percentage population increases expected from 2000 to 2100 for different regions. Population momentum is a major contributor to the expected population growth over the next century. These calculations assume a graduate demographic transition to replacement family size followed by no changes in age-specific birth and death rates, and consequently they are projections rather than predictions. (Data from Bongaarts and Bulatao 1999.)

1.5 billion people will be added to the population in the next 25 years because of population momentum. The questions that arise from these projections are, Is the world already overpopulated? If not, will it be overpopulated in 2050? How many humans can the biosphere support?

Carrying Capacity of the Earth

What is the carrying capacity of the Earth for humans? This question has been asked for more than 300 years by scientists interested in demography (Cohen 1995). As a first step, we can ask what range of estimates has so far been produced for how many people the Earth can support. **Figure 5** shows the ranges of the estimates for the carrying capacity of the Earth, beginning with the first estimate by Leeuwenhoek in 1679. Two aspects of this graph are striking. First, the estimates vary widely, from less than 1 billion to over 1000 billion; second, the estimates do not converge over time but seem to increase in variability, although the median of the estimates seems to stay around 10–15 billion. Why should these estimates of carrying capacity be so variable?

Carrying capacity is difficult to estimate, and the writers that produced the estimates in Figure 5 used quite different methods to get their answers. Some authors such as Raymond Pearl used curves like the logistic equation to predict the future maximum of the human population. Others generalized from existing "maximum"

population density and multiplied this by the area of land that could be inhabited. More scientific estimates were made by focusing on a single assumed population constraint such as food, but this promising approach is limited by the assumptions it must make about the amount of available land, average crop yields, the diet to be utilized (vegetarian or meat-eating), and the number of joules or calories to be provided to each person each day. The equation for carrying capacity used is

$$\text{Carrying capacity} = \frac{(\text{ha land})\left(\dfrac{\text{yield}}{\text{per ha}}\right)\left(\dfrac{\text{kJ per}}{\text{crop unit}}\right)}{\substack{\text{no. of kJ needed per} \\ \text{person per year}}} \quad (6)$$

This equation is a definition, and if we could quantify the terms in it we could identify the carrying capacity of the Earth. But because there are many crops grown on different soils, variable yields, losses to pests, and differences in assumed standards of living, it has been impossible to get anyone to agree on how these numbers should be constrained.

A more promising approach to estimating the carrying capacity of the Earth is to recognize that we have multiple constraints because we need food, fuel, wood, and other amenities like clothing and transportation. One approach is to express everything in the amount of land needed to support each activity (for example, wood production) and then to sum these requirements. But this approach also has limits because it is difficult to express energy requirements directly as land areas, and water requirements may be more of a constraint than land areas.

A recent advance in using multiple constraints to estimate carrying capacity is summarized in the concept of an **ecological footprint** (Wackernagel and Rees 1996). For each nation we can calculate the aggregate land and water area in various ecosystem categories that is appropriated by that nation to produce all the resources it consumes and to absorb all the waste it generates. Six types of ecologically productive areas are distinguished in calculating the ecological footprint: arable land, pasture, forest, ocean, built-up land, and fossil energy land. Fossil energy land is calculated on the basis of the land needed to absorb the CO_2 produced by burning fossil fuels. All measures are thus converted to land area per person. If we add up all the biologically productive land on the planet, we find there is about 2 ha of land per person alive in 2008. If we wish to reserve land for parks and conservation, we must reduce this to 1.7 ha per person of land available for human use. This is the benchmark for comparing the ecological footprints of nations. Because different countries differ in their agricultural capacity and resource base, the available ecological capacity for each country must be adjusted for its productivity in each of

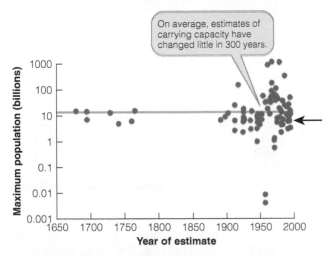

Figure 5 Estimates of the carrying capacity of the Earth for humans as suggested by various writers from 1679 to the present. Both minimum and maximum estimates are plotted. The median of the estimates is approximated by the blue line at 10–12 billion, and the arrow indicates the human population in 2008. (Data from Cohen 1995, Appendix 3.)

How to Calculate an Ecological Footprint

Ecological footprints are estimates of how much area is needed to support a given population. The easiest calculations of footprints are for nations because most data are compiled as national statistics. The first step in calculating an ecological footprint is to compile the total consumption of each commodity used by the population. Here we examine this approach for a single commodity, grain consumption, in Canada. A more complete example is given in Wackernagel et al. (1999) for all commodities for Italy.

1. Correct consumption data for trade imports and exports:

$$Consumption_{wheat} = production_{wheat} + imports_{wheat} - exports_{wheat}$$

For Canada in 1993, domestic production of wheat was 51,416,000 tons. Imports of wheat totaled 499,100 tons, and exports were 26,428,400 tons. Thus,

$$
\begin{aligned}
Consumption_{wheat} &= production_{wheat} + import_{wheat} - exports_{wheat} \\
&= 51,416,000 + 499,100 - 26,428,400 \\
&= 25,486,700 \text{ tons}
\end{aligned}
$$

2. Convert the consumption data into the land area required to produce the item:

$$a_{wheat} = \frac{c_{wheat}}{y_{wheat}}$$

where a_{wheat} = total ecological footprint of wheat in hectares of land

c_{wheat} = total consumption of wheat in kg

y_{wheat} = yield of wheat in kg per hectare

For Canada in 1993, the average yield of wheat was 2744 kg/ha. Thus,

$$
\begin{aligned}
a_{wheat} &= \frac{c_{wheat}}{y_{wheat}} \\
&= \frac{25,486,700}{2744} \\
&= 9,288,200 \text{ ha of arable land}
\end{aligned}
$$

3. Obtain the per capita ecological footprint by dividing the total ecological footprint by the human population (N):

$$
\begin{aligned}
f_{wheat} &= \frac{a_{wheat}}{N} \\
&= \frac{9,288,200}{28,817,000} \\
&= 0.32 \text{ ha}
\end{aligned}
$$

This is the per capita ecological footprint attributable to grain consumption for Canada in 1993. To obtain the total ecological footprint, this process would be repeated for each commodity and energy source that is consumed or used.

Most commodities that humans use can be converted into land area in a simple way. The exception is energy use, and to convert these into land area Wackernagel and Rees (1996) first converted all energy use into equivalent amounts of CO_2 production, and then calculated how much land would be needed to absorb this CO_2 in vegetation.

These calculations illustrate the ecological *demand* a country places on the biosphere. The last step is to compile the amount of productive land that a country has within its borders, the ecological *supply*. These values in land area are more readily compiled, and they are adjusted for the productivity of the country's ecosystems. Figure 6 illustrates some of the resulting ecological footprint values.

the six types of productive ecosystems. We end up with a simple comparison of ecological footprints and available ecological capacity. **Figure 6** illustrates these results for 13 countries. Two things are evident from this figure: the world in general was already in ecological deficit in 2003, and countries vary greatly in their individual footprint size and in their available capacity. The utility of these estimates of ecological footprints is that they can direct us toward sustainability in our use of the world's resources by targeting the specific areas in which particular countries overutilize resources.

The analysis of human impacts via ecological footprints suggests that the world is already slightly above its carrying capacity. Two other calculations can be made of human impacts, one for water and one for primary production. Freshwater is an important resource

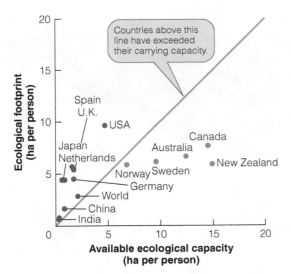

Figure 6 Ecological footprint in relation to available ecological capacity of several countries and the world in 2003. The ecological footprint expresses in hectares of land per person the current demand of global resources made by each country. The available ecological capacity measures in land area per person the resource base of each particular country. Countries in red above the diagonal are in an ecological deficit in 2003; countries in green below the diagonal still have resource surpluses. (Data from Halls 2006.)

for humans because it has no substitute and is difficult to transport more than a few hundred kilometers. The amount of water on the Earth and the volumes moving around are so large that we must discuss the water cycle in units of km³ (= 10¹² liters). If all the rainfall on land were evenly distributed, each weather station would record about 70 cm of rain per year (Schlesinger 1997). Freshwater constitutes only about 2.5% of the total volume of water on the Earth, and two-thirds of it is locked up in glaciers and ice.

Figure 7 shows the global water cycle. There is a net transport of water vapor from the oceans to the land that contributes about one-third of the rainfall on land areas. The mean residence time of a water molecule in the ocean is about 3100 years, a tribute to the large volumes of seawater in the ocean. The volume of groundwater is poorly known. It is fossil water because it cannot be used by plants, and it has a mean residence time of over 1000 years (Schlesinger 1997). We depend on freshwater flowing through the hydrological cycle through precipitation for our needs.

Precipitation on land takes one of two routes: evapotranspiration and runoff. The precipitation that is used for vegetative growth of forests, crops, and pastures, and that evaporates back into the atmosphere is called **evapotranspiration**. The remaining precipitation goes to runoff, which is the source of water to sustain cities, irrigated crops, industry, and aquatic ecosystems. Postel

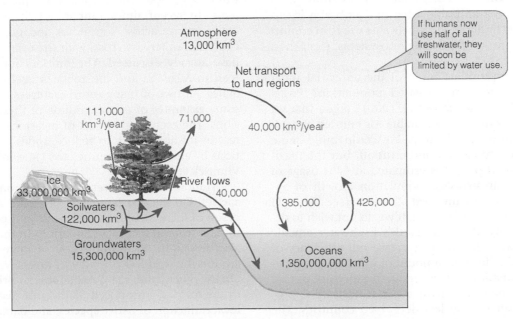

Figure 7 The global water cycle. Pools of water are in units of km³ (black), and flows of water are in km³ per year (red). About half of the pool of soil water is available to plants. Humans currently use about half of all the freshwater runoff. (Modified from Schlesinger 1997.)

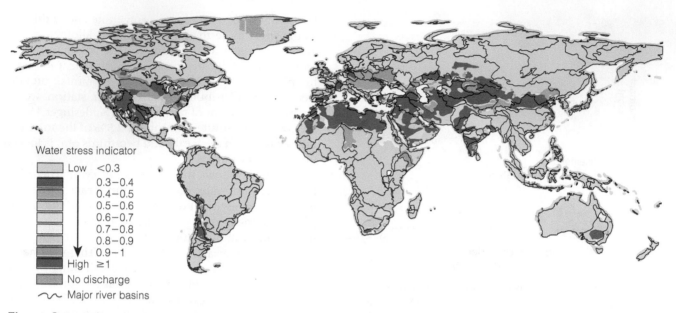

Figure 8 An indicator of water stress in the Earth's river catchments. Areas shown in red are at the crisis stage of water management. The most critical areas are the Murray-Darling Basin in Australia, the Yellow River Basin in China, the Orange River Basin in South Africa, and the Colorado River Basin in the United States. (From Smakhtin et al. 2004.)

et al. (1996) estimated that worldwide humans now use 26% of all evapotranspiration on land and 54% of freshwater runoff. Freshwater ecosystems require water to maintain their ecological integrity, and **Figure 8** shows an index of water stress in different catchments around the Earth. The water scarcity index shows what proportion of the utilizable water in river basins is currently used by humans and where this use is in conflict with the needs of freshwater ecosystems (Smakhtin et al. 2004).

Can we appropriate more of the terrestrial water supply to support an increased population? Postel et al. (1996) and Jackson et al. (2001) argue that we cannot because most land suitable for rain-fed agriculture is already in production. We could build more dams on rivers to capture more runoff, but the maximum this would permit is estimated at 64% usage of runoff. Given an expected population growth of approximately 33% in the next 25 years (see Table 1) something will have to change. If we do not wish to destroy completely all of the Earth's freshwater aquatic ecosystems, we will have to increase our efficiency of water use and reduce the amount of water used for irrigation. Desalination of seawater is an expensive option because it is energy intensive, and it is unlikely to provide a solution for the less developed countries. The message is similar to that we obtained from ecological footprints: humans are close to the carrying capacity of the Earth.

Invasive Species

Introduced species constitute a major impact on the biosphere. Humans are of course one of the main invasive species after their evolutionary origin in Africa, and from the earliest times have affected the Earth's ecosystems. In recent times human activities have moved many species across continents and oceans that they could not otherwise cross, with the resulting effects we have already discussed. The introduction of the cane toad to Australia and the zebra mussel to the United States are two of many more examples. **Table 2** gives some examples of the magnitude of these biotic invasions. The economic impact of invasions is one strong reason for attempting to reduce undesirable introductions by enacting quarantine laws (Ruesink et al. 1995; Vitousek et al. 1996).

Humans have a diverse set of responses to exotic species. On the one hand, the pet trade illustrates the desire of many people in temperate and polar climates to have tropical fish in an aquarium in their home. We also rejoice in the introduction of many crop plants around the world as one way of diversifying our diets. On the other hand, many exotic species bring problems in the form of introduced diseases or disease risk, the local extinction of native species, and damage to forests and crops. We begin with the understanding that we cannot automatically designate all introduced exotics as

Table 2 Biotic invasions of vascular plants, freshwater fish, and birds.

Taxa	Locale	Number of native species	Number of nonnative species	Percentage of nonnative species
Vascular plants	Germany	1718	429	20.0
	Finland	1006	221	18.0
	France	4200	438	9.4
	California	4844	1025	17.5
	Canada	9028	2840	23.9
	Greenland	427	86	16.8
	Australia	20,000	2000	10.0
	Tanzania	1940	19	1.0
	South Africa	20,263	824	3.9
	Bermuda	165	303	64.7
	Hawaii	956	861	47.4
	Fiji	1628	1000	38.1
	New Zealand	1790	1570	46.7
Freshwater fish	California	76	42	35.6
	Canada	177	9	4.8
	Australia	145	22	13.2
	South Africa	107	20	15.7
	Brazil	517	76	12.8
	Hawaii	6	19	76.0
	New Zealand	27	30	52.6
Birds	Europe	514	27	5.0
	South Africa	900	14	1.5
	Brazil	1635	2	0.1
	Hawaii	57	38	40.0
	New Zealand	155	36	18.8

Island habitats are particularly vulnerable to invasions, but continental areas have also been strongly affected.

SOURCE: Data from Vitousek et al. (1996).

"bad" and all native species as "good." These value judgments affect our perceptions and our actions. The job of the ecologist is to determine for any particular introduced species whether it is causing damage or will potentially cause damage. For some species like the fire ant and the zebra mussel, these conclusions are easy to draw because of past experience. For other introduced species the damage they cause may be difficult to determine (Brown and Sax 2004).

The process of invasion must go through five stages before an introduced species can become a pest (Richardson et al. 2000; Colautti and MacIsaac 2004).

Why Did the Pleistocene Megafauna Disappear?

About 50,000 years ago all the continents were populated with more than 150 genera of megafauna, animals larger than 44 kg. By 10,000 years ago more than 97% of these animals were extinct, one of the greatest extinctions of large animals known (Martin and Wright 1967; Barnosky et al. 2004). Since many of these losses of large animals occurred near the end of the Pleistocene Ice Age, they are often grouped as the *Pleistocene Extinctions*. Two characteristics of these extinctions stand out. First, they affected only large animals, and there was no unusual simultaneous loss of smaller animals or plants. Second, the losses did not occur at the same time on the different continents.

The size of the lost megafauna is difficult to comprehend: giant kangaroos weighing 250 kg and standing to 3 m, giant beavers weighing 200 kg and nearly 10 times the size of modern beavers, mastodons and mammoths up to 4 m at shoulder height and weighing up to 10 tons (**Figure 9**). Why did these large animals disappear? There are two broad hypotheses of the causes of these extinctions. The anthropogenic or "overkill" hypothesis suggests that early humans caused the extinction of these large animals by a combination of hunting and habitat changes brought about by burning. The alternative hypothesis suggests that rapid climate change doomed large animals to extinction. The majority of the evidence now favors the human factor as the dominating cause of these extinctions. **Figure 10** shows schematically how these extinctions occurred at different times on different continents, typically coinciding with the arrival of humans during the late Pleistocene. Australia's main extinction events occurred much earlier than that of the Northern Hemisphere (Johnson 2006). The picture in South America is less clear and more data are required.

As more details become available, the many different ways in which humans could have played a role in the extinction of the megafauna are being uncovered. Recent Australian data on the extinction of the giant flightless bird *Genyornis newtoni* indicate how indirect effects caused by early humans on ecosystem structure could result in megafaunal losses (Miller et al. 2005). The giant flightless emu-like bird *Genyornis newtoni* disappeared 50,000 years ago coincident with a rapid shift of vegetation in central Australia caused by fire. Fires set by early humans converted the vegetation from a drought-adapted mosaic of trees and shrubs to a fire-adapted grassland-shrubland complex that could not support the giant emu because of its dietary specialization. The Australian emu *Dromaius novaehollandiae* survived this fire transition by shifting its

(a)   (b)   (c)

1 m

Figure 9 A few of the megafauna taxa that were present in the late Pleistocene but then disappeared. (a) *Glyptodon*, an armadillo-type animal with bony plates weighing up to 2 tons, from fossil deposits in North and South America. (b) A wooly mammoth (*Mammuthus primigenius*) standing over 4 m at the shoulder and weighing up to 10 tons; an entire carcass was uncovered from the Siberian permafrost in 1806, at a time when extinction was thought to be a rare event. (c) The giant flightless bird *Genyornis newtoni* from Australia in comparison with the modern emu *Dromaius novaehollandiae*. *Genyornis* stood over 2 m tall and probably weighed more than 100 kg; emus are shorter and weigh less than one-third that amount. (*Glyptodon* image from Martin and Wright 1967, *Mammuthus* image from the Academy of Natural Sciences of Philadelphia, Thomas Jefferson Collection, and *Genyornis* from Johnson 2006.)

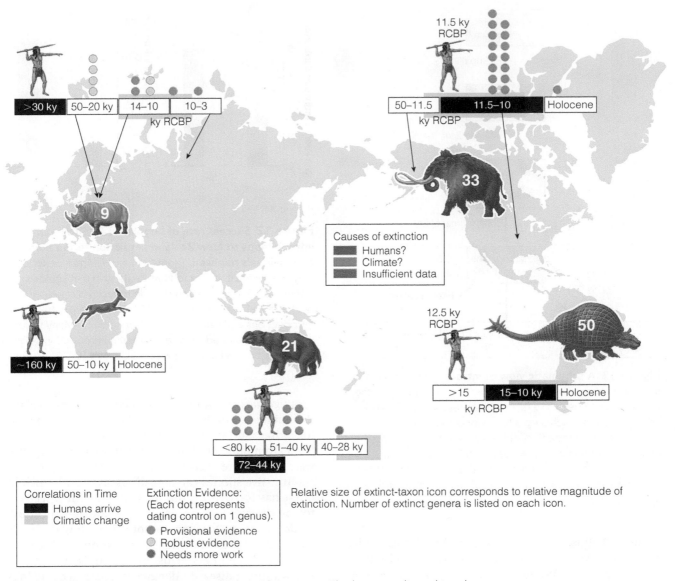

Figure 10 Megafauna extinctions in the late Pleistocene. The large numbers show the number of genera that went extinct on each continent, the strength of the paleontological data (provisional or robust), and a comparison of the timing of extinction with the timing of the arrival of humans and climate change. The recent extinction of the moas in New Zealand after the arrival of the Maori are not included here because they occurred during the last thousand years. (From Barnosky et al. 2004.)

opportunistic diet of a broad range of C_3 and C_4 plants to a diet composed almost entirely of C_3 plants. *Genyornis newtoni* was a dietary specialist on C_4 plants and did not survive the vegetation changes that occurred between 50,000 and 45,000 years ago. The key to this particular loss was dietary specialization rather than direct overharvesting of this large bird by human hunters.

The bottom line is that the large animals we see in the world today are a small subset of a much larger set of large animals that disappeared relatively recently due largely to human-caused changes in the biosphere. The circumstances now differ but it is a lesson we do not need to repeat again in the coming years.

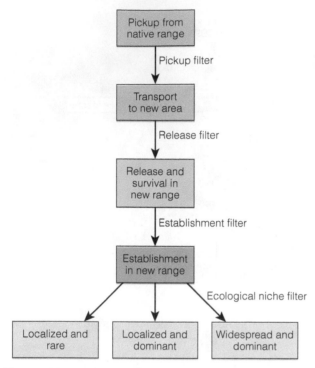

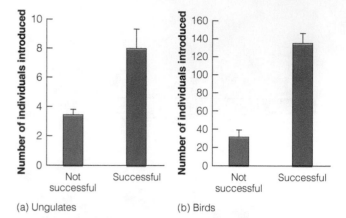

(a) Ungulates   (b) Birds

Figure 12 **Success rate of ungulate and bird introductions to New Zealand in relation to the number of individuals released.** From 1880 to 1930 many exotic ungulates were brought to New Zealand in a misguided attempt to change the countryside to resemble the Northern Hemisphere. From 1861 to 1885 there were 284 introductions of 43 species of passerine birds. Many fewer ungulates were needed for a successful introduction. (Data from Forsyth and Duncan 2001 and Duncan 1997.)

Figure 11 **A schematic view of introductions of alien species into a new region.** A series of steps lead from the transport of a species to its establishment and success in colonization. At each step a filter works that eliminates an approximate average 90% of the species moving down the chain, so that relatively few species are successful and relatively few become widespread and dominant pests. (Modified after Lockwood et al. 2005 and Colautti and MacIsaac 2004.)

Figure 11 illustrates these stages and indicates the filters that prevent many potential invasive species from becoming pests. Williamson and Fitter (1996) proposed the **tens-rule** to capture the expected probability of each filter being crossed as approximately 10%. The invasion cascade shown in Figure 11 can be separated into stages in which the key factors change (Williamson 2006):

- At import and release or escape, social and economic factors prevail.

- At establishment in the new area, ecology and biogeography come in.

- At success, spread, and impact, ecological and evolutionary factors prevail.

The exact importance of each of these factors is clear in some well-documented cases. Forsyth et al. (2004) found for introduced mammals in Australia that climatic suitability and introduction effort were the key factors predicting exotic species success. **Figure 12** shows that

for ungulates and for birds introduced to New Zealand, the number of individuals introduced was a key predictor of introduction success (Forsyth and Duncan 2001). For a given number of individuals introduced, ungulates were much more likely to succeed than birds.

Invasive species can spread diseases, and much interest has focused on the most prominent insect vectors of disease: mosquitoes. The Asiatic tiger mosquito (*Aedes albopictus*) began to spread worldwide in the 1970s thanks to the marine transport of tires and plants in containers. This mosquito is a vector of major human diseases: dengue, yellow fever, and West Nile virus (Eritja et al. 2005). *Aedes albopictus* is a treehole mosquito in Asia, but it is flexible enough to colonize water in old tires, cemetery flower pots, birdbaths, and soda cans. Adults in nature can fly only short distances, and all colonization has been by passive transport. **Figure 13** shows the area colonized by this mosquito since 1985 when it was first discovered in Texas. It now occupies 26 states and has been established (and eradicated) as far north as Chicago and Minnesota. Since its colonization, dengue fever has appeared in Texas in areas along the Mexican border (Brunkard et al. 2007). Mosquitoes such as the Asian tiger mosquito can create novel health threats or modify disease transmission for established viral diseases (Juliano and Lounibos 2005).

The key to invasive species management is to prevent the problem in the first place by good quarantine practices, and to have a rapid environmental response

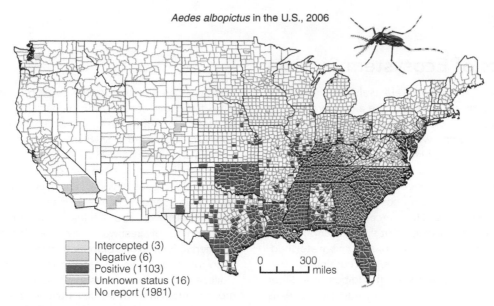

Aedes albopictus in the U.S., 2006

Intercepted (3)
Negative (6)
Positive (1103)
Unknown status (16)
No report (1981)

0   300 miles

Figure 13 Counties in the United States colonized by the Asiatic tiger mosquito, *Aedes albopictus*, in 2006. Areas presently colonized are shown in red, and blue areas have had established colonies but have been eradicated. Green areas show new areas of introduction outside the present range that have been intercepted. (Map courtesy of Chester G. Moore, Colorado State University, and the U.S. Centers for Disease Control and Prevention.)

team available to remove invasive species that manage to slip through quarantine before they spread and become a major pest.

Ecosystem Services

In our technological society we have lost touch with the numerous benefits ecosystems provide to human society. The term **ecosystem services** refers to all the processes through which natural ecosystems and the biodiversity they contain help sustain human life on this planet. The following is a list of a few ecosystem services we take for granted:

- Purification of air and water
- Mitigation of droughts and floods
- Generation and preservation of soils and soil fertility
- Detoxification and decomposition of wastes
- Pollination of crops and natural vegetation
- Dispersal of seeds
- Nutrient cycling
- Control of many agricultural pests by natural enemies

- Maintenance of biodiversity
- Protection of coastal shores from erosion
- Protection from ultraviolet rays
- Partial stabilization of climate
- Moderation of weather extremes
- Provision of aesthetic beauty

These ecosystem services are greatly undervalued by human society, as they have no dollar figure attached to them. But human life would cease to exist without these ecosystem services, and in this sense they are of immense value to us. Can we quantify this value?

When Robert Costanza and colleagues (1997) attempted to put a dollar figure on ecosystem services, they came up with an estimate of US$33 trillion per year, nearly twice as much as the gross national product of all the countries of the globe ($18 trillion). Many of the valuation techniques that are used in economic analyses of ecosystem services are based on the "willingness to pay" idea. If an individual owns a commercial forest, and ecosystem services provide a $50 increment to timber productivity, that individual should be willing to pay up to $50 for these services. The problem is that willingness to pay may not be a good measure if individuals are ill informed about the actual ecosystem services provided.

Economics of Ecosystem Services

The environment's services are valuable, everyone agrees. The water we drink and the air we breathe are available to us only because of ecosystem services that we take for granted. How can we enroll market forces in the conservation of ecosystems? How can we get corporations and governments to invest in natural capital? Two examples show how we might proceed to do this.

New York City's water supply comes from a watershed in the Catskill Mountains, and until recently water was purified by the natural processes of root systems and soil microorganisms, by filtration in the soils of this watershed, and by sedimentation in its streams. But continued additions of sewage to Catskill streams, in addition to fertilizer and pesticide use in local agriculture, had by the early 1990s degraded the Catskill water supplies to standards below those set for drinking water by the Environmental Protection Agency. In 1996 the city had two choices: (1) build and operate a water filtration plant to purify the water at a construction cost of over $6 billion to $8 billion and running costs of $300 million per year, or (2) restore the integrity of the Catskill ecosystem. The city chose to restore ecosystem integrity by buying land in and around the catchment area so that its use could be restricted, and by subsidizing the construction of better sewage treatment plants. By investing $1–1.5 billion in natural capital, New York City has saved a $6–8 billion investment in physical capital (Chichilnisky and Heal 1998).

Water is an ecosystem service that is both recognized as important by everyone but is also relatively easy to quantify in economic terms. Núñez et al. (2006) calculated the economic value of Chilean temperate forests to the provisioning of freshwater supplies for major cities in southern Chile. Exotic plantation forests of radiata pine or eucalypts cause a sharp drop in stream flows in southern Chile, and this difference is particularly striking in the summer months when water flows are minimal. Native forest cover allows more stream flow, with the result that one hectare of native forest provides a net benefit of $162 in summer and $61 in the remainder of the year, which translates into a per household savings of $21 per year for this part of Chile. The key point is that these ecosystem services do not in fact result in a direct payment of cash to each household, and so they are easily forgotten until they are lost by deforestation.

There are many uncertainties about this approach to quantifying the value of ecosystem services. The largest service contribution is from nutrient cycling, which makes up about half of the value of ecosystem services. The key point is that the value of ecosystem services is large, and if these services were actually paid for in our economic system, the global market system would be completely different. Many projects like large dams or irrigation projects would no longer be economical because their true cost would exceed social benefits.

Freshwater ecosystem services may be one of the simpler services to quantify, for humans need drinking water; they value lakes and rivers free of serious pollution (Wilson and Carpenter 1999). Specific indicators of water quality such as water clarity or the frequency of algal blooms are simple and are readily understood by the public as important indicators of ecosystem health. Most of the attempts to quantify freshwater ecosystem services have been site specific and cannot readily be extrapolated to larger areas like a state or country. For example, lakefront property values can be shown to decline substantially as a lake becomes more polluted and algal blooms become frequent. The challenge is to expand these very local economic evaluations to larger scales so that the value of ecosystem services can enter environmental policy debates.

Pollination is another ecosystem service that is both critically important and relatively easy to quantify. Pollination supplied by bees enhances the production of many crops. In the United States, managed honey bees have declined from over 4 million colonies in the 1970s to less that 2.4 million colonies in 2005 from parasitic mite infections and pesticide misuse. The key question arising from ecosystem services is how much wild bee populations could take over the pollination service formerly provided by commercial honey bees. Greenleaf and Kremen (2006) analyzed the impact of wild bees on tomato production in northern California. Bee pollination increases the number of tomatoes produced by sixfold over that of control plants from which all pollinators were excluded. The majority of the pollinators visiting their tomato plants were wild bees, and one species in particular, which made about one-third of the visits, was highly sensitive to the amount of natural habitat within range of the tomato patch. Few visits occurred if the tomatoes were more than 300 meters from natural shrubland or woodland. The practical recommendation is that keeping natural

Ecological Impacts of Biofuels

The impending shortage of oil and rising concern about climate change have fueled the search for ways to produce biofuels as a replacement for gasoline and diesel in the transport industry. Biofuels are typically ethanol produced from corn or sugarcane, or methane produced from decaying sewage or animal manures, but they include many forms of recycled petroleum products such as cooking oils. A detailed assessment of the environmental costs and benefits has now been provided by Zah et al. (2007).

Most efforts to evaluate the use of biofuels have concentrated on the relative reduction in greenhouse gas emissions from biofuels and the reduction in fossil fuel use. The key point is that nearly all biofuels reduce greenhouse gas emissions, although some like grass and wood are more effective than others (**Figure 14**). But this focus on greenhouse gases is too narrow, and we need as ecologists to consider the full environmental impacts of each particular biofuel. A key factor is whether native ecosystems are destroyed in order to produce the biofuels (Scharlemann and Laurance 2008). Sugarcane is a good example. If tropical rain forests are being cleared in order to grow sugarcane, greenhouse gas emissions increase from the land clearing, biodiversity is lost, and soil erosion increases. Some of the crops used for biofuels, like corn or canola, require nitrogen fertilizers, which release trace amounts of nitrous oxide, a powerful greenhouse gas that destroys ozone.

The problem is that each biofuel has specific benefits and specific costs. Zah et al. (2007) have boiled these costs and benefits into two criteria shown in Figure 14—greenhouse gas emissions and overall environmental impact. They found that almost all the biofuels reduce greenhouse gas emissions but about half of them have greater environmental costs than traditional fossil fuels. Ethanol production from grass and wood fare well in this comparison, but the use of crops like corn and potatoes have much more environmental costs than benefits (Pimentel et al. 2007).

The key message is that not all biofuels are beneficial when their full environmental costs are measured. In particular using corn and other foodstuffs for biofuels is perverse because government subsidies for these biofuels cause environmental degradation as well as increases in the price of food for people everywhere.

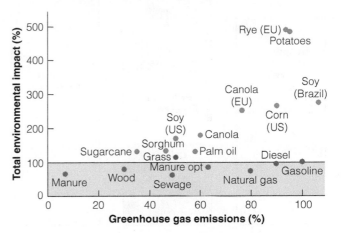

Figure 14 Greenhouse gas emissions in relation to overall environmental impacts of 17 transport fuels, scaled relative to gasoline as 100%. The origin of some of the biofuels produced is indicated by country codes: European Union (EU), United States (US). Palm oil estimate is from Malaysia. Fuels in the brown shaded area are considered more advantageous in both their overall environmental impacts and greenhouse gas emissions. (Data from Scharlemann and Laurance.)

One suggested solution to these problems comes from second-generation biofuels made from the breakdown of plant cellulose produced by nonfood plants like switchgrass (*Panicum virgatum*) and trees growing on marginal soils. But at present even these energy conversions operate at a net loss (Pimentel and Patzek 2005; Pimentel et al. 2007). Producing ethanol from corn requires 1.28 kcal of fossil fuel energy input for each 1 kcal ethanol energy yield, while at present to produce ethanol from switchgrass requires 1.50 kcal fossil fuel energy input for each 1 kcal ethanol yield. Sugarcane has a slightly positive energy balance. In Brazil 1.38 kcal of ethanol is produced for each 1 kcal of fossil energy expended in sugarcane cultivation and distillation (Pimentel and Patzek 2007). The ecological, economic, and social impacts of biofuels production should be important considerations in decisions about future energy paths to be followed.

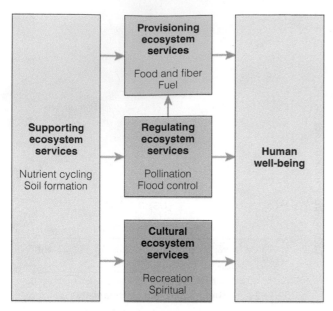

Figure 15 **Classification of ecosystem services from the Millennium Ecosystem Assessment (2005).** Agricultural lands are managed to maximize the provisioning services but in order to do so require many supporting services. While we know a great deal about the provisioning and regulating services, we know much less about the supporting services. (Modified after Zhang et al. 2007.)

habitats in a mosaic within agricultural crops can be one way to manage and increase wild pollinator populations (Kremen 2005).

The Millennium Ecosystem Assessment (2005) attempted to classify the ways in which ecosystem services influence human well-being (**Figure 15**). Assessments of ecosystem services demand some type of ecological indicators so that we can chart our progress over time in achieving our defined goals for conservation or land management. If we concentrate on the provisioning services made available to humans through agriculture, Zhang et al. (2007) pointed out that in addition to ecosystem services, we should recognize ecosystem dis-services that reduce productivity. These dis-services include problems already discussed—habitat loss, nutrient runoff, pesticide application to nontarget species, and weeds. Until we have a better understanding of the ecological machinery behind ecosystem services, we will not be able to maximize their utility for human well-being.

Sustainability

Sustainability is the mantra of our times, and unfortunately the word is misused more often than not. The concept of **sustainability** is to protect the biosphere

from degradation so that we pass to our children a planet unharmed by our presence. It is akin to the Hippocratic Oath in medicine to *do no harm* to the planet. Unfortunately, humans have not taken the notion of sustainability seriously, and global warming from greenhouse gases will give our children and grandchildren a different world than we have experienced. Neither agriculture, fisheries, nor forestry as currently practiced are sustainable, and we need to find out how we can set these critical industries on a more environmentally sustainable path.

One of the key questions of this century for humans is whether agriculture can become sustainable. Almost every agricultural ecologist will acknowledge that agriculture is not sustainable the way it is currently practiced. But a more fundamental question is whether soil fertility can be maintained in ecosystems with continuous cropping. Soil fertility will not decline if input equals output. Among the six macronutrients needed for plant growth, nitrogen and sulfur occur in gaseous form at normal temperatures, and nitrogen is also fixed by some plants and algae. Calcium, magnesium, and potassium are common constituents of many rocks and are released by weathering. Phosphorus, however, does not occur in gaseous form and is rare in most rocks, so its natural inputs to ecosystems are limited. Modern farming uses rock phosphate as fertilizer, but rock phosphate is a nonrenewable resource. What is the nutrient status of agricultural soils with and without artificial phosphate additions?

The most empirical approach to this question is to look at the productivity of a crop grown over a series of years on a site with no nutrient additions. Because rice has been grown for centuries in Southeast Asia, the broad historical overview would suggest that rice farming must be sustainable. **Figure 16** shows the yields of rice over 37 years of continuous cropping on land in the central Philippines with no added fertilizers. There is a slight downward trend in crop yields but in general the past 20 years have produced stable crop outputs, suggesting that rice cultivation in this region is sustainable. But the soils in the central Philippines are rich, volcanic soils and what other evidence do we have of sustainable yields on agricultural fields to which no artificial fertilizers have been added?

Newman (1997) analyzed data from a medieval farm in central England that kept good records of crops sown and the movements of grain sold to merchants in London so that he could obtain a phosphorus balance sheet for the whole farm, which grew mainly wheat and oats in a three-year rotation. The yield of wheat at the time averaged about 1 ton per ha per year, and Newman (1997) calculated that the farm was in deficit of phosphorus of about 0.7 to 0.9 kg/ha per year, about 4–5 times more than could be delivered by the weathering

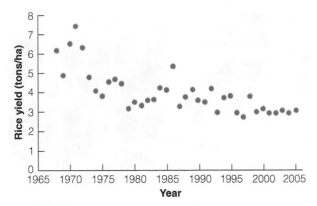

(a) Dry season

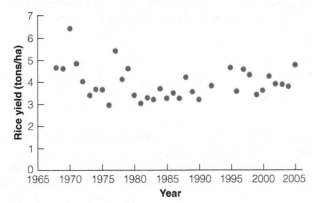

(b) Early wet season

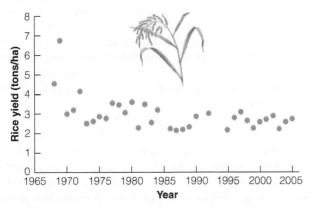

(c) Late wet season

Figure 16 Yields of rice planted continuously since 1968 on a single plot of land at the International Rice Research Institute at Los Banos, Philippines. No additions of nutrients in fertilizer were added to the soil on this plot, and three crops of rice were harvested each year. The nutrient removals from the crop had therefore to be made up from weathering or nitrogen fixation by algae. (Data provided by the International Rice Research Institute 2007.)

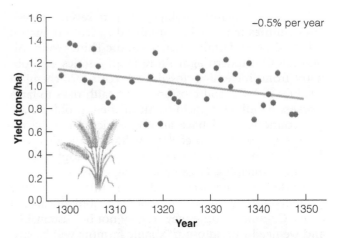

Figure 17 Wheat yield at the manor farm at Cuxham in Oxfordshire, England, from AD 1298 to 1347. Yields fluctuated annually but with an overall downward trend, possibly because the farm was in continuous phosphorus deficit from the harvesting and sale of the wheat. No outside source of fertilizer was available at this time to compensate for continual soil nutrient losses from cropping. (Data from Newman 1997.)

of soil parent materials. One consequence of this deficiency of phosphorus is shown in the wheat yields from AD 1298 to 1347 (**Figure 17**), which fluctuated with climatic conditions but overall declined at an average rate of 0.5% per year. This medieval farm with no source of artificial fertilizer was not operating in a sustainable manner.

Similar results were obtained for a set of plots planted in corn (maize) in central Illinois and maintained without artificial fertilizer input from 1876 to 1937. Corn yields fell on average 3% per year over this period as crop harvesting removed more phosphorus than could be released by natural weathering of soil elements (Newman 1997). Corn production in Illinois could not be sustained without artificial fertilizer additions to compensate for the phosphorus losses. Of all the farming systems examined by Newman (1997), only the grain grown in the fields along the Nile River in ancient Egypt were in long-term phosphorus balance, and this resulted from the annual Nile floods that deposited silt across the Nile delta each year.

There is much research currently under way to address the problems of sustainable agriculture. It is clear to all agricultural scientists that the current system of high external inputs of nitrogen (which is highly energy intensive to produce from oil) and phosphorus (a nonrenewable resource from rocks) is not sustainable. The world reserve of phosphate rock is currently estimated at 50 million tons, and at current rates of use for agricultural fertilizer, this will be completely depleted in 84 years (data in Jasinski 2007).

The sustainability problem is more severe in western countries that use industrialized agriculture than it is in traditional farming systems of the Third World (Altieri 2004). Modern agriculture faces a series of problems from increasing fossil fuel costs, to chemical residues in food, soil erosion, and health risks to farm workers handling pesticides. All of these problems are interconnected and have an economic and social dimension as well as an ecological framework. One way forward that has been suggested is to adopt organic farming principles. Organic farming excludes artificial pesticides and fertilizers, and maintains soil fertility with organic manures, crop rotation, and diversified crops. Organic farming may or may not be sustainable, and we need to find out if organic farming will be sustainable in the sense of producing high yields and being profitable, while at the same time protecting the environment and conserving soil nutrients.

There have been relatively few attempts to compare the results of organic farming with those of conventional farming with a wide variety of biological and economic indicators. Reganold et al. (2001) did this with apple production systems in the Yakima Valley of Washington. They planted Golden Delicious apples on four replicate plots of 1.7 ha each, and utilized organic farming, conventional farming with pesticides and herbicides, and integrated farming, which combined methods from both organic and conventional systems. Soils were sampled to look for any deterioration in soil nutrients, and the apples produced were sold on the open market to determine their economic value. They carried out their studies over five years with the results shown in **Figure 18**. All three agricultural systems gave similar apple yields, although organic apples were smaller than conventional apples. Over the five years, organic trees produced 222 tonnes per ha, while conventional trees produced 244 tonnes. Yields varied from year to year for each of the farming systems, showing the need for studies to extend over more than one year.

The organic farming system was most efficient when measured in terms of energy output per unit of energy

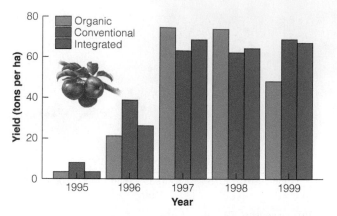

Figure 18 Yield of apples from orchards planted in 1994 in the Yakima Valley of Washington State. Organic farming used no artificial pesticides, herbicides, or fertilizers. Integrated farming used a mix of conventional and organic methods. The yields of apples fluctuated from year to year but overall were comparable for all three types of agriculture. (From Reganold et al. 2001.)

input, as well as in dollars returned, and because organic farming methods used less herbicides and pesticides, they scored much better in environmental impact (**Table 3**). The overall message was clear that organic apple farming was a win-win situation in which production could be achieved with reduced environmental impacts. This study of apple production has highlighted the need to analyze other agricultural cropping systems to find ways of changing to production methods that reduce environmental impacts without losing profitability. Many similar, small-scale changes in modern agriculture are happening to move it in the direction of sustainability, but there is a long way to go (Crews 2005).

One large-scale experiment illustrates all too well how little we understand the value of ecosystem services for sustainability. The Biosphere 2 in Oracle, Arizona, was an attempt to construct a closed experimental ecosystem covering 1.27 ha (**Figure 19**). A forest with soil and a miniature ocean were constructed inside

Table 3 A comparison of the economic and environmental value of organic apple farming and conventional apple farming in central Washington State.

	Organic farming	Conventional farming	Integrated farming
Energy ratio – output/input	1.18	1.11	1.13
Net return in 1999 ($/ha)	$9602	$5434	$4267
Environmental impact score	1.0	6.2	4.7

The energy ratio includes inputs from labor as well as fuel, herbicides, and insecticides. The environmental impact score is a relative measure of the potential adverse effects of all the agricultural chemicals used in each farming method; the higher the score the worse the impact.

SOURCE: Data from Reganold et al. (2001).

Figure 19 View of part of the Biosphere 2 site near Tucson, Arizona. The original aim was to construct a completely enclosed, functioning ecosystem that humans could occupy indefinitely, but it was a complete failure. Oxygen levels collapsed and many species went extinct, so the experiment had to be stopped after one year.

the airtight Biosphere, at a cost of over $200 million. From October 1991, eight people began a two-year stint living in isolation inside the Biosphere 2 ecosystem. Unlimited energy and technology were available from the outside to support the project. But the system failed, and the experiment had to be stopped after 15 months (Cohen and Tilman 1996). Atmospheric oxygen dropped to 14%, and CO_2 fluctuated irregularly. Most of the vertebrate species in Biosphere 2 went extinct, and all of the pollinators died out. Population explosions of pests such as cockroaches showed clearly that the ecosystem services were impaired even before the experiment was halted. The conclusion is clear: No one knows yet how to engineer a system that will provide humans with all the life-support services that natural ecosystems produce for free. We have no alternative but to maintain the health of ecosystems on the Earth, and sustainability is the ecological challenge for the twenty-first century.

Ecosystem Health

The news every day is filled with conflicts between "greenies" and developers, economists and environmentalists, over economic growth and the environment. One way to judge these controversies might be to obtain some index of the environment that we can call *ecosystem health*. The general notion of ecosystem health

is a useful one to provide for society and decision makers some indication of how the global environment is coping with stresses arising from physical factors such as hurricanes, fires, and floods; from population growth; and from human actions such as fossil fuel use. The problem is that these environmental issues have become polarized between two different worldviews (Costanza et al. 2000). For simplicity we will call these two the *technological optimist* view and the *technological skeptic* view. The technological optimist believes that through technological innovation humans will dominate nature and become independent of nature. All future challenges, whether water shortages or AIDS, will be overcome by further technological progress. This view is the dominant worldview now, and is based partly on the fact that this approach to life has produced the great progress in human health and living standards over the last 250 years. By contrast, the technological skeptic is dubious that the trends of the past can be extrapolated into the future. The basis for this skeptical view is that the human population has increased so much in recent years that we risk destroying essential natural systems via pollution, climate change, overharvesting, and habitat destruction (Rapport and Whitford 1999). Ecologists in general are technological skeptics, while many business economists are technological optimists. Because these are such different worldviews, it would be useful to move the debate to a new level to see if there could be some approach that could reconcile these two groups.

One suggestion is to view the Earth as an investment portfolio, and to ask how we ought to manage this portfolio, particularly in situations of uncertainty. For example, there is much uncertainty about future climate change. How can we accommodate to ecological and environmental uncertainty? One way is to use game theory to construct a payoff matrix of what might happen under these two views of the world (Costanza et al. 2000). The payoff matrix is an oversimplified, right-or-wrong kind of game that looks at our current beliefs and what will become future reality:

		Real state of the world	
		Optimists correct	Skeptics correct
Current policies to be used in environmental decisions	Technological optimists' policies	High payoff	Disaster
	Technological skeptics' policies	Good payoff	Very good payoff

575

We ask in the payoff matrix what will be the consequences of these two views of the world. If the optimists are correct, and we adopt their policies, and they work, we reach the highest payoff, and all is well. But if the skeptics are in fact correct about the future state of the world, and we continue with the optimist policies, we reach the worst payoff, disaster. By contrast, if we pursue the skeptics' policies and in fact the optimists are correct, we reach only a good outcome, because skeptical policies such as reducing carbon dioxide emissions and other pollutants are costly and will reduce economic growth. Finally, if the skeptics are correct about the future and we adopt skeptic policies, we have done the right thing. Which box do we wish to choose?

The payoff matrix basically asks whether we humans wish to gamble with the whole biosphere. The prudent person will not wish to risk disaster, and, even if you are a technological optimist, it would be sensible to adopt the skeptic's position to hedge your bets. The important point is that the skeptical viewpoint does not mean we should stifle new technology, but rather that we should adopt the *precautionary principle*. We need to acknowledge the great uncertainty we have about future events, and ask how we can manage our environmental affairs while acknowledging the uncertainty of our scientific information. The simple principle of *not to implement changes that cannot be reversed* is a good summary of the skeptics' position on environmental policies.

Business portfolio managers have long ago worked out how to manage financial affairs under uncertainty, and while ecologists talk too little to business managers it is useful to look at the four principles of financial management as they might be applied to the Earth's environment.

1. **Protect your capital.** The first rule of all financial managers is to live off interest and not erode capital. This principle focuses on the concept of natural capital and the services that humans obtain from the environment.

2. **Hedge your investments.** The simple rule of not to put all one's eggs in one basket applies to the environment as well. We should not assume that all the technological optimists' eggs are golden, and none will have adverse side effects on the environment. DDT is the classic example.

3. **Do not risk more than you can afford to lose.** Risk management is critical for all environmental policies, and the decisions that humans make about environmental actions must take into account the preservation of natural capital for future generations.

4. **Buy insurance.** All prudent persons protect their assets with insurance against unforeseen catastrophic events. Environmental insurance is obtained by setting aside national parks and marine reserves, and protecting biodiversity. Who knows what species will be useful in 100 years?

Environmental problems are often large-scale, global problems, and while we might agree on policies in one country, others will disagree so that it is difficult to achieve global action on issues like climate change. The principle of sustainability needs to be an accepted principle of all governments, and we humans need to begin managing our global environmental problems with an acknowledgement of the uncertainty of ecological knowledge and with the wisdom to be humble.

Summary

Human impacts on the Earth's ecosystems are rooted in the continuing increase in world population. Over 6.64 billion people now inhabit the Earth, and the population is growing by 211,000 people every day. The less developed countries of the globe contribute most of this increase, but much of the adverse effects of humans arises in the developed world, with its high demands for energy and materials.

The carrying capacity of the Earth is difficult to estimate. One approach is to estimate the ecological footprint of a nation in terms of the amount of land it utilizes to produce the commodities it consumes. By this measure the Earth is already beyond its carrying capacity, and many countries have exceeded their available ecological resources and are in deficit. By no measure can the Earth sustain its current population if all live at the high consumption lifestyle of the developed world.

Coupled with the growth of the human population has been the movement of exotic species over the Earth's surface, unleashing a large number of pest species. Not all introduced species become pests, and some like the crops we eat, are desirable. Fortunately most introduced species do not survive in their new domain, and they die out, causing no trouble. But a small number of serious pests have become invasive, resulting in serious damage to agriculture, forestry, and human health, while compromising many native species that have become endangered.

Ecosystems provide services in air and water purification, pollination, and nutrient cycling that are greatly undervalued by society because they are not traded in the marketplace. The Millennium Ecosystem Assessment (2005) put special emphasis on the need to respect nature's services lest their destruction lead to humans destroying their life support system.

Sustainability is the key word for this century, but all evidence shows without a doubt that our current lifestyles are not sustainable in the future. Agricultural systems are currently mining the natural capital of our soils and replacing the lost nutrients with artificial fertilizers that are very expensive to make (nitrogen from natural gas) or nonrenewable (phosphorus from rock phosphate). We do not yet know how to construct and engineer artificial ecosystems that can support human life. We had best take care of the Earth and safeguard its ecosystems, a task only slowly being learned by ourselves and by our political and business leaders.

Review Questions and Problems

1. Find two or three quotations in the current media that include "sustainability" and discuss the concept of sustainability, its origin in recent times, and the ways in which it is currently used in discussions of environmental problems.

2. Some authors have argued that biological invasions are nothing new and have occurred throughout evolutionary time. Imagine that you are an alien scientist visiting a new continent with no prior information about which species are native and which are exotic. How would you decide which species are "pests" and would your job be easier if someone gave you a list of which species are exotics and which are natives? Larson (2007) discusses this question.

3. Discuss how the payoff matrix might be used to evaluate decisions about the introduction of genetically modified crops to a new country. Is it possible to put a dollar figure on future risks?

4. While most evaluations of the benefits of ecosystem services give a positive value, Clinch (1999) estimated the value of water supply provisioning in an Irish temperate forest at minus $20 per hectare. Review the general issue of evaluating the dollar benefits of ecosystem services and discuss how some potential services might in fact have a negative value.

5. The concept of *ecosystem health* has been criticized as a concept that applies well to individuals but poorly to a whole ecosystem. Discuss the application of the idea of *health* to populations, communities, and ecosystems. Rapport et al. (1998) discuss this question.

6. Review the projections that have been given for the human population from the 1970s to the current time, and discuss how and why they have changed during the last 40 years. The United Nations, the Population Reference Bureau, and the U.S. Census Bureau provide Web sites. Keilman (1999) discusses the issue.

7. Nitrogen is often a critical nutrient limiting crop growth, and yet it occurs in abundance in the atmosphere. Discuss the nitrogen requirements of modern agriculture and the possible ways in which nitrogen requirements of crops can be provided in a sustainable manner. Crews (2005) discusses this problem.

8. Are there ecosystem services that cannot be evaluated in dollars? Is it necessary to place a monetary value on environmental concerns in order to obtain political action? Pearce et al. (2007) discuss this issue with respect to biodiversity.

9. If agricultural production is to provide for a growing human population, what components of agricultural production must increase in efficiency? Should we grow more food on less land or increase the land under cultivation? Is it feasible to produce more food on less land? Waggoner (1995) and Balmford et al. (2005) discuss the issue of land use for agriculture.

10. Discuss the economic and ecological meanings of the word "value." Is there a difference? What activities might an economist include in a valuation of the use of freshwater? Does economics value the nonuse of a resource like lake water? Wilson and Carpenter (1999) discuss these issues.

Overview Question

One suggested ecological response to help arrest increasing atmospheric carbon dioxide levels is to plant trees. Explain the mechanisms behind this recommendation, and discuss how you could calculate how many trees would need to be planted to achieve this policy goal.

Suggested Readings

- Barnosky, A. D., P. L. Koch, R. S. Feranec, S. L. Wing, and A. B. Shabel. 2004. Assessing the causes of Late Pleistocene extinctions on the continents. *Science* 306:70–75.

- Ehrlich, P. R., and A. H. Ehrlich. 2005. *One with Nineveh: Politics, Consumption, and the Human Future*. Washington, DC: Island Press.

- Hickey, G. M. 2008. Evaluating sustainable forest management. *Ecological Indicators* 8:109–114.

- Hoegh-Guldberg, O., L. Hughes, S. McIntyre, D. B. Lindenmayer, C. Parmesan, H. P. Possingham, and C. D. Thomas. 2008. Assisted colonization and rapid climate change. *Science* 321:345–346.

- Johnson, C. 2006. *Australia's Mammal Extinctions: A 50,000 year History*. Melbourne: Cambridge University Press.

- Kremen, C. 2005. Managing ecosystem services: What do we need to know about their ecology? *Ecology Letters* 8:468–479.

- Mack, R. N., D. Simberloff, W. M. Lonsdale, H. Evans, M. Clout, and F. A. Bazzaz. 2000. Biotic invasions: Causes, epidemiology, global consequences, and control. *Ecological Applications* 10:689–710.

- Newman, E. I. 1997. Phosphorus balance of contrasting farming systems, past and present. Can food production be sustainable? *Journal of Applied Ecology* 34:1334–1347.

- Pimentel, D., and T. Patzek. 2007. Ethanol production: Energy and economic issues related to U.S. and Brazilian sugarcane. *Natural Resources Research* 16:235–242.

- Wackernagel, M., L. Onisto, P. Bello, A. Callejas Linares, I. S. Lopez Falfan, J. Mendez Garcia, A. I. Suarez Guerrero, and M. G. Suarez Guerrero. 1999. National natural capital accounting with the ecological footprint concept. *Ecological Economics* 29:375–390.

- Yang, H. S. 2006. Resource management, soil fertility and sustainable crop production: Experiences of China. *Agriculture, Ecosystems & Environment* 116:27–33.

- Zhang, W., T. H. Ricketts, C. Kremen, K. Carney, and S. M. Swinton. 2007. Ecosystem services and dis-services to agriculture. *Ecological Economics* 64:253–260.

Credits

Illustration and Table Credits

Photo Credits

A Primer on Population Genetics

Population genetics is the algebraic description of evolution, of how allelic frequencies change over time. Here we consider a very abbreviated basic statement of this approach for diploid organisms. See Tamarin (2002, Chapters 19–20) for more details.

We begin with individuals distinguished by a single locus with two possible alleles (A, B). Every individual is thus one of three genotypes (AA, AB, BB). The most important theorem in population genetics is called the Hardy-Weinberg law. In 1908 a British mathematician, G. H. Hardy, and a German physician, G. Weinberg, independently discovered that an equilibrium will arise and be maintained in both allelic and genotypic frequencies in any diploid population that is large in size, undergoes random mating, has no mutation or migration of individuals, and is subject to no selection. We can illustrate this law simply with an example. Consider a hypothetical population with the following composition:

Genotype	Genotypic frequency (%)	No. individuals
AA	25	1000
AB	0	0
BB	75	3000

We define:

$$\text{Allelic frequency of } A = \frac{\text{Total number of } A \text{ alleles in population}}{\text{Total number of alleles in population}}$$

$$= \frac{2(1000)}{2(1000) + 2(3000)} = \frac{2000}{8000}$$

$$= 0.25$$

$$\text{Allelic frequency of } B = \frac{\text{Total number of } B \text{ alleles in population}}{\text{Total number of alleles in population}}$$

$$= \frac{2(3000)}{2(1000) + 2(3000)} = \frac{6000}{8000}$$

$$= 0.75$$

Under the given assumptions, the Hardy-Weinberg law states that (1) allelic frequencies will not change over time, and (2) genotypic frequencies will come into equilibrium within one generation and will be as follows:

$$\text{Genotypic frequency of } AA = (\text{allelic frequency of } A)^2$$

$$\text{Genotypic frequency of } BB = (\text{allelic frequency of } B)^2$$

$$\text{Genotypic frequency of } AB = 2(\text{allelic frequency of } A)$$
$$(\text{allelic frequency of } B)$$

For our hypothetical example:

$$\text{Genotypic frequency of } AA = (0.25)^2 = 0.0625$$

$$\text{Genotypic frequency of } BB = (0.75)^2 = 0.5625$$

$$\text{Genotypic frequency of } AB = 2(0.25)(0.75) = 0.3750$$

These frequencies sum to 1.00. If we apply the definition again for a second generation, a third generation, and so on, we find that none of these frequencies change. The Hardy-Weinberg law produces a globally stable equilibrium because no matter what the starting genotypic frequencies are, we always come to the same result (for one set of allelic frequencies). We can demonstrate this by doing these same calculations with genotype frequencies of $AA = 0\%$, $AB = 50\%$, and $BB = 50\%$.

The important conclusion is that gene frequencies, or *genetic variability*, in a population will be maintained over time without any change *unless outside forces are applied*. Evolution results from these outside forces. A brief comment on these outside forces follows.

Mutations   Genetic mutations are relatively rare events and do not typically cause shifts from Hardy-Weinberg equilibrium. Mutations are important as the source of genetic variation on which natural selection acts.

Migration   Migration, or movements of individuals between populations, can be a means of adding or subtracting alleles in a population. Migration can be critical in preventing or aiding adaptation in local populations.

Population size   In small populations, chance may be a critical element. For example, if two individuals in the previous example colonize an island, by chance both could be BB individuals. This is called random genetic drift and must be considered when populations are small in size.

From Appendix I of *Ecology: The Experimental Analysis of Distribution and Abundance*, Sixth Edition. Eugene Hecht.

Random mating   Individuals may mate on the basis of similarity (or dissimilarity) so that assortative mating would result. For example, *AA* individuals may prefer to mate with *AA* individuals. If the choice of mates involves relatives, then either inbreeding or outbreeding may also be involved. In either case, mating is not random, and the Hardy-Weinberg equilibrium is disturbed.

Natural selection   Selection produces adaptation by altering allelic frequencies by eliminating individuals that are less fit. We can introduce the idea of natural selection into our simple model population by defining the notion of *fitness*: the relative reproductive success of a given genotype. The following simple example shows how fitness can be calculated:

Genotype	*AA*	*AB*	*BB*
No. in generation 1	3000	6000	3000
No. in generation 2	2500	6000	3500

$$\text{Relative reproductive success of } AA = \frac{\text{No. in generation 2}}{\text{No. in generation 1}} = \frac{2500}{3000} = 0.83$$

$$\text{Relative reproductive success of } AB = \frac{6000}{6000} = 1.0$$

$$\text{Relative reproductive success of } BB = \frac{3500}{3000} = 1.17$$

By convention, the genotype with the highest relative reproductive success has fitness equal to 1.0, and we thus define *relative fitness* of the three genotypes as follows:

$$\text{Relative fitness of } BB = \frac{\text{Relative reproductive success of } BB}{\text{Highest relative reproductive success observed}}$$

$$= \frac{1.17}{1.17} = 1.0$$

$$\text{Relative fitness of } AB = \frac{1.00}{1.17} = 0.86$$

$$\text{Relative fitness of } AA = \frac{0.83}{1.17} = 0.71$$

We can now define the *selection coefficient* for each genotype:

$$\text{Selection coefficient for genotype } x = 1.0 - \text{relative fitness of genotype } x$$

The selection coefficient against the best genotype is always zero. Note that fitness (equivalent to relative fitness, relative Darwinian fitness, and adaptive value) is always *relative*, and evolution always deals with how fit one genotype is relative to other genotypes. If there is only one genotype in the population, one cannot define fitness.

Appendix

Instantaneous and Finite Rates

The concept of *rates* is critical for quantitative work in ecology, and students may find a brief review useful.

A *rate* is a numerical proportion between two sets of things. For example, the number of students failing an examination might be 27 of 350, a failure rate of 7.7%. In ecological usage, a rate is usually expressed with a standard time base. Thus if eight out of 12 seedlings die within one year, the mortality rate is 66.7% *per year*. If a population grows from 100 to 150 within one month, the rate of population increase will be 50% *per month*.

We usually think in terms of *finite rates*, which are simple expressions of observed values. Some ecological examples are

$$\text{Annual survival rate} = \frac{\text{No. alive at end of year}}{\text{No. alive at start of year}}$$

$$\frac{\text{Annual rate of}}{\text{population change}} = \frac{\text{Population size at end of year}}{\text{Population size at start of year}}$$

Rates can also be expressed as *instantaneous rates*, in which the time base becomes very short rather than a year or a month. The general relationship between finite rates and instantaneous rates is

$$\text{Finite rate} = e^{\text{instantaneous rate}}$$

$$\text{Instantaneous rate} = \log_e \text{finite rate}$$

where $e = 2.71828. \ldots$

The idea of an instantaneous rate can be explained most simply by the use of compound interest. Suppose we have a population of 100 organisms increasing at a *finite* rate of 10% per year. The population size at the end of year 1 will be

$$100\left(1 + \frac{1}{10}\right) = 110$$

At the end of year 2, it will be

$$110\left(1 + \frac{1}{10}\right) = 121$$

At the end of year 3, it will be

$$121\left(1 + \frac{1}{10}\right) = 133.1$$

In general, for an interest rate of $1/m$ carried out n times,

$$y_n = y_0\left(1 + \frac{1}{m}\right)^n$$

where y_n = Amount at end of the nth operation
y_0 = Amount at start

We can repeat these calculations for a finite interest rate of 5% per half year. Everyone who has invested money in a savings account knows that 5% interest per half year is a *better* interest rate than 10% per year. We can see this quite simply. At six months our population size will be

$$100\left(1 + \frac{1}{20}\right) = 105$$

At one year, it will be

$$105\left(1 + \frac{1}{20}\right) = 110.25$$

and similarly at two years, 121.55, and at three years, 134.01. Compare these values with those obtained earlier for 10% annual interest.

Biological systems often operate on a time schedule of hours and days, so we may be more realistic in using rates that are instantaneous, that divide a year into very many short time periods. Let us repeat the first calculation with an *instantaneous* rate of increase of 10% per year. If we divide the year into 1000 short time periods, each time period having a rate of increase of 0.10/1000, or 0.0001, for the first 1000th of the year we have:

$$100(1 + 0.0001) = 100.01$$

For the second 1000th of the year,

$$100.01(1 + 0.0001) = 100.020001$$

If we repeat this for all 1000 time intervals, we end with 110.5 organisms at the end of one year.

Instantaneous rates and finite rates are nearly complementary when rates are very small. The following table and figure show how they diverge as the rates become large and illustrate the change in size of a hypothetical population that starts at 100 organisms and

From Appendix II of *Ecology: The Experimental Analysis of Distribution and Abundance*, Sixth Edition. Eugene Hecht.

increases or decreases at the specified rate for one time period:

Change (%)	Finite rate	Instantaneous rate	Hypothetical population at end of one time period
−75	.25	−1.386	25
−50	.50	−0.693	50
−25	.75	−0.287	75
−10	.90	−0.105	90
−5	.95	−0.051	95
0	1.00	0.000	100
+5	1.05	0.049	105
+10	1.10	0.095	110
+25	1.25	0.223	125
+50	1.50	0.405	150
+75	1.75	0.560	175
+100	2.00	0.693	200
+200	3.00	1.099	300
+400	5.00	1.609	500
+900	10.00	2.303	1000

	Decreases	No change	Increases
Finite rates	0 to 1.00	1.00	1.00 to $+\infty$
Instantaneous rates	$-\infty$ to 0.00	0.00	0.00 to $+\infty$

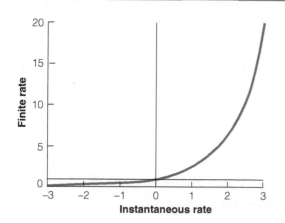

This illustrates one difference between finite rates (always positive or zero) and instantaneous rates (range from $-\infty$ to $+\infty$).

Mortality rates can be expressed as finite rates or as instantaneous rates. If the number of deaths in a short interval of time is proportional to the total population size at that time, then the rate of drop in numbers can be described by the geometric equation

$$\frac{dN}{dt} = iN$$

where N = population size
i = instantaneous mortality rate
t = time

In integral form, we have

$$\frac{N_t}{N_0} = e^{it}$$

where N_0 = starting population size
N_t = population size at time t^*

Taking logs, if $t = 1$ time unit, we obtain:

$$\log_e\left(\frac{N_t}{N_0}\right) = i$$

Since N_t/N_0 is the finite survival rate by definition, we have obtained

$$\log_e(\text{finite survival rate}) = \text{instantaneous mortality rate}$$

We thus obtain the following relationships for expressing mortality rates:

$$\text{Finite survival rate} = 1.0 - \text{finite mortality rate}$$

$$\log_e(\text{finite survival rate}) = \text{instantaneous mortality rate}$$

$$\text{Finite survival rate} = e^{\text{instantaneous mortality rate}}$$

$$\text{Finite mortality rate} = 1.0 - e^{\text{instantaneous mortality rate}}$$

Why do we need to use instantaneous rates? The principal reason is that instantaneous rates are easier to deal with mathematically. A simple example will illustrate this property. Suppose we have data on an insect population and know that the mortality rate is 50% in the egg stage and 90% in the larval stages. How can we combine these mortalities? If they are expressed as finite mortality rates, we cannot add them because a 50% loss followed by a 90% loss is obviously not 140% mortality, but only 95% mortality. If, however, the mortality is expressed as instantaneous rates, we can add them directly:

	Instantaneous mortality rate
Egg stage (50%)	−0.693
Larval stages (90%)	−2.303
Combined loss	−2.996

---

*Note that instantaneous rates are determined for a specific time base (per year, per month, etc.), even though the rate applies to a very short time interval.

We can convert back to a finite mortality rate by the formula given earlier:

$$\text{Finite mortality rate} = 1.0 - e^{\text{instantaneous mortality rate}}$$

$$= 1.0 - e^{-2.996}$$

$$= 0.950$$

and the combined mortality is seen to be 95%.

Four examples will illustrate some of these ideas, and students are referred to Ricker (1975, Ch. 1) for further discussion.

Example 1

A population increases from 73 to 97 within one year. This can be expressed as

a. Finite rate of population growth = 97/73 = 1.329 per head per year (or, the population grew 32.9% in one year).

b. Instantaneous rate of population growth = $\log_e(97/73) = 0.284$ per head per year.

Example 2

A population decreases from 67 to 48 within one month. This can be expressed as

a. Finite rate of population growth = 48/67 = 0.716 per head per month (or, the population decreased 28.4% over the month).

b. Instantaneous rate of population growth = $\log_e(48/67) = -0.333$ per head per month.

Example 3

A cohort of trees decreases in number from 24 to 19 within one year. This can be expressed as

a. Annual survival rate (finite) = 19/24 = 0.792.

b. Annual mortality rate (finite) = 1.0 − annual survival rate = 0.208.

c. Instantaneous mortality rate = $\log_e(19/24) = -0.234$ per year.

Example 4

A cohort of fish decreases in number from 350,000 to 79,000 within one year. This can be expressed as

a. Annual survival rate (finite) = 79,000/350,000 = 0.2257.

b. Annual mortality rate (finite) = 1.0 − 0.2257 = 0.7743.

c. Instantaneous mortality rate = $\log_e 0.2257 = -1.488$ per year.

Species Diversity Measures of Heterogeneity

There are several different measures of species diversity that are sensitive to both the number of species in the sample and the relative abundances of the species (Krebs 1999). Here we discuss only two of the most commonly used measures of heterogeneity.

The Shannon-Wiener function approaches the measure of species diversity through information theory. We ask the question: How difficult would it be to predict correctly the species of the next individual collected? This is the same problem faced by communication engineers interested in predicting correctly the name of the next letter in a message. This uncertainty can be measured by the Shannon-Wiener function[1]:

$$H = - \overset{s}{\underset{i=1}{a}} (p_i)(\log_2 p_i)$$

where H = information content of sample (bits/individual) = index of species diversity
S = number of species
p_i = proportion of total sample belonging to the i the species

Information content is a measure of the amount of uncertainty, so the larger the value of H, the greater the uncertainty. A message such as *bbbbbbb* has no uncertainty in it, and $H = 0$. For an example of two species of 99 and 1 individuals,

$$H = -[(p_1)(\log_2 p_1) + (p_2)(\log_2 p_2)]$$

$$= -[(0.99)(\log_2 0.99) + (0.01)(\log_2 0.01)]$$

$$= 0.81 \text{ bit/individual}$$

---

[1] This function was derived independently by Shannon and Wiener. It is sometimes mislabeled the Shannon-Weaver function.

For a sample of two species with 50 individuals in each,

$$H = -[(0.50)(\log_2 0.50) + (0.50)(\log_2 0.50)]$$

$$= 1.00 \text{ bit/individual}$$

This agrees with our intuitive feeling that the second sample is more diverse than the first sample.

Two components of diversity are combined in the Shannon-Wiener function: (1) number of species and (2) equitability or evenness of allotment of individuals among the species (Krebs 1999). A greater number of species increases species diversity, and a more even or equitable distribution among species will also increase species diversity measured by the Shannon-Wiener function. Equitability can be measured in several ways. The simplest approach is to ask, What would be the species diversity of this sample if all S species were equal in abundance? In this case,

$$H_{max} = -S\left(\frac{1}{S}\log_2\left[\frac{1}{S}\right]\right) = \log_2 S$$

where H_{max} = species diversity under conditions of maximal equitability
S = number of species in the community

Thus, for example, in a community with two species only,

$$H_{max} = \log_2 2 = 1 \text{ bit/individual}$$

as we observed earlier. Equitability can now be defined as the ratio:

$$E = \frac{H}{H_{max}}$$

where E = equitability (range 0–1)
H = observed species diversity
H_{max} = maximum species diversity = $\log_2 S$

The following table presents a sample calculation illustrating the use of these formulas.

Tree species	Proportional abundance (p_i)	$(p_i)(\log_2 p_i)a$
Hemlock	0.521	0.490
Beech	0.324	0.527
Yellow birch	0.046	0.204
Sugar maple	0.036	0.173
Black birch	0.026	0.137
Red maple	0.025	0.133
Black cherry	0.009	0.061
White ash	0.006	0.044
Basswood	0.00:	0.032
Yellow poplar	0.002	0.018
Magnolia	0.001	0.010
Total	1.000	$H = 1.829$

$H_{max} = \log_2 S = \log_2 11 = 3.459$

Sample calculations of species diversity and equitability through the use of the Shannon-Wiener function.

NOTES: Based on the composition of large trees (over 21.5 m tall) in a virgin forest in northwestern Pennsylvania. Note that there is no special theoretical reason to use $\log_2$ instead of $\log_e$ or $\log_{10}$. The $\log_2$ usage gives us information units in "bits" (binary digits) and is preferred by information theorists.
SOURCE: Hough (1936).

Other measures of species diversity can be derived from probability theory. Simpson (1949) suggested the following question: What is the probability that two specimens picked at random in a community of infinite size will be the same species? If a person went into the boreal forest in northern Canada and picked two trees at random, there is a fairly high probability that they would be the same species. If a person went into the tropical rain forest, by contrast, two trees picked at random would have a low probability of being the same species. We can use this approach to determine an index of diversity:

$$\text{Simpson's index of diversity} = \text{probability of picking two organisms at random that are different species}$$

$$= 1 - (\text{probability of picking two organisms that are the same species})$$

If a particular species i is represented in the community by p_i (proportion of individuals), the probability of picking two of these at random is the joint probability $[(p_i)(p_i)]$ or p_i^2. If we sum these probabilities for all the i species in the community, we get Simpson's diversity (D):

$$D = 1 - \overset{s}{\underset{i=1}{a}} (p_i)^2$$

where D = Simpson's index of diversity
p_i = proportion of individuals of species i in the community

For example, for our two-species community with 99 and 1 individuals,

$$D = 1 - [(0.99)^2 + (0.01)^2] = 0.02$$

Simpson's index gives relatively little weight to rare species and more weight to common species. It ranges in value from 0 (low diversity) to a maximum of $(1 - 1/S)$, where S is the number of species.

A more detailed discussion of these and other measures of species diversity is given in Magurran (2004) and Krebs (1999).

Credits

Illustration and Table Credits

T1 A. F. Hough, "A climax forest community on East Tionesta Creek in Northwestern Pennsylvania," *Ecology* 17:9–28, 1936.

Bibliography

Aanes, R., B.-E. Sæther, and N. A. Øritsland. 2000. Fluctuations of an introduced population of Svalbard reindeer: The effects of density dependence and climatic variation. *Ecography* 23:437–443.

Abrams, P. A. 1986. Adaptive responses of predators to prey and prey to predators: The failure of the arms-race analogy. *Evolution* 40:1229–1247.

Abrams, P. A. 1987. On classifying interactions between populations. *Oecologia* 73:272–281.

Abrams, P. A. 1998. High competition with low similarity and low competition with high similarity: Exploitative and apparent competition in consumer-resource systems. *American Naturalist* 152:114–128.

Acquaah, G. 2007. *Principles of Plant Genetics and Breeding*. Malden, MA: Blackwell Publishing. 569 pp.

Adams, B., A. White, and T. M. Lenton. 2004. An analysis of some diverse approaches to modelling terrestrial net primary productivity. *Ecological Modelling* 177:353–391.

Aguirre, A. A., and E. E. Starkey. 1994. Wildlife disease in U.S. National Parks: Historical and coevolutionary perspectives. *Conservation Biology* 8:654–661.

Ahlgren, I., T. Frisk, and L. Kamp-Nielsen. 1988. Empirical and theoretical models of phosphorus loading, retention and concentration vs. lake trophic status. *Hydrobiologia* 170:285–303.

Ainley, D. G., G. Ballard, and K. M. Dugger. 2006. Competition among penguins and cetaceans reveals trophic cascades in the western Ross Sea, Antarctica. *Ecology* 87:2080–2093.

Akçkaya, H. R., R. Arditi, and L. R. Ginzburg. 1995. Ratio-dependent predation: An abstraction that works. *Ecology* 76:995–1004.

Akselsson, C., O. Westling, H. Sverdrup, and P. Gundersen. 2007. Nutrient and carbon budgets in forest soils as decision support in sustainable forest management. *Forest Ecology and Management* 238:167–174.

Albani, M., D. Medvigy, G. C. Hurtt, and P. R. Moorcroft. 2006. The contributions of land-use change, CO_2 fertilization, and climate variability to the Eastern US carbon sink. *Global Change Biology* 12:2370–2390.

Alerstam, T. 1990. *Bird Migration*. Cambridge: Cambridge University Press. 420 pp.

Allee, W. C. 1931. *Animal Aggregations. A Study in General Sociology*. Chicago: University of Chicago Press. 431 pp.

Allen, A. P., and J. F. Gillooly. 2006. Assessing latitudinal gradients in speciation rates and biodiversity at the global scale. *Ecology Letters* 9:947–954.

Allen, K. R. 1980. *Conservation and Management of Whales*. Seattle: University of Washington Press. 107 pp.

Allen, M. F., W. Swenson, J. I. Querejeta, L. M. Egerton-Warburton, and K. K. Treseder. 2003. Ecology of mycorrhizae: A conceptual framework for complex interactions among plants and fungi. *Annual Review of Phytopathology* 41:271–303.

Alonso, D., R. S. Etienne, and A. J. McKane. 2006. The merits of neutral theory. *Trends in Ecology & Evolution* 21:451–457.

Altieri, M. A. 2004. Linking ecologists and traditional farmers in the search for sustainable agriculture. *Frontiers in Ecology and the Environment* 2:35–42.

Altieri, M. A., and C. I. Nicholls. 2004. *Biodiversity and Pest Management in Agroecosystems*. New York: Food Products Press. 236 pp.

Anderson, D. R. 2003. Index values rarely constitute reliable information. *Wildlife Society Bulletin* 31:288–291.

Anderson, G. R. V., A. H. Ehrlich, P. R. Ehrlich, J. D. Roughgarden, B. C. Russell, and F. H. Talbot. 1981a. The community structure of coral reef fishes. *American Naturalist* 117:476–495.

Anderson, G. S. 2001. Insect succession on carrion and its relationship to determining time since death. In *Forensic Entomology: The Utility of Arthropods in Legal Investigations*, (ed. J. H. Byrd and J. L. Castner), 143–175. Boca Raton, FL: CRC Press.

Anderson, K. G., H. Kaplan, and J. Lancaster. 1999. Paternal care by genetic fathers and stepfathers I: Reports from Albuquerque men. *Evolution and Human Behavior* 20:405–432.

Anderson, K. J. 2007. Temporal patterns in rates of community change during succession. *American Naturalist* 169:780–793.

Anderson, R. M. 1991. Populations and infectious diseases: Ecology or epidemiology? *Journal of Animal Ecology* 60:1–50.

Anderson, R. M., H. C. Jackson, R. M. May, and A. M. Smith. 1981b. Population dynamics of fox rabies in Europe. *Nature* 289:765–771.

Anderson, R. M., and R. M. May. 1978. Regulation and stability of host-parasite population interactions. *Journal of Animal Ecology* 47:219–247.

Anderson, R. M., and R. M. May. 1979. Population biology of infectious diseases: Part I. *Nature* 280:361–367.

Anderson, R. M., and R. M. May. 1980. Infectious diseases and population cycles of forest insects. *Science* 210:658–661.

Anderson, R. M., and R. M. May. 1982. Coevolution of hosts and parasites. *Parasitology* 85:411–426.

Anderson, R. M., and R. M. May. 1991. *Infectious Diseases of Humans: Dynamics and Control*. Oxford: Oxford University Press. 757 pp.

Anderson, S. 1985. The theory of range-size (RS) distributions. *American Museum Novitates* 2833:1–20.

Anderson, W. B., and G. A. Polis. 1998. Marine subsidies of island communities in the Gulf of California: evidence from stable carbon and nitrogen isotopes. *Oikos* 81: 75–80.

Andrewartha, H. G. 1961. *Introduction to the Study of Animal Populations*. Chicago: University of Chicago Press, 281 pp.

Andrewartha, H. G., and L. C. Birch. 1954. *The Distribution and Abundance of Animals*. Chicago: University of Chicago Press, 782 pp.

Anthony, R. G., E. D. Forsman, A. B. Franklin, D. R. Anderson, K. P. Burnham, G. C. White, C. J. Schwarz, et al. 2006. Status and trends in demography of northern spotted owls, 1985–2003. *Wildlife Monographs* 163:1–48.

Antonovics, J., A. D. Bradshaw, and R. G. Turner. 1971. Heavy metal tolerance in plants. *Advances in Ecological Research* 7:1–85.

Aplet, G. H., R. D. Laven, and F. W. Smith. 1988. Patterns of community dynamics in Colorado Engelmann spruce–subalpine fir forests. *Ecology* 69:312–319.

Applegate, V. C. 1950. Natural history of the sea lamprey, *Petromyzon marinus*, in Michigan. Special Scientific Report, Fisheries No. 55. Washington, DC: U.S. Fish and Wildlife Service.

Apps, C. D., B. N. McLennan, and J. G. Woods. 2006. Landscape partitioning and spatial inferences of competition between black and grizzly bears. *Ecography* 29:561–572.

Arditi, R., L. R. Ginzburg, and H. R. Akçakaya. 1991. Variation in plankton densities among lakes: A case for ratio-dependent predation models. *American Naturalist* 138:1287–1296.

Arita, H. T., P. Rodríguez, and E. Vázquez-Domínguez. 2005. Continental and regional ranges of North American mammals: Rapoport's rule in real and null worlds. *Journal of Biogeography* 32:961–971.

Armstrong, R. A., and R. McGehee. 1980. Competitive exclusion. *American Naturalist* 115:151–170.

Arrigo, K. R. 2005. Marine microorganisms and global nutrient cycles. *Nature* 437:349–355.

Arsenault, R., and N. Owen-Smith. 2002. Facilitation versus competition in grazing herbivore assemblages. *Oikos* 97:313–318.

Arthur, A., D. Ramsey, and M. Efford. 2004. Impact of bovine tuberculosis on a population of brushtail possums (*Trichosurus vulpecula* Kerr) in the Orongorongo Valley, New Zealand. *Wildlife Research* 31:389–395.

Ashley, R. M., D. Channon, D. Blackwood, and H. Smith. 2003. A new model for sewer rodent control. *International Pest Control* 45:183–188.

Ashmole, N. P. 1968. Body size, prey size, and ecological segregation in five sympatric tropical terns (Aves: Laridae). *Systematic Zoology* 17:292–304.

Åstrom, M., P. Lundberg, and S. Lundberg. 1996. Population dynamics with sequential density-dependencies. *Oikos* 75:174–181.

Atkinson, A., V. Siegel, E. Pakhomov, and P. Rothery. 2004. Long-term decline in krill stock and increase in salps within the Southern Ocean. *Nature* 432:100–103.

Attiwill, P. 1981. Energy, nutrient flow, and biomass. *Proceedings of the Australian Forest Nutrition Workshop*. Volume 1, CSIRO, Melbourne, pp. 131–144.

Austin, M. P. 1990. Community theory and competition in vegetation. In *Perspectives on Plant Competition*, (ed. J. B. Grace and D. Tilman), 215–238. San Diego: Academic Press.

Austin, M. P. 1999. A silent clash of paradigms: Some inconsistencies in community ecology. *Oikos* 86:170–178.

Bach, C. E. 1991. Direct and indirect interactions between ants (*Pheidole megacephala*), scales (*Coccus viridis*) and plants (*Pluchea indica*). *Oecologia* 87:233–239.

Bacon, P. J. 1985. A systems analysis of wildlife rabies epizootics. In *Population Dynamics of Rabies in Wildlife*, (ed. P. J. Bacon), 109–130. London: Academic Press.

Bailey, N. T. J. 1975. *The Mathematical Theory of Infectious Diseases and Its Application*. London: Griffin. 413 pp.

Bais, H. P., S.-W. Park, T. L. Weir, R. M. Callaway, and J. M. Vivanco. 2004. How plants communicate using the underground information superhighway. *Trends in Plant Science* 9:26–32.

Baker, C. S., and P. J. Clapham. 2004. Modelling the past and future of whales and whaling. *Trends in Ecology & Evolution* 19:365–371.

Balmford, A., R. E. Green, and J. P. W. Scharlemann. 2005. Sparing land for nature: Exploring the potential impact of changes in agricultural yield on the area needed for crop production. *Global Change Biology* 11:1594–1605.

Barbosa, P., and I. Castellanos. 2005. *Ecology of Predator-Prey Interactions*. New York: Oxford University Press, p. 394.

Barclay-Estrup, P., and C. H. Gimingham. 1969. The description and interpretation of cyclical processes in a heath community. I. Vegetational change in relation to the Calluna cycle. *Journal of Ecology* 57:737–758.

Barker, K. R. 2003. Integrated pest management toolbox. *Integrated Pest Management: Current and Future Strategies*. Ames, IA: Council for Agricultural Science and Technology, pp. 25–81.

Barlow, N. D. 1995. Critical evaluation of wildlife disease models. In *Ecology of Infectious Diseases in Natural Populations*, (ed. B. T. Grenfell and A. P. Dobson), 230–259. Cambridge, England: Cambridge University Press.

Barlow, N. D. 2000. Non-linear transmission and simple models for bovine tuberculosis. *Journal of Animal Ecology* 69:703–713.

Barnola, J. M., M. Anklin, J. Porcheron, D. Raynaud, J. Schwander, and B. Stauffer. 1995. CO_2 evolution during the last millennium as recorded by Antarctic and Greenland ice. *Tellus* 47B:264–272.

Barnosky, A. D., P. L. Koch, R. S. Feranec, S. L. Wing, and A. B. Shabel. 2004. Assessing the causes of Late Pleistocene extinctions on the continents. *Science* 306:70–75.

Bart, J., and E. D. Forsman. 1992. Dependence of northern spotted owls *Strix occidentalis caurina* on old-growth forests in the western USA. *Biological Conservation* 62:95–100.

Bast, M., and R. J. Reader. 2003. Regrowth response of young black spruce (*Picea mariana*) trees to meristem removal and resource addition. *Canadian Journal of Botany* 81:956–963.

Bayly, N. J. 2006. Optimality in avian migratory fuelling behaviour: A study of a trans-Saharan migrant. *Animal Behaviour* 71:173–182.

Bazzaz, F. A. 1996. *Plants in Changing Environments: Linking Physiological, Population, and Community Ecology.* Cambridge, England: Cambridge University Press. 320 pp.

Bazzaz, F. A., S. L. Bassow, G. M. Berntson, and S. C. Thomas. 1996. Elevated CO_2 and terrestrial vegetation: Implications for and beyond the global carbon budget. In *Global Change and Terrestrial Ecosystems*, (ed. B. Walker and W. Steffen), 43–76. Cambridge, England: Cambridge University Press.

Beaugrand, G., K. M. Brander, J. A. Lindley, S. Souissi, and P. C. Reid. 2003. Plankton effect on cod recruitment in the North Sea. *Nature* 426:661–664.

Behrenfeld, M. J., A. J. Bale, Z. S. Kolber, J. Aiken, and P. G. Falkowski. 1996. Confirmation of iron limitation of phytoplankton photosynthesis in the equatorial Pacific Ocean. *Nature* 383:508–511.

Behrenfeld, M. J., R. T. O'Malley, D. A. Siegel, C. R. McClain, J. L. Sarmiento, G. C. Feldman, A. J. Milligan, P. G. Falkowski, R. M. Letelier, and E. S. Boss. 2006. Climate-driven trends in contemporary ocean productivity. *Nature* 444:752–755.

Beier, P., and R. F. Noss. 1998. Do habitat corridors provide connec-

tivity? *Conservation Biology* 12:1241–1252.

Beissinger, S. R., and D. R. McCullough. 2002. *Population Viability Analysis.* Chicago: University of Chicago Press, p. 577.

Beitinger, T. L., W. A. Bennett, and R. W. McCauley. 2000. Temperature tolerances of North American freshwater fishes exposed to dynamic changes in temperature. *Environmental Biology of Fishes* 58:237–275.

Belayneh, Y. T. 2005. Acridid pest management in the developing world: A challenge to the rural population, a dilemma to the international community. *Journal of Orthoptera Research* 14:187–195.

Bell, G. 1980. The costs of reproduction and their consequences. *American Naturalist* 116:45–76.

Bell, G. 2000. The distribution of abundance in neutral communities. *American Naturalist* 155:606–617.

Bell, G. 2001. Neutral macroecology. *Science* 293:2413–2418.

Bell, R. H. V. 1971. A grazing ecosystem in the Serengeti. *Scientific American* 225:86–93.

Bellocq, M. I., and R. Gomez-Insausti. 2005. Raptorial birds and environmental gradients in the southern Neotropics: A test of species-richness hypotheses. *Austral Ecology* 30:900–906.

Belsky, A. J. 1986. Does herbivory benefit plants? A review of the evidence. *American Naturalist* 127:870–892.

Benedict, M. Q., and A. S. Robinson. 2003. The first releases of transgenic mosquitoes: An argument for the sterile insect technique. *Trends in Parasitology* 19:349–355.

Benkman, C. W. 1993. Adaptation to single resources and the evolution of crossbill (*Loxia*) diversity. *Ecological Monographs* 63:305–325.

Benkman, C. W., W. C. Holimon, and J. W. Smith. 2001. The influence of a competitor on the geographic mosaic of coevolution between crossbills and lodgepole pine. *Evolution* 55:282–294.

Bennett, W. A. 1990. Scale of investigation and the detection of competition: An example from the house sparrow and house finch intro-

ductions in North America. *American Naturalist* 135:725–747.

Bentley, M. D., and J. F. Day. 1989. Chemical ecology and behavioral aspects of mosquito oviposition. *Annual Review of Entomology* 34:401–421.

Berendse, F., and R. Aerts. 1987. Nitrogen-use-efficiency: A biologically meaningful definition? *Functional Ecology* 1:293–296.

Bergerud, A. T., W. J. Dalton, H. Butler, L. Camps, and R. Ferguson. 2007. Woodland caribou persistence and extirpation in relic populations on Lake Superior. *Rangifer Special Issue* 17:57–78.

Bergerud, A. T., and J. P. Elliot. 1998. Wolf predation in a multiple-ungulate system in northern British Columbia. *Canadian Journal of Zoology* 76:1551–1569.

Berryman, A. A. 1992. The origins and evolution of predator-prey theory. *Ecology* 73: 1530–1535.

Berryman, A. A. 2002. *Population Cycles: The Case for Trophic Interactions.* New York: Oxford University Press, 192 pp.

Berryman, A. A., M. L. Arce, and B. A. Hawkins. 2002. Population regulation, emergent properties, and a requiem for density dependence. *Oikos* 99:600–606.

Bertness, M. D., G. C. Trussell, P. J. Ewanchuk, and B. R. Silliman. 2002. Do alternate stable community states exist in the Gulf of Maine rocky intertidal zone? *Ecology* 83:3434–3448.

Berube, C. H., M. Festa-Bianchet, and J. T. Jorgenson. 1999. Individual differences, longevity, and reproductive senescence in bighorn ewes. *Ecology* 80:2555–2565.

Berumen, M., and M. Pratchett. 2006. Recovery without resilience: Persistent disturbance and long-term shifts in the structure of fish and coral communities at Tiahura Reef, Moorea. *Coral Reefs* 25:647–653.

Beurden, E. V. 1981. Bioclimatic limits to the spread of *Bufo marinus* in Australia: A baseline. *Proceedings of the Ecological Society of Australia* 1:143–149.

Beverton, R. J. H. 1962. Long-term dynamics of certain North Sea fish populations. In *The Exploitation of*

Natural Animal Populations, (ed. E. D. LeCren and M. W. Holdgate), 242–259. Oxford, England: Blackwell.

Beverton, R. J. H., and S. H. Holt. 1957. *On the Dynamics of Exploited Fish Populations*. London: H. M. Stationery Office, 533 pp.

Bhattarai, K. R., and O. R. Vetaas. 2006. Can Rapoport's rule explain tree species richness along the Himalayan elevation gradient, Nepal? *Diversity & Distributions* 12:373–378.

Biedermann, R. 2003. Body size and area-incidence relationships: Is there a general pattern? *Global Ecology & Biogeography* 12:381–387.

Biggins, D. E., J. L. Godbey, L. R. Hanebury, B. Luce, P. E. Marinari, R. Marc, and A. Vargas. 1998. The effect of rearing methods on survival of reintroduced black-footed ferrets. *Journal of Wildlife Management* 62:643–653.

Billings, W. D., J. O. Luken, D. A. Mortensen, and K. M. Peterson. 1983. Increasing atmospheric carbon dioxide: Possible effects on arctic tundra. *Oecologia* 58:286–289.

Billings, W. D., K. M. Peterson, J. O. Luken, and D. A. Mortensen. 1984. Interaction of increasing atmospheric carbon dioxide and soil nitrogen on the carbon balance of tundra microcosms. *Oecologia* 65:26–29.

Bininda-Emonds, O. R. P., M. Cardillo, K. E. Jones, R. D. E. MacPhee, R. M. D. Beck, R. Grenyer, S. A. Price, R. A. Vos, J. L. Gittleman, and A. Purvis. 2007. The delayed rise of present-day mammals. *Nature* 446:507–512.

Binkley, C. S., and R. S. Miller. 1983. Population characteristics of the whooping crane, *Grus americana*. *Canadian Journal of Zoology* 61:2768–2776.

Binkley, C. S., and R. S. Miller. 1988. Recovery of the whooping crane *Grus americana*. *Biological Conservation* 45:11–20.

Binkley, D., and P. Högberg. 1997. Does atmospheric deposition of nitrogen threaten Swedish forests? *Forest Ecology and Management* 92:119–152.

Binkley, D., Y. Son, and D. W. Valentine. 2000. Do forests receive occult inputs of nitrogen? *Ecosystems* 3:321–331.

Birch, L. C. 1948. The intrinsic rate of natural increase of an insect population. *Journal of Animal Ecology* 17:15–26.

Birch, L. C. 1953a. Experimental background to the study of the distribution and abundance of insects. I. The influence of temperature, moisture, and food on the innate capacity for increase of three grain beetles. *Ecology* 34:698–711.

Birch, L. C. 1953b. Experimental background to the study of the distribution and abundance of insects. III. The relations between innate capacity for increase and survival of different species of beetles living together on the same food. *Evolution* 7:136–144.

Birch, L. C. 1957. The meanings of competition. *American Naturalist* 91:5–18.

Björkman, O. 1975. Inaugural address. In: Marcell, R., ed. *Environmental and Biological Control of Photosynthesis*. The Hague: Dr. W. Junk Publishers, pp. 1–16.

Björkman, O., and J. Berry. 1973. High efficiency photosynthesis. *Scientific American* 229:80–93.

Black, C. C. 1971. Ecological implications of dividing plants into groups with distinct photosynthetic production capacities. *Advances in Ecological Research* 7:87–114.

Blackburn, T. M., P. Cassey, R. P. Duncan, K. L. Evans, and K. J. Gaston. 2004. Avian extinction and mammalian introductions on oceanic islands. *Science* 305:1955–1958.

Blake, S., S. Strindberg, P. Boudjan, C. Makombo, I. Bila-Isia, O. Ilambu, F. Grossmann, et al. 2007. Forest elephant crisis in the Congo Basin. *PLoS Biology* 5:0945–0953.

Blanc, J. J., R. F. W. Barnes, G. C. Craig, H. T. Dublin, C. R. Thouless, I. Douglas-Hamilton, and J. A. Hart. 2007. African Elephant Status Report 2007: An update from the African Elephant Database. *Occasional Paper Series of the IUCN Species Survival Commission* 33:1–276.

Blanchong, J. A., M. D. Samuel, D. R. Goldberg, D. J. Shadduck, and L. H. Creekmore. 2006. Wetland environmental conditions associated with the risk of avian cholera outbreaks and the abundance of *Pasteurella multocida*. *Journal of Wildlife Management* 70:54–60.

Blankenship, L. E., and L. A. Levin. 2007. Extreme food webs: Foraging strategies and diets of scavenging amphipods from the ocean's deepest 5 kilometers. *Limnology & Oceanography* 52:1685–1697.

Blossey, B. 1995. A comparison of various approaches for evaluating potential biological control agents using insects on *Lythrum salicaria*. *Biological Control* 5:113–122.

Blossey, B., L. C. Skinner, and J. Taylor. 2001. Impact and management of purple loosestrife (*Lythrum salicaria*) in North America. *Biodiversity and Conservation* 10:1787–1807.

Blumstein, D. T., J. C. Daniel, and B. P. Springett. 2004. A test of the multi-predator hypothesis: Rapid loss of antipredator behavior after 130 years of isolation. *Ethology* 110:919–934.

Bocherens, H., A. Argant, J. Argant, D. Billiou, E. Cregut-Bonnoure, B. Donat-Ayache, M. Philippe, and M. Thinon. 2004. Diet reconstruction of ancient brown bears (*Ursus arctos*) from Mont Ventoux (France) using bone collagen stable isotope biogeochemistry (^{13}C, ^{15}N). *Canadian Journal of Zoology* 82:576–586.

Boerema, L. K., and J. A. Gulland. 1973. Stock assessment of the peruvian anchovy (*Engraulis ringens*) and management of the fishery. *Journal of the Fisheries Research Board of Canada* 30:2226–2235.

Boisvenue, C., and S. W. Running. 2006. Impacts of climate change on natural forest productivity—evidence since the middle of the 20th century. *Global Change Biology* 12:862–882.

Bonesi, L., and D. W. Macdonald. 2004. Impact of released Eurasian otters on a population of American mink: A test using an experimental approach. *Oikos* 106:9–18.

Bonesi, L., R. Strachan, and D. W. Macdonald. 2006. Why are there fewer signs of mink in England? Considering multiple hypotheses. *Biological Conservation* 130:268–277.

Bongaarts, J., and R. A. Bulatao. 1999. Completing the demographic transition. *Population and Development Review* 25:515–529.

Borer, E. T., B. S. Halpern, and E. W. Seabloom. 2006. Asymmetry in community regulation: Effects of predators and productivity. *Ecology* 87:2813–2820.

Botkin, D. B. 1981. Causality and succession. In *Forest Succession*, (ed. D. C. West, H. H. Shugart, and D. B. Botkin), 36–55. New York: Springer-Verlag.

Bouzat, J. L., H. H. Cheng, H. A. Lewin, R. L. Westermeier, J. D. Brawn, and K. N. Paige. 1998a. Genetic evaluation of a demographic bottleneck in the greater prairie chicken. *Conservation Biology* 12:836–843.

Bouzat, J. L., H. A. Lewin, and K. N. Paige. 1998b. The ghost of genetic diversity past: Historical DNA analysis of the greater prairie chicken. *American Naturalist* 152:1–6.

Boveng, P. L., L. M. Hiruki, M. K. Schwartz, and J. L. Bengtson. 1998. Population growth of Antarctic fur seals: Limitation by a top predator, the leopard seal? *Ecology* 79:2863–2877.

Bowen, L., and D. Van Vuren. 1997. Insular endemic plants lack of defenses against herbivores. *Conservation Biology* 11:1249–1254.

Bowman, D. D. 2006. Successful and currently ongoing parasite eradication programs. *Veterinary Parasitology* 139:293–307.

Boyd, P. W., T. Jickells, C. S. Law, S. Blain, E. A. Boyle, K. O. Buesseler, K. H. Coale, et al. 2007. Mesoscale iron enrichment experiments 1993–2005: Synthesis and future directions. *Science* 315:612–617.

Bradford, J. B., J. A. Hicke, and W. K. Laurenroth. 2005. The relative importance of light-use efficiency modifications from environmental conditions and cultivation for estimation of large-scale net primary productivity. *Remote Sensing of Environment* 96:246–255.

Bradshaw, A. D., and K. Hardwick. 1989. Evolution and stress—genotypic and phenotypic components. *Biological Journal of the Linnean Society* 37:137–155.

Brady, K. U., A. R. Kruckeberg, and H. D. J. Bradshaw. 2005. Evolutionary ecology of plant adaptation to serpentine soils. *Annual Review of Ecology, Evolution, & Systematics* 36:243–266.

Brandt, J. P., H. F. Cerezke, K. I. Mallett, W. J. A. Volney, and J. D. Weber. 2003. Factors affecting trembling aspen (*Populus tremuloides* Michx.) health in the boreal forest of Alberta, Saskatchewan, and Manitoba, Canada. *Forest Ecology and Management* 178:287–300.

Brett, J. R. 1970. Temperature. In *Marine Ecology. Vol. I, Environmental Factors, Part I*, (ed. O. Kinne), 515–560. New York: Wiley-Interscience.

Bridgham, S. D., J. Pastor, C. A. McClaugherty, and C. J. Richardson. 1995. Nutrient-use efficiency: A litterfall index, a model, and a test along a nutrient-availability gradient in North Carolina peatlands. *American Naturalist* 145:1–21.

Brito, D., and C. Grelle. 2006. Estimating minimum area of suitable habitat and viable population size for the northern muriqui (*Brachyteles hypoxanthus*). *Biodiversity and Conservation* 15:4197–4210.

Brodie, E. D. III., and F. J. Janzen. 1995. Experimental studies of coral snake mimicry: Generalized avoidance of ringed snake patterns by free-ranging avian predators. *Functional Ecology* 9:186–190.

Brodie, E. D. III., and E. D. Brodie Jr. 1999. Predator-prey arms races. *BioScience* 49:557–568.

Bromham, L., and M. Cardillo. 2003. Testing the link between the latitudinal gradient in species richness and rates of molecular evolution. *Journal of Evolutionary Biology* 16:200–207.

Bronikowski, A. M., S. C. Alberts, J. Altmann, C. Packer, K. D. Carey, and M. Tatar. 2002. The aging baboon: Comparative demography in a non-human primate. *Proceedings of the National Academy of Sciences of the USA* 99:9591–9595.

Bronte, C. R., M. P. Ebener, D. R. Schreiner, D. S. Devault, M. M. Petzold, D. A. Jensen, C. Richards, and S. J. Lozano. 2003. Fish community change in Lake Superior, 1970–2000. *Canadian Journal of Fisheries & Aquatic Sciences* 60:1552–1574.

Brook, B. W., L. W. Traill, and C. J. A. Bradshaw. 2006. Minimum viable population sizes and global extinction risk are unrelated. *Ecology Letters* 9:375–382.

Brooks, J. L., and S. I. Dodson. 1965. Predation, body size, and composition of plankton. *Science* 150:28–35.

Brower, L. P. 1988. Avian predation on the monarch butterfly and its implications for mimicry theory. *American Naturalist* 131 Supp.: S4–S6.

Brown, G. P., B. L. Phillips, J. K. Webb, and R. Shine. 2006. Toad on the road: Use of roads as dispersal corridors by cane toads (*Bufo marinus*) at an invasion front in tropical Australia. *Biological Conservation* 133:88–94.

Brown, J. H. 1971. Mammals on mountaintops: Non-equilibrium insular biogeography. *American Naturalist* 105:467–478.

Brown, J. H. 1978. The theory of insular biogeography and the distribution of boreal birds and mammals. *Great Basin Naturalist Memoirs* 2:209–227.

Brown, J. H. 1981. Two decades of homage to Santa Rosalia: Toward a general theory of diversity. *American Zoologist* 21:877–888.

Brown, J. H. 1984. On the relationship between abundance and distribution of species. *American Naturalist* 124:255–279.

Brown, J. H. 1995. *Macroecology*. Chicago, Illinois: University of Chicago Press. 269 pp.

Brown, J. H., and D. W. Davidson. 1979. An experimental study of competition between seed-eating desert rodents and ants. *American Zoologist* 19:1129–1145.

Brown, J. H., J. F. Gillooly, A. P. Allen, V. M. Savage, and G. B. West. 2004. Toward a metabolic theory of ecology. *Ecology* 85:1771–1789.

Brown, J. H., and E. J. Heske. 1990. Control of a desert-grassland transition by a keystone rodent guild. *Science* 250:1705–1707.

Brown, J. H., and M. V. Lomolino. 1998. *Biogeography*. Sunderland, MA: Sinauer Associates. 691 pp.

Brown, J. H., and W. McDonald. 1995. Livestock grazing and conservation on southwestern rangelands. *Conservation Biology* 9:1644–1647.

Brown, J. H., and D. F. Sax. 2004. An essay on some topics concerning invasive species. *Austral Ecology* 29:530–536.

Brown, J. H., T. G. Whitham, S. K. M. Ernest, and C. A. Gehring. 2001. Complex species interactions and the dynamics of ecological systems: Long-term experiments. *Science* 293:643–650.

Brown, P. M. 2006. Climate effects on fire regimes and tree recruitment in Black Hills ponderosa pine forests. *Ecology* 87:2500–2510.

Brundrett, M., N. Bougher, B. Dell, T. Grove, and N. Malajczuk. 1996. *Working with Mycorrhizas in Forestry and Agriculture*. Canberra: Australian Centre for International Agricultural Research. 374 pp.

Brunkard, J. M., J. L. R. López, J. Ramirez, E. Cifuentes, S. J. Rothenberg, E. A. Hunsperger, C. G. Moore, et al. 2007. Dengue fever seroprevalence and risk factors, Texas–Mexico border, 2004. *Emerging Infectious Diseases* 13:1477–1483.

Brussard, P. F., and J. C. Tull. 2007. Conservation biology and four types of advocacy. *Conservation Biology* 21:21–24.

Buckle, A. P., and R. H. Smith. 1994. *Rodent Pests and Their Control*. Wallingford, England: CAB International, 405 pp.

Buckley, R. C. 1987. Interactions involving plants, homoptera, and ants. *Annual Review of Ecology and Systematics* 18:111–135.

Buckner, C. H., and W. J. Turnock. 1965. Avian predation on the larch sawfly, *Pristiphora erichsonii* (Htg.), (Hymenoptera: Tenthredinidae). *Ecology* 46:223–236.

Buesseler, K. O., J. E. Andrews, S. M. Pike, and M. A. Charette. 2004. The effects of iron fertilization on carbon sequestration in the Southern Ocean. *Science* 304:414–417.

Burbidge, A. A., and N. L. McKenzie. 1989. Patterns in the modern decline of Western Australia's vertebrate fauna: Causes and conservation implications. *Biological Conservation* 50:143–198.

Burger, C. V., K. T. Scribner, W. J. Spearman, C. O. Swanton, and D. E. Campton. 2000. Genetic contribution of three introduced life history forms of sockeye salmon to colonization of Frazer Lake, Alaska. *Canadian Journal of Fisheries and Aquatic Sciences* 57:2096–2111.

Burns, K. C. 2005. Is there limiting similarity in the phenology of fleshy fruits? *Journal of Vegetation Science* 16:617–624.

Busenberg, S. N., and K. L. Cooke. 1993. *Vertically Transmitted Diseases, Models and Dynamics*. Berlin: Springer-Verlag. 248 pp.

Buss, L. W. 1980. Competitive intransitivity and size-frequency distributions of interacting populations. *Proceedings of the National Academy of Science (USA)* 77:5355–5359.

Buss, L. W., and J. B. C. Jackson. 1979. Competitive networks: Nontransitive competitive relationships in cryptic coral reef environments. *American Naturalist* 113:223–234.

Cabana, G., and J. B. Rasmussen. 1994. Modelling food chain structure and contaminant bioaccumulation using stable nitrogen isotopes. *Nature* 372:255–257.

Cain, S. A. 1944. *Foundations of Plant Geography*. New York: Harper & Row, 556 pp.

Caley, M. J., and D. Schluter. 1997. The relationship between local and regional diversity. *Ecology* 78:70–80.

Caley, P. 2006. Bovine tuberculosis in brushtail possums: Models, dogma and data. *New Zealand Journal of Ecology* 30:25–34.

Caley, P., and J. Hone. 2004. Disease transmission between and within species, and the implications for disease control. *Journal of Applied Ecology* 41:94–104.

Callaghan, T. V., L. O. Björn, Y. Chernov, T. Chapin, T. R. Christensen, B. Huntley, R. A. Ims, M. Johansson, D. Jolly, S. Jonasson, N. Matveyeva, N. Panikov, W. Oechel, G. Shaver, and H. Henttonen. 2004. Effects on the structure of arctic ecosystems in the short- and long-term perspectives. *Ambio* 33:436–447.

Caltagirone, L. E., and R. L. Doutt. 1989. The history of the vedalia beetle importation to California and its impact on the development of biological control. *Annual Review of Entomology* 34:1–16.

Campbell, B. M. 1986. Plant spinescence and herbivory in a nutrient poor ecosystem. *Oikos* 47:168–172.

Canadell, J. G., C. Le Quere, M. R. Raupach, C. B. Field, E. T. Buitenhuis, P. Ciais, T. J. Conway, et al. 2007a. Contributions to accelerating atmospheric CO_2 growth from economic activity, carbon intensity, and efficiency of natural sinks. *Proceedings of the National Academy of Sciences of the USA* 104:18866–18870.

Canadell, J. G., D. E. Pataki, and L. F. Pitelka. 2007b. *Terrestrial Ecosystems in a Changing World*. Berlin: Springer, 336 pp.

Caputi, N., C. Chubb, R. Melville-Smith, A. Pearce, and D. Griffin. 2003. Review of relationships between life history stages of the western rock lobster, *Panulirus cygnus*, in Western Australia. *Fisheries Research* 65:47–61.

Carey, A. B., S. P. Horton, and B. L. Biswell. 1992. Northern spotted owls: Influence of prey-base and landscape character. *Ecological Monographs* 62:223–250.

Carey, C. 1996. Female reproductive energetics. In *Avian Energetics and Nutritional Ecology*, (ed. C. Carey), 324–374. London: Chapman and Hall.

Carey, J. R. 1993. *Applied Demography for Biologists With Special Emphasis on Insects*. New York: Oxford University Press.

Carey, J. R. 1995. Insect demography. In *Encyclopedia of Environmental Biology*, (ed. W. A. Nierenberg), 289–303. San Diego: Academic Press.

Carey, J. R., P. Liedo, D. Orozco, M. Tatar, and J. W. Vaupel. 1995. A male-female longevity paradox in medfly cohorts. *Journal of Animal Ecology* 64:107–116.

Cargill, S. M., and R. L. Jefferies. 1984. Nutrient limitation of primary production in a sub-arctic salt marsh. *Journal of Applied Ecology* 21:657–668.

Carpenter, F. L., D. C. Paton, and M. A. Hixon. 1983. Weight gain and adjustment of feeding territory size in migrant hummingbirds. *Proceedings of the National Academy of Sciences USA* 80:7259–7263.

Carpenter, S. R. 1996. Microcosm experiments have limited relevance for community and ecosystem ecology. *Ecology* 77:677–680.

Carpenter, S. R., N. F. Caraco, D. L. Correll, R. W. Howarth, A. N. Sharpley, and V. H. Smith. 1998. Nonpoint pollution of surface waters with phosphorus and nitrogen. *Ecological Applications* 8:559–568.

Carpenter, S. R., J. F. Kitchell, and J. R. Hodgson. 1985. Cascading trophic interactions and lake productivity. *BioScience* 35:634–639.

Carson, R. 1962. *Silent Spring*. Boston: Houghton Mifflin. 368 pp.

Carveth, C. J., A. M. Widmer, and S. A. Bonar. 2006. Comparison of upper thermal tolerances of native and nonnative fish species in Arizona. *Transactions of the American Fisheries Society* 135:1433–1440.

Casanova, C., and A. P. do Prado. 2002. Key-factor analysis of immature stages of *Aedes scapularis* (Diptera: Culicidae) populations in southeastern Brazil. *Bulletin of Entomological Research* 92:271–277.

Case, T. J. 1996. Global patterns in the establishment and distribution of exotic birds. *Biological Conservation* 78:69–96.

Caswell, H. 2001. *Matrix Population Models: Construction, Analysis, and Interpretation*. Sunderland, MA: Sinauer Associates. 722 pp.

Caughley, G. 1970. Eruption of ungulate populations, with emphasis on Himalayan thar in New Zealand. *Ecology* 51:53–72.

Caughley, G. 1976a. Wildlife management and the dynamics of ungu-

late populations. *Applied Biology* 1: 183–246.

Caughley, G. 1976b. Plant-herbivore systems. In *Theoretical Ecology*, (ed. R. M. May), 94–113. Philadelphia: Saunders.

Caughley, G. 1977. *Analysis of Vertebrate Populations*. London: Wiley. 234 pp.

Caughley, G. 1994. Directions in conservation biology. *Journal of Animal Ecology* 63:215–244.

Caughley, G., and A. Gunn. 1996. *Conservation Biology in Theory and Practice*. Oxford: Blackwell Science. 459 pp.

Caughley, G., and J. H. Lawton. 1981. Plant-herbivore systems. In *Theoretical Ecology*, (ed. R. M. May), 132–166. Oxford, England: Blackwell.

Caughley, G., N. Shepherd, and J. Short. 1987. *Kangaroos: Their Ecology and Management in the Sheep Rangelands of Australia*. Cambridge, England: Cambridge University Press. 253 pp.

Caughley, G., and A. R. E. Sinclair. 1994. *Wildlife Ecology and Management*. Boston: Blackwell Scientific Publications. 334 pp.

Celada, C., G. Bogliana, A. Gariboldi, and A. Maracci. 1994. Occupancy of isolated woodlots by the red squirrel *Sciurus vulgaris* L. in Italy. *Biological Conservation* 69:177–183.

Cerri, C. E. P., K. Paustian, M. Bernoux, R. L. Victoria, J. M. Melillo, and C. C. Cerri. 2004. Modeling changes in soil organic matter in Amazon forest to pasture conversion with the Century model. *Global Change Biology* 10:815–832.

Chadwick, O. A., L. A. Derry, P. M. Vitousek, B. J. Huebert, and L. O. Hedin. 1999. Changing sources of nutrients during four million years of ecosystem development. *Nature* 397:491–497.

Chapin, F. S., III. 1980. The mineral nutrition of wild plants. *Annual Review of Ecology and Systematics* 11:233–260.

Chapman, D. G. 1981. Evaluation of marine mammal population models. In *Dynamics of Large Mammal Populations*, (ed. C. W. Fowler and T. D. Smith), 277–296. New York: Wiley.

Chapman, R. N. 1928. The quantitative analysis of environmental factors. *Ecology* 9:111–122.

Charnov, E. L., and W. M. Schaffer. 1973. Life-history consequences of natural selection: Cole's result revisited. *American Naturalist* 107:791–793.

Chase, J. M., and M. A. Leibold. 2003. *Ecological Niches: Linking Classical and Contemporary Approaches*. Chicago: University of Chicago Press. 216 pp.

Chesson, P. L. 1986. Environmental variation and the coexistence of species. In *Community Ecology*, (ed. J. Diamond and T. J. Case), 240–256. New York: Harper & Row.

Chesson, P. L., and T. J. Case. 1986. Overview: Nonequilibrium community theories: Chance, variability, history, and coexistence. In *Community Ecology*, (ed. J. Diamond and T. J. Case), 229–239. New York: Harper & Row.

Chesson, P. L., and R. R. Warner. 1981. Environmental variability promotes coexistence in lottery competitive systems. *American Naturalist* 117:923–943.

Chetkiewicz, C.-L. B., C. C. St. Clair, and M. S. Boyce. 2006. Corridors for conservation: Integrating pattern and process. *Annual Review of Ecology, Evolution and Systematics* 37:317–342.

Chichilnisky, G., and G. Heal. 1998. Economic returns from the biosphere. *Nature* 391:629–630.

Chitty, D. 1960. Population processes in the vole and their relevance to general theory. *Canadian Journal of Zoology* 38:99–113.

Christou, P., T. Capell, A. Kohli, J. A. Gatehouse, and A. M. R. Gatehouse. 2006. Recent developments and future prospects in insect pest control in transgenic crops. *Trends in Plant Science* 11:302–308.

Chu, K. H., P. F. Tam, C. H. Fung, and Q. C. Chen. 1997. A biological survey of ballast water in container ships entering Hong Kong. *Hydrobiologia* 352:201–206.

Civeyrel, L., and D. Simberloff. 1996. A tale of two snails: Is the cure worse than the disease? *Biodiversity and Conservation* 5:1231–1252.

Clark, B. W., T. A. Phillips, and J. R. Coats. 2005. Environmental fate and effects of *Bacillus thuringiensis* (Bt) proteins from transgenic crops: A review. *Journal of Agricultural and Food Chemistry* 53:4643–4653.

Clark, C. W. 1990. *Mathematical Bioeconomics: The Optimal Management of Renewable Resources.* New York: Wiley, 386 pp.

Clark, C. W., and M. L. Rosenzweig. 1994. Extinction and colonization processes: Parameter estimates from sporadic surveys. *American Naturalist* 143:583–596.

Clark, J. S. 1998. Why trees migrate so fast: Confronting theory with dispersal biology and the paleorecord. *American Naturalist* 152:204–224.

Clark, J. S., M. Lewis, and L. Horvath. 2001. Invasion by extremes: Population spread with variation in dispersal and reproduction. *American Naturalist* 157:537–554.

Clark, R. G., and D. Shutler. 1999. Avian habitat selection: Pattern from process in nest-site use by ducks. *Ecology* 80:272–287.

Clarke, A. 1990. Temperature and evolution: Southern Ocean cooling and the antarctic marine fauna. In *Antarctic Ecosystems*, (ed. K. R. Kerry and G. Hempel), 9–22. Berlin: Springer-Verlag.

Clausen, C. P. 1956. *Biological Control of Insect Pests in the Continental United States.* Technical Bulletin No. 1139. Washington, DC: U.S. Department of Agriculture.

Clausen, J., D. D. Keck, and W. M. Hiesey. 1948. Experimental Studies on the Nature of Species. III. Environmental Responses of Climatic Races of *Achillea* Publication. No. 581. Washington, DC: Carnegie Institute of Washington.

Clavero, M., and E. Garciai-Berthou. 2005. Invasive species are a leading cause of animal extinctions. *Trends in Ecology & Evolution* 20:110.

Clements, F. E. 1916. *Plant Succession: An Analysis of the Development of Vegetation.* Washington, DC: Carnegie Institute of Washington, 512 pp.

Clements, F. E. 1936. Nature and structure of the climax. *Journal of Ecology* 24:252–284.

Clementz, M. T., K. A. Hoppe, and P. L. Koch. 2003. A paleoecological paradox: The habitat and dietary preferences of the extinct tethythere *Desmostylus*, inferred from stable isotope analysis *Paleobiology* 29:506–519.

Clinch, J. P. 1999. *Economics of Irish Forestry: Evaluating the Returns to Economy and Society.* Dublin: COFORD. 300 pp.

Clobert, J., E. Danchin, A. A. Dhondt, and J. D. Nichols. 2001. *Dispersal.* Oxford: Oxford University Press, 452 pp.

Clutton-Brock, T. H., S. D. Albon, and F. E. Guinness. 1989. Fitness costs of gestation and lactation in wild mammals. *Nature* 337:260–262.

Clutton-Brock, T. H., F. E. Guiness, and S. D. Albon. 1982. *Red Deer: Behavior and Ecology of Two Sexes.* Chicago: University of Chicago Press, 378 pp.

Coale, K. H., K. S. Johnson, F. P. Chavez, K. O. Buesseler, R. T. Barber, M. A. Brzezinski, W. P. Cochlan, et al. 2004. Southern Ocean Iron Enrichment Experiment: Carbon cycling in high- and low-Si waters. *Science* 304:408–414.

Coble, D. W., R. E. Brueswitz, T. W. Fratt, and J. W. Scheirer. 1990. Lake trout, sea lamprey, and overfishing in the upper Great Lakes: A review and reanalysis. *Transactions of the American Fishery Society* 119:985–995.

Cohen, A. A. 2004. Female post-reproductive lifespan: A general mammalian trait *Biological Reviews* 79:733–750.

Cohen, J. E. 1978. *Food Webs and Niche Space.* Princeton, NJ: Princeton University Press, 189 pp.

Cohen, J. E. 1994. Marine and continental food webs: Three paradoxes? *Philosophical Transactions of the Royal Society of London,* Series B 343:57–69.

Cohen, J. E. 1995a. *How Many People Can the Earth Support?* New York: W. W. Norton. 532 pp.

Cohen, J. E. 1995b. Population growth and earth's human carrying capacity. *Science* 269:341–346.

Cohen, J. E., and D. Tilman. 1996. Biosphere 2 and biodiversity: The lessons so far. *Science* 274:1150–1151.

Colautti, R. I., and H. J. MacIsaac. 2004. A neutral terminology to define 'invasive' species. *Diversity & Distributions* 10:135–141.

Cole, D. W., and M. Rapp. 1981. Elemental cycling in forest ecosystems. In *Dynamic Properties of Forest Ecosystems,* (ed. D. E. Reichle), 341–409. Cambridge, England: Cambridge University Press.

Cole, L. C. 1954. The population consequences of life history phenomena. *Quarterly Review of Biology* 29:103–137.

Cole, L. C. 1958. Sketches of general and comparative demography. *Cold Spring Harbor Symposia on Quantitative Biology* 22:1–15.

Coles, S., and E. Brown. 2007. Twenty-five years of change in coral coverage on a hurricane impacted reef in Hawaii: The importance of recruitment. *Coral Reefs* 26:705–717.

Coley, P. D. 1987. Interspecific variation in plant anti-herbivore properties: The role of habitat quality and rate of disturbance. *New Phytologist* 106:251–263.

Coley, P. D., J. P. Bryant, and F. S. Chapin III. 1985. Resource availability and plant antiherbivore defense. *Science* 230:895–899.

Colinvaux, P. 1973. *Introduction to Ecology.* New York: Wiley, 621 pp.

Collins, S. L., S. M. Glenn, and D. J. Gibson. 1995. Experimental analysis of intermediate disturbance and initial floristic composition: Decoupling cause and effect. *Ecology* 76:486–492.

Coltman, D. W., P. O'Donoghue, J. T. Jorgenson, J. T. Hogg, C. Strobeck, and M. Festa-Bianchet. 2003. Undesirable evolutionary consequences of trophy hunting. *Nature* 426:655–658.

Colvin, B. A., and W. B. Jackson. 1999. Urban rodent control programs for the 21st century. In *Ecologically-based Management of Rodent Pests,* (ed. G. R. Singleton, L. A. Hinds, H. Leirs, and Z. Zhang), 243–257. Canberra: Australian Centre for International Agricultural Research.

Condit, R. 1995. Research in large, long-term tropical forest plots.

Trends in Ecology and Evolution 10:18–22.

Condit, R., P. S. Ashton, H. Balslev, N. Brokaw, S. Bunyavejchewin, G. Chuyong, L. Co, et al. 2005. Tropical tree α-diversity: Results from a worldwide network of large plots. *Biologiske Skrifter* 55:565–582.

Connell, J. H. 1961a. Effects of competition, predation by *Thais lapillus*, and other factors on natural populations of the barnacle *Balanus balanoides*. *Ecological Monographs* 31:61–104.

Connell, J. H. 1961b. The influence of interspecific competition and other factors on the distribution of the barnacle *Chthamalus stellatus*. *Ecology* 42:710–732.

Connell, J. H. 1971. On the role of natural enemies in preventing competitive exclusion in some marine animals and in rain forest trees. In *Dynamics of Numbers in Populations*, (ed. P. J. den Boer and G. R. Gradwell), 298–312. Wageningen, Netherlands: Centre for Agricultural Publishing and Documentation.

Connell, J. H. 1978. Diversity in tropical rain forests and coral reefs. *Science* 199:1302–1310.

Connell, J. H. 1980. Diversity and the coevolution of competitors, or the ghost of competition past. *Oikos* 35:131–138.

Connell, J. H. 1987. Change and persistence in some marine communities. In *Colonization, Succession and Stability*, (ed. A. J. Gray, M. J. Crawley, and P. J. Edwards), 339–352. Oxford, England: Blackwell.

Connell, J. H. 1990. Apparent versus "real" competition in plants. In *Perspectives on Plant Competition*, (ed. J. B. Grace and D. Tilman), 9–26. San Diego: Academic Press.

Connell, J. H., T. P. Hughes, and C. C. Wallace. 1997. A 30-year study of coral abundance, recruitment, and disturbance at several scales in space and time. *Ecological Monographs* 67:461–488.

Connell, J. H., and R. O. Slatyer. 1977. Mechanisms of succession in natural communities and their role in community stability and organization. *American Naturalist* 111:1119–1144.

Connell, J. H., and W. P. Sousa. 1983. On the evidence needed to judge ecological stability or persistence. *American Naturalist* 121:789–824.

Connor, E. F., A. C. Courtney, and J. M. Yoder. 2000. Individuals-area relationships: The relationship between animal population density and area. *Ecology* 81:734–748.

Conover, D. O., and S. B. Munch. 2002. Sustaining fisheries yields over evolutionary time scales. *Science* 297:94–96.

Conroy, M. J., M. W. Miller, and J. E. Hines. 2002. Identification and synthetic modeling of factors affecting American black duck populations. *Wildlife Monographs*:1–64.

Cook, S. M., Z. R. Khan, and J. A. Pickett. 2007. The use of push-pull strategies in integrated pest management. *Annual Review of Entomology* 52:375–400.

Cooke, F., and C. S. Findlay. 1982. Polygenic variation and stabilizing selection in a wild population of lesser snow geese (*Anser caerulescens caerulescens*). *American Naturalist* 120:543–547.

Cooper, D. W., and E. Larsen. 2006. Immunocontraception of mammalian wildlife: Ecological and immunogenetic issues. *Reproduction* 132:821–828.

Copson, G., and J. Whinam. 1998. Response of vegetation on subantarctic Macquarie Island to reduced rabbit grazing. *Australian Journal of Botany* 46:15–24.

Copson, G., and J. Whinam. 2001. Review of ecological restoration programme on subantarctic Macquarie Island: Pest management progress and future directions. *Ecological Management and Restoration* 2:129–138.

Cory, J. S., and J. H. Myers. 2003. The ecology and evolution of insect baculoviruses. *Annual Review of Ecology, Evolution, and Systematics* 34:239–272.

Costanza, R., H. Daly, C. Folke, P. Hawken, C. S. Holling, A. J. McMichael, D. Pimentel, and D. J. Rapport. 2000. Managing our environmental portfolio. *BioScience* 50:149–155.

Costanza, R., R. Darge, R. de Groot, S. Farber, M. Grasso, B. Hannon, K. Limburg, et al. 1997. The value of the world's ecosystem services and natural capital. *Nature* 387:253–260.

Coupland, R. T. 1979. *Grassland ecosystems of the world: Analysis of grassland and their uses. International Biological Programme.* Cambridge, England: Cambridge University Press, 401 pp.

Cousins, S. 1985. Ecologists build pyramids again. *New Scientist* 106:50–54.

Cousins, S. 1987. The decline of the trophic level concept. *Trends in Ecology and Evolution* 2:312–316.

Cowles, H. C. 1899. The ecological relations of the vegetation on the sand dunes of Lake Michigan. *Botanical Gazette* 27:95–117, 167–202, 281–308, 361–391.

Craige, B. J. 2001. *Eugene Odum: Ecosystem Ecologist and Environmentalist.* Athens, GA: University of Georgia Press, 226 pp.

Cramer, W., D. W. Kicklighter, A. Bondeau, B. I. Moore, G. Churkina, B. Nemry, A. Ruimy, et al. 1999. Comparing global models of terrestrial net primary productivity (NPP): Overview and key results. *Global Change Biology* 5:1–15.

Crawley, M. J. 1986. The structure of plant communities. In *Plant Ecology*, (ed. M. J. Crawley), 1–50. Oxford, England: Blackwell.

Crawley, M. J. 1990. The population dynamics of plants. *Philosophical Transactions of the Royal Society of London*, Series B 330:125–140.

Crawley, M. J. 1997. *Plant Ecology.* Oxford: Blackwell Science, 717 pp.

Crews, T. E. 2005. Perennial crops and endogenous nutrient supplies. *Renewable Agriculture and Food Systems* 20:25–37.

Crombie, A. C. 1945. On competition between different species of graminivorous insects. *Proceedings of the Royal Society of London* 132:362–395.

Crouse, D. T., B. Crowder, and H. Caswell. 1987. A stage-based population model for loggerhead sea turtles and implications for conservation. *Ecology* 68:1412–1423.

Crow, J. F., and M. Kimura. 1970. *An Introduction to Population Genetics Theory.* New York: Harper & Row, 591 pp.

Crowcroft, P. 1991. *Elton's Ecologists: A History of the Bureau of Animal Population*. Chicago: University of Chicago Press. 177 pp.

Crowder, L. B., D. T. Crouse, S. S. Heppell, and T. H. Martin. 1994. Predicting the impact of turtle excluder devices on loggerhead sea turtle populations. *Ecological Applications* 4:437–445.

Cunnington, D. C., and R. J. Brooks. 1996. Bet-hedging theory and eigenelasticity: A comparison of the life histories of loggerhead sea turtles (*Caretta caretta*) and snapping turtles (*Chelydra serpentina*). *Canadian Journal of Zoology* 74:291–296.

Currie, D. J. 1991. Energy and large-scale patterns of animal- and plant-species richness. *American Naturalist* 137:27–49.

Currie, D. J., G. G. Mittelbach, H. V. Cornell, R. Field, J. F. Guégan, B. A. Hawkins, D. M. Kaufman, et al. 2004. Predictions and tests of climate-based hypotheses of broad-scale variation in taxonomic richness. *Ecology Letters* 7:1121–1134.

Currie, D. J., and V. Paquin. 1987. Large-scale biogeographical patterns of species richness of trees. *Nature* 329:326–327.

Cushing, D. H. 1990. Plankton production and year-class strength in fish populations: An update of the match/mismatch hypothesis. *Advances in Marine Biology* 26:249–292.

Cuthbert, R. 2002. The role of introduced mammals and inverse density-dependent predation in the conservation of Hutton's shearwater. *Biological Conservation* 108:69–78.

Cutler, A. 1991. Nested faunas and extinction in fragmented habitats. *Conservation Biology* 5:496–505.

Cyr, H., and M. L. Pace. 1993. Magnitude and patterns of herbivory in aquatic and terrestrial ecosystems. *Nature* 361:148–150.

Cyr, H., R. H. Peters, and J. A. Downing. 1997. Population density and community size structure: Comparison of aquatic and terrestrial systems. *Oikos* 80:139–149.

Dahl, K. 1919. Studies of trout and troutwaters in Norway. *Salmon and Trout Magazine*, 16–33.

Dale, V. H., C. M. Crisafulli, and F. J. Swanson. 2005. 25 years of ecological change at Mount St. Helens. *Science* 308:961–962.

Danell, K., and K. Huss-Danell. 1985. Feeding by insects and hares on birches earlier affected by moose browsing. *Oikos* 44:75–81.

Darlington, P. J., Jr. 1965. *Biogeography of the Southern End of the World*. Cambridge, MA: Harvard University Press, 236 pp.

Darwin, C. 1859. *The Origin of Species, by Means of Natural Selection or the Preservation of Favoured Races in the Struggle for Life*. London: John Murray. 458 pp.

Daubenmire, R. 1966. Vegetation: Identification of typal communities. *Science* 151:291–298.

Davidson, C., and R. A. Knapp. 2007. Multiple stressors and amphibian declines: Dual impacts of pesticides and fish on yellow-legged frogs. *Ecological Applications* 17:587–597.

Davis, M. B. 1986. Climatic instability, time lags, and community disequilibrium. In *Community Ecology*, (ed. J. Diamond and T. J. Case), 269–284. New York: Harper & Row.

Davis, S., M. Begon, L. De Bruyn, V. S. Ageyev, N. L. Klassovskly, S. B. Pole, H. Viljugrein, et al. 2004. Predictive thresholds for plague in Kazakhstan. *Science* 304:736–738.

Dawkins, R. 1982. *The Extended Phenotype*. Oxford, England: Oxford University Press, 307 pp.

Dawkins, R., and J. R. Krebs. 1979. Arms races between and within species. *Proceedings of the Royal Society of London*, Series B 205:489–511.

Dayan, T., and D. Simberloff. 2005. Ecological and community-wide character displacement: The next generation. *Ecology Letters* 8:875–894.

Dayton, P. K. 1971. Competition, disturbance, and community organization: The provision and subsequent utilization of space in a rocky intertidal community. *Ecological Monographs* 41:351–389.

Dayton, P. K. 2003. The importance of the natural sciences to conservation. *American Naturalist* 162:1–13.

Dayton, P. K., W. A. Newman, and J. Oliver. 1982. The vertical zonation of the deep-sea antarctic acorn barnacle, *Bathylasma corolliforme* (Hoek.): Experimental transplants from the shelf into shallow water. *Journal of Biogeography* 9:95–109.

de Blij, H. J., P. O. Muller, and R. S. Williams, Jr. 2004. *Physical Geography: The Global Environment*. New York: Oxford University Press. 702 pp.

de Graaff, M.-A., K.-J. van Groenigen, J. Six, B. Hungate, and C. van Kessel. 2006. Interactions between plant growth and soil nutrient cycling under elevated CO_2: A meta-analysis. *Global Change Biology* 12:2077–2091.

de Heij, M. E., P. J. van den Hout, and J. M. Tinbergen. 2006. Fitness cost of incubation in great tits (*Parus major*) is related to clutch size. *Proceedings of the Royal Society of London*, Series B 273:2353–2361.

de Lope, F., and A. P. Møller. 1993. Effects of ectoparasites on reproduction of their swallow hosts: A cost of being multi-brooded. *Oikos* 67:557–562.

de Wit, C. T. 1960. On competition. *Verslagen Landbouwkundige Onderzoekingen* 66:1–82.

DeAngelis, D. L. 1992. *Dynamics of Nutrient Cycling and Food Webs*. New York: Chapman & Hall. 270 pp.

DeAngelis, D. L., and J. C. Waterhouse. 1987. Equilibrium and nonequilibrium concepts in ecological models. *Ecological Monographs* 57:1–21.

Death, R. G., and M. J. Winterbourn. 1995. Diversity patterns in stream benthic invertebrate communities: The influence of habitat stability. *Ecology* 76:1446–1460.

DeBach, P. 1974. *Biological Control by Natural Enemies*. London: Cambridge University Press, 323 pp.

del Giorgio, P. A., J. J. Cole, N. F. Caraco, and R. H. Peters. 1999. Linking plankton biomass and metabolism to net gas fluxes in northern temperate lakes. *Ecology* 80:1422–1431.

del Moral, R. 2007. Limits to convergence of vegetation during early primary succession. *Journal of Vegetation Science* 18:479–488.

del Moral, R., and I. L. Lacher. 2005. Vegetation patterns 25 years after the eruption of Mount St. Helens, Washington, USA. *American Journal of Botany* 92:1948–1956.

del Moral, R., J. H. Titus, and A. M. Cook. 1995. Early primary succession on Mount St. Helens, Washington, USA. *Journal of Vegetation Science* 6:107–120.

del Moral, R., and D. M. Wood. 1993. Early primary succession on the volcano Mount St. Helens. *Journal of Vegetation Science* 4:223–234.

Delcourt, P. A., and H. R. Delcourt. 1987. *Long-Term Forest Dynamics of the Temperate Zone: A Case Study of Late-Quaternary Forest in Eastern North America.* New York: Springer-Verlag, 439 pp.

DeMott, W. R., and F. Moxter. 1991. Foraging on cyanobacteria by copepods: Responses to chemical defenses and resource abundance. *Ecology* 72:1820–1834.

Dennis, B., R. A. Desharnis, J. M. Cushing, and R. F. Costantino. 1997. Transitions in population dynamics: Equilibria to periodic cycles to aperiodic cycles. *Journal of Animal Ecology* 66:704–729.

Denoth, M., L. Frid, and J. H. Myers. 2002. Multiple agents in biological control: Improving the odds? *Biological Control* 24:20–30.

Dewar, R. C., and A. D. Watt. 1992. Predicted changes in synchrony of larval emergence and budburst under climatic warming. *Oecologia* 89:557–559.

Dhondt, A. A., F. Adriaensen, E. Matthysen, and B. Kempenaers. 1990. Nonadaptive clutch sizes in tits. *Nature* 348:723–725.

Diamond, J. 1986. Overview: Laboratory experiments, field experiments, and natural experiments. In *Community Ecology*, (ed. J. Diamond and T. J. Case), 3–22. New York: Harper and Row.

Diamond, J. 1989. Overview of recent extinctions. In *Conservation for the Twenty-first Century*, (ed. D. West-ern and M. Pearl), 37–41. New York: Oxford University Press.

Diamond, J. M. 1999. *Guns, Germs, and Steel: The Fates of Human Societies.* New York: W. W. Norton & Company. 480 pp.

Diamond, J. M., K. D. Bishop, and S. Van Balen. 1987. Bird survival in an isolated Javan woodland: Island or mirror? *Conservation Biology* 1:132–142.

Diaz, S., J. P. Grime, J. Harris, and E. McPherson. 1993. Evidence of a feedback mechanism limiting plant response to elevated carbon dioxide. *Nature* 364:616–617.

Dillon, P. J., and F. H. Rigler. 1975. A simple method for predicting the capacity of a lake for development based on lake trophic status. *Journal of the Fisheries Research Board of Canada* 32:1519–1531.

Dippery, J. K., D. T. Tissue, R. B. Thomas, and B. R. Strain. 1995. Effects of low and elevated CO_2 on C_3 and C_4 annuals. I. Growth and biomass allocation. *Oecologia* 101:13–20.

Dobson, A., and M. Meagher. 1996. The population dynamics of brucellosis in Yellowstone National Park. *Ecology* 77:1026–1036.

Dobson, A. P., A. D. Bradshaw, and A. J. M. Baker. 1997a. Hopes for the future: Restoration ecology and conservation biology. *Science* 277:515–522.

Dobson, A. P., J. P. Rodriguez, W. M. Roberts, and D. S. Wilcove. 1997b. Geographic distribution of endangered species in the United States. *Science* 275:550–553.

Dobzhansky, T. 1950. Evolution in the tropics. *American Scientist* 38:209–221.

Dodd, A.P. 1940. The biological campaign against prickly-pear. *Commonwealth Prickly Pear Board*, Brisbane.

Dodd, A. P. 1959. The biological control of prickly pear in Australia. In *Biogeography and Ecology in Australia*, (ed. A. Keast, R. L. Crocker, and C. S. Christian), 565–577. Den Haag: Dr. W. Junk.

Dodd, J. L., and W. K. Lauenroth. 1979. Analysis of the response of a grassland ecosystem to stress. In *Perspectives in Grassland Ecology*, (ed. N. R. French), 43–58. New York: Springer-Verlag.

Doering, P. H., C. A. Oviatt, B. L. Nowicki, E. G. Klos, and L. W. Reed. 1995. Phosphorus and nitrogen limitation of primary production in a simulated estuarine gradient. *Marine Ecology Progress Series* 124:271–287.

Doherty, P., and T. Fowler. 1994. An empirical test of recruitment limitation in a coral reef fish. *Science* 263:935–939.

Doka, S. E., D. K. McNicol, M. L. Mallory, I. Wong, C. K. Minns, and N. D. Yan. 2003. Assessing potential for recovery of biotic richness and indicator species due to changes in acidic deposition and lake pH in five areas of southeastern Canada. *Environmental Monitoring and Assessment* 88:53–101.

Doležal, J., M. Šrutek, T. Hara, A. Sumida, and T. Penttilä. 2006. Neighborhood interactions influencing tree population dynamics in non-pyrogenous boreal forest in northern Finland. *Plant Ecology* 185:135–150.

Donlan, C. J., J. Knowlton, D. F. Doak, and N. Biavaschi. 2005. Nested communities, invasive species and Holocene extinctions: Evaluating the power of a potential conservation tool. *Oecologia* 145:475–485.

Dormann, C. F., and S. J. Woodin. 2002. Climate change in the Arctic: Using plant functional types in a meta-analysis of field experiments. *Functional Ecology* 16:4–17.

Dornelas, M., S. R. Connolly, and T. P. Hughes. 2006. Coral reef diversity refutes the neutral theory of biodiversity. *Nature* 440:80–82.

Dornhaus, A., and L. Chittka. 2004. Why do honey bees dance? *Behavioral Ecology and Sociobiology* 55:395–401.

Downing, J. A., C. W. Osenberg, and O. Sarnelle. 1999. Meta-analysis of marine nutrient-enrichment experiments: Variation in the magnitude of nutrient limitation. *Ecology* 80:1157–1167.

Drenner, R. W., and A. Mazumder. 1999. Microcosm experiments have limited relevance for community and ecosystem ecology: Comment. *Ecology* 80:1081–1085.

Duan, K., L. G. Thompson, T. Yao, M. E. Davis, and E. Mosley-Thompson. 2007. A 1000 year history of atmospheric sulfate concentrations in southern Asia as recorded by a Himalayan ice core. *Geophysical Research Letters* 34(1): doi: 2006GL027456.

Dublin, L. I., and A. Lotka. 1925. On the true rate of natural increase as exemplified by the population of the United States, 1920. *Journal of the American Statistical Association* 20:305–339.

Dulvy, N. K., R. P. Freckleton, and N. V. C. Polunin. 2004. Coral reef cascades and the indirect effects of predator removal by exploitation. *Ecology Letters* 7:410–416.

Duncan, R., D. Forsyth, and J. Hone. 2007. Testing the metabolic theory of ecology: Allometric scaling exponents in mammals. *Ecology* 88:324–333.

Duncan, R. P. 1997. The role of competition and introduction effort in the success of Passeriform birds introduced to New Zealand. *American Naturalist* 149:903–915.

Dunne, J. A., R. J. Williams, and N. D. Martinez. 2004. Network structure and robustness of marine food webs. *Marine Ecology Progress Series* 273:291–302.

Durell, S. E. A., and R. T. Clarke. 2004. The buffer effect of non-breeding birds and the timing of farmland bird declines. *Biological Conservation* 120:375–382.

Dwyer, G., J. Dushoff, and S. H. Yee. 2004. The combined effects of pathogens and predators on insect outbreaks. *Nature* 430:341–345.

Dybas, C. L. 2006. On a collision course: Ocean plankton and climate change. *BioScience* 56:642–646.

Dyck, V. A., J. Hendrichs, and A. S. Robinson. 2005. *Sterile Insect Technique: Principles and Practice in Area-wide Integrated Pest Management*. Dordrecht, Netherlands: Springer, 787 pp.

Dyer, S. J., J. P. O'Neill, S. M. Wasel, and S. Boutin. 2001. Avoidance of industrial development by woodland caribou. *Journal of Wildlife Management* 65:531–542.

Ebenhard, T. 1988. Introduced birds and mammals and their ecological effects. *Swedish Wildlife Research* 13:1–107.

Eberhardt, L. L. 1988. Using age structure data from changing populations. *Journal of Applied Ecology* 25:373–378.

Eberhardt, L. L. 1998. Applying difference equations to wolf predation. *Canadian Journal of Zoology* 76:380–386.

Eberhardt, L. L. 2000. Reply: Predator-prey ratio dependence and regulation of moose populations. *Canadian Journal of Zoology* 78:511–513.

Ebert, D. 1998. Experimental evolution of parasites. *Science* 282:1432–1435.

Ebert, D. 1999. The evolution and expression of virulence. In *Evolution of Health and Disease*, (ed. S. C. Stearns), 161–172. Oxford: Oxford University Press.

Ebert, D., and J. J. Bull. 2003. Challenging the trade-off model for the evolution of virulence: Is virulence management feasible? *Trends in Microbiology* 11:15–20.

Echeverria, C., D. Coomes, J. Salas, J. M. Rey-Benayas, A. Lara, and A. Newton. 2006. Rapid deforestation and fragmentation of Chilean Temperate Forests. *Biological Conservation* 130:481–494.

Eckert, R., D. J. Randall, W. Burggren, and K. French. 1997. *Animal Physiology: Mechanisms and Adaptations*. New York: W. H. Freeman. 727 pp.

Edmondson, W. T. 1991. *The Uses of Ecology*. Seattle: University of Washington Press, 329 pp.

Egerton, F. N., III. 1968a. Ancient sources for animal demography. *Isis* 59:175–189.

Egerton, F. N., III. 1968b. Leeuwenhoek as a founder of animal demography. *Journal of the History of Biology* 1:1–22.

Egerton, F. N., III. 1968c. Studies of animal populations from Lamarck to Darwin. *Journal of the History of Biology* 1:225–259.

Egerton, F. N., III. 1969. Richard Bradley's understanding of biological productivity: A study of eighteenth-century ecological ideas. *Journal of the History of Biology* 2:391–410.

Egerton, F. N., III. 1973. Changing concepts of the balance of nature. *Quarterly Review of Biology* 48:322–350.

Egler, F. E. 1954. Vegetation science concepts. I. Initial floristic composition, a factor in old-field vegetation development. *Vegetatio* 14:412–417.

Ehleringer, J. R., T. E. Cerling, and M. D. Dearing. 2002. Atmospheric CO_2 as a global change driver influencing plant-animal interactions. *Integrative and Comparative Biology* 42:424–430.

Ehleringer, J. R., T. E. Cerling, and B. R. Helliker. 1997. C_4 photosynthesis, atmospheric CO_2 and climate. *Oecologia* 112:285–299.

Ehrlich, P. R., and A. Ehrlich. 1981. *Extinction: The Causes and Consequences of the Disappearance of Species*. New York: Random House. 305 pp.

Ehrlich, P. R., and A. H. Ehrlich. 2005. *One with Nineveh: Politics, Consumption, and the Human Future*. Washington, DC: Island Press. 465 pp.

Ehrlich, P. R., and P. H. Raven. 1964. Butterflies and plants: A study in coevolution. *Evolution* 18:586–608.

Elliott, J. M. 2004. Prey switching in four species of carnivorous stoneflies. *Freshwater Biology* 49:709–720.

Ellis, J. C., M. J. Shulman, M. Wood, J. D. Witman, and S. Lozyniak. 2007. Regulation of intertidal food webs by avian predators on New England rocky shores. *Ecology* 88:853–863.

Elser, J. J., M. E. S. Bracken, E. E. Cleland, D. S. Gruner, W. S. Harpole, H. Hillebrand, J. T. Ngai, et al. 2007. Global analysis of nitrogen and phosphorus limitation of primary producers in freshwater, marine and terrestrial ecosystems. *Ecology Letters* 10:1135–1142.

Elton, C. 1927. *Animal Ecology*. London: Sidgwick and Jackson. 209 pp.

Elton, C., and M. Nicholson. 1942. The ten-year cycle in numbers of the lynx in Canada. *Journal of Animal Ecology* 11:215–244.

Elton, C. S. 1958. *The Ecology of Invasions by Animals and Plants*. London: Methuen, 181 pp.

Elton, C. S. 1966. *The Pattern of Animal Communities*. London: Methuen, 432 pp.

Enright, J. T. 1976. Climate and population regulation: The biogeographer's dilemma. *Oecologia* 24:295–310.

Epstein, H. E., W. K. Lauenroth, I. C. Burke, and D. P. Coffin. 1997. Productivity patterns of C_3 and C_4 functional types in the U.S. Great Plains. *Ecology* 78:722–731.

Erickson, G. M., P. J. Currie, B. D. Inouye, and A. A. Winn. 2006. Tyrannosaur life tables: An example of nonavian dinosaur population biology. *Science* 313:213–217.

Eritja, R., R. Escosa, J. Lucientes, E. Marquès, D. Roiz, and S. Ruiz. 2005. Worldwide invasion of vector mosquitoes: Present European distribution and challenges for Spain. *Biological Invasions* 7:87–97.

Errington, P. L. 1963. *Muskrat Populations*. Ames: Iowa State University Press, 665 pp.

Essington, T. E., A. H. Beaudreau, and J. Wiedenmann. 2006. Fishing through marine food webs. *Proceedings of the National Academy of Sciences of the USA* 103:3171–3175.

Estes, J. A., and D. O. Duggins. 1995. Sea otters and kelp forests in Alaska: Generality and variation in a community ecological paradigm. *Ecological Monographs* 65:75–100.

Estes, J. A., M. T. Tinker, T. M. Williams, and D. F. Doak. 1998. Killer whale predation on sea otters linking oceanic and nearshore ecosystems. *Science* 282:473–476.

Ewald, P. W. 1995. The evolution of virulence: A unifying link between parasitology and ecology. *Journal of Parasitology* 81:659–669.

Ewers, R. M., A. D. Kliskey, S. Walker, D. Rutledge, J. S. Harding, and R. K. Didham. 2006. Past and future trajectories of forest loss in New Zealand. *Biological Conservation* 133:312–325.

Fagan, W. F., and L. E. Hurd. 1994. Hatch density variation of a generalist arthropod predator: Population consequences and community impact. *Ecology* 75:2022–2032.

Fahnestock, J. T., and J. K. Detling. 2002. Bison-prairie dog-plant interactions in a North American mixed-grass prairie. *Oecologia* 132:86–95.

Falkowski, P., R. T. Barber, and V. Smetacek. 1998. Biogeochemical controls and feedbacks on ocean primary production. *Science* 281:200–206.

Falkowski, P. G. 1997. Evolution of the nitrogen cycle and its influence on the biological sequestration of CO_2 in the ocean. *Nature* 387:272–275.

Fanshawe, S., G. R. Vanblaricom, and A. A. Shelly. 2003. Restored top carnivores as detriments to the performance of marine protected areas intended for fishery sustainability: A case study with red abalones and sea otters. *Conservation Biology* 17:273–283.

Fauth, J. E., and W. J. Resetarits, Jr. 1991. Interactions between the salamander *Siren intermedia* and the keystone predator *Notophthalmus viridescens*. *Ecology* 72:827–838.

Feeny, P. P. 1970. Seasonal changes in oak leaf tannins and nutrients as a cause of spring feeding by winter moth caterpillars. *Ecology* 51:565–581.

Feeny, P. P. 1976. Plant apparency and chemical defence. *Recent Advances in Phytochemistry* 10:1–40.

Feeny, P. P. 1992. The evolution of chemical ecology: Contributions from the study of herbivorous insects. In *Herbivores: Their Interactions with Secondary Plant Metabolites. Vol. II. Evolutionary and Ecological Processes*, (ed. G. A. Rosenthal and M. Berenbaum), 1–44. San Diego, CA: Academic Press.

Feinsinger, P. 1976. Organization of a tropical guild of nectivorous birds. *Ecological Monographs* 46:257–291.

Feller, M. C. 1980. Biomass and nutrient distribution in two eucalypt forest ecosystems. *Australian Journal of Ecology* 5:309–333.

Fenchel, T., and B. J. Finlay. 2004. The ubiquity of small species: Patterns of local and global diversity. *BioScience* 54:777–784.

Fenner, F., and K. Myers. 1978. Myxoma virus and myxomatosis in retrospect: The first quarter century of a new disease. In *Viruses and Environment*, (ed. E. Kurstak and K. Maramorosch), 539–570. New York: Academic Press.

Fenner, F., and F. N. Ratcliffe. 1965. *Myxomatosis*. Cambridge, England: Cambridge University Press. 379 pp.

Ferraz, G., J. D. Nichols, J. E. Hines, P. C. Stouffer, R. O. Bierregaard, Jr., and T. E. Lovejoy. 2007. A large-scale deforestation experiment: Effects of patch area and isolation on Amazon birds. *Science* 315:238–241.

Ferry, N., M. G. Edwards, J. Gatehouse, T. Capell, P. Christou, and A. M. R. Gatehouse. 2006. Transgenic plants for insect pest control: A forward looking scientific perspective. *Transgenic Research* 15:13–19.

Field, C. B., M. J. Behrenfeld, J. T. Randerson, and P. Falkowski. 1998. Primary production of the biosphere: Integrating terrestrial and oceanic components. *Science* 281:237–240.

Finegan, B. 1996. Pattern and process in neotropical secondary rain forests: the first 100 years of succession. *Trends in Ecology and Evolution* 11: 119–124.

Fischer, A. G. 1960. Latitudinal variations in organic diversity. *Evolution* 14:64–81.

Fisher, R. A., A. S. Corbet, and C. B. Williams. 1943. The relation between the number of species and the number of individuals in a random sample of an animal population. *Journal of Animal Ecology* 12:42–58.

Fleischner, T. L. 1994. Ecological costs of livestock grazing in western North America. *Conservation Biology* 8:629–644.

Fleishman, E., G. T. Austin, and A. D. Weiss. 1998. An empirical test of Rapoport's rule: Elevational gradients in montane butterfly communities. *Ecology* 79:2482–2493.

Flores, D., and J. W. Carlson. 2006. Biological control of giant salvinia in East Texas waterways and the impact on dissolved oxygen levels. *Journal of Aquatic Plant Management* 44:115–121.

Forero, M. G., J. L. Tella, K. A. Hobson, M. Bertellotti, and G. Blanco.

2002. Conspecific food competition explains variability in colony size: A test in Magellanic penguins. *Ecology* 83:3466–3475.

Forrester, G. E., R. R. Vance, and M. A. Steele. 2002. Simulating large-scale population dynamics using small-scale data. In *Coral Reef Fishes: Dynamics and Diversity in a Complex Ecosystem*, (ed. P. F. Sale), 275–301. Amsterdam: Elsevier.

Forsyth, D. M., and P. Caley. 2006. Testing the irruptive paradigm of large-herbivore dynamics. *Ecology* 87:297–303.

Forsyth, D. M., and R. P. Duncan. 2001. Propagule size and the relative success of exotic ungulate and bird introductions to New Zealand. *American Naturalist* 157:583–595.

Forsyth, D. M., R. P. Duncan, M. Bomford, and G. Moore. 2004. Climatic suitability, life-history traits, introduction effort, and the establishment and spread of introduced mammals in Australia. *Conservation Biology* 18:557–569.

Forsyth, D. M., J. Hone, J. P. Parkes, G. H. Reid, and D. Stronge. 2003. Feral goat control in Egmont National Park, New Zealand, and the implications for eradication. *Wildlife Research* 30:437–450.

Fowler, K., and L. Partridge. 1989. A cost of mating in female fruitflies. *Nature* 338:760–761.

Fox, C. W., D. A. Roff, and D. J. Fairbairn, eds. 2001. *Evolutionary Ecology: Concepts and Case Studies*. New York: Oxford University Press, 424 pp.

Fox, J. F. 1977. Alternation and coexistence of tree species. *American Naturalist* 111:69–89.

Fraenkel, G. S. 1959. The raison d'etre of secondary plant substances. *Science* 129:1466–1470.

Frank, D. A., and S. J. McNaughton. 1991. Stability increases with diversity in plant communities: Empirical evidence from the 1988 Yellowstone drought. *Oikos* 62:360–362.

Franklin, I. R. 1980. Evolutionary change in small populations. In *Conservation Biology: An Evolutionary-Ecological Perspective*, (ed. M. Soulé and B. A. Wilcox), 135–149. Sunderland, MA: Sinauer Associates.

Franzreb, K. E. 1997. Success of intensive management of a critically imperiled population of red-cockaded woodpeckers in South Carolina. *Journal of Field Ornithology* 68:458–470.

Fraser, L. H., and P. Keddy. 1997. The role of experimental microcosms in ecological research. *Trends in Ecology and Evolution* 12:478–481.

Fraser, R. H., and D. J. Currie. 1996. The species richness-energy hypothesis in a system where historical factors are thought to prevail: Coral reefs. *American Naturalist* 148:138–159.

Freed, L. A., R. L. Cann, M. L. Goff, W. A. Kuntz, and G. R. Bodner. 2005. Increase in avian malaria at upper elevation in Hawaii. *Condor* 107:753–764.

French, N. R., ed. 1979. *Perspectives in Grassland Ecology*. New York: Springer-Verlag, 204 pp.

French, N. R., R. K. Steinhorst, and D. M. Swift. 1979. Grassland biomass trophic pyramids. In: French, N. R., ed. *Perspectives in Grassland Ecology*. New York: Springer-Verlag, pp. 59–87.

Fretwell, S. D. 1972. *Populations in a Seasonal Environment*. Princeton, NJ: Princeton University Press. 217 pp.

Friäriksson, S. 1987. Plant colonization of a volcanic island, Surtsey, Iceland. *Arctic and Alpine Research* 19:425–431.

Fryxell, J. M., and A. R. E. Sinclair. 1988. Causes and consequences of migration by large herbivores. *Trends in Ecology and Evolution* 3:237–241.

Fuentes, M. 2004. Slight differences among individuals and the unified neutral theory of biodiversity. *Theoretical Population Biology* 66:199–203.

Futuyma, D. J. 2005. *Evolution*. Sunderland, MA: Sinauer Associates. 603 pp.

Gagan, M. K., L. K. Ayliffe, J. W. Beck, J. E. Cole, E. R. M. Druffel, R. B. Dunbar, and D. P. Schrag. 2000. New views of tropical paleoclimates from corals. *Quaternary Science Reviews* 19:45–64.

Gagan, M. K., L. K. Ayliffe, D. Hopley, J. A. Cali, G. E. Mortimer, J. Chappell, M. T. McCulloch, and M. J. Head. 1998. Temperature and surface-ocean water balance of the Mid-Holocene tropical western Pacific. *Science* 279:1014–1018.

Gaillard, J.-M., M. Festa-Bianchet, and N. G. Yoccoz. 1998. Population dynamics of large herbivores: Variable recruitment with constant adult survival. *Trends in Ecology and Evolution* 13:58–63.

Gaines, S. D., and J. Lubchenco. 1982. A unified approach to marine plant-herbivore interactions. II. Biogeography. *Annual Review of Ecology and Systematics* 13:111–138.

Garcia-Berthou, E., C. Alcaraz, Q. Pou-Rovira, L. Zamora, G. Coenders, and C. Feo. 2005. Introduction pathways and establishment rates of invasive aquatic species in Europe. *Canadian Journal of Fisheries & Aquatic Sciences* 62:453–463.

Gascoigne, J. C., and R. N. Lipcius. 2004. Allee effects driven by predation. *Journal of Applied Ecology* 41:801–810.

Gaston, K. 1994. Measuring geographic range sizes. *Ecography* 17:198–205.

Gaston, K. J. 1988. Patterns in the local and regional dynamics of moth populations. *Oikos* 53:49–59.

Gaston, K. J. 1990. Patterns in the geographical ranges of species. *Biological Reviews* 65:105–129.

Gaston, K. J. 1991. How large is a species' geographic range? *Oikos* 61:434–438.

Gaston, K. J. 2003. *The Structure and Dynamics of Geographic Ranges*. Oxford: Oxford University Press. 266 pp.

Gaston, K. J., and T. M. Blackburn. 1996. Global scale macroecology: Interactions between population size, geographic range size and body size in the Anseriformes. *Journal of Animal Ecology* 65:701–714.

Gaston, K. J., T. M. Blackburn, and J. H. Lawton. 1997. Interspecific abundance-range size relationships: An appraisal of mechanism. *Journal of Animal Ecology* 66:579–601.

Gaston, K. J., T. M. Blackburn, and J. L. Spicer. 1998. Rapoport's rule: Time for an epitaph. *Trends in Ecology and Evolution* 13:70–74.

Gaston, K. J., and J. L. Curnutt. 1998. The dynamics of abundance—range size relationships. *Oikos* 81:38–44.

Gaston, K. J., and R. A. Fuller. 2008. Commonness, population depletion and conservation biology. *Trends in Ecology & Evolution* 23:14–19.

Gause, G. F. 1932. Experimental studies on the struggle for existence. I. Mixed population of two species of yeast. *Journal of Experimental Biology* 9:389–402.

Gause, G. F. 1934. *The Struggle for Existence.* New York: Macmillan (Hafner Press), 163 pp.

Gavin, D. G., and F. S. Hu. 2006. Spatial variation of climatic and non-climatic controls on species distribution: The range limit of *Tsuga heterophylla. Journal of Biogeography* 33:1384–1396.

Gehring, C. A., and T. G. Whitham. 1994. Comparisons of ectomycorrhizae on pinyon pines (*Pinus edulis*; Pinaceae) across extremes of soil type and herbivory. *American Journal of Botany* 81:1509–1516.

Gelbard, A., C. Haub, and M. M. Kent. 1999. World population beyond six billion. *Population Bulletin* 54:1–44.

Gifford, R. M. 2003. Plant respiration in productivity models: Conceptualisation, representation and issues for global terrestrial carbon-cycle research. *Functional Plant Biology* 30:171–186.

Gilbert, G. S., S. P. Hubbell, and R. B. Foster. 1994. Density and distance-to-adult effects of a canker disease of trees in a moist tropical forest. *Oecologia* 98:100–108.

Gilchrist, H. G. 1999. Declining thick-billed murre *Uria lomvia* colonies experience higher gull predation rates: An inter-colony comparison. *Biological Conservation* 87:21–29.

Gill, B., and P. Martinson. 1991. *New Zealand's Extinct Birds.* Auckland, New Zealand: Random Century. 109 pp.

Gill, D. E. 1974. Intrinsic rate of increase, saturation density, and competitive ability. II. The evolution of competitive ability. *American Naturalist* 108:103–116.

Gilliam, F. S. 2006. Response of the herbaceous layer of forest ecosystems to excess nitrogen deposition. *Journal of Ecology* 94:1176–1191.

Gillson, L. 2004. Testing non-equilibrium theories in savannas: 1400 years of vegetation change in Tsavo National Park, Kenya. *Ecological Complexity* 1:281–298.

Gilpin, M. E., and F. J. Ayala. 1973. Global models of growth and competition. *Proceedings of the National Academy of Sciences of the USA* 70:3590–3593.

Gilpin, M. E., and M. E. Soulé. 1986. Minimum viable populations: Processes of species extinction. In *Conservation Biology*, (ed. M. E. Soulé), 19–34. Sunderland, MA: Sinauer Associates.

Gimeno, I., M. Vila, and P. E. Hulme. 2006. Are islands more susceptible to plant invasions than continents? A test using *Oxalis pes-caprae* L. in the western Mediterranean. *Journal of Biogeography* 33:1559–1565.

Gintis, H., S. Bowles, R. Boyd, and E. Fehr. 2003. Explaining altruistic behavior in humans. *Evolution and Human Behavior* 24:153–172.

Givnish, T. J. 1988. Adaptation to sun and shade: A whole-plant perspective. *Australian Journal of Plant Physiology* 15:63–92.

Glazier, D. S. 2005. Beyond the '3/4-power law': Variation in the intra- and interspecific scaling of metabolic rate in animals. *Biological Reviews* 80:611–662.

Glen, A. S., C. R. Dickman, M. E. Soulé, and B. G. Mackey. 2007. Evaluating the role of the dingo as a trophic regulator in Australian ecosystems. *Austral Ecology* 32:492–501.

Gliwicz, Z. M. 1990. Food thresholds and body size in cladocerans. *Nature* 343:638–640.

Goldman, C. R. 1968. Aquatic primary production. *American Zoologist* 8:31–42.

Goldman, C. R. 1988. Primary productivity, nutrients, and transparency during the early onset of eutrophication in ultra-oligotrophic Lake Tahoe, California-Nevada. *Limnology and Oceanography* 33:1321–1333.

Goldstein, J. R. 2002. Population momentum for gradual demographic transitions: An alternative approach. *Demography* 39:65–73.

Golley, F. B. 1961. Energy values of ecological materials. *Ecology* 42:581–584.

Gomez, J. M., and R. Zamora. 2002. Thorns as induced mechanical defense in a long-lived shrub (*Hormathophylla spinosa*, Cruciferae). *Ecology* 83:885–890.

Gonzalez Sagrario, M. A., E. Jeppesen, J. Goma, M. Sondergaard, J. P. Jensen, T. Lauridsen, and F. Landkildehus. 2005. Does high nitrogen loading prevent clear-water conditions in shallow lakes at moderately high phosphorus concentrations? *Freshwater Biology* 50:27–41.

Good, R. 1964. *The Geography of the Flowering Plants.* London: Longman's. 518 pp.

Gordon, H. S. 1954. The economic theory of a common property resource: The fishery. *Journal of Political Economics* 62:124–142.

Gotelli, N. J. 1998. *A Primer of Ecology.* Sunderland, MA: Sinauer Associates. 236 pp.

Gough, K. F., and G. I. H. Kerley. 2006. Demography and population dynamics in the elephants *Loxodonta africana* of Addo Elephant National Park, South Africa: Is there evidence of density dependent regulation? *Oryx* 40:434–441.

Gould, J. L. 1975. Honey bee recruitment: The dance-language controversy. *Science* 189:685–693.

Gould, S. J. 1981. Palaeontology plus ecology as palaebiology. In *Theoretical Ecology*, (ed. R. M. May), 295–317. Oxford, England: Blackwell.

Goulden, C. E., and L. L. Hornig. 1980. Population oscillations and energy reserves in planktonic Cladocera and their consequences to competition. *Proceedings of the National Academy of Sciences of the USA* 77:1716–1720.

Goward, S. N., C. J. Tucker, and D. G. Dye. 1985. North American vegetation patterns observed with the NOAA-7 advanced very high resolution radiometer. *Vegetatio* 64:3–14.

Gowda, J. H. 1996. Spines of *Acacia tortilis*: What do they defend and how? *Oikos* 77:279–284.

Gower, S. T., R. E. McMurtrie, and D. Murty. 1996. Aboveground net primary production decline with stand age: Potential causes. *Trends in Ecology & Evolution* 11:378–382.

Grace, J. B. 1995. In search of the Holy Grail: Explanations for the coexistence of plant species. *Trends in Ecology & Evolution* 10:263–264.

Graetz, R. D., R. Fisher, and M. Wilson. 1992. *Looking Back: The Changing Face of the Australian Continent 1972–1992*. Canberra, Australia: CSIRO, 159 pp.

Grafton-Cardwell, E. E., and P. Gu. 2003. Conserving vedalia beetle, *Rodolia cardinalis* (Mulsant) (Coleoptera: Coccinellidae), in citrus: A continuing challenge as new insecticides gain registration. *Journal of Economic Entomology* 96:1388–1398.

Graham, M. 1935. Modern theory of exploiting a fishery, and application to North Sea trawling. *Journal du Conseil Permanente International pour l'Exploration de la Mer* 10:264–274.

Graham, M. 1939. The sigmoid curve and the overfishing problem. *Rapport du Conseil International pour l'Exploration de la Mer* 110:15–20.

Granett, J., M. A. Walker, L. Kocsis, and A. D. Omer. 2001. Biology and management of grape phylloxera. *Annual Review of Entomology* 46:387–412.

Grant, B. R., and L. L. Wiseman. 2002. Recent history of melanism in American peppered moths. *Journal of Heredity* 93:86–90.

Grant, B. S. 2005. Industrial melanism. *Encyclopedia of Life Sciences.* Chichester, England: John Wiley & Sons, Ltd., pp. 1–9.

Grant, P. R. 1986. *Ecology and Evolution of Darwin's Finches*. Princeton, NJ: Princeton University Press. 484 pp.

Grant, P. R., and B. R. Grant. 2002. Adaptive radiation of Darwin's finches. *American Scientist* 90:130–139.

Grant, P. R., and B. R. Grant. 2006. Evolution of character displacement in Darwin's finches. *Science* 313:224–226.

Grant, P. R., and J. Weiner. 2000. *Ecology and Evolution of Darwin's Finches*. Princeton, NJ: Princeton University Press. 512 pp.

Gras, E. K., J. Read, C. T. Mach, G. D. Sanson, and F. J. Clissold. 2005. Herbivore damage, resource richness and putative defences in juvenile versus adult Eucalyptus leaves. *Australian Journal of Botany* 53:33–44.

Green, R. E. 1997. The influence of numbers released on the outcome of attempts to introduce exotic bird species to New Zealand. *Journal of Animal Ecology* 66:25–35.

Greene, H. W., and R. W. McDiarmid. 1981. Coral snake mimicry: does it occur? *Science* 213:1207–1212.

Greenleaf, S. S., and C. Kremen. 2006. Wild bee species increase tomato production and respond differently to surrounding land use in Northern California. *Biological Conservation* 133:81–87.

Grenfell, B. T., and A. P. Dobson. 1995. *Ecology of Infectious Diseases in Natural Populations*. Cambridge, England: Cambridge University Press, 521 pp.

Griffiths, D. 2006. The direct contribution of fish to lake phosphorus cycles. *Ecology of Freshwater Fish* 15:86–95.

Grim, T. 2006. The evolution of nestling discrimination by hosts of parasitic birds: Why is rejection so rare? *Evolutionary Ecology Research* 8:785–802.

Grime, J. P. 1973. Competitive exclusion in herbaceous vegetation. *Nature* 242:344–347.

Grime, J. P. 1979. *Plant Strategies and Vegetation Processes*. New York: John Wiley and Sons, 222 pp.

Grime, J. P. 1997a. Biodiversity and ecosystem function: The debate deepens. *Science* 277:1260–1261.

Grime, J. P. 1997b. Climate change and vegetation. In *Plant Ecology*, (ed. M. J. Crawley), 582–594. Oxford: Blackwell Science.

Grubb, P. J. 1987. Global trends in species-richness in terrestrial vegetation: A view from the Northern Hemisphere. In *Organization of Communities Past and Present*, (ed. J. H. R. Gee and P. S. Giller), 99–118. Oxford, England: Blackwell.

Guillet, C., and R. Bergström. 2006. Compensatory growth of fast-growing willow (*Salix*) coppice in response to simulated large herbivore browsing. *Oikos* 113:33–42.

Gulland, F. M. D. 1995. The impacts of infectious diseases on wild animal populations—a review. In *Ecology of Infectious Diseases in Natural Populations*, (ed. B. T. Grenfell and A. P. Dobson), 20–51. Cambridge, England: Cambridge University Press.

Gundersen, P., B. A. Emmett, O. J. Kjonaas, C. J. Koopmans, and A. Tietema. 1998. Impact of nitrogen deposition on nitrogen cycling in forests: A synthesis of NITREX data. *Forest Ecology and Management* 101:37–55.

Gunn, J. M., and K. H. Mills. 1998. The potential for restoration of acid-damaged lake trout lakes. *Restoration Ecology* 6:390–397.

Gurevitch, J., J. A. Morrison, and L. V. Hedges. 2000. The interaction between competition and predation: A meta-analysis of field experiments. *American Naturalist* 155:435–453.

Gurevitch, J., L. L. Morrow, A. Wallace, and J. S. Walsh. 1992. A meta-analysis of competition in field experiments. *American Naturalist* 140:539–572.

Gurr, G. M., S. D. Wratten, and M. A. Altieri. 2004. *Ecological Engineering for Pest Management: Advances in Habitat Manipulation for Arthropods*. Collingwood, Victoria: CSIRO Press, 232 pp.

Gwynne, M. D., and R. H. Bell. 1968. Selection of vegetation components by grazing ungulates in the Serengeti National Park. *Nature* 220:390–393.

Hairston, N. G. 1991. The literature glut: Causes and consequences (reflections of a dinosaur). *Bulletin of the Ecological Society of America* 72:171–174.

Hairston, N. G., F. E. Smith, and L. B. Slobodkin. 1960. Community structure, population control, and competition. *American Naturalist* 94:421–425.

Hairston, N. G. J., and N. G. S. Hairston. 1993. Cause-effect relationships in energy flow, trophic structure, and interspecific interactions. *American Naturalist* 142:379–411.

Hall, R. W., and L. E. Ehler. 1979. Rate of establishment of natural enemies in classical biological control. *Bulletin of the Entomological Society of America* 25:280–282.

Hall, R. W., L. E. Ehler, and B. Bisabri-Ershadi. 1980. Rate of success in classical biological control of arthropods. *Bulletin of the Entomological Society of America* 26:111–114.

Hall, S. J., and D. G. Raffaelli. 1993. Food webs: Theory and reality. *Advances in Ecological Research* 24:187–239.

Hall, S. J., and D. G. Raffaelli. 1997. Food-web patterns: What do we really know? In *Multitrophic Interactions in Terrestrial Systems,* (ed. A. C. Gange and V. K. Brown), 395–417. London: Blackwell Science.

Hall, S. R., J. B. Shurin, S. Diehl, and R. M. Nisbet. 2007. Food quality, nutrient limitation of secondary production, and the strength of trophic cascades. *Oikos* 116:1128–1143.

Halls, C. 2006. *Living Planet Report 2006.* Gland, Switzerland: World Wildlife Fund and Global Footprint Network, 44 pp.

Halpern, B. S. 2003. The impact of marine reserves: Do reserves work and does reserve size matter? *Ecological Applications* 13: S117–S137.

Hamilton, N. R., M. C. Sackville, and G. Lemaire. 1995. In defence of the –3/2 boundary rule: A re-evaluation of self-thinning concepts and status. *Annals of Botany* 76:569–577.

Hamilton, W. D. 1971. Geometry of the selfish herd. *Journal of Theoretical Biology* 31:295–311.

Hanski, I. 1982. Dynamics of regional distribution: The core and satellite species hypothesis. *Oikos* 38:210–221.

Hanski, I. 1998. Metapopulation dynamics. *Nature* 396:41–49.

Hanski, I., J. Koukli, and A. Halkka. 1993. Three explanations of the positive relationship between distribution and abundance of species. In *Species Diversity in Ecological Communities,* (ed. R. Ricklefs and D. Schulter), 108–116. Chicago: University of Chicago Press.

Hanski, I., M. Kuussaari, and M. Nieminen. 1994. Metapopulation structure and migration in the butterfly *Melitaea cinxia. Ecology* 75:747–762.

Hanski, I., A. Moilanen, T. Pakkala, and M. Kuussaari. 1996. The quantitative incidence function model and persistence of an endangered butterfly metapopulation. *Conservation Biology* 10:578–590.

Hanski, I., T. Pakkala, M. Kuussaari, and G. Lei. 1995. Metapopulation persistence of an endangered butterfly in a fragmented landscape. *Oikos* 72:21–28.

Harcourt, A. H., S. A. Coppeto, and S. A. Parks. 2005. The distribution-abundance (density) relationship: its form and causes in a tropical mammal order, Primates. *Journal of Biogeography* 32:565–579.

Hardin, G. 1960. The competitive exclusion principle. *Science* 131:1292–1297.

Hardin, G. 1968. The tragedy of the commons. *Science* 162:1243–1248.

Hardy, C. M., L. A. Hinds, P. J. Kerr, M. L. Lloyd, A. J. Redwood, G. R. Shellam, and T. Strive. 2006. Biological control of vertebrate pests using virally vectored immuno-contraception. *Journal of Reproductive Immunology* 71:102–111.

Harley, C. D. G., and B. S. T. Helmuth. 2003. Local- and regional-scale effects of wave exposure, thermal stress, and absolute versus effective shore level on patterns of intertidal zonation. *Limnology & Oceanography* 48:1498–1508.

Harley, D. K. P. 2006. A role for nest boxes in the conservation of Leadbeater's possum (*Gymnobelideus leadbeateri*). *Wildlife Research* 33:385–395.

Harper, J. L. 1977. *Population Biology of Plants.* New York: Academic Press. 892 pp.

Harper, J. L. 1981. The meanings of rarity. In *The Biological Aspects of Rare Plant Conservation,* (ed. H. Synge), 189–203. London: Wiley.

Harper, J. L., R. B. Rosen, and J. E. White. 1986. The growth and form of modular organisms. *Philosophical Transactions of the Royal Society of London,* Series B 313:1–250.

Harvell, C. D., C. E. Mitchell, J. R. Ward, S. Altizer, A. P. Dobson, R. S. Ostfeld, and M. D. Samuel. 2002. Climate warming and disease risks for terrestrial and marine biota. *Science* 296:2158–2162.

Harwood, J. 1989. Lessons from the seal epidemic. *New Scientist* 121:38–42.

Hastings, A. 1997. *Population Biology: Concepts and Models.* New York: Springer. 220 pp.

Hattersley, P. W. 1983. The distribution of C_3 and C_4 grasses in Australia in relation to climate. *Oecologia* 57:113–128.

Hawkes, C. V., and J. J. Sullivan. 2001. The impact of herbivory on plants in different resource conditions: A meta-analysis. *Ecology* 82:2045–2058.

Hawkins, B. A., and H. V. Cornell. 1999. *Theoretical Approaches to Biological Control.* Cambridge: Cambridge University Press, 412 pp.

Hawkins, B. A., and J. A. F. Diniz-Filho. 2006. Beyond Rapoport's rule: Evaluating range size patterns of New World birds in a two-dimensional framework. *Global Ecology & Biogeography* 15:461–469.

Hawkins, B. A., R. Field, H. V. Cornell, D. J. Currie, J.-F. Guégan, D. M. Kaufman, J. T. Kerr, G. G. Mittelbach, T. Oberdorff, E. M. O'Brien, E. E. Porter, and J. R. G. Turner. 2003a. Energy, water and broad-scale geographic patterns of species richness. *Ecology* 84:3105–3117.

Hawkins, B. A., E. E. Porter, and J. A. F. Diniz-Filho. 2003b. Productivity and history as predictors of the latitudinal diversity gradient of terrestrial birds. *Ecology* 84:1608–1623.

Haydon, D. 1994. Pivotal assumptions determining the relationship between stability and complexity: An analytical synthesis of the stability-complexity debate. *American Naturalist* 144:14–29.

Hayward, T. L. 1991. Primary production in the North Pacific Central Gyre: A controversy with important implications. *Trends in Ecology and Evolution* 6:281–284.

Hebblewhite, M., C. A. White, C. G. Nietvelt, J. A. McKenzie, T. E. Hurd, J. M. Fryxell, S. E. Bayley, and P. C. Paquet. 2005. Human activity mediates a trophic cascade caused by wolves. *Ecology* 86:2135–2144.

Hecky, R. E., P. Campbell, and L. L. Hendzel. 1993. The stochiometry of carbon, nitrogen and phosphorus in particulate matter of lakes and oceans. *Limnology & Oceanography* 38:709–724.

Hecnar, S. J., and R. T. M'Closkey. 1997. Changes in the composition of a ranid frog community following bullfrog extinction. *American Midland Naturalist* 137:145–150.

Hedenström, A., and T. Alerstam. 1997. Optimum fuel loads in migratory birds: Distinguishing between time and energy minimization. *Journal of Theoretical Biology* 189:227–234.

Heesterbeek, J. A. P., and M. G. Roberts. 1995. Mathematical models for microparasites of wildlife. In *Ecology of Infectious Diseases in Natural Populations*, (ed. B. T. Grenfell and A. P. Dobson), 90–122. Cambridge, England: Cambridge University Press.

Heide-Jorgensen, M.-P., and T. Harkonen. 1992. Epizootiology of the seal disease in the eastern North Sea. *Journal of Applied Ecology* 29:99–107.

Heinsohn, R., and S. Legge. 1999. The cost of helping. *Trends in Ecology & Evolution* 14:53–57.

Helfield, J., and R. Naiman. 2006. Keystone interactions: Salmon and bear in riparian forests of Alaska. *Ecosystems* 9:167–180.

Henry, H. A. L., and L. W. Aarssen. 1997. On the relationship between shade tolerance and shade avoidance strategies in woodland plants. *Oikos* 80:575–582.

Hickey, G. M. 2008. Evaluating sustainable forest management. *Ecological Indicators* 8:109–114.

Hickling, R., D. B. Roy, J. K. Hill, R. Fox, and C. D. Thomas. 2006. The distributions of a wide range of taxonomic groups are expanding poleward. *Global Change Biology* 12:450–455.

Hierro, J. L., J. L. Maron, and R. M. Callaway. 2005. A biogeographical approach to plant invasions: The importance of studying exotics in their introduced and native range. *Journal of Ecology* 93:5–15.

Hietala, T., P. Hiekkala, H. Rosenqvist, S. Laakso, L. Tahvanainen, and T. Repo. 1998. Fatty acid and alkane changes in willow during frost-hardening. *Phytochemistry* 47: 1501–1507.

Hilborn, R. 2007a. Defining success in fisheries and conflicts in objectives. *Marine Policy* 31:153–158.

Hilborn, R. 2007b. Moving to sustainability by learning from successful fisheries. *Ambio* 36:1–9.

Hilborn, R., J. Annala, and D. S. Holland. 2006. The cost of overfishing and management strategies for new fisheries on slow-growing fish: Orange roughy (*Hoplostethus atlanticus*) in New Zealand. *Canadian Journal of Fisheries & Aquatic Sciences* 63:2149–2153.

Hilborn, R., and C. J. Walters. 1992. *Quantitative Fisheries Stock Assessment: Choice, Dynamics, and Uncertainty*. New York: Chapman and Hall. 570 pp.

Hill, M. J., S. H. Roxburgh, J. O. Carter, and G. M. McKeon. 2005. Vegetation state change and consequent carbon dynamics in savanna woodlands of Australia in response to grazing, drought and fire: A scenario approach using 113 years of synthetic annual fire and grassland growth. *Australian Journal of Botany* 53:715–739.

Hinsley, S. A., P. E. Bellamy, I. Newton, and T. H. Sparks. 1996. Influences of population size and woodland area on bird species distributions in small woods. *Oecologia* 105:100–106.

Hixon, M. A., and M. S. Webster. 2002. Density dependence in reef fish populations. In *Coral Reef Fishes: Dynamics and Diversity in a Complex Ecosystem*, (ed. P. F. Sale), 303–325. Amsterdam: Elsevier.

Hjort, J. 1914. Fluctuations in the great fisheries of northern Europe, viewed in the light of biological research. *Rapport du Conseil International pour l'Exploration de la Mer* 20:1–228.

Hobson, K. A., J. F. Piatt, and J. Pitochelli. 1994. Using stable isotopes to determine seabird trophic relationships. *Journal of Animal Ecology* 63:786–798.

Hocker, H. W. J. 1956. Certain aspects of climate as related to the distribution of loblolly pine. *Ecology* 37:824–834.

Holdaway, R. N. 1989. New Zealand's pre-human avifauna and its vulnerability. *New Zealand Journal of Ecology* 12:11–25.

Holey, M. E., R. E. Rybicki, G. W. Eck, E. H. J. Brown, J. E. Marsden, D. S. Lavis, M. L. Toneys, et al. 1995. Progress toward lake trout restoration in Lake Michigan. *Journal of Great Lakes Research* 21:128–151.

Holland, H. D. 1995. Atmospheric oxygen and the biosphere. In *Linking Species and Ecosystems*, (ed. C. G. Jones and J. H. Lawton), 127–136. New York: Chapman and Hall.

Holling, C. S. 1959. The components of predation as revealed by a study of small mammal predation of the European pine sawfly. *Canadian Entomologist* 91:293–320.

Holt, R. D. 1977. Predation, apparent competition and the structure of prey communities. *Theoretical Population Biology* 28:181–208.

Holzapfel, C., and B. E. Mahall. 1999. Bidirectional facilitation and interference between shrubs and annuals in the Mojave Desert. *Ecology* 80:1747–1761.

Hone, J. 2007. *Wildlife Damage Control*. Collingwood, Victoria: CSIRO Publishing. 179 pp.

Hone, J., R. Pech, and P. Yip. 1992. Estimation of the dynamics and rate of transmission of classical swine fever (hog cholera) in wild pigs. *Epidemiology and Infection* 108:377–386.

Hone, J., and R. M. Sibly. 2002. Demographic, mechanistic and density-dependent determinants of population growth rate: A case study in an avian predator. *Philosophical Transactions of the Royal Society*, Series B 357:1171–1177.

Honnay, O., and H. Jacquemyn. 2007. Susceptibility of common and rare plant species to the genetic consequences of habitat fragmentation. *Conservation Biology* 21:823–831.

Hooper, D. U., and P. M. Vitousek. 1998. Effects of plant composition and diversity on nutrient cycling. *Ecological Monographs* 68:121–149.

Horn, H. S. 1975a. Forest succession. *Scientific American* 232:90–98.

Horn, H. S. 1975b. Markovian properties of forest succession. In *Ecology and Evolution of Communities*, (ed. M. L. Cody and J. M. Diamond), 196–211. Cambridge, MA: Harvard University Press.

Horn, H. S. 1981. Succession. In *Theoretical Ecology*, (ed. R. M. May), 253–271. Oxford, England: Blackwell.

Horne, A. J., and C. R. Goldman. 1994. *Limnology*. New York: McGraw-Hill. 576 pp.

Horppila, J., H. Peltonen, T. Malinen, E. Luokkanen, and T. Kairesalo. 1998. Top-down or bottom-up effects by fish: Issues of concern in biomanipulation of lakes. *Restoration Ecology* 6:20–28.

Hough, A. F. 1936. A climax forest community on East Tionesta Creek in northwestern Pennsylvania. *Ecology* 17:9–28.

Howarth, F. G. 1991. Environmental impacts of classical biological control. *Annual Review of Entomology* 36:485–509.

Howarth, R. W., and R. Marino. 2006. Nitrogen as the limiting nutrient for eutrophication in coastal marine ecosystems: Evolving views over three decades. *Limnology & Oceanography* 51:364–376.

Huang, Y., W. Zhang, W. Sun, and X. Zheng. 2007. Net primary production of Chinese croplands from 1950 to 1999. *Ecological Applications* 17:692–701.

Hubbell, S. P. 2001. *The Unified Neutral Theory of Biodiversity and Biogeography*. Princeton, NJ: Princeton University Press. 375 pp.

Hubbell, S. P., R. Condit, and R. B. Foster. 2005. Barro Colorado Forest Census Plot Data *http://ctfs.si.edu/datasets/bci*.

Hubbell, S. P., and R. B. Foster. 1986. Biology, chance, and history and the structure of tropical rain forest tree communities. In *Community Ecology*, (ed. J. Diamond and T. J.

Case), 314–329. New York: Harper & Row.

Huberty, A. F., and R. F. Denno. 2004. Plant water stress and its consequences for herbivorous insects: A new synthesis. *Ecology* 85:1383–1398.

Hudson, P. J., A. P. Dobson, and D. Newborn. 1998. Prevention of population cycles by parasite removal. *Science* 282:2256–2258.

Huffaker, C. B. 1958. Experimental studies on predation: Dispersion factors and predator-prey oscillations. *Hilgardia* 27:343–383.

Huffaker, C. B., K. P. Shea, and S. G. Herman. 1963. Experimental studies on predation: Complex dispersion and levels of food in an acarine predator-prey interaction. *Hilgardia* 34:305–330.

Hulme, P. E. 1996. Natural regeneration of yew (*Taxus baccata* L.): Microsite, seed or herbivore limitation? *Journal of Ecology* 84:853–861.

Humphreys, W. F. 1979. Production and respiration in animal populations. *Journal of Animal Ecology* 48:427–453.

Humphreys, W. F. 1984. Production efficiency in small mammal populations. *Oecologia* 62:85–90.

Hungate, B. A., D. W. Johnson, P. Dijkstra, G. Hymus, P. Stiling, J. P. Megonigal, A. L. Pagel, et al. 2006. Nitrogen cycling during seven years of atmospheric CO_2 enrichment in a scrub oak woodland. *Ecology* 87:26–40.

Huntley, B., R. E. Green, Y. C. Collingham, J. K. Hill, S. G. Willis, P. J. Bartlein, W. Cramer, et al. 2004. The performance of models relating species geographical distributions to climate is independent of trophic level. *Ecology Letters* 7:417–426.

Huntzinger, M., R. Karban, T. P. Young, and T. M. Palmer. 2004. Relaxation of induced indirect defenses of acacias following exclusion of mammalian herbivores. *Ecology* 85:609–614.

Husak, M. S., and A. L. Husak. 2003. Latitudinal patterns in range sizes of New World woodpeckers. *Southwestern Naturalist* 48:61–69.

Huse, G. 1998. Sex-specific life history strategies in capelin (*Mallotus villo-*

sus)? *Canadian Journal of Fisheries and Aquatic Sciences* 55:631–638.

Huston, M. 1979. A general hypothesis of species diversity. *American Naturalist* 113:81–101.

Huston, M., and T. Smith. 1987. Plant succession: Life history and competition. *American Naturalist* 130:168–198.

Huston, M. A. 1994. *Biological Diversity: The Coexistence of Species on Changing Landscapes*. Cambridge, England: Cambridge University Press. 681 pp.

Hutchings, M. J. 1983. Ecology's law in search of a theory. *New Scientist* 98:765–767.

Hutchins, L. W. 1947. The bases for temperature zonation in geographical distribution. *Ecological Monographs* 17:325–335.

Hutchinson, G. E. 1958. Concluding remarks. *Cold Spring Harbor Symposium on Quantitative Biology* 22:415–427.

Hutchinson, G. E. 1959. Homage to Santa Rosalia, or why are there so many kinds of animals? *American Naturalist* 93:145–159.

Hutchinson, G. E. 1961. The paradox of the plankton. *American Naturalist* 95:137–145.

Hyatt, L. A., M. S. Rosenberg, T. G. Howard, G. Bole, G. Wei Fang, J. Anastasia, K. Brown, et al. 2003. The distance dependence prediction of the Janzen-Connell hypothesis: A meta-analysis. *Oikos* 103:590–602.

Inbar, M., H. Doostdar, and R. T. Mayer. 2001. Suitability of stressed and vigorous plants to various insect herbivores. *Oikos* 94:228–235.

Ishibashi, Y., T. Saitoh, and M. Kawata. 1998. Social organization of the vole *Clethrionomys rufocanus* and its demographic and genetic consequences: A review. *Researches on Population Ecology* 40:39–50.

Ito, A., and T. Oikawa. 2004. Global mapping of terrestrial primary productivity and light-use efficiency with a process-based model. In *Global Environmental Change in the Ocean and on Land*, (ed. M. Shiyomi), 343–358. Tokyo: Terrapub.

Iverson, L. R., and A. M. Prasad. 1998. Predicting abundance of 80 tree species following climate change in the eastern United States. *Ecological Monographs* 68:465–485.

Iverson, L. R., A. M. Prasad, and M. W. Schwartz. 1999. Modeling potential future individual tree-species distributions in the Eastern United States under a climate change scenario: A case study with *Pinus virginiana*. *Ecological Modelling* 115:77–93.

Ives, A. R., and S. R. Carpenter. 2007. Stability and diversity of ecosystems. *Science* 317:58–62.

Jackson, R. B., S. R. Carpenter, C. N. Dahm, D. M. McKnight, R. J. Naiman, S. L. Postel, and S. W. Running. 2001. Water in a changing world. *Ecological Applications* 11:1027–1045.

Jahn, G. C., J. A. Litsinger, Y. Chen, and A. Barrion. 2007. Integrated pest management of rice: Ecological concepts. In *Ecologically Based Integrated Pest Management*, (ed. O. Koul and G. W. Cuperus), 315–366. CAB International.

Jahncke, J., D. M. Checkley, and G. L. Hunt. 2004. Trends in carbon flux to seabirds in the Peruvian upwelling system: Effects of wind and fisheries on population regulation. *Fisheries Oceanography* 13:208–223.

James, C. D., and R. Shine. 2000. Why are there so many coexisting species of lizards in Australian deserts? *Oecologia* 125:127–141.

Janes, S. W. 1985. Habitat selection in raptorial birds. In *Habitat Selection in Birds*, (ed. M. L. Cody), 159–188. Orlando: Academic Press.

Janmaat, A. F., and J. H. Myers. 2003. Rapid evolution and the cost of resistance to *Bacillus thuringiensis* in greenhouse populations of cabbage loopers, *Trichoplusia ni*. *Proceedings of the Royal Society of London*, Series B 270:2263–2270.

Janse, J. H. 1997. A model of nutrient dynamics in shallow lakes in relation to multiple stable states. *Hydrobiolgia* 342/343:1–8.

Janzen, D. H. 1966. Coevolution of mutualism between ants and acacias in Central America. *Evolution* 20:249–275.

Janzen, D. H. 1970. Herbivores and the number of tree species in tropical forests. *American Naturalist* 104:501–528.

Janzen, D. H. 1986. Chihuahuan desert nopaleras: Defaunated big mammal vegetation. *Annual Review of Ecology and Systematics* 17:595–636.

Jasinski, S. M. 2007. Phosphate rock. *US Geological Survey Minerals Yearbook—2006, Volume 1, Chapter 56, 10 pp.*

Jeffries, M.J., and J.H. Lawton. 1984. Enemy free space and the structure of ecological communities. *Biological Journal of the Linnean Society* 23: 269–286.

Jenkins, B., R. L. Kitching, and S. L. Pimm. 1992. Productivity, disturbance and food web structure at a local spatial scale in experimental container habitats. *Oikos* 65:249–255.

Jensen, C. X. J., and L. R. Ginzburg. 2005. Paradoxes or theoretical failures? The jury is still out. *Ecological Modelling* 188:3–14.

Jeppesen, E., M. Meerhoff, B. Jacobsen, R. Hansen, M. Søndergaard, J. Jensen, T. Lauridsen, et al. 2007a. Restoration of shallow lakes by nutrient control and biomanipulation—the successful strategy varies with lake size and climate. *Hydrobiologia* 581:269–285.

Jeppesen, E., M. Søndergaard, M. Meerhoff, T. Lauridsen, and J. Jensen. 2007b. Shallow lake restoration by nutrient loading reduction—some recent findings and challenges ahead. *Hydrobiologia* 584:239–252.

Jeschke, J. M., and D. L. Strayer. 2005. Invasion success of vertebrates in Europe and North America. *Proceedings of the National Academy of Sciences USA* 102:7198–7202.

Johns, B. 2005. Whooping Crane recovery—A North American success story. *Biodiversity* 6:2–6.

Johnson, C. 2006. *Australia's Mammal Extinctions: A 50,000 Year History*. Melbourne: Cambridge University Press. 278 pp.

Johnson, C. N. 1998. Rarity in the tropics: Latitudinal gradients in distribution and abundance in Australian mammals. *Journal of Animal Ecology* 67:689–698.

Johnson, C. N., M. Clinchy, A. C. Taylor, C. J. Krebs, P. J. Jarman, A. Payne, and E. G. Ritchie. 2001. Adjustment of offspring sex ratios in relation to the availability of resources for philopatric offspring in the common brushtail possum. *Proceedings of the Royal Society of London*, Series B 268:2001–2005.

Johnson, C. N., J. L. Isaac, and D. O. Fisher. 2007. Rarity of a top predator triggers continent-wide collapse of mammal prey: Dingoes and marsupials in Australia. *Proceeding of the Royal Society of London*, Series B 274:341–346.

Johnson, D. W. 2006. Progressive N limitation in forests: Review and implications for long-term responses to elevated CO_2. *Ecology* 87:64–75.

Johnson, D. W., T. Sogn, and S. Kvindesland. 2000. The nutrient cycling model: Lessons learned. *Forest Ecology and Management* 138:91–106.

Johnson, E. A., and K. Miyanishi. 2008. Testing the assumptions of chronosequences in succession. *Ecology Letters* 11:419–431.

Johnson, J., and P. Dunn. 2006. Low genetic variation in the Heath Hen prior to extinction and implications for the conservation of prairie-chicken populations. *Conservation Genetics* 7:37–48.

Johnson, J. A., M. R. Bellinger, J. E. Toepfer, and P. Dunn. 2004. Temporal changes in allele frequencies and low effective population size in greater prairie-chickens. *Molecular Ecology* 13:2617–2630.

Johnson, L. E., J. M. Bossenbroek, and C. E. Kraft. 2006. Patterns and pathways in the post-establishment spread of non-indigenous aquatic species: The slowing invasion of North American inland lakes by the zebra mussel. *Biological Invasions* 8:475–489.

Jolliffe, P. A. 1997. Are mixed populations of plant species more productive than pure stands? *Oikos* 80:595–602.

Jolliffe, P. A. 2000. The replacement series. *Journal of Ecology* 88:371–385.

Jones, G. P. 1990. The importance of recruitment to the dynamics of a coral reef fish population. *Ecology* 71:1691–1698.

Jones, M. 1997. Character displacement in Australian dasyurid carnivores: Size relationships and prey size patterns. *Ecology* 78:2569–2587.

Jonzen, N., A. R. Pople, G. C. Grigg, and H. P. Possingham. 2005. Of sheep and rain: Large-scale population dynamics of the red kangaroo. *Journal of Animal Ecology* 74:22–30.

Jordan, C. F., and R. Herrera. 1981. Tropical rain forests: Are nutrients really critical? *American Naturalist* 117:167–180.

Joshi, J., B. Schmid, M. C. Caldeira, P. G. Dimitrakopoulos, J. Good, R. Harris, A. Hector, et al. 2001. Local adaptation enhances performance of common plant species. *Ecology Letters* 4:536–544.

Juliano, S. A., and L. P. Lounibos. 2005. Ecology of invasive mosquitoes: Effects on resident species and on human health. *Ecology Letters* 8:558–574.

Kalisz, S. 1991. Experimental demonstration of seed bank age structure in the winter annual *Collinsia verna*. *Ecology* 72:575–585.

Kalmar, A., and D. J. Currie. 2006. A global model of island biogeography. *Global Ecology & Biogeography* 15:72–81.

Karban, R. E., and I. T. Baldwin. 1997. *Induced Responses to Herbivory*. Chicago: University of Chicago Press. 319 pp.

Kashiwagi, A., T. Kanaya, T. Yomo, and I. Urabe. 1998. How small can the difference among competitors be for coexistence to occur? *Researches on Population Ecology* 40:223–226.

Kaufman, D. M., and M. R. Willig. 1998. Latitudinal patterns of mammalian species richness in the New World: The effects of sampling method and faunal group. *Journal of Biogeography* 25:795–805.

Kautz, R., R. Kawula, T. Hoctor, J. Comiskey, D. Jansen, D. Jennings, J. Kasbohm, et al. 2006. How much is enough? Landscape-scale conservation for the Florida panther. *Biological Conservation* 130:118–133.

Kazimirov, N. I., and R. N. Morozova. 1973. *Biological Cycling of Matter in Spruce Forests of Karelia*. Leningrad: Nauka Publishing House.

Keddy, P. 2005. Putting the plants back into plant ecology: Six pragmatic models for understanding and conserving plant diversity. *Annals of Botany* 96:177–189.

Keeling, C. D., and T. P. Whorf. 2004. Atmospheric CO_2 records from sites in the Scripps Institute of Oceanography air sampling network *Trends: A Compendium of Data on Global Change*. Oak Ridge, TN: Carbon Dioxide Information Analysis Center, Oak Ridge National Laboratory, U.S. Department of Energy.

Keever, C. 1953. Present composition of some stands of the former oak-chestnut forest in the southern Blue Ridge Mountains. *Ecology* 34:44–54.

Keilman, N. 1999. How accurate are the United Nations' world population projections? In *Frontiers of Population Forecasting*, (ed. W. Lutz, J. W. Vaupel, and D. A. Ahlburg), 15–41. New York: Population Council.

Keith, H. 1997. Nutrient cycling in eucalypt ecosystems. In *Eucalypt Ecology: Individuals to Ecosystems*, (ed. J. E. Williams and J. C. Z. Woinarski), 197–226. Cambridge, England: Cambridge University Press.

Kelly, J. F. 2000. Stable isotopes of carbon and nitrogen in the study of avian and mammalian trophic ecology. *Canadian Journal of Zoology* 78:1–27.

Kerfoot, W. C. 1987. Predation: Direct and Indirect Impacts on Aquatic Communities. Hanover, NH: New England University Press, 386 pp.

Kershaw, K. A., and J. H. H. Looney. 1985. *Quantitative and Dynamic Plant Ecology*. London: Edward Arnold. 282 pp.

Ketterson, E. D., and V. Nolan, Jr. 1982. The role of migration and winter mortality in the life history of a temperate-zone migrant, the dark-eyed junco, as determined from demographic analyses of winter populations. *Auk* 99:243–259.

Keyfitz, N. 1971. On the momentum of population growth. *Demography* 8:71–80.

Khan, Z. R., J. A. Pickett, L. J. Wadhams, A. Hassanali, and C. A. O. Midega. 2006. Combined control

of *Striga hermonthica* and stemborers by maize-*Desmodium* spp. intercrops. *Crop Protection* 25:989–995.

Kilpatrick, A. M. 2006. Facilitating the evolution of resistance to avian malaria in Hawaiian birds. *Biological Conservation* 128:475–485.

Kilpatrick, A. M., L. D. Kramer, M. J. Jones, P. P. Marra, and P. Daszak. 2006. West Nile virus epidemics in North America are driven by shifts in mosquito feeding behavior. *PLoS Biology* 4: e82.

Kimberling, D. N. 2004. Lessons from history: Predicting successes and risks of intentional introductions for arthropod biological control. *Biological Invasions* 6:301–318.

Kingsland, S. 2004. Conveying the intellectual challenge of ecology: An historical perspective. *Frontiers in Ecology and the Environment* 2:367–374.

Kingsland, S. E. 1995. *Modeling Nature: Episodes in the History of Population Ecology*. Chicago: University of Chicago Press. 306 pp.

Kingsland, S. E. 2005. *The Evolution of American Ecology, 1890–2000*. Baltimore: Johns Hopkins University Press. 313 pp.

Kinnear, J. E., M. L. Onus, and N. R. Sumner. 1998. Fox control and rock-wallaby population dynamics. II. An update. *Wildlife Research* 25:81–88.

Kinnison, M., M. Unwin, N. Broustead, and T. Quinn. 1998. Population-specific variation in body dimensions of adult chinook salmon (*Oncorhynchus tshawytscha*) from New Zealand and their source population, 90 years after introduction. *Canadian Journal of Fisheries and Aquatic Sciences* 55:554–563.

Kira, T. 1975. Primary production of forests. In *Photosynthesis and Productivity in Different Environments*, (ed. J. P. Cooper), 5–40. London: Cambridge University Press.

Kirchner, J. W. 2002. The Gaia Hypothesis: Fact, theory, and wishful thinking. *Climatic Change* 52:391–408.

Kirkpatrick, J. B., and J. J. Scott. 2002. Change in undisturbed vegetation on the coastal slopes of subantarctic Macquarie Island, 1980–1995.

Arctic, Antarctic, and Alpine Research 34:300–307.

Kitching, J. A., and F. J. Ebling. 1967. Ecological studies at Lough Inc. *Advances in Ecological Research* 4:197–291.

Kleiman, D. G. 1977. Monogamy in mammals. *Quarterly Review of Biology* 52:39–69.

Klomp, H. 1970. The determination of clutch size in birds: A review. *Ardea* 58:1–124.

Klun, J. A. 1974. Biochemical basis of resistance of plants to pathogens and insects: Insect hormone mimics and selected examples of other biologically active chemicals derived from plants. In *Proceedings of the Summer Institute on Biological Control of Plant Insects and Diseases*, (ed. F. G. Maxwell and F. A. Harris), 463–484. Jackson: University Press of Mississippi.

Knapp, A., C. Burns, R. Fynn, K. Kirkman, C. Morris, and M. Smith. 2006. Convergence and contingency in production–precipitation relationships in North American and South African C_4 grasslands. *Oecologia* 149:456–464.

Knobe, D. L., S. Cleaveland, P. G. Coleman, E. M. Favre, M. I. Meltzer, M. E. Miranda, A. Shaw, et al. 2005. Re-evaluating the burden of rabies in Africa and Asia. *Bulletin of the World Health Organization* 83:360–368.

Knoll, A. H. 1986. Patterns of change in plant communities through geological time. In *Community Ecology*, (ed. J. Diamond and T. J. Case), 126–141. New York: Harper & Row.

Knox, E. A. 1970. Antarctic marine ecosystems. In *Antarctic Ecology*, (ed. M. W. Holdgate), 69–96. London: Academic Press.

Koch, G. W., S. C. Sillett, G. M. Jennings, and S. D. Davis. 2004. The limits to tree height. *Nature* 428:851–854.

Koch, R. L., R. C. Venette, and W. D. Hutchison. 2006. Predicted impact of an exotic generalist predator on monarch butterfly (Lepidoptera: Nymphalidae) populations: A quantitative risk assessment. *Biological Invasions* 8:1179–1193.

Kodric-Brown, A., and J. H. Brown. 1978. Influence of economics, interspecific competition, and sexual dimorphism on territoriality of migrant rufous hummingbirds. *Ecology* 59:285–296.

Kokkinn, M. J., and A. R. Davis. 1986. Secondary production: Shooting a halcyon for its feathers. In *Limnology in Australia*, (ed. P. De Decker and W. D. Williams), 251–261. Dordrecht, Netherlands: Dr. W. Junk.

Körner, C. 1998. A re-assessment of high elevation treeline positions and their explanation. *Oecologia* 115:445–459.

Körner, C. 2006. Plant CO_2 responses: An issue of definition, time and resource supply. *New Phytologist* 172:393–411.

Körner, C., and J. Paulsen. 2004. A world-wide study of high altitude treeline temperatures. *Journal of Biogeography* 31:713–732.

Korzukhin, M. D., S. D. Porter, L. C. Thompson, and S. Wiley. 2001. Modeling temperature-dependent range limits for the fire ant *Solenopsis invicta* (Hymenoptera: Formicidae) in the United States. *Environmental Entomology* 30:645–655.

Kotler, B. P., J. S. Brown, and O. Hasson. 1991. Factors affecting gerbil foraging behaviour and rates of owl predation. *Ecology* 72:2249–2260.

Kozhov, M. 1963. Lake Baikal and its life. *Monographiae Biologicate* 11:1–344.

Kozlowski, T. T., S. G. Pallardy, and P. J. Kramer. 1997. *The Physiology of Woody Plants*. San Diego: Academic Press. 411 pp.

Krause, J., and J.-G. J. Godin. 1995. Predator preferences for attacking particular prey group sizes: Consequences for predator hunting success and prey predation risk. *Animal Behaviour* 50:465–473.

Krebs, C. J. 1999. *Ecological Methodology*. Menlo Park, CA: Addison Wesley Longman Inc. 620 pp.

Krebs, C. J. 2002. Two complementary paradigms for analyzing population dynamics. *Philosophical Transactions of the Royal Society of London*, Series B 357:1211–1219.

Krebs, C. J. 2006. Ecology after 100 years: Progress and pseudo-progress. *New Zealand Journal of Ecology* 30:3–11.

Krebs, C. J., S. Boutin, and R. Boonstra. 2001. *Ecosystem Dynamics of the Boreal Forest: The Kluane Project*. New York: Oxford University Press, 511 pp.

Krebs, J. R., and N. B. Davies. 1993. *An Introduction to Behavioural Ecology*. Oxford: Blackwell Scientific Publications. 420 pp.

Krebs, J. R., J. T. Erichsen, M. I. Webber, and E. L. Charnov. 1977. Optimal prey selection in the great tit (*Parus major*). *Animal Behaviour* 25:30–38.

Krebs, J. W., E. J. Mandel, D. L. Swerdlow, and C. E. Rupprecht. 2005. Rabies surveillance in the United States during 2004. *Journal of the American Veterinary Medicine Association* 227:1912–1925.

Krebs, J. W., J. S. Smith, C. E. Rupprecht, and J. E. Childs. 1998. Rabies surveillance in the United States during 1997. *Journal of the American Veterinary Medical Association* 213: 1713–1728.

Krebs, J. W., M. L. Wilson, and J. E. Childs. 1995. Rabies—epidemiology, prevention, and future research. *Journal of Mammalogy* 76:681–694.

Kremen, C. 2005. Managing ecosystem services: What do we need to know about their ecology? *Ecology Letters* 8:468–479.

Krkosek, M., J. S. Ford, A. Morton, S. Lele, R. A. Myers, and M. A. Lewis. 2007. Declining wild salmon populations in relation to parasites from farm salmon. *Science* 318:1772–1775.

Krueger, C. C., M. L. Jones, and W. W. Taylor. 1995. Restoration of lake trout in the Great Lakes: Challenges and strategies for future management. *Journal of Great Lakes Research* 21:547–558.

Kruuk, H. 1972. *The Spotted Hyaena: A Study of Predation and Social Behavior*. Chicago: University of Chicago Press, 335 pp.

Kubanek, J., M. K. Hicks, J. Naar, and T. A. Villareal. 2005. Does the red tide dinoflagellate *Karenia brevis* use allelopathy to outcompete other phytoplankton? *Limnology & Oceanography* 50:883–895.

Kuparinen, A., and J. Merila. 2007. Detecting and managing fisheries-induced evolution. *Trends in Ecology & Evolution* 22:652–659.

Kuussaari, M., I. Saccheri, M. Camara, and I. Hanski. 1998. Allee effect and population dynamics in the Glanville fritillary butterfly. *Oikos* 82:384–392.

Lack, D. 1947a. The significance of clutch size. *Ibis* 89:302–352.

Lack, D. 1954. *The Natural Regulation of Animal Numbers*. Oxford. 343 pp.

Lack, D. L. 1947b. *Darwin's Finches*. Cambridge: Cambridge University Press. 208 pp.

Lacy, R. C. 2000. Structure of the VORTEX simulation model for population viability analysis. Ecological Bulletin. In *The Use of Population Viability Analyses in Conservation Planning*, (ed. P. S. Gulve and T. Ebenhard), 191–203. Oxford: Blackwell Publishing.

LaHaye, W. S., and R. J. Gutierrez. 1999. Nest sites and nesting habitat of the northern spotted owl in northwestern California. *Condor* 101:324–330.

Lambert, B., and M. Peferoen. 1992. Insecticidal promise of *Bacillus thuringiensis*: Facts and mysteries about a successful biopesticide. *BioScience* 42:112–121.

Lambin, X., J. Aars, and S. B. Piertney. 2001. Dispersal, intraspecific competition, kin competition and kin facilitation: A review of the empirical evidence. In *Dispersal*, (ed. J. Clobert, E. Danchin, A. A. Dhondt, and J. D. Nichols), 110–122. Oxford: Oxford University Press.

Lande, R. 1993. Risks of population extinction from demographic and environmental stochasticity and random catastrophes. *American Naturalist* 142:911–927.

Landsberg, J. H. 2002. The effects of harmful algal blooms on aquatic organisms. *Reviews in Fisheries Science* 10:113–390.

Larkin, P. A. 1977. An epitaph for the concept of maximum sustained yield. *Transactions of the American Fisheries Society* 106:1–11.

Larkin, P. A., B. Scott, and A. W. Trites. 1990. The red king crab fishery of the southeastern Bering Sea. *Fisheries Management Foundation* 78.

Larson, B. 2007. An alien approach to invasive species: Objectivity and society in invasion biology. *Biological Invasions* 9:947–956.

Larsson, S. 1989. Stressful times for the plant stress-insect performance hypothesis. *Oikos* 56:277–283.

Lauenroth, W. K. 1979. Grassland primary production: North American grasslands in perspective. In *Perspectives in Grassland Ecology*, (ed. N. R. French), 3–24. New York: Springer-Verlag.

Laurance, W. F., H. E. M. Nascimento, S. G. Laurance, A. Andrade, J. E. L. S. Ribeiro, J. P. Giraldo, T. E. Lovejoy, et al. 2006. Rapid decay of tree-community composition in Amazonian forest fragments. *Proceedings of the National Academy of Sciences of the USA* 103:19010–19014.

Law, R. 1979. The cost of reproduction in annual meadow grass. *American Naturalist* 113:3–16.

Lawlor, T. 1998. Biogeography of Great Basin mammals: Paradigm lost? *Journal of Mammology* 79:1111–1130.

Laws, R. M. 1970. Elephants as agents of habitat and landscape change in East Africa. *Oikos* 21:1–15.

Lawton, J. H. 1987. Are there assembly rules for successional communities? In *Colonization, Succession and Stability*, (ed. A. J. Gray, M. J. Crawley, and P. J. Edwards), 225–244. Oxford, England: Blackwell.

Le Galliard, J.-F., P. S. Fitze, R. Ferriere, and J. Clobert. 2005. Sex ratio bias, male aggression, and population collapse in lizards. *Proceedings of the National Academy of Sciences of USA* 102:18231–18236.

Le Treut, H., R. Somerville, U. Cubasch, Y. Ding, C. Mauritzen, A. Mokssit, T. Peterson, and M. Prather. 2007. Historical Overview of Climate Change. In *Climate Change 2007: The Physical Science Basis. Contribution of Working Group I to the Fourth Assessment Report of the Intergovernmental Panel on Climate Change*, (ed. Solomon et al.). Cambridge: Cambridge University Press.

Leach, M. K., and T. J. Givnish. 1996. Ecological determinants of species loss in remnant prairies. *Science* 273:1555–1558.

Leader-Williams, N. 1988. *Reindeer on South Georgia: The Ecology of an Introduced Population*. Cambridge, England: Cambridge University Press. 319 pp.

Lefkovitch, L. P. 1965. The study of population growth in organisms grouped by stages. *Biometrics* 21:1–18.

Lertzman, K. P. 1992. Patterns of gap-phase replacement in a subalpine, old-growth forest. *Ecology* 73:657–669.

Lesica, P., R. Yurkewycz, and E. E. Crone. 2006. Rare plants are common where you find them. *American Journal of Botany* 93:454–459.

Leslie, P. H. 1945. On the use of matrices in certain population mathematics. *Biometrika* 33:183–212.

Leslie, P. H. 1966. The intrinsic rate of increase and the overlap of successive generations in a population of guillemots (*Uria aalge* Pont.). *Journal of Animal Ecology* 35:291–301.

Leslie, P. H., and R. M. Ranson. 1940. The mortality, fertility, and rate of natural increase of the vole (*Microtus agrestis*) as observed in the laboratory. *Journal of Animal Ecology* 9:27–52.

Lesser, M. P. 2007. Coral reef bleaching and global climate change: Can corals survive the next century? *Proceedings of the National Academy of Sciences of the USA* 104:5259–5260.

Letcher, B. H., J. A. Priddy, J. R. Walters, and L. B. Crowder. 1998. An individual-based, spatially-explicit simulation model of the population dynamics of the endangered red-cockaded woodpecker, *Picoides borealis*. Biological Conservation 86:1–14.

Lever, C. 1985. *Naturalized Mammals of the World*. London: Longman. 487 pp.

Lever, C. 1996. *Naturalized Fishes of the World*. San Diego: Academic Press. 408 pp.

Levin, D. A. 1976. Alkaloid-bearing plants: An ecogeographic perspective. *American Naturalist* 110:261–284.

Levin, S., and D. Pimentel. 1981. Selection of intermediate rates of increase in parasite-host systems. *American Naturalist* 117:308–315.

Levin, S. A. 1970. Community equilibria and stability, and an extension of the competitive exclusion principle. *American Naturalist* 104:413–423.

Levin, S. A. 1974. Dispersion and population interactions. *American Naturalist* 108:207–228.

Lewis, J. R. 1972. *The Ecology of Rocky Shores*. London: English Universities Press, 323 pp.

Lewontin, R. C. 1965. Selection of colonizing ability. In *The Genetics of Colonizing Species*, (ed. H. G. Baker, and G. L. Stebbins), 77–94. New York: Academic Press.

Li, J., W. A. Loneragan, J. A. Duggin, and C. D. Grant. 2004. Issues affecting the measurement of disturbance response patterns in herbaceous vegetation—A test of the intermediate disturbance hypothesis. *Plant Ecology* 172:11–26.

Libralato, S., V. Christensen, and D. Pauly. 2006. A method for identifying keystone species in food web models. *Ecological Modelling* 195:153–171.

Lichter, J. 1998. Primary succession and forest development on coastal Lake Michigan sand dunes. *Ecological Monographs* 68:487–510.

Lichter, J. 2000. Colonization constraints during primary succession on coastal Lake Michigan sand dunes. *Journal of Ecology* 88:825–839.

Liebhold, A., and N. Kamata. 2000. Are population cycles and spatial synchrony a universal characteristic of forest insect populations? *Population Ecology* 42:205–209.

Likens, G. E. 1972. *Nutrients and Eutrophication: The Limiting Nutrient Controversy*. Lawrence, KS: American Society of Limnology and Oceanography, 328 pp.

Likens, G. E., C. T. Driscoll, and D. C. Buso. 1996. Long-term effects of acid rain: Response and recovery of a forest ecosystem. *Science* 272:244–245.

Likens, G. E., R. F. Wright, J. N. Galloway, and T. J. Butler. 1979. Acid rain. *Scientific American* 241:43–51.

Lima, S. L. 1998. Nonlethal effects in the ecology of predator-prey interactions. *BioScience* 48:25–34.

Lindenmayer, D., and J. Fischer. 2006. *Habitat Fragmentation and Landscape Change: An Ecological and Conservation Synthesis*. Washington, DC: Island Press. 344 pp.

Lindenmayer, D., and M. McCarthy. 2006. Evaluation of PVA models of arboreal marsupials: Coupling models with long-term monitoring data. *Biodiversity and Conservation* 15:4079–4096.

Lindenmayer, D. B. 1996. *Wildlife and Woodchips: Leadbeater's Possum as a Test Case of Ecologically Sustainable Forestry*. Sydney, New South Wales: University Press. 156 pp.

Lindenmayer, D. B., C. I. MacGregor, R. B. Cunningham, R. D. Incoll, M. Crane, D. Rawlins, and D. R. Michael. 2003. The use of nest boxes by arboreal marsupials in the forests of the Central Highlands of Victoria. *Wildlife Research* 30:259–264.

Litzgus, J. D. 2006. Sex differences in longevity in the spotted turtle (*Clemmys guttata*). *Copeia* 2006:281–288.

Lloyd, A. H. 2005. Ecological histories from Alaskan tree lines provide insight into future change. *Ecology* 86:1687–1695.

Lobon-Cervia, J., and E. Mortensen. 2006. Two-phase self-thinning in stream-living juveniles of lake-migratory brown trout *Salmo trutta* L. Compatibility between linear and non-linear patterns across populations? *Oikos* 113:412–423.

Lockwood, J. L., P. Cassey, and T. Blackburn. 2005. The role of propagule pressure in explaining species invasions. *Trends in Ecology and Evolution* 20:223–228.

Lockwood, J. L., D. Simberloff, M. L. McKinney, and B. Von Holle. 2001. How many, and which, plants will invade natural areas? *Biological Invasions* 3:1–8.

Lodge, D. M., G. Cronin, E. van Donk, and A. J. Froelich. 1998. Impact of herbivory on plant standing crop: Comparisons among biomes, between vascular and nonvascular plants, and among freshwater herbivore taxa. Ecological Studies. In *The Structuring Role of Submerged Macrophytes in Lakes*, (ed. E. Jeppesen, M. Sondergaard, M. Sondergaard, and K. Christoffersen), 149–174. New York: Springer-Verlag.

Loehr, J., J. Carey, M. Hoefs, J. Suhonen, and H. Ylonen. 2007. Horn growth rate and longevity: implications for natural and artificial selection in thinhorn sheep (*Ovis dalli*). *Journal of Evolutionary Biology* 20:818–828.

Loher, T., D. A. Armstrong, and B. G. Stevens. 2001. Growth of juvenile red king crab (*Paralithodes camtschaticus*) in Bristol Bay (Alaska) elucidated from field sampling and analysis of trawl-survey data. *Fishery Bulletin* 99:572–587.

Long, J. L. 1981. *Introduced Birds of the World*. London: Davis and Charles, 528 pp.

Long, S. P., E. A. Ainsworth, A. D. B. Leakey, J. Nosberger, and D. R. Ort. 2006. Food for thought: Lower-than-expected crop yield stimulation with rising CO_2 concentrations. *Science* 312:1918–1921.

Long, Z. T., and P. J. Morin. 2005. Effects of organism size and community composition on ecosystem functioning. *Ecology Letters* 8:1271–1282.

Lotka, A. J. 1907. Studies on the mode of growth of material aggregates. *American Journal of Science* 24:199–216.

Lotka, A. J. 1913. A natural population norm. *Journal of the Washington Academy of Science* 3: 241–248 and 289–293.

Lotka, A. J. 1922. The stability of the normal age distribution. *Proceedings of the National Academy of Sciences of the USA* 8:339–345.

Lotka, A. J. 1925. *Elements of Physical Biology*. New York (reprint): Dover Publications, 465 pp.

Lovelock, J. 1988. *The Ages of Gaia: A Biography of Our Living Earth*. New York: Norton, 252 pp.

Lubchenco, J. 1978. Plant species diversity in a marine intertidal community: Importance of herbivore food preference and algal competitive abilities. *American Naturalist* 112:23–39.

Lubchenco, J. 1986. Relative importance of competition and predation: Early colonization by seaweeds in New England. In *Community Ecology*, (ed. J. Diamond and T. J. Case), 537–555. New York: Harper & Row.

Lubina, J. A., and S. A. Levin. 1988. The spread of a reinvading species: Range expansion in the California sea otter. *American Naturalist* 131:526–543.

Lucas, J. A. 1998. *Plant Pathology and Plant Pathogens*. Malden, MA: Blackwell Science. 274 pp.

Ludwig, D., R. Hilborn, and C. Walters. 1993. Uncertainty, resource exploitation, and conservation: Lessons from history. *Science* 260:17, 36.

Ludwig, D., M. Mangel, and B. Haddad. 2001. Ecology, conservation, and public policy. *Annual Review of Ecology and Systematics* 32:481–517.

Ludwig, D., and C. J. Walters. 1985. Are age-structured models appropriate for catch-effort data? *Canadian Journal of Fisheries and Aquatic Sciences* 42:1066–1072.

Ma, Y. Q. 2005. Allelopathic studies of common wheat (*Triticum aestivum* L.). *Weed Biology & Management* 5:93–104.

MacArthur, R., and E. O. Wilson. 1967. *The Theory of Island Biogeography*. Princeton, NJ: Princeton University Press. 203 pp.

MacArthur, R. H. 1958. Population ecology of some warblers of northeastern coniferous forests. *Ecology* 39:599–619.

MacArthur, R. H. 1965. Patterns of species diversity. *Biological Reviews* 40:510–533.

MacArthur, R. H. 1972. *Geographical Ecology*. New York: Harper and Row. 269 pp.

MacCracken, M. C. 1985. Carbon dioxide and climate change: Background and overview. In *Projecting the Climatic Effects of Increasing Carbon Dioxide*, (ed. M. C. MacCracken and F. M. Luther), 1–23. Washington, DC: U.S. Department of Energy.

Mace, P. 2001. A new role for MSY in single-species and ecosystem approaches to fisheries stock assessment and management. *Fish and Fisheries* 2:2–32.

MacIsaac, H. 1996. Potential abiotic and biotic impacts of zebra mussels on the inland waters of North America. *American Zoologist* 36:287–299.

Mack, M. C., E. A. G. Schuur, M. S. Bret-Harte, G. R. Shaver, and F. S. Chapin, III. 2004. Ecosystem carbon storage in Arctic tundra reduced by long-term nutrient fertilization. *Nature* 431:440.

Mack, R. N., D. Simberloff, W. M. Lonsdale, H. Evans, M. Clout, and F. A. Bazzaz. 2000. Biotic invasions: Causes, epidemiology, global consequences, and control. *Ecological Applications* 10:689–710.

MacLean, D. A., and R. W. Wein. 1977. Nutrient accumulation for post fire jack pine and hardwood successional patterns in New Brunswick. *Canadian Journal of Forest Research* 7:562–578.

Macnair, M. R. 1987. Heavy metal tolerance in plants: A model evolutionary system. *Trends in Ecology and Evolution* 2:254–259.

Madigan, M. T., J. M. Martinko, and T. D. Brock. 2006. *Biology of Microorganisms*. Upper Saddle River, NJ: Pearson Prentice Hall. 992 pp.

Madsen, T., B. Ujvari, R. Shine, and M. Olsson. 2006. Rain, rats and pythons: Climate-driven population dynamics of predators and prey in tropical Australia. *Austral Ecology* 31:30–37.

Magurran, A. E. 2004. *Measuring Biological Diversity*. Oxford: Blackwell Publishing. 256 pp.

Magurran, A. E., B. H. Seghers, G. R. Carvalho, and P. W. Shaw. 1992. Behavioural consequences of an artificial introduction of guppies (*Poecilia reticulata*) in N. Trinidad: Evidence for the evolution of antipredator behaviour in the wild. *Proceeding of the Royal Society of London*, Series B 248:117–122.

Majerus, M. E. N. 1998. *Melanism: Evolution in Action*. Oxford: Oxford University Press. 338 pp.

Makarieva, A. M., V. G. Gorshkov, and B.-L. Li. 2005. Energetics of the smallest: Do bacteria breathe at the same rate as whales? *Proceeding of the Royal Society of London*, Series B 272:2219–2224.

Mangin, S., M. Gauthier-Clerc, Y. Frenot, J.-P. Gendner, and Y. L. Maho. 2003. Ticks *Ixodes uriae* and the breeding performance of a colonial seabird, king penguin *Aptenodytes patagonicus*. *Journal of Avian Biology* 34:30–34.

Manne, L. L., S. L. Pimm, J. M. Diamond, and T. M. Reed. 1998. The form of the curves: A direct evaluation of MacArthur & Wilson's classic theory. *Journal of Animal Ecology* 67:784–794.

Mansson, B. A., and J. M. McGlade. 1993. Ecology, thermodynamics and H. T. Odum's conjectures. *Oecologia* 93:582–596.

Mantua, N. J., S. R. Hare, Y. Zhang, J. M. Wallace, and R. C. Francis. 1997. A Pacific interdecadal climate oscillation with impacts on salmon production. *Bulletin of the American Meteorological Society* 78:1069–1079.

Margules, C. R., and R. L. Pressey. 2000. Systematic conservation planning. *Nature* 405:243–253.

Marks, P. L. 1983. On the origin of the field plants of the northeastern United States. *American Naturalist* 122:210–228.

Marples, N. M., D. J. Kelly, and R. J. Thomas. 2005. The evolution of warning coloration is not paradoxical. *Evolution* 59:933–940.

Marra, P. P., S. Griffing, C. Caffrey, A. M. Kilpatrick, R. McLean, C. Brand, E. Saito, et al. 2004. West Nile virus and wildlife. *BioScience* 54:393–402.

Marrs, R. H., S. W. Johnson, and M. G. Le Duc. 1998. Control of bracken and restoration of the heathland. VI. The response of bracken fronds to 18 years of continued bracken control or 6 years of control followed by recovery. *Journal of Applied Ecology* 35:479–490.

Martin, J. H., R. M. Gordon, and S. E. Fitzwater. 1990. Iron in antarctic waters. *Nature* 345:156–158.

Martin, P. S., and H. E. Wright. 1967. *Pleistocene Extinctions: The Search for a Cause*. New Haven, CT: Yale University Press, 453 pp.

Martin, T. E. 1995. Avian life history evolution in relation to nest sites, nest predation and food. *Ecological Monographs* 65:101–127.

Martinez, N. D. 1991. Artifacts or attributes? Effects of resolution on the Little Rock Lake food web. *Ecological Monographs* 61:367–392.

Marzal, A., F. Lope, C. Navarro, and A. P. Møller. 2005. Malarial parasites decrease reproductive success: An experimental study in a passerine bird. *Oecologia* 142:541–545.

Masaki, T., Y. Kominami, and T. Nakashizuka. 1994. Spatial and seasonal patterns of seed dissemination of *Cornus controversa* in a temperate forest. *Ecology* 75:1903–1910.

Matsuda, H., and P. A. Abrams. 2006. Maximal yields from multispecies fisheries systems: Rules for systems with multiple trophic levels. *Ecological Applications* 16:225–237.

Matter, W. J., and R. W. Mannan. 2005. How do prey persist? *Journal of Wildlife Management* 69:1315–1320.

May, R. M. 1973. *Stability and Complexity in Model Ecosystems*. Princeton, NJ: Princeton University Press, 265 pp.

May, R. M. 1974. Biological populations with nonoverlapping generations: Stable points, stable cycles, and chaos. *Science* 186:645–647.

May, R. M. 1975. Patterns of species abundance and diversity. In *Ecology and Evolution of Communities*, (ed. M. L. Cody and J. M. Diamond), 81–120. Cambridge, MA: Belknap Press, Harvard University.

May, R. M. 1981. Models for single populations. In *Theoretical Ecology*, (ed. R. M. May), 5–29. Oxford: Blackwell.

Mayewski, P. A., W. B. Lyons, M. J. Spencer, M. Twickler, W. Dansgaard, B. Koci, C. I. Davidson, and R. E. Honrath. 1986. Sulfate and nitrate concentrations from a south Greenland ice core. *Science* 232:975–977.

Maynard Smith, J. 1968. *Mathematical Ideas in Biology*. New York: Cambridge University Press, 152 pp.

Mayr, E. 1982. *The Growth of Biological Thought*. Cambridge, MA: Belknap Press, Harvard University. 974 pp.

McCallum, H., N. Barlow, and J. Hone. 2001. How should pathogen transmission be modelled? *Trends in Ecology and Evolution* 16:295–300.

McCann, K. S. 2000. The diversity-stability debate. *Nature* 405:228–233.

McCann, K. S., A. Hastings, and D. R. Strong. 1998. Trophic cascades and trophic trickles in pelagic food webs. *Proceedings of the Royal Society of London*, Series B 265:205–209.

McCarthy, S. A. 1996. Effects of temperature and salinity on survival of toxigenic *Vibrio cholerae* 01 in seawater. *Microbial Ecology* 31:167–175.

McGill, B. J. 2003. A test of the unified neutral theory of biodiversity. *Nature* 422:881–885.

McGowan, D. P. J., and D. L. Otis. 1998. Population demographics of two local South Carolina mourning dove populations. *Journal of Wildlife Management* 62:1443–1451.

McGowan, J. A., and P. W. Walker. 1979. Structure in the copepod community of the North Pacific Central Gyre. *Ecological Monographs* 49:195–226.

McGowan, J. A., and P. W. Walker. 1985. Dominance and diversity maintenance in an oceanic ecosystem. *Ecological Monographs* 55:103–118.

McIntosh, R. P. 1985. *The Background of Ecology: Concept and Theory*. Cambridge: Cambridge University Press. 383 pp.

McLeod, S. R. 1997. Is the concept of carrying capacity useful in variable environments? *Oikos* 79:529–542.

McMillan, C. 1956. The edaphic restriction of Cupressus and Pinus in the Coast Ranges of Central California. *Ecological Monographs* 26:177–212.

McNaughton, S. J. 1976. Serengeti migratory wildebeest: Facilitation of energy flow by grazing. *Science* 191:92–94.

McNaughton, S. J., M. Oesterheid, D. A. Frank, and K. J. Williams. 1989. Ecosystem-level patterns of primary productivity and herbivory in terrestrial habitats. *Nature* 341:142–144.

Mduma, S. A. R., A. R. E. Sinclair, and R. Hilborn. 1999. Food regulates the Serengeti wildebeest: A 40 year record. *Journal of Animal Ecology* 68:1101–1122.

Meagher, M., and M. E. Meyer. 1994. On the origin of brucellosis in bison of Yellowstone National Park: A review. *Conservation Biology* 8:645–653.

Menge, B. A., and J. P. Sutherland. 1987. Community regulation: Variation disturbance, competition, and predation in relation to environmental stress and recruitment. *American Naturalist* 130:730–757.

Menzel, D. W., and J. H. Ryther. 1961. Nutrients limiting the production of phytoplankton in the Sargasso Sea, with special reference to iron. *Deep Sea Research* 7:276–281.

Mertz, D. B. 1970. Notes on methods used in life-history studies. In *Readings in Ecology and Ecological Genetics*, (ed. J. H. Connell, D. B. Mertz, and W. W. Murdoch), 4–17. New York: Harper and Row.

Meskhidze, N., A. Nenes, W. L. Chameides, C. Luo, and N. Mahowald. 2007. Atlantic Southern Ocean productivity: Fertilization from above or below? *Global Biogeochemical Cycles* 21.

Messier, F., and D. O. Joly. 2000. Comment: Regulation of moose populations by wolf predation. *Canadian Journal of Zoology* 78:506–510.

Mezquida, E. T., S. J. Slater, and C. W. Benkman. 2006. Sage-grouse and indirect interactions: Potential implications of coyote control on sage-grouse populations. *Condor* 108:747–759.

Miles, J. 1987. Vegetation succession: Past and present perceptions. In *Colonization, Succession and Stability*, (ed. A. J. Gray, M. J. Crawley, and P. J. Edwards), 1–29. Oxford, England: Blackwell.

Miller, G. H., M. L. Fogel, J. W. Magee, M. K. Gagan, S. J. Clarke, and B. J. Johnson. 2005. Ecosystem collapse in Pleistocene Australia and a human role in megafaunal extinction. *Science* 309:287–290.

Mills, J. N. 1999. The role of rodents in emerging human disease: examples from the hantaviruses and arenaviruses. In: Singleton, G. R., L. Hinds, H. Leirs, and Z. Zhang, eds. *Ecologically-based Management of Rodent Pests*. Canberra: Australian Centre for International Agricultural Research, pp. 134–160.

Mills, L. S., M. E. Soulé, and D. F. Doak. 1993. The keystone-species concept in ecology and conservation. *BioScience* 43:219–224.

Mills, N. J. 2006. Accounting for differential success in the biological control of homopteran and lepidopteran pests. *New Zealand Journal of Ecology* 30:61–72.

Millspaugh, J. J., and J. M. Marzluff. 2001. *Radio Tracking and Animal Populations.* San Diego, CA: Academic Press, 474 pp.

Moles, A. T., and M. Westoby. 2004. Seedling survival and seed size: A synthesis of the literature. *Journal of Ecology* 92:372–383.

Monaghan, P., and R. G. Nager. 1997. Why don't birds lay more eggs? *Trends in Ecology and Evolution* 12:270–273.

Mondor, E. B., M. N. Tremblay, C. S. Awmack, and R. L. Lindroth. 2004. Divergent pheromone-mediated insect behaviour under global atmospheric change. *Global Change Biology* 10:1820–1824.

Monson, R. K. 1989. On the evolutionary pathways resulting in C_4 photosynthesis and crassulacean acid metabolism (CAM). *Advances in Ecological Research* 19:57–110.

Monteith, J. L. 1972. Solar radiation and productivity in tropical ecosystems. *Journal of Applied Ecology* 9:747–766.

Montevecchi, W. A., and D. A. Kirk. 1996. Great auk: *Pinguinus impennis*. *Birds of North America* 260:1–20.

Moore, P. D., and S. B. Chapman. 1986. *Methods in Plant Ecology.* Oxford: Blackwell Scientific Publications, 589 pp.

Morato, T., R. Watson, T. J. Pitcher, and D. Pauly. 2006. Fishing down the deep. *Fish & Fisheries* 7:24–34.

Morgan, N. C. 1980. Secondary production. In *The Functioning of Freshwater Ecosystems*, (ed. E. D. LeCren and R. H. Lowe-McDonnell), 247–340. Cambridge, England: Cambridge University Press.

Morris, C. D., V. L. Larson, and L. P. Lounibos. 1991. Measuring mosquito dispersal for control programs. *Journal of the American Mosquito Control Association* 7:608–615.

Morris, L. A., and R. E. Miller. 1994. Evidence for long-term productivity change as provided by field trials. In *Impacts of Forest Harvesting on Long-term Site Productivity*, (ed. W. J. Dyck, D. W. Cole, and N. B. Comerford), 41–80. London: Chapman and Hall.

Morris, R. F. 1957. The interpretation of mortality data in studies on population dynamics. *Canadian Entomologist* 89:49–69.

Morris, R. F. 1963. The dynamics of epidemic spruce budworm populations. *Memoirs of the Entomological Society of Canada* 31:1–332.

Moss, B. 2002. *The Broads: The People's Wetland.* London: Harper Collins, 392 pp.

Moss, B. 2007. The art and science of lake restoration. *Hydrobiologia* 581:15–24.

Moss, R., and A. Watson. 2001. Population cycles in birds of the grouse family (Tetraonidae). *Advances in Ecological Research* 32:53–111.

Moss, R., A. Watson, and J. Ollason. 1982. *Animal Population Dynamics.* London: Chapman-Hall, 80 pp.

Mougeot, F., S. M. Redpath, R. Moss, J. Matthiopoulos, and P. J. Hudson. 2003. Territorial behaviour and population dynamics in red grouse *Lagopus lagopus scoticus*. I. Population experiments. *Journal of Animal Ecology* 72:1073–1082.

Moulton, M. P., and S. L. Pimm. 1986. Species introductions to Hawaii. In *Ecology of Biological Invasions of North America and Hawaii*, (ed. H. A. Mooney and J. A. Drake), 231–249. New York: Springer-Verlag.

Moutia, L. A., and R. Mamet. 1946. A review of twenty-five years of economic entomology in the islands of Mauritius. *Bulletin of Entomological Research* 36:439–472.

Muirhead-Thomson, R. C. 1951. *Mosquito Behaviour in Relation to Malaria Transmission and Control in the Tropics.* London: Edward Arnold, 219 pp.

Murdoch, W. W. 1966. Community structure, population control, and competition: A critique. *American Naturalist* 100:219–226.

Murdoch, W. W., and C. J. Briggs. 1996. Theory for biological control: Recent developments. *Ecology* 77:2001–2013.

Murdoch, W. W., C. J. Briggs, and R. M. Nisbet. 2003. *Consumer-resource Dynamics.* Princeton, NJ: Princeton University Press. 456 pp.

Murdoch, W. W., C. J. Briggs, and S. Swarbrick. 2005. Host suppression and stability in a parasitoid-host system: Experimental demonstration. *Science* 309:610–613.

Murdoch, W. W., and A. Oaten. 1975. Predation and population stability. *Advances in Ecological Research* 9:1–131.

Murdoch, W. W., and A. Sih. 1978. Age-dependent interference in a predatory insect. *Journal of Animal Ecology* 47:581–592.

Murdoch, W. W., and A. Stewart-Oaten. 1989. Aggregation by parasitoids and predators: Effects on equilibrium and stability. *American Naturalist* 134:288–310.

Murdoch, W. W., J. Chesson, and P. L. Chesson. 1985. Biological control in theory and practice. *The American Naturalist* 125:344–366.

Murdoch, W. W., S. L. Swarbrick, and C. J. Briggs. 2006. Biological control: Lessons from a study of California red scale. *Population Ecology* 48:297–305.

Murdoch, W. W., S. L. Swarbrick, R. F. Luck, S. Walde, and D. S. Yu. 1996. Refuge dynamics and metapopulation dynamics: An experimental test. *American Naturalist* 147:424–444.

Murphy, E. J., J. L. Watkins, P. N. Trathan, K. Reid, M. P. Meredith, S. E. Thorpe, N. M. Johnston, et al. 2007. Spatial and temporal operation of the Scotia Sea ecosystem: A review of large-scale links in a krill centred food web. *Philosophical Transactions of the Royal Society B: Biological Sciences* 362:113–148.

Murray, B. G. J. 1999. Can the population regulation controversy be buried and forgotten? *Oikos* 84:148–152.

Mutze, G., P. Bird, J. Kovaliski, D. Peacock, S. Jennings, and B. Cooke. 2002. Emerging epidemiological patterns in rabbit haemorrhagic disease, its interaction with myxomatosis, and their effects on rabbit populations in South Australia. *Wildlife Research* 29:577–590.

Myers, J. H. 1988. Can a general hypothesis explain population cycles of forest Lepidoptera? *Advances in Ecological Research* 18:179–242.

Myers, J. H. 1993. Population outbreaks in forest Lepidoptera. *American Scientist* 81:240–250.

Myers, J. H. 2000. Population fluctuations of the western tent caterpillar in southwestern British Columbia. *Population Ecology* 42:231–242.

Myers, J. H., and D. Bazely. 1991. Thorns, spines, prickles, and hairs: Are they stimulated by herbivory and do they deter herbivores? In *Phytochemical Induction by Herbivores*, (ed. D. W. Tallamy and M. J. Raupp), 325–344. New York: John Wiley and Sons.

Myers, J. H., J. Monro, and N. Murray. 1981. Egg clumping, host plant selection and population regulation in *Cactoblastis cactorum* (Lepidoptera). *Oecologia* 51:7–13.

Myers, J. H., D. Simberloff, A. M. Kuris, and J. R. Carey. 2000a. Eradication revisited: Dealing with exotic species. *Trends in Ecology and Evolution* 15:316–320.

Myers, K., I. D. Marshall, and F. Fenner. 1954. Studies in epidemiology of infectious myxomatosis of rabbits. III. Observations on two succeeding epizootics in Australian wild rabbits on the Riverine Plain of south-eastern Australia. *Journal of Hygiene* 52:337–360.

Myers, N., R. A. Mittermeier, C. G. Mittermeier, G. A. B. da Fonseca, and J. Kent. 2000b. Biodiversity hotspots for conservation priorities. *Nature* 403:853–858.

Myers, R. A. 2002. Recruitment: Understanding density-dependence in fish populations. In *Handbook of Fish Biology and Fisheries*, (ed. P. J. B. Hart and J. D. Reynolds), 123–148. Oxford: Blackwell Publishing.

Myers, R. A., and N. J. Barrowman. 1996. Is fish recruitment related to spawner abundance? *Fishery Bulletin* 94:707–724.

Mysak, L. A. 1986. El Niño, interannual variability and fisheries in the Northeast Pacific Ocean. *Canadian Journal of Fisheries and Aquatic Sciences* 43:464–497.

Mysterud, A. 2006. The concept of overgrazing and its role in management of large herbivores. *Wildlife Biology* 12:129–141.

Nagy, K. A. 1989. Field bioenergetics: Accuracy of models and methods. *Physiological Zoology* 62:237–252.

Nagy, K. A. 2005. Field metabolic rate and body size. *Journal of Experimental Biology* 208:1621–1625.

Nakai, S., A. N. Halliday, and D. K. Rea. 1993. Provenance of dust in the Pacific Ocean. *Earth and Planetary Science Letters* 119:143–157.

National Center for Health Statistics. 2006. United States Life Tables, 2003. *National Vital Statistics Reports* 54:1–40.

Nedelman, J., J. A. Thompson, and R. J. Taylor. 1987. The statistical demography of whooping cranes. *Ecology* 68:1401–1411.

Nee, S. 2005. The neutral theory of biodiversity: Do the numbers add up? *Functional Ecology* 19:173–176.

Needham, J. 1986. *Science and Civilisation in China, Vol. 6: Part 1. Biology and Biological Technology: Botany*. Cambridge, England: Cambridge University Press.

Nemani, R. R., C. D. Keeling, H. Hashimoto, W. M. Jolly, S. C. Piper, C. J. Tucker, R. B. Myneni, and S. W. Running. 2003. Climate-driven increases in global terrestrial net primary production from 1982 to 1999. *Science* 300:1560–1563.

Newman, E. I. 1997. Phosphorus balance of contrasting farming systems, past and present. Can food production be sustainable? *Journal of Applied Ecology* 34:1334–1347.

Newmark, W. D. 1985. Legal and biotic boundaries of Western North American National Parks: A problem of congruence. *Biological Conservation* 33:197–208.

Newmark, W. D. 1995. Extinction of mammal populations in Western North American national parks. *Conservation Biology* 9:512–526.

Newton, I. 1972. *Finches*. London: Collins, 288 pp.

Newton, P. F. 2006. Asymptotic size-density relationships within self-thinning black spruce and jack pine stand-types: Parameter estimation and model reformulations. *Forest Ecology and Management* 226:49–59.

Nicholls, A. O., and C. R. Margules. 1993. An upgraded reserve selection algorithm. *Biological Conservation* 64:165–169.

Nichols, J. D., M. J. Conroy, D. R. Anderson, and K. P. Burnham. 1984. Compensatory mortality in waterfowl populations: A review of the evidence and implications for research and management. *Transactions of the North American Wildlife and Natural Resources Conference* 49:535–554.

Nicholson, A. J., and V. A. Bailey. 1935. The balance of animal populations. Part 1. *Proceedings of the Zoological Society of London* 3:551–598.

Nicklas, K. J., B. H. Tiffney, and A. H. Knoll. 1980. Apparent changes in the diversity of fossil plants. *Evolutionary Biology* 12:1–89.

Nielsen, E. S., and E. A. Jensen. 1957. Primary oceanic production. *Galathea Report* 1:49–136.

Niinemets, U., and F. Valladares. 2006. Tolerance to shade, drought, and waterlogging of temperate Northern Hemisphere trees and shrubs. *Ecological Monographs* 76:521–547.

Nilssen, A. C., O. Tenow, and H. Bylund. 2007. Waves and synchrony in *Epirrita autumnata/Operophtera brumata* outbreaks. II. Sunspot activity cannot explain cyclic outbreaks. *Journal of Animal Ecology* 76:269–275.

Noble, I. R. 1981. Predicting successional change. In: Mooney, H. A., ed. *Fire Regimes and Ecosystem Properties*. U.S. Department of Agriculture, Forest Service, General Technical Report WO-26, pp. 278–300.

Noble, J. C. 1997. *The Delicate and Noxious Scrub: CSIRO Studies on Native Tree and Shrub Proliferation in the Semi-arid Woodlands of Eastern Australia*. Canberra: CSIRO Wildife and Ecology. 137 pp.

Noble, J. C., D. S. Hik, and A. R. E. Sinclair. 2007. Landscape ecology of the burrowing bettong: Fire and marsupial biocontrol of shrubs in semi-arid Australia. *Rangeland Journal* 29:107–119.

Norby, R. J., C. A. Gunderson, S. D. Wullschleger, E. G. O'Neill, and M. K. McCracken. 1992. Productivity and compensatory responses of yellow-poplar trees in elevated CO_2. *Nature* 357:322–324.

Noss, R. F. 1987. Corridors in real landscapes: A reply to Simberloff and Cox. *Conservation Biology* 1:159–164.

Noy-Meir, I. 1975. Stability of grazing systems: An application of predator-prey graphs. *Journal of Ecology* 63:459–481.

Núñez, D., L. Nahuelhual, and C. Oyarzun. 2006. Forests and water: The value of native temperate forests in supplying water for human consumption. *Ecological Economics* 58:606–616.

O'Connor, R. J. 2000. Why ecology lags behind biology. *The Scientist* 14:35.

O'Donoghue, M., S. Boutin, C. J. Krebs, and E. J. Hofer. 1997. Numerical responses of coyotes and lynx to the snowshoe hare cycle. *Oikos* 80:150–162.

O'Donoghue, M., S. Boutin, C. J. Krebs, G. Zuleta, D. L. Murray, and E. J. Hofer. 1998. Functional responses of coyotes and lynx to the snowshoe hare cycle. *Ecology* 79:1193–1208.

O'Dowd, D. J., P. T. Green, and P. S. Lake. 2003. Invasional 'meltdown' on an oceanic island. *Ecology Letters* 6:812–817.

Odum, E. P. 1963. *Ecology.* New York: Holt, Rinehart and Winston. 152 pp.

Odum, H. T. 1983. *Systems Ecology: An Introduction.* New York: John Wiley and Sons. 644 pp.

Oechel, W. C., S. Cowles, N. Grulke, S. J. Hastings, B. Lawrence, T. Prudhomme, G. Riechers, B. Strain, D. Tissue, and G. Vourlitis. 1994. Transient nature of CO_2 fertilization in arctic tundra. *Nature* 371:500–503.

Oechel, W. C., G. L. Vourlitis, S. J. Hastings, R. C. Zulueta, L. Hinzman, and D. Kane. 2000. Acclimation of ecosystem CO_2 exchange in the Alaskan Arctic in response to decadal climate warming. *Nature* 406:978–981.

Oerke, E.-C. 2006. Crop losses to pests. *Journal of Agricultural Science* 144:31–43.

Oksanen, L., and T. Oksanen. 1992. Long-term microtine dynamics in north Fennoscandian tundra: The vole cycle and the lemming chaos. *Ecography* 15:226–236.

Olin, M., M. Rask, J. Ruuhijarvi, J. Keskitalo, J. Horppila, P. Tallberg, T. Taponen, et al. 2006. Effects of biomanipulation on fish and plankton communities in ten eutrophic lakes of southern Finland. *Hydrobiologia* 553:67–88.

Osmond, C. H., and J. Monro. 1981. Prickly pear. In *Plants and Man in Australia*, (ed. D. J. Carr and S. G. M. Carr), 194–222. New York: Academic Press.

Osterhaus, A. D. M. E., and E. J. Vedder. 1988. Identification of virus causing recent seal deaths. *Nature* 335:20.

Ostfeld, R. S. 1997. The ecology of Lyme-disease risk. *American Scientist* 85:338–346.

Owen, D. F., and R. G. Wiegert. 1981. Mutualism between grasses and grazers: An evolutionary hypothesis. *Oikos* 36:376–378.

Pac, H. I., and K. Frey. 1991. Some population characteristics of the Northern Yellowstone bison herd during the winter of 1988–89. *Montana Department of Fish, Wildlife and Parks, Bozeman, Montana.*

Pacala, S. W., and J. Roughgarden. 1985. Population experiments with the *Anolis* lizards of St. Maarten and St. Eustatius. *Ecology* 66:129–141.

Pace, M. L., S. E. G. Findlay, and D. Fischer. 1998. Effects of an invasive bivalve on the zooplankton community of the Hudson River. *Freshwater Biology* 39:103–116.

Packer, C., L. Herbst, A. E. Pusey, J. D. Bygott, J. P. Hanby, S. J. Cairns, and M. B. Mulder. 1988. Reproductive success in lions. In *Reproductive Success*, (ed. T. H. Clutton-Brock), 363–383. Chicago: University of Chicago Press.

Packer, C., R. Hilborn, A. Mosser, B. Kissui, M. Borner, G. Hopcraft, J. Wilmshurst, S. Mduma, and A. R. E. Sinclair. 2005. Ecological change, group territoriality, and population dynamics in Serengeti lions *Science* 307:390–393.

Pagel, M. D., R. M. May, and A. R. Collie. 1991. Ecological aspects of the geographical distribution and diversity of mammalian species. *American Naturalist* 137:791–815.

Paine, R. T. 1966. Food web complexity and species diversity. *American Naturalist* 100:65–75.

Paine, R. T. 1969. A note on trophic complexity and community stability. *American Naturalist* 104:91–93.

Paine, R. T. 1974. Intertidal community structure: Experimental studies on the relationship between a dominant competitor and its principal predator. *Oecologia* 15:93–120.

Paine, R. T. 1984. Ecological determinism in the competition for space. *Ecology* 65:1339–1348.

Paine, R. T. 1992. Food-web analysis through field measurement of per capita interaction strength. *Nature* 355:73–75.

Paine, R. T. 2002. Advances in ecological understanding: By Kuhnian revolution or conceptual evolution? *Ecology* 83:1553–1559.

Paine, R. T., M. J. Tegner, and E. A. Johnson. 1998. Compounded perturbations yield ecological surprises. *Ecosystems* 1:535–545.

Palmer, T. M., M. L. Stanton, T. P. Young, J. R. Goheen, R. M. Pringle, and R. Karban. 2008. Breakdown of an ant-plant mutualism follows the loss of large herbivores from an African savanna. *Science* 319:192–195.

Pandolfi, J. M., J. B. C. Jackson, and H. Hillebrand. 2006. Ecological persistence interrupted in Caribbean coral reefs. *Ecology Letters* 9:818–826.

Parer, I., D. Conolly, and W. R. Sobey. 1985. Myxomatosis: The effects of annual introductions of an immunizing strain and a highly virulent strain of myxoma virus into rabbit populations at Urana, N. S. W. *Australian Wildlife Research* 12:407–423.

Park, S. E., L. R. Benjamin, and A. R. Watkinson. 2003. The theory and application of plant competition models: An agronomic perspective. *Annals of Botany* 92:741–748.

Park, T., P. H. Leslie, and D. B. Mertz. 1964. Genetic strains and competition in populations of *Tribolium*. *Physiological Zoology* 37:97–162.

Parmesan, C., and G. Yohe. 2003. A globally coherent fingerprint of climate change impacts across natural systems. *Nature* 421:37–42.

Parry, M. L., O. F. Canziani, J. P. Palutikof, P. J. van der Linden, and C. E. Hanson. 2007. *Climate Change 2007: Impacts, Adaptation and Vulnerability. Contribution of Working Group II to the Fourth Assessment Report of the Intergovernmental Panel on Climate Change.* Cambridge: Cambridge University Press, 976 pp.

Pascual, M., J. A. Ahumada, L. F. Chaves, X. Rodo, and M. Bouma. 2006. Malaria resurgence in the East African highlands: Temperature trends revisited. *Proceedings of the National Academy of Sciences of the USA* 103:5829–5834.

Pastor, J., and S. D. Bridgham. 1999. Nutrient efficiency along nutrient availability gradients. *Oecologia* 118:50–58.

Pastor, J., R. H. Gardner, V. H. Dale, and W. M. Post. 1987. Successional changes in nitrogen availability as a potential factor contributing to spruce declines in boreal North America. *Canadian Journal of Forest Research* 17:1394–1400.

Pastor, J., and W. M. Post. 1988. Response of northern forests to CO_2—induced climate change. *Nature* 334:55–58.

Pastoret, P. P., and B. Brochier. 1999. Epidemiology and control of fox rabies in Europe. *Vaccine* 17:1750–1754.

Patten, B. C. 1993. Toward a more holistic ecology, and science: The contribution of H. T. Odum. *Oecologia* 93:597–602.

Patterson, B. D. 1987. The principle of nested subsets and its implications for biological conservation. *Conservation Biology* 1:323–334.

Patterson, R. S., D. E. Weidhaas, H. R. Ford, and C. S. Lofgren. 1970. Suppression and elimination of an island population of *Culex pipiens quinquefasciatus* with sterile males. *Science* 168:1368–1370.

Pauly, D., J. Alder, E. Bennett, V. Christensen, P. Tyedmers, and R. Watson. 2003. The future for fisheries. *Science* 302:1359–1361.

Pauly, D., and V. Christensen. 1995. Primary production required to sustain global fisheries. *Nature* 374:255–257.

Pauly, D., V. Christensen, J. Dalsgaard, R. Froese, and F. Torres Jr. 1998. Fishing down marine food webs. *Science* 279:860–863.

Pearce, D., S. Hecht, and P. Vorhies. 2007. What is biodiversity worth? Economics as a problem and a solution. In *Key Topics in Conservation Biology,* (ed. D. W. Macdonald and K. Service), 35–45. Oxford: Blackwell Publishing.

Pearcy, R. W. 1983. The light environment and growth of C3 and C4 tree species in the understory of a Hawaiian forest. *Oecologia* 58:19–25.

Pearcy, R. W., and J. Ehleringer. 1984. Comparative ecophysiology of C_3 and C_4 plants. *Plant Cell and Environment* 7:1–13.

Pearl, R. 1922. *The Biology of Death.* Philadelphia: Lippincott, 275 pp.

Pearl, R. 1927. The growth of populations. *Quarterly Review of Biology* 2:532–548.

Pearl, R. 1928. *The Rate of Living.* New York: Knopf, 185 pp.

Pearson, D. L. 1977. A pantropical comparison of bird community structure on six lowland forest sites. *Condor* 79:232–244.

Pech, R., G. M. Hood, J. McIlroy, and G. Saunders. 1997. Can foxes be controlled by reducing their fertility? *Reproduction, Fertility and Development* 9:41–50.

Pech, R. P., A. R. E. Sinclair, and A. E. Newsome. 1995. Predation models for primary and secondary prey species. *Wildlife Research* 22:55–64.

Peckham, S. D., J. W. Chipman, T. M. Lillesand, and S. I. Dodson. 2006. Alternate stable states and the shape of the lake trophic distribution. *Hydrobiologia* 571:401–407.

Pemberton, R. W., and H. Liu. 2007. Control and persistence of native *Opuntia* on Nevis and St. Kitts 50 years after the introduction of *Cactoblastis cactorum. Biological Control* 41:272–282.

Penn, D. 1999. Explaining the human demographic transition. *Trends in Ecology and Evolution* 14:32–33.

Peters, R. H. 1977. Unpredictable problems with tropho-dynamics. *Environmental Biology of Fishes* 2:97–101.

Peters, R. H. 1983. *The Ecological Implications of Body Size.* New York: Cambridge University Press. 329 pp.

Peterson, C. H. 1984. Does a rigorous criterion for environmental identity preclude the existence of multiple stable states? *American Naturalist* 124:127–133.

Peterson, R. O. 1999. Wolf-moose interaction on Isle Royale: The end of natural regulation? *Ecological Applications* 9:10–16.

Petit, J. R., J. Jouzel, D. Raynaud, N. I. Barkov, J. M. Barnola, I. Basile, M. Bender, et al. 1999. Climate and atmospheric history of the past 420,000 years from the Vostok Ice Core, Antarctica. *Nature* 399:429–436.

Petraitis, P. S., and S. R. Dudgeon. 2004. Detection of alternative stable states in marine communities. *Journal of Experimental Marine Biology and Ecology* 300:343–371.

Petraitis, P. S., and S. R. Dudgeon. 2005. Divergent succession and implications for alternative states on rocky intertidal shores. *Journal of Experimental Marine Biology and Ecology* 326:14–26.

Petraitis, P. S., and E. T. Methratta. 2006. Using patterns of variability to test for multiple community states on rocky intertidal shores. *Journal of Experimental Marine Biology and Ecology* 338:222–232.

Petrides, G. A., and W. G. Swank. 1966. Estimating the productivity and energy relations of an African elephant population. *Proceedings of the Ninth International Grassland Congress, Sao Paulo, Brazil:*831–842.

Petrusewicz, K., and A. MacFadyen. 1970. *Productivity of Terrestrial Animals: Principles and Methods.* Oxford, England: Blackwell. 190 pp.

Pettifor, R. A. 1993. Brood-manipulation experiments. I. The number of offspring surviving per nest in blue tits (*Parus caeruleus*). *Journal of Animal Ecology* 62:131–144.

Pettifor, R. A., C. M. Perrins, and R. H. McCleery. 2001. The individual optimization of fitness: Variation in reproductive output, including clutch size, mean nestling mass and offspring recruitment, in manipulated broods of great tits *Parus major. Journal of Animal Ecology* 70:62–79.

Pham, A. T., L. De Grandpre, S. Gauthier, and Y. Bergeron. 2004. Gap dynamics and replacement patterns in gaps of the northeastern boreal forest of Quebec. *Canadian Journal of Forest Research* 34:353–364.

Phillips, B. L., and R. Shine. 2006a. Allometry and selection in a novel predator-prey system: Australian snakes and the invading cane toad. *Oikos* 112:122–130.

Phillips, B. L., and R. Shine. 2006b. An invasive species induces rapid adaptive change in a native predator: Cane toads and black snakes in Australia. *Proceedings of the Royal Society of London,* Series B 273:1545–1550.

Phillips, J. 1934–1935. Succession, development, the climax, and the complex organism: An analysis of concepts. *Journal of Ecology* 22:554–571; 23:210–246, 488–508.

Phillips, O. L., Y. Malhi, N. Higuchi, W. F. Laurance, P. V. Nunez, R. M. Vasquez, S. G. Laurance, L. V. Ferreira, M. Stern, S. Brown, and J. Grace. 1998. Changes in the carbon balance of tropical forests: Evidence from long-term plots. *Science* 282:439–441.

Pianka, E. R. 1994. *Evolutionary Ecology.* New York: Harper Collins. 486 pp.

Pianka, E. R., and W. S. Parker. 1975. Age-specific reproductive tactics. *American Naturalist* 109:453–464.

Pickering, S. 1917. The effect of one plant on another. *Annals of Botany* 31:181–187.

Pickett, S. T. A., and P. S. White. 1985. *The Ecology of Natural Disturbance and Patch Dynamics.* Orlando, FL: Academic Press, 472 pp.

Pielou, E. C. 1969. *An Introduction to Mathematical Ecology.* New York: Wiley. 286 pp.

Pielou, E. C. 1979. *Biogeography.* New York: John Wiley and Sons. 351 pp.

Pielou, E. C. 1991. *After the Ice Age: The Return of Life to Glaciated North America.* Chicago: University of Chicago Press. 366 pp.

Pieper, R. D. 1988. Rangeland vegetation productivity and biomass. In *Vegetation Science Applications for Rangeland Analysis and Management,* (ed. P. T. Tueller), 449–467. Dordrecht, Netherlands: Kluwer Academic Publishers.

Pierson, E. A., and R. M. Turner. 1998. An 85-year study of saguaro (*Carnegiea gigantea*) demography. *Ecology* 79:2676–2693.

Pimentel, D. 2002. *Encyclopedia of Pest Management.* New York: Marcel Dekker, 929 pp.

Pimentel, D., H. Acquay, M. Biltonen, P. Rice, M. Silva, J. Nelson, V. Lipner, S. Giordano, A. Horowitz, and M. D'Amore. 1992. Assessment of environmental and economic impacts of pesticide use. In: D. Pimentel and H. Lehman, eds. *The Pesticide Question: Environment Economics and Ethics.* New York: Chapman and Hall, pp. 47–83.

Pimentel, D., L. Lach, R. Zuniga, and D. Morrison. 2000. Environmental and economic costs of nonindigenous species in the United States. *BioScience* 50:53–65.

Pimentel, D., and T. Patzek. 2007. Ethanol production: Energy and economic issues related to U.S. and Brazilian sugarcane. *Natural Resources Research* 16:235–242.

Pimentel, D., T. Patzek, and G. Cecil. 2007. Ethanol production: Energy, economic, and environmental losses. *Reviews of Environmental Contamination and Toxicology* 189:25–41.

Pimentel, D., and T. W. Patzek. 2005. Ethanol production using corn, switchgrass, and wood; biodiesel production using soybean and sunflower. *Natural Resources Research* 14:65–76.

Pimm, S. L. 1982. *Food Webs.* London: Chapman and Hall, 219 pp.

Pimm, S. L. 1984. The complexity and stability of ecosystems. *Nature* 307:321–326.

Pimm, S. L. 1991. *The Balance of Nature?* Chicago: University of Chicago Press. 434 pp.

Pimm, S. L., L. Dollar, and O. L. Bass. 2006. The genetic rescue of the Florida panther. *Animal Conservation* 9:115–122.

Pimm, S. L., J. H. Lawton, and J. E. Cohen. 1991. Food web patterns and their consequences. *Nature* 350:669–674.

Platt, J. R. 1964. Strong inference. *Science* 146:347–353.

Platt, T., C. Fuentes-Yaco, and K. T. Frank. 2003. Spring algal bloom and larval fish survival. *Nature* 423:398–399.

Platt, T., S. Sathyendranath, O. Ulloa, W. G. Harrison, N. Hoepffner, and J. Goes. 1992. Nutrient control of phytoplankton photosynthesis in the western North Atlantic. *Nature* 356:229–231.

Polis, G. A. 1999. Why are parts of the world green? Multiple factors control productivity and the distribution of biomass. *Oikos* 86:3–15.

Pollock, K. H., J. D. Nichols, C. Brownie, and J. E. Hines. 1990. Statistical inference for capture-recapture experiments. *Wildlife Monographs* 107:1–97.

Popper, K. R. 1963. *Conjectures and Refutations: The Growth of Scientific Knowledge.* London: Routledge and Kegan Paul. 412 pp.

Population Reference Bureau. 2007. 2007 World Population Data Sheet. http://www.prb.org/.

Possingham, H. P., D. B. Lindenmayer, and G. N. Tuck. 2002. Decision theory for population viability. In *Population Viability Analysis,* (ed. S. R. Beissinger and D. R. McCullough), 470–489. Chicago: University of Chicago Press.

Postel, S. L., G. C. Daily, and P. R. Ehrlich. 1996. Human appropriation of renewable fresh water. *Science* 271:785–788.

Pough, F. H. 1988. Mimicry of vertebrates: Are the rules different? *American Naturalist* 131 Supp.: S67–S102.

Poulson, T. L., and W. J. Platt. 1996. Replacement patterns of beech and sugar maple in Warren Woods, Michigan. *Ecology* 77: 1234–1253.

Powell, J. A., and N. E. Zimmermann. 2004. Multiscale analysis of active seed dispersal contributes to resolving Reid's Paradox. *Ecology* 85:490–506.

Power, M. E. 1990. Effects of fish in river food webs. *Science* 250:811–814.

Power, M. E. 1992a. Habitat heterogeneity and the functional significance of fish in river food webs. *Ecology* 73:1675–1688.

Power, M. E. 1992b. Top-down and bottom-up forces in food webs: Do plants have primacy? *Ecology* 73:733–746.

Power, M. E., D. Tilman, J. A. Estes, B. A. Menge, W. J. Bond, L. S. Mills, G. Daily, J. C. Castilla, J. Lubchenco, and R. T. Paine. 1996. Challenges in the quest for keystones. *BioScience* 46:609–620.

Pöysä, H. 2004. Ecological basis of sustainable harvesting: Is the prevailing paradigm of compensatory mortality still valid? *Oikos* 104:612–615.

Pratt, D. M. 1943. Analysis of population development in Daphnia at different temperatures. *Biological Bulletin* 85:116–140.

Preston, F. W. 1948. The commonness and rarity of species. *Ecology* 29:254–283.

Preston, F. W. 1962. The canonical distribution of commonness and rarity. *Ecology* 43:185–215, 410–432.

Price, D. 1999. Carrying capacity reconsidered. *Population and Environment* 21:5–26.

Price, P. W. 1991. The plant vigor hypothesis and herbivore attack. *Oikos* 62:244–251.

Price, T., M. Kirkpatrick, and S. J. Arnold. 1988. Directional selection and the evolution of breeding date in birds. *Science* 240:798–799.

Prins, H. H. T., and F. J. Weyerhaeuser. 1987. Epidemics in populations of wild ruminants: Anthrax and impala, rinderpest and buffalo in Lake Manyara National Park, Tanzania. *Oikos* 49:28–38.

Prober, S. M., and A. H. D. Brown. 1994. Conservation of the Grassy White Box Woodlands: Population genetics and fragmentation of *Eucalyptus albens*. *Conservation Biology* 8:1003–1013.

Probst, J. R., D. M. Donner, C. I. Bocett, and S. Sjogren. 2003. Population increase in Kirtland's warbler and summer range expansion to Wisconsin and Michigan's

Upper Peninsula, USA. *Oryx* 37:365–373.

Proches, S. 2001. Back to the sea: Secondary marine organisms from a biogeographical perspective. *Biological Journal of the Linnean Society* 74:197–203.

Pulliam, H. R. 1988. Sources, sinks, and population regulation. *American Naturalist* 132:652–661.

Queller, D. C., and J E. Strassmann. 1998. Kin selection and social insects. *BioScience* 48:165–175.

Quinn, J. F., and A. E. Dunham. 1983. On hypothesis testing in ecology and evolution. *American Naturalist* 122:602–617.

Quinn, R. M., K. J. Gaston, and D. B. Roy. 1998. Coincidence in the distributions of butterflies and their foodplants. *Ecography* 21:279–288.

Rabalais, N. N., R. E. Turner, and D. Scavia. 2002. Beyond science into policy: Gulf of Mexico hypoxia and the Mississippi River. *BioScience* 52:129–142.

Rabinowitz, D. 1981. Seven forms of rarity. In *The Biological Aspects of Rare Plant Conservation*, (ed. H. Synge), 205–217. London: John Wiley and Sons.

Raffaelli, D. 2000. Trends in research on shallow water food webs. *Journal of Experimental Marine Biology and Ecology* 250:223–232.

Raffel, T. R., J. R. Rohr, J. M. Kiesecker, and P. J. Hudson. 2006. Negative effects of changing temperature on amphibian immunity under field conditions. *Functional Ecology* 20:819–828.

Ramsey, D. S. L., and J. C. Wilson. 1997. The impact of grazing by macropods on coastal foredune vegetation in southeast Queensland. *Australian Journal of Ecology* 22:288–297.

Rankin, D. J., and H. Kokko. 2007. Do males matter? The role of males in population dynamics. *Oikos* 116:335–348.

Rapport, D. J., R. Costanza, and A. J. McMichael. 1998. Assessing ecosystem health. *Trends in Ecology and Evolution* 13:397–402.

Rapport, D. J., and W. G. Whitford. 1999. How ecosystems respond to stress. *BioScience* 49:193–203.

Raupach, M. R., G. Marland, P. Ciais, C. Le Quere, J. G. Canadell,

G. Klepper, and C. B. Field. 2007. Global and regional drivers of accelerating CO_2 emissions. *Proceedings of the National Academy of Sciences of the USA* 104:10288–10293.

Réale, D., A. G. McAdam, S. Boutin, and D. Berteaux. 2003. Genetic and plastic responses of a northern mammal to climate change. *Proceedings of the Royal Society of London*, Series B 270:591–596.

Recher, H. F. 1969. Bird species diversity and habitat diversity in Australia and North America. *American Naturalist* 103:75–80.

Reddingius, J., and P. J. den Boer. 1970. Simulation experiments illustrating stabilization of animal numbers by spreading of risk. *Oecologia* 5:240–284.

Redpath, S. M., F. Mougeot, F. M. Leckie, D. A. Elston, and P. J. Hudson. 2006. Testing the role of parasites in driving the cyclic population dynamics of a gamebird. *Ecology Letters* 9:410–418.

Reeve, J. D., and W. W. Murdoch. 1986. Biological control by the parasitoid *Aphytis melinus*, and population stability of the California red scale. *Journal of Animal Ecology* 55:1069–1082.

Reganold, J. P., J. D. Glover, P. K. Andrews, and H. R. Hinman. 2001. Sustainability of three apple production systems. *Nature* 410:926–930.

Reid, C. 1899. *The Origin of the British Flora*. London: Dulau, 191 pp.

Reid, K., J. P. Croxall, D. R. Briggs, and E. J. Murphy. 2005. Antarctic ecosystem monitoring: Quantifying the response of ecosystem indicators to variability in Antarctic krill. *ICES Journal of Marine Science* 62:366–373.

Reid, W. V. 1998. Biodiversity hotspots. *Trends in Ecology and Evolution* 13:275–280.

Reisewitz, S., J. Estes, and C. Simenstad. 2006. Indirect food web interactions: Sea otters and kelp forest fishes in the Aleutian archipelago. *Oecologia* 146:623–631.

Rejmankova, E., D. R. Roberts, S. Manguin, K. O. Pope, J. Komarek, and R. A. Post. 1996. *Anopheles albimanus* (Diptera: Culicidae) and

Cyanobacteria: An example of larval habitat selection. *Environmental Entomology* 25:1058–1067.

Reznick, D., M. Bryant, and D. Holmes. 2006. The evolution of senescence and post-reproductive lifespan in guppies (*Poecilia reticulata*). PLoS Biology 4: e7.

Rhoades, D. F., and R. G. Cates. 1976. Towards a general theory of plant antiherbivore chemistry. *Recent Advances in Phytochemistry* 19:168–213.

Rhyan, J. C., T. Gidlewski, T. J. Roffe, K. Aune, L. M. Philo, and D. R. Ewalt. 2001. Pathology of brucellosis in bison from Yellowstone National Park. *Journal of Wildlife Diseases* 37:101–109.

Rice, E. L. 1984. *Allelopathy*. New York: Academic Press, 422 pp.

Richards, J. D., and J. Short. 2003. Reintroduction and establishment of the western barred bandicoot *Perameles bougainville* (Marsupialia: Peramelidae) at Shark Bay, Western Australia. *Biological Conservation* 109:181–195.

Richards, O. W. 1928. Potentially unlimited multiplication of yeast with constant environment and the limiting of growth by changing environment. *Journal of General Physiology* 11:525–538.

Richards, O. W. 1939. An American textbook. (Review of A. S. Pearse, Animal Ecology.). *Journal of Animal Ecology* 8:387–388.

Richardson, D. M., P. Pysek, M. Rejmanek, M. G. Barbour, F. D. Panetta, and C. J. West. 2000. Naturalization and invasion of alien plants: Concepts and definitions. *Diversity & Distributions* 6:93–107.

Richardson, R. H., J. J. Ellison, and W. W. Averhoff. 1982. Autocidal control of screwworms in North America. *Science* 215:361–370.

Ricker, W. E. 1975. Computation and interpretation of biological statistics of fish populations. *Fisheries Research Board of Canada Bulletin* 191:1–382.

Ricklefs, R. E., and D. Schluter. 1993. *Species Diversity in Ecological Communities: Historical and Geographical Perspectives*. Chicago: University of Chicago Press, 414 pp.

Ridgway, M. S., J. B. Pollard, and D. V. C. Weseloh. 2006. Density-dependent growth of double-crested cormorant colonies on Lake Huron. *Canadian Journal of Zoology* 84:1409–1420.

Rijnsdorp, A. D., and R. S. Millner. 1996. Trends in population dynamics and exploitation of North Sea plaice (*Pleuronectes platessa* L.) since the late 1800s. *ICES Journal of Marine Science* 53:1170–1184.

Riney, T. 1964. The impact of introductions of large herbivores on the tropical environment. *International Union for the Conservation of Nature Publications,* New Series 4:261–273.

Ripple, W. J., and R. L. Beschta. 2006. Linking a cougar decline, trophic cascade, and catastrophic regime shift in Zion National Park. *Biological Conservation* 133:397–408.

Ripple, W. J., and R. L. Beschta. 2007. Restoring Yellowstone's aspen with wolves. *Biological Conservation* 138:514–519.

Robbins, C. S., J. R. Sauer, R. S. Greenberg, and S. Droege. 1989. Population declines in North American birds that migrate to the Neotropics. *Proceedings of the National Academy of Sciences of the USA* 86:7658–7662.

Roberts, C. M., J. P. Hawkins, and F. Gell. 2005. The role of marine reserves in achieving sustainable fisheries. *Philosophical Transactions of the Royal Society of London,* Series B 360:123–132.

Robinson, W. D. 1999. Long-term changes in the avifauna of Barro Colorado Island, Panama, a tropical forest isolate. *Conservation Biology* 13:85–97.

Rodda, G. H., T. H. Fritts, and D. Chiszar. 1997. The disappearance of Guam's wildlife. *BioScience* 47:565–574.

Roff, D. A. 1992. *The Evolution of Life Histories: Theory and Analysis*. New York: Chapman and Hall. 535 pp.

Rogers, H. H., J. F. Thomas, and G. E. Bingham. 1983. Response of agronomic and forest species to elevated atmospheric carbon dioxide. *Science* 220:428–429.

Rohde, K. 1992. Latitudinal gradients in species diversity: The search for the primary cause. *Oikos* 65:514–527.

Rohde, K. 1998. Latitudinal gradients in species diversity. Area matters, but how much? *Oikos* 82:184–190.

Rohner, C. 1997. Non-territorial 'floaters' in great horned owls: Space use during a cyclic peak of snowshoe hares. *Animal Behavior* 53:901–912.

Rohner, C., and D. W. Ward. 1997. Chemical and mechanical defense against herbivory in two sympatric species of desert Acacia. *Journal of Vegetation Science* 8:717–726.

Rolstad, J. 1991. Consequences of forest fragmentation for the dynamics of bird populations: Conceptual issues and the evidence. *Biological Journal of the Linnean Society* 42:149–163.

Room, P. M. 1990. Ecology of a simple plant-herbivore system: Biological control of Salvinia. *Trends in Ecology & Evolution* 5:74–78.

Root, R. B. 1967. The niche exploitation pattern of the blue-gray gnatcatcher. *Ecological Monographs* 37:317–350.

Root, R. B. 1973. Organization of a plant-arthropod association in simple and diverse habitats: The fauna of collards (*Brassica oleracea*). *Ecological Monographs* 43:95–124.

Root, T. 1988. Energy constraints on avian distributions and abundances. *Ecology* 69:330–339.

Root, T. L., D. Liverman, and C. Newman. 2007. Managing biodiversity in the light of climate change: Current biological effects and future impacts. In *Key Topics in Conservation Biology*, (ed. D. W. Macdonald and K. Service), 85–104. Oxford: Blackwell Publishing.

Root, T. L., J. T. Price, K. R. Hall, S. H. Schneider, C. Rosenzweig, and J. A. Pounds. 2003. Fingerprints of global warming on wild animals and plants. *Nature* 421:57–60.

Rosenzweig, M. L. 1968. Net primary productivity of terrestrial communities: Prediction from climatological data. *American Naturalist* 102:67–74.

Rosenzweig, M. L. 1985. Some theoretical aspects of habitat selection. In *Habitat Selection in Birds*, (ed. M. L. Cody), 517–540. New York: Academic Press.

Rosenzweig, M. L. 1995. *Species Diversity in Space and Time*. Cambridge: Cambridge University Press. 436 pp.

Rosenzweig, M. L., and R. H. Macarthur. 1963. Graphical representation and stability conditions of predator-prey interactions. *American Naturalist* 97:209–223.

Ross, J., and A. M. Tittensor. 1986. Influence of myxomatosis in regulating rabbit numbers. *Mammal Review* 16:163–168.

Ross, R. 1911. *The Prevention of Malaria*. London: Waterloo, 669 pp.

Ross, R. M., and L. B. Quetin. 1986. How productive are Antarctic krill? *BioScience* 36:264–269.

Roughgarden, J. 1974. Niche width: Biogeographic patterns among *Anolis* lizard populations. *American Naturalist* 108:429–442.

Roughgarden, J. 1986. A comparison of food-limited and space-limited animal competition communities. In *Community Ecology*, (ed. J. Diamond and T. J. Case), 492–516. New York: Harper & Row.

Roughgarden, J., S. D. Gaines, and S. W. Pacala. 1987. Supply side ecology: The role of physical transport processes. In *Organization of Communities: Past and Present*, (ed. J. H. R. Gee and P. S. Giller), 491–518. Oxford, England: Blackwell.

Rowley, I., and E. Russell. 1997. *Fairy-wrens and Grasswrens, Maluridae*. Oxford: Oxford University Press. 274 pp.

Royama, T. 1970. Factors governing the hunting behaviour and selection of food by the great tit (*Parus major* L.). *Journal of Animal Ecology* 39:619–668.

Royama, T. 1996. A fundamental problem in key factor analysis. *Ecology* 77:87–93.

Royama, T., W. E. MacKinnon, E. G. Kettela, N. E. Carter, and L. K. Hartling. 2005. Analysis of spruce budworm outbreak cycles in New Brunswick, Canada, since 1952. *Ecology* 86:1212–1224.

Rueffler, C., T. J. M. Van Dooren, O. Leimar, and P. A. Abrams. 2006. Disruptive selection and then what? *Trends in Ecology & Evolution* 21:238–245.

Ruesink, J. L., I. M. Parker, M. J. Groom, and P. M. Kareiva. 1995. Reducing the risks of nonindigenous species introductions. *BioScience* 45:465–477.

Ruimy, A., L. Kergoat, and A. Bondeau. 1999. Comparing global models of terrestrial net primary productivity (NPP): Analysis of differences in light absorption and light-use efficiency. *Global Change Biology* 5:56–64.

Ruiz, G. M., J. T. Carlton, E. D. Grosholz, and A. H. Hines. 1997. Global invasions of marine and estuarine habitats by non-indigenous species: Mechanisms, extent, and consequences. *American Zoologist* 37:621–632.

Rundle, H., L. Nagel, J. W. Boughman, and D. Schluter. 2000. Natural selection and parallel speciation in sympatric sticklebacks. *Science* 287:306–309.

Running, S. W., R. R. Nemani, F. A. Heinsch, M. Zhao, M. Reeves, and H. Hashimoto. 2004. A continuous satellite-derived measure of global terrestrial primary production. *BioScience* 54:547–560.

Russell, E. S. 1931. Some theoretical considerations on the "overfishing" problem. *Journal du Conseil Permanente International pour l'Exploration de la Mer* 6:3–27.

Russell, G. J., J. M. Diamond, S. L. Pimm, and T. M. Reed. 1995. A century of turnover: Community dynamics at three timescales. *Journal of Animal Ecology* 64:628–641.

Russell, P. F., and T. R. Rao. 1942. On the relation of mechanical obstruction and shade to ovipositing of *Anopheles culicifacies*. *Journal of Experimental Zoology* 91:303–329.

Ruttimann, J. 2006. Sick seas. *Nature* 442:978–980.

Ryther, J. H. 1963. Geographic variation in productivity. In *The Sea*, (ed. M. N. Hill), 347–380. New York: Wiley-Interscience.

Ryther, J. H., and W. M. Dunstan. 1971. Nitrogen, phosphorus, and eutrophication in the coastal marine environment. *Science* 171:1008–1013.

Sæther, B.-E., S. Engen, F. Filli, R. Aanes, W. Schröder, and R. Andersen. 2002. Stochastic population dynamics of an introduced Swiss population of the ibex. *Ecology* 83:3457–3465.

Sæther, B.-E., T. H. Ringsby, O. Bakke, and E. J. Solberg. 1999. Spatial and temporal variation in demography of a house sparrow metapopulation. *Journal of Animal Ecology* 68:628–637.

Sage, R. F. 2004. The evolution of C_4 photosynthesis. *New Phytologist* 161:341–379.

Sage, R. F., and D. S. Kubien. 2003. Quo vadis C_4?* An ecophysiological perspective on global change and the future of C_4 plants. *Photosynthesis Research* 77:209–225.

Sage, R. F., and A. D. McKown. 2006. Is C_4 photosynthesis less phenotypically plastic than C_3 photosynthesis? *Journal of Experimental Botany* 57:303–317.

Sagoff, M. 2005. Do non-native species threaten the natural environment? *Journal of Agricultural and Environmental Ethics* 18:215–236.

Sala, O. E., W. J. Parton, L. A. Joyce, and W. K. Lauenroth. 1988. Primary production of the central grassland region of the United States. *Ecology* 69:40–45.

Sale, P. 2002. *Coral Reef Fishes. Dynamics and diversity in a complex ecosystem*. San Diego, CA: Academic Press, 549 pp.

Sale, P. F. 1977. Maintenance of high diversity in coral reef fish communities. *American Naturalist* 111:337–359.

Sale, P. F. 1982. Stock-recruit relationships and regional coexistence in a lottery competitive system: A simulation study. *American Naturalist* 120:139–159.

Salt, G., and F. S. J. Hollick. 1944. Studies of wireworm populations. I. A census of wireworms in pasture. *Annals of Applied Biology* 31:52–64.

Samelius, G. 2004. Foraging behaviours and population dynamics of arctic foxes. *Arctic* 57:441–443.

Sanchez, P. A., D. E. Bandy, J. H. Villachica, and J. J. Nicholaides. 1982. Amazon basin soils: Management for continuous crop production. *Science* 216:821–827.

Sandercock, B. K., K. Martin, and S. J. Hannon. 2005. Life history strategies in extreme environments: Comparative demography of arctic and alpine ptarmigan. *Ecology* 86:2176–2186.

Sanders, H. L. 1968. Marine benthic diversity: A comparative study. *American Naturalist* 102:243–282.

Sanderson, E. W. 2006. How many animals do we want to save? The many ways of setting population target levels for conservation. *BioScience* 56:911–922.

Sang, J. H. 1950. Population growth in *Drosophila* cultures. *Biological Reviews* 25:188–219.

Saporito, R. A., M. A. Donnelly, H. M. Garraffo, T. F. Spande, and J. W. Daly. 2006. Geographic and seasonal variation in alkaloid-based chemical defenses of *Dendrobates pumilio* from Bocas del Toro, Panama. *Journal of Chemical Ecology* 32:795–814.

Sauer, J. R., J. E. Hines, and J. Fallon. 2005. The North American Breeding Bird Survey, Results and Analysis 1966–2005. Version 6.2.2006. Laurel, Maryland: USGS Patuxent Wildlife Research Center.

Savory, A. 1988. *Holistic Resource Management*. Washington, DC: Island Press, 564 pp.

Schall, J. J. 1983. Lizard malaria: Cost to vertebrate host's reproductive success. *Parasitology* 87:1–6.

Scharlemann, J. P. W., and W. F. Laurance. 2008. How green are biofuels? *Science* 319:43–44.

Scheffer, M., and S. R. Carpenter. 2003. Catastrophic regime shifts in ecosystems: Linking theory to observation. *Trends in Ecology & Evolution* 18:648–656.

Schiegg, K., S. J. Daniels, J. R. Walters, J. A. Priddy, and G. Pasinelli. 2006. Inbreeding in red-cockaded woodpeckers: Effects of natal dispersal distance and territory location. *Biological Conservation* 131:544–552.

Schimel, D. S. 2006. Climate change and crop yields: beyond Cassandra. *Science* 312:1889–1890.

Schindler, D. W. 1977. Evolution of phosphorus limitation in lakes. *Science* 195:260–262.

Schindler, D. W. 1990. Experimental perturbations of whole lakes as tests of hypotheses concerning ecosystem structure and function. *Oikos* 57:25–41.

Schindler, D. W. 1997. Liming to restore acidified lakes and streams: A typical approach to restoring damaged ecosystems? *Restoration Ecology* 5:1–6.

Schindler, D. W. 1998. Replication versus realism: The need for ecosystem-scale experiments. *Ecosystems* 1:323–334.

Schindler, D. W. 2006. Recent advances in the understanding and management of eutrophication. *Limnology & Oceanography* 51:356–363.

Schlesinger, W. H. 1997. *Biogeochemistry: An Analysis of Global Change*. San Diego, CA: Academic Press. 588 pp.

Schluter, D., and J. D. McPhail. 1992. Ecological character displacement and speciation in sticklebacks. *American Naturalist* 140:85–108.

Schneider, S. S., G. DeGrandi-Hoffman, and D. R. Smith. 2004. The African honey bee: Factors contributing to a successful biological invasion. *Annual Review of Entomology* 49:351–376.

Schoener, T. W. 1986. Resource partitioning. In *Community Ecology: Pattern and Process*, (ed. J. Kikkawa and D. J. Anderson), 91–126. Melbourne: Blackwell.

Schoenly, K. 1992. A statistical analysis of successional patterns in carrion-arthropod assemblages: Implications for forensic entomology and determination of the postmortem interval. *Journal of Forensic Sciences* 37:1489–1513.

Schoenly, K., and J. E. Cohen. 1991. Temporal variation in food web structure: 16 empirical cases. *Ecological Monographs* 61:267–298.

Schoenly, K., and W. Reid. 1987. Dynamics of heterotrophic succession in carrion arthropod assemblages: Discrete seres or a continuum of change? *Oecologia* 73:192–202.

Schultz, A. M. 1969. A study of an ecosystem: The arctic tundra. In *The Ecosystem Concept in Natural Resource Management*, (ed. G. Van Dyne), 77–93. New York: Academic Press.

Scott, J. A., N. R. French, and J. W. Leetham. 1979. Patterns of consumption in grasslands. In *Perspectives in Grassland Ecology*, (ed. N. R. French), 89–105. New York: Springer-Verlag.

Scott, J. M., J. L. Rachlow, R. T. Lackey, A. B. Pidgorna, J. L. Aycrigg, G. R. Feldman, L. K. Svancara, et al. 2007. Policy advocacy in science: Prevalence, perspectives, and implications for conservation biologists. *Conservation Biology* 21:29–35.

Seber, G. A. F. 1982. *The Estimation of Animal Abundance*. London: Charles Griffin and Company. 654 pp.

Seehausen, O. 2006. Conservation: Losing biodiversity by reverse speciation. *Current Biology* 16: R334–R337.

Seehausen, O., F. Witte, E. F. Katunzi, J. Smits, and N. Bouton. 1997. Patterns of the remnant cichlid fauna in southern Lake Victoria. *Conservation Biology* 11:890–904.

Seeley, T. D. 1985. *Honeybee Ecology: A Study of Adaptation in Social Life*. Princeton, NJ: Princeton University Press. 201 pp.

Seghers, B. H. 1974. Schooling behavior in the guppy (*Poecilia reticulata*): An evolutionary response to predation. *Evolution* 28:486–489.

Service, M. W. 1997. Mosquito (Diptera: Culicidae) dispersal—the long and short of it. *Journal of Medical Entomology* 34:579–588.

Shaffer, M. L. 1981. Minimum population sizes for species conservation. *BioScience* 31:131–134.

Sherman, G., and P. K. Visscher. 2002. Honeybee colonies achieve fitness through dancing. *Nature* 419:920.

Sherman, P. W. 1977. Nepotism and the evolution of alarm calls. *Science* 197:1246–1253.

Short, J. 1998. The extinction of rat-kangaroos (Marsupialia: Potoroidae) in New South Wales, Australia. *Biological Conservation* 86:365–377.

Short, J., J. E. Kinnear, and A. Robley. 2002. Surplus killing by introduced predators in Australia—evidence for ineffective anti-predator adaptations in native prey species? *Biological Conservation* 103:283–301.

Short, J., and B. Turner. 2000. Reintroduction of the burrowing bettong *Bettongia lesueur* (Marsupialia: Potoroidae) to mainland Australia. *Biological Conservation* 96:185–196.

Shugart, H. H. 1984. *A Theory of Forest Dynamics: The Ecological Implications of Forest Succession Models*. New York: Springer-Verlag, 278 pp.

Shurin, J. B., D. S. Gruner, and H. Hillebrand. 2006. All wet or dried up? Real differences between aquatic and terrestrial food webs. *Proceedings of the Royal Society of London*, Series B 273:1–9.

Sibly, R. M., D. Barker, M. C. Denham, J. Hone, and M. Pagel. 2005. On the regulation of populations of mammals, birds, fish, and insects. *Science* 309:607–610.

Sibly, R. M., J. Hone, and T. H. Clutton-Brock. 2003. *Wildlife Population Growth Rates*. Cambridge: Cambridge University Press. 362 pp.

Siegel, V. 2005. Distribution and population dynamics of *Euphausia superba*: Summary of recent findings. *Polar Biology* 29:1–22.

Signorini, S. R., R. G. Murtugudde, C. R. McClain, J. R. Christian, J. Picaut, and A. J. Busalacchi. 1999. Biological and physical signatures in the tropical and subtropical Atlantic. *Journal of Geophysical Research, C. Oceans* 104:18367–18382.

Sih, A. 1991. Reflections on the power of a grand paradigm. *Bulletin of the Ecological Society of America* 72:174–178.

Sih, A., P. Crowley, M. McPeek, J. Petranka, and K. Strohmeier. 1985. Predation, competition, and prey communities: A review of field experiments. *Annual Review of Ecology and Systematics* 16:269–311.

Sih, A., G. Englund, and D. Wooster. 1998. Emergent impacts of multiple predators on prey. *Trends in Ecology and Evolution* 13:350–355.

Sillero-Zubiri, C., D. Gottelli, and D. W. Macdonald. 1996. Male philopatry, extra-pair copulations and inbreeding avoidance in Ethiopian wolves (*Canis simensis*). *Behavioral Ecology and Sociobiology* 38:331–340.

Silva, M., J. H. Brown, and J. A. Downing. 1997. Differences in population density and energy use between birds and mammals: A macroecological perspective. *Journal of Animal Ecology* 66:327–340.

Silvertown, J., P. Poulton, E. Johnston, G. Edwards, M. Heard, and P. M. Biss. 2006. The Park Grass Experiment 1856–2006: Its contribution to ecology. *Journal of Ecology* 94:801–814.

Silvertown, J. W., and D. Charlesworth. 2001. *Introduction to Plant Population Biology*. Oxford: Blackwell Science. 347 pp.

Simberloff, D. 2005. Non-native species do threaten the natural environment! *Journal of Agricultural and Environmental Ethics* 18:595–607.

Simberloff, D. 2006. Invasional meltdown 6 years later: Important phenomenon, unfortunate metaphor, or both? *Ecology Letters* 9:912–919.

Simberloff, D., and J. Cox. 1987. Consequences and costs of conservation corridors. *Conservation Biology* 1:63–71.

Simberloff, D., and T. Dayan. 1991. The guild concept and the structure of ecological communities. *Annual Review of Ecology and Systematics* 22:115–143.

Simberloff, D., and P. Stiling. 1996. Risks of species introduced for biological control. *Biological Conservation* 78:185–192.

Simberloff, D. S., and W. Boecklen. 1981. Santa Rosalia reconsidered: Size ratios and competition. *Evolution* 35:1206–1228.

Simpson, G. G. 1964. Species density of North American recent mammals. *Systematic Zoology* 13:57–73.

Simpson, G. G. 1969. The first three billion years of community evolution. *Brookhaven Symposium in Biology* 22:162–177.

Sinclair, A. R. E. 1977. *The African Buffalo: A Study of Resource Limitation of Populations*. Chicago: University of Chicago Press, 355 pp.

Sinclair, A. R. E. 1989. Population regulation in animals. In *Ecological Concepts*, (ed. J. M. Cherrett), 197–241. Oxford: Blackwell Scientific.

Sinclair, A. R. E. 1997. Fertility control of mammal pests and the conservation of endangered marsupials. *Reproduction, Fertility and Development* 9:1–16.

Sinclair, A. R. E., and P. Arcese. 1995. Serengeti II: Dynamics, management, and conservation of an ecosystem. Chicago: University of Chicago Press, 665 pp.

Sinclair, A. R. E., and M. Norton-Griffiths. 1982. Does competition or facilitation regulate migrant ungulate populations in the Serengeti? A test of hypotheses. *Oecologia* 53:364–369.

Sinclair, A. R. E., C. J. Krebs, J. M. Fryxell, R. Turkington, S. Boutin, R. Boonstra, P. Seccomb-Hett, P. Lundberg, and L. Oksanen. 2000. Testing hypotheses of trophic level interactions: A boreal forest ecosystem. *Oikos* 89:313–328.

Sinclair, A. R. E., R. P. Pech, C. R. Dickman, D. Hik, P. Mahon, and A. E. Newsome. 1998. Predicting effects of predation on conservation of endangered prey. *Conservation Biology* 12:564–575.

Sinclair, A. R. E., and J. N. M. Smith. 1984. Do plant secondary compounds determine feeding preferences of snowshoe hares? *Oecologia* 61:403–410.

Singer, F. J., A. Harting, K. K. Symonds, and M. B. Coughenour. 1997. Density dependence, compensation, and environmental effects on elk calf mortality in Yellowstone National Park. *Journal of Wildlife Management* 61:12–25.

Singleton, G. R., L. Hinds, H. Leirs, and Z. Zhang. 1999. *Ecologically-based Management of Rodent Pests*. Canberra, Australia: Australian Centre for International Agricultural Research, 494 pp.

Sitar, S. P., and J. X. He. 2006. Growth and maturity of hatchery and wild lean lake trout during population recovery in Michigan waters of Lake Superior. *Transactions of the American Fisheries Society* 135:915–923.

Skaggs, R. W., and W. J. Boecklen. 1996. Extinctions of montane mammals reconsidered: Putting a global-warming scenario on ice. *Biodiversity and Conservation* 5:759–778.

Skellam, J. G. 1951. Random dispersal in theoretical populations. *Biometrika* 38:196–218.

Skellam, J. G. 1955. The mathematical approach to population dynamics. In *The Numbers of Man and Animals*, (ed. J. B. Cragg and N. W. Pirie), 31–46. Edinburgh: Oliver and Boyd.

Skutch, A. F. 1967. Adaptive limitation of the reproductive rate of birds. *Ibis* 109:579–599.

Smakhtin, V. U., C. Revenga, and P. Doll. 2004. Taking into account environmental water requirements in global-scale water resources assessments. *Research Report of the CGIAR Comprehensive Assessment of Water Management in Agriculture. No. 2, International Water Management Institute*:1–24.

Smith, F. E. 1970. Analysis of ecosystems. In: Reichle, D., ed. *Analysis of Temperate Forest Ecosystems*. Berlin: Springer-Verlag, pp. 7–18.

Smith, J. N. M. 1988. Determinants of lifetime reproductive success in the song sparrow. In *Reproductive Success*, (ed. T. H. Clutton-Brock), 154–172. Chicago: University of Chicago Press.

Smith, T. J., III. 1987. Seed predation in relation to tree dominance and distribution in mangrove forests. *Ecology* 68:266–273.

Smith, V. H. 1982. The nitrogen and phosphorous dependence of algal biomass in lakes: An empirical and theoretical analysis. *Limnology and Oceanography* 27:1101–1112.

Smith, V. H. 2006. Responses of estuarine and coastal marine phytoplankton to nitrogen and phosphorus enrichment. *Limnology & Oceanography* 51:377–384.

Smith, V. H., S. B. Joye, and R. W. Howarth. 2006. Eutrophication of freshwater and marine ecosystems. *Limnology & Oceanography* 51:351–355.

Snaydon, R. W. 1991. The productivity of C_3 and C_4 plants: A reassessment. *Functional Ecology* 5:321–330.

Snyder, G. K., and W. W. Weathers. 1975. Temperature adaptations in amphibians. *American Naturalist* 109:93–101.

Sokal, R. R., and F. J. Rohlf. 1995. *Biometry*. New York: W. H. Freeman. 887 pp.

Solomon, M. E. 1949. The natural control of animal populations. *Journal of Animal Ecology* 18:1–35.

Solomon, S., D. Qin, M. Manning, R. B. Alley, T. Berntsen, N. L. Bindoff, Z. Chen, et al. 2007. Technical Summary. In *Climate Change 2007: The Physical Science Basis. Contribution of Working Group I to the Fourth Assessment Report of the Intergovernmental Panel on Climate Change*, (ed. S. Solomon et al.). Cambridge: Cambridge University Press.

Solomon, S., D. Qin, M. Manning, Z. Chen, M. Marquis, K. B. Averyt, M. Tignor, and H. L. Miller. 2007. *Climate Change 2007: The Physical Science Contribution of Working Group I to the Fourth Assessment Report of the Intergovernmental Panel on Climate Change.* Cambridge: Cambridge University Press, 996 pp.

Somero, G. N. 2002. Thermal physiology and vertical zonation of intertidal animals: Optima, limits, and costs of living. *Integrative and Comparative Biology* 42:780–789.

Soulé, M., and L. S. Mills. 1992. Conservation genetics and conservation biology: A troubled marriage. In *Conservation of Biodiversity for Sustainable Development*, (ed. O. T. Sandlund, K. Hindar, and A. H. D. Brown), 55–69. Oslo, Norway: Scandinavian University Press.

Soulé, M. E. 1987. *Viable Populations for Conservation*. Cambridge, England: Cambridge University Press, 189 pp.

Sousa, W. P. 1985. Disturbance and patch dynamics on rocky intertidal shores. In *The Ecology of Natural Disturbance and Patch Dynamics*, (ed. S. T. A. Pickett and P. S. White), 101–124. Orlando: Academic Press.

Sousa, W. P., and J. H. Connell. 1985. Further comments on the evidence for multiple stable points in natural communities. *American Naturalist* 125:612–615.

Southwood, T. R. E., and P. A. Henderson. 2000. *Ecological Methods*. Oxford: Blackwell Science. 575 pp.

Spiller, D. A., and T. W. Schoener. 2007. Alteration of island food-web dynamics following major disturbance by hurricanes. *Ecology* 88:37–41.

Srivastava, D. S. 1999. Using local-regional richness plots to test for species saturation: Pitfalls and potential. *Journal of Animal Ecology* 68:1–16.

Srivastava, D. S., J. Kolasa, J. Bengtsson, A. Gonzalez, S. P. Lawler, T. E. Miller, P. Munguia, et al. 2004. Are natural microcosms useful model systems for ecology? *Trends in Ecology and Evolution* 19:379–384.

Stamp, N. 2003. Out of the quagmire of plant defense hypotheses. *Quarterly Review of Biology* 78:23–55.

Stamp, N. 2005. The problem with the messages of plant-herbivore interactions in ecology textbooks. *Bulletin of the Ecological Society of America* 86:27–31.

Stapley, L. 1998. The interaction of thorns and symbiotic ants as an effective defence mechanism of swollen-thorn acacias. *Oecologia* 115:401–405.

Steffan, W. L., P. D. Tyson, J. Jaeger, P. A. Matson, B. I. Moore, E. Oldfield, K. Richardson, et al. 2005. *Global Change and the Earth System: A Planet Under Pressure*. New York: Springer. 336 pp.

Steinberg, C. E. W., and R. F. Wright. 1994. *Acidification of Aquatic Ecosystems: Implications for the Future*. Chichester, England: John Wiley and Sons, 404 pp.

Stenseth, N. C. 1981. How to control pest species: Application of models from the theory of island biogeography in formulating pest control strategies. *Journal of Applied Ecology* 18:773–794.

Stephens, D. W., and J. R. Krebs. 1986. *Foraging Theory*. Princeton, NJ: Princeton University Press, 247 pp.

Stephenson, N. L. 1990. Climatic control of vegetation distribution:

The role of water balance. *American Naturalist* 135:649–670.

Sterner, R. W. 2004. A one-resource "stoichiometry"? *Ecology* 85:1813–1816.

Sterner, R. W., J. Clasen, W. Lampert, and T. Weisse. 1998. Carbon: phosphorus stoichiometry and food chain production. *Ecology Letters* 1:146–150.

Sterner, R. W., and D. O. Hessen. 1994. Algal nutrient limitation and the nutrition of aquatic herbivores. *Annual Review of Ecology and Systematics* 25:1–29.

Stevens, G. C. 1989. The latitudinal gradient in geographical range: How so many species coexist in the tropics. *American Naturalist* 133:240–256.

Stevens, G. C., and J. F. Fox. 1991. The causes of treeline. *Annual Review of Ecology and Systematics* 22:177–191.

Stiling, P. 2002. Potential non-target effects of a biological control agent, prickly pear moth, *Cactoblastis cactorum* (Berg) (Lepidoptera: Pyralidae), in North America, and possible management actions. *Biological Invasions* 4:273–281.

Stiling, P., and T. Cornelissen. 2005. What makes a successful biocontrol agent? A meta-analysis of biological control agent performance. *Biological Control* 34:236–246.

Stiling, P., and T. Cornelissen. 2007. How does elevated carbon dioxide (CO_2) affect plant-herbivore interactions? A field experiment and meta-analysis of CO_2-mediated changes on plant chemistry and herbivore performance. *Global Change Biology* 13:1823–1842.

Stiling, P., D. Moon, and D. Gordon. 2004. Endangered cactus restoration: Mitigating the non-target effects of a biological control agent (*Cactoblastis cactorum*) in Florida. *Restoration Ecology* 12:605–610.

Strengborn, J., A. Nordin, T. Nasholm, and L. Ericson. 2001. Slow recovery of boreal forest ecosystem following decreased nitrogen input. *Functional Ecology* 15:451–457.

Stromayer, K. A. K., and R. J. Warren. 1997. Are overabundant deer herds in the eastern United States creating alternate stable states in forest plant communities? *Wildlife Society Bulletin* 25:227–234.

Strong, D. R. 1984. Density-vague ecology and liberal population regulation in insects. In *A New Ecology*, (ed. P. W. Price, C. N. Slobodchikoff, and W. S. Gaud), 313–327. New York: John Wiley and Sons.

Strong, D. R. 1986. Density-vague population change. *Trends in Ecology and Evolution* 1:39–42.

Stuart, S. N., J. S. Chanson, N. A. Cox, B. E. Young, A. S. L. Rodrigues, D. L. Fischman, and R. W. Waller. 2004. Status and trends of amphibian declines and extinctions worldwide. *Science* 306:1783–1786.

Sturm, M., J. Schimel, G. Michaelson, J. M. Welker, S. F. Oberbauer, G. E. Liston, J. Fahnestock, and V. E. Romanovsky. 2005. Winter biological processes could help convert arctic tundra to shrubland. *BioScience* 55:17–26.

Sugihara, G. 1980. Minimal community structure: An explanation of species abundance patterns. *American Naturalist* 116:770–787.

Sutherland, J. P. 1990. Perturbations, resistance, and alternative views of the existence of multiple stable points in nature. *American Naturalist* 136:270–275.

Sutherland, W. J. 1996. *From Individual Behaviour to Population Ecology*. New York: Oxford University Press. 213 pp.

Sutherland, W. J. 2006. *Ecological Census Techniques: A Handbook*. Cambridge: Cambridge University Press, 432 pp.

Sutherst, R. W., R. B. Floyd, and G. F. Maywald. 1995. The potential geographical distribution of the cane toad, *Bufo marinus* L. in Australia. *Conservation Biology* 9:294–299.

Swinton, J., J. Harwood, B. T. Grenfell, and C. A. Gilligan. 1998. Persistence thresholds for phocine distemper virus infection in harbour seal *Phoca vitulina* metapopulations. *Journal of Animal Ecology* 67:54–68.

Tabashnik, B. E., N. L. Cushing, N. Finson, and M. W. Johnson. 1990. Field development of resistance to *Bacillus thuringiensis* in diamondback moth (Lepidoptera: Plutellidae). *Journal of Economic Entomology* 83:1671–1676.

Tabashnik, B. E., Y.-B. Liu, T. Malvar, D. G. Heckel, L. Masson, and J. Ferre. 1998. Insect resistance to *Bacillus thuringiensis*: uniform or diverse? *Phiolsophical Transactions of the Royal Society of London, Series B* 353:1751–1756.

Tabashnik, B. E., J. A. Fabrick, S. Henderson, R. W. Biggs, C. M. Yafuso, M. E. Nyboer, N. M. Manhardt, et al. 2006. DNA screening reveals pink bollworm resistance to *Bt* cotton remains rare after a decade of exposure. *Journal of Economic Entomology* 99:1525–1530.

Tabor, K. L., C. C. Brewster, and R. D. Fell. 2004. Analysis of the successional patterns of insects on carrion in southwest Virginia. *Journal of Medical Entomology* 41:785–795.

Takasu, F. 1998. Modeling the arms race in avian brood parasitism. *Evolutionary Ecology* 12:969–987.

Talbot, F. H., B. C. Russell, and G. R. V. Anderson. 1978. Coral reef fish communities: Unstable, high-diversity systems. *Ecological Monographs* 48:425–440.

Tamarin, R. H. 1999. *Principles of Genetics*. New York: McGraw Hill, 683 pp.

Tamarin, R. H. 2002. *Principles of Genetics*, 7th ed. New York: McGraw Hill, 788 pp.

Tanner, J. E., T. P. Hughes, and J. H. Connell. 1996. The role of history in community dynamics: A modelling approach. *Ecology* 77:108–117.

Tannerfeldt, M., B. Elmhagen, and A. Angerbjörn. 2002. Exclusion by interference competition? The relationship between red and arctic foxes. *Oecologia* 132:213–220.

Tansley, A. G. 1939. *The British Islands and Their Vegetation*. Cambridge, England: Cambridge University Press. 930 pp.

Tape, K., M. Sturm, and C. Racine. 2006. The evidence for shrub expansion in northern Alaska and the Pan-Arctic. *Global Change Biology* 12:686–702.

Taylor, E. B., J. W. Boughman, M. Groenenboom, M. Sniatynski, D. Schluter, and J. L. Gow. 2006.

Speciation in reverse: Morphological and genetic evidence of the collapse of a three-spined stickleback (*Gasterosteus aculeatus*) species pair. *Molecular Ecology* 15:343–355.

Teeri, J. A., and L. G. Stowe. 1976. Climatic patterns and the distribution of C_4 grasses in North America. *Oecologia* 23:1–12.

Temple, S. A. 1987. Do predators always capture substandard individuals disproportionately from prey populations? *Ecology* 68:669–674.

Templeton, A. R. 1986. Coadaptation and outbreeding depression. In *Conservation Biology*, (ed. M. E. Soulé), 105–116. Sunderland, MA: Sinauer Associates.

Thomas, A. S. 1960. Changes in vegetation since the advent of myxomatosis. *Journal of Ecology* 48:287–306.

Thomas, A. S. 1963. Further changes in vegetation since the advent of myxomatosis. *Journal of Ecology* 51:151–183.

Thompson, J. 1999. What we know and do not know about coevolution: Insect herbivores and plants as a test case. In *Herbivores: Between Plants and Predators*, (ed. H. Olff, V. K. Brown, and R. H. Drent), 7–30. Oxford: Blackwell Science.

Thompson, J. N. 1994. *The Coevolutionary Process*. Chicago: University of Chicago Press. 376 pp.

Thompson, J. N., O. J. Reichman, P. J. Morin, G. A. Polis, M. E. Power, R. W. Sterner, C. A. Couch, et al. 2001. Frontiers of ecology. *BioScience* 51:15–24.

Thompson, R. M., M. Hemberg, B. M. Starzomski, and J. B. Shurin. 2007. Trophic levels and trophic tangles: The prevalence of omnivory in real food webs. *Ecology* 88:612–617.

Thornthwaite, C. W. 1948. An approach toward a rational classification of climate. *Goegraphical Review* 38:55–94.

Tiessen, H., E. Cuevas, and P. Chacon. 1994. The role of soil organic matter in sustaining soil fertility. *Nature* 371:783–785.

Tilghman, N. G. 1989. Impacts of white-tailed deer on forest regeneration in northwestern Pennsylvania. *Journal of Wildlife Management* 53:524–532.

Tilman, D. 1977. Resource competition between planktonic algae: An experimental and theoretical approach. *Ecology* 58:338–348.

Tilman, D. 1982. *Resource Competition and Community Structure*. Princeton, NJ: Princeton University Press, 296 pp.

Tilman, D. 1985. The resource-ratio hypothesis of plant succession. *American Naturalist* 125:827–852.

Tilman, D. 1986a. Evolution and differentiation in terrestrial plant communities: The importance of the soil resource: Light gradient. In *Community Ecology*, (ed. J. Diamond and T. J. Case), 359–380. New York: Harper and Row.

Tilman, D. 1986b. Resources, competition and the dynamics of plant communities. In *Plant Ecology*, (ed. M. J. Crawley), 51–75. Oxford: Blackwell.

Tilman, D. 1987. The importance of the mechanisms of interspecific competition. *American Naturalist* 129:769–774.

Tilman, D. 1990. Mechanisms of plant competition for nutrients: The elements of a predictive theory of competition. In *Perspectives on Plant Competition*, (ed. J. B. Grace and D. Tilman), 117–141. San Diego, CA: Academic Press.

Tilman, D., J. HilleRisLambers, S. Harpole, R. Dybzinski, J. Fargione, C. Clark, and C. Lehman. 2004. Does metabolic theory apply to community ecology? It's a matter of scale. *Ecology* 85:1797–1799.

Tilman, D., S. S. Kilham, and P. Kilham. 1982. Phytoplankton community ecology: The role of limiting nutrients. *Annual Reviews of Ecological Systematics* 13:349–372.

Tilman, D., J. Knops, D. Wedin, P. Reich, M. Ritchie, and E. Siemann. 1997. The influence of functional diversity and composition on ecosystem processes. *Science* 277:1300–1302.

Tilman, D., P. B. Reich, J. Knops, D. Wedin, T. Mielke, and C. Lehman. 2001. Diversity and productivity in a long-term grassland experiment. *Science* 294:843–845.

Tilman, D., P. B. Reich, and J. M. H. Knops. 2006. Biodiversity and ecosystem stability in a decade-long grassland experiment. *Nature* 441:629–632.

Tilman, D., D. Wedin, and J. Knops. 1996. Productivity and sustainability influenced by biodiversity in grassland ecosystems. *Nature* 379:718–720.

Tinbergen, J. M., and J. J. Sanz. 2004. Strong evidence for selection for larger brood size in a great tit population. *Behavioral Ecology* 15:525–533.

Tinbergen, N. 1963. On aims and methods of ethology. *Zeitschrift für Tierpsychologie* 20:410–433.

Tittler, R., L. Fahrig, and M.-A. Villard. 2006. Evidence of large-scale source-sink dynamics and long-distance dispersal among wood thrush populations. *Ecology* 87:3029–3036.

Tollrian, R., and C. D. Harvell. 1999. *The Ecology and Evolution of Inducible Defenses*. Princeton, NJ: Princeton University Press, 383 pp.

Tomlinson, G. H. 2003. Acidic deposition, nutrient leaching and forest growth. *Biogeochemistry* 65:51–81.

Török, J., G. Hegyi, L. Tóth, and R. Könczey. 2004. Unpredictable food supply modifies costs of reproduction and hampers individual optimization. *Oecologia* V141:432–443.

Torti, S. D., P. D. Coley, and T. A. Kursar. 2001. Causes and consequences of monodominance in tropical lowland forests. *American Naturalist* 157:141–153.

Toth, G. B., O. Langhamer, and H. Pavia. 2005. Inducible and constitutive defenses of valuable seaweed tissues: Consequences for herbivore fitness. *Ecology* 86:612–618.

Towns, D. R., and K. G. Broome. 2003. From small Maria to massive Campbell: Forty years of rat eradications from New Zealand islands. *New Zealand Journal of Zoology* 30:377–398.

Trenberth, K. E., P. D. Jones, P. Ambenje, R. Bojariu, D. Easterling, A. K. Tank, D. Parker, et al. 2007. Observations: Surface and atmospheric climate change. In

Climate Change 2007: The Physical Science Basis. Contribution of Working Group I to the Fourth Assessment Report of the Intergovernmental Panel on Climate Change, (ed. S. Solomon et al.). Cambridge: Cambridge University Press.

Trout, R. C., J. Ross, A. M. Tittensor, and A. P. Fox. 1992. The effect on a British wild rabbit population (*Oryctolagus cuniculus*) of manipulating myxomatosis. *Journal of Applied Ecology* 29:679–686.

Tsurumi, M. 2003. Diversity at hydrothermal vents. *Global Ecology & Biogeography* 12:181–190.

Turchin, P. 1999. Population regulation: A synthetic view. *Oikos* 84:153–159.

Turchin, P. 2003. *Complex Population Dynamics: A Theoretical/Empirical Synthesis*. Princeton, NJ: Princeton University Press. 456 pp.

Turesson, G. 1922. The genotypical response of the plant species to the habitat. *Hereditas* 3:211–350.

Turesson, G. 1925. The plant species in relation to habitat and climate. *Hereditas* 6:147–236.

Turesson, G. 1930. The selective effect of climate upon the plant species. *Hereditas* 14:99–152.

Turkington, R., E. John, S. Watson, and P. Seccomb-Hett. 2002. The effects of fertilization and herbivory on the herbaceous vegetation of the boreal forest in north-western Canada: A 10-year study. *Journal of Ecology* 90:325–337.

Turner, J. R., Jr., I. K. M. Liu, D. R. Flanagan, A. T. Rutberg, and J. F. Kirkpatrick. 2007. Immunocontraception in wild horses: One inoculation provides two years of infertility. *Journal of Wildlife Management* 71:662–667.

Turner, J. R. G. 2004. Explaining the global biodiversity gradient: Energy, area, history and natural selection. *Basic and Applied Ecology* 5:435–448.

Turner, J. R. G., J. J. Lennon, and J. A. Lawrenson. 1988. British bird species distributions and the energy theory. *Nature* 335:539–541.

Turner, M. G., W. L. Baker, C. J. Peterson, and R. K. Peet. 1998. Factors influencing succession: Lessons from large, infrequent natural disturbances. *Ecosystems* 1:511–523.

Twigg, L. E., and C. K. Williams. 1999. Fertility control of overabundant species: Can it work for feral rabbits? *Ecology Letters* 2:281–285.

Underwood, A. J., and E. J. Denley. 1984. Paradigms, explanations, and generalizations in models for the structure of intertidal communities on rocky shores. In *Ecological Communities*, (ed. D. R. J. Strong, D. Simberloff, L. G. Abele, and A. B. Thistle), 153–180. Princeton, NJ: Princeton University Press.

Underwood, A. J., and P. G. Fairweather. 1989. Supply-side ecology and benthic marine assemblages. *Trends in Ecology and Evolution* 4:16–20.

Valkama, J., E. Korpimäki, B. Arroyo, P. Beja, V. Bretagnolle, E. Bro, R. Kenward, S. Manosa, S. M. Redpath, S. Thirgood, and J. Vinuela. 2005. Birds of prey as limiting factors of gamebird populations in Europe: A review. *Biological Reviews* 80:171–203.

van Asch, M., and M. E. Visser. 2007. Phenology of forest caterpillars and their host trees: The importance of synchrony. *Annual Review of Entomology* 52:37–55.

Van de Koppel, J., J. Huisman, R. Van der Wal, and H. Olff. 1996. Patterns of herbivory along a productivity gradient: An empirical and theoretical investigation. *Ecology* 77:736–745.

van den Bosch, R., P. S. Messenger, and A. P. Gutierrez. 1982. *An Introduction to Biological Control*. New York: Plenum. 247 pp.

van der Heijden, M. G. A., and I. R. Sanders, eds. 2002. *Mycorrhizal Ecology*. Berlin: Springer-Verlag, 469 pp.

van Langevelde, F., C. A. D. M. van de Vijver, K. Lalit, J. van de Koppel, N. De Ridder, J. Van Andel, A. K. Skidmore, et al. 2003. Effects of fire and herbivory on the stability of savanna ecosystems. *Ecology* 84:337–350.

Van Lenteren, J. C., J. Bale, F. Bigler, H. M. T. Hokkanen, and A. J. M. Loomans. 2006. Assessing risks of releasing exotic biological control agents of arthropod pests. *Annual Review of Entomology* 51:609–634.

Van Riper III, C., S. G. Van Riper, M. L. Goff, and M. Laird. 1986. The epizootiology and ecological significance of malaria in Hawaiian land birds. *Ecological Monographs* 56:327–344.

Van Valen, L. 1973. A new evolutionary law. *Evolutionary Theory* 1:1–30.

Vander Zanden, J. M., and W. W. Fetzer. 2007. Global patterns of aquatic food chain length. *Oikos* 116:1378–1388.

Vanderwerf, E. 1992. Lack's clutch size hypothesis: An examination of the evidence using meta-analysis. *Ecology* 73:1699–1705.

Vanni, M. J. 2002. Nutrient cycling by animals in freshwater ecosystems. *Annual Review of Ecology & Systematics* 33:341–370.

Vanni, M. J., K. K. Arend, M. T. Bremigan, D. B. Bunnell, J. E. Garvey, M. J. Gonzalez, W. H. Renwick, P. A. Soranno, and R. A. Stein. 2005. Linking landscapes and food webs: Effects of omnivorous fish and watersheds on reservoir ecosystems. *BioScience* 55:155–167.

Vargas-Teran, M., H. C. Hofmann, and N. E. Tweddle. 2005. Impact of screwworm eradication programmes using the sterile insect technique. In *Sterile Insect Technique: Principles and Practice in Area-Wide Integrated Pest Management*, (ed. V. A. Dyck, J. Hendrichs, and A. S. Robinson), 629–650. Dordrecht, Netherlands: Springer.

Varley, G. C., and G. R. Gradwell. 1960. Key factors in population studies. *Journal of Animal Ecology* 29:399–401.

Varley, G. C., G. R. Gradwell, and M. P. Hassell. 1973. *Insect Population Ecology: An Analytical Approach*. Oxford. 212 pp.

Vitousek, P. M. 1982. Nutrient cycling and the nutrient use efficiency. *American Naturalist* 119:553–572.

Vitousek, P. M. 1984. Litterfall, nutrient cycling, and nutrient limitation in tropical forests. *Ecology* 65:285–298.

Vitousek, P. M. 2004. *Nutrient Cycling and Limitation: Hawaii as a Model System*. Princeton, NJ: Princeton University Press. 223 pp.

Vitousek, P. M., J. Aber, R. W. Howarth, G. E. Likens, P. A. Matson, D. W. Schindler, W. H. Schlesinger, and D. G. Tilman. 1997. Human alteration of the global nitrogen cycle: Causes and consequences. *Ecological Applications* 7:737–750.

Vitousek, P. M., C. M. D'Antonio, L. L. Loope, and R. Westbrooks. 1996. Biological invasions as global environmental change. *American Scientist* 84:468–478.

Voigt, W., J. Perner, A. J. Davis, T. Eggers, J. Schumacher, R. Bährmann, B. Fabian, et al. 2003. Trophic levels are differentially sensitive to climate. *Ecology* 84:2444–2453.

Vollenweider, R. A. 1974. *A Manual on Methods of Measuring Primary Production in Aquatic Environments*. Oxford, England: Blackwell, 225 pp.

Volterra, V. 1926. Fluctuations in the abundance of a species considered mathematically. *Nature* 118:558–560.

von Frisch, K. 1967. *The Dance Language and Orientation of Bees*. Cambridge, MA: Harvard University Press. 566 pp.

Wackernagel, M., L. Onisto, P. Bello, A. Callejas Linares, I. S. Lopez Falfan, J. Mendez Garcia, A. I. Suarez Guerrero, and M. G. Suarez Guerrero. 1999. National natural capital accounting with the ecological footprint concept. *Ecological Economics* 29:375–390.

Wackernagel, M., and W. E. Rees. 1996. *Our Ecological Footprint: Reducing Human Impact on the Earth*. Gabriola Island, BC: New Society Publishers. 160 pp.

Waggoner, P. E. 1995. How much land can ten billion people spare for nature? Does technology make a difference? *Technology in Society* 17:17–34.

Walker, B. H. 1992. Biodiversity and ecological redundancy. *Conservation Biology* 6:18–23.

Walker, L. R., and F. S. Chapin, III. 1987. Interactions among processes controlling successional change. *Oikos* 50:131–135.

Walker, L. R., and R. del Moral. 2003. *Primary Succession and Ecosystem Rehabilitation*. Cambridge, England: Cambridge University Press. 442 pp.

Wallace, A. R. 1878. *Tropical Nature and Other Essays*. London: Macmillan, 356 pp.

Walls, J. G. 1994. *Jewels of the Rainforest: Poison Frogs of the Family Dendrobatidae*. Neptune City, NJ: T. F. H. Publications Inc., 288 pp.

Walters, C., and J. Korman. 1999. Linking recruitment to trophic factors: Revisiting the Beverton-Holt recruitment model from a life history and multispecies perspective. *Reviews in Fish Biology and Fisheries* 9:187–202.

Walters, C. J., V. Christensen, S. J. Martell, and J. F. Kitchell. 2005. Possible ecosystem impacts of applying MSY policies from single-species assessment. *ICES Journal of Marine Science* 62:558–568.

Walters, C. J., and S. J. D. Martell. 2004. *Fisheries Ecology and Management*. Princeton, NJ: Princeton University Press. 448 pp.

Walters, C. J., C. E. Robinson, and T. G. Northcote. 1990. Comparative population dynamics of *Daphnia rosea* and *Holopedium gibberum* in four oligotrophic lakes. *Canadian Journal of Fisheries and Aquatic Sciences* 47:401–409.

Walters, J. R. 1991. Application of ecological principles to the management of endangered species: The case of the red-cockaded woodpecker. *Annual Review of Ecology and Systematics* 22:505–523.

Walther, B. A., and P. W. Ewald. 2004. Pathogen survival in the external environment and the evolution of virulence. *Biological Reviews* 79:849–869.

Waples, R. S. 2002. Definition and estimation of effective population size in the conservation of endangered species. In *Population Viability Analysis*, (ed. S. R. Beissinger and D. R. McCullough), 147–168. Chicago: University of Chicago Press.

Ward, J. K., and B. R. Strain. 1999. Elevated CO_2 studies: Past, present and future. *Tree Physiology* 19:211–220.

Ward, J. P. J., R. J. Gutierrez, and B. R. Noon. 1998. Habitat selection by northern spotted owls: The consequences of prey selection and distribution. *Condor* 100:79–92.

Ward, N. L., and G. J. Masters. 2007. Linking climate change and species invasion: An illustration using insect herbivores. *Global Change Biology* 13:1605–1615.

Wardle, P., and M. C. Coleman. 1992. Evidence for rising upper limits of four native New Zealand forest trees. *New Zealand Journal of Botany* 30:303–314.

Waring, R. H., and W. H. Schlesinger. 1985. *Forest Ecosystems: Concepts and Management*. Orlando, FL: Academic Press. 340 pp.

Warming, E. 1896. *Lehrbuch der Okologischen Pflanzengeographie*. Berlin: Gebruder Borntraeger. 442 pp.

Warming, E. 1909. *Oecology of Plants: An Introduction to the Study of Plant Communities*. Oxford: Clarendon Press. 422 pp.

Warner, R. E. 1968. The role of introduced diseases in the extinction of the endemic Hawaiian avifauna. *Condor* 70:101–120.

Warren, M. S. 1994. The UK status and suspected metapopulation structure of a threatened European butterfly, the marsh fritillary *Eurodryas aurinia*. *Biological Conservation* 67:239–249.

Waters, T. F. 1957. The effects of lime application to acid bog lakes in northern Michigan. *Transactions of the American Fisheries Society* 86:329–344.

Watkinson, A. R. 1997. Plant population dynamics. In *Plant Ecology*, (ed. M. J. Crawley), 359–400. Oxford: Blackwell Science.

Watson, J. E. M., R. J. Whittaker, and D. Freudenberger. 2005. Bird community responses to habitat fragmentation: How consistent are they across landscapes? *Journal of Biogeography* 32:1353–1370.

Watt, A. S. 1947a. Contributions to the ecology of bracken (*Pteridium aquilinum*). IV. The structure of the community. *New Phytologist* 47:97–121.

Watt, A. S. 1947b. Pattern and process in the plant community. *Journal of Ecology* 35:1–22.

627

Watt, A. S. 1955. Bracken versus heather, a study in plant sociology. *Journal of Ecology* 43:490–506.

Way, M. J., and K. L. Heong. 1994. The role of biodiversity in the dynamics and management of insect pests of tropical irrigated rice—a review. *Bulletin of Entomological Research* 84:567–587.

Wayne, P. M., and F. A. Bazzaz. 1995. Seedling density modifies the growth responses of yellow birch maternal families to elevated carbon dioxide. *Global Change Biology* 1:315–324.

Webb, T. J., D. Noble, and R. P. Freckleton. 2007. Abundance-occupancy dynamics in a human dominated environment: Linking interspecific and intraspecific trends in British farmland and woodland birds. *Journal of Animal Ecology* 76:123–134.

Webster, M. S., K. A. Tarvin, E. M. Tuttle, and S. Pruett-Jones. 2004. Reproductive promiscuity in the splendid fairy-wren: Effects of group size and auxiliary reproduction. *Behavioral Ecology* 15:907–915.

Wedin, D. A., and D. Tilman. 1996. Influence of nitrogen loading and species composition of the carbon balance of grasslands. *Science* 274:1720–1723.

Weeks, J. R. 1996. *Population: An Introduction to Concepts and Issues*. San Francisco: Wadsworth Publishing Company, 676 pp.

Weidenhamer, J. D. 1996. Distinguishing resource competition and chemical interference: Overcoming the methodological impasse. *Agronomy Journal* 88:866–875.

Weladji, R. B., A. Mysterud, Ø. Holand, and D. Lenvik. 2002. Age-related reproductive effort in reindeer (*Rangifer tarandus*): Evidence of senescence. *Oecologia* 131:79–82.

Weller, D. E. 1987. A re-evaluation of the −3/2 power rule of plant self-thinning. *Ecological Monographs* 57:23–43.

Weller, D. E. 1991. The self-thinning rule: Dead or unsupported?—a reply to Lonsdale. *Ecology* 72:747–750.

West, G. B., and J. H. Brown. 2005. The origin of allometric scaling laws in biology from genomes to ecosystems: Towards a quantitative unifying theory of biological structure and organization. *Journal of Experimental Biology* 208:1575–1592.

Westemeier, R. L., J. D. Brawn, S. A. Simpson, T. L. Esker, R. W. Jansen, J. W. Walk, E. L. Kershner, J. L. Bouzat, and K. N. Paige. 1998. Tracking the long-term decline and recovery of an isolated population. *Science* 282:1695–1697.

Westoby, M. 1984. The self-thinning rule. *Advances in Ecological Research* 14:167–225.

Westoby, M. 1998. A leaf-height-seed (LHS) plant ecology strategy scheme. *Plant and Soil* 199: 213–227.

Westoby, M., D. S. Falster, A. T. Moles, P. A. Vesk, and I. J. Wright. 2002. Plant ecological strategies: Some leading dimensions of variation between species. *Annual Review of Ecology and Systematics* 33:125–159.

Westoby, M., B. Walker, and I. Noy-Meir. 1989. Opportunistic management for rangelands not at equilibrium. *Journal of Range Management* 42:266–274.

Weston, L. A. 1996. Utilization of allelopathy for weed management in agroecosystems. *Agronomy Journal* 88:860–866.

Wethey, D. S. 2002. Biogeography, competition, and microclimate: The barnacle *Chthamalus fragilis* in New England. *Integrative and Comparative Biology* 42:872–880.

Wetzel, R. C. 2001. *Limnology: Lake and River Ecosystems* San Diego, CA: Academic Press. 1,006 pp.

White, C. R., and R. S. Seymour. 2003. Mammalian basal metabolic rate is proportional to body mass $^{2/3}$. *Proceedings of the National Academy of Sciences of the USA* 100:4046–4049.

White, C. R., and R. S. Seymour. 2005. Allometric scaling of mammalian metabolism. *Journal of Experimental Biology* 208:1611–1619.

White, G. G. 1981. *Current status of prickly pear control by Cactoblastis cactorum in Queensland*. Proceedings of the 5th International Symposium on the Biological Control of Weeds. Brisbane, 1980:609–616.

White, T. C. R. 1993. *The Inadequate Environment: Nitrogen and the Abundance of Animals*. New York: Springer-Verlag. 425 pp.

Whitfield, S. M., K. E. Bell, T. Philippi, M. Sasa, F. Bolanos, G. Chaves, J. M. Savage, and M. A. Donnelly. 2007. Amphibian and reptile declines over 35 years at La Selva, Costa Rica. *Science* 104:8352–8356.

Whitham, T. G., and S. Mopper. 1985. Chronic herbivory: Impacts on architecture and sex expression of pinyon pine. *Science* 228:1089–1091.

Whitmore, T. C. 1998. *An Introduction to Tropical Rain Forests*. Oxford: Oxford University Press. 282 pp.

Whittaker, R. H. 1953. A consideration of climax theory: The climax as a population and pattern. *Ecological Monographs* 23:41–78.

Whittaker, R. H. 1975. *Communities and Ecosystems*. New York: Macmillan, 385 pp.

Whittaker, R. H., and P. P. Feeny. 1971. Allelochemics: Chemical interactions between species. *Science* 171:757–770.

Whittaker, R. J., M. B. Bush, and K. Richards. 1989. Plant recolonization and vegetation succession on the Krakatau Islands, Indonesia. *Ecological Monographs* 59:59–123.

Wiens, J. A. 1977. On competition and variable environments. *American Scientist* 65:590–597.

Wiens, J. A. 1984. On understanding a non-equilibrium world: Myth and reality in community patterns and processes. In *Ecological Communities*, (ed. D. R. Strong Jr., D. Simberloff, L. G. Abele, and A. B. Thistle), 439–457. Princeton, NJ: Princeton University Press.

Wiens, J. A. 1989. *The Ecology of Bird Communities Volume 2: Processes and Variations*. Cambridge: Cambridge University Press, 539 pp.

Wiens, J. A., J. F. Addicott, T. J. Case, and J. Diamond. 1986. Overview: The importance of spatial and temporal scale in ecological investigations. In *Community Ecology*, (ed. J. Diamond and T. J. Case),

145–153. New York: Harper & Row.

Wiles, G. J., J. Bart, R. E. Beck, and C. F. Aguon. 2003. Impacts of the brown tree snake: Patterns of decline and species persistence in Guam's avifauna. *Conservation Biology* 17:1350–1360.

Wilkinson, D. M. 2003. Catastrophes on Daisyworld. *Trends in Ecology & Evolution* 18:266–268.

Williams, B. L., E. D. J. Brodie, and E. D. I. Brodie. 2003. Coevolution of deadly toxins and predator resistance: Self-assessment of resistance by garter snakes leads to behavioral rejection of toxic newt prey. *Herpetologica* 59:155–163.

Williams, C. B. 1964. *Patterns in the Balance of Nature*. London: Academic Press. 324 pp.

Williams, C. K., R. S. Lutz, and R. D. Applegate. 2004a. Winter survival and additive harvest in northern bobwhite coveys in Kansas. *Journal of Wildlife Management* 68:94–100.

Williams, G. D. 1966. *Adaptation and Natural Selection*. Princeton, NJ: Princeton University Press. 307 pp.

Williams, T. M., J. A. Estes, D. F. Doak, and A. M. Springer. 2004b. Killer appetites: Assessing the role of predators in ecological communities. *Ecology* 85:3373–3384.

Williamson, M. 2006. Explaining and predicting the success of invading species at different stages of invasion. *Biological Invasions* 8:1561–1568.

Williamson, M., and A. Fitter. 1996a. The varying success of invaders. *Ecology* 77:1661–1666.

Williamson, M. H., and A. Fitter. 1996b. The character of successful invaders. *Biological Conservation* 78:163–170.

Willig, M. R., D. M. Kaufman, and R. D. Stevens. 2003. Latitudinal gradients of biodiversity: Pattern, process, scale, and synthesis. *Annual Review of Ecology, Evolution and Systematics* 34:273–309.

Wills, C. 1996. *Yellow Fever, Black Goddess: The Coevolution of People and Plagues*. Reading, MA: Addison Wesley. 324 pp.

Wills, C., R. Condit, R. B. Foster, and S. P. Hubbell. 1997. Strong density- and diversity-related effects help to maintain tree species diversity in a neotropical forest. *Proceedings of the National Academy of Sciences of the USA* 94:1252–1257.

Wills, C., K. E. Harms, R. Condit, D. King, J. Thompson, F. He, H. C. Muller-Landau, et al. 2006. Nonrandom processes maintain diversity in tropical forests. *Science* 311:527–531.

Wilson, E. O. 1994. *Naturalist*. Washington, DC: Island Press. 380 pp.

Wilson, J. B. 1999. Guilds, functional types, and ecological groups. *Oikos* 86:507–522.

Wilson, J. B., H. Gitay, S. H. Roxburgh, W. M. King, and R. S. Tangney. 1992. Egler's concept of 'Initial floristic composition' in succession—ecologists citing it don't agree what it means. *Oikos* 64:591–593.

Wilson, J. B., and W. G. Lee. 2000. C-S-R triangle theory: Community-level predictions, tests, evaluations of criticisms, and relation to other theories. *Oikos* 91:77–96.

Wilson, J. W. I. 1974. Analytical zoogeography of North American mammals. *Evolution* 28:124–140.

Wilson, M. A., and S. R. Carpenter. 1999. Economic valuation of freshwater ecosystem services in the United States: 1971–1997. *Ecological Applications* 9:772–783.

Winkler, D. W., and J. R. Walters. 1983. The determination of clutch size in precocial birds. *Current Ornithology* 1:33–68.

Winters, G. H., and J. P. Wheeler. 1985. Interaction between stock area, stock abundance, and catchability coefficient. *Canadian Journal of Fisheries and Aquatic Sciences* 42:989–998.

Wires, L. R., and F. J. Cuthbert. 2006. Historic populations of the double-crested cormorant (*Phalacrocorax auritus*): Implications for conservation and management in the 21st century. *Waterbirds* 29:9–37.

Wirsing, A. J., T. D. Steury, and D. L. Murray. 2002. Relationship between body condition and vulnerability to predation in red squirrels and snowshoe hares. *Journal of Mammalogy* 83:707–715.

Witte, F., B. S. Msuku, J. H. Wanink, O. Seehausen, E. F. B. Katunz, P. C. Goudswaard, and T. Goldschmidt. 2000. Recovery of cichlid species in Lake Victoria: An examination of factors leading to differential extinction. *Reviews in Fish Biology and Fisheries* 10:233–241.

Wittmer, H. U., A. R. E. Sinclair, and B. N. McLellan. 2005. The role of predation in the decline and extirpation of woodland caribou. *Oecologia* 144:257–267.

Wolda, H., and B. Dennis. 1993. Density dependence tests, are they? *Oecologia* 95:581–591.

Wolff, E. W., H. Fischer, F. Fundel, U. Ruth, B. Twarloh, G. C. Littot, R. Mulvaney, et al. 2006. Southern Ocean sea-ice extent, productivity and iron flux over the past eight glacial cycles. *Nature* 440:491–496.

Wolff, J. O. 1997. Population regulation in mammals: An evolutionary perspective. *Journal of Animal Ecology* 66:1–13.

Wolff, J. O., W. D. Edge, and G. Wang. 2002. Effects of adult sex ratios on recruitment of juvenile gray-tailed voles, *Microtus canicaudus*. *Journal of Mammalogy* 83:947–956.

Wolff, J. O., and D. W. Macdonald. 2004. Promiscuous females protect their offspring. *Trends in Ecology & Evolution* 19:127–134.

Wong, K. T. M., J. H. W. Lee, and I. J. Hodgkiss. 2007. A simple model for forecast of coastal algal blooms. *Estuarine, Coastal and Shelf Science* 74:175–196.

Wood, D. M., and R. del Moral. 1987. Mechanisms of early primary succession in subalpine habitats on Mount St. Helens. *Ecology* 68:780–790.

Woodward, G., D. C. Speirs, and A. G. Hildrew. 2005. Quantification and resolution of a complex, size-structured food web. *Advances in Ecological Research* 36:85–135.

Wootton, J. T. 1998. Effects of disturbance on species diversity: A multitrophic perspective. *American Naturalist* 152:803–825.

Wootton, J. T., M. S. Parker, and M. E. Power. 1996. Effects of disturbance on river food webs. *Science* 273:1558–1561.

Worm, B., and R. A. Myers. 2003. Meta-analysis of cod-shrimp interactions reveals top-down control in oceanic food webs. *Ecology* 84:162–173.

Wright, D. H. 1983. Species-energy theory: An extension of species-area theory. *Oikos* 41:496.

Wright, D. H., B. D. Patterson, G. M. Mikkelson, A. Cutler, and W. Atmar. 1998. A comparative analysis of nested subset patterns of species composition. *Oecologia* 113:1–20.

Wright, S. 1931. Evolution in Mendelian populations. *Genetics* 16:97–159.

Wrona, F. J., T. D. Prowse, J. D. Reist, J. E. Hobbie, L. M. J. Levesque, and W. F. Vincent. 2006. Climate change effects on aquatic biota, ecosystem structure and function. *Ambio* 35:359–369.

Wu, H., J. Pratley, D. Lemerle, and T. Haig. 2000. Laboratory screening for allelopathic potential of wheat (*Triticum aestivum*) accessions against annual ryegrass (*Lolium rigidum*). *Australian Journal of Agricultural Research* 51:259–266.

Wu, L., A. D. Bradshaw, and D. A. Thurman. 1975. The potential for evolution of heavy metal tolerance in plants. III. The rapid evolution of copper tolerance in *Agrostis stolonifera*. *Heredity* 34:165–187.

Wyatt, J. L., and M. R. Silman. 2004. Distance-dependence in two Amazonian palms: Effects of spatial and temporal variation in seed predator communities. *Oecologia* 140:26–35.

Yang, H. S. 2006. Resource management, soil fertility and sustainable crop production: Experiences of China. *Agriculture, Ecosystems & Environment* 116:27–33.

Yaremych, S. A., R. E. Warner, P. C. Mankin, J. D. Brawn, A. Raim, and R. Novak. 2004. West Nile virus and high death rate in American crows. *Emerging Infectious Diseases* 10:709–711.

Yoda, K., T. Kira, H. Ogawa, and K. Hozumi. 1963. Self-thinning in overcrowded pure stands under cultivated and natural conditions. *Journal of the Institute of Polytechnics, Osaka City University*, Series D 14:107–129.

Yodzis, P. 1986. Competition, mortality, and community structure. In *Community Ecology*, (ed. J. Diamond and T. J. Case), 480–491. New York: Harper & Row.

Young, A., T. Boyle, and T. Brown. 1996. The population genetic consequence of habitat fragmentation for plants. *Trends in Ecology & Evolution* 11:413–418.

Young, B. E. 1996. An experimental analysis of small clutch size in tropical house wrens. *Ecology* 77:472–488.

Young, T. P. 1990. Evolution of semelparity in Mount Kenya lobelias. *Evolutionary Ecology* 4:157–171.

Zah, R., H. Böni, M. Gauch, R. Hischier, M. Lehmann, and P. Wäger. 2007. *Ökobilanz von Energieprodukten: Ökologische Bewertung von Biotreibstoffen*. St. Gallen, Switzerland: EMPA, Abteilung Technologie und Gesellschaft. Download: *http://www.bfe.admin.ch*.

Zar, J. H. 1999. *Biostatistical Analysis*. London: Prentice-Hall. 663 pp.

Zhang, W., T. H. Ricketts, C. Kremen, K. Carney, and S. M. Swinton. 2007. Ecosystem services and disservices to agriculture. *Ecological Economics* 64:253–260.

Zhang, X., K. Harvey, W. Hogg, and T. Yuzyk. 2001. Trends in Canadian streamflow. *Water Resources Research* 37:987–998.

Zheng, D., S. Prince, and R. Wright. 2003. Terrestrial net primary production estimates for 0.5° grid cells from field observations—a contribution to global biogeochemical modeling. *Global Change Biology* 9:46–64.

Zhu, Y., H. Chen, J. Fan, Y. Wang, Y. Li, J. Chen, J. Fan, et al. 2000. Genetic diversity and disease control in rice. *Nature* 406:718–722.

Zillio, T., and R. Condit. 2007. The impact of neutrality, niche differentiation and species input on diversity and abundance distributions. *Oikos* 116:931–940.

Index